The Concise Encyclopedia of Genomic Diseases

The Concise Encyclopedia of Genomic Diseases

The Concise Encyclopedia of Genomic Diseases

The Concise Encyclopedia of Genomic Diseases

Preface

The Concise Encyclopedia of Genomic Diseases is a single-source reference for students (of medicine, biology, molecular biology, genomics, bioinformatics, etc.), physicians, and biological scientists. The major objective of this work is to educate the afore-mentioned medical professionals for the sake of genomic disease prevention, as well as for understanding the genomic basics of clinical knowledge. I am writing for medical specialists because they should be aware of the 6.2 billion base pairs of DNA to prevent diseases and certain traits. They can, thus, provide a better perspective for generations to come. The book points out that those future generations will be able to decrease the number of diseases and unwanted traits, depending on initiatives that can be taken by lawmakers, decision makers, and religious groups.

My hope in writing the book is that my approach encourages trainers of medical students and decision makers for medical school programs to first educate students to have a deep knowledge of genomics. Then, they can obtain the clinical knowledge to interpret the clinical syndrome. The current medical programs require students to do exactly the opposite. By this, I mean that medical students first receive a depth of clinical knowledge, and are later taught about genomics. Thus, in this book I suggest that medical students should start by understanding that the genomic changes that produce each disease have happened over a period of hundreds to thousands of years, depending on the type of the disease. Then, the students should have access to the clinical descrip-tion of each disease, in terms of the recent-ly acquired genomic knowledge.

This method will allow students to better understand the discipline by consid-ering all the clinical syndromes instead of focusing solely on a small amount of data for each test section.

Acknowledgments

My first and deepest expression of gratitude goes to the memory of my father Luis Oscar Salavaggione, who often quot-ed a famous line from the Talmud: "have a child, plant a tree, and write a book." I am grateful to my wife Rafika Zahrouni for her constant assistance, and my son Idris Salavaggione Zahrouni. I am also grateful to the rest of my family, particularly my sister Andrea Lorena Salavaggione and my mother Luisa Soplan de Salavaggione. This project could not have taken place without the valuable training that I re-ceived early on in my career, especially with both Dr. Richard Weinshilboum at Mayo Clinic and Dr. Roberto Diez at Bue-nos Aires University. I would like to thank other formers advisors Dr. Eric Wieben, Dr. Howard McLeod, Dr. Jeffrey Mil-brandt and Dr. Mark Watson. I also would like to thank Dr. Pedro Politi, Dr. Norberto Terragno, Dr. Juan Carlos Romero, and all physicians and scientists with whom I ex-changed early ideas for this project. Spe-cial thanks to Kaleb Demerew who proof-read this work and assisted me with the project. I would also like to thank all my friends, especially those from my hometown Juan B. Alberdi in the province of Buenos Aires in Argentina.

Oreste Ezequiel Salavaggione, M.D.

Introduction

This book attempts to simplify the highly technical research conducted in genomics for the medical specialist who may be unfamiliar with this field. The book will enable the medical specialist to understand: the reasons why genomic diseases emerge, the demographic groups in which they prevail, and the strategies needed to prevent them. Moreover, this study aims to make a distinction between diseases that can be considered genomically transmitted and those that are thought to be of genomic origin.

My aim in writing this book is to point out a possibility for a tradition of awareness concerning the science of genomics. This tradition should be built up in every community throughout the world to make it a better place. So far, the data sources that account for the bulk of my research span over many years of projects from around the world. The progress in genomics started more than two centuries ago and has continued thanks to the contributions of countless scientists such as Mendel and the laws of inheritance, the DNA structure discovery by Watson and Crick 60 years ago, and further with the launch of the Human Genome Project 25 years ago. However, the significant developments of genomics have not yet reached out to all populations.

Currently, no book describes all the 4,000 clinical and genomic diseases. This book contains over 2,000 diseases, from a clinical perspective, as well as from a genomic and biological point of view. The number of genes is sometimes higher than the number of diseases because there may be several genes that cause the same disease. On the other hand, the number of diseases may be higher if the same gene causes several diseases (see table 1.12). Currently, the only existing alternate sources to my book are the databases that are highly technical and hard to understand without the necessary training. This book aims to help the medical specialist to use these databases in order to be able to check and read the original data.

The market evolved over 12 years ago when J.C Venter and E.S Lander published the first two papers that described the draft of the human genome. At the time, the full DNA sequence could be obtained only from a handful of individuals. Now, with the advance of next generation sequencing, the full DNA sequence data can be obtained within a few hours, although the final analysis and interpretation of the 4,000 diseases is not yet fully accurate. The reasons for this are many. Among the 4,000 diseases, two hundred will form the foundation of my book, since they can be identified in terms of the genomic change and the resulting phenotype. However, we struggle to completely understand and recognize more than 90% of the diseases either because they are not very common or because the genotype-phenotype is variable and our understanding of it is less than desirable. Therefore, this book combines the clinical information with the genomic research data to create a comprehensive way of classifying disease and genomic information by providing the structure of each.

The Concise Encyclopedia of Genomic Diseases describes most known genomic diseases from a clinical perspective. Next, it examines them from genomic and biological points of view. Then, it analyzes the correlation between clinical and genomic data, which contributes towards creating accurate knowledge in the Personalized Medicine field. The book explains to the specialist reader the incidence of genomic diseases and the ways to prevent their prevalence in generations to come.

The book will provide an update on genomics throughout its rapid evolution. While the clinical field is old and the genomic field is recent, the integration of the two of them is partially available only in highly technical databases. Physicians trained in the last 20 years would have received some data and would have some knowledge of the genomic and Personalized Medicine field. However, physicians who were trained before the mid-1980s did not receive this knowledge since the Human Genome Project started after 1986. Therefore, medical specialists will gain a comprehensive and analytical view of the correlation between Clinical and Genomic Medicine, altogether forming a broad scope of science, and suggesting new ways of rethinking the Personalized Medicine field.

This book has 5 chapters. The first chapter describes Genomics in the past, the present and the future. The goal of the book is to adapt Mendel's law of inheritance into the contemporary Genomic field to prevent Genomic diseases in the future.

The second chapter analyzes the strategies of understanding the "pillars of the Genomic origin" in diseases, using not only the correlation between alteration of DNA information and disease, but also the criteria of inclusion and exclusion. It is sometime hard to determine the level of genomic component in a disease, but several studies (family, twin etc...) have tried to address this question. Sometimes, environment and behavior provide a better explanation for the disease than Genomics.

The third chapter examines the classical method of understanding clinical syndrome and Genomics. Most of the diseases, from susceptibility to infection, to cancer or cardiovascular disorders, are rooted in our genes. One of the most important objectives was to look at all the genomic information and the clinical syndrome in order to see the entire pattern to comprehensively understand the clinical Genomic aspect of diseases. This chapter introduces a new way of thinking about genomic diseases, with respect to their corresponding genomic data, and their correlation with other diseases. The chapter has two main parts. The first part discusses in detail the disease groups and classification methods for a select group of diseases, based on an established set of criteria. The second part of the chapter concisely describes all known genomic diseases and their traits. For instance, by studying neurology, one can focus only on the neurologic disease. However, by overlooking the genomic information that is in the same locus, the traditional method will miss important clinical information. Instead, the new method will avoid dropping important clinical and genomic information. In considering the genomic diseases and conditions in this chapter, it is important to reference back to the OMIM database, which is the cornerstone for this chapter, for more detailed descriptions and characterizations. However, some information in this database is slightly outdated, since many genes have been renamed and the database hasn't been updated. The rules and coding mechanisms that I have used to help the reader better engage with and understand the disease classifications are presented in the section "Classification Codes".

The fourth chapter illustrates the reverse method of understanding Genomics and clinical syndrome. It provides ample evidence to support the formation of different groups of disease. Several groups of diseases such as SNP syndrome, Subtelomeric syndromes, Trinucleotide syndromes, Insertion syndrome, Deletion syndrome and *de novo* syndrome are based on DNA alteration. Since not all the SNPs or mutations produce a disease, this chapter focuses on gene alteration rather than

SNP or mutation. Because there are 2,300 Gene Producing Diseases and 54,000,000 SNPs the only manageable study to conduct at this point is the data from Gene Producing Disease, when available.

The fifth chapter describes the biological method of understanding Genomics and clinical syndrome affecting the organelles.

This book provides information to understand how to prevent human genomic diseases. It draws on multiple perspectives, including geographical, historical, statistical, ethnic, national, and religious factors. Considering these aspects will help to better understand genomic diseases that are prevalent today.

Genomic Disease Perspectives:

0.1 Historical

In genomics, there are three separate groups of traits or alleles that could explain contemporary diseases. These traits and diseases are divided according to the time when their genomic alteration occurred. First, the oldest or ancestral genomic alterations are polymorphic in all people around the world. Table 1.1 (below) illustrates different traits according to the time of origin, the disease and the gene involved, and the starting age of the manifestation of these traits. Examples of ancestral traits are the ABO blood group, RH blood group, Apolipoprotein E, etc. Second, the polymorphism of the Early Migration traits can be present or absent at different levels in all eth-nic groups. These traits explain many current diseases. Third, the de novo traits appear in present generations. But this subchapter focuses on the traits and diseases of the Early Migration. Table 1.3 provides further details about the diseases, the genes, people self-referring, and the alteration of DNA estimated in years ago.

Table 1.1 Ancestral, Early Migration and *de novo* Trait and Disease

	Disease	Gene	Allele or Alteration	Clinical presenta-tion in
Ancestral trait	Hyperlipoproteinemia type II	*APOE*	E2 (Cys112 and Cys158)	Adulthood
Ancestral trait	Alzheimer Disease	*APOE*	E4 (Arg112 and Arg158)	Elderly
Early Migration trait	Cystic fibrosis	*CFTR*	ΔF508	Childhood
Early Migration trait	Huntington disease	*HTT*	CAG repeat expansion	Adulthood
de novo trait	Fragile X syndrome	*FMR1*	CGG repeat expansion	Childhood
de novo trait	Neurofibromatosis type 2	*NF2*	Several mutations or de-letions	Childhood

According to this chart, the ancestral traits are less symptomatic. On the other hand, the traits that are able to explain *de novo* diseases have clinical manifestations since early childhood.

An examination of the current diseases requires an understanding of the historical estimates of world population, the scientific theory of human migration, and a concise description of the history of clinical and genomic knowledge. First, as illustrated on Table 1.2, many events that occurred in history explain changes in the size of the human population. At the beginning of the agricultural times (10,000 years ago) many genomic traits were already present. When the population started to grow exponentially, so did the traits. 2,000 years ago, there were 170 Million people and it took 1,500 years for this size to be doubled. Today, the human population has increased to seventeen times the level that it was less than 500 years ago.

Table 1.2 Historical Estimates of World Population

Years ago	Population number	Events in history
200,000	Few persons	
74,000	2,000-15,000	Post "volcanic winter"
10,000	4 Millions	Agriculture times
2,000	170 Millions	
500	400 Millions	New world times
110	1.6 Billions	Twenty century
Now	7 Billions	Contemporary

Second, I would like to describe the scientific theory of human migration in order to understand the origins of genomic diseases, as well as their current prevalence. About 200,000 years ago *Homo sapiens* emerged in Africa; some of them travelled to the Near East around 125,000 years ago. Then, these people spread east to South Asia. Modern humans were on the verge of disappearance 71,000 to 74,000 years ago. During the volcanic winter event a genetic bottleneck was created, at which time only 2,000 to 15,000 people survived. Then, the modern humans were likely in Asia and Australia 50,000 and 40,000 years ago respectively. Europe began to be populated around 43,000 years ago. East Asia was inhabited 30,000 years ago. The America migration took place around 18,000 years ago. The genomic alterations in these places of the world that happened hundreds, or even thousands of years and/or generations ago is of a great interest in today's medicine.

In order to prevent genomic diseases, multiple measures should be taken: collecting data about backgrounds of genomic diseases, and integrating both the historical and geographical facts according to when the genomic alteration events took place. Chart 1.3 exhibits the disease, the genes, and the estimated age when the genomic alteration event took place in each group of people. This is a key factor to understanding the founder effect phenomena. For example, some genomic alterations events took place before the development of the agricultural times (more than 10,000 years ago). Nowadays, these traits produce diseases such as, Cystic fibrosis, Factor V Leiden, and Friedreich ataxia. In fact, these traits and diseases are present in many ethnic groups. A more recent example of alteration is Mucolipidosis II or I-cell disease, which has been described in more restricted ethnic groups. Among these groups, French Canadians are affected by the actual DNA alteration that happened in the Mediterranean region over 2,000 years ago. Additionally, the chart illustrates that many genomic alterations that happened in the last few hundred years are unlikely to be present today in many ethnic groups. For example, Spanish Gypsies suffer from Charcot-Marie-Tooth disease 4C and Amish-Mennonites suffer from Sitosterolemia. Most of these diseases are described in Eurasian and African people, with the exception of Amyotrophic lateral sclerosis that is present in Amerindians due to SOD1 mutations. Sometimes, information may be available when it comes to specific geographical areas (Southwest Norway or North Sweden) and religious background (Jews from Lithuania and Byelorussia). Despite these conclusions, the table 1.3 has its own limitations. The origin of the genomic alteration event is limited to one variant only. For most of these conditions I am able to obtain hundreds of different DNA variant that explain the disease, but the origin is limited to one variant. For instance, Cystic fibrosis has thousands of variants, but only one is included in this table.

Table 1.3 Estimated Age of Genomic Alteration Events

Disease	Gene	People with higher prevalence	Alteration of DNA estimated in years ago
Familial hyperphosphatemic tumoral calcinosis	GALNT3	Druze people	88-200
Charcot-Marie-Tooth disease-4C	SH3TC2	Spanish Gypsies	225
Sitosterolemia or Phytosterolemia	ABCG8	Amish-Mennonites	250
Hereditary myopathy with lactic acidosis	ISCU	Swedish (North)	300
Hereditary hemorrhagic telangiectasia	ACVRL1	French	300
Torsion dystonia, early-onset	TOR1A	Jews (Lithuania and Byelorussia)	350
Lipodystrophy	BSCL2	Norwegians (Southwest)	400
Congenital chloride diarrhea	SLC26A3	Finnish	400
Glanzmann thrombasthenia	ITGA2B	Gypsies	300-400
Xeroderma pigmentosum group C	XPC	Mediterraneans	300-500
Amyloidosis	TTR	Portuguese and Swedish	375-750
Neuronal ceroid lipofuscinosis-5 or Finnish Late Infantile Variant	CLN5	Finnish	400-750
Amyotrophic lateral sclerosis	SOD1	Amerindians	400-600
Sjögren-Larsson syndrome	ALDH3A2	Swedish (North)	600
Susceptibility/resistance to HIV, hepatitis C, and West nile virus	CCR5	Europeans (Northeast)	700

The Concise Encyclopedia of Genomic Diseases

Table 1.3. Continued.

Disease	Gene	People with higher prevalence	Alteration of DNA estimated in years ago
Susceptibility/resistance to HIV, hepatitis C, and West nile virus	CCR5	Europeans (North-east)	700
Congenital myasthenic syndrome	CHRNE	North Africans	700
Breast and ovarian cancer syndrome	BRCA2	Ashkenazi Jews	580-725
Breast and ovarian cancer syndrome	BRCA1	Ashkenazi Jews	920-1150
Arthrogryposis, renal dysfunction, and cholestasis-1 or ARC syndrome	VPS33B	Pakistanis	900-1,000
Acute intermittent porphyria	HMBS	Europeans	1,000
Mental retardation, truncal obesity, retinal dystrophy, and micropenis	INPP5E	Pakistanis (Salt Range of Punjab)	1,000
Achalasia-Addisonianism-Alacrimia syndrome or Allgrove syndrome	AAAS	North Africans	1,000-1,175
Hemoglobin E	HBB	Southeast Asians	1,240-4,440
Sickle cell anemia, Hemoglobin S or Bantu	HBB	Equatorial Africans	2,000
Mucolipidosis II or I-cell disease	GNPTAB	Mediterraneans	2,063
Stargardt disease-1	ABCA4	Europeans	2,400-3,000
Familial adenomatous polyposis-1	APC	Ashkenazi Jews, Sephardi Jews and Arabs	2,200-2,950
Parkinson disease-8	LRRK2	Chinese	2,500
Smith-Lemli-Opitz syndrome	DHCR7	Europeans (Northwest)	3,000
Leber congenital amaurosis and Cone-rod dystrophy	GUCY2D	Finnish	3,000-3,750
Citrullinemia-2	SLC25A13	Chinese	3,000-7,000
Cartilage-hair hypoplasia, metaphyseal dysplasia, and hypotrichosis	RMRP	Finnish	3,900-4,800
Huntington disease*	HTT	Europeans	4,700-10,000
Hemoglobin C	HBB	Africans (Central West)	5,000
Hypercholesterolemia	APOB	Europeans	6,000
Emphysema and liver disease, due to alpha-1 antitrypsin	SERPINA1	Europeans (North)	6,000
Friedreich ataxia*	FXN	Europeans	9,000-24,000

The Concise Encyclopedia of Genomic Diseases

Table 1.3. Continued.

Disease	Gene	People with higher prevalence	Alteration of DNA estimated in years ago
Cystic fibrosis	CFTR	Europeans, the Middle Easterns, and Western Asians	10,000-52,000
Chuvash polycythemia or familial erythrocytosis-2	VHL	Chuvash people (Mid-Volga river region)	14,000-62,000
Oculocutaneous albinism-4	SLC45A2	Eastern Asians and Caucasians	15,000-35,000
Deafness-1A	GJB2	Europeans	14,000-17,000
Factor V Leiden	F5	Europeans	21,340
Thrombophilia	F2	Europeans	23,720
Axonal neuropathy with hoarse-ness and vocal cord paresis	GDAP1	Iberians	33,000
Hypertrophic cardiomyopathy-4	MYBPC3	Inhabitants of Indi-an subcontinent	33,000
Ataxia-telangiectasia or Boder-Sedgwick syndrome	ATM	Europeans, the Middle Easterns, and Western Asians	50,000

* In this case I am considering an allele that matches with the trinucleotide expansion.

Third, I would like to present a concise history of the clinical and genomic knowledge over the last 220 years, and provide additional information to better understand the recent history of clinical and genomic knowledge in terms of discovery date, identification either of the disease (Italicized), or of any biological discovery (Underlined), description, the contributor to the clinical or genomic description if known, and the quote.

1798 *Daltonism, color blindness or deuteranopia*, John Dalton:

"My vision was not like that of other persons; and, at the same time, that the difference between day-light and candle-light, on some colours, was indefinitely more perceptible to me than to other." Dalton endeavors to craft a theory explaining his particular type of color blindness.

1803 *Hemophilia*, John C. Otto:

"About seventy or eighty years ago, a woman by name of Smith, settled in the vicinity of Plymouth, New Hampshire, and transmitted the following idiosyncrasy to her descendants. It is one, she observed, to which her family is unfortunately subject, and had been the source not only of great solicitude, but frequently the cause of death. If the least scratch is made on the skin of some of them, as mortal a hemorrhagy will eventually ensue as if the largest wound is inflicted. So assured are the members of this family of the terrible consequences of the least wound, that they will not suffer themselves to be bled on any consideration, having lost a relation by not being able to stop the discharge occasioned by this operation."

18th century Literature from Germany and Switzerland, *Cystic fibrosis or mucoviscidosis*:

"Woe to the child who tastes salty from a kiss on the brow, for he is cursed and soon must die"

1856 *Congenital deafness and prolongation of the QT interval*, Meissner. (In a textbook on deaf-mutism) A young girl was called before the director of her school for a minor offense and fell instantly dead. The parents were not surprised, having lost 2 other "deaf-mute" children under similar circumstances of fright and rage.

1859 The Origin of Species, Charles Darwin.

It introduced the scientific theory that populations evolve over the course of generations through a process of natural selection. It presented a body of evidence that the diversity of life arose by common descent through a branching pattern of evolution.

1865 and 1866 Mendelian inheritance, Gregor Johann Mendel.

This theoretical framework of scientific theory of how hereditary characteristics are passed from parent organisms to their offspring; it underlies much of genetics.

1872 *Huntington disease or chorea*, George Huntington:

"Of its hereditary nature. When either or both the parents have shown manifestations of the disease one or more of the offspring almost invariably suffer from the disease. But if by any chance these children go through life without it, the thread is broken and the grandchildren and great-grandchildren of the original shakers may rest assured that they are free from the disease."

1881-1887 *Tay–Sachs disease*, Waren Tay and Bernard Sachs.

1900 "Re-discovered Mendel's theories" by three European scientists, Hugo de Vries, Carl Correns and Erich von Tschermak.

1900 *ABO blood group*, Landsteiner.

1902-1903 Chromosome theory of inheritance Walter Sutton and Theodor Boveri.

This theory identifies chromosomes as the carriers of genetic material. It correctly explains the mechanism underlying the laws of Mendelian inheritance by identifying chromosomes with the paired factors required by Mendel's laws.

1906 *Tay–Sachs disease or Amaurotic Idiocy*:

"It is a curious fact that amaurotic family idiocy, a rare and fatal disease of children, occurs mostly among Jews. The largest number of cases has been observed in the United States—over thirty in number. It was at first thought that this was an exclusively Jewish disease, because most of the cases at first reported were among Russian and Polish Jews; but recently there have been reported a few cases occurring in non-Jewish children. The chief characteristics of the disease are progressive mental and physical enfeeblement; weakness and paralysis of all the extremities; and marasmus, associated with symmetrical changes in the macula lutea. On investigation of the reported cases it has been found that neither consanguinity nor syphilitic, alcoholic, or nervous antecedents in the family history are factors in the etiology of the disease. No preventive measures have as yet been discovered, and no treatment has been of any benefit, all the cases having terminated fatally".

1911 *Daltonism, color blindness or deuteranopia*.

Color blindness was the foundation to start mapping the human genome (When red-green color blindness was assigned to the X chromosome). This was based upon the observation that color blindness is passed on from mothers to their sons.

The Concise Encyclopedia of Genomic Diseases

1910–11 Heritability of ABO blood groups, Ludwik Hirszfeld and E. von Dungern.

1915 The Mechanism of Mendelian Heredity, Thomas Hunt Morgan.

It demonstrated that genes are carried on chromosomes and are the mechanical bases of heredity.

1928 The significance of pneumococcal types, Griffith F.

1943 *Fragile X syndrome,* J. Purdon Martin and Julia Bell.

1953 Molecular structure of nucleic acids, Watson JD, Crick FH.

1969 *Tay–Sachs disease*, enzyme defect; the patients could be diagnosed by an assay of hexosaminidase A activity.

1977 Developed the chain termination method for sequencing DNA, Fred Sanger.

1978 The duplicated human alpha globin lie close together in cellular DNA, Orkin, S. H.

1983 A polymorphic DNA marker genetically linked to *Huntington's disease.*

1985 PKD1 locus is closely linked to the alpha-globin locus on 16p, Reeders, S. T.

1986 "Specific enzymatic amplification of DNA in vitro: the polymerase chain reaction." Mullis KB.

1986 Human Genome Initiative. It was founded by the Department of Energy to develop critical resources and technologies.

1989 "Identification of the cystic fibrosis gene: chromosome walking and jumping"

1993 A novel gene containing a trinucleotide repeat that is expanded and unstable on *Huntington's disease* chromosomes.

1998 Celera Genomics started to sequence much of human genome in 3 years using HGP-generated resources.

1999 First Human Chromosome Completely Sequenced.

2003 Human Genome Project Declared Complete.

0.2 Statistical

While statistical genomics is the study of different alleles' frequency in a group of individuals, this study is concerned with alleles in common and rare diseases. There is neither a single, commonly accepted definition for rare diseases (as each country defines them differently) nor an established cutoff. The United States defines it strictly in terms of prevalence; specifically any condition that affects less than 200,000 individuals (about 1 in every 1,500 people) is rare according to the Rare Disease Act. Japan considers a rare disease to be one that occurs in less than 50,000 Japanese (about 1 in every 2,500 people). The present section of the book takes less than 1 affected individual in every 2,000 people as a definition for a rare disease. On the other hand, it defines the term common disease as one that appears in more than 1 individual in 2,000 people. Table 1.4 describes the frequency of traits, the alleles, the diseases, the product of conception alterations, and the set of people who suffer from the listed conditions. The expectancy of alleles and common disorders in a different set of people are established by using the principle of Hardy–Weinberg, as shall be summarized below.

0.2.1 Diseases in Recessive Syndrome

In order to understand recessive disorder it is important to know the prevalence of the allele frequency for a pathogenic DNA alteration. For example in people from Ireland, one in every eighteen persons has a pathogenic trait for Cystic fibrosis (Allele frequency of 1/37). Statistically, if 1 Irish man (Allele frequency of 1/37) x 1 Irish woman (Allele frequency of

1/37) or 1/ $(37)^2$ it makes 1/ 1,353 persons, who are at the risk of having Cystic fibrosis in Ireland. While certain diseases are often present in the same specific ethnic group, the color of the skin or any general characteristic of an ethnic group is not the best predictor of the prevalence of Cystic fibrosis. For example a neighboring people in Finland have a 20 fold less occurring frequency of this disease. This conclusion emphasizes that although the Finns and the Irish are white Caucasians, they do not have a similar prevalence of Cystic fibrosis.

Moreover, it is worth noting that any trait in any set of people who have an allele frequency of at least one percent will be defined as polymorphism in each specific group. Statistically, when 1 person in 50 has a pathogenic allele, (Allele frequency of 1%), subsequent generations with this background will eventually result in the frequency 1/10,000 persons suffering from the condition. To put it differently: (1/100 allele frequency) x (1/100 allele frequency) or 1/ $(100)^2$. On the other hand, any trait that has an allele frequency of less than one percent will be defined as mutation. As for common diseases, any recessive trait has to be present in less than 1/22 person which is the same as an allele frequency of 1/44. Numerically, a prevalence of 1/44 x 1/44, or 1/ $(44)^2$, corresponds to 1/ 1,936 persons who suffer from these recessive conditions. This number is generally used by nations to organize diseases by frequency. However, from a point of view of Personalized medicine, these rare diseases become more common if we are aware of the heterozygosis, specifically pathogenic allele for the same gene, in both parents. Indeed, the frequency of a disease as Mendel predicted turns out to be common in that 1 in 4 people will go to suffer from the condition. With regard to table 1.4 only Sickle cell anemia, Hemochromatosis, and Cystic fibrosis (Irish people) are considered common in the ethnic groups described. Besides the information that the table provides, there are many other examples of diseases, such as Sickle cell anemia that are common in a selected group of people (1/625 African Americans suffer from the condition), but rare in a nationality (1/5,000 persons in the United States).

0.2.2 Diseases in Dominant Syndrome

In order to understand dominant disorders, it is important to know that every allele produces the disease. For any dominant trait to become a common disease, it has to be present in less than 1/2,000 persons. As shown on table 1.4, only von Willebrand disease type 1, Adult polycystic kidney disease 1, and Familial hypercholesterolemia are common. Furthermore, if both parents have diseases such as Familial hypercholesterolemia and Von Willebrand disease, and they have offspring, one of every four children is going to have the full symptomatic disease. Arithmetically, for Familial hypercholesterolemia due to LDLR, the frequency will be 1/1,000 x 1/1,000 or 1/ $(1,000)^2$. Consequently 1/1,000,000 person is going to suffer from the full severity disease. Also, statistically for Von Willebrand disease, 1/200 x 1/200 or 1/ $(200)^2$ will result in 1/40,000 person, who are going to suffer from the full severity of the disease.

The Concise Encyclopedia of Genomic Diseases

Table 1.4 Expectancy of Common and Rare Disorders

Disease	Trait Frequency	Allele Frequency	Disease Frequency	People with higher prevalence	Product of conception Alteration
Sickle cell anemia	1 in 3.5	1 in 7	1 in 50	Equatorial Africans	AR
Hemochromatosis	1 in 10	1 in 20	1 in 400	Inhabitants of Utah	AR
Cystic fibrosis	1 in 18.5	1 in 37	1 in 1,353	Irish	AR + KDN
Cystic fibrosis	1 in 25	1 in 50	1 in 2,500	Caucasians (USA)	AR + KDN
Severe combined immunodeficiency	1 in 25	1 in 50	1 in 2,500	Navajo and Apache peoples (Athabaskan)	AR
Tay-Sachs disease	1 in 27	1 in 54	1 in 2,916	Ashkenazi Jews	AR
Cystic fibrosis	1 in 79	1 in 158	1 in 25,000	Finnish	AR + KDN
Tay-Sachs disease	1 in 110	1 in 220	1 in 48,400	Sephardic Jews	AR

The above chart reflects the expectancy of diseases. For example, the actual frequency of Tay-Sachs disease in Ashkenazi Jewish communities is 10 fold less than it appears on the chart. This is mainly due to education and encouragement of disassortative mating.

0.3 Nationality

For different reasons countries around the world have invested money and resources to better serve their own countrymen. Table 1.5 reflects the total population and the contribution of each selected country (dependent territories are classified separate) in the world in terms of genomic research. This section illustrates how disparities in genomic research differ from one country to another on the basis of key examples. The top 10 countries that conduct genomic research are the United States, United Kingdom, Japan, Italy, Germany, France, Israel, Netherlands, Finland, and Saudi Arabia. Most of these countries are developed and fully aware of many inherited diseases. Some of these diseases result from both founder effect and endogamy. This section is also based on a selection of a few countries, or territories including Oman, Faroe Islands (Denmark), Réunion Islands (France), and Philippines with similar genomic knowledge regarding their own countrymen. It is important to note that there is incongruity between the sizes of population and the depth of knowledge that these countries have in genomics. For example, while the Philippines is 2,000 times more heavily populated than Faroe Islands, the two have similar genomic knowledge. A few countries in the Middle East also have developed genomic knowledge. However, Sub-Saharan Africa and Latin America have little knowledge regarding the genomic information of their own countrymen.

Table 1.5 Contribution of Selected Country in Genomic Research

Ranking in overall research	Country	Population	World Genomic Research Contribution (%)
1	United States	313,053,000	24.275
2	United Kingdom	62,300,000	5.778
3	Japan	127,770,000	4.158
4	Italy	60,757,278	3.366
5	Germany	81,796,000	3.157
6	France	65,350,000	3.04
7	Israel	7,836,000	3.017
8	Netherlands	16,727,255	2.772
9	Finland	5,404,750	2.726
10	Saudi Arabia	27,136,977	2.679
11	Canada	33,476,688	2.586
12	Spain	46,196,278	1.98
13	Sweden	9,482,855	1.98
14	Turkey	74,724,269	1.736
15	Australia	22,837,348	1.701
16	India	1,210,193,422	1.514
17	Ireland	4,581,269	1.223
18	China	1,347,350,000	1.211
19	Denmark	5,579,204	1.142
20	Brazil	192,376,496	1.13
53	Oman	2,773,479	0.28
54	Faroe Islands (D)	48,738	0.28
65	Réunion (France)	816,364	0.21
66	Philippines	94,013,200	0.198
78	Åland Islands (F)	28,007	0.175
79	Malaysia	28,334,135	0.163

(D) Denmark and (F) Finland

0.4 Ethnicity, Race and Population

Ethnicity is the cultural identity of a group of people from a nation state (The Greeks used this term 5 centuries BC). Throughout human history, a group of persons always identifies itself as distinct from other groups. In contrast, the term race dates back to the 17th century. For example, in the year 1684 the book "New Division of Earth by the Different Species or Races which Inhabit it" was published.

Race is a colonial European concept without any scientific evidence; in fact new DNA information has caused some legal and social concepts such as race to be inappropriate. In other words, race has acquired pejorative connotations. Race also focuses on physical characteristics of visible traits such as skin color, eye shape and hair texture, which are not biologically or genomically relevant. Whereas the use of some other terms such as Caucasian (which is a race) has been established due to the abundance of information that is available about it, the author of this book will still use such terms despite the fact that they are not appropriate.

Population or sexual biological population is a set of people in which any pair of members can procreate together. There is only 1 population in the world, i.e. *Homo sapiens*, since the importance is just to exchange gametes in order to produce fertile offspring. In this book the term population is going to be used in a limited way since many people use the term subpopulation that is as inappropriate as race. Instead of the term population, the author of this book prefers to use people "self-referring as," or ethnic group.

0.5 Ethnic, Religious, and Geographical

Ethnic, religious, and geographical perspectives are going to be described altogether because in some cases the three factors correlate with a disease. As Table 1.6 describes, most of the genomic traits and diseases, which prevail around the world are determined by the aforementioned factors. The chart also includes examples of diseases that are several folds more prevalent in certain groups than the

worldwide expectancy. In addition to the worldwide frequency and the description of these diseases in these groups of people, the table includes the genes and the product of conception alteration. For example, Familial hypercholesterolemia is seven times more common in these groups than in the world. Moreover, in Ashkenazi Jews and in Lebanese Christians 1 disease occurs in every 69 or 71 individuals, respectively, whereas the worldwide prevalence is 1 Familial hypercholesterolemia every 500 individuals. Another example of disease is Maple syrup urine that is 470 times more prevalent in Old Order Mennonite than in worldwide people. This disease is present in 1 every 380 individuals in Old Order Mennonite whereas the worldwide prevalence occurs in one patient every 180,000 individuals. Conclusions that may be drawn from Table 1.6 indicate that some diseases such as Familial hypercholesterolemia, Tay-Sachs disease, Maple syrup urine disease are shared among different ethnic and religious groups. These diseases are defined as non-ethnic specific ones. However, there are few diseases that are highly related to a small group of people, or almost absent in any other group of individuals. These diseases are defined as ethnic specific ones.

Table 1.6 Ethnic, Religious, and Geographical Factors of Genomic Diseases

Disease	Gene	Product of conception altered	Worldwide Frequency	Prevalent Group
Familial hypercholesterolemia	LDLR	AD + de novo	2 in 1,000	Ashkenazi Jews, Afrikaners, French Canadians, Lebanese Christians and Finnish
Tay-Sachs disease	HEXA	AR	3 in 1,000,000	Ashkenazi Jews, French Canadians, Cajuns, Irish Americans, also in lesser extent in Sephardic Jews

The Concise Encyclopedia of Genomic Diseases

Table 1.6. Continued.

Disease	Gene	Product of conception altered	Worldwide Frequency	Prevalent Group
Microcephaly, Amish type	SLC25A19	AR	<1 in 1,000,000	1 in 500 Amish-Christians
Polygamist Down's	FH	AR	<1 in 1,000,000	Inhabitants of Colorado City (Arizona) and Hildale (Utah)-Christians)
Familial hypercholesterolemia	LDLR	AD + AD = AR	1 in 1,000,000	Ashkenazi Jews, Afrikaners, French Canadians, Lebanese Christians and Finnish
Familial Mediterranean fever	MEFV	AR	1-9 in 1,000,000	1-4 in 1,000 in Armenian people, Sephardi Jews, Oriental Jews, Greeks, Turkish and Arabs
Krabbe disease	GALC	AR	1 in 100,000	15 in 100.000 Druze communities in Israel
Propionic acidemia	PCCB	AR	3 in 100,000	Saudi Arabians, Amish and Mennonites
Primary hyperoxaluria type 1	AGXT	AR	1-3 in 1,000,000	Tunisians and Mediterraneans
Maple syrup urine disease-1A or branched-chain ketoaciduria	BCKDHA	AR	6 in 1,000,000	1 in 380 Old Order Mennonites also high in Amish, Gypsies of Portugal and Ashkenazi Jews
Biotin-responsive basal ganglia disease	SLC19A3	AR	<1 in 1,000,000	Saudi Arabians, Syrians and Yemenis
Majeed syndrome	LPIN2	AR	<1 in 1,000,000	Middle Easterns
Dubin-Johnson syndrome	ABCC2	AR	<1 in 1,000,000	7 in 10.000 Iranian Jews
Creutzfeldt-Jakob disease	PRNP	AD + de novo	6-15 in 10,000,000	4 in 100.000 in Sephardic Jews

Religious

I chose to study the Ashkenazi and the Amish groups because there is valuable genomic information that describes them. This section focuses first on the Ashkenazi group that is an urban community with a tradition of genomic research awareness. Table 1.7 describes the diseases, the genes, the frequency of traits, alleles, and diseases, and the product of conception alteration regarding this group. There are common recessive traits that are less severe diseases for the Ashkenazis such as Nonclassical 21 OH deficiency, Hemophilia C, and Gaucher disease type 1. (Gaucher Type 1 can in very rare cases, be more severe) There are also common dominant diseases such as Breast and Ovarian cancer, and Familial hypercholesterolemia. Additionally, the table describes several highly symptomatic rare diseases that are inherited in an autosomal recessive pattern and could be prevented. In fact, the religious organization Dor Yeshorim advocates for anonymous genomic testing in the first 9 rare syndromes (**).

The Concise Encyclopedia of Genomic Diseases

These diseases include Tay–Sachs disease, Familial dysautonomia, Cystic fibrosis, Canavan disease, Glycogen Storage Disease type 1A, Fanconi anemia C, Bloom syndrome, Niemann–Pick type A, and Mucolipidosis IV. Dor Yeshorim also provides a genomic test for Gaucher disease only by request (*).

Table 1.7 Common and Rare Genomic Disorders in Ashkenazi Jews

Disease	Gene	Trait frequency	Allele frequency	Disease frequency	Product of Conception Alteration
Tay–Sachs disease	HEXA	1 in 27	1 in 54	1 in 3,000	AR**
Familial dysautonomia	IKBKAP	1 in 30	1 in 60	1 in 3,600	AR**
Cystic fibrosis	CFTR	1 in 25	1 in 50	1 in 2,500	AR + KDN**
Canavan disease	ASPA	1 in 60	1 in 120	1 in 14,400	AR**
Glycogen Storage Disease type 1A	G6PC	1 in 71	1 in 142	1 in 20,000	AR**
Fanconi anemia C	FANCC	1 in 90	1 in 180	1 in 32,400	AR**
Bloom syndrome	BLM	1 in 100	1 in 200	1 in 40,000	AR**
Niemann–Pick type A	SMPD1	1 in 90	1 in 180	1 in 32,000	AR**
Mucolipidosis IV	MCOLN1	1 in 110	1 in 220	1 in 48,000	AR**
Gaucher disease type 1	GBA	1 in 11	1 in 450	1 in 450	AR*
Breast cancer and ovarian cancer-2	BRCA2	1 in 75	1 in 150	1 in 75	AD + de novo
Breast cancer and ovarian cancer-1	BRCA1	1 in 100	1 in 200	1 in 100	AD + de novo
Congenital deafness	GJB2 or GJB6	1 in 25	1 in 50	1 in 2,500	AR and Digenetic
Familial hypercholesterolemia monoallelic	LDLR	1 in 69	1 in 138	1 in 69	AD + de novo
Familial hypercholesterolemia biallelic	LDLR	1 in 69	1 in 138	1 in 19,000	AD + AD = AR
Familial hyperinsulinemic hypoglycemia	ABCC8	1 in 52	1 in 104	1 in 11,000	AR
Usher syndrome I	PCDH15	1 in 72	1 in 144	1 in 21,000	AR
Nonclassical 21 OH deficiency	CYP21	1 in 6	1 in 12	1 in 144	AR
Hemophilia C	F11	1 in 12	1 in 24	1 in 576	AR

Among these disorders, the only one that seems to be ethnic specific is the Familial dysautonomia; for example, Tay–Sachs disease is also prevalent among Cajuns, French Canadians, and Irish Americans.

The Amish group is another example that allows the study of correlation between religion with genomics. This Christian group has a few pathogenic traits that are both highly prevalent and symptomatic. For example, the pathogenic allele for Microcephaly, Amish type is present in 1 every 11 individuals. Statistically, 1 Amish man (Allele frequency of 1/22) x 1 Amish woman (Allele frequency of 1/22), or $1/(22)^2$, corresponds to 1/484 persons who are at risk of the disease in the Amish group. In all the diseases described in Table 1.8, the product of conception is altered in autosomal recessive patterns. The table includes the biological functions of genes and displays the different ways in which these genes and proteins can be altered. Unlike other diseases that do not carry the name of any group of individuals, the term Amish appears in the original name of the disease to describe it (Amish infantile epilepsy syndrome, Microcephaly, Amish type, etc.).

Table 1.8 Diseases in Amish People, Genes, and Biological Function Altered

Disease	Gene	Biological Function
Amish infantile epilepsy syndrome	ST3GAL5	Golgi apparatus protein
Nemaline myopathy-5 or Amish nemaline myopathy	TNNT1	Microfilaments
Microcephaly, Amish type	SLC25A19	Solute carrier
Syndromic multisystem autoimmune disease	ITCH	E3 ubiquitin ligase
Familial hyperreninemic hypoaldosteronism 2	CYP11B2	Mitochondrial inner membrane
Amish Brittle Hair Brain syndrome or BIDS syndrome	MPLKIP	DNA repair
Familial hypercholanemia	BAAT	Enzyme-Acyltransferase
Pitt-Hopkins like syndrome -1	CNTNAP2	Cell adhesion molecule
McKusick-Kaufman syndrome	MKKS	Ciliary proteins
Ellis-van Creveld syndrome-1	EVC	Transmembrane
Sudden infant death with dysgenesis of the testes syndrome	TSPYL1	Nucleolus

Geography

There are diseases that are more prevalent in one geographical area than another. According to Table 1.9, there are many diseases that have been named with reference to a geographical place. The reference could be a city (e.g Bombay and Karak), a province (e.g Quebec, Avellino, and Newfoundland), an island (e.g Tangier, Cayman, and Åland), a river (Haw), or a country (e.g Finland, Norway, Britain, Denmark, and France). Also the chart shows the corresponding genes for the diseases, and different, specific geographical regions, which could be countries, islands, states, provinces, or continents, etc. Stating the place is important to avoid confusion about places in the world that have the same name. For example, Tangier disease exists in Virginia, not Morocco, as shown on the table.

The Concise Encyclopedia of Genomic Diseases

Table 1.9 Diseases, Genes, and Geography

Disease	Gene	Place
Megaloblastic anemia 1, Finnish type	CUBN	Finland
Megaloblastic anemia 1, Norwegian type	AMN	Norway
Quebec platelet disorder	PLAU	Canada
Ocular albinism 2 or Åland Island eye disease or Forsius-Eriksson syndrome	CACNA1F	Åland Island
Sialuria, Finnish type or Salla disease	SLC17A5	Finland
Sialuria, French type	GNE	France
Tangier disease	ABCA1	Virginia (USA)
Amyloidosis, Finnish type	GSN	Finland
Familial Mediterranean fever	MEFV	Mediterranean area
Dentatorubro-pallidoluysian atrophy or Haw River Syndrome	ATN1	North Carolina (USA)
Cerebellar ataxia, Cayman type	ATCAY	Cayman islands
Spastic ataxia, Charlevoix-Saguenay type	SACS	Canada
Karak syndrome	PLA2G6	Jordan
Familial British dementia	ITM2B	United kingdom
Familial Danish dementia	ITM2B	Denmark
Leigh syndrome, French-Canadian type	LRPPRC	Canada
Corneal dystrophy, Avellino type	TGFBI	Italy
Newfoundland rod-cone dystrophy	RLBP1	Canada
Bombay phenotype	FUT1	India
Hemochromatosis-4 or African iron overload	SLC40A1	African descent

0.6 Ecology

Based on Charles Lindbergh's concept regarding ecology in terms of being the custodian rather than the absolute owner, it is useful to adapt this ecologist viewpoint to the field of genomics. By this I mean that our genomic information should consider that each generation becomes the custodian rather than the absolute owner of our DNA material, and each generation has the responsibility to pass the best inheritance on to the future.

0.7 Miscellaneous Factors

Unlike the genomics information which describes how identical almost all human beings are to each other, there are key factors that differentiate people in terms of language, culture, and environment. This section considers several pathogenic DNA traits and attempts to avoid senseless use of physical traits that do not correlate with the diseases. According to Table 1.10, it is significant to note that the DNA's of any two random individuals in the world are 99.9% identical. Genomically speaking, a man and a woman are 2.5 % different at the DNA level. This corresponds to a 25 fold difference between any 2 most diverse males in the world. Instead, actual differences among groups arise from differences in history, beliefs, and cultures, rather than from genomics. Learning about genomic and environmental factors helps to approach diseases in new ways. In other words, the table below looks at diseases in ways that consider both scientific and non-scientific facts.

The Concise Encyclopedia of Genomic Diseases

Table 1.10 Miscellaneous Scientific Facts

	Number
Population of interest in this book	1 (*Homo sapiens*)
DNA Identity between distant related humans	99.90%
DNA Identity between close related males and females	97.50%
No. of Chromosomes	46
Countries	>200
Ethnic groups	*>1,000*
Race	*Social concept not scientific*
Religious groups	>1,000
Genes Producing Disease	2,300
Genomic disease	>4,000
Genes encoding proteins	>20,000
Y-chromosomal Adam	140.000 YA
Mitochondrial Eve	200,000 YA
Size of diploid genome	6.2 Billion bp*
Total human population	over 7 billion
Number of SNPs in (dbSNP)	54,000,000
Number of SNPs detected in the 1000 Genomes Project	15,000,000
Number of indels detected in the 1000 Genomes Project	1,000,000
Structural variants detected in the 1000 Genomes Project	20,000
Loss-of-function variants in each individual	250-300
Variants in each individual implicated in inherited disorders	50-100
CpG content in genome	0.80%
SNPs occurring in CpG sites	35%

*bp Base Pairs, Years Ago (YA), single nucleotide polymorphisms (SNPs) and short insertions and deletions (indels)

Chapter 1 Genomics: Past, Present and Future

Genomics is the art of understanding the 200,000 years of human history and the correlation between DNA alteration and diseases. The purpose of this chapter is to adapt the description of genomics from the past based on an exemplary contribution from Mendel. It also aims to better understand genomics in the present, to create strategies for better quality of life, and to avoid diseases in future generations.

1.1 Adapting Mendel Law of Inheritance into the Contemporary Genomic Field

This analysis provides a room to rethink Mendel's law of inheritance theory in order to adapt it to the contemporary genomic field, particularly to include de novo alterations in the DNA that explain de novo diseases. A reconsideration of the principle of Mendel theory does not alter the principle itself. Instead, it modifies the interpretation of Mendel's work by adding a new variable, de novo alteration in the DNA, which includes the product of conception alteration. This revision modernizes the classical contribution to science given by Gregor Mendel, and allows it to be adapted to post-genomic sequencing times. This research will open up expanded vistas of studies, which go beyond the limited domain of clinical and genomic information due to the separation of the two fields. Hopefully this chapter breaks new grounds regarding the relationship between the two fields reflecting a more comprehensive view.

1.1.1 Modernized Mendel's Inheritance Laws, or Post-genomic Revolution Laws of Product of Conception

Before the 1900s, there were many hypotheses about how a pregnant woman could control the sex of her baby. One of them was linked to what she eats: if she wanted a boy she would eat bananas, peaches, and red meat. If she wanted a girl she would eat fresh fish, vegetables, and sesame seed. Nowadays, we could say that there is a gene (SRY) on the chromosome Y that determines the sex of the baby. This is to say that a hypothesis has to be modernized to adapt to the current scientific knowledge.

While Mendel's inheritance law applies to monogenic traits in individuals, there is new genomic knowledge that should be imparted to include a larger set of genomic and non-genomic diseases. Approaching genomics in this way will allow us to understand the disease in terms of clinical, genomic, and biological view points. This will help to better understand diseases in the present and possibly prevent them from happening in the future. For example, using time and location of historical DNA alterations to identify a specific group of people will provide an insight into each person's DNA. We can then focus on pathogenic alleles to avoid diseases in next generations.

According to Table 1.11, the Mendelian law differs from the post genomic revolution hypotheses in that the first uses a genetic approach while the second uses a genomic methodology. This means, instead of scrutinizing the functioning of a single gene, the post genomic revolution hypotheses examine the whole pathway, including the interrelationship of clinical, genomic, and biological aspects of diseases in order to identify their combined effects in diseases.

Table 1.11 Comparison between Mendelian Theory and Post Genomic Revolution Hypotheses

	Mendelian	Post genomic revolution
Scientific Approach	Genetic	Genomic
Defined by	Pattern of inheritance	Product of conception alteration
Main subject to be studied	Plant phenotype	Human disorder or trait
de novo alteration	Not known in the mid-1800s	Known and included (e.g. mutation, imprinting, nondisjunction in the chromosome)

1.2 Classification of Product of Conception Alteration and Clinical Examples

It is virtually impossible to classify all the human genomic diseases under the same category since the common DNA alterations that define a disease are quite different at the clinical, genomic, and biological levels. The classification of all the diseases has redundancy, overlapping, and ambiguity at those three levels. Ambiguity at the genomic level for example, arises from multiple instances where the same gene is implicated in different diseases, or the same disease implicates different genes.

As shown in Table 1.12, there are several instances where the same gene (LMNA) could be involved in different diseases (e.g. Hutchinson-Gilford progeria, Dilated Cardiomyopathy, etc), an occurence defined as allelic disorder, or Polysyndromic Gene. By the same token, there are cases where the same disease is caused by several different genes (e.g. Brugada syndrome caused by SCN5A, GPD1L, etc). This is defined as Polygenomic Disease.

Table 1.12 Polysyndromic Genes and Polygenomic Diseases

Disease	Gene
LADD syndrome	FGFR3
Crouzon syndrome	FGFR3
Muenke syndrome	FGFR3
Thanatophoric dysplasia	FGFR3
Achondroplasia	FGFR3
Hypochondroplasia	FGFR3
Camptodactyly, tall stature, scoliosis, and hearing loss	FGFR3
Hutchinson-Gilford progeria	LMNA
Familial partial lipodystrophy	LMNA
Charcot-Marie-Tooth disease	LMNA
Emery-Dreifuss muscular dystrophy	LMNA
Limb-girdle Muscular dystrophy	LMNA
Dilated Cardiomyopathy	LMNA
Mandibuloacral dysplasia	LMNA
Agammaglobulinemia-1	IGHM
Agammaglobulinemia-2	IGLL1
Agammaglobulinemia-3	CD79A
Agammaglobulinemia-4	BLNK
Agammaglobulinemia-5	LRRC8A
Agammaglobulinemia-6	CD79B
Brugada syndrome-1	SCN5A
Brugada syndrome-2	GPD1L
Brugada syndrome-3	CACNA1C
Brugada syndrome-4	CACNB2
Brugada syndrome-5	SCN1B
Brugada syndrome-6	KCNE3
Brugada syndrome-7	SCN3B
Brugada syndrome-8	HCN4

1.2.1 Product of Conception Alteration

The product of conception alteration will be explained by the sum of patterns of inheritance (Mendel laws of inheritance) plus *de novo* alteration (the post genomic revolution hypotheses of product of conception). Table 1.13 displays the product of conception alteration, the disorders, and genes involved in the affected people. Additionally, it defines the product of conception alteration by means of integrating the whole pathway. This method of classification may also be clinically relevant. For example in autosomal dominant syndromes, the ascending percentage of *de novo* alteration in general is related to a more severe phenotype. Based on this classification, many disorders belong to more than one group such as Familial hypercholesterolemia and von Willebrand disease. There are at least 28 different categories in which the patterns of inheritance plus de novo alterations are important to know. Examples of diseases and product of conception alterations shall be summarized below.

1. Product of conception alteration explained by an autosomal recessive inheritance plus known de novo alterations.

These products of conception alterations are defined as occasional *de novo* autosomal recessive syndrome or autosomal recessive syndrome plus known de novo altera-tions (AR + KDN). Most of the syndromes described below have been extensively studied. Examples of (AR + KDN) are: Cystic fibrosis, Spinal muscular atrophy-1 or Werdnig-Hoffmann disease, Alpha-thalassemia, Beta-thalassemia, Phenylketonuria, Gaucher disease, Maple syrup urine disease-2 or branched-chain ketoaciduria, Fanconi anemia complementation group A and Shwachman–Bodian–Diamond syndrome.

2. Product of conception alteration explained by an autosomal recessive inheritance plus unknown de novo alterations. These products of conception alterations are defined as classic autosomal recessive syndrome (AR). Examples of (AR) are: Sickle cell anemia, Hemochromatosis and Tay-Sachs disease.

3. Product of conception alteration explained by an autosomal recessive inheritance plus anticipation. These products of conception alterations are defined as autosomal recessive syndrome plus anticipation (AR + Ant). Example of (AR + Ant) is: Friedreich ataxia.

4. Product of conception alteration explained by an autosomal recessive inheritance in known symptomatic autosomal dominant trait in both parents. These products of conception alterations are defined as autosomal recessive syndrome, where both parents possess known autosomal dominant traits in their alleles and both parents provide the faulty copy to descendants (AD + AD = AR). Examples of (AD + AD = AR) are: Familial hypercholesterolemia IIA (LDLR mutation), and von Willebrand disease type 3. While these examples are uncommon, they are clinically important since they display the most extreme phenotypes.

5. Product of conception alteration explained by an autosomal dominant inheritance plus a low frequency of de novo alteration. These products of conception alterations are defined as classic autosomal dominant syndrome (AD). Example of (AD) is: Factor V Leiden.

6. Product of conception alteration explained by an autosomal dominant inheritance plus 10% de novo alteration. These products of conception alterations are defined as autosomal dominant syndrome plus 10% (AD + 10%). Examples of (AD + 10%) are: Li-Fraumeni syndrome and Multiple endocrine neoplasia 1 (Wermer syndrome).

7. Product of conception alteration explained by an autosomal dominant inheritance plus 15-30% de novo alteration. These products of conception alterations are defined as autosomal dominant syndrome plus 25% (AD + 25%). Examples of (AD + 25%) are: Marfan syndrome and von Hippel-Lindau syndrome.

8. Product of conception alteration explained by an autosomal dominant inheritance plus 60-70% de novo alteration. These products of conception alterations are defined as autosomal dominant syndrome plus 65% (AD + 65%). Examples of (AD + 65%) are: Osteogenesis imperfecta type 1 and Tuberous sclerosis 1.

9. Product of conception alteration explained by an autosomal dominant inheritance plus 80-95% de novo alteration. These products of conception alterations are defined as autosomal dominant plus 90% (AD + 90%). Examples of (AD + 90%) are: Neurofibromatosis type 2, Beckwith-Wiedemann syndrome, Williams-Beuren syndrome, Chromosome 22q11.2 deletion syndrome (DiGeorge syndrome), and Dravet syndrome.

10. Product of conception alteration explained by an autosomal dominant inheritance plus over 95% *de novo* alterations. These products of conception altera-

tions are defined as autosomal dominant syndrome plus 99% (AD + 99%). Examples of (AD + 99%) are: CHARGE syndrome, Cornelia de Lange syndrome 1.

11. Product of conception alteration explained by an autosomal dominant inheritance plus anticipation. These products of conception alterations are defined as autosomal dominant syndrome plus anticipation (AD + Ant). Examples of (AD + Ant) are: Huntington Disease and Myotonic Dystrophy.

12. Product of conception alteration explained by an X-linked recessive inher-itance plus a low frequency of de novo alteration. These products of conception alterations are defined as classic X-linked recessive syndrome (XR). Example of (XR) is: Glucose-6-phosphate dehydrogenase deficiency.

13. Product of conception alteration explained by an X-linked recessive inher-itance plus 25-35% de novo alteration. These products of conception alterations are defined as X-linked recessive syndrome plus 30% (XR + 30%). Examples of (XR + 30%) are: Duchenne muscular dystrophy, Hemophilia A, and Hemophilia B.

14. Product of conception alteration explained by an X-linked dominant inher-itance plus anticipation. These products of conception alterations are defined as classic X-linked dominant syndrome plus anticipation (XD + Ant). Example of (XD + Ant) is: Fragile X syndrome.

15. Product of conception alteration explained by an X-linked dominant inheritance plus over 95% de novo alterations. These products of conception alterations are defined as classic X-linked dominant syndrome plus 99% (XD + 99%). Example of (XD + 99%) is: Rett syndrome.

16. Product of conception alteration explained by a Y-linked inheritance plus over 95% de novo alterations. These products of conception alterations are defined as classic Y-linked syndrome plus 99% (Y + 99%). Example of (Y + 99%) is: Swyer syndrome.

17. Product of conception alteration explained by a translocation (recombination) of part of the Y chromosome containing the SRY gene to the X chromosome. These products of conception alterations are defined as chromosome XY recombination (XY + Rec). Example of (XY + Rec) is: XX male syndrome (XY + Rec) or de la Chapelle syndrome.

18. Product of conception alteration explained by an autosomal recessive inheritance plus an environmental factor, which produce the disease in adulthood. These products of conception alterations are defined as autosomal recessive and environmental syndrome (AR + Env). Example of (AR + Env) is: Hemochromatosis-1.

19. Product of conception alteration explained by mitochondrial DNA alterations (Mit). These products of conception alterations are defined as Mitochondrial DNA alterations (Mit). Examples of (Mit) are: NARP syndrome, MELAS syndrome and MERRF syndrome.

20. Product of conception alteration explained by maternal DNA alterations (MAT). These products of conception al-terations are defined as maternal DNA alterations (MAT). Examples of (MAT) are: Angelman syndrome, Pseudo-hypo-parathyroidism-1, and Birk-Barel Mental retardation dysmorphism syndrome.

21. Product of conception alteration explained by paternal DNA alterations (PAT). These products of conception alterations are defined as paternal DNA alterations (PAT). Examples of (PAT) are: Prader-Willi syndrome and Myoclonic dystonia-11.

22. Product of conception alteration explained by monoallelic alterations

in the pseudoautosomal region on X chromosome. These products of conception alterations are defined as pseudoautosomal dominant on X chromosome (PADX). Examples of (PADX) are: Léri-Weill dyschondrosteosis and Madelung deformity.

23. Product of conception alteration explained by monoallelic alterations in the pseudoautosomal region on Y chromosome. These products of conception alterations are defined as pseudoautosomal dominant on Y chromosome (PADY). Examples of (PADY) are: Léri-Weill dyschondrosteosis and Madelung deformity.

24. Product of conception alteration explained by a pseudoautosomal recessive inheritance in known symptomatic pseu-doautosomal dominant traits in both par-ents. These products of conception altera-tions are defined as pseudoautosomal re-cessive syndrome, where both parents possess known pseudoautosomal dominant traits in their alleles and both parents provide the faulty copy to descendants. These product of conception alterations are defined as pseudoautosomal chromosome XY (PADX + PADY = PAR). Example of (PADX + PADY = PAR) is: Langer mesomelic dysplasia.

25. Product of conception alteration explained by an X-linked semidominant inheritance or incomplete dominance. These products of conception alterations are defined as X-linked sem-

idominant (XsD). Examples of (XsD) are: X-linked mental retardation-41/48, X-linked retinitis pigmentosa-3, X-linked Lissencephaly-1 and Nance-Horan syndrome.

26. Product of conception alteration explained by concomitant monoallelic al-teration in 2 separate genes. These products of conception alterations are defined as digenic or digenomic monoallelic (DigM). Examples of (DigM) are: Alpha-thalassemia, Hereditary coproporphyria and Waardenburg syndrome/ocular albinism.

27. Product of conception alteration explained by concomitant biallelic altera-tion in 2 separate genes. These products of conception alterations are defined as digenic or digenomic biallelic (DigB). Examples of (DigB) are: Alpha-thalassemia, Pendred syndrome, Cystinuria and Rotor syndrome.

28. Product of conception alteration explained by 2 copies of a chromosome, or part of a chromosome, from 1 parent, and no copies from the other parent. In this case, the product of conception is altered in both parents either by missing 1 chromosome or having an extra copy. These products of conception alterations are defined as uniparental disomy (UniPD). Examples of (UniPD) are: Cystic Fibrosis, Silver-Russell syndrome, Beck-with-Wiedemann syndrome, Prader-Willi syndrome and Angelman syndrome.

Table 1.13 Classification of Product of Conception Alteration and Clinical Examples.

Product of Conception Alteration	Disorder	People self-reporting	Gene alteration
AR + KDN	Cystic fibrosis	Caucasians	CFTR
AR	Sickle cell anemia	Bantu Peoples	HBB
AR + Ant	Friedreich ataxia	French-Canadians	FXN
AD + AD = AR	Familial hypercholesterolemia IIA	Afrikaners, Ashkenazi Jews and Lebanese Christians	LDLR
AD	Factor V Leiden	Caucasians	F5
AD + 10%	Li-Fraumeni syndrome	Worldwide	TP53
AD + 25%	Marfan syndrome	Worldwide	FBN1
AD + 65%	Tuberous sclerosis-1	Worldwide	TSC1
AD + 90%	DiGeorge syndrome	Worldwide	TBX1
AD + 99%	Cornelia de Lange syndrome-1	Worldwide	NIPBL
AD + Ant	Huntington Disease	Caucasians	HTT
XR	Glucose-6-phosphate dehydrogenase deficiency	Caucasians*	G6PD
XR + 30%	Hemophilia A	Worldwide	F8
XD + Ant	Fragile X syndrome	Worldwide	FMR1
XD + 99%	Rett syndrome	Worldwide	MECP2
Y + 99%	Swyer syndrome	Worldwide	SRY
XY + Rec	De la Chapelle syndrome	Worldwide	SRY
AR + Env	Hemochromatosis-1	Caucasians	HFE
Mit	NARP syndrome	Worldwide	MT-ATP6
MAT	Angelman syndrome	Worldwide	UBE3A
PAT	Prader-Willi syndrome	Worldwide	NDN
PADX	Léri-Weill dyschondrosteosis	Worldwide	SHOX
PADY	Léri-Weill dyschondrosteosis	Worldwide	SHOXY
PADX + PADY = PAR	Langer mesomelic dysplasia	Worldwide	SHOX/SHOXY
XsD	Nance-Horan syndrome	Worldwide	NHS
DigM	Alpha-thalassemia	African-Americans	HBA1/HBA2
DigB	Alpha-thalassemia	Southeastern Asians	HBA1/HBA2
UniPD	Cystic fibrosis		CFTR

(*) Caucasians in USA or Europe

1.2.2 Parent of Origin of the *de novo* Alteration

The main reason for studying the product of conception alterations is the fact that parents, due to alteration in oogenesis or spermatogenesis, are the main sources of *de novo* alteration in DNA.

These changes in DNA could produce diseases. Table 1.14 shows how the parents of origin are the sources of these diseases or the main alleles that alter most of the product of conception. The chart also displays the gene name when available. There are *de novo* diseases that are traced mainly in the maternal alleles or chromosomes, including Albright hereditary osteodystrophy, Spinocerebellar ataxia 8, Fragile X syndrome, Beckwith-Wiedemann syndrome, Neurofibromatosis type 2, Hyperphenylalaninemia, Angelman syndrome, Wilms tumor, Down syndrome, Long QT syndrome, IMAGE syndrome, Orofacial cleft-2, and Kearns-Sayre syndrome. For example, Down syndrome is a *de novo* disease in which 88 % of the syndromes are traced back to maternal alleles due to alteration in oogenesis during meiosis I. There are *de novo* diseases that are sourced mainly in the paternal alleles or chromosomes, including Léri-Weill dyschondrosteosis, Spinal and bulbar muscular atrophy of Kennedy, Dentatorubro-pallidoluysian atrophy, Spinocerebellar ataxia 10, Machado-Joseph disease, Huntington disease, Cerebral palsy, Psoriatic arthritis, Pheochromocytoma, Charcot-Marie-Tooth disease type 1A,

Retinoblastoma, Myoclonic Dystonia, Spinocerebellar ataxia 17, Multiple endocrine neoplasia 2B, Diabetes mellitus, insulin-dependent-1, Basal cell nevus syndrome, and Myotonic dystrophy 1. For instance, Huntington disease can be inherited from the mother or the father, but the main phenomenon of anticipation in this case is due to CAG expansion during spermatogenesis (paternal allele). There are also *de novo* diseases that are sourced in either parent. Moreover, from a genomic point of view there are alterations in genes that can produce different diseases that can be traced back to either parent. For example, in the RET gene mutations, the paternal allele produces Multiple endocrine neoplasia 2B and the maternal allele mainly produces Hirschsprung disease. For further information on this subject see chapter 4 on genomics.

Table 1.14 Diseases Related to Genes, and Parents of Origin

Disease	Gene	Product of conception affected by allele from
Albright hereditary osteo-dystrophy	*GNAS*	Maternal
Spinocerebellar ataxia-8	*ATXN8OS*	Maternal
Fragile X syndrome	*FMR1*	Maternal
Beckwith-Wiedemann syndrome	*H19*	Maternal
Neurofibromatosis type 2	*NF2*	Maternal
Hyperphenylalaninemia	*PAH*	Maternal
Angelman syndrome	*UBE3A*	Maternal
Wilms tumor	*WT1*	Maternal
Down syndrome		Maternal
Long QT syndrome	*KCNQ1*	Maternal
IMAGE syndrome	*CDKN1C*	Maternal
Orofacial cleft-2	*TGFA*	Maternal

Table 1.14. Continued.

Disease	Gene	Product of conception affected by allele from
Spinocerebellar ataxia 2	ATXN2	Both
Neurofibromatosis type 1	NF1	Both
Thrombocytopenia-absent radius syndrome	RBM8A	Both
Amyotrophic lateral sclerosis	SOD1	Both
Prader-Willi syndrome	NDN	Both
Hirschsprung disease	RET	Mostly maternal
Spinocerebellar ataxia 1	ATXN1	Mostly paternal
Spinocerebellar ataxia 7	ATXN7	Mostly paternal
Attention deficit-hyperactivity disorder		Mostly paternal
Léri-Weill dyschondrosteosis	SHOXY	Paternal
Spinal and bulbar muscular atrophy of Kennedy	AR	Paternal
Dentatorubro-pallidoluysian atrophy	ATN1	Paternal
Spinocerebellar ataxia 10	ATXN10	Paternal
Machado-Joseph disease	ATXN3	Paternal
Huntington disease	HTT	Paternal
Cerebral palsy, spastic quadriplegic 2	KANK1	Paternal
Psoriatic arthritis	LTA	Paternal
Pheochromocytoma	MAX	Paternal
Charcot-Marie-Tooth disease type 1A	PMP22	Paternal
Retinoblastoma	RB1	Paternal
Myoclonic Dystonia	SGCE	Paternal
Spinocerebellar ataxia 17	TBP	Paternal
Multiple endocrine neoplasia 2B	RET	Paternal
Diabetes mellitus, insulin-dependent-1		Paternal
Basal cell nevus syndrome	PTCH1	Paternal
Myotonic dystrophy-1	DMPK	Paternal in alleles of 40 to 80 repeats

1.3 Prevention of Genomic Diseases

The present section of the book focuses on the strategy for the prevention of genomic diseases. However, it is important to first describe the clinical severity and the variability of these diseases. The clinical severity of a disease is orientative in that it can be explained in terms of when the speculated symptoms of prognosis can be diagnosed after a product of conception is formed. The clinical severity does not necessarily lead to death, but it can either produce changes in quality of life, or lead to reduction in the lifespan. Table 1.15 takes genes, and product of conception alterations into account. The chart describes mainly intracellular diseases except for Alzheimer disease due to APOE gene alteration, since the extracellular diseases are more likely to be successfully treated. The chart catalogues diseases in a decreasing order of clinical severity. For example, the most severe are Chromosome 16 Trisomy, Tay-Sachs disease, and I-cell disease. The least severe are Sickle cell anemia, Triple X Chromosome, and Alzheimer disease. The chart also includes Duchenne and Becker muscular dystrophy in which the same gene (DMD) is altered, but the first one is more symptomatic and occurs at an early age.

Table 1.15 Clinical Severities of Genomics Diseases, Genes, and Product of Conception Alteration

	Disease	Quality of life or Lifespan	Gene	Product of Conception Alteration
1	Chromosome 16 Trisomy	Lethal during pregnancy		*de novo*
2	Tay-Sachs disease	Lethal before 5 years of age	*HEXA*	AR
3	I-cell disease	Lethal between 5 and 10 years of age	*GNPTAB*	AR
4	Familial hypercholesterolemia	Lethal between 10 and 30 years of age	*LDLR*	AD + AD = AR
5	Duchenne muscular dystrophy	Lethal between 20 and 30 years of age	*DMD*	XR + 30%
6	Cystic Fibrosis	Lethal between 30 and 50 years of age	*CFTR*	AR + KDN
7	Becker muscular dystrophy	Lethal between 40 and 50 years of age	*DMD*	XR + 30%
8	Down syndrome	Lethal between 45 and 55 years of age		*de novo*
9	Sickle cell anemia	Lethal between 45 and 55 years of age	*HBB*	AR
10	Triple X Chromosome	Undistinguishable		*de novo*
11	Alzheimer disease	Quality of life reduced	*APOE*	AD and Polygenomic

Besides the clinical severity, it is important to include the clinical variability concerning the starting signs and symptoms in particular disease. Table 1.16 also includes the genes and the product of conception alterations. This chart shows that the clinical variability could be large. For instance, Gaucher is a disease that takes a mild form of the disease in most patients, but with a huge variability regarding the prognosis. In other diseases such as in Huntington the starting symptoms could came to a variable ages (5-80). In some other diseases, such as von Willebrand could be explained by the product of conception alteration (AD + de novo or AD + AD = AR). In disease, the ABO blood group can alter the phenotype. Variability can also be explained in terms of the contribution of the environment and the sex of the person as can be exemplified in Hemochromatosis.

Table 1.16 Variability of Clinical Presentation, Genes, and Product of Conception Alteration

Disease	Variability in sign or symptoms	Gene	Product of Conception Alteration
Chromosome 16 Trisomy	Almost always lethal during pregnancy except in mosaicism		*de novo*
Familial hyper-cholesterole-mia	LDL-cholesterol greater than 500 mg per dl	*LDLR*	AD +AD = AR
Familial hyper-cholesterole-mia	LDL-cholesterol 250-450 mg per dl	*LDLR*	AD + de novo
Huntington disease	Symptoms starting at 5 to 80 years of age	*HTT*	AD + Ant
Gaucher disease type 1	Severe form patient has an enlarged liver and spleen with easily bruising. Most of the patients have a mild form of the disease	*GBA*	AR + KDN
Von Willebrand disease	As in Multimeric syndrome there are several factors such as more severity in 0 blood type	*VWF*	AD + AD = AR
Von Willebrand disease	Asymptomatic or mild symptoms	*VWF*	AD + de novo
Hemochroma-tosis	Depending on sex and environment	*HFE*	AR + Env

There are many different ways to prevent genomic diseases including first mate selection, which is a screening program in which any couples with any genomic background can avoid homozygosis in lethal allele for any of the over 3,000 autosomal recessive inherited diseases. The number of genes may be more or less depending on whether a gene causes multiple diseases or a disease is caused by multiple genes. Of course this could prevent inherited diseases, but not *de novo* diseases. Mate selection at the current time is generally the only method that is of no ethical dilemma since the basic principle is conscious reprogenomics. Many ethical committees or councils of health organization often approve mating selection. Table 1.17 includes the diseases, the genes, and the countries that have implemented a mandatory premarital screening. However, the chart indicates that the Orthodox Jewish organization takes some measures to decrease nine autosomal recessive diseases

(* The table below mentions only one disease, but the other eight diseases are in table 1.7).

The second way consists in avoiding overcompensation of new children mainly in cases of homozygosis in lethal allele, such as Tay-Sachs disease. Preimplantation genomic diagnosis is a third technique that makes it possible to test the product of conception for the inherited diseases plus the *de novo* alterations prior to an embryo implantation. This method requires a retrieval of the mother's oocytes for in-vitro fertilization. While this technique can prevent inherited and *de novo* diseases, it might present an ethical dilemma. People have diverging views regarding the discarding of unhealthy embryos. However, people disagree less on healthy embryos that are selected and transferred into the mother's uterus. The fourth way to prevent genomic diseases is the prenatal diagnosis such as Chorionic villus sampling and Amniocentesis, which are tests done to decide for a possible   abortion.

Table 1.17 Premarital Screening of Diseases and Genes around the World

Country or Organization	Disease	Gene
Cyprus	Beta-thalassemia	*HBB*
Iran	Beta-thalassemia	*HBB*
Orthodox Jewish	Tay-Sachs*	*HEXA*
Bahrain	Sickle cell anemia	*HBB*
Bahrain	Beta-thalassemia	*HBB*
Saudi Arabia	Sickle cell anemia	*HBB*
Saudi Arabia	Beta-thalassemia	*HBB*

1.3.1 Background Information on Genomics: Inherited and *de novo* Sources of Diseases

To prevent human genomic diseases, it is crucial to acquire basic knowledge in this field. A great deal of this information was accomplished by virtue of the invaluable contribution of the Human Genome Project. As a result of this progress in genomics and medicine, conscious reprogenomic programs can be established and help to decrease autosomal recessive diseases, such as Cystic fibrosis and Sickle cell anemia. There are many causes of genomic diseases, which derive from three major sources. These are endogamy, founder effect, and advance parental age.

Firstly, endogamy is about reproducing a product of conception within a specific ethno-religious group. Secondly, the founder effect occurs many generations ago when a small number of individuals separate from a large group of people. The small new group is characterized by a low genomic variation and increased inbreeding rate. Typical examples of the founder effect are the people, who live in islands such as Faroe Island, Iceland, and The Åland Islands. In other words, the founder effect is the product of a limited gene pool. Thirdly, the advanced parental age effect refers to the statistical relationships between the parents' age and abnormalities in the gametogenesis; consequently, an increase in the prevalence of *de novo* diseases occurs.

Besides acquiring knowledge it is important to decrease the endogamy based on reducing the inbreeding rate in village or communities and to increase our own knowledge of genomic background, particularly, in relation to known diseases or traits. It is also essential to revise our own perception that physical characteristics are not necessarily of genomic importance to understand diseases. Every individual has between 5 and 10 hypothetically deadly polymorphisms or mutations in every diploid genome, but in general these traits do not manifest as diseases since most of them require the two alleles (maternal and paternal) to be altered. There is a strong likelihood of passing on a genomic disease when breeding occurs between closely related people.

Conscious reprogenomic programs have been targets of criticism due to certain atrocities in the 20th century such as compulsory sterilization, enforced abortions, enforced pregnancies, marriage restrictions, segregation, and genocide. Unlike the 20th century approach, according to which marriage restrictions were based on race, this new approach of conscious reprogenomics encourages disassortative mating based on known disease traits only. It also eradicates the occurrence of homozygosis in lethal alleles. This approach will cost one generation for each discouraged trait to be reduced to half. The Human Genome Project (1986 to 2003) was the preparatory genomic information to be used as a tool to prevent disease. Nowadays, by using a refined roadmap of the DNA alteration we can prevent diseases and traits. While this book displays that the frequency of a disease is 1 in 2,500 individuals or 1 disease in 40,000 individuals, the expected probability is as Mendel predicted 1 disease every 4 individuals if both parents are heterozygous for a specific trait.

1.3.2 Reproduction

One of every ten marriages worldwide is between first or second cousins. These are predominantly common in the Middle East, where they account for over half of all marriages in some nations. Of course, these recessive syndromes are more common in this geographical area. On the other hand, it is harder to decrease the prevalence of polygenetic or complex diseases, since the identification of genomic risk factors is still limited, and does not follow a clear-cut pattern of inheritance.

Table 1.18 consists of 4 examples of diseases regarding endogamy, founder effect, and advance parental age. For instance, Cystic fibrosis is present mainly due to founder effect and endogamy. The whole community where this trait prevails has to be educated to understand the basic reprogenomic factors in order to prevent this disease. Each individual in this world has at least 5 to 10 lethal or pathogenic genes as heterozygous traits. If in the next generation of people any of these traits appear in homozygosity, recessive diseases prevail. On the contrary, if the entire community where these traits are present

is not educated, it will then not understand how to avoid homozygosity in pathogenic alleles or recessive diseases. Moreover, the founder effect can be seen in Huntington disease, which is highly prevalent in the lake region of Maracaibo, Venezuela. Currently, there are 18,000 patients in that area, all of whom are descendants from the same gene pool. There are some diseases, including Huntington disease and Multiple Endocrine Neoplasia type 1 in which the late onset of symptoms does not usually affect reproduction. On the other hand, Multiple endocrine neoplasia 2B is both an early onset of symptoms and a *de novo* syndrome caused by advanced parental age. The prevention of this disease depends on decreasing the average age of both parents at conception. The chart also illustrates the allele frequency of these 4 above diseases. For example, the Cystic fibrosis pathogenic allele is at least 1,000 fold higher than the others.

1.3.3 Avoidance of the Advanced Parental Age

Avoiding the advanced parental age can be considered as one of the three most important ways to prevent genomic diseases. In most modern societies, no measures have been taken to avoid late parenthood. For example, in the United States, the generations are getting longer at least in the last 30 years, in which the average age first pregnancy has been delayed for 4 years, but the data concerning this issue is limited. The data regarding parental age should be broader in the sense that it should not only focus on the first pregnancy but more on the age of the last pregnancy. Data regarding the male parent is equally or perhaps more important than data on the mother since males often reproduce at an older age. Unfortunately data records of the age of male parents are limited. This study suggests that data on males' age at the time of reproduction should be taken into account in *de novo* genomic diseases studies.

Table 1.18 Alterations in the Product of Conception, Differences, and Similarities in Caucasian Americans

Product of Conception Alteration	AR + KDN	AD + Ant	AD + 10%	AD + 50%
Examples	Cystic fibrosis	Huntington Disease	Multiple endocrine neoplasia 1	Multiple endocrine neoplasia 2B
Endogamy	Yes	No	No	No
Founder effect	Yes	Yes	No	No
Parental advancing age	Unknown	Yes	Unknown	Yes
Number of allele in 100,000 people	5,000	4-10	3	1
Number of disease in 100,000 people	40	4-10	3	1
Symptomatic at age	Newborn	40 years	40 years	Newborn

1.3.4 Reprogenomics in the Sub-Saharan Africa

Theoretically speaking, if the power of reprogenomics is used in the Sub-Saharan Africa, many endemic diseases, including Malaria, alpha-Thalassemia, Sickle cell anemia, and Glucose-6-phosphate dehydrogenase deficiency will be prevented in future generations. Note that by using conscious reprogenomics programs this book hopes to prevent widespread diseases. When these programs are applied for sickle cell traits, Malaria and Sickle Cell Disease will be targeted at the same time; subsequently both of them could be eradicated. Indeed, the reproduction of more than 400,000 newborn infants with Sickle cell anemia can be avoided yearly if reprogenomics programs are put into practice. Furthermore, since these programs will stimulate to have a higher prevalence of Sickle cell traits, people in this area will be more resistant to Malaria. This goal can be achieved if education is promoted and disassortative mating is encouraged.

Chapter 2 Strategies to Understand the Pillars of the Genomic Origin in Diseases

The second chapter examines the strategies to understand whether a given disease has a genomic origin based on various factors. A selection of these factors is based on clinical, genomic, and biological information. This section examines all the diseases that are of genomic origin unless there is clear evidence for being related to environmental factors. A discussion of the origin of genomic diseases raises many questions that will be summarized in 8 enquiries. Some of these questions concern the criteria of inclusion and exclusion in order to discern whether a disease is genomic or not. These assessments depend on various factors that should be used to identify the clinical, genomic, and biological knowledge of each disease. I take these 8 factors as the pillars that enable the researcher to carry out genomic diagnosis. In other words, these pillars have the ability to define if the characteristics of a disease can be traced to genomic origin.

1) Correlation between Alteration of DNA Information and Disease
2) Set of People where Variation in DNA Correlates with a Disease
3) Being Able to Dissect the Mechanisms of Disease
4) Being Able to Classify the Diseases in Groups
5) Being Able to Identify the Alteration in the Product of Conception
6) Being Able to Establish the Origin of the Disease
7) Making use of Exclusion Criteria
8) Making use of Inclusion Criteria

2.1 Correlation between Alteration of DNA Information and Disease

The relationship between genomic alterations and diseases needs close examination. To account for the complexity of this relationship, there are diseases that are genomic, without having alterations in DNA. At the same time, not all the alterations of the DNA produce a disease. This is mainly due to the fact that there are 50 million genomic alterations and just a few thousand genomic diseases. Moreover, 98.5 % of the genome does not encode for protein, and has repetitive sequence; for example the genome contains Line 17 % and Alu 11% repetitive sequences. Furthermore, each of us had millions of genomic alterations (SNP, mutation, indels); most of them do not matter, since they do not contribute to diseases or response to drugs. At the current times, however, we are not sure about what all the pathogenic DNA alterations are due to shortage of studies in the genomic field.

This study aims to describe the different aspects of genomic diseases by considering first whether they belong to the germline, somatic, or germline-somatic DNA category.

An emphasis will be put on the germline diseases since they are inherited in the offspring, including the inherited and *de novo* aspects. The somatic or acquired mutation disease will occupy less importance in this section of the book, because they are not inherited in the offspring. Instead they are acquired, regardless of linking one generation to the other. For this reason I consider somatic or acquired genomic disease unable to provide the tools to better understand the origin of genomic disorder, a most pertinent example being cancer syndrome. This argument will be discussed in depth on the basis of genomic disease characteristics.

What make an examination of germline diseases more useful to this study is that they are inherited from your parents or are a product of a *de novo* event, which is crucial to understanding genomics disorder. A consideration of the fact that there are 4,000 genomic diseases, whose origins are mainly germline and have been subject of research, provides the tools to conduct this study.

None the less genomic diseases are not always regarded as germline or somatic. These 2 aspects could overlap, producing a third category, germline-somatic. For example a typical germline disease is Huntington, and in some nucleus of the brain you are going to have mitochondria with somatic changes.

2.2 Set of People where Variation in DNA Correlates with a Disease

The relationship between variations in DNA and diseases is covered mainly in the perspective where I consider multiple factors, some of which are geographical, historical, statistical, ethnic, national, and religious. Each of us has many genomic differences that make us unique. But in this book we are going to focus in DNA changes that correlate with diseases; in other words we emphasize the pathogenic DNA changes.

2.3 Being Able to Dissect the Mechanisms of Disease

In every single genomic disease we need to understand the mechanism of the disease. To achieve this objective we use two methods, the genomic and the biological. These are describe in chapter four and five, respectively. The mechanism has to be classified depending on the number of genes involved. For example cystic fibrosis is caused by only one gene, hemochromatosis is caused by more than one gene, and cancer or hypertension is caused by polygenetic factors. Genomic diseases also have to be studied according to the main macromolecule that is affected: the DNA, the RNA, or the protein. The disease could also be caused by a combination of all the macromolecules mentioned above. Nondisjunction of the DNA is the case of Down syndrome where there is an extra chromosome; surely the RNA and protein would also be affected. Other diseases's mechanisms include disturbing the RNA. This could be caused by different mechanisms such as changing the gene-expression, splicing, editing and many more tools. A similar mechanism also affects the functionality of the protein. As shown in table 2.1 the mechanisms of diseases correlate with their clinical significance, as is the case with Cystic Fibrosis. This mechanism involves mainly a dysfunction in the CFTR protein, but also could be due to promoter alteration and alternative splicing. The same chart includes the different dbSNP rs#, the minor allele frequency, the codon change, and the DNA alteration. Other mechanisms of disease are: Wilson disease in Sardinian people is due to an alteration in the promoter of ATP7B, and reduces gene expression. Patient with increased risk of venous thrombus has a genomic variation in the 3-prime untranslated region of the prothrombin gene (F2). This is a gain-of-function mutation, causing increased cleavage site recognition, increased 3-prime end processing, and increased mRNA accumulation and protein synthesis. In most of the diseases we could have many different DNA alterations but at the end, the altered allele could be catalogue as pathogenic or not.

Table 2.1 CFTR Clinical Significance - Loss of Function Disease

dbSNP rs#	Minor Allele Frequency	Clinical Significance	Codon Change	DNA Alteration
rs1800073	0.0022	Unknown	Arg31Cys	Missense
rs1800076	0.023	Probable-non-pathogenic	Arg75Gln	Missense
rs75961395		Pathogenic	Gly85Glu	Missense
rs121908803		Untested	Pro205Ser	Missense
rs213950	0.4539	Non-pathogenic	Val470Met	Missense
rs121909001	0.03	Pathogenic	Phe508Del	Deletion
rs1800094	0.0259	Non-pathogenic	Glu527Glu	Synonymous
rs77010898		Pathogenic	Trp1282Ter	Nonsense

2.4 Being Able to Classify the Diseases in Groups

There are three groups, the clinical, the genomics, and the biological point of view. This will be covered in the third, fourth, and fifth chapters respectively. To better understand the genomics discipline, the book has created groups of diseases based on different categories. For the clinical point of view, I use information based on similar data as the ICD-10 code to describe 2,000 – 4,000 genomics diseases, depending on how you do the counting. For the genomic point of view, I consider the time of origin, the structure of the gene and/or genes, structure of the chromosome, function of the gene, product of conception alteration, protein sequence alteration, genomic silencing, and multimeric protein dysfunction. For the Biological point of view, I discuss how the affected organelles correlate with the genomics and clinical syndrome. In this chapter I will discuss the protein that has a key function in the different organelles, including the nucleus, the biological membranes, the cytoskeleton, the extracellular matrix, other organelles such as endoplasmic reticulum, golgi, mitochondrial, vesicles, lysosomes, and peroxisomes, as well as other miscellaneous such as oncogene, proto-oncogene, tumor suppressor, ciliary protein, ribosomal, enzymes, and cell adhesion.

2.5 Being Able to Identify the Product of Conception Alteration (pattern of inheritance and *de novo* alterations).

Genomic alteration disease has to be defined as a disorder that can be detected as soon as an ovule is fertilized. In other words, as soon as a product of conception is formed, an immediate genomic diagnosis can theoretically be made even in the absence of a clinical syndrome. At this point, the product of conception has the confluence of the two major sources of disease, the inherited and *de novo* diseases. To be specific, inherited genomic disorders are passed down from the parents' genes. Even when *de novo* is a genomic alteration that neither parent possesses, they may sometimes transmit this DNA alteration. This said, genomic diseases may emerge at different times of a patient's lifespan. All the genomic DNA alteration that is present as a product of conception has the particularity of being inherited in the following generation. The product of conception is thus where the diagnosis should be made. That's where the money is. Waiting until after birth, childhood or adulthood is just to wait for the inevitable. Of course, in preventative individualized medicine, we will try to avoid the inevitable, and of course this is just applicable to many genomic diseases. Genomic alteration disease is a model to predict, and combined information of alterations in DNA can be used to diagnose a disease before the appearance of symptoms.

In genomics, we always have room for improvement: to learn new diseases and to better characterize the diseases in different groups by analyzing different patterns of history, beliefs, and cultures, even with 99.9 % identity between individuals at the DNA level.

This book is made to utilize the knowledge in genomics, to make you think, and come up with new ideas, to impart the knowledge of genomics and disorders. Genomic disease is a broad term where almost all the disease should be included; there are some diseases that have other factors as psychological or environmental causes. While there are many diseases that originate from non-genomic factors such as psychological and environmental influences, the present study examines how almost all types of diseases are genomic-based.

2.5.1 Background information in pregnancy

The causes of losing a pregnancy are many, and include genomic alteration (inherited plus *de novo*), uterine or hormonal abnormalities, reproductive tract infections. Genomic alterations are extremely common, and occur in around one in every 10 pregnancies. There are some cases where pregnancy cannot survive genomic alterations.

Chromosome 16 trisomy and Turner syndrome are the two most common (3-10% of pregnancies; in Turner syndrome, only 1% survive the pregnancy, making it 2-5 cases/10,000). This also explains the higher sex ratio of males to females (1.05 in USA) at birth.

Pregnancies with fetuses that have strong genomic alterations do not even survive more than a trimester (Strong genomic alterations do not correlate well with genomic DNA size alterations). Besides the cases discussed earlier, Homozygous Achondroplasia (1 nucleotides altered in each parent), or Rett syndrome in males (1 nucleotide altered) are also relevant diseases to strong genomic alterations.

Diseases like Triple X syndrome, Klinefelter syndrome, XYY syndrome are less common (0.1-0.2% of pregnancies). They usually have no distinguishable clinical difference, except the presence of karyotype abnormalities, which make it less likely that the individual can reproduce.

2.6 Being Able to Establish the Origin of the Disease

While most of this information was covered earlier in the Genomic Disease Perspectives section, I would like to remind the reader the importance of the genomic alteration origin. We need to have a sense of when this DNA changes happened. They could have occured hundreds, or even thousands of years or generations ago. But it is important to know them, since they explain today's genomic diseases. These scientific items have their own limitations, since the information is scattered and limited to only a few variants. (See table in historical perspective)

2.7-8 Exclusion and Inclusion criteria

These are the standard criteria that this book proposes to hypothetically determine whether a disease should be considered of environmental or genomic origin. The use of exclusion and inclusion criteria is to either: exclude a disease as having a genomic origin (consequently, the origin would be related to environmental factors) or to include the disease as having a genomic origin. The exclusion and inclusion criteria, is not meant to discourage the study of diseases. In fact, the main goal is completely opposite; the purpose involves the creation of a new subgroup to better fit the genomic or environmental origins. Many times, the decision regarding each particular disease is inconclusive, in which we have many inclusion and exclusion criteria altogether, without a clear definition in the genomic or environmental origin of a particular disease. Inclusion and exclusion criteria should be studied together. For example, if we have a disease that has a high correlation between twins, this is an inclusion criterion, whereas on the other hand if the correlation is low, it is an exclusion criterion. In other words, there are two main source of disease: the environment and the genes rooted in our genome. For example, there are small communities such as the Amish that are at an increased risk for a number of monogenomic traits and disorders. But interestingly enough, these communities have a lower frequency of polygenetic and environmental diseases (from cancers to even suicide) due to healthier life styles (there is very little tobacco or alcohol use and limited sexual partners). The healthy life style creates an almost perfect control group to better understand environmental and genomic diseases. As Albert Einstein puts it: "Many of the things you can count, don't count. Many of the things you can't count, really count." There are occasions when there are twin correlations, such as in type 2 Diabetes Mellitus, which has almost perfect correlation. However, the main causes of type 2 Diabetes Mellitus are related to nongenomic factors, such as life style, and western society diet. It is hard to study the criteria of inclusion and exclusion since there are two possibilities to understand the accurate factors that produce diseases. When these possibilities overlap, they can confound the genomic and environmental factors.

2.7 Making use of Exclusion Criteria

There are many factors that exclude certain diseases from being of Genomic origin. These include environmental factors, psychological factors, and differences in the prevalence of diseases which affect males and females, and they cannot be explained by genomics and developmental factors. There are also risk factors that are not of genomic origin. Certain factors cannot contribute to a disease being considered as one of genomic origin. For example, diseases due to urbanized life or industrialization, or diseases whose incidence of this disorder has been significantly decreased or increased in historical comparison.

Environmental or psychological factors or any other factor that better explain the disease will be considered in the clinical chapter. There are large differences between males and females in genomic diseases, most of them explained by the presence of a Y or X chromosome, respectively. There are several examples of differences in activation and inactivation of genes, in a sex specific pattern. For example, the male phenotype is determined by the SRY gen in the Y-Chromosome, but then it requires the action of multiple genes to fully determine the sex. Most of these genes are located in autosomal chromosome. The frequency of every autosomal disease should be almost identical between males and females, to be of an accurate genomic origin. If the frequency of a disease is not similar, it should be considered as an exclusion criterion. There are diseases linked to chromosome X and Y, also there are linked to autosomal chromosome in which activation and inactivation of genes, in a sex specific pattern are the exceptions.

2.8 Making use of Inclusion Criteria

There are many factors that include certain disease as being of Genomic origin. These include twin study diseases, ethnic diseases, familial aggregation diseases etc. Certain factors do contribute to a disease being considered as one of genomic origin. (e.g. family history of genomic diseases, parental consanguinity, and first-degree relatives)

2.8.1 Twin studies

Most of the twin studies were popular a few years ago, when the genetic information was not available. But it is always a good idea to dust off these books, and think about the genomic contributions in this condition.

Most of this information is from a small set of samples, since these natural occurring experimental conditions are not common (Twin plus condition). Table 2.2 shows the concordance and discordance of diseases in monozygotic twins (MZ) and dizygotic twins (DZ) twin studies. Most of the conditions studied and review in this book are thought to be of polygenetic causes (Major depressive disorder and Diabetes mellitus, insulin-dependent, etc.). There are many environmental factors (e.g. alcohol) that make this analysis even less consistent.

One of the major purposes of this subchapter is to show the twin studies results. In a perfect setting of autosomal dominant inheritance condition, the results are going to be as follows. There is a 100% of correlation between MZ and 50% of correlation between DZ twins. The table below shows only a few diseases including Huntington disease, Tourette syndrome, Fetal alcohol syndrome and Essential tremor have more than 90% of correlation between MZ twins. Most of the remaining conditions are still rooted in our genes, but the genomic information falls short. Even so, a DNA origin trend could be implicated. With this information we could consider that there are several conditions that have a strong genomic basis, mainly when the correlation between MZ and DZ twin is high, such as in Multiple sclerosis, Narcolepsy, Psoriasis, Anorexia nervosa, Schizophrenia, Celiac disease, etc. Also, this analysis points out that certain conditions do not seem to have a strong genomic basis such as late onset Parkinson disease. Sometimes monozygotic twins could be phenotypically discordant as a result of skewed pattern of X inactivation or postzygotic genomic alterations.

Table 2.2 Correlations of Diseases in Twin Studies

Twin studies	Concordance in MZ twins (%)	Concordance in DZ twins (%)
Alcoholism	55	28
Fetal alcohol syndrome	100	63.6
Huntington disease	100	
Major depressive disorder	36-70	
Breast cancer	11.1	5.2
Diabetes mellitus, insulin-dependent	30-50	6
Anorexia nervosa	52-56	5-11
Polycystic ovary syndrome	71	38
Temporal lobe Epilepsy	22.7	0
Narcolepsy	25-31	0
Multiple sclerosis	25.9	2.3
Schizophrenia	46	14
Parkinson disease, late-onset	2.3	0
Tune deafness	67	44
Essential tremor	93	29
Systemic lupus erythematosus	57	
Psoriasis	65	15
Celiac disease	70	30
Strabismus	73	35
Tourette syndrome	89-94	
Bilateral renal agenesis	16.6	
Malpositioning or ectopic placement of teeth	28.6	

2.8.2 First-degree relatives and familial risk

The risk of an individual suffering the same condition as the parents is one of the most important circumstances in genomic counseling. In autosomal genomic disorders, the risk is clear. For example a typical autosomal dominant disorder such as Huntington disease has a 50% chance of being inherited if either parent is heterozygous for pathogenic mutations in the HTT gene. Another example could be a typical autosomal recessive disorder such as Cystic Fibrosis, which has a 25% chance of being inherited, if both parents are heterozygous for pathogenic mutations in the CFTR gene. In polygenomic disorders, it is a great challenge to really calculate the familial risk.

The first-degree relatives and familial risk have been presented in different forms. Table 2.3 shows the information in 2 different ways to interpret the data, sometimes there is information regarding the relative risk (RR). For example, in Ovarian cancer the relative risk is 3 times, but if BRCA 2 or 1 is mutated the RR is 20 times. Another is just the estimated risk in first-degree relatives. For example, if patients with type I diabetes have descendency, the risk in the child to develop the disease is 3.4% if the father had diabetes and 1.8% if the mother had diabetes. This statistic does not fit typical recessive or dominant patterns, but can only be described as a familial increased risk related to inheritance.

Table 2.3 Disease Risk in First-degree Relatives

Disease	Relative risk (RR)	Estimated risk in first-degree relatives
Lung cancer	1.3	
Cervical cancer	1.45	
Ovarian cancer	3	
Schizophrenia	4.2	
Bilateral breast cancer	5.3	
Polycystic ovary syndrome	10	
Wolff-Parkinson-White syndrome	10	
Unilateral renal agenesis	15	
Ovarian cancer (BRCA 1 or 2 mutated)	40	
Prostate cancer	1.72-2.62	
Migraine with or without aura	1.88-2.5	
Thyroid cancer	1.93-4.10	
Acute myeloid leukemia (Down syndrome)	10-18	
Acute myeloid leukemia	2-3	
Major depressive disorder	2-4	
Diabetes mellitus, insulin-dependent	1	3.4% from father and 1.8% from mothers
Psoriasis	1	8-23%

2.8.3 Wrong Assumptions

There are many wrong assumptions regarding inheritance and genomics; one of the most common assumptions is that the traits are mainly dominant. This is based mainly in common sense information: "if your father had a heart attack, or your mother had breast cancer, you have the same probability of suffering those conditions." These assumptions are generally wrong.

First dominant pattern of inheritance are less common than recessive. And the true fact is that the cause of the disease in the previous generation could be different. With your parents, you only have a 50% chance of having the same polymorphic information.

So even if you have to treat a female patient with breast cancer, you have to consider that the cause is not the fact that the mother had this same disease. This could be easily be coming from the father, or from the mother and father combined as a recessive or digenetic or polygenetic disorder. Another wrong assumption is the one that holds that a genomic DNA alteration produces a disease. There are millions of SNP and mutations but the number of diseases is in thousands.

Chapter 3 Clinical Syndrome and Genomics

This chapter has been further divided into 2 sections; the first section deals with the approach to creating a comprehensive database and the second section deals with clinical descriptions of genomic diseases.

The methodology for creating this database required the annotation of all known diseases. I personally refer to this subject as the "Sickenome Project" or sick genome, beginning the project by obtaining the clinical and genomic information in all the known diseases annotated in the OMIM databases (around 4,000). The information that I retrieved in each disease involved disease name, OMIM # reference, gene the official symbol (causing this disorder), chromosome mapping information (range), clinical classification (based in a similar information than the ICD-10 code), product of conception alteration (inheritance plus de novo), biological function of gene associated and localization of subcellular proteins. Many Clinical syndromes are not part of the OMIM database, since they do not follow a Mendelian inheritance. The Clinical syndromes were added to fill a gap by correlating the genomic and clinical aspect of a disease. For example, the sickenome and the clinical laboratories can detect mutation in factor II and V Leiden of coagulation, but cannot detect Deep vein thrombosis. This study is based on collecting data, using two different approaches, which are genomic and genetic based. As for the genomic approach, most of the information was first obtained from a crude annotation of all the genomic diseases, then for an iterative process, I was able to better set some rules to understand the genomic knowledge. In calling this subchapter the "Sickenome Project", I mean that all the diseases have been annotated and catalogued to create a map of the disease in the human genome using public databases. This provides a starting point for a comprehensive analysis of all the human diseases as well as a detailed clinical and genomic classification of the human genome disease. First, I identify all the known diseases (around 4,000) and 2,300 gene-producing diseases which is the annotation stage. All the diseases have been assigned to a clinical group according to a similar methodology as the one used for ICD-10 nomenclature (described later in this section). The second approach is using the genetic data from clinical laboratories to classify diseases. For this approach, the data was obtained from the most commonly investigated diseases in clinical laboratories around the world. The findings display that while 2200 diseases form the subject of research in these laboratories, only 147 diseases are clinically investigated in at least 20 laboratories. Table 3.0.2 displays a more detailed list of these diseases. Despite the fact that this approach allows us to identify these diseases, it is not capable of considering the diseases in a comprehensive way. In other words, these approaches examine each disease gene by gene. The genetic approach gained acceptance a few years ago until the emergence of the next generation sequencing approach, which concentrates on studying all the diseases in 1 round. This is regarded as a promising approach for future individualized medicine. I argue that these approaches are all important, but there are some diseases that are going to be studied in a more efficient way in terms of budget and data quality based on the next generation sequencing approach. The third approach is a combination of the previous approaches, but with a focus in obtaining clinical and genomic knowledge. The above-mentioned approaches are by no means able to provide satisfying information for all the diseases..

3.0.1 Foundation of this book

The first part of this chapter includes a selected group of diseases that form the core or foundation of all the genomic diseases. All human disorders and traits that follow the Mendelian principle are listed in a major resource "Online Mendelian Inheritance of Man" OMIM. These groups of diseases were selected based on the following criteria: a) Diseases that have at least 50 or more references on the OMIM databases; b) Diseases for which there are at least 20 or more clinical laboratories around the world with a set up assay; c) Diseases that are either common or of clinical significance but have not been included in the OMIM databases; and/or d) Diseases that have an unusual mechanism.

a) Diseases that have at least 50 or more reference on the OMIM databases. Table 3.0.1 (Number of References in Selected Disorders) displays all genomic diseases that have 50 or more OMIM database references and the subchapters in the book that cover them, along with the corresponding number of references, OMIM numbers, genes, and product of conception

alterations. There are at least 87 conditions that form part of this highly cited group.

b) Diseases for which there are at least 20 or more clinical laboratories around the world with a set up assay. Table 3.0.2 (Number of Clinical Laboratories in Selected Disorders) displays all genomic diseases that have undergone clinical testing in at least 20 laboratories, along with the number of laboratories that conduct tests on each, their corresponding genes, and their product of conception alterations. There are at least 147 conditions that form part of this group.

c) Diseases that are either common or of clinical significance but have not been included in the OMIM database. There is a large set of conditions that have not been included in the database due to their non-Mendelian nature. These conditions mainly include nondisjunction diseases such as Down syndrome, Triple X syndrome and Klinefelter syndrome.

d) Diseases that have an unusual mechanism such as chimera mechanism, subtelomeric syndromes, pericentromeric syndrome, imprinting syndrome, RNA editing conditions, severe conditions and microRNA conditions.

There are almost 200 genomic diseases that satisfy the above mentioned criteria and consequently form the foundation of this book

Table 3.0.1 Number of References in Selected Disorders

Number of refer- ences	Disease	OMIM	Gene	Product of Conception Alteration	Subchapter where De- scribed
196	Familial adenomatous polyposis or Gardner syndrome	175100	APC	AD + 25%	3.02.01
96	Lynch Syndrome	120435	MSH2	AD + de novo	3.02.01
96	Lynch Syndrome	120435	MLH1	AD + de novo	3.02.01
130	Hereditary breast and ovarian cancer syndrome 2	114480	BRCA2	AD + de novo	3.02.04
173	Hereditary breast and ovarian cancer syndrome 1	604370	BRCA1	AD + de novo	3.02.04

The Concise Encyclopedia of Genomic Diseases

Table 3.0.1. Continued.

Number of references	Disease	OMIM	Gene	Product of Conception Alteration	Subchapter where Described
118	Multiple endocrine neoplasia 1	131100	*MEN1*	AD + 10%	3.02.11
104	Multiple endocrine neoplasia 2A	171400	*RET*	AD + 10%	3.02.11
399	G6PD deficiency	305900	*G6PD*	XR	3.03.2
77	Sickle cell anemia	603903	*HBB*	AR	3.03.2
433	Alpha-thalassemia	141800	*HBA1-HBA2*	Digenic de novo	3.03.2
132	Fanconi anemia, complementation group A	227650	*FANCA*	AR + KDN	3.03.3
153	Hemophilia A	306700	*F8*	XR + 30%	3.03.4
124	Hemophilia B	306900	*F9*	XR + 30%	3.03.4
103	Von Willebrand disease	193400	*VWF*	AD or AD + AD = AR	3.03.4
66	Factor V Leiden	227400	*F5*	AD or AD + AD = AR	3.03.4
66	Thrombophilia	613679	*F2*	AD or AD + AD = AR	3.03.4
107	Chronic granulomatous disease or Bridges–Good syndrome	306400	*CYBB*	XR + de novo	3.03.5
176	Chromosome 22q11.2 deletion syndrome or DiGeorge syndrome	181500	*TBX1*	AD + 85%	3.03.6
100	Severe combined immunodeficiency	102700	*ADA*	AR	3.03.6
94	Wiskott-Aldrich syndrome-1	301000	*WAS*	XR	3.03.6
161	Congenital adrenal hyperplasia	201910	*CYP21A2*	AR + KDN	3.04.05
197	Complete androgen insensitivity syndrome	300068	*AR*	XR + de novo	3.04.07
178	Phenylketonuria	261600	*PAH*	AR + KDN	3.04.09
89	Maple syrup urine disease 1A or Branched-chain ketoaciduria	248600	*BCKDHA*	AR	3.04.09
159	Cerebral adrenoleukodystrophy and Adrenomyeloneuropathy	300100	*ABCD1*	XR + de novo	3.04.09.3
92	Ornithine transcarbamylase deficiency	311250	*OTC*	XR + de novo	3.04.09.6
84	Glycogen storage disease 2 or Pompe disease	232300	*GAA*	AR	3.04.10.1
56	Galactosemia	230400	*GALT*	AR	3.04.10.1
173	Tay-Sachs disease or GM2-gangliosidosis	272800	*HEXA*	AR	3.04.11.1
135	Fabry disease	301500	*GLA*	XR	3.04.11.1
119	Gaucher Disease	606463	*GBA*	AR + KDN	3.04.11.1

Table 3.0.1. Continued.

Number of references	Disease	OMIM	Gene	Product of Conception Alteration	Subchapter where Described
106	Mucopolysaccharidosis 2 or Hunter syndrome	309900	IDS	XR	3.04.12.1
137	Familial hypercholesterolemia	143890	LDLR	AD or AD + AD = AR	3.04.13
318	Familial dysbetalipoproteinemia or hyperlipoproteinemia 3	107741	APOE	AD or AR	3.04.13
103	Menkes disease	309400	ATP7A	XR	3.04.16.1
125	Wilson disease	277900	ATP7B	AR	3.04.16.1
188	Hemochromatosis	235200	HFE	AR + Env	3.04.16.2
282	Cystic fibrosis	219700	CFTR	AR + KDN	3.04.17
101	Familial mediterranean fever	249100	MEFV	AR	3.04.18.1
141	Alpha-1-antitrypsin deficiency	107400	SERPINA1	AR	3.04.20
55	Major depressive disorder	608516	TOR1A	AD	3.05.4
84	Autism susceptibility 1	300425	NLGN3	XR	3.05.9
116	Rett syndrome	312750	MECP2	XD de novo	3.05.9
96	Gilles de la Tourette syndrome	137580	SLITRK1	AD	3.05.10
389	Huntington disease	143100	HTT	AD + Ant	3.06.02
140	Friedreich ataxia	229300	FXN	AR + Ant	3.06.02.1
199	Ataxia-telangiectasia or Boder-Sedgwick syndrome or Louis–Bar syndrome	208900	ATM	AR	3.06.02.1
122	Spinocerebellar ataxia-1	164400	ATXN1	AD + Ant	3.06.02.1
124	Spinocerebellar ataxia-3	109150	ATXN3	AD + Ant	3.06.02.1
99	Spinal muscular atrophy 1 or Werdnig-Hoffmann disease	253300	SMN1	AR + KDN	3.06.02.2
122	Amyotrophic lateral sclerosis, due to SOD1 deficiency	105400	SOD1	AD + de novo	3.06.02.2
129	Parkinson disease	168600	SNCA	AD or AR	3.06.03
212	Alzheimer disease 1	104300	APP	AD or AR	3.06.04
82	Charcot-Marie-Tooth disease 1B	118200	MPZ	AD	3.06.08.1
77	Charcot-Marie-Tooth Neuropathy Type 1A	118220	PMP22	AD + 10%	3.06.08.1
217	Duchenne muscular dystrophy	310200	DMD	XR + 30%	3.06.09.3
274	Myotonic dystrophy 1 or Steinert disease	160900	DMPK	AD + Ant	3.06.09.3
81	Age-related macular degeneration 1	603075	HMCN1	AD or AR	3.07.05
131	Leber hereditary optic neuropathy (LHON)	535000	MT-ND1	Mitochondria	3.07.07

Table 3.0.1. Continued.

Number of references	Disease	OMIM	Gene	Product of Conception Alteration	Subchapter where De-scribed
92	Essential hypertension	145500	*ADD1*	Polygenic disorders	3.09.2
91	Familial hypertrophic cardiomyopathy	192600	*CAV3*	AD	3.09.5.2
79	Long QT syndrome-1	192500	*KCNQ1*	AD + de novo	3.09.5.3
114	Hereditary hemorrhagic telangiectasia-1	187300	*ENG*	AD	3.09.7
85	Asthma	600807	*HNMT*	Polygenic disorders	3.10.2
102	Inflammatory bowel disease 1	266600	*NOD2*	AD or AR	3.11.2
142	Susceptibility to systemic lupus erythematosus	152700	*DNASE1*	AD or AR	3.12.7
52	ABO hemolytic disease of the newborn	110300	*ABO*	Digenic	3.16.3
84	Familial hyperinsulinemic hypoglycemia	125853	*ABCC8*	AD or AD + AD = AR	3.16.4
113	Adult polycystic kidney disease 1	173900	*PKD1*	AD + de novo	3.17.7.2
118	Achondroplasia	100800	*FGFR3*	AD + 75%	3.17.8.10
115	Osteogenesis imperfecta type 1	166200	*COL1A1/2*	AD + 65%	3.17.8.11
104	Ichthyosis	308100	*STS*	XR	3.17.9.1
295	Neurofibromatosis type 1 or von Recklinghausen disease	162200	*NF1*	AD + 50%	3.17.9.4
79	Neurofibromatosis type 2	101000	*NF2*	AD + 90%	3.17.9.4
124	Tuberous sclerosis 1	191100	*TSC1*	AD + 65%	3.17.9.4
129	Von Hippel-Lindau syndrome	193300	*VHL*	AD + 25%	3.17.9.4
222	Prader-Willi syndrome	176270	*NDN*	Paternal de novo/imprinting	3.17.9.6.2
117	Noonan syndrome-1	163950	*PTPN11*	AD + de novo	3.17.9.6.2
124	Cornelia de Lange syndrome-1	122470	*NIPBL*	AD + 99%	3.17.9.6.2
172	Beckwith-Wiedemann syndrome	130650	*KCNQ1*	AD + 85%	3.17.9.6.4
143	Marfan syndrome	154700	*FBN1*	AD + 25%	3.17.9.6.5
111	Alport syndrome	301050	*COL4A5*	XR + de novo	3.17.9.6.7
101	Kartagener syndrome or primary ciliary dyskinesia 1, with or without situs inversus	244400	*DNAI1*	AR	3.17.9.6.8
123	Angelman syndrome	105830	*UBE3A*	Maternal de novo	3.17.10.2

The Concise Encyclopedia of Genomic Diseases

Table 3.0.1. Continued.

Number of references	Disease	OMIM	Gene	Product of Conception Alteration	Subchapter where De-scribed
156	Williams-Beuren syndrome	194050	*ELN*	AD + 90%	3.17.10.2
207	Fragile X syndrome, Martin–Bell syndrome or Escalante syndrome	300624	*FMR1*	XD + Ant	3.17.10.5

Table 3.0.2 Number of Clinical Laboratories in Selected Disorders

Number of Laboratories	Disease	Gene	Product of Conception Alteration
131	Cystic fibrosis	*CFTR*	AR + KDN
125	Fragile X syndrome	*FMR1*	XD + Ant
121	Factor V Leiden Thrombophilia	*F5*	AD or AD + AD = AR
120	Angelman Syndrome	*UBE3A*	Mat de novo
118	Prothrombin Thrombophilia	*F2*	AD or AD + AD = AR
117	Prader-Willi Syndrome	*SNRPN*	Pat de novo/imprinting
107	Hemochromatosis	*HFE*	AR + Env
74	Huntington Disease	*HTT*	AD + Ant
70	Spinal Muscular Atrophy	*SMN1*	AR + KDN
65	Rett syndrome	*MECP2*	XD de novo
63	22q11.2 Deletion Syndrome	*TBX1*	AD + 85%
62	Medium Chain Acyl-Coenzyme A Dehydrogenase Deficiency	*ACADM*	AR
61	Fabry Disease	*GLA*	XR
60	Multiple Endocrine Neoplasia Type 2	*RET*	AD + 10%
59	Gaucher Disease	*GBA*	AR + KDN
55	Hereditary Breast and Ovarian Cancer	*BRCA1*	AD + de novo
53	Hereditary Breast and Ovarian Cancer	*BRCA2*	AD + de novo
53	Y Chromosome Infertility	*USP9Y*	Y + 99%
53	Tay-Sachs disease	*HEXA*	AR
52	Lynch Syndrome	*MSH2*	AD + de novo
52	Myotonic Dystrophy Type 1	*DMPK*	AD + Ant
52	Canavan Disease	*ASPA*	AR
52	Phenylketonuria	*PAH*	AR + KDN
50	Achondroplasia	*FGFR3*	AD + 75%
50	Duchenne and Becker muscular dystrophy	*DMD*	XR + 25%
50	Lynch Syndrome	*MLH1*	AD + de novo

Table 3.0.2. Continued.

Number of Laboratories	Disease	Gene	Product of Conception Alteration
48	Adenomatous polyposis coli or Gardner syndrome	APC	AD + 25%
47	Friedreich Ataxia	FXN	AR + Ant
46	Noonan Syndrome	PTPN11	AD + de novo
44	Marfan Syndrome	FBN1	AD + 25%
43	Von Hippel-Lindau Syndrome	VHL	AD + 25%
42	Lynch Syndrome	MSH6	AD + de novo
41	Hypochondroplasia	FGFR3	AD + 75%
41	Li-Fraumeni Syndrome	TP53	AD + 10%
40	Leber Hereditary Optic Neuropathy	Mit	Mitochondria
39	Dilated Cardiomyopathy	LMNA	AD
39	Glycogen Storage Disease Type II or Pompe Disease	GAA	AR
38	Familial Mediterranean Fever	MEFV	AR
38	Mitochondrial encephalomyopathy, lactic acidosis, and stroke-like episodes	Mitochondria	
38	Spinocerebellar Ataxia Type 1	ATXN1	AD + Ant
38	Charcot-Marie-Tooth Neuropathy Type 1A	PMP22	AD + 10%
37	Spinal and Bulbar Muscular Atrophy	AR	XR
37	Galactosemia	GALT	AR
36	Homocystinuria	MTHFR	AR
36	Attenuated familial adenomatous polyposis	MUTYH	AR
36	Dilated Cardiomyopathy	DMD	XR + 25%
36	Myoclonic epilepsy with ragged-red fibers	Mitochondria	
36	Spinocerebellar Ataxia Type 2	CACNA1A	AD
36	Spinocerebellar Ataxia Type 6	SLC1A3	AD
36	Hamartoma Tumor Syndrome	PTEN	AD + de novo
36	Neurofibromatosis type 1	NF1	AD + 50%
36	Biotinidase Deficiency	BTD	AR
36	Ornithine Transcarbamylase Deficiency	OTC	XR + de novo
35	Leigh Syndrome and NARP	Mit	Mitochondria
35	Sickle cell anemia	HBB	AR
35	Spinocerebellar Ataxia Type 3	ATXN3	AD + Ant
35	Spinocerebellar Ataxia Type 7	ATXN7	AD + Ant
35	Very Long Chain Acyl-Coenzyme A Dehydrogenase Deficiency	ACADVL	AR
34	Familial Hypercholesterolemia	LDLR	AD or AD + AD = AR
34	Wilson Disease	ATP7B	AR
33	Familial Dysautonomia	IKBKAP	AR
33	Beckwith-Wiedemann Syndrome	KCNQ1	AD + 85%

Table 3.0.2. Continued.

Number of Laboratories	Disease	Gene	Product of Conception Alteration
32	Alpha1-Antitrypsin Deficiency	SERPINA1	AR
32	Cerebral arteriopathy with subcortical infarcts and leukoencephalopathy	NOTCH3	AD
32	CHARGE Syndrome	CHD7	AD + 99%
32	Dentatorubro-pallidoluysian atrophy	ATN1	AD + Ant
32	Gilbert Syndrome	UGT1A1	AD or AR
32	Noonan Syndrome-4	SOS1	AD + de novo
32	Early infantile epileptic encephalopathy 2 or Ohtahara syndrome	CDKL5	XR + de novo
31	Congenital Adrenal Hyperplasia	CYP21A2	AR + KDN
31	Familial Hypercholesterolemia Type B	APOB	AD or AD + AD = AR
31	Mental retardation	ARX	XR + de novo
31	Noonan Syndrome-3	KRAS	AD + de novo
31	Torsion dystonia	TOR1A	AD
31	Hereditary Neuropathy with Liability to Pressure Palsies	PMP22	AD + 10%
31	Sotos Syndrome	NSD1	AD + de novo
30	Charcot-Marie-Tooth Neuropathy X Type 1	GJB1	XD
30	Apert syndrome or Craniosynostosis	FGFR2	AD + 80%
30	Crouzonodermoskeletal syndrome	FGFR3	AD + de novo
30	Multiple Endocrine Neoplasia Type 1	MEN1	AD + 10%
29	Cardiovascular Disease Risk Factor	APOE	AD or AR
29	Charcot-Marie-Tooth Neuropathy Type 1B	MPZ	AD
29	Hemophilia A	F8	XR + 30%
28	Amyotrophic Lateral Sclerosis	SOD1	AD + de novo
28	Loeys-Dietz Syndrome	TGFBR1	AD + 75%
28	Osteogenesis Imperfecta	COL1A1/2	AD + 65%
28	Paraganglioma-Pheochromocytoma Syndrome	SDHB	AD
28	Bloom's Syndrome	BLM	AR
27	Charcot-Marie-Tooth Neuropathy Type 1E	PMP22	AD
27	Glycogen Storage Disease Type Ia	G6PC	AR
27	Hearing Loss and Deafness	Mitochondria	
27	Loeys-Dietz Syndrome	TGFBR2	AD + 75%
27	POLG-Related Disorders	POLG	AR
27	Smith-Lemli-Opitz Syndrome	DHCR7	AR
27	Thanatophoric Dysplasia	FGFR3	AD + de novo
27	Adrenoleukodystrophy, X-Linked	ABCD1	XR + de novo
26	Alpha-thalassemia	HBA1/ HBA2	Digenic de novo

Table 3.0.2. Continued.

Number of Laboratories	Disease	Gene	Product of Conception Alteration
26	LEOPARD Syndrome-1	PTPN11	AD + de novo
26	Noonan Syndrome-5	RAF1	AD + de novo
26	Paraganglioma-Pheochromocytoma Syndrome	SDHD	AD
26	Pendred Syndrome	SLC26A4	AR
26	Spinocerebellar Ataxia Type 17	TBP	AD + Ant
26	Familial Transthyretin Amyloidosis	TTR	AD
25	Familial hypertrophic Cardiomyopathy 1	MYH7	AD
25	Fanconi Anemia	FANCC	AR
25	Hemoglobin SC	HBB	AR
25	Peutz-Jeghers Syndrome	STK11	AD + de novo
25	Russell-Silver Syndrome	H19	AD + de novo
25	Fructose Intolerance	ALDOB	AR
25	Alagille Syndrome	JAG1	AD + de novo
24	Brugada Syndrome	SCN5A	AD
24	Emery-Dreifuss Muscular Dystrophy	LMNA	XR
24	Long QT Syndrome 2	KCNH2	AD
24	Long QT Syndrome 5	KCNE1	AD
24	Optic Atrophy Type 1	OPA1	AD
24	Dopa-responsive dystonia with or without hyperphenylalainemia	GCH1	AR
24	Isolated Aniridia	PAX6	AD + de novo
23	Cutaneous Malignant Melanoma	CDKN2A	AD + de novo
23	Dilated Cardiomyopathy	MYH7	AD
23	Lynch Syndrome	PMS2	AD + de novo
22	Charcot-Marie-Tooth Neuropathy Type 2A2	MFN2	AD
22	Hereditary Diffuse Gastric Cancer	CDH1	AD
22	Hypertrophic Cardiomyopathy	MYBPC3	AD
22	Intellectual disability and Duplication Syndrome	MECP2	XD de novo or XR
22	Long QT Syndrome 1	KCNQ1	AD + de novo
22	Long QT Syndrome 6	KCNE2	AD
22	Mitochondrial DNA Deletion Syndromes	Mit	Mitochondria
22	Paraganglioma-Pheochromocytoma Syndrome	SDHC	AD
21	Arrhythmogenic Right Ventricular Dysplasia/Cardiomyopathy 9	PKP2	AD
21	Calcium-sensing receptor disease	CASR	AR or AD
21	Charcot-Marie-Tooth Neuropathy Type 2B1	LMNA	AD + de novo
21	Costello Syndrome	HRAS	AD + de novo
21	Hereditary Hemorrhagic Telangiectasia 1	ENG	AD

Table 3.0.2. Continued.

Number of Laboratories	Disease	Gene	Product of Conception Alteration
21	Mucolipidosis IV	MCOLN1	AR
21	Niemann-Pick Disease Type C1	NPC1	AR
21	Spinal Muscular Atrophy	SMN2	AR + KDN
21	Androgen Insensitivity Syndrome	AR	XR + de novo
21	Hemophilia B	F9	XR + 30%
21	Hereditary Hemorrhagic Telangiectasia 2	ACVRL1	AD
21	Neurofibromatosis type 2	NF2	AD + 90%
21	Seizure Disorders	SCN1A	AD + 90%
20	Alzheimer Disease Risk Factor	APOE	AD or AR
20	Craniosynostosis	FGFR1	AD + de novo
20	Glucose Transporter Type 1 Deficiency Syndrome	SLC2A1	AD + de novo
20	Muscle Diseases	FKRP	AR
20	Muscle Diseases	LMNA	AD
20	Spastic Paraplegia 4	SPAST	AD

3.0.1.1 Selection Criteria of Diseases for this Book

As mentioned before, there are 4 base criteria that are used to determine whether a disease will be described in this part of the book. Accordingly, Table 3.0.03 (Select Group of Diseases and Corresponding Sub-Chapters) displays all diseases that have either: a) been referenced at least 50 times, b) undergone clinical studies in at least 20 laboratories, c) have exceptionally common occurrence rates, and/or d) have unusual mechanisms. The corresponding sub-chapters where the diseases are discussed are also listed on the table, along with their corresponding genes (if available).

At the end of each group of diseases in this chapter, there will be at least one table. Sometimes, there are more conditions that are described than there are in the table. Other times, the tables may contain more diseases than were originally described in the section. This is because there are some diseases that have very little inclusion criteria for genomic origin, and others that have very little clinical characteristics that need to be specifically explained.

Table 3.0.03 Select Group of Diseases and Corresponding Sub-Chapters

Subchapter where Described	Subchapter Name	Gene	Disease
3.02.01	Malignant Neoplasms of Digestive Organs	APC	Familial adenomatous polyposis-1
3.02.01	Malignant Neoplasms of Digestive Organs	MUTYH	Familial adenomatous polyposis-2

Table 3.0.03. Continued.

Subchapter where Described	Subchapter Name	Gene	Disease
3.02.01	Malignant Neoplasms of Digestive Organs	MSH6	Lynch Syndrome
3.02.01	Malignant Neoplasms of Digestive Organs	PMS2	Lynch Syndrome
3.02.01	Malignant Neoplasms of Digestive Organs	MLH1	Lynch Syndrome
3.02.01	Malignant Neoplasms of Digestive Organs	MSH2	Lynch Syndrome
3.02.03	Malignant Neoplasms of Skin	CDKN2A	Cutaneous Malignant Melanoma
3.02.04	Malignant Neoplasms of Breast and Female Genital Organs	BRCA1	Hereditary breast and ovarian cancer syndrome
3.02.04	Malignant Neoplasms of Breast and Female Genital Organs	BRCA2	Hereditary breast and ovarian cancer syndrome
3.02.04	Malignant Neoplasms of Breast and Female Genital Organs	CDH1	Hereditary breast cancer syndrome
3.02.06	Malignant Neoplasms of Urinary Organs	WT1	Wilms tumor
3.02.07	Malignant Neoplasms of Eye, Brain and Central Nervous System	RB1	Retinoblastoma
3.02.08	Malignant Neoplasms of Endocrine Glands and Related Structures	SDHB	Paraganglioma-Pheochromocytoma Syndrome
3.02.08	Malignant Neoplasms of Endocrine Glands and Related Structures	SDHD	Paraganglioma-Pheochromocytoma Syndrome
3.02.08	Malignant Neoplasms of Endocrine Glands and Related Structures	SDHC	Paraganglioma-Pheochromocytoma Syndrome
3.02.11	Neoplasms of Uncertain or Unknown Behavior	MEN1	Multiple endocrine neoplasia 1
3.02.11	Neoplasms of Uncertain or Unknown Behavior	RET	Multiple endocrine neoplasia 2A and 2B
3.03.2	Hemolytic Anemias	G6PD	G6PD deficiency
3.03.2	Hemolytic Anemias	HBA1	Alpha-thalassemia
3.03.2	Hemolytic Anemias	HBA2	Alpha-thalassemia
3.03.2	Hemolytic Anemias	LCRA	Alpha-thalassemia
3.03.2	Hemolytic Anemias	HBB	Sickle cell anemia
3.03.2	Hemolytic Anemias	HBB	Hemoglobin C
3.03.3	Aplastic and other Anemias	FANCA	Fanconi anemia
3.03.3	Aplastic and other Anemias	FANCC	Fanconi anemia
3.03.4	Coagulation Defects, Purpura and other Hemorrhagic Conditions	F8	Hemophilia A

Table 3.0.03. Continued.

Subchapter where Described	Subchapter Name	Gene	Disease
3.03.4	Coagulation Defects, Purpura and other Hemorrhagic Conditions	F9	Hemophilia B
3.03.4	Coagulation Defects, Purpura and other Hemorrhagic Conditions	VWF	Von Willebrand disease
3.03.4	Coagulation Defects, Purpura and other Hemorrhagic Conditions	F5	Factor V Leiden
3.03.4	Coagulation Defects, Purpura and other Hemorrhagic Conditions	F2	Thrombophilia
3.03.5	Other Diseases of Blood and Blood-forming Organs	CYBB	Chronic granulomatous disease or Bridges–Good syndrome
3.03.6	Certain Disorders Involving the Immune Mechanism	ADA	Severe combined immunodeficiency
3.03.6	Certain Disorders Involving the Immune Mechanism	WAS	Wiskott-Aldrich syndrome
3.03.6	Certain Disorders Involving the Immune Mechanism	TBX1	Chromosome 22q11.2 deletion syndrome or DiGeorge syndrome
3.04.01	Thyroid Gland / Thyroid Hormone	SLC26A4	Pendred syndrome and Enlarged vestibular aqueduct
3.04.02	Pancreas / Insulin, Glucagon	Mitochondrial	Maternally inherited Diabetes mellitus and deafness or Ballinger-Wallace syndrome
3.04.03	Parathyroid Gland / PTH	CASR	Familial isolated hypoparathyroidism
3.04.05	Adrenal Gland / Aldosterone, Cortisol, Epinephrine and Norepinephrine	CYP21A2	Congenital adrenal hyperplasia, due to 21-hydroxylase deficiency
3.04.07	Other Endocrine Diseases	AR	Complete androgen insensitivity syndrome
3.04.08	Nutritional Diseases	BTD	Biotinidase deficiency
3.04.08	Nutritional Diseases		Obesity
3.04.09	Amino-acids Disorders	PAH	Phenylketonuria
3.04.09	Amino-acids Disorders	GCH1	Dopa-responsive dystonia with or without hyperphenylalainemia
3.04.09	Amino-acids Disorders	BCKDHA	Maple syrup urine disease 1A or branched-chain ketoaciduria
3.04.09.2	Disorders of Branched-chain Amino-acid Metabolism and Fatty-acid Metabolism	MTHFR	Homocystinuria

Table 3.0.03. Continued.

Subchapter where De- scribed	Subchapter Name	Gene	Disease
3.04.09.3	Disorders of Fatty-acid Metabolism	ABCD1	Cerebral adrenoleu-kodystrophy and Adre-nomyeloneuropathy
3.04.09.3	Disorders of Fatty-acid Metabolism	ACADM	Medium chain acyl-CoA dehydrogenase deficiency
3.04.09.3	Disorders of Fatty-acid Metabolism	ACADVL	Very long chain acyl-CoA dehydrogenase deficiency
3.04.09.6	Disorders of Urea Cycle and Ornithine Metabolism	OTC	Ornithine transcarbamyl-ase deficiency
3.04.10.1	Glycogen Storage Disease	G6PC	Glycogen storage disease 1A or von Gierke disease
3.04.10.1	Glycogen Storage Disease	GAA	Glycogen storage disease 2 or Pompe disease
3.04.10.2	Disorders of Fructose Metabolism	ALDOB	Fructose intolerance
3.04.10.3	Disorders of Galactose Metabolism	GALT	Classic Galactosemia
3.04.11.1	Disorders of Sphingolipid Metabolism and other Lipid Storage Disorders	HEXA	Tay-Sachs disease or GM2-gangliosidosis
3.04.11.1	Disorders of Sphingolipid Metabolism and other Lipid Storage Disorders	MCOLN1	Mucolipidosis-4
3.04.11.1	Disorders of Sphingolipid Metabolism and other Lipid Storage Disorders	ASPA	Canavan disease
3.04.11.1	Disorders of Sphingolipid Metabolism and other Lipid Storage Disorders	GLA	Fabry disease
3.04.11.1	Disorders of Sphingolipid Metabolism and other Lipid Storage Disorders	GBA	Gaucher Disease
3.04.11.1	Disorders of Sphingolipid Metabolism and other Lipid Storage Disorders	NPC1	Niemann-Pick disease C1
3.04.12.1	Disorders of Glycosaminoglycan Me-tabolism	IDS	Mucopolysaccharidosis-2 or Hunter syndrome
3.04.13	Disorders of Lipoprotein Metabolism and other Lipidemias	LDLR	Familial hypercholesterol-emia
3.04.13	Disorders of Lipoprotein Metabolism and other Lipidemias	APOB	Hypercholesterolemia, due to ligand-defective apo B
3.04.13	Disorders of Lipoprotein Metabolism and other Lipidemias	APOE	Familial dysbetalipopro-teinemia or Hyperlipopro-teinemia 3
3.04.15.4	Disorder of Bilirubin Metabolism	UGT1A1	Gilbert Syndrome
3.04.16.1	Disorders of Copper Metabolism	ATP7A	Menkes disease
3.04.16.1	Disorders of Copper Metabolism	ATP7B	Wilson disease
3.04.16.2	Disorders of Iron Metabolism	HFE	Hemochromatosis
3.04.17	Cystic Fibrosis	CFTR	Cystic fibrosis

Table 3.0.03. Continued.

Subchapter where Described	Subchapter Name	Gene	Disease
3.04.18.1	Heredofamilial Amyloidosis	MEFV	Familial Mediterranean fever
3.04.18.1	Heredofamilial Amyloidosis	TTR	Hereditary Amyloidosis
3.04.19	Disorders of Fluid, Electrolyte and Acid-base Balance	Mitochondrial	Mitochondrial DNA Depletion Syndrome
3.04.20	Other Metabolic Disorders (not elsewhere classified)	SERPINA1	Alpha1-Antitrypsin Deficiency
3.04.20.3	Mitochondrial Metabolism Disorders	Mitochondrial	MELAS syndrome
3.04.20.3	Mitochondrial Metabolism Disorders	Mitochondrial	MERRF syndrome
3.04.20.4	Other Specified Metabolic Disorders	Mitochondrial	Leigh Syndrome
3.05.1	Organic Mental Disorders (including symptomatic)	NOTCH3	Cerebral arteriopathy with subcortical infarcts and leukoencephalopathy
3.05.10	Behavioral and Emotional Disorders with onset usually occurring in Childhood and Adolescence	SLITRK1	Gilles de la Tourette syndrome
3.05.3	Schizophrenia, Schizotypal and Delusional Disorders	COMT	Chromosome 22q11.21 deletion syndrome or Susceptibility to Schizophrenia
3.05.4	Mood (affective) Disorders	TOR1A	Major depressive disorder
3.05.8	Mental Retardation	ARX	X-linked mental retardation-43
3.05.9	Disorders of Psychological Development	NLGN3	Autism susceptibility 1
3.05.9	Disorders of Psychological Development	MECP2	Rett syndrome
3.06.02	Systemic Atrophies Primarily Affecting the Central Nervous System	HTT	Huntington disease
3.06.02.1	Hereditary Ataxia	Mitochondrial	NARP syndrome
3.06.02.1	Hereditary Ataxia	ATN1	Dentatorubro-pallidoluysian atrophy
3.06.02.1	Hereditary Ataxia	FXN	Friedreich ataxia
3.06.02.1	Hereditary Ataxia	ATM	Ataxia-telangiectasia or Boder-Sedgwick syndrome or Louis–Bar syndrome
3.06.02.1	Hereditary Ataxia	SPAST	Spastic paraplegia-4
3.06.02.1	Hereditary Ataxia	ATXN1	Spinocerebellar ataxia-1
3.06.02.1	Hereditary Ataxia	TBP	Spinocerebellar ataxia-17
3.06.02.1	Hereditary Ataxia	ATXN3	Spinocerebellar ataxia-3
3.06.02.1	Hereditary Ataxia	ATXN7	Spinocerebellar ataxia-7

Table 3.0.03. Continued.

Subchapter where Described	Subchapter Name	Gene	Disease
3.06.02.1	Hereditary Ataxia	ATXN2	Spinocerebellar Ataxia-2
3.06.02.1	Hereditary Ataxia	CACNA1A	Spinocerebellar Ataxia-6
3.06.02.2	Spinal Muscular Atrophy and Related Syndromes	SMN2	Spinal Muscular Atrophy
3.06.02.2	Spinal Muscular Atrophy and Related Syndromes	SMN1	Spinal muscular atrophy 1 or Werdnig-Hoffmann disease
3.06.02.2	Spinal Muscular Atrophy and Related Syndromes	AR	Spinal and bulbar muscular atrophy of Kennedy
3.06.02.2	Spinal Muscular Atrophy and Related Syndromes	SOD1	Amyotrophic lateral sclerosis, due to SOD1 deficiency
3.06.03	Extrapyramidal and Movement Disorders	SNCA	Parkinson disease
3.06.04	Other Degenerative Diseases of the Nervous System	APP	Alzheimer disease-1
3.06.04	Other Degenerative Diseases of the Nervous System	APOE	Familial dysbetalipoproteinemia - Alzheimer Disease Risk Factor
3.06.06.1	Epilepsy	SCN1A	Early infantile epileptic encephalopathy-6 or Dravet syndrome
3.06.06.1	Epilepsy	CDKL5	Early infantile epileptic encephalopathy-2
3.06.08.1	Hereditary and Idiopathic Neuropathy	LMNA	Charcot-Marie-Tooth disease
3.06.08.1	Hereditary and Idiopathic Neuropathy	PMP22	Charcot-Marie-Tooth disease-1A
3.06.08.1	Hereditary and Idiopathic Neuropathy	MPZ	Charcot-Marie-Tooth disease-1B
3.06.08.1	Hereditary and Idiopathic Neuropathy	MFN2	Charcot-Marie-Tooth disease-2A2
3.06.08.1	Hereditary and Idiopathic Neuropathy	GJB1	Charcot-Marie-Tooth disease X-Linked
3.06.08.3	Other Hereditary and Idiopathic Neuropathies	IKBKAP	Hereditary sensory and autonomic neuropathy type 3 or Familial dysautonomia
3.06.09.3	Muscular dystrophy	DMD	Duchenne muscular dystrophy
3.06.09.3	Muscular dystrophy	LMNA	Emery-Dreifuss muscular dystrophy-2 and 3
3.06.09.3	Muscular dystrophy	FKRP	Muscle disease or muscular dystrophy

Table 3.0.03. Continued.

Subchapter where Described	Subchapter Name	Gene	Disease
3.06.09.3	Muscular dystrophy	DMPK	Myotonic dystrophy-1 or Steinert disease
3.06.11	Other Disorders of the Nervous System	SLC2A1	Glucose Transporter Type 1 Deficiency Syndrome or De Vivo disease
3.07.05	Disorders of Choroid and Retina	HMCN1	Age-related macular degeneration-1
3.07.07	Disorders of Optic Nerve and Visual Pathways	Mitochondrial	Leber hereditary optic neuropathy
3.07.07	Disorders of Optic Nerve and Visual Pathways	OPA1	Optic atrophy-1 or Kjer optic atrophy
3.07.08	Disorders of Ocular Muscles, Binocular Movement, Accommodation and Refraction	POLG	POLG-Related Disorders
3.09.2	Hypertensive Diseases		Essential hypertension
3.09.5.2	Cardiomyopathy	MYBPC3	Cardiomyopathy
3.09.5.2	Cardiomyopathy	DMD	Dilated Cardiomyopathy
3.09.5.2	Cardiomyopathy	MYH7	Familial hypertrophic cardiomyopathy-1
3.09.5.2	Cardiomyopathy	CAV3	Familial hypertrophic Cardiomyopathy
3.09.5.3	Other Cardiomyopathy	PKP2	Arrhythmogenic right ventricular dysplasia-9
3.09.5.3	Other Cardiomyopathy	SCN5A	Brugada syndrome-1
3.09.5.3	Other Cardiomyopathy	KCNQ1	Long QT syndrome-1
3.09.5.3	Other Cardiomyopathy	KCNH2	Long QT syndrome-2
3.09.5.3	Other Cardiomyopathy	KCNE1	Long QT Syndrome-5
3.09.5.3	Other Cardiomyopathy	KCNE2	Long QT Syndrome-6
3.09.7	Diseases of Arteries, Arterioles and Capillaries	TGFBR1	Loeys-Dietz syndrome-1A and 2A
3.09.7	Diseases of Arteries, Arterioles and Capillaries	TGFBR2	Loeys-Dietz syndrome-1B and 2B
3.09.7	Diseases of Arteries, Arterioles and Capillaries	ENG	Hereditary hemorrhagic telangiectasia-1
3.09.7	Diseases of Arteries, Arterioles and Capillaries	ACVRL1	Hereditary hemorrhagic telangiectasia-2
3.10.2	Chronic Lower Respiratory Diseases		Asthma
3.11.2	Noninfective Enteritis and Colitis	NOD2	Inflammatory bowel disease
3.12.7	Other Disorders of the Skin and Subcutaneous Tissue	DNASE1	Systemic lupus erythematosus

Table 3.0.03. Continued.

Subchapter where Described	Subchapter Name	Gene	Disease
3.14.6	Diseases of Male Genital Organs	USP9Y	Chromosome Yq11 microdeletion syndrome or Spermatogenic failure
3.16.3	Hemorrhagic and Hematological Disorders of Fetus and Newborn	ABO	ABO hemolytic disease of the newborn
3.16.4	Transitory Endocrine and Metabolic Disorders Specific to Fetus and Newborn	ABCC8	Diabetes mellitus
3.17.2.4	Congenital Malformations of Anterior Segment of Eye	PAX6	Aniridia
3.17.5.5	Congenital Malformations of Gallbladder, Bile Ducts and Liver	JAG1	Alagille syndrome-1
3.17.7.2	Cystic Kidney Disease	PKD1	Adult polycystic kidney disease-1
3.17.8.10	Osteochondrodysplasia with Defects of Growth of Tubular Bones and Spine	FGFR3	Thanatophoric dysplasia
3.17.8.10	Osteochondrodysplasia with Defects of Growth of Tubular Bones and Spine	FGFR3	Achondroplasia
3.17.8.10	Osteochondrodysplasia with Defects of Growth of Tubular Bones and Spine	FGFR3	Hypochondroplasia
3.17.8.11	Other Osteochondrodysplasias	COL1A1	Osteogenesis imperfecta type I
3.17.8.8.1	Craniosynostosis	FGFR1	Craniosynostosis
3.17.8.8.2	Craniofacial Dysostosis	FGFR3	Crouzono-dermoskeletal syndrome
3.17.9.1	Congenital Ichthyosis	STS	X-linked ichthyosis
3.17.9.3	Other congenital Malformations of Skin	BLM	Bloom syndrome
3.17.9.4	Phakomatoses (not elsewhere classified)	NF1	Neurofibromatosis type 1 or von Recklinghausen disease
3.17.9.4	Phakomatoses (not elsewhere classified)	NF2	Neurofibromatosis type 2
3.17.9.4	Phakomatoses (not elsewhere classified)	TSC1	Tuberous sclerosis-1
3.17.9.4	Phakomatoses (not elsewhere classified)	TSC2	Tuberous sclerosis-2
3.17.9.4	Phakomatoses (not elsewhere classified)	STK11	Peutz-Jeghers syndrome
3.17.9.4	Phakomatoses (not elsewhere classified)	VHL	von Hippel-Lindau syndrome
3.17.9.6.1	Congenital Malformation Syndromes predominantly Affecting Facial Appearance	FGFR2	Apert syndrome or Craniosynostosis

Table 3.0.03. Continued.

Subchapter where Described	Subchapter Name	Gene	Disease
3.17.9.6.2	Congenital Malformation Syndromes predominantly Associated with Short Stature	SNRPN	Prader-Willi syndrome
3.17.9.6.2	Congenital Malformation Syndromes predominantly Associated with Short Stature	PTPN11	Noonan syndrome-1
3.17.9.6.2	Congenital Malformation Syndromes predominantly Associated with Short Stature	KRAS	Noonan syndrome-3
3.17.9.6.2	Congenital Malformation Syndromes predominantly Associated with Short Stature	SOS1	Noonan syndrome-4
3.17.9.6.2	Congenital Malformation Syndromes predominantly Associated with Short Stature	RAF1	Noonan syndrome-5
3.17.9.6.2	Congenital Malformation Syndromes predominantly Associated with Short Stature		Russell-Silver Syndrome
3.17.9.6.2	Congenital Malformation Syndromes predominantly Associated with Short Stature	DHCR7	Smith-Lemli-Opitz syndrome
3.17.9.6.2	Congenital Malformation Syndromes predominantly Associated with Short Stature	NIPBL	Cornelia de Lange syndrome-1
3.17.9.6.4	Congenital Malformation Syndromes Involving Early Overgrowth	KCNQ1	Beckwith-Wiedemann syndrome
3.17.9.6.4	Congenital Malformation Syndromes Involving Early Overgrowth	NSD1	Sotos Syndrome
3.17.9.6.5	Marfan's Syndrome	FBN1	Marfan syndrome
3.17.9.6.7	Other Specified Congenital Malformation Syndromes (not elsewhere classified)	COL4A5	Alport syndrome
3.17.9.6.7	Other Specified Congenital Malformation Syndromes (not elsewhere classified)	CHD7	CHARGE syndrome
3.17.9.6.7	Other Specified Congenital Malformation Syndromes (not elsewhere classified)	HRAS	Costello syndrome
3.17.9.6.7	Other Specified Congenital Malformation Syndromes (not elsewhere classified)	MIR17HG	Feingold syndrome-2
3.17.9.6.7	Other Specified Congenital Malformation Syndromes (not elsewhere classified)	PTPN11	LEOPARD syndrome-1

The Concise Encyclopedia of Genomic Diseases

Table 3.0.03. Continued.

Subchapter where Described	Subchapter Name	Gene	Disease
3.17.9.6.8	Other Congenital Malformations (not elsewhere classified)	DNAI1	Primary ciliary dyskinesia-1 with or without situs inversus or Kartagener syndrome
3.17.10.1	Autosomal Trisomy		Chromosome 21 trisomy or Down syndrome
3.17.10.1	Autosomal Trisomy		Chromosome 16 trisomy
3.17.10.2	Microduplication and Microdeletion Syndrome (not elsewhere classified)	UBE3A	Angelman syndrome
3.17.10.2	Microduplication and Microdeletion Syndrome (not elsewhere classified)	MECP2	Intellectual disability and Duplication Syndrome
3.17.10.2	Microduplication and Microdeletion Syndrome (not elsewhere classified)	ELN	Williams-Beuren syndrome
3.17.10.3	Monosomy		Turner syndrome or Ullrich-Turner syndrome
3.17.10.4	Sex Chromosome Trisomy		Triple X syndrome
3.17.10.4	Sex Chromosome Trisomy		Klinefelter syndrome, or 47, XXY or XXY syndrome
3.17.10.4	Sex Chromosome Trisomy		XYY syndrome
3.17.10.5	Unclassified Miscellaneous Syndrome and Mechanism	SRY	XX male syndrome or de la Chapelle syndrome
3.17.10.5	Unclassified Miscellaneous Syndrome and Mechanism	FMR1	Fragile X syndrome, Martin–Bell syndrome or Escalante syndrome
3.17.10.5	Unclassified Miscellaneous Syndrome and Mechanism		Uniparental disomy
3.18.8	Persons with Potential Health Hazards Related to Family and Personal History and certain Conditions influencing Health Status	TP53	Li-Fraumeni syndrome

3.0.1.2 New Classification Mechanisms for Diseases

When I started annotating the diseases, I noticed a trend for different types of diseases to fall under one or more new groups of classification. This led me to devise a new mechanism for disease classification, based on the clearest trend that the diseases display. This new mechanism will be discussed in detail in Chapters 4 and 5.

Table 3.0.04 (Classification Mechanisms for Select Groups of Diseases) contains a sampling of diseases that have been classified using a new mechanism. It displays the new classification categories for select diseases, according to the groups to which they belong. The corresponding genes are also listed if available, along with the corresponding subchapter where the diseases are discussed.

Table 3.0.04 – Classification Mechanisms for Select Groups of Diseases

Subchapter where Described	Gene	Disease	Classification
3.04.13	APOE	Familial dysbetalipoproteinemia or Hyperlipoproteinemia 3	Ancestral
3.16.3	ABO	ABO hemolytic disease of the newborn	Ancestral
3.17.10.5	SRY	XX male syndrome or de la Chapelle syndrome	Chromosome chimera
3.04.01	SLC26A4/ FOXI1	Pendred syndrome and Enlarged vestibular aqueduct	Digenic
3.03.4	F5	Factor V Leiden	Gain-of-function
3.03.4	F2	Thrombophilia	Gain-of-function
3.17.10.2	UBE3A	Angelman syndrome	Imprinting
3.17.9.6.2	SNRPN	Prader-Willi syndrome	Imprinting
3.17.9.6.2	NDN	Prader-Willi syndrome	Imprinting
3.17.9.6.2		Russell-Silver Syndrome	Imprinting
3.17.9.6.4	KCNQ1	Beckwith-Wiedemann syndrome	Imprinting
3.03.6	TBX1	Chromosome 22q11.2 deletion syndrome or DiGeorge syndrome	Microdeletion
3.17.10.2		Chromosome 7 microdeletion or Williams-Beuren syndrome	Microdeletion
3.06.08.1	PMP22	Charcot-Marie-Tooth disease 1A	Microinsertion
3.17.9.6.7	MIR17HG	Feingold syndrome 2	microRNA
3.17.10.3		Turner syndrome or Ullrich-Turner syndrome	Monosomy
3.03.4	VWF	Von Willebrand disease	Multimeric
3.03.6	TBX1	Chromosome 22q11.2 deletion syndrome or DiGeorge syndrome	Pericentromeric
3.04.08		Obesity	Polygenomic
3.09.2		Essential hypertension	Polygenomic
3.10.2		Asthma	Polygenomic
3.11.2		Inflammatory bowel disease	Polygenomic
3.12.7		Systemic lupus erythematosus	Polygenomic
3.16.4		Diabetes mellitus 2	Polygenomic
3.03.2	G6PD	G6PD deficiency	Subtelomeric
3.03.2	HBA1	Alpha-thalassemia	Subtelomeric
3.03.2	HBA2	Alpha-thalassemia	Subtelomeric
3.03.2	LCRA	Alpha-thalassemia	Subtelomeric
3.03.4	F8	Hemophilia A	Subtelomeric
3.04.09.3	ABCD1	Cerebral adrenoleukodystrophy and Adrenomyeloneuropathy	Subtelomeric
3.04.11.1	GLA	Fabry disease	Subtelomeric
3.04.18.1	MEFV	Familial mediterranean fever	Subtelomeric

The Concise Encyclopedia of Genomic Diseases

Table 3.0.04. Continued.

Subchapter where Described	Gene	Disease	Classification
3.05.9	*MECP2*	Rett syndrome	Subtelomeric
3.06.02	*HTT*	Huntington disease	Subtelomeric
3.09.5.3	*KCNQ1*	Long QT syndrome 1	Subtelomeric
3.09.7	*ENG*	Hereditary hemorrhagic telangiectasia 1	Subtelomeric
3.17.10.2	*MECP2*	Intellectual disability and Duplication Syndrome	Subtelomeric
3.17.10.5	*SRY*	XX male syndrome or de la Chapelle syndrome	Subtelomeric
3.17.7.2	*PKD1*	Adult polycystic kidney disease 1	Subtelomeric
3.17.8.10	*FGFR3*	Thanatophoric dysplasia	Subtelomeric
3.17.8.10	*FGFR3*	Achondroplasia	Subtelomeric
3.17.8.10	*FGFR3*	Hypochondroplasia	Subtelomeric
3.17.8.8.2	*FGFR3*	Crouzono-dermoskeletal syndrome	Subtelomeric
3.17.9.4	*TSC1*	Tuberous sclerosis 1	Subtelomeric
3.17.9.4	*TSC2*	Tuberous sclerosis 2	Subtelomeric
3.17.9.4	*STK11*	Peutz-Jeghers syndrome	Subtelomeric
3.17.9.6.4	*KCNQ1*	Beckwith-Wiedemann syndrome	Subtelomeric
3.16.3	*ABO*	ABO hemolytic disease of the newborn	Subtelomeric
3.06.02	*HTT*	Huntington disease	Trinucleotide
3.06.02.1	*ATN1*	Dentatorubro-pallidoluysian atrophy	Trinucleotide
3.06.02.1	*FXN*	Friedreich ataxia	Trinucleotide
3.06.02.1	*ATXN1*	Spinocerebellar ataxia 1	Trinucleotide
3.06.02.1	*TBP*	Spinocerebellar ataxia 17	Trinucleotide
3.06.02.1	*ATXN3*	Spinocerebellar ataxia 3	Trinucleotide
3.06.02.1	*ATXN7*	Spinocerebellar ataxia 7	Trinucleotide
3.06.02.1	*ATXN2*	Spinocerebellar Ataxia 2	Trinucleotide
3.06.02.1	*CACNA1A*	Spinocerebellar Ataxia 6	Trinucleotide
3.17.10.5	*FMR1*	Fragile X syndrome, Martin–Bell syndrome or Escalante syndrome	Trinucleotide
3.17.10.1		Chromosome 21 trisomy or Down syndrome	Trisomy
3.17.10.1		Chromosome 16 trisomy	Trisomy
3.17.10.4		Triple X syndrome	Trisomy
3.17.10.4		Klinefelter syndrome, or 47, XXY or XXY syndrome	Trisomy
3.17.10.4		Chromosome XYY syndrome	Trisomy

The basis of classification is different for different diseases. Additionally, since they have been annotated in different databases, the information available for each of them is different but still categorizable. But, as seen on the table, some pieces of data such as the gene involved are not available for some diseases.

3.0.1.3 Classification Codes

In order to annotate and categorize the different genomic diseases, I used a coding mechanism based on signs, symptoms and circumstances. This coding mechanism is similar to the one used by The International Statistical Classification of Diseases and Related Health Problems, 10th Revision. This database, also known as "ICD-10", is maintained by the World Health Organization (WHO).

- Furthermore, some of the diseases in this chapter have either complex names or alternate names. Along those lines, a general rule to follow is that the connecting word "and" in a disease usually signifies that what follows is part of the disease's complex name, whereas "or" signifies that a different name with which the disease is also known as will follow.

- In considering the disease classifications, the general rule to follow is that the diseases are caused by autosomal DNA alterations, unless otherwise stated as "X chromosome", "Y chromosome", or "mitochondrial". Mitochondrial diseases of mitochondrial DNA origin can be recognized by the prefix MT- in the name of the gene and by an OMIM number that begins with "5".

- Some of the diseases discussed in this chapter have multiple associated genes, which each have their own characterizations. In order to clearly distinguish between the different genes producing the disease, the symbol ;; (double semi-colon) will be used before each new gene. The exceptions to this rule are digenic and trigenic diseases, which can be identified by the use of the phrase "concomitant mutations", and generally are not separated using the ;; symbol.

- For long disease names, quotation marks are used, especially in sentences where the long name of a disease might cause confusion regarding verb tense.

- When describing the conditions, I will generally name the protein that is encoded by the causatory gene for the condition without specifying that it is a protein. This is because the word "protein" is not part of the official name. However, there are instances where the word "protein" is specifically used to enhance readability.

- For diseases that have acronyms, the letters in the acronym will be capitalized and underlined on the full name of the disease that will be the title of the section. At the end of this chapter, there will be a description of acronym diseases.

- For diseases that have non-English names, their original name and characters are used in this book.

- Finally, regarding the terms locus and loci, I use "locus" only when the region in the gene where the encoded protein is located is well known and identified. Otherwise and more often, I will use the term "loci" as both singular and plural to signify a large area within the gene for which our understanding of the mechanisms producing the disease is insufficient.

3.0.1.4 Databases

The miscellaneous information that forms part of this book is based on the databases that any student or physician might find beneficial. My first step was to retrieve information from the National Center for Biotechnology Information or NCBI (http://www.ncbi.nlm.nih.gov), which I found extremely helpful. The second step of my research was to retrieve essential information for this project from the Gene Information

(http://www.ncbi.nlm.nih.gov/gene/), the Online Mendelian Inheritance in Man or OMIM (http://www.ncbi.nlm.nih.gov/omim), and the SNP variation databases (http://www.ncbi.nlm.nih.gov/snp). Third, the code for the clinical information was derived mainly from the ICD-10, which is mentioned above. The information *per se* was obtained using all the previous databases. Finally, other databases that I found useful for this project are:
http://genome.ucsc.edu/
http://www.reactome.org/Reactome

http://hapmap.ncbi.nlm.nih.gov/
http://ensembl.org/
http://www.orpha.net/
http://www.sanger.ac.uk/genetics/CGP/cosmic/

The following Table (3.0.05) displays a list of all genomic disease groups that are covered in this chapter, and the subchapters that cover them. In each subchapter, diseases that are caused by environmental factors, genomic factors, and an interaction of the two will be described.

Table 3.0.05 List of Genomic Disorder Groups

Chapter	Subchapter where Described	Title
I	3.01.1	Infectious and Parasitic Diseases
II	3.02.1-11	Neoplasms
III	3.03.1-7	Diseases of the Blood and Blood-forming Organs and Disorders Involving the Immune Mechanism
IV	3.04.1-21	Endocrine, Nutritional and Metabolic Diseases
V	3.05.1-10	Mental and behavioral Disorders
VI	3.06.1-12	Diseases of the Nervous System
VII	3.07.1-10	Diseases of the Eye and Adnexa
VIII	3.08.1	Diseases of the Ear and Mastoid process
IX	3.09.1-9	Diseases of the Circulatory System
X	3.10.1-4	Diseases of the Respiratory System
XI	3.11.1-6	Diseases of the Digestive System
XII	3.12.1-7	Diseases of the Skin and Subcutaneous Tissue
XIII	3.13.1-9	Diseases of the Musculoskeletal System and Connective Tissue
XIV	3.14.1-8	Diseases of the Genitourinary System
XV	3.15.1-3	Pregnancy, Childbirth and the Puerperium
XVI	3.16.1-4	Conditions Originating in the Perinatal Period
XVII	3.17.1-10	Congenital Malformations, Deformations and Chromosomal Abnormalities
XVIII	3.18.1-8	Disease (not elsewhere Classified)

3.01.1 Subchapter I

Infectious and Parasitic Diseases:

A group of diseases where the signs and/or symptoms of the disease are caused by pathogenic biological agents. This chapter explains the hosts' genomic factors which are involved in infectious disease including the transmission of pathogenic agents, as well as the tools used to prevent infectious diseases, and immunities dealing with contagious agents. This chapter is devoted to the genomics of human beings in relationship to susceptibility and resistance to infection. For this reason, I will not consider the genomics of microorganisms in this chapter. The only people considered here are those with intrinsic genomic susceptibility or resistance and the concomitant presence of an infectious agent.

Tuberculosis: A disease caused by Mycobacterium tuberculosis. There are several genes that have been linked to tuberculosis susceptibility and resistance including *MC3R, IFNG, TIRAP, IRGM, SLC11A1, IL12RB1, CD209, CCL2, CISH, SP110, IFNGR1, IL12B, STAT1, IKBKG* and *IFNGR2*. There are several loci that have been identified, but not all the genes have been characterized. There is not a clear ethnic group in which DNA changes correlate with disease. Tuberculosis infection is common; between 90–95% of infections remain asymptomatic. Tuberculosis is a disease stemming from poverty, ill-health and malnutrition, rather than from a genomic disease. Furthermore, it has been concluded for a fact that each year this disease sees more than 8 million new cases worldwide and more than 1 million resultant deaths, mostly in developing countries. The understanding of the host's genomic factors in tuberculosis is limited since it is hard to study such a complex polygenetic disease. There are many exclusion criteria for genomic origin, mainly because there are many risk factors that are not of genomic origin: people with silicosis have a 30-fold higher risk; those with diabetes mellitus and low body weight (BMI below 18.5) have a 3-fold higher risk. Other risk factors that are not of genomic origin include: chronic renal failure, hemodialysis, cancer, etc… One of the most interesting inclusion criteria of genomic origin was seen in twin studies in the 1940s, which showed that susceptibility to tuberculosis is heritable.

Leprosy or Hansen disease: A chronic granulomatous disease of the peripheral nerves and mucosa of the upper respiratory tract caused by the bacteria Mycobacterium leprae. This disease has one of the highest social stigmas - a major obstacle to self-reporting and early treatment. There is not a clear ethnic group in which DNA changes correlate with disease. In the 4,000 year history of this disease, forced quarantine and segregation of patients have been implemented, but this was almost unnecessary since approximately 95% of people are naturally immune. There are at least 6 loci associated with vulnerability to leprosy. Polymorphisms in the *TLR2, LTA* and *TLR1* genes have been associated with leprosy susceptibility. Most of the cases worldwide are in India and Brazil.

Meningococcal disease: An infection caused by Neisseria meningitides. There is not a clear ethnic group in which DNA changes correlate with disease. However, many different patients could be at risk. The most striking risk factor, increasing risk by 7,000-fold, is the presence of deficiencies in the complement pathway. Most of the meningococcal diseases are in under-developed countries, particularly in Sub-Saharan Africa.

The Concise Encyclopedia of Genomic Diseases

Human immunodeficiency virus: An infection that causes acquired immunodeficiency syndrome which progresses to life-threatening opportunistic infections and cancers. There are several genes that have been linked to HIV susceptibility/resistance including *DEFA1, DEFA3, APOBEC3G, CEBPB, CD4, CXCR4, CCNT1, CCL5, KIR3DL1, HFE, ZBTB7A, OTUD4, AGFG1, AGFG2, HTATSF1, KAT5, HTATIP2, KIR2DL1, KIR2DL2, KIR2DL3, MAP3K5, ELANE, PPIA, CCL5, IFNG, CX3CR1, CXCL12, CCL3, CCL11, CCL2, SERPINA1, IL4R, CXCR1, CD209, CCL3L1, IL10,HLA-C, IL6, CCR5, TARDBP, TARBP2, TLR7, TRIM32*, and *TRIM5* genes. There is not a clear ethnic group in which DNA changes correlate with the disease (except the CCR5-Δ32 below), since this infection occurs through the transfer of bodily fluids (blood, semen, vaginal fluid, breast milk) rather than as a genomic disease. Furthermore, more than 30 million people are currently living with HIV. The estimated number of AIDS-related deaths is more than 1.5 million people per year. Sub-Saharan Africa is the worst affected area. It is well known that the viral mechanisms of disease in this RNA genome consist of 9 or 10 genes (gag, pol, env, tat, rev, nef, vif, vpr, vpu, and sometimes tev), encoding 19 proteins. General understanding of the host's genomic factors is more limited. This includes regulation in the apoptotic genes that down-regulate CD4, and in MHC class I and class II molecules. These considerations can help us to understand the ability of HIV to infect cells, replicate and cause disease. Also limited is the identification of the pattern of inheritance, or the ability to establish the genomic origin of the disease. One clear genomic example is the gene *CCR5*. The CCR5-Δ32 is a deletion mutation that has a specific impact on the function of T cells. The CCR5-Δ32 alleles originated from a single mutation event. This mutation event probably took place at least 700 years ago in northeastern Europe. CCR5-Δ32 biallelic individuals appear to be protected against smallpox and HIV. On the other hand, these individuals have a disproportionately higher risk of West Nile virus infection than controls. There are many exclusion criteria for genomic origin, mainly the transmission of HIV which is primarily via unprotected sexual intercourse, contaminated blood transfusions and hypodermic needles, etc. One of the most interesting inclusion criteria of genomic origin is some kind of natural immunity. In a low proportion of patients, the disease does not progress to classical AIDS. Instead, these patients become long-term non-progressors.

Malaria: A parasitic disease caused by microorganisms of the genus Plasmodium. There are several genes that have been linked to malaria susceptibility, resistance, and to susceptibility to cerebral malaria and placental malaria infection including *HBA1, HP, SPTA1, G6PD, HMOX1, CAPN2, IL12A, IL4, CAPN1, CAPNS1, KIR2DS4, HBG1, FLT1, FUT9, TAS2R16, TIRAP, HLA-DRB1, PKLR, GYPC, GYPA, NOS2, CR1, GYPB, TNF, ICAM1, NCR3* and *TCN2* genes. This disease has a clear correlation amongst DNA traits, such as sickle cell disease, thalassemias, glucose-6-phosphate dehydrogenase deficiency and resistance to malaria. Malaria correlates well with poverty and geographical areas. Furthermore, it has been concluded for a fact that each year this disease has more than 200 million new cases worldwide, and almost 1 million deaths. One clear example of genomic interest is the *HBB* gene, specifically the sickle cell trait, which is due to a change of glutamic acid 6 with valine in the beta-hemoglobin. This variant originated across the waist of Africa before the Bantu expansion which occurred about 2,000 years ago. Heterozygous people for this trait are resistant to malaria, while people with the homozygous variants are sensitive to malaria and have sickle cell disease. Finally, the homozygous wild-type are sensitive to malaria. The sickle cell trait or allele was expanded via the selective pressure of malaria. Malaria has many exclusion criteria for genomic origin, but it is mainly the fact that it is widely spread in tropical and subtropical regions, containing mosquito-borne infections occurring predominantly in Sub-Saharan Africa, Asia and the Americas. One of the most interesting inclusion criteria for genomic origin is the fact that other species genomically close to humans do not get the disease. For example, Plasmodium falciparum infects humans but does not infect chimpanzees. On the other hand, Plasmodium reichenowi affects chimpanzees, but not humans.

3.02.00 Subchapter II

Neoplasms

Malignant Neoplasia: An abnormal proliferation of tumoral cells. Cancerous cells divide and grow uncontrollably, forming malignant tumors, and invade other parts of the body. An examination of neoplasia reveals two main possibilities. The first suggests that 5-10% of patients with cancers is entirely related to familial, inherited or genomic origins. Second, the remaining 90-95% of patients have cancers which are still rooted in their genes. However, the main determining causes of patients with neoplasia are complex and mostly related to non-genomic factors. In order to describe each cancer disease, it is important to consider the criteria of exclusion and inclusion in several cancers. My aim is to show how understanding genomic factors and environmental causes of neoplasia in individuals helps to prevent

such diseases. Most of the patients with neoplasia have strong exclusion criteria, which mean that the genomic origin is not accurate. Instead, there are strong environmental factors which increase the risk of cancer, including tobacco use, certain infections, radiation, lack of physical activity, obesity and environmental pollutants. In addition to environmental factors, the age variable conveys that longevity can also increase the probability of cancer incidence. There are also historical comparisons that make use of exclusion criteria of genomic origin. These have been described, for instance, in lung cancer. An analysis of neoplasia also reveals that many patients have strong inclusion criteria for being of genomic origin. Besides the inherited and de novo alterations, which are examples of genomic or familial origins of cancer, there are twin studies, family history and first-degree relatives information that contribute to making each specific disease genomic based. Malignancy will be briefly considered in this subchapter, since the main reason for cancer is not inheritance or de novo germline alteration. Instead, the different types of cancer are related to de novo somatic mutations, which are not covered in this book. New DNA information has helped scientists to rethink the classification of cancer. For example, some genomic disorders have been split into multiple entities while other genomic disorders were merged into one condition. At the current time, 1 in every 100 persons is a carrier of a genomic alteration, which has a large effect on cancer risk. Nonetheless, in developed countries, the risk of developing cancer in the average individual's lifetime is 1 neoplasia in every 3 persons. Table 3.02.00.1 describes different types of Neoplasia, while Table 3.02.00.2 shows the prevalence of traits and disorders associated with Neoplasia.

Table 3.02.00.1 Types of Neoplasia

3.02.00-11	Neoplasms	
	3.02.01	Malignant Neoplasms of Digestive Organs
	3.02.02	Malignant Neoplasms of Respiratory System and Intrathoracic Organs
	3.02.03	Malignant Neoplasms of Skin
	3.02.04	Malignant Neoplasms of Breast and Female Genital Organs
	3.02.05	Malignant Neoplasms of Male Genital Organs
	3.02.06	Malignant Neoplasms of Urinary Organs
	3.02.07	Malignant Neoplasms of Eye, Brain and Central Nervous System
	3.02.08	Malignant Neoplasms of Endocrine Glands and Related Structures
	3.02.09	Malignant Neoplasms Stated or Presumed to be Primary, of Lymphoid, Hematopoietic, and Related Tissue
	3.02.10	Benign Neoplasms
	3.02.11	Neoplasms of Uncertain or Unknown Behavior

Table 3.02.00.2 Prevalence of Traits and Disorders Associated with Risk of Neoplasia

Disease	Gene altered	Incidence every 100,000 persons
Ataxia telangiectasia monoallelic	ATM	630-1,000
Down syndrome	*	140
Hereditary breast and ovarian cancer syndrome	BRCA1 and BRCA2	125-250
Lynch syndrome	MSH2 and MLH1	220
Familial adenomatous polyposis	APC	6-12
Wilms tumor	WT1	10
Retinoblastoma	RB1	5
Multiple endocrine neoplasia-1	MEN1	3-5
Multiple endocrine neoplasia-2	RET	2-3
Juvenile intestinal polyposis	SMAD4	1-6
Li-Fraumeni syndrome	TP53 and CHEK2	1-2
Basal cell nevus syndrome or Gorlin syndrome	PTCH1, PTCH2 and SUFU	1

(*) Trisomy

3.02.01 Malignant Neoplasms of Digestive Organs

Colorectal cancer: A neoplasia that arises in the colon and/or rectum, from polyps in most cases. The lifetime risk of colorectal cancer in the United States is: 7% in patients without any known gene mutations; 80% in patients suffering from Lynch syndrome or Hereditary nonpolyposis colorectal cancer; and almost 100% by the age of 40 in patients suffering from Familial adenomatous polyposis, due to mutations in the *APC* gene. There are also several exclusion criteria for genomic origin in colorectal cancer. There are clear geographical considerations for colorectal cancer, including mainly industrialized nations. These nations have many environmental factors that increase the risk compared to less developed nations. Unhealthy diet is the major risk factor, particularly diets high in red meat and low in fiber, fresh fruit, vegetables, poultry and fish. There are also other risks such as smoking, heavy drinking of alcoholic beverages, physical inactivity and inflammatory diseases. There are several inclusions criteria in colorectal cancer considered to be of genomic origin: for example, the product of conception alteration in the autosomal dominant pattern with certain degrees of de novo syndrome, such as Familial adenomatous polyposis and Hereditary nonpolyposis colorectal cancer. Products of conception altered in autosomal recessive patterns include: Turcot syndrome and Familial adenomatous polyposis, due to mutations in the *MUTYH* gene. Polyps of the colon, principally adenomatous polyps, are a risk factor for colon cancer. First degree relatives have a 2-3-fold larger risk, such as in patients with family history of colon cancer, particularly colorectal cancer in a family member that is diagnosed before the age of 55, or if colon cancer occurs in multiple relatives. Table 3.02.01 displays the diseases, the genes, and their biological function. It also includes the product of conception alteration. The biological functions include tumor suppression, DNA damage repair, histone acetyltransferase, enzyme-linked receptors, etc.

Lynch syndrome or Hereditary nonpolyposis colorectal cancer: A condition which has a high risk of developing into colorectal and endometrium cancer. There are also other cancers associated with this condition, such as those of the ovary, stomach, small intestine, etc... The combined risk for all cancers at age 70 is 91% for males and 69% for females. Male patients with mutations in any gene listed below have a 74% lifelong risk of devel-

oping colorectal cancer, while females have a significantly lower 30% risk. Among females, the largest risk is endometrium cancer at around 42%. The increased risk for these cancers is due to inherited monoallelic mutations that impair DNA mismatch repair mechanisms. Lynch syndrome I is associated with colon cancer, and Lynch syndrome II is associated with extracolonic cancer, predominantly carcinoma of the endometrium stomach, biliary and pancreatic system, and urinary tract. Muir-Torre syndrome, which is associated with sebaceous skin tumors, is another form of Lynch syndrome II. The genes related to these conditions are shared, and include proteins related to DNA mismatch repair *MSH2, MLH1, MSH6, PMS2, PMS1* and *MLH3*. There are other genes related to these conditions such as *TGFBR2* and *EPCAM*. There are several underlying causes of this disorder with at least 8 autosomal genomic loci: Lynch syndrome I or Hereditary nonpolyposis colorectal cancer-1 is caused by monoallelic mutations in the *MSH2* gene which encodes the mutS homolog 2 protein, this form may account for as many as 60% of Lynch syndrome cases;; Hereditary nonpolyposis colorectal cancer-2 is caused by monoallelic mutations in the *MLH1* gene which encodes the mutL homolog 1 protein: this form may account for as many as 30% of Lynch syndrome cases;; Hereditary nonpolyposis colorectal cancer-3 is caused by mutations in the *PMS1* gene which encodes the PMS1 postmeiotic segregation increased 1 protein;; Hereditary nonpolyposis colorectal cancer-4 is caused by mutations in the PMS2 gene which encodes the *PMS2* postmeiotic segregation increased 2 protein;; Hereditary nonpolyposis colorectal cancer-5 is caused by mutations in the *MSH6* gene which encodes the mutS homolog 6 protein;; Hereditary nonpolyposis colorectal cancer-6 is caused by mutations in the *TGFBR2* gene which encodes the transforming growth factor (beta receptor II);; Hereditary nonpolyposis colorectal cancer-7 is caused by mutations in the *MLH3* gene which encodes the mutL homolog 3 protein;; and Hereditary nonpolyposis colorectal cancer-8 is caused by epigenetic silencing of MSH2 triggered by deletion of 3-prime exons of the *EPCAM* gene and intergenic regions directly upstream of the MSH2 gene. Lynch syndrome accounts for around 1-3% of colorectal cancer cases, and 0.8-1.4% of endometrial cancers. Lynch syndrome is estimated to affect 220 in every 100,000 individuals.

Mismatch repair cancer syndrome: A disorder characterized by childhood cancer syndrome with 4 main tumor types: colorectal tumors and multiple intestinal polyps, hematologic malignancies, brain/central nervous system tumors, and other malignancies including embryonic tumors and rhabdomyosarcoma. Many patients display signs reminiscent of Neurofibromatosis type 1, predominantly multiple cafe-au-lait macules. The product of conception is altered when both parents are heterozygous and both parents provide the faulty copy to descendants (AD + AD = AR).

Turcot syndrome: A Mismatch repair cancer syndrome characterized by the combination of colorectal polyposis and primary tumors of the central nervous system. This inherited condition is uncommon and highly severe. The product of conception is altered when both parents are heterozygous and both parents provide the faulty copy to descendants (AD + AD = AR). Based on the original description provided by Turcot, we assume that the product of conception was altered in both alleles since the parents of the first 2 patients with this condition were third cousins. Turcot syndrome is caused by biallelic mutations in any of the DNA mismatch repair genes. Turcot syndrome is estimated to affect less than 1 in every 1,000,000 individuals.

Familial adenomatous polyposis: A condition characterized by the development of hundreds of polyps in the large intestine; cancerous transformation is then expected by the 4th decade of life. There are several underlying causes of this disorder with at least 2 autosomal genomic loci: Familial adenomatous polyposis-1 is caused by monoallelic loss-of-function mutations in the *APC* gene which encodes the adenomatosis polyposis coli, a tumor suppressor. These *APC* gene mutations can either be inherited mutations in 75% of all cases or de novo mutations in the remaining cases. In Ashkenazi Jews, Sephardi Jews and Arabs at least one pathogenic allele appears 2,200-2,950 years ago;; and Familial adenomatous polyposis-2 is caused by biallelic mutations in the *MUTYH* gene which encodes the mutY homolog protein (A/G-specific adenine DNA glycosylase). Familial adenomatous polyposis is a rare disorder affecting 6-12 patients in every 100,000 individuals. Familial adenomatous polyposis accounts for less than 1% of colorectal cancer cases. Gardner syndrome is characterized by the combination of polyposis, osteomas, fibromas and sebaceous cysts. Nowadays, it has been merged into Familial adenomatous polyposis, since the gene (*APC*) involved is the same.

Table 3.02.01 Neoplasms of Digestive Organs

Disease	OMIM	Gene	Biological Function	Product of Conception Alteration
Familial adenomatous polyposis-1	175100	APC	Tumor suppressor	AD + 25%
Gardner syndrome	175100	APC	Tumor suppressor	AD + 25%
Attenuated familial adenomatous polyposis-2	608456	MUTYH	DNA damage repair	AR
Hereditary mixed polyposis syndrome 2	120435	BMPR1A	Enzyme-linked receptor	AD
Hereditary nonpolyposis colorectal cancer-1	120435	MSH2	DNA mismatch repair	AD + de novo
Hereditary nonpolyposis colorectal cancer-2	609310	MLH1	DNA mismatch repair	AD + de novo
Hereditary nonpolyposis colorectal cancer-3	600258	PMS1	DNA mismatch repair	AD
Hereditary nonpolyposis colorectal cancer-4	600259	PMS2	DNA mismatch repair	AD
Hereditary nonpolyposis colorectal cancer-5	600678	MSH6	DNA mismatch repair	AD
Hereditary nonpolyposis colorectal cancer-6	190182	TGFBR2	Enzyme-linked receptor	AD
Hereditary nonpolyposis colorectal cancer-7	604395	MLH3	DNA mismatch repair	AD
Hereditary nonpolyposis colorectal cancer-8	613244	EPCAM	Cell adhesion molecule	AD
Mismatch repair cancer syndrome	120436	MLH1	DNA mismatch repair	AD + AD = AR
Mismatch repair cancer syndrome	609309	MSH2	DNA mismatch repair	AD + AD = AR
Mismatch repair cancer syndrome	600678	MSH6	DNA mismatch repair	AD + AD = AR
Mismatch repair cancer syndrome	600259	PMS2	DNA mismatch repair	AD + AD = AR
Turcot syndrome	276300		DNA mismatch repair	AD + AD = AR
Colonic adenoma recurrence	114500	ODC1	Decarboxylase	AD
Colorectal cancer	114500	EP300	Histone acetyltransferase	AD
Colorectal cancer	114500	TP53	DNA repair	AD
Colorectal cancer	114500	NRAS	Oncogene	AD
Colorectal cancer	114500	AXIN2	Regulation of G-protein signaling	AD
Colorectal cancer	114500	BUB1B	Serine/threonine kinase	AD
Colorectal cancer	114500	PDGFRL	Tumor suppressor	AD
Colorectal cancer	114500	PLA2G2A	Phospholipase	AD

Table 3.02.01. Continued.

Disease	OMIM	Gene	Biological Function	Product of Conception Alteration
Oligodontia-colorectal cancer syndrome	608615	AXIN2	Regulation of G-protein signaling	AD
Susceptibility to Colorectal cancer	114500	CCND1	Cyclins	
Susceptibility to Colorectal cancer	114500	TLR2	Receptor Other/ungrouped	
Susceptibility to Colorectal cancer	114500	TLR4	Receptor Other/ungrouped	
Susceptibility to Colorectal cancer	114500	AURKA	Serine/threonine protein kinase	

Hepatocellular carcinoma: A malignant primary liver neoplasm which seems to be due mainly to environmental factors such as aflatoxin, hepatitis B or C infection, and alcoholic cirrhosis. There are genomic causes of Hepatocellular carcinoma, such as Glycogen storage disease type IA, Hemochromatosis and Wilson's disease. There are several monoallelic mutations in the *CTNNB1* gene which encodes the catenin beta 1 (cadherin-associated protein), *PDGFRL* gene which encodes the platelet-derived growth factor receptor-like protein, and the *TP53* gene associated with hepatocellular carcinoma and hepatoblastoma. Hepatocellular carcinoma, childhood type, can be caused by mutations in the *MET* gene which encodes the met proto-oncogene (hepatocyte growth factor receptor).

Pancreatic cancer: A malignant neoplasm of the pancreas. There are several exclusion criteria in pancreatic cancer to be considered of genomic origin, such as smoking, diet, obesity and diabetes mellitus. There are several inclusion criteria in pancreatic cancer to be considered of genomic origin. There are several underlying causes of this disorder with at least 4 autosomal genomic loci: Pancreatic cancer-1 can be caused by mutations in the *PALLD* gene which encodes the palladin, a cytoskeletal associated protein;; Pancreatic cancer-2 can be caused by mutations in the *BRCA2* gene;; Pancreatic cancer-3 can be caused by mutations in the *PALB2* gene which encodes the partner and localizer of BRCA2 protein;; and Pancreatic cancer-4 can be caused by mutations in the *BRCA1* gene. Several familial cancer syndromes increase the risk of pancreatic cancer including: Hereditary nonpolyposis colon cancer syndrome; Peutz-Jeghers syndrome caused by monoallelic mutations in the *STK11* gene which encodes the serine/threonine protein kinase 11; the Melanoma-pancreatic cancer syndrome caused by monoallelic mutations in the *CDKN2A* gene which encodes the cyclin-dependent kinase inhibitor 2A; von Hippel-Lindau syndrome caused by monoallelic mutations in the *VHL* gene which encodes the von Hippel-Lindau tumor suppressor (E3 ubiquitin protein ligase); Ataxia-telangiectasia caused by biallelic mutations in the *ATM* gene which encodes the ataxia telangiectasia mutated protein; and Juvenile polyposis syndrome caused by monoallelic mutations in the *SMAD4* gene which encodes the SMAD family member 4 protein. Other genomic factors include family histories of pancreatic cancer.

3.02.02 Malignant Neoplasms of Respiratory System and Intrathoracic Organs

Lung cancer: A leading cause of cancer deaths worldwide. The 2 major forms of lung cancer are small cell lung cancer (SCLC) and nonsmall cell lung cancer (NSCLC), which account for 15% and 85% of all lung cancers, respectively. There are 3 major histologic subtypes in NSCLC: squamous cell carcinoma, adenocarcinoma and large cell lung cancer. Smoking causes all types of lung cancer, but is most strongly linked with SCLC and squamous cell carcinoma. In nonsmoking patients, Adenocarcinoma is most common. There are several exclusion criteria in lung cancer to be considered of genomic origin. Many environmental factors, such as long-term exposure to tobacco smoke, radon gas, asbestos, air pollution and secondhand smoking. Historical comparison in the lung cancer incidence is also an exclusion criterion for genomic origin in this condition. There are several inclusion criteria in lung cancer to be considered of genomic origin. For instance, there are germline mutations identified in the *EGFR* and *TP53* genes. Lung cancer susceptibility loci have been mapped to chromosome 6q23-q25, 5p15, 6p21 and 3q28. There are ethnic differences in lung cancer; for instance Japanese and Han Chinese individuals have a reduced risk of lung cancer than other ethnic groups, believed to be due to a more prevalent deletion in *CYP2A6* gene which encodes the cytochrome P450 2A6 protein, and the *CASP8* gene which encodes the caspase 8 protein (apoptosis-related cysteine peptidase). There are smoking individuals with reduced risk of lung cancer, associated with SNPs in *MPO* gene which encodes the myeloperoxidase. Nonsmokers account for 15% of lung cancer cases. Family history of lung cancer in first-degree relatives contributes to a risk of 1.45 times control.

3.02.03 Malignant Neoplasms of Skin

Melanoma: A malignant tumor of melanocytes. This neoplasia predominantly occurs in the skin, but is also found in other parts of the body, such as the eyes and bowel. 3 out of 4 deaths related to skin cancer are caused by melanoma. There are several exclusion criteria for genomic origin in melanoma. The most relevant environmental factor is sun exposure, particularly the exposure to ultraviolet radiation. There are socio-economic conditions that increase the risk of this condition. The incidence of melanoma has been increasing in recent years: this fact points to non-genomic factors. There are several inclusion criteria in melanoma to be considered of genomic origin. For instance, there are melanoma-prone families, mainly caused by mutations in the *CDKN2A* gene. Family history, mainly in ethnically fair and red-headed people, also contributes to the risk for this condition. Furthermore, individuals with multiple atypical nevi or dysplastic nevi, and individuals born with giant congenital melanocytic nevi are also at increased risk for melanoma.

Familial cutaneous malignant melanoma: A cutaneous condition characterized by susceptibility to familial cutaneous cancer. There are several underlying causes of this disorder with at least 9 autosomal genomic loci: Familial cutaneous malignant melanoma-1 has been mapped but the gene has not been characterized;; Familial cutaneous malignant melanoma-2 is caused by mutations in the *CDKN2A* gene which encodes the cyclin-dependent kinase inhibitor 2A;; Familial cutaneous malignant melanoma-3 is caused by mutations in the *CDK4* gene which encodes the cyclin-dependent protein kinase 4;; Familial cutaneous malignant melanoma-4 has been mapped but the gene has not been characterized;; Familial cutaneous malignant melanoma-5 is caused by mutations in the *MC1R* gene which encodes the melanocortin 1 receptor (alpha melanocyte stimulating hormone receptor);; Familial cutaneous malignant melanoma-6 is caused by mutations in the *XRCC3* gene which encodes the X-ray repair complementing defective repair in Chinese hamster cells 3;; Familial cutaneous malignant melanoma-7 has been mapped but the gene has not been characterized;; Familial cutaneous malignant melanoma-8 is caused by mutations in the *MITF* gene which encodes the microphthalmia-associated transcription factor protein;; and Familial cutaneous malignant melanoma-9 is caused by mutations in the *TERT* gene which encodes the telomerase reverse transcriptase protein. Familial cutaneous malignant melanoma accounts for around 3-15% of malignant melanoma cases.

3.02.04 Malignant Neoplasms of Breast and Female Genital Organs

Breast cancer: A condition that has a lifetime risk of 12% in women in the developed world. There are several exclusion criteria for genomic origin in breast cancer. Breast cancer varies across all the different ethnic backgrounds, affecting mainly African-American women, who are more likely to die of breast cancer than any other group, followed by Caucasian, Hispanic, American Indian/Alaska Native and Asian/Pacific Islander women. There are several non-genomic sources of breast cancer such as lack of childbearing or breastfeeding, higher hormone levels, smoking, high-fat diet, economic status, alcohol intake, lack of physical activity and maintaining an unhealthy weight. In other words, the number of cases of breast

cancer has increased worldwide mainly due to modern lifestyles. The primary risk factors for breast cancer are developmental since it is a disease that has an overwhelming majority of cases in women; in fact, breast cancer is more than 100 times more common in women than in men. Currently, most of the clinical and genomic information from the patient is obtained almost exclusively from the maternal allele. But the paternal allele could in many cases contribute to the cancer since most of the genes highly associated with breast cancer are mapped on autosomal chromosomes. There are several inclusion criteria in breast cancer to be considered of genomic origin. For instance, there is evidence that mutations at more than one locus can be involved in familial breast cancer. These familial forms can be caused by monoallelic mutations in the *BRCA1* gene which encodes the breast cancer susceptibility protein 1(in Ashkenazi Jews at least one pathogenic allele appears 920-1150 years ago), *BRCA2* gene which encodes the breast cancer susceptibility protein 2 (in Ashkenazi Jews at least one pathogenic allele appears 580-725 years ago), *BRCA3* gene which encodes the breast cancer susceptibility protein 3, *TP53* gene which encodes the tumor protein p53, *RAD51* gene which encodes the RAD51 homolog protein, *CHEK2* gene which encodes the checkpoint kinase 2, *BARD1* gene which encodes the BRCA1 associated RING domain 1 protein, and the *CDH1* gene which encodes the cadherin 1 protein. The product of conception alteration in Hereditary breast-ovarian cancer syndromes follows autosomal dominant patterns with certain degrees of de novo DNA syndrome. A twin study shows a two-fold increase in the risk for breast cancer in monozygotic twins, compared to dizygotic twins. The heritability component for breast cancer has been calculated at 30-40%. Family history of breast cancer is particularly important in bilateral breast cancer cases. Mutations in genes responsible for several forms of Fanconi anemia have been recognized as susceptibility factors for breast cancer. These include *BRCA2, BRIP1, PALB2* and *RAD51C*.

Hereditary breast-ovarian cancer syndromes: Are familial disorders which have higher than normal levels of breast cancer and ovarian cancer. The mutated genes associated with hereditary breast-ovarian cancer syndromes usually increase the risk 5-fold for breast cancer and 40-fold for ovarian cancer. In other words, the lifetime risk of breast cancer in women increases from 12-60% in the case of BRCA mutations. The lifetime risk for ovarian cancer in women is 1.6% but increases to 60-85 % in the case of BRCA mutations. In the United States, mutations in these BRCA genes account for 10-20% of all ovarian cancers and only 2-3% of all breast cancers. Table 3.02.04 displays the diseases, the OMIM, the genes, the biological functions and the product of conception alterations. Biological functions include tumor suppressor and DNA repair protein, while subcellular localization includes intranuclear localization. The product of conception alteration is mainly due to autosomal dominant inheritance and de novo alterations, except in Ataxia telangiectasia which is autosomal recessive. Table 3.02.04 displays the most common genes associated with breast-ovarian cancer; *BRCA1* and *BRCA2* genes can produce very high rates of breast and ovarian cancer, as well as increased rates of other cancers. There are also other genes related to this condition, such as *TP53, PTEN, CDH1, STK11, CHEK2,* and *ATM*. Li–Fraumeni syndrome occurs with a variety of tumor types including soft tissue sarcomas and osteosarcomas, breast cancer, brain tumors, leukemia and adrenocortical carcinoma. Li–Fraumeni syndrome pa-

tients have a 25-fold higher risk of developing cancer. The last syndrome is caused by monoallelic mutations in *TP53* gene, which responds to diverse cellular stresses to regulate target genes that induce cell cycle arrest, apoptosis, senescence, DNA repair or changes in metabolism. Li–Fraumeni syndrome is inherited in 80-93% of the cases, and occurs de novo in the remaining cases. Li–Fraumeni syndrome is a common cause of breast cancer in women under 30 years of age. Li–Fraumeni syndrome affects 1-2 patients in every 100,000 individuals. Patients with Li-Fraumeni simil syndrome have a variety of tumor types including sarcoma, breast cancer and brain tumors, and also an increased risk of colon cancer and prostate cancer. Li-Fraumeni syndrome-2 is caused by mutations in *CHEK2* gene. Cowden syndrome patients suffer from skin growths, hamartomas in the colon, and increased risk for many cancers. Cowden syndrome-1 is caused by monoallelic mutations *PTEN* gene. Lobular breast cancer and gastric cancer is caused by mutations in *CDH1* gene. Patients with Peutz–Jeghers syndrome or hereditary intestinal polyposis syndrome have a predisposition to breast cancer, pancreatic cancer, hamartomatous polyps, and hyperpigmented macules on the lips and oral mucosa. Peutz–Jeghers syndrome is caused by mutations in *STK11* gene which encodes the serine/threonine protein kinase 11. Peutz–Jeghers syndrome affects 1-6 in every 100,000 individuals. Patients with Ataxia telangiectasia syndrome have ataxia and telangiectasia, and increased incidences of lymphomas and leukemia. While the mode of inheritance for Ataxia telangiectasia has been classically described as autosomal recessive or biallelic, the product of conception alteration is considered as (AD + AD = AR) in this chapter, since both parents provide the faulty copy to descendents. In fact, female heterozygous or monoallelic carriers have double the normal risk of developing breast cancer. 1% of people are heterozygous carriers for Ataxia telangiectasia. Ataxia telangiectasia syndrome is caused by biallelic mutations in *ATM* gene which encodes the ataxia telangiectasia mutated protein. There are several cases of hereditary breast-ovarian cancer syndromes in which either the genes involved have not been identified, or there are multiple genes. Hereditary breast-ovarian cancer syndromes type 1 and 2 affect 12-25 patients in every 10,000 individuals. There is a higher mutation rate for *BRCA1* and *BRCA2* in Ashkenazi Jews.

Table 3.02.04 Neoplasms of Breast and Female Genital Organs

Disorder	OMIM	Gene	Biological Function	Product of Conception Alteration
Hereditary breast and ovarian cancer syndrome-1	604370	BRCA1	Repairing damaged DNA	AD + de novo
Hereditary breast and ovarian cancer syndrome-2	612555	BRCA2	Repairing damaged DNA	AD + de novo
Hereditary breast and ovarian cancer syndrome-3	613399	RAD51C	DNA repair protein	AD
Breast cancer	605100	PPM1D	Phosphatase magnesium-dependent	AD
Breast cancer	191170	TP53	Tumor suppressor	AD + de novo
Breast cancer	133430	ESR1	Transcription factor diseases Zinc finger	AD
Early-onset breast cancer	605882	BRIP1	Tumor suppressor	AD
Invasive ductal breast cancer	603615	RAD54L	DNA repair and recombination	AD
Lobular breast cancer	192090	CDH1	Cytoskeleton	AD
Protection against breast cancer	601763	CASP8	Caspase	AD
Susceptibility to breast cancer	607585	ATM	DNA repair MRN complex	
Susceptibility to breast cancer	601593	BARD1	Tumor suppressor	
Susceptibility to breast cancer	604373	CHEK2	DNA repair protein	
Susceptibility to breast cancer	600936	HMMR	Cell motility	
Susceptibility to breast cancer	160998	NQO2	Dehydrogenase	
Susceptibility to breast cancer	610355	PALB2	Intranuclear localization	
Susceptibility to breast cancer	176705	PHB	Tumor suppressor	
Susceptibility to breast cancer	179617	RAD51	Recombinase	
Susceptibility to breast cancer	600675	XRCC3	DNA repair protein	
Susceptibility to breast cancer in males	600185	BRCA2	Tumor suppressor	AD
Familial endometrial cancer	608089	MLH3	DNA mismatch repair	AD
Familial endometrial cancer	608089	MSH6	DNA mismatch repair	AD

Cervical cancer: A malignant neoplasm of the cervix, the lower part of the uterus. The main factor associated with this condition is the human papillomavirus (HPV) infection. Genomic factors include a family history of cervical cancer, which causes a 1.8-fold increase in the risk among female members compared to the control group.

Ovarian cancer: A neoplasia arising from the ovary. There are hereditary forms already described such as those caused by mutations in *BRCA1* gene, *BRCA2* gene, *MSH2* gene which encodes the mutS homolog 2, and *MLH1* gene which encodes the mutL homolog 1. Non-genomic factors include the fact that the more children a woman has, the lower her risk of ovarian cancer. Genomic factors include a lifetime risk 3 times higher in affected first-degree relatives compared to the control group.

3.02.05 Malignant Neoplasms of Male Genital Organs

Prostate cancer: A form of cancer that develops in the prostate. There are exclusion criteria for genomic origin such as diet, obesity and geographical factors. The primary risk factors for prostate cancer are developmental since only men are affected. There are several inclusion criteria in prostate cancer to be considered of genomic origin such as family history, large differences among ethnic groups, mutated genes, etc. In the United States, Prostate cancer is more prevalent in African-Americans than in Caucasians or Hispanic Americans, and is also more deadly in African-Americans. There are several underlying causes of this disorder with at least 15 autosomal genomic loci and 1 X-linked genomic locus. No single gene is responsible for prostate cancer; many different genes have been involved. There are sev-

eral genes which predispose one to this cancer, including *EHBP1* gene which encodes the EH domain binding protein 1, *MSMB* gene which encodes the beta-microseminoprotein, *EPHB2* gene which encodes the EH domain binding protein 2, *HIP1* gene which encodes the huntingtin interacting protein 1, *RNASEL* gene which encodes the ribonuclease L protein (2',5'-oligoisoadenylate synthetase-dependent), *CDH1* gene which encodes the cadherin 1 protein, *CD82* gene which encodes the CD82 molecule, *ELAC2* gene which encodes the elaC homolog 2 protein, *ZFHX3* gene which encodes the zinc finger homeobox 3, *MXI1* gene which encodes the MAX interactor 1 (dimerization protein), and *AR* gene which encodes the androgen receptor. There are several genes which predispose one to hereditary Prostate cancer, including *MSR1* gene which encodes the macrophage scavenger receptor 1, *BRCA1* gene, *BRCA2* gene, and *CHEK2* gene which encodes the checkpoint kinase 2. There is also a locus associated with Prostate cancer aggressiveness on chromosome 19. Unlike in the case of breast cancer, the clinical information from the patient is obtained almost exclusively from the paternal allele. However, the maternal allele in many cases could be contributing to this form of cancer, since most of the genes highly related with prostate cancer are mapped on autosomal chromosome except for the Androgen receptor gene (*AR*). The risk for prostate cancer seems to be larger for men with an affected brother than for men with an affected father.

Testicular cancer: A cancer that arises in the testicles. The lifetime risk of testicular cancer is 1 in every 250 male individuals. There are several exclusion criteria for genomic origin in testicular cancer. The major risk factor for the development of testicular cancer is cryptorchidism, which can be either of environmental or genomic origin. Other non-genomic risk factors for Testicular cancer include: inguinal hernia, mumps orchitis, lack of physical activity and sedentary lifestyle. The fact that worldwide incidence has doubled since the 1960's also points to non-genomic factors. An inclusion criterion for testicular cancer is the fact that Caucasian men commonly have this tumor, while it is rare among African-American men. Patients with certain genomic disorders such as Down syndrome, Peutz-Jeghers syndrome, Androgen insensitivity, Aromatase excess syndrome and Male Precocious puberty have risk several times higher than individuals without these conditions.

3.02.06 Malignant Neoplasms of Urinary Organs

Renal cell carcinoma or Hypernephroma: A kidney cancer that arises in the lining of the proximal convoluted tubule. There are several exclusion criteria for genomic origin in hypernephroma. The strongest known risk factors are cigarette smoking, hypertension and obesity. Patients in dialysis due to cystic disease of the kidney revealed a risk 30 times greater than in the general populace for the development of hypernephroma. The incidence of hypernephroma has been rising progressively in the last 35 years. Renal cell cancer is 1.6 times more common in men than women. There are several inclusion criteria in Hypernephroma to be considered of genomic origin, such as other genomic conditions, large differences in ethnic groups, mutated genes, etc. Patients with certain genomic disorders such as von Hippel-Lindau disease and Birt-Hogg-Dubé syndrome display a greater risk of Hypernephroma. Inclusion criteria for Hypernephroma include the fact that African-Americans have this tumor more commonly than Caucasians. There are several gene mutations which predispose one to this cancer, including: *HNF1A* gene which encodes the HNF1 homeobox A, *RNF139* gene which encodes the ring finger protein 139, *DIRC2* gene which encodes the disrupted in renal carcinoma 2 protein, and *HNF1B* gene which encodes the HNF1 homeobox B.

Wilms' tumor or Nephroblastoma: A cancer of the kidneys that typically occurs in children. There are several underlying causes of this disorder with at least 5 autosomal genomic loci: Wilms tumor-1 can be caused by mutations in the *WT1* gene which encodes the Wilms tumor 1 protein, a transcription factor essential to the normal development of the urogenital system;; Wilms tumor-2 can be caused by mutations of the imprinted *H19* gene, a maternally imprinted expressed transcript (non-protein coding). This non-coding RNA functions as a tumor suppressor;; Wilms tumor-3 and 4 have been mapped, but the genes have not been characterized;; and Wilms tumor-5 can be caused by mutations in the *POU6F2* gene which encodes the POU class 6 homeobox 2 protein. There are several gene mutations which predispose one to this cancer, including *BRCA2* gene and the non-protein coding *WT1-AS* gene (WT1 antisense RNA). Wilms tumor affects 1 child in every 10,000 and accounts for 8% of childhood cancers.

3.02.07 Malignant Neoplasms of Eye, Brain and Central Nervous System

Retinoblastoma: A rapidly developing cancer in the retina. Retinoblastoma affects approximately 1 in 15,000-20,000 live births, but it is the most commonly inherited childhood malignancy. Retinoblastoma can be caused by mutations on the *RB1* gene gene which encodes the retinoblastoma 1 protein. The defective *RB1* gene can be inherited from either parent, or occurs as a *de novo* event. Most of the de novo mutations in the *RB1* gene seem to originate in the paternal allele.

Meningioma: A benign neoplasia that affects the meninges. There are several exclusion criteria for genomic origin. For example, patients who have undergone radiation to the scalp are at higher risk for developing meningioma. There are familial cases of meningioma. There are several gene mutations which predispose one to the familial form of meningioma, including *PDGFB* gene which encodes the platelet-derived growth factor beta polypeptide, *SMARCB1* gene which encodes the SWI/SNF related matrix associated actin dependent regulator of chromatin (subfamily b member 1), and *MN1* gene which encodes the meningioma 1 protein (disrupted in balanced translocation). On the contrary, de novo forms of meningioma are caused by de novo mutations in the *NF2* gene which is predisposed to this condition. This *de novo* product of conception alteration originates from the maternal allele.

Medulloblastoma: An extremely malignant primary brain tumor that affects children. Medulloblastoma originates in the cerebellum or posterior fossa. Medulloblastoma accounts for 15-20% of all pediatric brain tumors. There are several exclusion criteria for genomic origin such as the fact that the incidence is higher in males (62%) than in females (38%). Patients with certain genomic disorders such as Gorlin syndrome, Turcot syndrome and leukoencephalopathy display a higher risk of medulloblastomas.

Neuroblastoma: A condition that is both the most common cancer in infancy and the most common extracranial solid cancer in childhood. The etiological factors associated with neuroblastoma are not well understood, but there are familial predispositions to neuroblastoma. For example, Familial neuroblastoma can be triggered by germline mutations in the *ALK* gene which encodes the anaplastic lymphoma kinase. There are also mutations (in the *NME1* gene which encodes the NME/NM23 nucleoside diphosphate kinase 1, the *KIF1B* gene which encodes the kinesin family member 1B protein, and the *PHOX2B* gene which encodes the paired-like homeobox 2b) which are highly associated with neuroblastoma predisposition.

3.02.08 Malignant Neoplasms of Endocrine Glands and Related Structures

Medullary thyroid cancer: A tumor that originates in the parafollicular cells or C cells. 1 in every 4 of these tumors is genomic in essence, principally caused by mutations in the *RET* proto-oncogene. Autosomal dominant or monoallelic mutations in the *RET* proto-oncogene cause Familial medullary thyroid cancer. When this familial tumor coexists with hyperplasia of the parathyroid gland and pheochromocytoma, it is called Multiple endocrine neoplasia type 2. The product of conception alteration in the *RET* proto-oncogene involves inheritance, and/or de novo alterations originating mainly in the paternal allele, especially in multiple endocrine neoplasia 2B type.

Pheochromocytoma: A neuroendocrine tumor originating in the chromaffin cells of the adrenal glands, or extra-adrenal chromaffin tissue. This tumor secretes excessive amounts of catecholamines displaying hyperactivity of the sympathetic nervous system. Pheochromocytoma was classically described as the 10% tumor. Nowadays, the inheritance that correlates with this tumor is higher: 1 in 4 of all Pheochromocytomas cases may be of genomic origin. Familial pheochromocytoma is caused by monoallelic mutations in: the *VHL* gene which encodes the von Hippel-Lindau tumor suppressor (E3 ubiquitin protein ligase), *RET* gene which encodes the ret proto-oncogene, *NF1* gene which encodes the neurofibromin 1, *SDHB* gene which encodes the succinate dehydrogenase complex (subunit B), *SDHC* gene which encodes the succinate dehydrogenase complex (subunit C), or the *SDHD* gene which encodes the succinate dehydrogenase complex (subunit D). Pheochromocytoma is a tumor of the Multiple endocrine neoplasia syndrome type IIA and type IIB, caused by mutations in the *RET* gene.

Paraganglioma: A generally benign neuroendocrine neoplasm. The familial forms of Paragangliomas are caused by mutations in the *SDHD* gene, *SDHB* gene, *SDHC* gene, or the *SDHAF2* gene which encodes the succinate dehydrogenase complex assembly factor 2. Paragangliomas might also occur in Multiple endocrine neoplasia syndrome type IIA and type IIB, caused by mutations in the *RET* gene.

3.02.09 Malignant Neoplasms Stated or Presumed to be Primary, of Lymphoid, Hematopoietic, and Related Tissue

Lymphoma: A type of blood cancer. Nowadays, there are over 80 different forms of lymphomas in 4 broad groups. This classification is based on somatic mutations in the Cluster of Differentiation CD 19, CD 20, CD5, etc. This book refers to inheritance, not to somatic mutations, so the preference here is to use the traditional classification "Hodgkin" vs. "Non-Hodgkin".

Hodgkin's lymphoma: A type of lymphoma characterized by the orderly spread of disease from one lymph node group to another. This condition was named after Thomas Hodgkin in 1832. There are several exclusion criteria such as the fact that the incidence of Hodgkin's lymphoma is increasing due to the higher prevalence of patients with weakened immune systems, including those infected with HIV; this association is particularly different from other lymphomas since it affects patient with higher CD4 T cell counts. The exposure to exotoxins, such as Agent Orange, also increases the incidence of Hodgkin's lymphoma. The familial forms can be caused by mutations in the *KLHDC8B* gene which encodes the kelch domain containing protein 8B. It is generally more common in males, though the nodular sclerosis variant is marginally more common in females.

Non-Hodgkin's lymphoma: A type of lymphoma characterized by significant variations in their severity, from indolent to very aggressive. The fact that certain types of non-Hodgkin's lymphoma such as Kaposi's sarcoma are AIDS-defining cancers is an example of exclusion criteria for genomic origin. There are familial forms which can be caused by mutations in the *PRF1* gene which encodes the perforin 1 (pore forming protein). Patients with certain genomic disorders such as Lymphoproliferative syndrome, Ataxia telangiectasia syndrome, Klinefelter syndrome, Chédiak-Higashi syndrome, and Cartilage-hair hypoplasia display a greater risk of non-Hodgkin's lymphoma.

Acute lymphoblastic leukemia: A form of leukemia characterized by excess lymphoblasts. Acute lymphoblastic leukemia is the most common type of childhood cancer. Acute lymphoblastic leukemia accounts for approximately 70% of all childhood leukemia cases. There are several exclusion criteria for genomic origin such as the association between radiation and leukemia. An inclusion criterion for genomic origin in childhood leukemia is the fact Caucasians more commonly have this form of leukemia than African-Americans, Asians or Hispanics. Patients with certain genomic disorders such as Down syndrome, Fanconi anemia, Bloom syndrome, Ataxia telangiectasia, X-linked agammaglobulinemia and Severe combined immunodeficiency display a greater risk of Acute lymphoblastic leukemia. Males are slightly more commonly affected than females.

Acute myeloid leukemia: A form of leukemia characterized by an excess of abnormal white blood cells. There are several exclusion criteria for genomic origin in Acute myeloid leukemia. For example, there are clear correlations for Acute myeloid leukemia with chemical exposures (anti-cancer chemotherapy, benzene, etc...), ionizing radiation and blood disorders. Acute myeloid leukemia is 1.3 times more common in males than in females. There is a hereditary or familial risk for Acute myeloid leukemia especially when multiple members of a family develop this condition. First-degree relatives of patients with this condition have a relative risk 3 times higher than control. Several congenital conditions may increase the risk of this form of leukemia, including the common (Down syndrome) and the rare (Fanconi anemia), platelet disorder with associated myeloid malignancy, telomere-related pulmonary fibrosis and/or bone marrow failure. Germline mutations in the *GATA2* gene which encodes the GATA binding protein 2 (transcription factor),

TERC gene which encodes the telomerase RNA component, and *TERT* gene which encodes the telomerase reverse transcriptase are associated with susceptibility to the development of Acute myeloid leukemia.

Chronic myelogenous leukemia or Chronic granulocytic leukemia: A clonal myeloproliferative disorder of a pluripotent stem cell with a specific somatic-cytogenetic chromosomal translocation known as the Philadelphia chromosome. There are few cases of Chronic myelogenous leukemia in successive generations, which points to a familial or genomic mechanism. The only well-described risk factor for Chronic myelogenous leukemia is exposure to ionizing radiation.

3.02.10 Benign Neoplasms

Juvenile Polyposis Syndrome: A disorder characterized by the appearance of hamartomatous polyps in the digestive tract. There are familial cases of Juvenile Polyposis Syndrome associated with mutations in the *BMPR1A* gene which encodes the bone morphogenetic protein receptor (type 1A), and *SMAD4* gene which encodes the SMAD family (member 4) protein. There are also other genomic syndromes such as Cowden syndrome or other phenotypes of the PTEN hamartoma tumor syndrome that also course with Juvenile Polyposis Syndrome.

Hepatocellular adenoma: An uncommon benign liver tumor which is associated with the use of hormonal contraception with high estrogen content. 90% of hepatic adenomas arise in women. Familial hepatic adenomas can be caused by biallelic loss-of-function mutations in the *HNF1A* gene which encodes the HNF1 homeobox

A, a transcription factor. Hepatic adenomas also occur in high frequency with Glycogen storage disease (Type Ia) or von Gierke disease, which is caused by biallelic mutations in the *G6PC* gene.

Cavernous angioma or cavernous hemangioma: A vascular disorder with features of hamartomatous tumor. The product of conception in this vascular disorder appears to exhibit monoallelic mutations either by inheritance or de novo alterations. There are several underlying causes of this disorder with at least 4 autosomal genomic loci: Cerebral cavernous malformations-1 can be caused by monoallelic mutations in the *KRIT1* gene which encodes the ankyrin repeat-containing protein KRIT1;; Cerebral cavernous malformations-2 can be caused by monoallelic mutations in the *CCM2* gene which encodes the cerebral cavernous malformation 2 protein;; Cerebral cavernous malformations-3 can be caused by monoallelic mutations in the *PDCD10* gene which encodes the programmed cell death 10 protein;; and Cerebral cavernous malformations-4 locus has been identified but the gene has not yet been characterized. "Cerebral cavernous malformations" is estimated to affect around 1 in every 1,000 individuals.

Atrial myxoma: A benign tumor found in the atria of the heart. There are familial myxomas which account for 10% of all mixomas. Atrial myxoma can be associated with mutations in the *PRKAR1A* gene which encodes the cAMP-dependent protein kinase regulatory alpha subunit (type 1). Myxomas are more common in females than in males.

Pallister-Hall syndrome: A disorder characterized by development disturbance of many parts of the body. This condition is named after Judith Hall and Philip Pal-

lister. Pallister-Hall syndrome is caused by autosomal dominant mutations in the *GLI3* gene which encodes the GLI family zinc finger 3 protein. The product of conception alterations could exhibit either inheritance or de novo DNA alterations.

Gorlin syndrome or Basal cell nevus syndrome: A condition that can cause unusual facial appearances and a predisposition for basal cell carcinoma. Gorlin syndrome also involves defects in multiple body systems such as the skin, nervous system, eyes, endocrine system and bones. Affected individuals are predominantly prone to developing a common and typically non-life-threatening form of non-melanoma skin cancers. Roughly 10% of individuals with the condition do not develop basal cell carcinomas. This condition is named after Robert J. Gorlin. The product of conception alterations in Gorlin syndrome appears to exhibit autosomal dominant or monoallelic mutations, either by inheritance in 60% of cases or de novo alterations in 40% of cases. Most of the de novo mutations in Gorlin syndrome seem to originate in the paternal allele and there is evidence of advanced paternal age effect in de novo mutations. Gorlin syndrome is caused by mutations in the *PTCH1* gene which encodes the patched 1 protein, *PTCH2* gene which encodes the patched 2 protein, and *SUFU* gene which encodes the suppressor of fused homolog protein. Gorlin syndrome affects 1 in every 57,000-256,000 individuals.

3.02.11 Neoplasms of Uncertain or Unknown Behavior

Carney complex or LAMB and NAME syndrome: A disorder characterized by lentiginosis, endocrine overactivity, and myxomas of the heart and skin. About 7% of all cardiac myxomas are associated with

Carney complex. There are several underlying causes of this disorder with at least 2 autosomal genomic loci: Carney complex type 1 can be caused by monoallelic mutations in the *PRKAR1A* gene which encodes the cAMP-dependent protein kinase regulatory subunit (type 1 alpha). Type 1 is the most common form;; and Carney complex type 2 has been mapped but the gene has not yet been characterized. Carney complex has been reported in association with several conditions including: Isolated primary pigmented nodular adrenocortical disease-1 and Isolated cardiac myxoma. There are also patients with Carney complex and Distal arthrogryposis, which are caused by mutations in the MYH8 gene.

Multiple endocrine neoplasia type 1 or Wermer syndrome: A disorder that affects the endocrine system. These individuals exhibit multiple endocrine neoplasia affecting the parathyroid, pancreas and pituitary glands. The product of conception alteration in Wermer syndrome appears to exhibit either autosomal dominant inheritance in 90% of cases or de novo alterations in the remaining 10% of cases. Multiple endocrine neoplasia type 1 is caused by monoallelic mutations in the MEN1 gene which encodes the menin, a putative tumor suppressor. Multiple endocrine neoplasia 1 is estimated to affect 3-5 in every 100,000 individuals.

Multiple endocrine neoplasia 2: A disorder that affects the endocrine system. The features include Pheochromocytoma and Medullary thyroid carcinoma. Most cases of MEN2 derive from gain-of-function monoallelic mutations in the *RET* gene which encodes the *RET* proto-oncogene, a receptor tyrosine kinase that transduce signals for cell growth and differentiation. The encoded protein is specific for cells of neural crest origin. The product of conception in Multiple endocrine neoplasia 2 appears to exhibit autosomal dominant or monoallelic mutations, either by inheritance or de novo alterations. Multiple endocrine neoplasia 2 is estimated to affect 2-3 in every 100,000 individuals.

Multiple endocrine neoplasia 2A or Sipple syndrome: Besides the tumor in thyroid and adrenal glands, patients are additionally characterized by the presence of parathyroid hyperplasia or tumor. Most cases are inherited but 5.6-9% of cases are caused by de novo alterations.

Multiple endocrine neoplasia 2B or 3 or Mucosal neuromata with endocrine tumors: Besides the tumor in thyroid and adrenal glands, patients are additionally characterized by the presence of mucocutaneous neuroma, gastrointestinal symptoms, muscular hypotonia and Marfanoid habitus. Multiple endocrine neoplasia 2B is the most severe type of multiple endocrine neoplasia. The product of conception alteration in this condition is equally due to autosomal dominant inheritance and de novo events. The *de novo* mutated *RET* gene is almost exclusively paternal, predominantly from older fathers. The gender ratio is also uneven: sons are 2 times more likely than daughters to develop Multiple endocrine neoplasia 2B.

3.03.00 Subchapter III

3.03.00.1 Diseases of the Blood and Blood-forming Organs and Disorders Involving the Immune Mechanism

Table 3.03.00-6 Diseases of the Blood and Blood-forming Organs and Disorders Involving the Immune Mechanism

3.03.00-6	Diseases of the Blood and Blood-forming Organs and Disorders Involving the Immune Mechanism	
	3.03.1	Nutritional Anemias
	3.03.2	Hemolytic Anemias
	3.03.3	Aplastic and other Anemias
	3.03.4	Coagulation Defects, Purpura and other Hemorrhagic Conditions
	3.03.5	Other Diseases of Blood and Blood-forming Organs
	3.03.6	Certain Disorders Involving the Immune Mechanism
3.03.7	Dissasortative Mating in Ethnics Groups and Disease Prevention	

The primary goal of the subchapter on Blood and blood-forming organs and disorders involving the immune mechanism is to provide a sense of the prevalence of these diseases. The first section will address altogether the set of people where variations in DNA correlate with a disease. Table 3.03.00.1 displays the diseases, the worldwide frequencies of conditions, and the set of people with higher prevalence (if available). This section illustrates common and rare diseases. With regard to the table, only von Willebrand disease, Factor V Leiden, Thrombophilia Factor II, Hemolytic anemia (G6PD), Alpha-thalassemia, Beta-thalassemia, Immunodeficiency (complements deficiency C9), Thrombophilia, due to protein C and S deficiency, Sickle cell anemia, Elliptocytosis and Spherocytosis are considered common in the nationalities and/or ethnic groups described. The rest of these diseases are considered to be rare, occuring in less than 1 in every 2,000 individuals. In the following subsections, there are many common diseases not found in Table 3.03.00.2, since they are caused by a combination of genomic and environmental factors (e.g. Nutritional anemia).

The Concise Encyclopedia of Genomic Diseases

Table 3.03.00.2 Diseases of the Blood and Blood-forming Organs and Disorders Involving the Immune Mechanism in Different Ethnic Groups

Disease	Frequency of condition	People with higher prevalence (if available)
von Willebrand disease monoallelic	1 in 100	Worldwide and Åland Islanders
von Willebrand disease biallelic	1 in 40,000*	Worldwide and Åland Islanders
Factor V Leiden monoallelic	1-5 in 100	Caucasians
Factor V Leiden biallelic	1 in 2,000*	Caucasians
Thrombophilia Factor II monoallelic	2 in 100	Caucasians
Hemolytic anemia, due to G6PD	2-260 in 1,000	Africans, Middle Easterns, and Southern Asians
Alpha-thalassemia trait, Hemoglobin H disease and *hydrops fetalis* syndrome	1-300 in 1,000	Equatorial Africans and South-eastern Asians
Beta-thalassemia intermediate and major	1-50 in 10,000	Mediterraneans, Middle Easterns, Transcaucasians, Central Asians, Indians, Pakistanis and Bangladeshis
Immunodeficiency, due to complements deficiency C9	1 in 1,000	Japanese and Koreans
Thrombophilia, due to protein C deficiency	2 in 1,000	Caucasians
Thrombophilia, due to protein S deficiency	2 in 1,000	Caucasians
Hemophilia A	1 in 6,000 Males	Worldwide
Sickle cell anemia	1-200 in 10,000	Sub-Saharan Africans, Indians and Middle Easterns
Elliptocytosis	3-2,000 in 10,000	Malaysians and Equatorial Africans
Spherocytosis	1-5 in 10,000	Northern Europeans
Platelet glycoprotein IV deficiency	2-400 in 10,000	Japanese
Thrombophilia, due to congenital antithrombin deficiency	3-5 in 10,000	Caucasians
Thrombophilia, due to heparin cofactor 2 deficiency	3-5 in 10,000	Caucasians
DiGeorge syndrome	3 in 10,000	Unknown
Common variable immunodeficiency	2 in 100,000	Unknown
Hemophilia B	3-5 in 100,000 Males	Worldwide
Hemophilia C	1 in 100,000	Ashkenazi Jews and French Basques
Factor VII deficiency or hypoproconvertinemia	3 in 1,000,000	Oriental Jews
Immunodeficiency, due to complements deficiency C5toC8	1-9 in 100,000	Unknown
Hereditary Angioedema 1 and 2	2 in 100,000	Unknown

The Concise Encyclopedia of Genomic Diseases

Table 3.03.00.2. Continued.

Disease	Frequency of condition	People with higher prevalence (if available)
Agammaglobulinemia, X-linked or Bruton type agammaglobulinemia	1 in 100,000 Males	Unknown
Severe combined immunodeficiency	1-2 in 100,000	Unknown
Wiskott-Aldrich syndrome	4-10 in 1,000,000	Unknown
Severe combined immunodeficiency, Athabaskan type	<1 in 1,000,000	Navajo and Apache people
Immunodeficiency with hyper IgM type 1	2 in 1,000,000 Males	Unknown
Bernard-Soulier syndrome	1 in 1,000,000	Unknown
Agammaglobulinemia	<1 in 1,000,000	Unknown
Autoimmune lymphoproliferative syndrome 1A	<1 in 1,000,000	Unknown
Hemorrhagic diathesis, due to alpha2-plasmin inhibitor deficiency	<1 in 1,000,000	Unknown
ATRUS syndrome or Radioulnar synostosis with amegakaryocytic thrombocytopenia	<1 in 1,000,000	Unknown
Combined deficiency of Factor V and factor VIII	<1 in 1,000,000	Sephardic Jews
Dyserythropoietic anemia with thrombocytopenia	<1 in 1,000,000	Unknown
Factor XII deficiency	<1 in 1,000,000	Unknown
Factor XIIIA deficiency	<1 in 1,000,000	Unknown
Factor XIIIB deficiency	<1 in 1,000,000	Unknown
Familial platelet disorder with associated myeloid malignancy	<1 in 1,000,000	Unknown
Fletcher factor deficiency	<1 in 1,000,000	Unknown
Glanzmann thrombasthenia	<1 in 1,000,000	Arabs in Israel and French Gypsies (Manouche)
Hemorrhagic diathesis, due to P2RY12 defect	<1 in 1,000,000	Unknown
Hemorrhagic diathesis, due to plasminogen activator inhibitor 1	<1 in 1,000,000	Amish and Chinese
Macrothrombocytopenia	<1 in 1,000,000	Unknown
Growth hormone insensitivity with immunodeficiency	<1 in 1,000,000	Unknown
Hyper-IgE recurrent infection syndrome	<1 in 1,000,000	Unknown
Immunodeficiency-centromeric instability-facial anomalies syndrome	<1 in 1,000,000	Unknown
Immunodysregulation, polyendocrinopathy, and enteropathy	<1 in 1,000,000	Unknown
Lymphoproliferative syndrome	<1 in 1,000,000	Unknown
RIDDLE syndrome	<1 in 1,000,000	Unknown
WHIM syndrome or Warts-hypogammaglobulinemia-infections-myelokathexis	<1 in 1,000,000	Unknown

3.03.1 Nutritional Anemias

Nutritional anemia: while a common environmental condition, has causes attributed to nutritional genomic disorders which are uncommon, occurring in less than 1 in 1,000,000 individuals. Most of these conditions follow autosomal recessive monogenetic traits with loss-of-function. Examples of nutritional anemia of genomic origin are: Atransferrinemia, Hypochromic microcytic anemia, Iron-refractory iron deficiency anemia, Congenital intrinsic factor deficiency, Megaloblastic anemia (Finnish and Norwegian types), Transcobalamin 2 deficiency and Hereditary folate malabsorption.

Table 3.03.1 Recognizable Nutritional Anemias

Disorder	OMIM	Gene	Biological Function	Product of Conception Alteration
Atransferrinemia	209300	*TF*	Secreted	AR
Hypochromic microcytic anemia	206100	*SLC11A2*	Solute carrier	AR
Iron-refractory iron deficiency anemia	206200	*TMPRSS6*	Transmembrane protease	AR
Congenital intrinsic factor deficiency	261000	*GIF*	Secreted	AR
Megaloblastic anemia 1, Finnish type	261100	*CUBN*	Receptor for intrinsic factor-vitamin B12	AR
Megaloblastic anemia 1, Norwegian type	261100	*AMN*	Transmembrane protein	AR
Transcobalamin 2 deficiency	275350	*TCN2*	Secreted	AR
Hereditary folate malabsorption	229050	*SLC46A1*	Solute carrier	AR
Thiamine-responsive megaloblastic anemia syndrome	249270	*SLC19A2*	Solute carrier	AR
Elevated Adenosine triphosphate of erythrocytes	102900	*PKLR*	Pyruvate kinase	AD
Erythrocyte lactate transporter defect	245340	*SLC16A1*	Solute carrier	AR

3.03.2 Hemolytic Anemias

Hemolytic anemia: A disorder characterized by hemolysis, either intravascular or extravascular. Hemolytic anemia has numerous possible causes including those of genomic or acquired origin. Hereditary hemolytic anemia can be due to: Defective red cell metabolism, such as in Glucose-6-phosphate dehydrogenase deficiency and disorders of glycolytic enzymes and gluta-thione metabolism; defects in hemoglobin production, such as in Sickle-cell disease, thalassemia, and Congenital dyserythropoietic anemia; and defects of red blood cell membrane production, such as in Hereditary spherocytosis and Hereditary elliptocytosis.

Glucose-6-phosphate dehydrogenase deficiency or Favism: A trait characterized by a hemolytic reaction to the consumption of fava beans. These individuals might suffer from nonimmune hemolytic anemia in response to the consumption of fava beans, infection, or exposure to certain medications (primaquine, chloroquine, sulfonamides, etc.) or chemicals. Glucose-6-phosphate dehydrogenase deficiency is the most common human enzyme defect, affecting more than 400 million people worldwide. This disorder affects mostly male individuals, since it is an X-linked recessive syndrome. This trait confers protection against malaria caused by Plasmodium falciparum. This trait and malaria also share the same endemic areas, such as sub-Saharan Africa, the Middle East and areas of South Asia. Glucose-6-phosphate dehydrogenase deficiency is caused by X-linked recessive mutations in *G6PD* gene which encodes the glucose-6-phosphate dehydrogenase. This gene is located on the subtelomeric long arm of the X chromosome, on band Xq28.

3.03.2.1 Defects in hemoglobin production

Alpha-thalassemia: A disorder characterized by a reduction in the amount of hemoglobin in the red blood cells. This form of thalassemia involves 2 subtelomeric genes: the *HBA1* gene which encodes the alpha 1hemoglobin, and the *HBA2* gene which encodes the alpha 2 hemoglobin. Alpha-thalassemia is a digenomic disorder characterized by impaired production of 1, 2, 3, or 4 alpha globin chains. Consequently, there is a relative excess of beta globin chains in red blood cells. The clinical features of this condition depend on how many alpha chains are non-functional, due either to mutation or deletion. There are two main different genomic subtypes. The first subtype is common in sub-Saharan Africa, the Mediterranean Basin and the Middle East. It is characterized by the single monoallelic deletion of an alpha-globin gene. The Alpha-thalassemia minor affects 30% of African-Americans (one alpha globin chain deleted). Also, 3% of this ethnic group have alpha-thalassemia trait which is characterized by single biallelic deletion of a alpha-globin gene (two alpha globin chains deleted); in this homozygous form, patients exhibit a mild microcytic anemia. The second subtype is rare, affecting individuals from Southeast Asia, producing serious clinical disorders of Hemoglobin H disease (three alpha globin chains deleted) and *hydrops fetalis* syndrome. The last form is characterized by double biallelic deletion of an alpha-globin gene. In Southeast Asians, the biallelic deletion trait frequency is estimated to be 1 in 25 individuals, which means the frequency of this condition is 1 in 2,500 individuals or $1/(50)^2$. These genes are located on the subtelomeric short arm of the 16 chromosome.

Beta-thalassemias: A form of thalassemia characterized by reduced or absent synthesis of the beta chains of hemoglobin. There are variable phenotypes, ranging from asymptomatic individuals to those with severe anemia. This disorder appears after 6 months of age, occuring with severe anemia, poor growth and skeletal abnormalities. This form of thalassemia is caused by autosomal recessive mutations in the *HBB* gene which encodes the beta hemoglobin. Beta-thalassemia is highly prevalent in people around the Mediterranean Sea, Middle East, Transcaucasus, Central Asia, Indian subcontinent, and Southeast Asia. Beta-thalassemia is also relatively common in people of African ancestry. The highest carrier incidences are reported in Cyprus (14%), Sardinia (12%) and Southeast Asia. In Cyprus, the carrier frequency for beta-thalassemia traits is 1 in 7 individuals, which means the frequency of this disorder is about 1 in 200 individuals or $1/(14)^2$. There are several countries in the Middle East that use disassortative mating to decrease the prevalence of the most severe form of anemia: the thalassemia major.

Sickle cell anemia or Drepanocytosis: A form of anemia characterized by hemoglobin polymerization, leading to erythrocyte rigid with a sickle shape and resulting in vasoocclusion. The clinical presentation of sickle cell anemia usually starts during childhood. The sickling phenomenon occurs since there is mutation in the hemoglobin beta gene. Sickle cell anemia occurs more commonly in individuals from Equatorial Africa; it also affects individuals from the Mediterranean area, India and the Middle East. In other words, the geographical area of Sickle cell anemia is also shared with malaria. There is a fitness advantage in carrying the sickle cell trait, since these individuals are more tolerant to malarial infection and consequently show less severe symptoms when infected. The sickle cell trait originated around 2,000 years ago. Sickle cell anemia is caused by biallelic polymorphism in the HBB gene that causes HbS. The sickle cell trait is referred as "HbAS". There are other, less common forms of sickle cell anemia, which include sickle-hemoglobin C disease (HbSC), sickle beta-zero-thalassemia (HbS/β0), and sickle beta-plus-thalassemia (HbS/β+). There are different geographical origins in the sickle cell gene alteration: these variants are known as Benin, Bantu, Cameroon, Saudi-Asian and Senegal. The genomic disorder is due to the polymorphism of a single nucleotide, from a GAG to GTG codon mutation, resulting in glutamic acid being substituted by valine at position 6 of the beta-globin protein (Glu6Val). In the United States, it is a rare syndrome since it affects 1 in 5,000 individuals. In African-Americans, the carrier frequency for sickle cell trait is 1 in 12.5 individuals, which means the frequency of

this condition is 1 in 625 individuals or $1/(25)^2$.

Hemoglobin E disease: A disorder characterized by a mild hemolytic anemia and mild splenomegaly. Hemoglobin E disease is caused by biallelic polymorphism (Glu26Lys) in the HBB gene that causes HbE. Hemoglobin E is most prevalent in Southeast Asia; this allele appears 1,240-4,440 years ago.

Hemoglobin C disease: A disorder charac-terized by a mild hemolytic anemia. Hemoglobin C disease is caused by biallelic polymorphism (Glu6Lys) in the HBB gene that causes HbC. Hemoglobin C is most prevalent in Central West Africa; this allele appears 5,000 years ago.

Hereditary persistence of fetal hemoglobin: A trait characterized by a substantial elevation of fetal hemoglobin in adult red blood cells. Affected individuals do not display other phenotypic or hematologic manifestations. Hereditary persistence of fetal hemoglobin is caused by autosomal recessive point mutations in the promoter regions of either the *HBG1* or the *HBG2* gene which encode the gamma hemoglobin (1 and 2, respectively).

3.03.2.2 Other Hereditary Hemolytic Anemias

Hereditary spherocytosis: A disorder characterized by the production of erythrocytes, that are sphere-shaped rather than bi-concave (disk shaped) and consequently make an individual more prone to hemolysis. The product of conception alteration in hereditary spherocytosis appears to exhibit autosomal dominant or monoallelic mutations which are either inherited in 75% of cases or de novo in 25% of cases. Hereditary spherocytosis type 1 is caused by mutations in the *ANK1* gene which encodes the ankyrin protein;; Hereditary spherocytosis type 2 is caused by mutations in the *SPTB* gene which encodes the beta spectrin protein, which along with ankyrin, play a role in cell membrane organization and stability;; Hereditary spherocytosis type 3 is caused by mutations in the *SPTA1* gene which encodes the alpha 1 spectrin protein;; Hereditary spherocytosis type 4 is caused by mutations in the *SLC4A1* gene which encodes the solute carrier (family 4 member A1), an anion exchanger or erythrocyte membrane protein band 3;; and Hereditary spherocytosis type 5 is caused by mutations in the *EPB42* gene which encodes the erythrocyte membrane protein band 4.2. Hereditary spherocytosis affects 1 in 2,000 individuals from Northern European countries and Japanese families. The prevalence of hereditary spherocytosis in people of other ethnic ancestries is unknown.

Hereditary elliptocytosis or Ovalocytosis: A disorder characterized by elliptical erythrocytes rather than the typical biconcave disc shape. In its severe forms, this disorder predisposes one to hemolytic anemia. Less than 10% of those with this disorder are thought to fall into the symptomatic group of people. Hereditary elliptocytosis-1 is caused by mutations in the *EPB41* gene;; Hereditary elliptocytosis-2 is caused by mutations in the *SPTA1* gene;; Hereditary elliptocytosis-3 is caused by mutations in the *SPTB* gene;; and Hereditary elliptocytosis-4 is caused by mutations in the *SLC4A1* gene. The highest estimated incidence is 1,500-2,000 in 10,000 in Malayan natives; in equatorial Africa its incidence is estimated at 60-160 in 10,000; in the United States, it is a rare syndrome affecting 3-5 in 10,000 individuals, but is more common among those of African-American and Mediterranean descent. There are subtypes of Hereditary elliptocytosis which can confer resistance to Malaria, and are prevalent in regions where malaria is endemic. The product of conception alterations in Hereditary elliptocytosis appears to exhibit autosomal dominant inheritance. For instance, both sexes are consequently at equal risk of having the condition. The most important exception to this rule of autosomal dominance is for a subtype called hereditary pyropoikilocytosis, which is inherited in a severe form when both parents are heterozygous and both parents provide the faulty copy to descendants (AD + AD = AR). South-east Asian ovalocytosis or Stomatocytic elliptocytosis occurs in those individuals with mild hemolytic anemia, and has increased resistance to malaria. It affects individuals of Southeast Asian descent, such as Indonesian, Filipino, Malaysian, Melanesian and New Guinean individuals. Spherocytic elliptocytosis or Hemolytic ovalocytosis typically affects individuals of European descent; Elliptocytes and Spherocytes are concurrently present in their blood.

Hereditary Pyropoikilocytosis: An erythrocytes disorder characterized by marked instability at even mildly elevated temperatures. Pyropoikilocytosis is habitually found in burn victims. These individuals have life-threateningly severe hemolytic anemia with micropoikilocytosis. Patients tend to suffer severe hemolysis and anemia at infancy that progressively improves, and then typically develop elliptocytosis later in life. Hereditary pyropoikilocytosis typically affects individuals of African descent. Hereditary pyropoikilocytosis can be caused by biallelic mutations in either the *SPTA1* gene which encodes the alpha-spectrin protein, or the *SPTB* gene which encodes the beta-spectrin protein.

The Concise Encyclopedia of Genomic Diseases

Table 3.03.2.2 Recognizable Hereditary Hemolytic Anemias

Disorder	OMIM	Gene	Biological Function	Product of Conception Alteration
Glucose-6-phosphate dehydrogenase deficiency	305900	G6PD	Dehydrogenase	XR
Hemolytic anemia, due to gamma-glutamylcysteine synthetase deficiency	230450	GCLC	Glutathione synthesis	AR
Hemolytic anemia, due to glucose phosphate isomerase deficiency	613470	GPI	Isomerase	AR
Hemolytic anemia, due to hexokinase deficiency	235700	HK1	Hexokinase	AR
Hemolytic anemia, due to tri-osephosphate isomerase deficiency	190450	TPI1	Isomerase	AR
Hemolytic anemia, due to phospho-glycerate kinase 1 deficiency	300653	PGK1	Kinase	XR
Hemolytic anemia, due to pyruvate ki-nase deficiency	266200	PKLR	Pyruvate kinase	AR
Hemolytic anemia, due to adenylate kinase deficiency	612631	AK1	Kinase	AR
Hemolytic Anemia, due to UMPH1 deficiency	266120	NT5C3	Cytosolic nucleo-tidase	AR
Alpha-thalassemia	604131	HBA1-HBA2	Hemoglobin	DigM or DigB or AR + KDN
Alpha-thalassemia	141800	LCRA	Regulatory region	AR
Alpha-thalassemia/Mental retardation syndrome	301040	ATRX	Chromatin remod-eling	XR
Sickle cell anemia	603903	HBB	Hemoglobin	AR
Beta-thalassemias	604131	HBB	Hemoglobin	AR + KDN
Delta-beta thalassemia	141749	HBB	Hemoglobin	AD
Hispanic gamma-delta-beta-thalassemia	604131	LCRB	Regulatory region	AD
Hereditary persistence of fetal hemo-globin	141749	HBG1-HBG2	Hemoglobin	AR
Hereditary persistence of fetal hemo-globin	613566	KLF1	Transcription factor Zinc finger	AD
Hereditary persistence of fetal hemo-globin	141749	HBB	Hemoglobin	AD
Hereditary spherocytosis-1	182900	ANK1	Membrane cyto-skeletal	AD + 25%
Hereditary spherocytosis-2	182870	SPTB	Membrane cyto-skeletal	AD
Hereditary spherocytosis-3	130600	SPTA1	Membrane cyto-skeletal	AD + de novo
Hereditary spherocytosis-4	109270	SLC4A1	Solute carrier	AD + de novo
Hereditary spherocytosis-5	612690	EPB42	Membrane cyto-skeletal	AD

Table 3.03.2.2. Continued.

Disorder	OMIM	Gene	Biological Function	Product of Conception Alteration
Hereditary elliptocytosis-1	611804	*EPB41*	Membrane cyto-skeletal	AD
Hereditary elliptocytosis-2	130600	*SPTA1*	Membrane cyto-skeletal	AD + de novo
Hereditary elliptocytosis-3	182870	*SPTB*	Membrane cyto-skeletal	AD
Hereditary elliptocytosis-4	109270	*SLC4A1*	Solute carrier	AD + de novo
Pyropoikilocytosis	266140	*SPTA1*	Membrane cyto-skeletal	AD + AD = AR

Table 3.03.02.3 Hereditary Hemolytic Anemias in Different Ethnic Groups

Disease	Frequency of condition	People with higher prevalence (if available)
Hemolytic anemia, due to G6PD	2-260 in 1,000	Africans, Middle Easterns and South Asians
Alpha-thalassemia trait, Hemoglobin H disease and hydrops fetalis syndrome	1-300 in 1,000	Equatorial Africans and Southeastern Asians
Beta-thalassemia intermediate and major	1-50 in 10,000	Mediterraneans, Middle Easterns, Transcaucasians, Central Asians, Indians, Pakistanis and Bangladeshis
Sickle cell anemia	1-200 in 10,000	Sub-Saharan Africans, Indians and Middle Easterns
Elliptocytosis	3-2,000 in 10,000	Malaysians and Equatorial Africans
Spherocytosis	1-5 in 10,000	Northern Europeans

3.03.3 Aplastic and other Anemias

Diamond–Blackfan anemia or Inherited erythroblastopenia: A congenital erythroid aplasia with decreased erythroid progenitors in the bone marrow that generally occurs at infancy. It is characterized by anemia with normal platelets and white blood cell counts. Diamond-Blackfan anemia is associated in 30-40% of patients with other congenital anomalies, predominantly of the upper limb and craniofacial regions. The product of conception in Diamond-Blackfan anemia appears to exhibit de novo alteration in 55-60% of cases and autosomal dominant inheritance in 40-45% of cases. Patients have a risk of developing leukemia. Diamond-Blackfan anemia is caused by either monoallelic or biallelic mutations in the ribosomal protein gene including the *RPS19, RPS26, RPS24, RPS17, RPL35A, RPL5, RPL11, RPS7* and *RPS10* genes. Mutations in the *RPS19* gene account for 25% of all Diamond-Blackfan anemia cases. This condition does not show an ethnic predilection and both sexes are equally affected. In Europeans, Diamond-Blackfan anemia is estimated to affect 1 in every 150,000 individuals.

Shwachman–Bodian–Diamond syndrome: A congenital disorder characterized by bone marrow dysfunction, resulting primarily in neutropenia. It is also accompanied by exocrine pancreatic insufficiency (with normal sweat electrolytes and no respiratory difficulties), skeletal abnormalities and short stature. Shwachman–Bodian–Diamond syndrome is caused by biallelic mutations in the *SBDS* gene which encodes the Shwachman-Bodian-Diamond syndrome, a ribosome maturation protein. The product of conception alterations are defined as autosomal recessive syndrome plus known *de novo* alterations (AR + KDN). The de novo mutation in the *SBDS* gene appears to be the result of gene conversion. Shwachman–Bodian–Diamond syndrome is estimated to affect 1 in every 76,000 individuals.

Congenital dyserythropoietic anemia: A disorder characterized by macrocytic anemia, ineffective erythropoiesis and secondary hemochromatosis. There are several underlying causes of this disorder with at least 4 autosomal genomic loci: Congenital dyserythropoietic anemia-1 is caused by autosomal recessive mutations in the *CDAN1* gene which encodes the codanin 1 protein;; Congenital dyserythropoietic anemia-2 is caused by autosomal recessive mutations in the *SEC23B* gene which encodes the Sec23 homolog B protein. This protein is part of a protein complex found in the ribosome-free transitional face of the endoplasmic reticulum and associated vesicles;; Congenital dyserythropoietic anemia-3 has been mapped but the gene has not been characterized;; and Congenital dyserythropoietic anemia-4 is caused by autosomal dominant mutations in the the *KLF1* gene which encodes the Kruppel-like factor 1 protein (erythroid).

Congenital dyserythropoietic anemia is estimated to affect 1 in every 100,000 individuals.

Majeed Syndrome: A disorder characterized by congenital dyserythropoietic anemia, chronic recurrent multifocal osteomyelitis and inflammatory dermatosis. Majeed syndrome is caused by autosomal recessive mutations in the *LPIN2* gene which encodes the lipin 2 protein. Majeed syndrome is estimated to affect less than 1 in every 1,000,000 individuals. This disorder is prevalent in consanguineous families.

Fanconi anemia: A disorder that affects all blood cell lines resulting in pancytopenia. Characteristic clinical features include early-onset bone marrow failure, and a high predisposition to cancer (most frequently Acute myeloid leukemia). Individuals with Fanconi anemia suffer from developmental abnormalities in major organ systems, resulting commonly in short stature and abnormalities of the skin, eyes, kidneys, arms and head. Patients also have some form of endocrine problem. The disease is named after Guido Fanconi. The cellular hallmark of Fanconi anemia is genomic instability and hypersensitivity to DNA crosslinking agents. There are several underlying causes of this disorder with at least 16 autosomal genomic loci and 1 X-linked genomic locus. This condition is caused due to a defect in DNA repair, and presently there are 17 Fanconi anemia genes: *FANCA, FANCB, FANCC, BRCA2, FANCD2, FANCE, FANCF, FANCG, FANCI, FANCJ, FANCL, FANCM, FANCN, FANCP, RAD51C, SLX4* and *ERCC4*. The product of conception alteration in Fanconi anemia is primarily due to autosomal recessive inheritance, except *FANCB* which is X-linked recessive. Fanconi anemia complementation group A can be caused by biallelic mutations in the

FANCA gene which encodes the Fanconi anemia complementation group A protein;; Fanconi anemia complementation group B can be caused by X-linked recessive mutations in the *FANCB* gene which encodes the Fanconi anemia complementation group B protein;; Fanconi anemia complementation group C can be caused by biallelic mutations in the *FANCC* gene which encodes the Fanconi anemia complementation group C protein;; Fanconi anemia complementation group D1 can be caused by biallelic mutations in the *BRCA2* gene which encodes the breast and ovarian cancer susceptibility protein 2;; Fanconi anemia complementation group D2 can be caused by biallelic mutations in the *FANCD2* gene which encodes the Fanconi anemia complementation group D2 protein;; Fanconi anemia complementation group E can be caused by biallelic mutations in the *FANCE* gene which encodes the Fanconi anemia complementation group E protein;; Fanconi anemia complementation group F can be caused by biallelic mutations in the *FANCF* gene which encodes the Fanconi anemia complementation group F protein;; Fanconi anemia complementation group G can be caused by biallelic mutations in the *FANCG* gene which encodes the Fanconi anemia complementation group G protein;; Fanconi anemia complementation group I can be caused by biallelic mutations in the *FANCI* gene which encodes the Fanconi anemia complementation group I protein;; Fanconi anemia complementation group J can be caused by biallelic mutations in the *BRIP1* gene which encodes the BRCA1 interacting protein C-terminal helicase 1;; Fanconi anemia complementation group L can be caused by biallelic mutations in the *FANCL* gene which encodes the Fanconi anemia complementation group L protein;; Fanconi anemia complementation group M can be caused by biallelic mutations in the

FANCM gene which encodes the Fanconi anemia complementation group M protein;; Fanconi anemia complementation group N can be caused by biallelic mutations in the *PALB2* gene which encodes the partner and localizer of BRCA2 protein;; Fanconi anemia complementation group O can be caused by biallelic mutations in the *RAD51C* gene which encodes the RAD51 paralog C protein, a DNA repair protein;; Fanconi anemia complementation group P can be caused by biallelic mutations in the *SLX4* gene which encodes the SLX4 structure-specific endonuclease subunit;; and Fanconi anemia complementation group Q can be caused by biallelic mutations in the *ERCC4* gene which encodes the excision repair cross-complementing rodent repair deficiency (complementation group 4). The prevalence of Fanconi anemia is 1 in 350,000 births worldwide. There are two groups of people, Ashkenazi Jews and Afrikaners in South Africa, where such incidences are higher than average. In Ashkenazi Jews, the carrier frequency for *FANCC* gene trait is estimated to be 1 in 90 individuals, which means the frequency of this condition is about 1 in 32,400 individuals or $1/(180)^2$. The current frequency among Ashkenazi Jews is probably lower since they have implemented programs of disassortative mating.

Aplastic anemia: A disorder characterized by a bone marrow which produces insufficient new blood cells (red, white and platelets) to replenish blood cells. There are several exclusion criteria for genomic origin in aplastic anemia, since many environmental factors such as benzene, chloramphenicol, antiseizure medication, ionizing radiation, acute viral hepatitis and parvovirus B19 infection produce aplastic anemia. There are inclusion criteria to consider aplastic anemia of genomic origin, such as the fact that patients with alterations in the *SBDS, IFNG, NBS1, PRF1, SBDS, TERT* and *TERC* genes are predisposed to being affected with this condition.

Pearson syndrome or Sideroblastic anemia and exocrine pancreas dysfunction: A disorder caused by a deletion in mitochondrial DNA. Pearson syndrome is extremely rare. Sometimes, patients develop symptoms of Kearns-Sayre syndrome.

Sideroblastic anemia or Sideroachrestic anemia: A disorder characterized by a bone marrow that does not produce healthy erythrocytes. Instead, it produces ringed sideroblasts. In affected individuals, the bone marrow has accessible iron, but cannot incorporate it into hemoglobin. There are several underlying causes of this disorder with at least 2 X-linked genomic loci and 2 autosomal genomic loci: Sideroblastic anemia can be caused by X-linked mutations in the *ALAS2* gene which encodes the delta-aminolevulinate synthase 2 protein;; Sideroblastic anemia with spinocerebellar ataxia can be caused by X-linked mutations in the *ABCB7* gene which encodes the ATP-binding cassette (sub-family B member 7);; Autosomal recessive pyridoxine-refractory sideroblastic anemia is caused by biallelic mutations in the *SLC25A38* gene which encodes the solute carrier (family 25 member A38);; and Autosomal recessive pyridoxine-refractory sideroblastic anemia is caused by biallelic mutations in the *GLRX5* gene which encodes the glutaredoxin 5 protein. There is also a Pyridoxine-responsive sideroblastic anemia but the gene/loci have not been characterized.

Table 3.03.3.1 Recognizable Aplastic and other Anemias

Disorder	OMIM	Gene	Biological Function	Product of Conception Alteration
Diamond-Blackfan anemia-1	105650	*RPS19*	Ribosomal protein	AR or AD
Diamond-blackfan anemia-3	610629	*RPS24*	Ribosomal protein	AD
Diamond-Blackfan anemia-4	612527	*RPS17*	Ribosomal protein	AD
Diamond-Blackfan anemia-5	612528	*RPL35A*	Ribosomal protein	AD
Diamond-Blackfan anemia-6	612561	*RPL5*	Ribosomal protein	AD
Diamond-Blackfan anemia-7	612562	*RPL11*	Ribosomal protein	AD
Diamond-Blackfan anemia-8	612563	*RPS7*	Ribosomal protein	AD
Diamond-Blackfan anemia-9	613308	*RPS10*	Ribosomal protein	AD
Diamond-Blackfan anemia-10	613309	*RPS26*	Ribosomal protein	AD
Shwachman–Bodian–Diamond syndrome	260400	*SBDS*	Ribosome maturation protein	AR + KDN
Congenital dyserythropoietic anemia-1	224120	*CDAN1*	Nuclear envelope integrity	AR
Congenital dyserythropoietic anemia-2	224100	*SEC23B*	Vesicle budding from the ER	AR
Congenital dyserythropoietic anemia-4	613673	*KLF1*	Transcription factor Zinc finger	AD
Majeed syndrome	609628	*LPIN2*	Phosphatase	AR
Fanconi anemia, complementation group A	227650	*FANCA*	DNA repair, defective DNA repair	AR + KDN

Table 3.03.3.1. Continued.

Disorder	OMIM	Gene	Biological Function	Product of Conception Alteration
Fanconi anemia, complementation group B	300514	FANCB	DNA repair, defective DNA repair	XR
Fanconi anemia, complementation group C	227645	FANCC	DNA repair, defective DNA repair	AR
Fanconi anemia, complementation group D1	605724	BRCA2	Repairing damaged DNA	AR
Fanconi anemia, complementation group D2	227646	FANCD2	DNA repair, defective DNA repair	AR
Fanconi anemia, complementation group E	600901	FANCE	DNA repair, defective DNA repair	AR
Fanconi anemia, complementation group F	603467	FANCF	DNA repair, defective DNA repair	AR
Fanconi anemia, complementation group I	609053	FANCI	DNA repair, defective DNA repair	AR
Fanconi anemia, complementation group J	609054	BRIP1	RecQ DEAH helicase	AR
Fanconi anemia, complementation group N	610832	PALB2	Intranuclear localization	AR
Fanconi anemia, complementation group O	613390	RAD51C	DNA repair protein	AR
Fanconi anemia, complementation group P	613951	SLX4	Endonucleases mismatch repair complex and telomere binding	AR
Fanconi anemia, complementation group Q	615272	ERCC4	DNA repair Nucleotide excision repair	AR
Pearson syndrome or Pearson marrow-pancreas syndrome	557000	MT-Del	Mitochondrial deletion	Mit
Sideroblastic anemia with ataxia	301310	ABCB7	ABC-transporter	XR
X-linked Sideroblastic anemia	300751	ALAS2	Mitochondrial protein	XR
Pyridoxine-refractory sideroblastic anemia	205950	GLRX5	Mitochondrial iron homeostasis	AR
Pyridoxine-refractory sideroblastic anemia	205950	SLC25A3 8	Solute carrier	AR
Exocrine pancreatic insufficiency, dyserythropoietic anemia, and calvarial hyperostosis	612714	COX4I2	Mitochondrial protein	AR
Hoyeraal-Hreidarsson syndrome	300240	DKC1	Nucleolar ribonucleoproteins	XR + de novo
Idiophatic aplastic anemia	609135	TERC	Telomerase	
Susceptibility to aplastic anemia	609135	IFNG	Secreted	AR
Susceptibility to aplastic anemia	609135	TERC	Telomerase	AD
Susceptibility to aplastic anemia	609135	TERT	Telomerase	AD

Table 3.03.03.2 (Aplastic and other Anemias in Different Ethnic Groups) displays a group of anemic disorders, their corresponding frequencies, and the ethnic groups with highest prevalence rates of each disease (if available).

Table 3.03.03.2 Aplastic and other Anemias in Different Ethnic Groups

Disease	Frequency of condition	People with higher prevalence (if available)
Shwachman–Bodian–Diamond syndrome	1 in 76,000	Unknown
Congenital dyserythropoietic anemia	1 in 100,000	Unknown
Diamond-Blackfan anemia	1 in 150,000	Europeans
Fanconi anemia	1 in 350,000	Ashkenazi Jews and Afrikaners
Majeed Syndrome	<1 in 1,000,000	Unknown

3.03.4 Coagulation Defects, Purpura and other Hemorrhagic Conditions

Table 3.03.4 Bleeding Disorders

Disease	Gene	Bleeding	Product of Conception Alteration
von Willebrand disease monoallelic	VWF	Mild	AD + de novo
von Willebrand disease biallelic	VWF	Mild-to-severe	AD + AD = AR
Hemophilia A	F8	Mild-to-severe	XR + 30 %
Hemophilia B	F9	Mild-to-severe	XR + 30 %
Deficiency of Vitamin K-dependent clotting factors	GGCX	Reversed by administration of vitamin K	AR
Deficiency of Vitamin K-dependent clotting factors	VKORC1	Reversed by administration of vitamin K	AR
Dysfibrinogenemia, alpha-type, causing bleeding diathesis	FGA	Moderate	AD
Dysfibrinogenemia, beta-type	FGB	Moderate	AD
Dysfibrinogenemia, gamma type	FGG	Moderate	AD
Factor V deficiency	F5	Mild-to-severe	AR
Combined deficiency of Factor V and factor VIII	LMAN1	Mild-to-moderate	AR
Combined deficiency of Factor V and factor VIII	MCFD2	Mild-to-moderate	AR
Factor VII deficiency or hypoproconvertinemia	F7	Mild-to-severe	AR
Hemophilia C	F11	Mild	AD + de novo or AR
Factor XII deficiency	F12	Moderate	AR

Table 3.03.4. Continued.

Disease	Gene	Bleeding	Product of Conception Alteration
Factor XIIIA deficiency	*F13A1*	Diathesis associated with spontaneous abortions	AR
Factor XIIIB deficiency	*F13B*	Diathesis associated with spontaneous abortions	AR
Platelet-type bleeding disorder	*P2RY12*	Mild-to-moderate	AR
Plasminogen activator inhibitor-1 deficiency	SERPINE1	Moderate	AR or AD
Alpha2-plasmin inhibitor deficiency	SERPINF2	Moderate-to-severe	AR

3.03.4.1 Coagulation Defects

There are many genomic disorders that course with coagulation defects. von Willebrand disease is the only condition that is common worldwide. Most of the disorder described below are rare or produce only mild symptoms. The are many known non-genomic factors that augment the bleeding risk such as vitamin K deficiency, Disseminated intravascular coagulation, thrombocytopenia, liver failure, uremia, and use of warfarin, heparin and aspirin.

Hemophilia: A disorder characterized by lower blood plasma clotting factor levels than are required for a normal clotting process. Consequently, patients bleed for a much longer time. Patients with severe hemophilia have 20 to 30 yearly episodes of spontaneous or excessive bleeding after minor trauma, predominantly in joints and muscles.

Hemophilia A: A rare disorder, it represents 80% of all the hemophilia cases. This condition is the most prevalent form of the disorder, affecting 1 in 6,000 male births worldwide. Hemophilia A has a large variable clinical severity, according to the plasmatic levels of coagulation factor VIII: severe, with levels less than 1% of normal; moderate, with levels 2-5% of normal; and mild, with levels 6-30% of normal. Hemophilia A is caused by X-linked recessive mutations in the *F8* gene which encodes the coagulation factor VIII. These mutations are either inherited (in 70% of cases) or de novo (in the remaining 30%). The gene is located on chromosome X subtelomeric long arm. Hemophilia A belongs to different groups including Subtelomeric disorder, *de novo* disorder, and X-linked recessive disorder.

Hemophilia B: A condition with similar clinical features to those of Hemophilia A. Type B affects 1 in about 30,000 male births worldwide. Hemophilia B is caused by X-linked recessive mutations in the *F9* gene which encodes the coagulation factor IX. These mutations are either inherited in 70% of cases or de novo in the remaining 30%. There is evidence of advanced maternal age effect in de novo mutations particularly is more prominent for transversions.

Hemophilia C or Factor XI deficiency: An autosomal bleeding disorder characterized by reduced levels of factor XI in plasma. Bleeding occurs mainly after trauma or surgery. Most patients suffer from a mild form of hemophilia. Hemophilia C is caused by either autosomal recessive or autosomal dominant mutations in the *F11* gene which encodes the coagulation factor XI. The alteration of the gene F11 could be inherited or occur as a de novo event. Hemophilia C is prevalent in Ashkenazi Jews and French Basques. In Ashkenazi Jews there is a factor XI deficiency which is commonly inherited as a recessive trait. However, in French Basques, it follows a dominant pattern of the product of conception alteration, due to a dominant-negative effect in the dimeric structure of plasmatic F11.

von Willebrand disease: A condition that is the most common hereditary coagulation defect. It is caused by qualitative or quantitative deficiency of von Willebrand factor - a multimeric protein that is essential for platelet adhesion. von Willebrand disease belongs to different groups including digenomic disorder, multimeric disorder, severe disorder, etc. This condition was named after Erik Adolf von Willebrand in 1926. There are at least 3 inherited types of this condition and there are various subtypes: von Willebrand disease types 1 and 2 are caused by monoallelic mutations in the *VWF* gene which encodes the von Willebrand factor. This protein is crucial to the hemostasis process, since functions as both an antihemophilic factor carrier and a platelet-vessel wall mediator in the blood coagulation system. The product of conception in von Willebrand disease types 1 and 2 appear to either exhibit de novo alteration in a few cases or autosomal dominant inheritance in most of the cases. Type 1 is the most common type, accounting for 60-80% of all cases, and is a quantitative defect of the disorder. Type 2 accounts for 20-30% of all cases,

and is a qualitative defect of the disorder. Affected individuals display average levels of vWF, but the multimers are structurally anomalous. von Willebrand disease types 3 is caused by biallelic mutations in the *VWF* gene; von Willebrand type 3 is inherited in a severe form when both parents are heterozygous and both parents provide the faulty copy to descendants (AD + AD = AR). The prevalence of von Willebrand is about 1 in 100 individuals; most of these people do not display symptoms. Sometimes individuals display mild symptoms, as in the case of females who have increased bleeding tendency during menstruation. von Willebrand is a digenomic disorder, since it may be more severe in people with blood type O. On average, the carrier frequency for von Willebrand disease type 1 and 2 trait is estimated to be 1 in 100 individuals, which means the frequency of von Willebrand disease type 3 is about 1 in 40,000 individuals or $1/(200)^2$.

3.03.4.2 Hereditary Deficiency of other Clotting Factors

Familial dysfibrinogenemia: A coagulation disorder characterized by a bleeding tendency due to a functional anomaly of circulating fibrinogen. Most patients with dysfibrinogenemia are asymptomatic. Affected individuals may have mild bleeding symptoms or even thrombosis. Familial dysfibrinogenemia is caused by monoallelic mutations in the *FGA* gene which encodes the fibrinogen alpha chain, or *FGB* gene which encodes the fibrinogen beta chain, or the *FGG* gene which encodes the fibrinogen gamma chain.

Combined deficiency of vitamin K-dependent clotting factors: A disorder appearing during the first weeks of life characterized by episodes of intracranial hemorrhage. These bleedings occasionally result in a fatal outcome. Deficiency of all vitamin K-dependent clotting factors leads to a bleeding tendency that is usually reversed by oral administration of vitamin K. There are several underlying causes of this disorder with at least 2 autosomal genomic loci: Combined deficiency of vitamin K-dependent clotting factors-1 is caused by biallelic mutations in the *GGCX* gene which encodes the gamma-glutamyl carboxylase;; and Combined deficiency of vitamin K-dependent clotting factors-2 is caused by biallelic mutations in the *VKORC1* gene which encodes the vitamin K epoxide reductase.

Congenital factor V deficiency: A bleeding disorder that is caused by reduced plasma levels of factor V and is characterized by mild to severe bleeding symptoms. Affected individuals show epistaxis, bruising, mucosal bleeding, soft tissue bleeding and hemarthrosis. Congenital factor V deficiency is caused by biallelic mutations in the *F5* gene which encodes the coagulation factor 5. Congenital factor V deficiency is estimated to affect 1 in every 1,000,000 individuals. Both sexes are equally affected.

Combined deficiency of factor V and factor VIII: A bleeding disorder that is caused by low levels of factors V and VIII and characterized by mild-to-moderate bleeding symptoms. There are several underlying causes of this disorder with at least 2 autosomal genomic loci: Combined deficiency of factor V and factor VIII type 1 can be caused by autosomal recessive mutations in the *LMAN1* gene which encodes the mannose-binding lectin-1 protein;; and Combined deficiency of factor V and factor VIII type 2 can be caused by autosomal recessive mutations in the *MCFD2* gene which encodes the multiple coagulation factor deficiency 2 protein.

Combined deficiency of factor V and factor VIII is estimated to affect 1 in every 100,000-1,000,000 individuals. There is a higher prevalence among people living in the Mediterranean area and among Sephardic Jews.

Factor VII deficiency or hypoproconvertinemia: A bleeding disorder that is caused by low levels of factors VII and characterized by severe bleeding symptoms. Affected individuals might manifest intracerebral hemorrhages or repeated hemarthroses. Factor VII deficiency is caused by autosomal recessive mutations in the *F7* gene which encodes the coagulation factor VII. Factor VII deficiency is estimated to affect 1 in every 300,000 individuals. There is a higher prevalence among Oriental Jews.

Factor XII deficiency: A disorder characterized by slight to moderate bleeding tendency and a high incidence of cerebral apoplexy occurring at a relatively early age. Other features include episodes of local edema, severe headache, abdominal pain, and various forms of allergy. Factor XII deficiency is caused by biallelic mutations in the *F12* gene which encodes the coagulation factor XII.

Factor XIII deficiency: A bleeding disorder that is caused by reduced plasma levels of factor XIII and characterized by hemorrhagic diathesis frequently associated with spontaneous abortions and defective wound healing. There are several underlying causes of this disorder with at least 2 autosomal genomic loci: Factor XIIIA deficiency is caused by biallelic mutations in the *F13A1* gene which encodes the coagulation factor XIII (subunit A). Factor XIIIB deficiency is caused by biallelic mutations in the *F13B* gene which encodes the coagulation factor XIII (subunit B). Factor XIII deficiency is one of the most rare coagulation factor deficiencies. Factor XIII deficiency is estimated to affect 1 in every 2,000,000 individuals. Both sexes are equally affected.

Fletcher factor deficiency: A coagulation defect trait characterized by prolonged activated partial thromboplastin time and delayed thromboplastin generation but normal prothrombin time, and no abnormal bleeding tendency. Fletcher factor deficiency is caused by autosomal recessive mutations in the *KLKB1* gene which encodes the plasma kallikrein B (Fletcher factor).

Plasminogen activator inhibitor-1 deficiency: A disorder characterized by a premature lysis of hemostatic clots and a moderate bleeding syndrome. Affected individuals might manifest frequent bruising, severe menstrual bleeding, and intracranial and joint bleeding after mild trauma. Plasminogen activator inhibitor-1 deficiency is caused by either autosomal recessive or autosomal dominant mutations in the *SERPINE1* gene which encodes the serpin peptidase inhibitor (clade E member 1). Most cases have been reported in Amish, Japanese and Chinese individuals.

Alpha-2-plasmin inhibitor deficiency: A disorder characterized by hemorrhagic diathesis. Affected patients exhibit prolonged bleeding and ecchymoses after minor trauma, spontaneous joint hemorrhage and hemothorax. Alpha-2-plasmin inhibitor deficiency is caused by autosomal recessive mutations in the *SERPINF2* gene which encodes the serpin peptidase inhibitor (clade F member 2).

3.03.4.3 Primary Thrombophilia

Thrombophilia or hypercoagulability: An abnormality of blood coagulation that increases the risk of thrombosis. Most of the patients only develop thrombosis in the presence of an additional risk factor, such as immobilization, autoimmune disease, cancer, nephrotic syndrome, inflammatory bowel disease, pregnancy, elevated estrogens levels, obesity, etc. The ABO system should also be considered a risk factor; in fact, those with blood type O have lower levels of the blood protein factor for von Willebrand as well as factor VIII, which confer protection from thrombosis. The disorders described below will be grouped as DNA alteration with gain or loss of function.

Thrombophilia caused by gain-of-function disorder: A condition that results from overactivity of coagulation factors, including the factor V Leiden and F II due to prothrombin polymorphism. The genes altered are *F5* and *F2* respectively. Gains of function disorders are extremely common polymorphisms; for instance in Caucasians, factor V Leiden is present in 1-5% and F II or Prothrombin G20210A is present in 2%. These 2 disorders belong to different groups; besides being part of a gain-of-function disorder; they also include early migration disorders and severe disorders, due to inheritance. The last group occurs when both parents are heterozygous and both parents provide the faulty copy to descendants (AD + AD = AR). In fact, 1 copy of factor V Leiden mutation increases the chance of developing a clot (deep vein thrombosis or pulmonary embolism) by 4-8 times. People who inherit 2 copies of the factor V Leiden mutation may have up to 80 times the usual risk of developing a clot. The prevalence of clinically severe risk for deep vein thrombosis or pulmonary embolism is 1 in 1,600 since the allele frequency is 1 in 40. Prothrombin G20210A single nucleotide polymorphism increases the clotting risk in patient by 2-3 fold.

Thrombophilia caused by loss-of-function disorder: A condition that results from loss of activity of anticoagulation factors including: Antithrombin III deficiency, Protein C deficiency, Protein S deficiency, Heparin cofactor II deficiency, Histidine-rich glycoprotein deficiency, Familial hyperfibrinolysis, Sickle-cell disease and Dysfibrinogenemia (factor I or fibrinogen). The corresponding mutated genes are the *SERPINC1, PROC, PROS1, SERPIND1, HRG, PLAT, HBB, FGA, FGB* and *FGG* genes, respectively. Fibrinogen is encoded for the last 3 genes. Protein C deficiency and protein S deficiency increase the risk of clotting in patients by 5-10-fold.

3.03.4.4 Purpura and other Hemorrhagic Conditions

Thrombocytopenia: A condition characterized by a decrease of platelets in blood, generally to below 50,000 platelets per microliter, resulting in bleeding complications. There are several non-genomic causes of thrombocytopenia including medication-induced, autoimmune disease, infectious disease, leukemia, vitamin B12, folic acid deficiency, etc… Table 3.03.4.1 describes the clinical features of diseases that course with thrombocytopenia, including the genes related to these conditions.

Hereditary nonsyndromic thrombocytopenia: A disorder characterized by decreased numbers of platelets and increase in bleeding tendency. There are several underlying causes of this disorder with at least 1 X-linked genomic locus and 3 autosomal genomic loci: X-linked thrombocytopenia-1 is caused by X-linked mutations in the *WAS* gene which encodes the Wiskott-Aldrich syndrome, a protein involved in transduction of signals from receptors on the cell surface to the actin cytoskeleton;; Thrombocytopenia-2 is caused by monoallelic mutations in either the *ANKRD26* gene which encodes the ankyrin repeat domain 26 protein, or the *MASTL* gene which encodes the microtubule associated serine/threonine kinase-like protein;; Thrombocytopenia-3 a possible autosomal recessive form, but the gene/loci have not yet been characterized;; and Thrombocytopenia-4 is caused by monoallelic mutations in the *CYCS* gene which encodes the cytochrome c, a central component of the electron transport chain in mitochondria.

Epstein syndrome: A disorder characterized by thrombocytopenia, giant platelets, nephritis, and deafness. Epstein syndrome is caused by monoallelic mutations in the *MYH9* gene which encodes the nonmuscle myosin heavy chain 9 protein.

May-Hegglin anomaly: A disorder characterized by the triad of thrombocytopenia, giant platelets, and Dohle body-like inclusions in peripheral blood leukocytes. Around 25-50% of affected individuals have mild to moderate episodic bleeding. May-Hegglin anomaly is caused by monoallelic mutations in the *MYH9* gene.

Fechtner syndrome: A disorder characterized by May-Hegglin anomaly and other features including nephritis, hearing loss (especially in the high frequency range) and eye abnormalities, typically cataracts. Fechtner syndrome is caused by monoallelic mutations in the *MYH9* gene.

Sebastian syndrome: A disorder characterized by the triad of thrombocytopenia, giant platelets and inclusions in peripheral blood leukocytes. Sebastian syndrome is caused by monoallelic mutations in the *MYH9* gene.

Macrothrombocytopenia and progressive sensorineural deafness: A platelet disorder appearing during early childhood characterized by hereditary macrothrombocytopenia. In the 3rd decade of life, affected individuals manifest hearing impairment progressing to severe-to-profound bilateral hearing loss. "Macrothrombocytopenia and progressive sensorineural deafness" is caused by monoallelic mutations in the *MYH9* gene.

Macrothrombocytopenia: A disorder characterized by thrombocytopenia due to peripheral destruction. Macrothrombocytopenia is caused by autosomal dominant mutations in the *TUBB1* gene which encodes the beta 1 class VI tubulin.

3.03.4.5 Qualitative Platelet Defects

Bernard–Soulier syndrome or Hemorrhagiparous thrombocytic dystrophy: A bleeding disorder caused by a deficiency of glycoprotein Ib, which functions as the platelet membrane von Willebrand factor receptor complex. This glycoprotein is composed of 4 subunits encoded by 4 separate genes: *GP1BA, GP1BB, GP9* and *GP5*. Bernard-Soulier syndrome type A1 can be caused by autosomal recessive mutations in the *GP1BA* gene;; Bernard-Soulier syndrome type A2 can be caused by autosomal dominant mutations in the *GP1BA* gene;; Bernard-Soulier syndrome type B can be caused by autosomal recessive mutations in the *GP1BB* gene;; and Bernard-Soulier syndrome type C can be caused by autosomal recessive mutations in the *GP9* gene. Bernard-Soulier syndrome is estimated to affect less than 1 in every 1,000,000 individuals.

Glanzmann's thrombasthenia: An abnormality of the platelets characterized by failure of platelet aggregation and by absent or diminished clot retraction. There are several underlying causes of this disorder with at least 2 autosomal genomic loci: Glanzmann thrombasthenia can be caused by autosomal recessive mutations either in the *ITGA2B* gene which encodes the platelet glycoprotein alpha-IIb, or the *ITGB3* gene which encodes the platelet glycoprotein IIIa. Glanzmann thrombasthenia is an extremely rare coagulopathy, except among Arabs living in Israel and French Gypsies Manouche and is estimated to affect less than 1 in every 1,000,000 individuals. In Gypsies this pathogenic allele appears 300-400 years ago.

Platelet-type bleeding disorder-5 or Quebec platelet disorder: A bleeding disorder caused due to a gain-of-function defect in fibrinolysis mechanism. Affected individuals show delayed onset bleeding after occurrences such as surgery. The feature of this condition is strikingly increased PLAU levels within platelets, which causes intraplatelet plasmin generation and secondary degradation of alpha-granule proteins. Quebec platelet disorder is caused by monoallelic tandem duplication of the *PLAU* gene which encodes the urokinase plasminogen activator.

Platelet-type bleeding disorder-8: A disorder characterized by mild to moderate mucocutaneous bleeding, and excessive posttraumatic and postsurgical bleeding. The defect is due to the inability of ADP to induce platelet aggregation. Platelet-type bleeding disorder-8 is caused by autosomal recessive mutations in the *P2RY12* gene which encodes the G-protein coupled purinergic receptor (P2Y12). Platelet-type bleeding disorder is estimated to affect less than 1 in every 1,000,000 individuals.

Platelet-type bleeding disorder-9: A disorder characterized by thrombocytopenia, easy bruising, and absence of bleeding after surgery or tooth extraction. Platelet-type bleeding disorder-9 is caused by autosomal dominant mutations in the *ITGA2* gene which encodes the alpha 2 integrin (CD49B, alpha 2 subunit of VLA-2 receptor).

Platelet-type bleeding disorder-11 or Glycoprotein VI deficiency: A disorder characterized by mild to moderate bleeding disorder caused by defective platelet activation and aggregation in response to collagen. Platelet-type bleeding disorder-11 is caused by autosomal recessive mutations in the *GP6* gene which encodes the glycoprotein VI.

Platelet-type bleeding disorder-12: A disorder characterized by mildly increased bleeding due to a platelet defect. Platelet-type bleeding disorder-12 is caused by autosomal dominant mutations in the *PTGS1* gene which encodes the prostaglandin-endoperoxide synthase 1 (prostaglandin G/H synthase and cyclooxygenase). This enzyme catalyzes the formation of prostaglandin G2 and prostaglandin H2 from arachidonic acid, and the downstream formation of thromboxane A2 and prostacyclin.

Susceptibility to platelet-type bleeding disorder-13: A disorder characterized by mild mucocutaneous bleeding. Susceptibility to this platelet-type bleeding disorder-13 is caused by autosomal dominant mutations in the *TBXA2R* gene which encodes the thromboxane A2 receptor.

Platelet-type bleeding disorder-15: A disorder characterized by macrothrombocytopenia. Affected individuals usually show mild bleeding tendency, such as epistaxis. Platelet-type bleeding disorder-15 is caused by autosomal dominant mutations in the *ACTN1* gene which encodes the alpha 1 actinin.

Scott syndrome: A mild platelet-type bleeding disorder characterized by defective surface exposure of procoagulant phosphatidylserine, resulting in impaired thrombin formation. Scott syndrome can be caused by autosomal recessive mutations in the *ANO6* gene which encodes the anoctamin 6, a small-conductance calcium-activated nonselective cation channel.

Platelet glycoprotein IV deficiency or CD36 deficiency: A common trait that can be associated with thrombocytopenia. Platelet glycoprotein IV deficiency is caused by autosomal recessive mutations in the *CD36* antigen gene which encodes the thrombospondin receptor. Platelet glycoprotein IV deficiency can be divided into 2 subgroups: The Type I phenotype is characterized by no cells expressing the CD36 antigen; The Type II phenotype lacks the surface expression of CD36 in platelets, but expression in monocytes/macrophages is near normal. Platelet glycoprotein IV deficiency is present in 2-3% of Japanese, sub-Saharan African and Thai individuals, but in less than 0.3% of individuals of Caucasian descent.

Familial platelet disorder: A disorder characterized by moderate thrombocytopenia, abnormal platelet function and the propensity to develop myeloid malignancies. Familial platelet disorder is caused by autosomal dominant mutations in the *RUNX1* gene which encodes the runt-related transcription factor 1.

Thromboxane synthase deficiency: A disorder characterized by petechiae, bruises, nosebleeds hematuria and gastrointestinal bleeding. Affected individuals manifest a defective aggregation of platelets.

Thromboxane synthase deficiency is caused by autosomal dominant mutations in the *TBXAS1* gene which encodes the thromboxane A synthase 1 (platelet).

3.03.4.6 Other Primary Thrombocytopenia

Wiskott–Aldrich syndrome: This disease is discussed under Diseases of the blood and blood-forming organs and certain disorders involving the immune mechanism in the subchapter "Immunodeficiency Associated with other Major Defects" (3.03.6.3).

Thrombocytopenia-absent radius syndrome: This disease is discussed under congenital malformations, deformations and chromosomal abnormalities in the subchapter "Congenital Malformation Syndromes predominantly Involving Limbs" (3.17.9.6.3).

Radioulnar synostosis with amegakaryocytic thrombocytopenia: This disease is discussed under congenital malformations, deformations and chromosomal abnormalities in the subchapter "Other Congenital Malformations of Limb(s)" (3.17.8.7).

3.03.4.7 Thrombocytopenic Purpura

Autoimmune thrombocytopenic purpura: A disorder characterized by a low platelet count, normal bone marrow, and the absence of other causes of thrombocytopenia. Affected individuals manifest an increased platelet destruction mediated by autoantibodies to platelet-membrane antigens. Autoimmune thrombocytopenic purpura can be caused by mutations in the *FCGR2C* gene which encodes the low affinity immunoglobulin gamma Fc region receptor II-c.

Familial thrombotic thrombocytopenic purpura: A disorder characterized by thrombocytopenia, hemolytic anemia with fragmentation of erythrocytes, fever, diffuse and nonfocal neurologic findings, and decreased renal function. Familial thrombotic thrombocytopenic purpura is caused by mutations in the *ADAMTS13* gene which encodes the von Willebrand factor cleaving protease.

Table 3.03.4.1 Recognizable Coagulation Defects

Disorder	OMIM	Gene	Biological Function	Product of Conception Alteration
Hemophilia A	306700	*F8*	Coagulation factor	XR + 30%
Hemophilia B	306900	*F9*	Coagulation factor	XR + 30%
Hemophilia C	612416	*F11*	Coagulation factor	AR or AD
Von Willebrand disease mono-allelic	193400	*VWF*	von Willebrand factor	AD + de novo
Von Willebrand disease biallelic	277480	*VWF*	von Willebrand factor	AD + AD = AR
Dysfibrinogenemia, alpha type, causing bleeding diathesis	134820	*FGA*	Fibrinogen	AD
Dysfibrinogenemia, beta type	134830	*FGB*	Fibrinogen	AD
Dysfibrinogenemia, gamma type	134850	*FGG*	Fibrinogen	AD

Table 3.03.4.1. Continued.

Disorder	OMIM	Gene	Biological Function	Product of Conception Alteration
Fibrinogen Milano 12 digenic phenotype FGA/FGG	134850	FGA/FGG	Fibrinogen	DigM
Combined deficiency of vitamin K-dependent clotting factors-1	277450	GGCX	Posttranslational modification of vitamin K	AR
Combined deficiency of vitamin K-dependent clotting factors-2	607473	VKORC1	Epoxide reductase	AR
Factor V deficiency	227400	F5	Coagulation factor	AR
Combined deficiency of factor V and factor VIII type 1	227300	LMAN1	Mannose-binding lectin-1	AR
Combined deficiency of factor V and factor VIII type 2	613625	MCFD2	Multiple coagulation factor protein 2	AR
Factor VII deficiency or hypo-proconvertinemia	227500	F7	Coagulation factor	AR
Factor XII deficiency	234000	F12	Coagulation factor	AR
Factor XIIIA deficiency	613225	F13A1	Coagulation factor	AR
Factor XIIIB deficiency	613235	F13B	Coagulation factor	AR
Fletcher factor deficiency	612423	KLKB1	Plasma kallikrein	AR
Hemorrhagic diathesis, due to plasminogen activator inhibitor 1	613329	SERPINE1	Serpin proteinase inhibitor	AR
Hemorrhagic diathesis, due to α2-plasmin inhibitor deficiency	262850	SERPINF2	Serpin proteinase inhibitor	AR
Thrombophilia, due to factor V Leiden	188055	F5	Coagulation factor	AD or AD + AD = AR
Thrombophilia	613679	F2	Coagulation factor	AD or AD + AD = AR
Thrombophilia, due to elevated HRG	613116	HRG	Prothrombotic effect	AD
Thrombophilia, due to factor IX defect	300807	F9	Coagulation factor	XR
Hereditary thrombophilia, due to congenital antithrombin deficiency	613118	SERPINC1	Serpin proteinase inhibitor	AD
Thrombophilia, due to protein C deficiency	612304	PROC	Inactivator of coagulation factors Va and VIIIa	AR
Thrombophilia, due to protein C deficiency	176860	PROC	Inactivator of coagulation factors Va and VIIIa	AD
Thrombophilia, due to protein S deficiency	614514	PROS1	Anticoagulant protease	AR
Thrombophilia, due to protein S deficiency	612336	PROS1	Anticoagulant protease	AD

Table 3.03.4.1. Continued.

Disorder	OMIM	Gene	Biological Function	Product of Conception Alteration
Thrombophilia, due to heparin cofactor 2 deficiency	612356	SERPIND1	Serpin proteinase inhibitor	AD
Thrombophilia, due to HRG deficiency	613116	HRG	Prothrombotic effect	AD
Familial thrombophilia, due to decreased release of PLAT	612348	PLAT	Plasminogen activator	AD
Thrombocytopenia-1	313900	WAS	Cytoskeleton and signal transduction	XR
Thrombocytopenia-2	188000	MASTL	Microtubule associated serine/threonine kinase	AD
Thrombocytopenia-4	612004	CYCS	Mitochondrial protein	AD
Congenital amegakaryocytic thrombocytopenia	604498	MPL	Cell surface receptor JAK-STAT	AR
Epstein syndrome	153650	MYH9	Cytoskeletal microfilaments	AD + de novo
Fechtner syndrome	155100	MYH9	Cytoskeletal microfilaments	AD + de novo
Fechtner syndrome	153640	MYH9	Cytoskeletal microfilaments	AD + de novo
Sebastian syndrome	605249	MYH9	Cytoskeletal microfilaments	AD + de novo
Macrothrombocytopenia and progressive sensorineural Deafness	600208	MYH9	Cytoskeletal microfilaments	AD + de novo
Macrothrombocytopenia	300367	GATA1	Transcription factor Zinc finger	XR
Thrombocytopenia with beta-thalassemia	314050	GATA1	Transcription factor Zinc finger	XR
Dyserythropoietic anemia with thrombocytopenia	300367	GATA1	Transcription factor Zinc finger	XR
Macrothrombocytopenia	613112	TUBB1	Cytoskeleton microtubules	AD
Gray platelet syndrome	139090	NBEAL2	Megakaryocyte biogenesis	AR
Bernard-Soulier syndrome	153670	GP1BA	Cell surface receptor Other/ungrouped	AD
Bernard-Soulier syndrome A	231200	GP1BA	Cell surface receptor Other/ungrouped	AR
Bernard-Soulier syndrome B	231200	GP1BB	Cell surface receptor Other/ungrouped	AR
Bernard-Soulier syndrome C	231200	GP9	Cell surface receptor Other/ungrouped	AR

Table 3.03.4.1. Continued.

Disorder	OMIM	Gene	Biological Function	Product of Conception Alteration
Glanzmann thrombasthenia	273800	ITGA2B	Cell surface receptor Other/ungrouped	AR
Platelet-type bleeding disorder-5 or Quebec platelet disorder	601709	PLAU	Plasminogen activator	
Platelet-type bleeding disorder-8	609821	P2RY12	Cell surface receptor G protein	AR
Platelet-type bleeding disorder-9	614200	ITGA2	Alpha 2 integrin	AD
Platelet-type bleeding disorder-11	614201	GP6	Glycoprotein VI	AR
Platelet-type bleeding disorder-12	605735	PTGS1	Prostaglandin G/H synthase	AD
Platelet-type bleeding disorder-13	614009	TBXA2R	Cell surface receptor G protein	AD
Platelet-type bleeding disorder-15	615193	ACTN1	Alpha 1 actinin	AD
Scott syndrome	262890	ANO6	Cation channel	AR
Platelet glycoprotein IV deficiency	608404	CD36	Cell surface receptor Other/ungrouped	
Familial platelet disorder with associated myeloid malignancy	601399	RUNX1	Transcription factor β-Scaffold factors	AD
Thromboxane synthase deficiency	614158	TBXAS1	Thromboxane A synthase	AD
Autoimmune thrombocytopenic purpura	188030	FCGR2C	Cell surface receptor Enzyme-linked receptor	
Familial thrombotic thrombocytopenic purpura	274150	ADAMTS13	von Willebrand factor cleaving protease	AR

The Concise Encyclopedia of Genomic Diseases

Table 3.03.4.2 Coagulation Defects in Different Ethnic Groups

Disease	Frequency of condition	People with higher prevalence (if available)
von Willebrand disease monoallelic	1 in 100	Worldwide and Åland Islanders
von Willebrand disease biallelic	1 in 40,000*	Worldwide and Åland Islanders
Factor V Leiden monoallelic	1-5 in 100	Caucasians
Factor V Leiden biallelic	1 in 2,000*	Caucasians
Thrombophilia Factor II monoallelic	2 in 100	Caucasians
Thrombophilia, due to protein C deficiency	2 in 1,000	Caucasians
Thrombophilia, due to protein S deficiency	2 in 1,000	Caucasians
Hemophilia A	1 in 6,000 Males	Worldwide
Platelet glycoprotein IV deficiency	2-400 in 10,000	Japanese, Thais and sub-Saharan Africans
Thrombophilia, due to congenital antithrombin deficiency	3-5 in 10,000	Caucasians
Thrombophilia, due to heparin cofactor 2 deficiency	3-5 in 10,000	Caucasians
Hemophilia B	3-5 in 100,000 Males	Worldwide
Hemophilia C	1 in 100,000	Ashkenazi Jews and French Basques
Factor VII deficiency or hypoproconvertinemia	1 in 300,000	Oriental Jews
Combined deficiency of Factor V and factor VIII	1-10 in 1,000,000	Sephardic Jews
Bernard-Soulier syndrome	1 in 1,000,000	Unknown
Congenital factor V deficiency	1 in 1,000,000	Unknown
Hemorrhagic diathesis, due to alpha2-plasmin inhibitor deficiency	<1 in 1,000,000	Unknown
Dyserythropoietic anemia with thrombocytopenia	<1 in 1,000,000	Unknown
Factor XII deficiency	<1 in 1,000,000	Unknown
Factor XIIIA deficiency	<1 in 1,000,000	Unknown
Factor XIIIB deficiency	<1 in 1,000,000	Unknown
Familial platelet disorder with associated myeloid malignancy	<1 in 1,000,000	Unknown
Fletcher factor deficiency	<1 in 1,000,000	Unknown

3.03.5 Other Diseases of Blood and Blood-forming Organs

3.03.5.1 Agranulocytosis

Severe congenital neutropenia: A disorder characterized by early onset of severe bacterial infections. Patients suffer from an arrest in granulopoiesis maturation at the level of promyelocytes. Severe congenital neutropenia is a heterogeneous disorder of hematopoiesis characterized by showing autosomal dominant, autosomal recessive and X-linked inheritance.

Autosomal dominant severe congenital neutropenia-1: A disorder caused by autosomal monoallelic mutations in the *ELANE* gene which encodes the neutrophil expressed elastase, a serine protease that hydrolyze many proteins in addition to elastin. Mutations in the *ELANE* gene cause 60% of severe congenital neutropenia cases in individuals of European and Middle Eastern ancestry.

Cyclic neutropenia or Cyclical neutropenia: An allelic disorder, less symptomatic than Severe congenital neutropenia. Cyclic neutropenia is characterized by regular 3 week cyclic fluctuations in the number of blood neutrophils. These forms of neutropenia tend to last 3-6 days. Recurrent severe neutropenia causes patients to experience periodic symptoms of fever, oral ulceration, skin infection and malaise. The product of conception in cyclic neutropenia appears to exhibit de novo alteration in a few cases and autosomal dominant inheritance in most of the cases. Cyclic neutropenia is caused by autosomal monoallelic mutations in the *ELANE* gene.

Nonimmune chronic idiopathic neutropenia of adults: A disorder characterized by a relatively mild form of neutropenia.

There is a subset of patients with high predisposition to leukemia. Nonimmune chronic idiopathic neutropenia of adults is caused by mutations in the *GFI1* gene which encodes the growth factor independent 1 transcription repressor.

Autosomal dominant severe congenital neutropenia-2: An allelic disorder resulting from nonimmune chronic idiopathic neutropenia of adults.

Autosomal recessive severe congenital neutropenia-3 or Kostmann syndrome: A disorder caused by biallelic mutations in the *HAX1* gene which encodes the HCLS1 associated protein X-1. This protein is associated with hematopoietic cell-specific Lyn substrate 1 which is a substrate of Src family tyrosine kinases.

Dursun syndrome: A disorder characterized by leukopenia including intermittent neutropenia, monocytosis, lymphopenia and anemia. Patients also suffer from familial pulmonary arterial hypertension and cardiac abnormalities including atrial septal defect. Dursum syndrome is caused by mutations in the *G6PC3* gene which encodes the glucose 6 phosphatase, third subunit.

Autosomal recessive severe congenital neutropenia-4: An allelic disorder from Dursum syndrome.

X-linked Severe congenital neutropenia: An allelic disorder from Wiskott–Aldrich syndrome. It is caused by mutations in the *WAS* gene which encodes the Wiskott-Aldrich syndrome, a protein involved in transduction of signals from receptors on the cell surface to the actin cytoskeleton.

3.03.5.2 Functional Disorders of Polymorphonuclear Neutrophils

Chronic granulomatous disease or Bridges–Good syndrome: A disorder characterized by difficulty with forming the reactive oxygen compounds in certain cells of the immune system. This condition results from an inability of phagocytes to kill microbes. Affected children exhibit severe recurrent and chronic nonspecific infections. Other features include suppurative lymphadenitis, pulmonary infiltrates, hepatosplenomegaly and eczematoid dermatitis, with findings of granulomas in biopsy. There are several underlying causes of this disorder with at least 1 X-linked genomic locus and 4 autosomal genomic loci: X-linked Chronic granulomatous disease is caused by mutations in the *CYBB* gene which encodes the cytochrome b-245, beta polypeptide. The product of conception in this condition appears to exhibit X-linked recessive inheritance in most of the affected male cases and de novo alteration in a few cases. Chronic granulomatous disease can be caused by autosomal recessive mutations in any 1 of 4 genes encoding structural or regulatory subunits of the phagocyte NADPH oxidase complex;; Chronic granulomatous disease cytochrome b-negative is caused by autosomal recessive mutations in the *CYBA* gene;; Chronic granulomatous disease cytochrome b-positive type I is caused by autosomal recessive mutations in the *NCF1* gene;; Chronic granulomatous disease cytochrome b-positive type II is caused by autosomal recessive mutations in the *NCF2* gene;; and Chronic granulomatous disease cytochrome b-positive type III is caused by autosomal recessive mutations in the *NCF4* gene. Chronic granulomatous disease affects about 1 in 200,000 live births worldwide; the ratio of male to female patients is 6.6/1.

3.03.5.3 Genomic Anomalies of Leukocytes

Poikiloderma with neutropenia: A genodermatosis appearing during the 1st year of life. Affected individuals start manifesting papular erythematous rash on the extremities. This condition progressively spreads centripetally and as the papular rash resolves, hypo- and hyper-pigmentation result with development of telangiectasias. Affected individuals exhibit other features including pachyonychia and recurrent pneumonias due to neutropenia. Poikiloderma with neutropenia is caused by biallelic mutations in the *USB1* gene which encodes the U6 snRNA biogenesis 1. In the United States, this condition has been reported among Navajo families. This condition has been also reported in Turkish and Moroccan families.

Leukocyte adhesion deficiency: A disorder characterized by recurrent infections, impairment in the leukocyte adhesion process and marked leukocytosis. There are several underlying causes of this disorder with at least 3 autosomal genomic loci: Leukocyte adhesion deficiency-I is characterized by life-threatening, recurrent bacterial infections. It is caused by autosomal recessive mutations in the *ITGB2* gene which encodes the integrin beta 2 (complement component 3 receptor 3 and 4 subunit);; Leukocyte adhesion deficiency-II presents with leukocytosis, recurrent infections, severe growth delay and intellectual deficit. It is caused by mutations in the *SLC35C1* gene which encodes the solute carrier (family 35 member C1), a GDP-fucose transporter;; and Leukocyte adhesion deficiency-III is characterized by both severe bacterial infections and severe bleeding disorders. It is caused by mutations in the *FERMT3* gene which encodes the fermitin family member 3.

Pelger-Huet anomaly: A trait characterized by hyposegmented neutrophils. In these patients, neutrophils are clinically normal but may be mistaken for immature cells. Pelger-Huet anomaly can be caused by mutations in the *LBR* gene which encodes the lamin B receptor. This nuclear envelope inner membrane protein anchors the lamina and the heterochromatin to the membrane. This trait is uncommon worldwide except in two areas: Vasterbotten County in northern Sweden and the mountain village of Gelenau in southeastern Germany. Most of the hyposegmented neutrophils are inherited in autosomal dominant fashion. There is a more severe disorder that occurs when both parents are heterozygous and both parents provide the faulty copy to descendents (AD + AD = AR), in which case the inheritance is autosomal recessive. These homozygous individuals have epilepsy, developmental delay, skeletal abnormalities and ovoid neutrophil nuclei.

3.03.5.4 Methemoglobinemia

Methemoglobinemia: A disorder characterized by abnormally high levels of methemoglobin in the blood. After hemoglobin is oxidized to methemoglobin, there are reductions in the ability to release oxygen to tissues. Consequently, patients exhibit cyanosis and hypoxia. There are several underlying causes of this disorder with at least 4 autosomal genomic loci: Hereditary methemoglobinemia types 1 and 2 are caused by autosomal recessive mutations in the *CYB5R3* gene, which encodes the methemoglobin reductase;; Hereditary methemoglobinemia type 4 is caused by autosomal recessive mutations in the *CYB5A* gene which encodes the cytochrome b5. Another form of hereditary methemoglobinemia is seen in patients with certain hemoglobin variants such as hemoglobin M or H, which are caused by mutations in the *HBB, HBA1* and *HBA2* genes. There are also congenital methemoglobinemia cases caused due to pyruvate kinase deficiency and Glucose-6-phosphate dehydrogenase deficiency. Both types of deficiencies are caused by mutations in the *PKLR* and *G6PD* genes respectively. There are also several non-genomic causes of acquired methemoglobinemia, including exposure to exogenous oxidizing drugs and their metabolites, antibiotics, local anesthetics, ingestion of compounds containing nitrates, etc…

3.03.5.5 Familial Erythrocytosis

Primary erythrocytosis: A disorder characterized by increases in the proportion of red blood cell mass and hemoglobin concentrations in blood. There is no increase in leukocytes or platelets and the disorder does not progress to leukemia. There are several underlying causes of this disorder with at least 4 autosomal genomic loci: Primary erythrocytosis-1 or Familial erythrocytosis-1 is caused by mutations in the *EPOR* gene which encodes the erythropoietin receptor;; Primary erythrocytosis-2 (or Chuvash polycythemia) is a clear example of polycythemia of genomic origin. This form of polycythemia affects the ethnic Chuvash people; this pathogenic allele appears 14,000-62,000 years ago. It is caused by autosomal recessive polymorphisms in the *VHL* gene which encodes the von Hippel-Lindau tumor suppressor protein;; Primary erythrocytosis-3 is caused by gain-of-function mutations in *EGLN1* gene which encodes the egl nine homolog protein 1. This is a transcriptional complex that plays a role in oxygen homeostasis. This protein functions as a cellular oxygen sensor. Patients with primary erythrocytosis-3 also suffer from pulmonary hypertension;; and Primary erythrocytosis-4 is caused by loss-of-function mutations of the *EPAS1* gene which encodes the endothelial PAS domain protein 1. Polycythemia Vera is a form of primary erythrocytosis caused by somatic mutations in the *JAK2* gene.

3.03.5.6 Other Specified Diseases with Participation of Lymphoreticular and Reticulohistiocytic Tissue

Familial Hemophagocytic lymphohistiocytosis or Hemophagocytic syndrome: A disorder characterized by fever, lymphadenopathy, jaundice, hepatosplenomegaly and rash. There are several underlying causes of this disorder with at least 5 autosomal genomic loci: Familial Hemophagocytic lymphohistioctosis-1 has been mapped but the gene has not been characterized;; Familial Hemophagocytic lymphohistioctosis-2 is caused by autosomal recessive mutations in the *PRF1* gene which encodes the perforin 1 (pore forming protein);; Familial Hemophagocytic lymphohistioctosis-3 is caused by autosomal recessive mutations in the *UNC13D* gene which encodes the unc-13 homolog D protein;; Familial Hemophagocytic lymphohistioctosis-4 is caused by autosomal recessive mutations in the *STX11* gene which encodes the syntaxin 11, a protein that may regulate protein transport among late endosomes and the trans-Golgi network.;; and Familial Hemophagocytic lymphohistioctosis-5 is caused by autosomal recessive mutations in the *STXBP2* gene which encodes the syntaxin binding protein 2. Familial Hemophagocytic lymphohistioctosis is more prevalent in areas of higher parental consanguinity. This condition affects 1 in 50,000 and since it is autosomal, has equal gender distribution.

Table 3.03.5.1 Other Recognizable Diseases of Blood and Blood-forming Organs

Disorder	OMIM	Gene	Biological Function	Product of Conception Alteration
Cyclic neutropenia	162800	*ELANE*	Serine proteases	AD + de novo
X-linked severe congenital neutropenia	300299	*WAS*	Cytoskeleton and signal transduction	XR
Severe congenital neutropenia-1	202700	*ELANE*	Serine proteases	AD + de novo
Severe congenital neutropenia-2	613107	*GFI1*	Zinc finger protein	AD
Severe congenital neutropenia-3	610738	*HAX1*	Substrate of Src family tyrosine kinases	AR
Severe congenital neutropenia-4	612541	*G6PC3*	Endoplasmic reticulum protein	AR
Dursun syndrome	612541	*G6PC3*	Endoplasmic reticulum protein	AR
Nonimmune chronic idiopathic neutropenia of adults	607847	*GFI1*	Zinc finger protein	AD
Chronic granulomatous disease or Bridges–Good syndrome	306400	*CYBB*	Cytosolic NADPH oxidase	XR + de novo

Table 3.03.5.1. Continued.

Disorder	OMIM	Gene	Biological Function	Product of Conception Alteration
Chronic granulomatous disease or Bridges–Good syndrome	233690	CYBA	Cytosolic NADPH oxidase	AR
Chronic granulomatous disease or Bridges–Good syndrome	233700	NCF1	Cytosolic NADPH oxidase	AR
Chronic granulomatous disease or Bridges–Good syndrome	233710	NCF2	Cytosolic NADPH oxidase	AR
Chronic granulomatous disease or Bridges–Good syndrome	613960	NCF4	Cytosolic NADPH oxidase	AR
Hereditary neutrophilia	162830	CSF3R	Cell surface receptor Other/ungrouped	AD
Poikiloderma with neutropenia	604173	USB1	snRNA biogenesis 1	AR
Leukocyte adhesion deficiency I	116920	ITGB2	Cell surface receptor Other/ungrouped	AR
Leukocyte adhesion deficiency II	266265	SLC35C1	Solute carrier	AR
Leukocyte adhesion deficiency III	612840	FERMT3	Protein-protein interaction	AR
Pelger-Huet anomaly	169400	LBR	Nuclear envelope protein	AD
Methemoglobinemia-1/2	250800	CYB5R3	Endoplasmic reticulum protein	AR
Methemoglobinemia-4	250790	CYB5A	Endoplasmic reticulum protein	AR
Primary erythrocytosis-1	133100	EPOR	Erythropoietin receptor	AD
Primary erythrocytosis-2 or Chuvash polycythemia	263400	VHL	Tumor suppressor	AR
Primary erythrocytosis-3	609820	EGLN1	Cellular oxygen sensor	AD
Primary erythrocytosis-4	611783	EPAS1	Transcription factor	AD
Familial hemophagocytic lymphohistiocytosis 2	603553	PRF1	Pore forming protein	AR
Familial hemophagocytic lymphohistiocytosis 3	608898	UNC13D	Vesicle maturation during exocytosis	AR
Familial hemophagocytic lymphohistiocytosis 4	603552	STX11	Trafficking Vesicle fusion	AR
Familial hemophagocytic lymphohistiocytosis 5	613101	STXBP2	Trafficking from the golgi apparatus to the plasma membrane	AR

3.03.6 Certain Disorders Involving the Immune Mechanism

The only common disorder, affecting 1 in 333-600 individuals, is selective immunoglobulin A deficiency, which is a milder form of primary immunodeficiency. Most of the conditions described below are uncommon, with incidences between 1 in 100,000-2,000,000 individuals.

3.03.6.1 Immunodeficiency with Predominantly Antibody Defects

In these primary antibody deficiencies, one or more isotypes of immunoglobulin are impaired or decreased. These are rare syndromes, except for immunoglobulin A deficiency, which is common.

X-linked agammaglobulinemia or Bruton's agammaglobulinemia: A disorder characterized by a complete lack of antibodies in the bloodstream. Affected children are unusually prone to bacterial infection but not to viral infection. This condition is named after Ogden Bruton. Bruton's agammaglobulinemia is caused by X-linked recessive mutations in the *BTK* gene which encodes the Bruton tyrosine kinase. This protein produces an arrest in B cell development. Bruton's agammaglobulinemia is the most prevalent form of agammaglobulinemia affecting 1 in 100,000 male births worldwide. The X-linked form accounts for about 85-90% of cases of the disorder. Bruton's agammaglobulinemia belongs to different groups including de novo and X-linked recessive.

Autosomal Agammaglobulinemia: A disorder characterized by a low number or absence of both circulating B cells and serum antibodies. Affected children exhibit severe infections in the 1st years of life. There are several underlying causes of this disorder with at least 6 autosomal genomic loci: Agammaglobulinemia-1 is caused by autosomal recessive mutations in the *IGHM* gene which encodes the immunoglobulin heavy constant mu;; Agammaglobulinemia-2 is caused by autosomal recessive mutations in the *IGLL1* gene which encodes the immunoglobulin lambda-like polypeptide 1;; Agammaglobulinemia-3 is caused by autosomal recessive mutations in the *CD79A* gene which encodes the CD79a molecule (immunoglobulin-associated alpha);; Agammaglobulinemia-4 is caused by autosomal recessive mutations in the *BLNK* gene which encodes the B-cell linker, an adapter containing a Src homology 2 domain protein;; Agammaglobulinemia-5 is caused by autosomal recessive disruption of the *LRRC8* gene which encodes the leucine rich repeat containing protein 8A;; and Agammaglobulinemia-6 is caused by autosomal recessive mutations in the *CD79B* gene which encodes the CD79a molecule (immunoglobulin-associated beta). The autosomal forms account for approximately 10-15% of cases of the disorder. These conditions affect less than 1 child in every 1,000,000 individuals.

Immunodeficiency with hyper IgM: A disorder with a clinical course similar to Bruton-type agammaglobulinemia. Immunodeficiency with hyper-IgM is characterized by recurrent infections, low or absent IgG, IgE and IgA levels, and normal or elevated levels of IgM and IgD. There are several underlying causes of this disorder with at least 1 X-linked genomic locus and 4 autosomal genomic loci: X-linked immunodeficiency with hyper-IgM type 1 is caused by X-linked recessive mutations in the *CD40LG* gene which encodes the CD40 ligand, a protein expressed on the surface of T cells that regulates B cell function by engaging CD40 on the B cell surface;; Immunodeficiency with hyper-

IgM type 2 is caused by mutations in the *AICDA* gene which encodes the activation-induced cytidine deaminase, a RNA-editing deaminase that is involved in class-switch recombination of immunoglobulin genes, somatic hypermutation and gene conversion;; Immunodeficiency with hyper-IgM type 3 is caused by mutations in the *CD40* gene which encodes the CD40 molecule (TNF receptor superfamily member 5);; "Immunodeficiency with hyper-IgM type 4 with a B lymphocyte-intrinsic selective deficiency in Ig class-switch recombination" gene/loci have not been characterized;; and Immunodeficiency with hyper-IgM type 5 is caused by mutations in the *UNG* gene which encodes the uracil-DNA glycosylase. This protein prevents mutagenesis by eliminating uracil from DNA molecules by cleaving the N-glycosylic bond and initiating the base-excision repair pathway. X-linked immunodeficiency with hyper-IgM type 1 is estimated to affect 1 in every 500,000 males.

Immunoglobulin A deficiency: A common disorder characterized by deficiency in IgA, which are antibodies that protect against infections of the mucous membranes. The pattern of inheritance in this condition does not strictly follow Mendel's law; in fact, this condition is more common in males than in females. Immunoglobulin A deficiency can be caused by mutations either in the *MSH5* gene which encodes the mutS homolog protein 5, or the *TNFRSF13B* gene which encodes the tumor necrosis factor receptor superfamily member 13B. Selective immunoglobulin A deficiency is estimated to affect 1 in every 333-600 individuals.

3.03.6.2 Combined Immunodeficiencies

Severe Combined T and B–cell immunodeficiency or Thymic alymphoplasia: A disorder characterized by vulnerability to viral, fungal and bacterial infections, atrophy of the thymus, lack of delayed hypersensitivity, and lack of benefit from gamma globulin administration. Affected children manifest dysfunctional (or a decreased number of both) T lymphocytes and B lymphocytes, which are the regulators of adaptive immunity. Patients suffering from severe Combined T and B–cell immunodeficiency are sometimes referred to as having "bubble boy" disease. There are several possible genes which may affect the adaptive immune system, including those that are autosomal and X-linked. Severe Combined T and B–cell immunodeficiency can be divided into 2 main classes: those with B lymphocytes (B+) and those without (B-). This condition can be further divided according to the presence or absence of T lymphocytes (T) and presence or absence of NK cells (NK). There are several underlying causes of this disorder with at least 1 X-linked genomic locus and 11 autosomal genomic loci: Severe Combined T and B–cell immunodeficiency T-, B+, NK- is caused by X-linked recessive mutations in the *IL2RG* gene which encodes the interleukin 2 receptor gamma chain. The IL2RG forms complexes with at least 6 unique cytokine-specific interleukin receptor including: IL2RA, IL4RA, IL7RA, IL9RA, IL15RA and IL21RA. This is the most common form as it accounts for 46-70% of all cases. The products of conception in X-linked Severe Combined T and B–cell immunodeficiency appear to either exhibit X-linked recessive inheritance in most of the cases or de novo alteration in a few cases;; Severe Combined T and B–cell immunodeficiency T-, B-, NK- is caused by autosomal recessive mutations in the *ADA* which encodes the adenosine deaminase, an enzyme that catalyzes the hydrolysis of adenosine to inosine. Mutations in the *ADA* gene account for approximately 20% of all cases of Severe Combined T and B–cell immunodeficiency;; Severe Combined T and B–cell immunodeficiency T-, B-, NK+ is caused by autosomal recessive mutations in the *RAG1* gene which encodes the recombination activating gene 1 (a protein involved in activation of immunoglobulin V-D-J recombination), and the *RAG2* gene which encodes the recombination activating gene 2;; Severe Combined T and B–cell immunodeficiency T-, B-, NK+ with sensitivity to ionizing radiation and Athabaskan-type are caused by autosomal recessive mutations in the *DCLRE1C* gene which encodes the DNA cross-link repair 1C protein. This disease is ethnic specific. In Athabaskan people (Navajo and Apache), the carrier frequency for Severe Combined T and B–cell immunodeficiency trait due to *DCLRE1C* gene alteration is estimated to be 1 in 25 individuals, which means the frequency of this condition is about 1 in 2,500 individuals or $1/(50)^2$. This genomic condition is a major cause of illness and death among Navajo and Apache children;; Severe combined immunodeficiency T-, B-, NK+ with microcephaly, growth retardation and sensitivity to ionizing radiation is caused by autosomal recessive mutations in the *NHEJ1* gene which encodes the nonhomologous end-joining factor 1 protein;; Severe Combined T and B–cell immunodeficiency T-, B-, NK+ with sensitivity to ionizing radiation can be caused by autosomal recessive mutations in the *LIG4* gene which encodes the DNA ligase 4. This enzyme joins single-strand breaks in a double-stranded polydeoxynucleotide in an ATP-dependent reaction;; Severe Combined T and B–cell immunodeficiency T-, B+, NK- is caused by autosomal recessive mutations in the *JAK3* gene which encodes the Janus kinase 3, a tyrosine kinase;; Severe Combined T and B–

cell immunodeficiency T-, B+, NK+ is caused by autosomal recessive mutations in the *IL7R* gene which encodes the interleukin 7 receptor;; Severe Combined T and B–cell immunodeficiency T-, B+, NK+ is caused by autosomal recessive mutations in the *PTPRC* gene which encodes the protein tyrosine phosphatase (receptor type C);; Severe Combined T and B–cell immunodeficiency T-, B+, NK+ is caused by autosomal recessive mutations in the *CD3E* gene which encodes the CD3e molecule epsilon (CD3-TCR complex);; and Severe Combined T and B–cell immunodeficiency T-, B+, NK+ is caused by autosomal recessive mutations in the *CD3D* gene which encodes the CD3d molecule delta (CD3-TCR complex). Severe Combined T and B–cell immunodeficiency is estimated to affect 1 in every 54,000 children of Swiss origin and 1 in every 200,000 children of Japanese origin. The overall prevalence of all types of Severe Combined T and B–cell immunodeficiency is approximately 1 in 75,000 births

Omenn syndrome: A form of severe Combined T and B–cell immunodeficiency. Affected individuals exhibit failure to thrive, chronic diarrhea, erythroderma, desquamation, alopecia, lymphadenopathy and hepatosplenomegaly. Omenn syndrome is caused by autosomal recessive mutations either in the *RAG1* gene which encodes the recombination activating gene 1, or *RAG2* gene which encodes the recombination activating gene 2. These recombination-activating proteins are involved in activation of immunoglobulin V-D-J recombination. Omenn syndrome is also caused by autosomal recessive mutations in the *DCLRE1C* gene. The occurrence prevalence of Omenn syndrome is difficult to establish.

Immune dysfunction with T-cell inactivation due to calcium entry defect: A disorder appearing during the neonatal period characterized by intermittent fever and aphthous stomatitis. Affected children show failure to thrive and muscular hypotonia. There are several underlying causes of this disorder with at least 2 autosomal genomic loci: Immune dysfunction with T-cell inactivation due to calcium entry defect-1 is caused by autosomal recessive mutations in the *ORAI1* gene which encodes the ORAI calcium release-activated calcium modulator 1, a membrane calcium channel subunit;; and Immune dysfunction with T-cell inactivation due to calcium entry defect-2 is caused by autosomal recessive mutations in the *STIM1* gene which encodes the stromal interaction molecule 1, a transmembrane protein that mediates calcium influx.

Bare lymphocyte syndrome: A disorder classified in 2 groups according to the HLA class deficiency: Bare lymphocyte syndrome I and II. Type I patients do not express HLA class I and exhibit chronic bacterial infections affecting the respiratory tract in late childhood while Bare lymphocyte syndrome II patients do not express HLA class II and exhibit similar features to severe Combined T and B–cell immunodeficiency. Affected individuals show impairment in the function in genes of the major histocompatibility complex. There are several underlying causes of this disorder with at least 7 autosomal genomic loci: Bare lymphocyte syndrome I (or HLA class I deficiency) is caused by autosomal recessive mutations in the *TAP2* gene which encodes the ATP-binding cassette transporter 2, *TAP1* gene which encodes the ATP-binding cassette transporter 1, or *TAPBP* gene which encodes the TAP binding protein (tapasin). These 3 genes are mapped in the HLA region on chromosome 6p21.3. On the contrary, in Bare

lymphocyte syndrome II, the defects are not in the major histocompatibility complex II genes themselves. Bare lymphocyte syndrome II is caused by autosomal recessive mutations in the *CIITA* gene which encodes the major histocompatibility complex class II transactivator, *RFX5* gene which encodes the regulatory factor X 5 (influences HLA class II expression), *RFXAP* gene which encodes the regulatory factor X-associated protein, and *RFXANK* gene which encodes the regulatory factor X-associated ankyrin-containing protein. These genes are located at: *CIITA* on chromosome 16p13, *RFX5* on chromosome 1q21, *RFXAP* on chromosome 13q14, and *RFXANK* on chromosome 19p12.

Severe immunodeficiency due to deficiency of Interleukin-2 receptor, alpha chain: A disorder characterized by decreased numbers of peripheral T cells displaying abnormal proliferation but normal B-cell development. Severe immunodeficiency due to deficiency of Interleukin-2 receptor, alpha chain is caused by autosomal recessive mutations in the *IL2RA* gene which encodes the interleukin-2 receptor (alpha chain).

Selective T-cell defect: A disorder characterized by severe immunodeficiency associated with absence of CD8+ T lymphocytes. This intrathymic developmental disorder displays a defective T-cell selection. Affected patients peripheral circulating T cells exclusively expressed CD4, CD3, and T-cell receptor-alpha/beta, but not CD8 molecules on their surface. Selective T-cell defect is caused by autosomal recessive mutations in the *ZAP70* gene which encodes the zeta-chain (TCR) associated protein kinase 70kDa.

Purine nucleoside phosphorylase deficiency: A disorder characterized predominantly by decreased T-cell function. Some patients also have neurologic impairment. Immunodeficiency due to purine nucleoside phosphorylase deficiency is caused by autosomal recessive mutations in *PNP* gene which encodes the purine nucleoside phosphorylase.

WHIM syndrome (Warts, Hypogammaglobulinaemia, Infections and Myleokathexis): A disorder characterized by chronic noncyclic neutropenia. There are several underlying causes of this disorder with at least 2 autosomal genomic loci: WHIM syndrome is caused by either autosomal dominant or autosomal recessive mutations in the *CXCR4* gene which encodes the chemokine (C-X-C motif) receptor 4, a seven-transmembrane-segment receptor;; and WHIM syndrome can be caused by dysfunction in the *ADRBK2* gene which encodes the beta-adrenergic receptor kinase 2. WHIM syndrome is estimated to affect less than 1 in every 1,000,000 individuals.

3.03.6.3 Immunodeficiency Associated with other Major Defects

There are several conditions that will be referenced here, but the detailed descriptions are included in the other corresponding chapters.

22q11.2 deletion syndrome or DiGeorge syndrome or Velo-cardio-facial syndrome: A syndrome caused by the microdeletion on chromosome 22, q11.2. A common acronym used to recall the condition is CATCH22. Affected patients exhibit Cardiac Abnormality (especially tetralogy of Fallot), Abnormal facies, Thymic aplasia, Cleft palate, Hypocalcemia and Chromosome 22 q11.2 microdeletion. Other features include learning disabilities, recurrent infections, autoimmune disorders, kidney abnormalities, and psychiatric

illnesses - particularly an increased risk of schizophrenia. The syndrome was named after Angelo DiGeorge in 1968. DiGeorge syndrome is caused by a 1.5 to 3.0-Mb microdeletion of chromosome 22q11.2 or can be caused by point mutations in the *TBX1* gene which encodes the T-box 1. This gene is a transcription factor involved in the regulation of developmental processes. This gene maps to the center of the DiGeorge syndrome on chromosome 22, q11.2. The mechanism that causes DiGeorge syndrome features might involve migration defects of neural crest-derived tissues, principally disturbing development of the 3rd – 4th branchial pouches. The product of conception in DiGeorge's syndrome appears to exhibit de novo alteration in 72-94% of cases and autosomal dominant inheritance in the remaining cases. DiGeorge syndrome is estimated to affect 2-3 in every 10,000 individuals.

Wiskott–Aldrich syndrome: A disorder characterized by eczema, thrombocytopenia, and immune deficiency. Affected children, mostly males, also suffer from splenomegaly, autoimmune disorder, recurrent bacterial infections, bloody diarrhea and malignancies. The syndrome is named after Robert Anderson Aldrich. There are several underlying causes of this disorder with at least 1 X-linked genomic locus and 1 autosomal genomic locus: Wiskott–Aldrich syndrome-1 is caused by X-linked recessive mutations in the *WAS* gene which encodes the Wiskott-Aldrich syndrome protein. The encoded protein is involved in transduction of signals from receptors on the cell surface to the actin cytoskeleton. The product of conception in this condition appears to exhibit de novo alteration in a few cases and X-linked recessive inheritance in most of the cases;; and Wiskott-Aldrich syndrome-2 is caused by autosomal recessive mutations in the *WIPF1* gene which encodes the

WAS/WASL interacting protein family member 1. Wiskott-Aldrich syndrome is estimated to affect 4-10 in every 1,000,000 live births worldwide.

Hyperimmunoglobulin E syndrome or Job syndrome: A disorder characterized by very high concentrations of the serum antibody IgE, severe lung infections, periodic staphylococcal infections, and eczema-like skin rashes. A common mnemonic used to recall the symptoms is FATED. Affected patients exhibit: coarse Facies, cold staph Abscesses, retained primary Teeth, increase in IgE, Eczema, and Dermatologic problems. Death usually occurs in the 2nd-3rd decade of life due to severe pulmonary disease. The products of conception in hyper IgE syndrome appear to exhibit either autosomal dominant or autosomal recessive patterns of inheritance. There are several underlying causes of this disorder with at least 2 autosomal genomic loci: Hyperimmunoglobulin E syndrome can be caused by autosomal dominant mutations in the *STAT3* gene which encodes the signal transducer and activator of transcription 3 (acute-phase response factor);; and Hyperimmunoglobulin E syndrome can be caused by autosomal recessive mutations in the *DOCK8* gene which encodes the dedicator of cytokinesis protein 8. Hyperimmunoglobulin E syndrome is estimated to affect less than 1 in every 1,000,000 individuals.

Chédiak-Higashi syndrome: This disease is discussed under endocrine, nutritional and metabolic diseases in the subchapter "Disorders of Aromatic Amino-acid Metabolism" (3.04.22.1).

Griscelli syndrome type 2: This disease is discussed under diseases of the skin and subcutaneous tissue in the subchapter "Other Disorders of Pigmentation" (3.12.6.1).

Nijmegen breakage syndrome: A disorder characterized by chromosomal instability, immunodeficiency, predisposition to cancer, microcephaly, growth retardation and hypersensitivity to ionizing radiation. Nijmegen breakage syndrome is caused by autosomal recessive mutations in the *NBN* gene which encodes the nibrin, a member of the MRE11/RAD50 double-strand break repair complex. There is a phenotypically indistinguishable condition - the Berlin breakage syndrome. Sometimes Nijmegen breakage syndrome is referred to as Ataxia-telangiectasia variant-1. There is another syndrome with resemblance to Nijmegen breakage syndrome; due to similar clinical features, it is referred to as LIG4 syndrome. This last condition is caused by mutations in the *LIG4* gene which encodes the DNA ligase 4.

Immunodeficiency-centromeric instability-facial anomalies syndrome: A disorder characterized by facial dysmorphism, immunoglobulin deficiency, and branching of chromosomes 1, 9 and 16 after culturing lymphocytes with phytohemagglutinin. There are several underlying causes of this disorder with at least 2 autosomal genomic loci: Immunodeficiency-centromeric instability-facial anomalies syndrome-1 is caused by autosomal recessive mutations in the *DNMT3B* gene which encodes the DNA methyltransferase-3B, a protein that localize mainly in the nucleus and its expression is developmentally regulated;; and Immunodeficiency-centromeric instability-facial anomalies syndrome-2 is caused by mutations in the *ZBTB24* gene which encodes the zinc finger and BTB domain containing protein 24.

Ataxia telangiectasia: This disease is discussed under neurological diseases in the subchapter "Hereditary Ataxia" (3.06.03.1).

Bloom syndrome: This disease is discussed under congenital malformations, deformations and chromosomal abnormalities in the subchapter "Other Congenital Malformations of Skin" (3.17.9.3).

RIDDLE syndrome: This disease is discussed under congenital malformations, deformations and chromosomal abnormalities in the subchapter "Other Osteochondrodysplasias" (3.17.8.11).

Hermansky-Pudlak syndrome type 2: This disease is discussed under endocrine, nutritional and metabolic diseases in the subchapter "Disorders of Aromatic Amino-acid Metabolism" (3.04.09.1).

XL-dyskeratosis congenita or Hoyeraal-Hreidarsson syndrome: This disease is discussed under congenital malformations, deformations and chromosomal abnormalities in the subchapter "Other Congenital Malformations of Skin" (3.17.9.3).

Autoimmune polyendocrine syndrome type 1 or APECED Syndrome: This disease is discussed under endocrine, nutritional and metabolic diseases in the subchapter "Other Endocrine Diseases" (3.04.07.1).

IPEX or Immunodysregulation Polyendocrinopathy Enteropathy X-linked syndrome: This disease is discussed under endocrine, nutritional and metabolic diseases in the subchapter "Other Endocrine Diseases" (3.04.07.1).

Laron syndrome with immunodeficiency: This disease is discussed under endocrine, nutritional and metabolic diseases in the subchapter "Other Endocrine Diseases" (3.04.07.1).

3.03.6.4 Common Variable Immunodeficiency

Common variable immunodeficiency or Acquired hypogammaglobulinemia: A disorder where only B cells are affected. Common variable immunodeficiency is characterized by hypogammaglobulinemia with increased susceptibility to bacterial and viral infections. Most cases of common variable immunodeficiency are sporadic while the cause of these cases is unclear. Sporadic cases possibly result from a complex interaction between environmental and genomic factors. There are several underlying causes of this disorder with at least 8 autosomal genomic loci: Common variable immunodeficiency-1 is caused by autosomal recessive mutations in the *ICOS* gene which encodes the inducible T-cell co-stimulator protein;; Common variable immunodeficiency-2 is caused by autosomal recessive mutations in the *TNFRSF13B* gene which encodes the tumor necrosis factor receptor superfamily member 13B;; Common variable immunodeficiency-3 is caused by autosomal recessive mutations in the *CD19* gene which encodes the CD19 molecule;; Common variable immunodeficiency-4 is caused by autosomal recessive mutations in the *TNFRSF13C* gene which encodes the tumor necrosis factor receptor superfamily member 13C;; Common variable immunodeficiency-5 is caused by autosomal recessive mutations in the *MS4A1* gene which encodes the membrane-spanning 4-domains (subfamily A member 1);; Common variable immunodeficiency-6 is caused by autosomal recessive mutations in the *CD81* gene which encodes the CD81 molecule;; Common variable immunodeficiency-7 is caused by autosomal recessive mutations in the *CR2* gene which encodes the complement receptor 2;; and

Common variable immunodeficiency-8 is caused by autosomal recessive mutations in the *LRBA* gene which encodes the beach and anchor containing lipopolysaccharide-responsive vesicle trafficking, a protein that associate with protein kinase A and may be involved in leading intracellular vesicles to activated receptor complexes. Common variable immunodeficiency is estimated to affect 1 in every 25,000-50,000 individuals. Males and females are equally affected.

3.03.6.5 Other Immunodeficiencies

Complement deficiencies: Are examples related to diseases that are either activators/enzymes or inhibitors. Examples of activators/enzymes include the mutations of *C1QA/C1QB/C1QC/C1R/C1S* genes associated with infections, lupus-like syndromes, and rheumatoid diseases. Mutations in *CFP* and *CFB* genes are associated with Neisserial infections. There are autoimmune conditions mainly related to deficiencies in C1, 2, and 4. Mutations in the *C3* gene have been associated with recurrent pyogenic infections. Deficiencies of the membrane attack complex C6, C7, C8, C9 are predisposed to both infections (particularly Neisseria meningitidis) and autoimmune diseases. Recurrent Neisserial infection is more common due to C8 deficiency. C9 deficiency seems to be common among Japanese individuals. Examples of inhibitor conditions include: Hereditary angioedema due to mutations of *SERPING1* gene, Age-related macular degeneration, and Atypical hemolytic uremic syndrome due to SNPs in the *CFH* gene.

Hereditary angioedema: A disorder characterized by local swelling in subcutaneous tissues and submucosal edema comprising the upper respiratory and gastrointestinal tracts. There are several underlying causes of this disorder with at least 2 autosomal genomic loci: Hereditary angioedema type I and II are caused by monoallelic mutations in the *SERPING1* gene which encodes the serpin peptidase inhibitor clade G member 1 (C1 inhibitor). Other designations for this gene are: C1 esterase inhibitor or complement component 1 inhibitor. The product of conception in Hereditary angioedema types 1 and 2 appears to exhibit de novo alteration in a few cases and autosomal dominant inheritance in most of the cases. The 2 types are clinically indistinguishable. Type I is the most common, occurring in 85% of patients. Patients with type I display 35% or less of C1NH (serum levels) whereas in type II, the levels of C1NH are normal or elevated, but the protein is nonfunctional;; and Hereditary angioedema type III is caused by monoallelic mutations in the *F12* gene which encodes the coagulation factor XII (Hageman factor). Other causes of episodic angioedema include the deficiency in Carboxypeptidase N, a serum alpha globulin that inactivates C3a, C4a, C5a, bradykinin, kallidin and fibrinopeptides. These conditions are also familial and linked with hay fever/asthma, angioedema/chronic urticaria or both. Hereditary angioedema is estimated to affect 1 in every 50,000 individuals.

Anhidrotic and Hypohidrotic ectodermal dysplasia: This disease is discussed under congenital malformations, deformations and chromosomal abnormalities in the subchapter "Other Congenital Malformations of Skin" (3.17.9.3).

Chronic mucocutaneous candidiasis: A disease characterized by chronic infections with Candida that are limited to mucosal surfaces, skin and nails. This condition is caused by an impaired response of T cells. There are several underlying causes of this disorder with at least 7 autosomal genomic loci following mostly autosomal dominant patterns: Chronic mucocutaneous candidiasis-1 has been mapped, but the gene has not been characterized;; Chronic mucocutaneous candidiasis-2 is caused by mutations in the *CARD9* gene which encodes the caspase recruitment domain family (member 9);; Chronic mucocutaneous candidiasis-3, a restricted form affecting the nails of the hands and feet, has been mapped but the gene has not been characterized;; Chronic mucocutaneous candidiasis-4, caused by biallelic mutations in the *CLEC7A* gene which encodes the C-type lectin domain family 7 (member A);; Chronic mucocutaneous candidiasis-5, caused by mutations in the *IL17RA* gene which encodes the interleukin 17 receptor A;; Chronic mucocutaneous candidiasis-6, caused by mutations in the *IL17F* gene which encodes the interleukin 17F;; and Chronic mucocutaneous candidiasis-7, caused by mutations in the *STAT1* gene which encodes the signal transducer and activator of transcription 1.

X-linked pure immunodeficiency: A disorder characterized by a specific pattern of infectious susceptibility and immunodeficiency. Patients exhibit usual dentition, hair pattern and perspiration. X-linked pure immunodeficiency is caused by X-linked recessive mutations in the *IKBKG* gene which encodes the inhibitor of kappa light polypeptide gene enhancer in B-cells, kinase gamma. The *IKBKG* gene is located on the subtelomeric long arm of chromosome Xq28.

Autoinflammatory disorder: A group of disorders that are susceptible to excessive inflammation rather than infections. These periodic fever syndromes lead to amyloid deposition.

Familial Mediterranean fever: This disease is discussed under endocrine, nutritional and metabolic diseases in the subchapter "Amyloidosis" (3.04.18.1).

Muckle-Wells syndrome: This disease is discussed under endocrine, nutritional and metabolic diseases in the subchapter "Amyloidosis" (3.04.18.1).

Familial cold autoinflammatory syndrome: This disease is discussed under diseases of the skin and subcutaneous tissue in the subchapter "Urticaria and Erythema" (3.12.4).

Blau syndrome: A disorder characterized by familial granulomatous arthritis, iritis and skin granulomas. Patients also suffer from cranial neuropathies, exanthema, camptodactyly and Crohn's disease. This genomic condition has overlapping features with both sarcoidosis and granuloma annulare. Blau syndrome is associated with monoallelic mutations in the *NOD2* gene which encodes the nucleotide-binding oligomerization domain containing protein 2.

Syndromic multisystem autoimmune disease: This disease is discussed under diseases of the musculoskeletal system and connective tissue in the subchapter "Systemic Connective Tissue Disorders" (3.13.3).

Vitiligo: This disease is discussed under diseases of the skin and subcutaneous tissue in the subchapter "Other Disorders of the Skin and Subcutaneous Tissue" (3.12.7).

Hyperimmunoglobulinemia D with recurrent fever: A periodic fever syndrome due to mutations in the *MVK* gene which encodes the mevalonate kinase. More description is provided under Mevalonic aciduria which is discussed under endocrine, nutritional and metabolic diseases in the subchapter "Other Specified Metabolic Disorders" (3.04.20.4).

3.03.6.6 Sarcoidosis

Sarcoidosis or Besnier-Boeck-Schaumann disease: A disorder characterized by unusual collections of granulomas in multiple organs - mainly in the lungs or the lymph nodes, but practically any organ can be compromised. The mechanism might involve a genomic susceptibility after exposure to an environmental factor. There are a few candidate genes that have been confirmed. The most interesting candidate genes are located in the HLA region on chromosome 6p21.3; this includes the genes *BTNL2* and several *HLA-DR*. In patients suffering persistent sarcoidosis, the HLA haplotype HLA-B7-DR15 has strong genomic association. In non-persistent sarcoidosis there is a strong genomic association with the HLA haplotype DR3-DQ2. There might be another locus of susceptibility for sarcoidosis on chromosome 10q22-q23, which may be associated with variations in the *ANXA11* gene. First-degree relatives have a relative risk of 5-6.6% of developing the disease. This condition is slightly more common in females than in males. There are large geographical differences in the prevalence of this condition. People from Northern European countries (Caucasians from Sweden and Iceland), have the highest annual incidence of 60 in every 100,000. On the contrary, in the United States, sarcoidosis is more prevalent in African-Americans than Caucasians, with annual incidences of 35.5 and 10.9 in every 100,000, respectively. In other words, there is a 6-fold difference between Caucasians from Northern European countries and those from the United States, which is hard to explain

with a genomic mechanism. Sarcoidosis has a higher prevalence in non-smokers with a relative risk of 8.7 times.

3.03.6.7 Other Specified Disorders Involving the Immune Mechanism (not elsewhere classified)

Lymphoproliferative disorders: A group of disorders that refer to several genomic and non-genomic conditions in which lymphocytes are produced in excessive quantities. There are several conditions that course with lymphoproliferative disorders, including lymphoma, leukemia, multiple myeloma, Wiskott-Aldrich syndrome, etc... Lymphoproliferative disorders can be classified according to the pattern of inheritance in autosomal and X-linked. There are several underlying causes of this disorder with at least 2 X-linked genomic loci and 6 autosomal genomic.

X-linked lymphoproliferative disease: A disorder characterized by an inadequate immune response to infection with the Epstein-Barr virus, or by similar symptoms in the absence of infection. X-linked lymphoproliferative syndrome-1 is caused by mutations in the *SH2D1A* gene which encodes the SLAM-associated protein;; X-linked lymphoproliferative syndrome-2 is caused by mutations in the *XIAP* gene which encodes the X-linked inhibitor of apoptosis protein. These conditions have been associated with this T cell and NK cell lymphoproliferative disorder. The product of conception alteration in this condition appears to exhibit either X-linked recessive inheritance or de novo alteration. X-linked lymphoproliferative syndrome is estimated to affect 1 in every 1,000,000 individuals.

Autosomal lymphoproliferative disorder or Autoimmune lymphoprolifera-

tive syndrome: A lymphoproliferative disorder characterized by defective lymphocyte apoptosis, which results in the accumulation of autoreactive lymphocytes. Clinically, patients with this condition have lymphadenopathy and splenomegaly associated with autoimmune hemolytic anemia and thrombocytopenia. Autoimmune lymphoproliferative syndrome type IA is caused by autosomal dominant mutations in the *FAS* gene which encodes the Fas (TNF receptor superfamily, member 6);; Autoimmune lymphoproliferative syndrome type IB is caused by autosomal dominant mutations in the *FASLG* gene which encodes the FAS ligand;; Autoimmune lymphoproliferative syndrome type IIA is caused by mutations in the *CASP10* gene which encodes the caspase-10;; Autoimmune lymphoproliferative syndrome type IIB is caused by mutations in the *CASP8* gene which encodes the caspase-8;; Autoimmune lymphoproliferative syndrome type III has not been characterized;;

and Autoimmune lymphoproliferative syndrome type IV is caused by mutations in the *NRAS* gene which encodes the neuroblastoma RAS viral (v-ras) oncogene homolog.

EBV-associated autosomal lymphoproliferative syndrome: A lymphoproliferative disorder associated with EBV infection. This condition is characterized by generalized lymphadenopathy, hepatosplenomegaly with impaired liver function, recurrent episodes of fever, nodular interstitial pulmonary infiltrates, anemia, thrombocytopenia, hypogammaglobulinemia, and pleural and pericardial effusion. EBV-associated autosomal lymphoproliferative syndrome is caused by autosomal recessive mutations in the *ITK* gene which encodes the IL2-inducible T-cell kinase. This disorder has been reported in two sisters born of consanguineous Turkish parents who died in childhood.

Table 3.03.6.1 Recognizable Disorders Involving the Immune Mechanism

Disorder	OMIM	Gene	Biological Function	Product of Conception Alteration
X-linked Agammaglobulinemia or Bruton's agammaglobulinemia	300755	BTK	Tyrosine kinase	XR
Agammaglobulinemia-1	601495	IGHM	Surface receptor	AR
Agammaglobulinemia-2	613500	IGLL1	Surface receptor	AR
Agammaglobulinemia-3	613501	CD79A	Cell surface receptor Other/ungrouped	AR
Agammaglobulinemia-4	613502	BLNK	B cell receptor signalling	AR
Agammaglobulinemia-5	613506	LRRC8A	Transmembrane protein	AD
Agammaglobulinemia-6	612692	CD79B	Cell surface receptor Other/ungrouped	AR
Immunodeficiency with hyper IgM type 1	308230	CD40LG	Cell surface receptor Other/ungrouped	XR
Immunodeficiency with hyper-IgM type 2	605258	AICDA	RNA-editing deaminase	AR

Table 3.03.6.1. Continued.

Disorder	OMIM	Gene	Biological Function	Product of Conception Alteration
Immunodeficiency with hyper-IgM type 3	606843	CD40	Cell surface receptor Other/ungrouped	AR
Immunodeficiency with hyper IgM type 5	608106	UNG	Mitochondrial protein	AR
X-linked severe combined immunodeficiency	300400	IL2RG	Cell surface receptor JAK-STAT	XR
Severe combined immunodeficiency	102700	ADA	Adenosine deaminase	AR
Severe combined immunodeficiency	601457	RAG1	Recognition of the DNA substrate	AR
Severe combined immunodeficiency	601457	RAG2	Stable binding and cleavage	AR
Severe combined immunodeficiency, Athabascan type	602450	DCLRE1C	Exonuclease	AR
Severe combined immunodeficiency	611291	NHEJ1	DNA repair factor	AR
Severe combined immunodeficiency	602450	LIG4	DNA ligase	AR
Severe combined immunodeficiency	600802	JAK3	Tyrosine kinase	AR
Severe combined immunodeficiency	608971	IL7R	Cell surface receptor JAK-STAT	AR
Severe combined immunodeficiency	608971	PTPRC	Tyrosine phosphatase	AR
Severe combined immunodeficiency	608971	CD3D	Cell surface receptor Other/ungrouped	AR
Severe combined immunodeficiency	608971	CD3E	Cell surface receptor Other/ungrouped	AR
Severe combined immunodeficiency	603554	RAG1	Recognition of the DNA substrate	AR
Severe combined immunodeficiency	603554	RAG2	Stable binding and cleavage	AR
Severe combined immunodeficiency	603554	DCLRE1C	Exonuclease	AR
Immune dysfunction with T-cell inactivation due to calcium entry defect 1	612782	ORAI1	Membrane calcium channel	AR
Immune dysfunction, with T-cell inactivation due to calcium entry defect 2	612783	STIM1	Transmembrane protein	AR
Bare lymphocyte syndrome I	604571	TAP1	ABC-transporter protein	AR
Bare lymphocyte syndrome I	604571	TAP2	ABC-transporter protein	AR

Table 3.03.6.1. Continued.

Disorder	OMIM	Gene	Biological Function	Product of Conception Alteration
Bare lymphocyte syndrome I	604571	TAPBP	Transport of antigenic peptides across the endoplasmic reticulum membrane	AR
Bare lymphocyte syndrome II	209920	RFX5	Nuclear protein	AR
Bare lymphocyte syndrome II	209920	RFXANK	Nuclear protein	AR
Bare lymphocyte syndrome II	209920	RFXAP	Nuclear protein	AR
Bare lymphocyte syndrome II	209920	CIITA	Major histocompatibility complex transactivator	AR
Severe immunodeficiency, due to deficiency of Interleukin-2 receptor, alpha chain	606367	IL2RA	Cell surface receptor JAK-STAT	AR
Selective T-cell defect	269840	ZAP70	Associated protein kinase	AR
Immunodeficiency, due to purine nucleoside phosphorylase deficiency	613179	PNP	Phosphorolysis of purine nucleosides	AR
Immunodeficiency, due to defect in CD3-zeta	610163	CD247	Cell surface receptor Other/ungrouped	AR
Immunodeficiency, due to defect in MAPBP-interacting protein	610798	LAMTOR2	Endosomal/lysosomal protein	AR
WHIM syndrome or Warts-hypogammaglobulinemia-infections-myelokathexis	193670	CXCR4	Cell surface receptor G protein	AD
Chromosome 22q11.2 deletion syndrome or DiGeorge syndrome	188400	TBX1	Nuclear transcription factor	AD + 90%
Wiskott-Aldrich syndrome-1	301000	WAS	Cytoskeleton and signal transduction	XR
Wiskott-Aldrich syndrome-2	614493	WIPF1	WAS/WASL interacting protein	AR
Hyper-IgE recurrent infection syndrome	147060	STAT3	Transcription factor β-Scaffold factors	AD
Hyper-IgE recurrent infection syndrome	243700	DOCK8	Intracellular signaling networks	AR
Nijmegen breakage syndrome	251260	NBN	DNA repair MRN complex	AR
Nijmegen breakage syndrome-like disorder	613078	RAD50	DNA repair protein	AR
Immunodeficiency-centromeric instability-facial anomalies syndrome-1	242860	DNMT3B	DNA methyltransferase	AR

The Concise Encyclopedia of Genomic Diseases

Table 3.03.6.1. Continued.

Disorder	OMIM	Gene	Biological Function	Product of Conception Alteration
Immunodeficiency-centromeric instability-facial anomalies syndrome-2	614069	ZBTB24	Nuclear zinc finger protein	AR
Common variable immunodeficiency-1	607594	ICOS	Surface receptor	AR
Common variable immunodeficiency-2	240500	TNFRSF13B	Cell surface receptor TNF receptor	AR
Common variable immunodeficiency-3	613493	CD19	Cell surface receptor Other/ungrouped	AR
Common variable immunodeficiency-4	613494	TNFRSF13C	Cell surface receptor TNF receptor	AR
Common variable immunodeficiency-5	613495	MS4A1	Transmembrane protein	AR
Common variable immunodeficiency-6	613496	CD81	Cell surface receptor Other/ungrouped	AR
Common variable immunodeficiency-7	614699	CR2	Complement receptor 2	AR
Common variable immunodeficiency-8	614700	LRBA	Vesicle trafficking	AR
C9 deficiency with dermatomyositis	613825	C9	Complement factor	AR or AD
Immunodeficiency, due to complements deficiency C9	613825	C9	Complement factor	AR or AD
Immunodeficiency, due to complements deficiency C5	609536	C5	Complement factor	AR
Immunodeficiency, due to complements deficiency C6	612446	C6	Complement factor	
Immunodeficiency, due to complements deficiency C7	610102	C7	Complement factor	
Immunodeficiency, due to complements deficiency C8A	613790	C8A	Complement factor	
Immunodeficiency, due to complements deficiency C8B	613789	C8B	Complement factor	
Partial deficiency of complement component 4	120790	SERPING1	Serpin proteinase inhibitor	AD
Hereditary Angioedema-1 and 2	106100	SERPING1	Serpin proteinase inhibitor	AD + de novo
Hereditary Angioedema-3	610618	F12	Coagulation factor	AD
Isolated immunodeficiency	300584	IKBKG	Inhibitor of kappaB kinase	XR + de novo
Blau syndrome	186580	NOD2	Nucleotide oligomerization domain receptors	AD

Table 3.03.6.1. Continued.

Disorder	OMIM	Gene	Biological Function	Product of Conception Alteration
Growth hormone insensitivity with immunodeficiency	245590	STAT5B	Transcription factor β-Scaffold factors	AR
T-cell immunodeficiency, congenital alopecia, and nail dystrophy or Winged helix deficiency	601705	FOXN1	Transcription factor Helix-turn-helix domains	AR
Autoimmune lymphoproliferative syndrome-1A	601859	FAS	Cell surface receptor TNF receptor	AD
Autoimmune lymphoproliferative syndrome-1B	601859	FASLG	Fas ligand	AD
Autoimmune lymphoproliferative syndrome-2A	603909	CASP10	Caspase	AD
Autoimmune lymphoproliferative syndrome-2B	607271	CASP8	Caspase	AR
Autoimmune lymphoproliferative syndrome-4	614470	NRAS	Oncogene	AD
EBV-associated lymphoproliferative syndrome	613011	ITK	Tyrosine kinase	AR

Table 3.03.6.2 Disorders Involving the Immune Mechanism in Different Ethnic Groups

Disease	Frequency of condition	People with higher prevalence (if available)
Immunodeficiency, due to complements deficiency C9	1 in 1,000	Japanese and Koreans
Chromosome 22q11.2 deletion syndrome or DiGeorge syndrome	2-3 in 10,000	Worldwide
Common variable immunodeficiency	2-4 in 100,000	Worldwide
Immunodeficiency, due to complements deficiency C5toC8	1-9 in 100,000	Unknown
Hereditary Angioedema	2 in 100,000	Unknown
X-linked Agammaglobulinemia	1 in 100,000 Males	Worldwide
Severe combined immunodeficiency	1-4 in 200,000	Worldwide
Wiskott-Aldrich syndrome	4-10 in 1,000,000	Unknown
Severe combined immunodeficiency, Athabaskan type	<1 in 1,000,000	Navajo and Apache peoples
Immunodeficiency with hyper IgM type 1	2 in 1,000,000 Males	Unknown
X-linked lymphoproliferative disease	1 in 1,000,000	Unknown
Autosomal Agammaglobulinemia	<1 in 1,000,000	Unknown
Autoimmune lymphoproliferative syndrome 1A	<1 in 1,000,000	Unknown
Growth hormone insensitivity with immunodeficiency	<1 in 1,000,000	Unknown

Table 3.03.6.2. Continued.

Disease	Frequency of condition	People with higher prevalence (if available)
Hyper-IgE recurrent infection syndrome	<1 in 1,000,000	Unknown
Immunodeficiency-centromeric instability-facial anomalies syndrome	<1 in 1,000,000	Unknown
Immunodysregulation, polyendocrinopathy, and enteropathy	<1 in 1,000,000	Unknown
Lymphoproliferative syndrome	<1 in 1,000,000	Unknown
RIDDLE syndrome	<1 in 1,000,000	Unknown
WHIM syndrome or Warts-hypogammaglobulinemia-infections-myelokathexis	<1 in 1,000,000	Unknown

3.03.7 Dissasortative Mating in Ethnics Groups and Disease Prevention

Table 3.03.7 displays some relevant examples of autosomal recessive diseases that require the implementation of some methodology to decrease the prevalence of the conditions. Besides the diseases, the table also includes the gene associated with the conditions, the incidence of expected dissasortative mating frequency, and the ethnic group with a high prevalence of the conditions. For example, 1 in every 13 families of Sub-Saharan African ancestry are at risk of having a child with sickle cell anemia because both parents are carriers of a polymorphism in the *HBB* gene. In the United States, 1 in every 157 families of African American ancestry are at risk of having a child with sickle cell anemia because both parents are carriers of a polymorphism in the *HBB* gene. Most of the autosomal recessive disorders decrease in prevalence in children of interethnic marriages. In this chapter, an exception to this rule is the Hemoglobin H disease form of Alpha-thalassemia when one of the parents is of Southeastern Asian descent and the other is African-American. This disease affects 1 in 100 of these interethnic marriages. 1 in every 100 families of Mediterranean, Middle Eastern, Transcaucasus, Central Asian, and Indian subcontinent ancestry is at the risk of having a child with intermediate and/or major Beta-thalassemia. In such locations both parents are carriers of mutations in the *HBB* gene. Several Middle Eastern countries have implemented premarital dissasortative mating to decrease the frequency of this condition. 1 in every 625 families of Navajo and Apache ancestry is at risk of having a child with Severe combined immunodeficiency, Athabaskan type because both parents are carriers of mutations in the *DCLRE1C* gene. 1 in every 8,000 families of Ashkenazi Jewish ancestry is at risk of having a child with Fanconi anemia because both parents are carriers of mutations in the *FANCC* gene. Each of these families has a 1 out of 4 risk of having a child with any of these conditions. Each of these families should receive counseling to prevent these autosomal recessive diseases.

The Concise Encyclopedia of Genomic Diseases

Table 3.03.7 Frequency of Dissasortative Mating in Multiple Ethnic Groups

Disease	Gene	Dissasortative mating frequency	Individuals with higher prevalence
Sickle cell anemia	*HBB*	1 in 13	Sub-Saharan Africans
Alpha-thalassemia	*HBA1-HBA2*	1 in 100	Southeastern Asians/African-Americans
Beta-thalassemia	*HBB*	1 in 100	Mediterraneans, Middle Easterns, Transcaucasians, Central Asians, Indians, Pakistanis and Bangladeshis
Sickle cell anemia	*HBB*	1 in 157	African-Americans
Factor V Leiden biallelic	*F5*	1 in 500	Caucasians
Severe combined immunodeficiency, Athabaskan type	*DCLRE1C*	1 in 625	Navajo and Apache peoples
Alpha-thalassemia	*HBA1-HBA2*	1 in 625	Southeastern Asians
Factor II Leiden biallelic	*F2*	1 in 2,500	Caucasians
Fanconi anemia	*FANCC*	1 in 8,000	Ashkenazi Jews
von Willebrand disease biallelic	*VWF*	1 in 10,000	Worldwide

3.04.00.1 Subchapter IV

Endocrine, Nutritional and Metabolic Diseases

Table 3.04.00.1 Endocrine, Nutritional and Metabolic Diseases

3.04.00-07	Endocrine diseases	
	3.04.01	Thyroid gland / Thyroid Hormone
	3.04.02	Pancreas / Insulin, Glucagon
	3.04.03	Parathyroid gland / PTH
	3.04.04	Pituitary gland / ADH, Oxytocin, GH, ACTH, TSH, LH, FSH and Prolactin
	3.04.05	Adrenal gland / Aldosterone, Cortisol, Epinephrine and Norepinephrine
	3.04.06	Gonads / Estrogen, Androgens, Testosterone, etc.
	3.04.07	Other Endocrine Diseases
3.04.08	Nutritional Diseases	
3.04.09-20	Metabolic diseases	
	3.04.09	Amino-acids Disorders
	3.04.10	Carbohydrates Disorders
	3.04.11	Lipids Disorders
	3.04.12	Disorders of Glycosaminoglycan Metabolism
	3.04.13	Disorders of Lipoprotein Metabolism and other Lipidemias
	3.04.14	Other Metabolic Disorders
	3.04.15	Disorders of Porphyrin and Bilirubin Metabolism
	3.04.16	Disorders of Mineral Metabolism
	3.04.17	Cystic Fibrosis
	3.04.18	Amyloidosis
	3.04.19	Disorders of Fluid, Electrolyte and Acid-base Balance
	3.04.20	Other Metabolic Disorders (not elsewhere classified)
3.04.21	Disease Prevention for Disassortative Mating in Ethnic Groups	

The first goal of the chapter for endocrine, nutritional and metabolic diseases is to have sense of the prevalence of these diseases. The first section will address altogether the set of people where variation in DNA correlates with a disease. Table shows the diseases, the worldwide frequencies of conditions, and the set of people with higher prevalence if available. This section illustrates common and rare diseases. With regard to table only Fructose malabsorption, renal hypouricemia, Familial hypercholesterolemia, Hemochromatosis 1, Renal glucosuria, and Cystic fibrosis (Irish people) are considered common in the nationality and/or ethnic groups described. The rest of these diseases are considered as rare, which means less than a 1 disease in 2,000 individuals. In these subsection there are many common diseases that are caused by a combination of genomic and environmental factors (Hashimoto's thyroiditis, Type 2 diabetes), they will be include in the specific   subsection not in this table.

Table 3.04.00.2 Endocrine, Nutritional and Metabolic Diseases in Different Ethnic Groups

Disease	Frequency of condition	People with higher prevalence (if available)
Fructose malabsorption	1-300 in 1,000	Westerners and Africans
Renal hypouricemia	1-7 in 1,000	Unknown
Familial hypercholesterolemia	2 in 1,000	Ashkenazi Jews, Afrikaner, French Canadians, Lebanese Christians, and Finnish
Hemochromatosis-1	1 in 1.000	Northern Europeans
Renal glucosuria	1 in 1,000	Unknown
Cystic fibrosis	3-80 in 100,000	Irish
Phenylketonuria	1-40 in 100,000	Turkish
Tay-Sachs disease	3-300 in 1,000,000	Ashkenazi Jews, French Canadians, Cajuns, Irish-Americans, and Sephardic Jews
Obesity, due to melanocortin-4 receptor deficiency	5 in 10,000	Unknown
Acute intermittent porphyria	1-5 in 10,000	Northern Europeans
Emphysema and COPD, due to alpha-1 antitrypsin Pittsburgh mutation	1-5 in 10,000	West Europeans
Isolated growth hormone deficiency and Kowarski syndrome	3 in 10,000	Unknown
Congenital nongoitrous hypothyroidism	2-3 in 10,000	Unknown
Congenital sucrase-isomaltase deficiency	2 in 10,000	Greenlanders, Alaskans, and Canadians
Fructose intolerance	3-5 in 100,000	Unknown
Premature ovarian failure	1-10 in 10,000	Unknown
Combined pituitary hormone deficiency	1 in 10,000	Unknown
Cystinuria	1 in 10,000	Sephardic Jews
Hypogonadotropic hypogonadism	1 in 10,000	Unknown
Mitochondrial complex 1 deficiency	1 in 10,000	Unknown
Kallmann syndrome	1 in 10,000 Males	Unknown
Waardenburg syndrome	5-10 in 100,000	Unknown
Congenital adrenal hyperplasia	5-20 in 100,000	Yupik Eskimos (West Alaska)
Pendred syndrome and enlarged vestibular aqueduct	1-10 in 100,000	Unknown
Histidinemia or histidinuria	1-11 in 100,000	Unknown
Sarcosinemia or hypersarcosinemia	3-35 in 1,000,000	Unknown
Hypomagnesemia with secondary hypocalcemia	5 in 100,000	Unknown
Hypophosphatemic rickets	5 in 100,000	Unknown

The Concise Encyclopedia of Genomic Diseases

Table 3.04.00.2. Continued.

Disease	Frequency of condition	People with higher prevalence (if available)
Cerebral adrenoleukodystrophy and Adrenomyeloneuropathy	5 in 100,000 Males	Unknown
Congenital adrenal hypoplasia with hypo-gonadotropic hypogonadism	8 in 100,000	Unknown
Acatalasemia or Takahara disease	1-9 in 100,000	Unknown
Adenine phosphoribosyltransferase deficiency or Urolithiasis, 2,8-dihydroxyadenine	1-9 in 100,000	Unknown
Choreoathetosis, hypothyroidism, and neonatal respiratory distress	1-9 in 100,000	Unknown
Dysbetalipoproteinemia, hyperlipoproteinemia, Lipoprotein glomerulopathy, and Alzheimer disease	1-9 in 100,000	Unknown
Fructose-1,6-biphosphatase deficiency	1-9 in 100,000	Unknown
Hartnup disorder and Iminoglycinuria	1-9 in 100,000	Unknown
Iminoglycinuria digenic phenotype	1-9 in 100,000	Unknown
Methylcrotonylglycinuria	1-9 in 100,000	Unknown
Glycogen storage disease 1A or von Gierke disease	1-2 in 100,000	Unknown
Glycogen storage disease 6 or Hers disease	1-2 in 100,000	Unknown
Isobutyryl-CoA dehydrogenase deficiency or valinuria	1-2 in 100,000	Unknown
Metachromatic leukodystrophy	1-2 in 100,000	Oriental Jews
Methylmalonic acidemia	1-2 in 100,000	Unknown
Ornithine transcarbamylase deficiency	1-2 in 100,000	Unknown
Hypocalciuric hypercalcemia 1	1-2 in 100,000	Scottish (West)
Ocular albinism	1-2 in 100,000 Males	Unknown
Systemic primary Carnitine deficiency	1-3 in 100,000	Faroese (Faroe Islands)
Very long chain acyl-CoA dehydrogenase deficiency	1-3 in 100,000	Unknown
Mucopolysaccharidosis 6 or Maroteaux-Lamy syndrome	1-4 in 1,000,000	Unknown
Porphyria cutanea tarda and hepatoerythropoietic porphyria	1-5 in 100,000	Unknown
Long-chain hydroxyacyl-CoA dehydrogenase deficiency, and Maternal HELLP syndrome of pregnancy	3 in 100,000	Unknown
Propionic acidemia	3 in 100,000	Saudi Arabians, Amish, and Mennonites
Fish odor syndrome or Trimethylaminuria	2-3 in 100,000	Unknown
Oculocutaneous albinism	2-3 in 100,000	Sephardic Jews
Fabry disease	2-3 in 100,000 Males	Unknown
Thyroid dyshormonogenesis	2-4 in 100,000	Unknown

The Concise Encyclopedia of Genomic Diseases

Table 3.04.00.2. Continued.

Disease	Frequency of condition	People with higher prevalence (if available)
Wilson disease	2-4 in 100,000	Sardinians (Italy) and Canarians (Spain)
Glycogen storage disease 0 or liver glycogen synthase deficiency	2-5 in 100,000	Unknown
Myeloperoxidase deficiency and Susceptibility to Alzheimer disease	2-70 in 100,000	Unknown
Medium chain acyl-CoA dehydrogenase deficiency	6 in 100,000	Northern Europeans
Oculocutaneous albinism 2	6 in 100,000	African Americans, Native American groups, and sub-Saharan Africans
Trehalase deficiency	6 in 100,000	Greenlanders
Cystathioninuria	7 in 100,000	Unknown
Biotinidase deficiency	2 in 100,000	Unknown
Cerebrotendinous xanthomatosis or cerebral cholesterosis	2 in 100,000	Sephardic Jews
Citrullinemia	2 in 100,000	Unknown
Classic Galactosemia	2 in 100,000	Unknown
Diabetes mellitus, Leucine-sensitive hypoglycemia, and hyperinsulinemic hypoglycemia	2 in 100,000	Saudi Arabians, Finnish, and Ashkenazi Jews
Gaucher disease	2 in 100,000	Ashkenazi Jews
Hyperinsulinemic hypoglycemia	2 in 100,000	Unknown
Hypomyelinating leukodystrophy 3	2 in 100,000	Unknown
Pyruvate dehydrogenase deficiency and Leigh syndrome	4 in 100,000	Unknown
Epimerase-deficiency galactosemia	1 in 100,000	Pakistanis
Galactokinase deficiency with cataracts	1 in 100,000	Unknown
Glycogen storage disease	1 in 100,000	Unknown
Glycogen storage disease 3 or Cori Disease or glycogen debranching enzyme deficiency	1 in 100,000	Sephardic Jews
Glycogen storage disease 5 or McArdle disease	1 in 100,000	Unknown
Glycogen storage disease 9	1 in 100,000	Unknown
Holocarboxylase synthetase deficiency	1 in 100,000	Unknown
Hypophosphatasia and Odontohypophosphatasia	1 in 100,000	Unknown
Krabbe disease	1 in 100,000	Druze and Arab communities in Israel
Mucopolysaccharidosis	1 in 100,000	Unknown
Tyrosinemia	1 in 100,000	French Canadians (Saguenay-Lac-Saint-Jean)

The Concise Encyclopedia of Genomic Diseases

Table 3.04.00.2. Continued.

Disease	Frequency of condition	People with higher prevalence (if available)
Porphyria variegata	1 in 100,000	Afrikaners and Europeans
Glutaricaciduria	1-20 in 1,000,000	Caucasians
Nephropathic cystinosis and ocular nonnephropathic cystinosis	5-10 in 1,000,000	French and British
Glycogen storage disease 2 or Pompe disease	7-16 in 1,000,000	Unknown
Essential fructosuria or hepatic fructokinase deficiency or ketohexokinase deficiency	8 in 1,000,000	Unknown
Sandhoff disease	8 in 1,000,000	Unknown
4-hydroxybutyricaciduria or succinic semialdehyde dehydrogenase deficiency	1-9 in 1,000,000	Unknown
Acrodermatitis enteropathica	1-9 in 1,000,000	Unknown
Alpha-mannosidosis	1-9 in 1,000,000	Unknown
Argininosuccinic aciduria or argininosuccinic acidemia	1-9 in 1,000,000	Unknown
Citrullinemia 2	1-9 in 1,000,000	Unknown
Combined lipase deficiency	1-9 in 1,000,000	Unknown
Congenital lipodystrophy	1-9 in 1,000,000	Unknown
Familial combined hyperlipidemia or familial apoprotein C2 deficiency	1-9 in 1,000,000	Unknown
Familial Mediterranean fever	1-9 in 1,000,000	Armenians, Sephardi Jews, Oriental Jews, Greeks, Turkish and Arabs
Glycine encephalopathy	1-9 in 1,000,000	Northern Finnish
Glycogen storage disease 4 or Andersen disease or Glycogen branching enzyme deficiency	1-9 in 1,000,000	Ashkenazi Jews
Homocystinuria	1-9 in 1,000,000	Qataris and Norwegians
Laron dwarfism	1-9 in 1,000,000	Semitic and Mediterraneans
Neurohypophyseal diabetes insipidus	1-9 in 1,000,000	Unknown
ALPHAlpha-mannosidosis	1-9 in 1,000,000	Unknown
Hermansky-Pudlak syndrome	1-2 in 1,000,000	Northwest Puerto Ricans, Japanese, and Swiss
Polycystic lipomembranous osteodysplasia with sclerosing leukoencephalopathy or Nasu-Hakola disease	1-2 in 1,000,000	Japanese and Finnish
Wolfram syndrome	1-2 in 1,000,000	Unknown
Primary hyperoxaluria type 1	1-3 in 1,000,000	Tunisians and Mediterraneans

The Concise Encyclopedia of Genomic Diseases

Table 3.04.00.2. Continued.

Disease	Frequency of condition	People with higher prevalence (if available)
Mucopolysaccharidosis 4A or Morquio syndrome A	1-7 in 1,000,000	Uruguayans
Alkaptonuria or ochronosis	4-10 in 1,000,000	Slovaks and Dominicans
Isovaleric acidemia	4-10 in 1,000,000	Europeans
Menkes disease	4-10 in 1,000,000	Unknown
Mucopolysaccharidosis 3 or Sanfilippo syndrome	4-17 in 1,000,000	West Australians and Dutch
Maple syrup urine disease 1A or branched-chain ketoaciduria	6 in 1,000,000	Old Order Mennonites, Amish, Gypsies of Portugal, and Ashkenazi Jews
Autoimmune polyendocrinopathy type 1	2 in 1,000,000	Finnish
Oculocerebrorenal syndrome and Dent disease	2 in 1,000,000	Unknown
Wolman disease or Cholesteryl ester storage disease	2 in 1,000,000	Unknown
Werner syndrome	2-6 in 1,000,000	Sardinians and Japanese (North)
Congenital adrenal hyperplasia	3 in 1,000,000	Sephardic Jews
Lesch-Nyhan syndrome	3 in 1,000,000	Unknown
Homocystinuria, and Hyperhomocysteinemic thrombosis	4 in 1,000,000	Unknown
Mucopolysaccharidosis 2 or Hunter syndrome	4 in 1,000,000	Unknown
Mucopolysaccharidosis 7 or Sly syndrome	4 in 1,000,000	Unknown
Niemann-Pick disease	4 in 1,000,000	Ashkenazi Jews
Pyruvate carboxylase deficiency	4 in 1,000,000	Unknown
Sjögren-Larsson syndrome	4 in 1,000,000	Swedish
Crigler-Najjar syndrome	1 in 1,000,000	Marylanders (South)
Familial combined hyperlipidemia or Buerger-Gruetz syndrome or lipoprotein lipase deficiency	1 in 1,000,000	French Canadians
Abetalipoproteinemia	<1 in 1,000,000	Unknown
Achalasia-Addisonianism-Alacrimia syndrome or Allgrove syndrome	<1 in 1,000,000	Unknown
Acquired partial lipodystrophy or Barraquer–Simons syndrome	<1 in 1,000,000	Unknown
Adenosine monophosphate deaminase deficiency type 1 or Myoadenylate deaminase deficiency	<1 in 1,000,000	Unknown
Adenylosuccinate lyase deficiency	<1 in 1,000,000	Unknown
Adult-onset leukodystrophy	<1 in 1,000,000	Irish-Americans
AICA-ribosiduria	<1 in 1,000,000	Unknown
Alexander disease	<1 in 1,000,000	Unknown

Table 3.04.00.2. Continued.

Disease	Frequency of condition	People with higher prevalence (if available)
Allan-Herndon-Dudley syndrome	<1 in 1,000,000	Unknown
Alopecia, neurologic defects and endocrinopathy	<1 in 1,000,000	Unknown
Amyloidosis and Carpal tunnel syndrome	<1 in 1,000,000	Unknown
Amyloidosis or Meretoja syndrome	<1 in 1,000,000	Finnish
Arginine:glycine amidinotransferase deficiency	<1 in 1,000,000	Unknown
Argininemia	<1 in 1,000,000	Unknown
Aromatase deficiency and excess syndrome	<1 in 1,000,000	Unknown
Aspartylglucosaminuria	<1 in 1,000,000	Finnish
Bamforth-Lazarus syndrome	<1 in 1,000,000	Unknown
Bartter syndrome	<1 in 1,000,000	Unknown
Beta-mannosidosis	<1 in 1,000,000	Unknown
Beta-ureidopropionase deficiency	<1 in 1,000,000	Unknown
Brunner syndrome	<1 in 1,000,000	Dutch
Canavan disease	<1 in 1,000,000	Ashkenazi Jews and East Europeans
Central precocious puberty and hypogonadotropic hypogonadism	<1 in 1,000,000	Unknown
Chylomicron retention disease	<1 in 1,000,000	Unknown
Combined oxidative phosphorylation deficiency	<1 in 1,000,000	Unknown
Congenital adrenal hyperplasia and Junctional epidermolysis bullosa	<1 in 1,000,000	Brazilians
Congenital brain dysgenesis due to glutamine synthetase deficiency	<1 in 1,000,000	Unknown
Congenital disorder of glycosylation	<1 in 1,000,000	Unknown
Congenital erythropoietic porphyria or Gunther disease	<1 in 1,000,000	Unknown
Congenital generalized lipodystrophy	<1 in 1,000,000	Unknown
Congenital lactase deficiency	<1 in 1,000,000	Finnish
Corticosteroid-binding globulin deficiency	<1 in 1,000,000	Unknown
Cortisone reductase deficiency	<1 in 1,000,000	Unknown
Creatine deficiency syndrome	<1 in 1,000,000	Unknown
Cystic leukoencephalopathy without megalencephaly	<1 in 1,000,000	Unknown
D-glyceric aciduria	<1 in 1,000,000	Serbians, Turkish and Mexicans
Dihydropyrimidinuria	<1 in 1,000,000	Unknown
Dimethylglycine dehydrogenase deficiency	<1 in 1,000,000	Unknown
Dubin-Johnson syndrome	<1 in 1,000,000	Iranian Jews
Endocrine-cerebroosteodysplasia	<1 in 1,000,000	Unknown

Table 3.04.00.2. Continued.

Disease	Frequency of condition	People with higher prevalence (if available)
Ethylmalonic encephalopathy	<1 in 1,000,000	Mediterraneans and Arabs
Familial Glucocorticoid deficiency syndrome	<1 in 1,000,000	Unknown
Familial hypercholanemia	<1 in 1,000,000	Unknown
Familial hypoalphalipoproteinemia and Tangier disease	<1 in 1,000,000	Virginians
Familial isolated hypoparathyroidism	<1 in 1,000,000	Unknown
Familial short stature	<1 in 1,000,000	Unknown
Farber lipogranulomatosis	<1 in 1,000,000	Unknown
Fatal infantile lactic acidosis, and Mitochondrial DNA depletion syndrome	<1 in 1,000,000	Unknown
Fish-eye disease and Norum disease	<1 in 1,000,000	Unknown
Fucosidosis	<1 in 1,000,000	Unknown
Fumaric aciduria or Polygamist Down's	<1 in 1,000,000	People in Arizona and Utah
Galactosialidosis	<1 in 1,000,000	Japanese and Mexicans
Gamma aminobutyric acid transaminase deficiency	<1 in 1,000,000	Unknown
Glucose in galactose malabsorption	<1 in 1,000,000	Unknown
Glutamate formiminotransferase deficiency	<1 in 1,000,000	Japanese
Glycerol kinase deficiency or hyperglycerolemia	<1 in 1,000,000	French-Canadians
Glycine N-methyltransferase deficiency	<1 in 1,000,000	Italians
Glycogen storage disease 0 or muscle glycogen synthase deficiency	<1 in 1,000,000	Unknown
Glycogen storage disease 10 or muscle phosphoglycerate mutase deficiency	<1 in 1,000,000	Unknown
Glycogen storage disease 11 or lactate dehydrogenase A deficiency	<1 in 1,000,000	Japanese
Glycogen storage disease 12 or Aldolase A deficiency	<1 in 1,000,000	Unknown
Glycogen storage disease 13 or Enolase 3 deficiency	<1 in 1,000,000	Unknown
Glycogen storage disease 14	<1 in 1,000,000	Unknown
Glycogen storage disease 1C	<1 in 1,000,000	Unknown
Glycogen storage disease 2B or Danon disease	<1 in 1,000,000	Unknown
Glycogen storage disease 7 or Tarui disease or muscle phosphofructokinase deficiency	<1 in 1,000,000	Unknown
Glycogen storage disease of heart, Cardiomyopathy, and Wolff-Parkinson-White syndrome	<1 in 1,000,000	Unknown
Glycosylphosphatidylinositol deficiency	<1 in 1,000,000	Unknown

The Concise Encyclopedia of Genomic Diseases

Table 3.04.00.2. Continued.

Disease	Frequency of condition	People with higher prevalence (if available)
GM1-gangliosidosis	<1 in 1,000,000	Japanese and Italians
GM2-gangliosidosis variant	<1 in 1,000,000	Unknown
GRACILE, Leigh, and Björnstad syndrome	<1 in 1,000,000	Finnish
Growth delay due to IGF-I resistance	<1 in 1,000,000	African Pygmies
Growth retardation with Deafness and Mental retardation	<1 in 1,000,000	Unknown
Guanidinoacetate methyltransferase deficiency	<1 in 1,000,000	Unknown
Gyrate atrophy of choroid and retina with or without ornithinemia	<1 in 1,000,000	Finnish
Hemochromatosis 2	<1 in 1,000,000	Unknown
Hemochromatosis 3	<1 in 1,000,000	Unknown
Hemochromatosis 4	<1 in 1,000,000	Unknown
Hemolytic anemia due to glutathione synthetase deficiency	<1 in 1,000,000	Unknown
Hepatocerebral, spinocerebellar ataxia, and Progressive external ophthalmoplegia	<1 in 1,000,000	Unknown
Hyperinsulinism-hyperammonemia syndrome	<1 in 1,000,000	Unknown
Hyperornithinemia-hyperammonemia-homocitrullinemia syndrome or HHH syndrome	<1 in 1,000,000	Saskatchewanians (Canada)
Hyperprolinemia	<1 in 1,000,000	Unknown
Hyperprolinemia 2	<1 in 1,000,000	Unknown
Insulin-resistant diabetes mellitus	<1 in 1,000,000	Unknown
Isolated follicle-stimulating hormone deficiency	<1 in 1,000,000	Unknown
Kanzaki and Schindler disease	<1 in 1,000,000	Unknown
L-2-hydroxyglutaric aciduria	<1 in 1,000,000	Iranians
Leukodystrophy, spastic paraplegia, and dystonia	<1 in 1,000,000	Unknown
Leukoencephalopathy with brain stem and spinal cord involvement and lactate elevation	<1 in 1,000,000	Unknown
Lipoid congenital adrenal hyperplasia	<1 in 1,000,000	Japanese, Koreans and Palestinian Arabs
Lysinuric protein intolerance	<1 in 1,000,000	Finnish
Lysosomal acid phosphatase deficiency	<1 in 1,000,000	Unknown
Malonyl-CoA decarboxylase deficiency	<1 in 1,000,000	Unknown
Maple syrup urine disease 1B or branched-chain ketoaciduria	<1 in 1,000,000	Ashkenazi Jews
Maple syrup urine disease 2 or branched-chain ketoaciduria	<1 in 1,000,000	Japanese and Ashkenazi Jews

The Concise Encyclopedia of Genomic Diseases

Table 3.04.00.2. Continued.

Disease	Frequency of condition	People with higher prevalence (if available)
Megalencephalic leukoencephalopathy with subcortical cysts	<1 in 1,000,000	Libyan Jews
Mevalonic aciduria and Hyper-IgD syndrome	<1 in 1,000,000	Unknown
Mitochondrial complex 2 deficiency	<1 in 1,000,000	Unknown
Mitochondrial complex 3 deficiency	<1 in 1,000,000	Unknown
Mitochondrial complex 4 deficiency	<1 in 1,000,000	Unknown
Mitochondrial DNA depletion syndrome 3 (hepatocerebral type)	<1 in 1,000,000	Navajo
Mitochondrial neurogastrointestinal encephalomyopathy syndrome	<1 in 1,000,000	Unknown
Mucolipidosis 3 gamma	<1 in 1,000,000	Unknown
Mucolipidosis 4	<1 in 1,000,000	Ashkenazi Jews
Mucolipidosis or I-cell disease	<1 in 1,000,000	French Canadians (Saguenay-Lac-Saint-Jean)
Mucopolysaccharidosis 4B or Morquio syndrome B	<1 in 1,000,000	Japanese and French
Mucopolysaccharidosis 9	<1 in 1,000,000	Unknown
Multiple sulfatase deficiency	<1 in 1,000,000	Unknown
N-acetylglutamate synthase deficiency	<1 in 1,000,000	Unknown
Neonatal diabetes mellitus with congenital hypothyroidism	<1 in 1,000,000	Unknown
Nestor-Guillermo progeria syndrome	<1 in 1,000,000	Unknown
Neuronal ceroid lipofuscinosis	<1 in 1,000,000	Costa Ricans, South Americans, Portuguese and British
Neuronal ceroid lipofuscinosis	<1 in 1,000,000	Turkish
Neuronal ceroid lipofuscinosis 1 or Santavuori-Haltia disease	<1 in 1,000,000	Finnish
Neuronal ceroid lipofuscinosis 10	<1 in 1,000,000	Unknown
Neuronal ceroid lipofuscinosis 2 or Jansky-Bielschowsky disease	<1 in 1,000,000	Unknown
Neuronal ceroid lipofuscinosis 4B or Kufs disease	<1 in 1,000,000	Unknown
Neuronal ceroid lipofuscinosis 5 or Finnish Late Infantile Variant	<1 in 1,000,000	Finnish
Neuronal ceroid lipofuscinosis 7	<1 in 1,000,000	Unknown
Neutral lipid storage disease or Dorfman-Chanarin syndrome	<1 in 1,000,000	Unknown
Neutral lipid storage disease with myopathy	<1 in 1,000,000	Unknown
Niemann-pick disease C2	<1 in 1,000,000	French Acadians (Nova Scotia)
Nonphotosensitive trichothiodystrophy 1	<1 in 1,000,000	Amish

The Concise Encyclopedia of Genomic Diseases

Table 3.04.00.2. Continued.

Disease	Frequency of condi-tion	People with higher preva-lence (if available)
Ocular albinism 2 and Cone-rod dystro-phy	<1 in 1,000,000	Åland Islanders
Oculocutaneous albinism 3	<1 in 1,000,000	Africans and New Guineans
Oculocutaneous albinism 4	<1 in 1,000,000	Japanese, Koreans, and Germans
Ovarioleukodystrophy and Leukoenceph-alopathy with vanishing white matter	<1 in 1,000,000	Cree and Chippewayan people (Quebec and Mani-toba)
Oxoglutaric aciduria	<1 in 1,000,000	Algerians and Portuguese
Pentosuria	<1 in 1,000,000	Ashkenazi Jews
Phosphoglycerate dehydrogenase defi-ciency	<1 in 1,000,000	Dutch and Turkish
Phosphoserine aminotransferase defi-ciency	<1 in 1,000,000	Unknown
Primary hyperoxaluria type 2	<1 in 1,000,000	Unknown
Primary hyperoxaluria type 3	<1 in 1,000,000	Unknown
Primary hypomagnesemia	<1 in 1,000,000	Unknown
Prolidase deficiency	<1 in 1,000,000	Unknown
Pseudohypoaldosteronism	<1 in 1,000,000	Unknown
Pseudohypoaldosteronism, Bronchiecta-sis, and elevated sweat chloride	<1 in 1,000,000	Unknown
Pyruvate dehydrogenase E2 deficiency	<1 in 1,000,000	Unknown
Saccharopinuria or Hyperlysinemia	<1 in 1,000,000	Unknown
Sialidosis	<1 in 1,000,000	Italians
Sialuria, and myopathy	<1 in 1,000,000	Oriental Jews
Sialuria, Finnish type or Salla disease	<1 in 1,000,000	Finnish
Sitosterolemia or Phytosterolemia, and Gallbladder disease	<1 in 1,000,000	Unknown
Transaldolase deficiency	<1 in 1,000,000	Turkish and Arabs
Urbach–Wiethe disease or Lipoid pro-teinosis	<1 in 1,000,000	Unknown
Wolcott-Rallison syndrome	<1 in 1,000,000	Unknown
Woodhouse-Sakati syndrome	<1 in 1,000,000	Unknown

3.04.00-07 Endocrine Diseases

3.04.01 Thyroid Gland / Thyroid Hormone

There are two common disorders, Hashimoto's thyroiditis and Graves' disease, both of which are caused by a combination of genomic and environmental factors. The rest of the conditions are rare, and caused by genomic variations. Table 3.04.01.1 displays the diseases, the genes and their biological functions; it also includes the product of conception alteration. The biological function includes transcription factors, transporters, enzymes, thyroid hormone receptor, receptor G protein, etc...

Congenital hypothyroidism: A condition characterized by the underproduction of thyroid hormone. Affected individuals often do not develop specific symptoms until several months of age. Affected children exhibit increased sleep, decreased activity, feeding difficulty, prolonged jaundice, constipation, macroglossia, myxedematous facies, a distended abdomen with umbilical hernia and hypotonia. Approximately, 1 in 4,000 Caucasian-American newborns have a severe deficiency of thyroid function. This condition has a lower prevalence in African-Americans: around 1 in 20,000. This condition is more prevalent amongst Asian, Native American, and Hispanic infants. If this condition is untreated for several months after birth, it can lead to mental retardation and growth failure. In underdeveloped countries, the most common cause of congenital hypothyroidism is iodine deficiency, but in most of the developed world the causes, when known, are due to genomics alteration. Commonly there is a defect of development of the thyroid gland itself. Among other specific defects, the table below includes the diseas-

es, the genes, the biological functions, and the product of conception alterations, if they are known. In these diseases, there are very few examples of *de novo* mutation.

Congenital nongoitrous hypothyroidism: A disorder characterized by increased levels of plasma TSH and low levels of thyroid hormone. Most commonly this form of hypothyroidism is caused by an absent or underdeveloped gland. Congenital nongoitrous hypothyroidism is the most prevalent inborn endocrine disorder; it accounts for most cases of congenital hypothyroidism. There are several underlying causes of this disorder with at least 6 autosomal genomic loci: Congenital nongoitrous hypothyroidism-1 is caused by mutations in the *TSHR* gene which encodes the thyroid-stimulating hormone receptor;; Congenital nongoitrous hypothyroidism-2 is caused by mutations in the *PAX8* gene which encodes the paired box 8;; Congenital nongoitrous hypothyroidism-3 has been mapped but the gene has not been characterized;; Congenital nongoitrous hypothyroidism-4 is caused by mutations in the *TSHB* gene which encodes the thyroid-stimulating hormone (subunit beta);; Congenital nongoitrous hypothyroidism-5 is caused by mutations in the *NKX2-5* gene which encodes the NK2 homeobox 5;; and Congenital nongoitrous hypothyroidism-6 is caused by mutations in the *THRA* gene which encodes the thyroid hormone receptor alpha. Congenital nongoitrous hypothyroidism is estimated to affect 1 in every 3,000 to 4,000 live births.

Bamforth–Lazarus syndrome: A condition that results in thyroid dysgenesis. Affected individuals show congenital hypothyroidism, bilateral choanal atresia, bifid epiglottis, cleft palate and spiky hair. Bamforth-Lazarus syndrome is caused by autosomal recessive mutations in the *FOXE1* gene which encodes the forkhead box E1 (thyroid transcription factor 2). Bamforth-Lazarus syndrome is estimated to affect less than 1 in every 1,000,000 individuals.

Thyroid dyshormonogenesis: A disorder characterized by the develop hypothyroidism with a goiter. Thyroid dyshormonogenesis accounts for 10-15% of the causes of congenital hypothyroidism. There is one particular familial form which is associated with sensorineural deafness or Pendred's syndrome. There are several underlying causes of this disorder with at least 7 autosomal genomic loci: Thyroid dyshormonogenesis-1 is caused by mutations in the *SLC5A5* gene which encodes the solute carrier (family 5 member A5), a sodium/iodide symporter;; Thyroid dyshormonogenesis-2A is caused by mutations in the *TPO* gene which encodes the thyroid peroxidase;; Thyroid dyshormonogenesis-2B is caused by mutations in the *SLC26A4* gene which encodes the solute carrier (family 26 member A4), a sodium-independent chloride/iodide transporter. Mutations in the *SLC26A4* gene cause Pendred syndrome, which is a form of thyroid hormone dysgenesis associated with deafness;; Thyroid dyshormonogenesis-3 is caused by mutations in the *TG* gene which encodes the thyroglobulin, a glycoprotein homodimer produced mainly by the thyroid gland;; Thyroid dyshormonogenesis-4 is caused by mutations in the *IYD* gene which encodes the tyrosine deiodinase;; Thyroid dyshormonogenesis-5 is caused by mutations in the *DUOXA2* gene which encodes the dual oxidase maturation factor 2, an endoplasmic reticulum protein that is necessary for proper cellular localization and maturation of functional dual oxidase 2;; Thyroid dyshormonogenesis-6 is caused by mutations in the *DUOX2* gene which encodes the dual oxidase 2. Thyroid dyshormonogenesis is estimated to affect 1 in every 40,000-60,000 live births. There is a higher carrier frequency for thyroid dyshormonogenesis-2A trait amongst individuals from Slovenia, Bosnia, and Slovakia.

Pendred syndrome: A disorder characterized by developmental abnormalities of the cochlea, congenital bilateral hearing loss and goiter with occasional hypothyroidism. Pendred syndrome is the most common syndromal form of deafness. This condition is named after Vaughan Pendred. It accounts for 7.5% of all cases of congenital deafness, and is caused by autosomal recessive mutations in the gene *SLC26A4*. Another allelic disorder associated with this gene is Enlarged vestibular aqueduct syndrome, a form of congenital deafness. Pendred syndrome is estimated to affect 1 in every 10,000 to 100,000 live births, depending on geographic location.

Thyroid hormone resistance: A disorder characterized by thyroid hormone levels is elevated but the thyroid stimulating hormone (TSH) level is not suppressed. Thyroid hormone resistance accounts for less than 10% of all the cases of congenital hypothyroidism. Thyroid hormone resistance is caused by either autosomal dominant or recessive mutations in the *THRB* gene which encodes the thyroid hormone receptor (beta).

Allan–Herndon–Dudley syndrome: A disorder characterized by a defect in brain development leading to moderate to severe mental retardation and problems with movement. Most affected male children exhibit hypotonia and muscle hypoplasia, and in general are not capable of walking independently, often becoming wheelchair-bound by adulthood. This condition is characterized by disruption of the nervous development before birth. This condition is named after William Allan, C. Nash Herndon, and Florence C. Dudley. Allan–Herndon–Dudley syndrome is caused by X-linked recessive mutations in the *SLC16A2* gene which encodes the solute carrier (family 16 member A2), a thyroid hormone transporter. This protein transports thyroxin and/or triiodothyronine hormone into nerve cells in the developing brain. Allan–Herndon–Dudley syndrome is estimated to affect less than 1 in every 1,000,000 individuals.

Hashimoto's thyroiditis or chronic lymphocytic thyroiditis: A common condition which affects between 1-1.5% of all adults in Western countries. In this disease, the thyroid gland is progressively destroyed by immune processes. This condition was named after Hashimoto Hakaru in 1912. The cause of this disease is a combination of genomic and environmental factors. The genomic predispositions to develop this thyroid disorder include: association to the HLA-DR5 haplotype and alteration in the *CTLA4* gene. There is also susceptibility mapped on 8q23-q24, but the gene has not been characterized. Studies on monozygotic twins also show the strong hereditary component of this disease, with a concordance rate of 38-55%. There are some chromosomal disorders such as Turner, Down's and Klinefelter's syndromes, which are associated with an increased incidence in Hashimoto's thyroiditis. There are also environmental factors that contribute to this disease, including tobacco smoke, high iodine intake, etc. There are differences of prevalence between males and females; Hashimoto's thyroiditis occurs 10-20 times more often in women than in men. These differences do not seem to be explained by genomic factors, since the genes/loci with susceptibility to developing Hashimoto's thyroiditis are mapped on autosomal chromosome.

Graves disease: A common disorder characterized by an overproduction of thyroid hormones. Graves disease accounts for 60-90% of all hyperthyroidism cases. This autoimmune disorder is caused by thyroid autoantibodies (TSHR-Ab) that activate the TSH-receptor (TSHR). This condition results from a constitutive activation of the receptor, stimulating thyroid hormone synthesis, secretion and thyroid growth. 25-50% of affected patients display ophthalmopathy with protrusion of one or both eyes. The cause of Graves' disease is a combination of genomic and environmental factors. There are several loci/genes associated with susceptibility to Graves disease which have been mapped on 14q31, 20q13, 4q12, 2q33 and Xp11. There are several influential hereditary components, mainly related to the genes or haplotypes *HLA-DR3, CG* and *CTLA4*. There are also environmental factors such as cigarette smoke, which for example increases the incidence of Graves' ophthalmopathy by 7.7-fold, but doesn't increase the thyroid risks. This disease is also 5-10 times more common in females than in males. This difference does not seem to be explained by genomic factors, since most of the susceptibility loci/genes for Graves' disease mapped on autosomal chromosome.

The Concise Encyclopedia of Genomic Diseases

Table 3.04.01.1 Recognizable Thyroid Disorders

Disorder	OMIM	Gene	Biological Function	Product of Conception Alteration
Congenital nongoitrous hypo-thyroidism-1	275200	TSHR	Cell surface receptor G protein	AR
Congenital nongoitrous hypo-thyroidism-2	218700	PAX8	Transcription factor Helix-turn-helix domains	AR
Congenital nongoitrous hypo-thyroidism-4	275100	TSHB	Thyroid stimulating hormone	AR
Congenital nongoitrous hypo-thyroidism-5	225250	NKX2-5	Transcription factor	AD
Congenital nongoitrous hypo-thyroidism-6	614450	THRA	Thyroid hormone receptor	AD + de novo
Bamforth-Lazarus syndrome	241850	FOXE1	Transcription factor Helix-turn-helix domains	AR
Thyroid dyshormonogenesis-1	274400	SLC5A5	Solute carrier	AR
Thyroid dyshormonogenesis-2A	274500	TPO	Thyroid peroxidase	AR
Thyroid dyshormonogenesis-2B or Pendred syndrome	274600	SLC26A4	Solute carrier	AR
Thyroid dyshormonogenesis-3	274700	TG	Thyroglobulin	AR
Thyroid dyshormonogenesis-4	274800	IYD	Deiodinase	AR
Thyroid dyshormonogenesis-5	274900	DUOXA2	Endoplasmic reticulum protein	AR
Thyroid dyshormonogenesis-6	607200	DUOX2	Oxidase/peroxidase	AR
Pendred syndrome	274600	SLC26A4	Solute carrier	AR
Pendred syndrome and enlarged vestibular aqueduct	274600	SLC26A4/ FOXI1	Solute carrier/Transcription factor	DigB
Thyroid hormone resistance	188570	THRB	Transcription factor Zinc finger	AD
Thyroid hormone resistance	274300	THRB	Transcription factor Zinc finger	AR
Allan-Herndon-Dudley syndrome	300523	SLC16A2	Solute carrier	XR
Thyroxine-binding globulin deficiency	314200	SERPINA7	Serpin proteinase inhibitor	
Neonatal diabetes mellitus with congenital hypothyroidism	610199	GLIS3	Zinc finger protein	AR

Table 3.04.01.1. Continued.

Disorder	OMIM	Gene	Biological Function	Product of Conception Alteration
Choreoathetosis, hypothyroidism, and neonatal respiratory distress	610978	NKX2-1	Transcription factor	AD
Short stature-delayed bone age due to thyroid hormone metabolism deficiency	609698	SECISBP2	Nuclear protein	AR
Familial gestational hyperthyroidism	603373	TSHR	Cell surface receptor G protein	
Nonautoimmune hyperthyroidism	609152	TSHR	Cell surface receptor G protein	AD + de novo
Selective pituitary thyroid hormone resistance	145650	THRB	Transcription factor Zinc finger	AD

Table 3.04.01.2 Thyroid Disorders in Different Ethnic Groups

Disease	Frequency of condition	People with higher prevalence (if available)
Congenital nongoitrous hypothyroidism	1 in 3,000-4,000	Asians, Native Americans, and Hispanics
Pendred syndrome	1-10 in 100,000	Worldwide
Thyroid dyshormonogenesis	1 in 40,000-60,000	Slovenians, Bosnians, and Slovaks
Bamforth–Lazarus syndrome	<1 in 1,000,000	Unknown
Allan–Herndon–Dudley syndrome	<1 in 1,000,000	Unknown

3.04.02 Pancreas / Insulin, Glucagon

3.04.02.1 Diabetes Mellitus

There are several syndromes described in this subchapter; the only common one is type 2 diabetes mellitus. This subchapter includes several conditions of polygenomic causes of disease such as diabetes mellitus type 1 and 2. Table 3.04.02.1 displays the diseases, the genes, the biological functions and the product of conception alterations. The biological function includes transcription factors, transporters, enzymes, mitochondrial protein, enzyme-linked receptors, etc. The table is trying to emphasize that there are a few diseases, such as Donohue syndrome and "Maturity onset diabetes of the young" that could be inherited in a severe form when both parents are heterozygous and both parents provide the faulty copy to descendants.

Diabetes mellitus: A metabolic disorder in which a person has high glycemia, either because the body does not produce enough insulin, or because cells do not respond to the insulin that is produced. It is characterized by symptoms of polyuria, polydipsia and polyphagia. About 2.8% of people worldwide have diabetes. In the United States, 90-95% of diabetic patients suffer from Type 2 diabetes. There are several types of diabetes including:

Type 1 diabetes or insulin-dependent diabetes mellitus: A disorder characterized by pancreatic failure to produce insulin. This autoimmune disorder is dependent on environmental and genomic factors. It is associated with a combination of both inherited factors and environmental factors, such as infections by Coxsackie B4 virus. There is also an individual susceptibility linked to HLA haplotypes. The alleles for HLA, B8-B15-DR3-DR4 seem to have a synergistic effect on the predisposition to insulin-dependent diabetes mellitus. This condition exhibits a correlation rate of 30-50% in monozygotic twins. The average risk of insulin-dependent diabetes mellitus among the siblings of an affected patient is 6%. The incidence of insulin-dependent diabetes mellitus varies from 1 in 100,000 in Japan and China, which are the lowest rates, to a high of about 35 in 100,000 in Scandinavia.

Type 2 diabetes: A disorder characterized by inability of the cells to use insulin properly, occasionally combined with an absolute insulin deficiency. Type 2 diabetes is primarily due to a combination of lifestyle and genomics factors. There are certain ethnicities with a high incidence of diabetes such as Pima Indians. This ethnic group has among the highest prevalence in first-degree relatives as well as high concordance in identical twins, which provides strong evidence that genomic factors trigger susceptibility to type 2 Diabetes mellitus. There are several genes associated with susceptibility to type 2 Diabetes mellitus including: *ABCC8, NEUROD1, KCNJ11, MAPK8IP1, WFS1, LIPC, HMGA1, GCGR, GCK, GPD2, CDKAL1, IGF2BP2, TCF7L2, IRS1, IRS2, ENPP1, RETN, SLC2A4, SLC30A8, PDX1, HNF4A, HNF1A* and *HNF1B* genes. Diabetes mellitus occurs throughout the world, but is more common in more de-veloped countries. This condition has among the largest increases in prevalence in recent years, mainly due to changes in lifestyle.

Neonatal diabetes mellitus: This disease is discussed under conditions originating in the perinatal period in the subchapter "Transitory Endocrine and Metabolic Disorders Specific to Fetus and Newborn" (3.16.4).

Gestational diabetes: A disorder affecting women characterized by the development of diabetes after the 20th week of pregnancy. This condition may precede the development of type 2 Diabetes. Gestational diabetes is associated with mutations in the *GCK* gene which encodes the glucokinase (hexokinase 4). This enzyme phosphorylates glucose to produce glucose-6-phosphate, the 1st step in most glucose metabolism pathways.

Monogenic diabetes: A disorder characterized by Mendelian single gene alteration. There are several forms of monogenic diabetes, including congenital diabetes, cystic fibrosis-related diabetes, hemochromatosis-related diabetes, maturity onset diabetes of the young, defects in proinsulin conversion, insulin gene mutations, insulin receptor mutations, etc. There is also a form of diabetes caused by mitochondrial DNA mutations.

Wolcott-Rallison syndrome: This disease is discussed under congenital malformations, deformations and chromosomal abnormalities in the subchapter "Osteochondrodysplasia with Defects of Growth of Tubular Bones and Spine" (3.17.8.10).

Woodhouse–Sakati syndrome: A multisystemic disorder characterized by malformations throughout the body, associated with deficiencies affecting the endocrine system. Affected individuals exhibit hypogonadism, alopecia, diabetes mellitus, mental retardation and extrapyramidal syndrome. Woodhouse-Sakati syndrome can be caused by autosomal recessive mutations in the *DCAF17* gene which encodes the DDB1 and CUL4 associated factor 17, a nuclear transmembrane protein. Woodhouse-Sakati syndrome is estimated to affect less than 1 in every 1,000,000 individuals.

Maternally inherited Diabetes mellitus and deafness or Ballinger-Wallace syndrome: A disorder characterized by onset of sensorineural hearing loss and diabetes in adulthood. Affected individuals may have additional features, including myopathy, cardiomyopathy, pigmentary retinopathy, renal problems, ptosis, and neuropsychiatric symptoms. Ballinger-Wallace syndrome can be caused by mutations in several mitochondrial genes, including *MT-TL1, MT-TE* and *MT-TK* genes. The association of diabetes and deafness is detected in nuclear genome diseases such as Wolfram syndrome, Rogers syndrome, and Herrmann syndrome. But, most of these disorders have other clinical manifestations. Ballinger-Wallace syndrome accounts for 0.2-0.3% of all cases of diabetes. Ballinger-Wallace syndrome is estimated to affect 1-10 in every 1,000,000 individuals.

Wolfram syndrome or DIDMOAD syndrome (Diabetes Insipidus, Diabetes Mellitus, Optic Atrophy and Deafness): A disorder characterized by diabetes insipidus, diabetes mellitus, optic atrophy and deafness. This disease affects mainly the brain stem and central nervous system. Wolfram syndrome is believed to be triggered by both a malfunction of the myelination and mitochondria. This condition was named after Don J. Wolfram in 1938. There are several underlying causes of this

disorder with at least 2 autosomal genomic loci: Wolfram syndrome-1 is caused by autosomal recessive mutations in the *WFS1* gene which encodes the Wolfram syndrome 1 protein (wolframin), a transmembrane protein, which is located primarily in the endoplasmic reticulum;; and Wolfram syndrome-2 is caused by mutations in the *CISD2* gene which encodes the CDGSH iron sulfur domain-containing protein 2, a zinc finger protein that localizes to the endoplasmic reticulum. There is also a possible mitochondrial form of Wolfram syndrome. Wolfram syndrome is estimated to affect 1-10 in every 1,000,000 individuals.

Diseases caused by alteration in INSR gene: A group of 3 conditions which are allelic disorders including: Donohue syndrome, Insulin-resistant diabetes mellitus with acanthosis nigricans and Rabson–Mendenhall syndrome.

Donohue syndrome or leprechaunism: A disorder characterized by extreme examples of insulin-resistance syndromes. The affected children exhibit cessation of growth at about the seventh month of gestation, uncharacteristic facies creating a gnome-like appearance, and severe endocrine disturbance in the ovaries and pancreas. Donohue syndrome is caused by autosomal recessive mutations in the *INSR* gene which encodes the insulin receptor. The affected children often are a product of consanguineous parents. The product of conception is altered in both parent alleles (AD + AD = AR), so the child inherits both parents' faulty genes. Donohue syndrome is estimated to affect less than 1 in every 1,000,000 individuals.

Insulin-resistant diabetes mellitus with acanthosis nigricans: A disorder characterized by two forms: type A, a syndrome of younger females with signs of virilization; and type B, a syndrome in older females with signs of an immunologic disease and antibodies against the insulin receptor. Insulin-resistant diabetes mellitus with acanthosis nigricans is caused by either autosomal dominant or autosomal recessive mutations in the *INSR* gene.

Rabson–Mendenhall syndrome: A disorder characterized by severe insulin resistance, developmental abnormalities and acanthosis nigricans. These infants exhibit fasting hypoglycemia, postprandial hyperglycemia and hyperinsulinemia, which develop into permanent hyperglycemia and recurrent diabetic ketoacidosis. Rabson–Mendenhall syndrome is caused by mutations in the *INSR* gene.

Maturity onset diabetes of the young: A group of disorder characterized by monogenic diabetes. The severity of the different types varies considerably, but most commonly Maturity onset diabetes of the young acts like a very mild version of type 1 diabetes. However, the two have much in common, and are often studied together. There are several underlying causes of this disorder with at least 11 autosomal genomic loci: Maturity onset diabetes of the young-1 is caused by autosomal dominant mutations in the *HNF4A* gene which encodes the hepatocyte nuclear factor-4-alpha;; Maturity onset diabetes of the young-2 is caused by autosomal dominant mutations in the *GCK* gene which encodes the glucokinase;; Maturity onset diabetes of the young-3 is caused by autosomal dominant mutations in the *HNF1A* gene which encodes the hepatocyte nuclear factor-1 (alpha);; Maturity onset diabetes of the young-4 is caused by autosomal dominant mutations in the *PDX1* gene which encodes the pancreas/duodenum homeobox protein-1;; Maturity onset diabetes of the young-5 is caused by autosomal dominant mutations in the *HNF1B* gene which

encodes the HNF1 homeobox B;; Maturity onset diabetes of the young-6 is caused by autosomal dominant mutations in the *NEUROD1* gene which encodes the neuronal differentiation protein 1;; Maturity onset diabetes of the young-7 is caused by autosomal dominant mutations in the *KLF11* gene which encodes the Kruppel-like factor 11;; Maturity onset diabetes of the young-8 or diabetes-pancreatic exocrine dysfunction syndrome, is caused by autosomal dominant mutations in the *CEL* gene (frameshift deletions in the VNTR) which encodes the carboxyl ester lipase;; Maturity onset diabetes of the young-9 is caused by autosomal dominant mutations in the *PAX4* gene which encodes the paired box 4;; Maturity onset diabetes of the young-10 is caused by autosomal dominant mutations in the *INS* gene which encodes the insulin;; and Maturity onset diabetes of the young-11 is caused by autosomal dominant mutations in the *BLK* gene which encodes the B lymphoid tyrosine kinase. Maturity onset diabetes of the young is estimated to affect 1 in every 9,000-14,500 individuals; it has an unusually high prevalence in African Americans and people from Romania. The prevalence of this condition is about 5% of type 2 diabetes patients in most ethnics groups. Maturity onset diabetes of the young- 2 and 3 are the most common forms in Caucasians. Maturity onset diabetes of the young 2 and 4 could be inherited in a severe form when both parents are heterozygous and both parents provide the faulty copy to descendants (AD + AD = AR) as neonatal diabetes.

3.04.02.2 Other Disorders of Glucose Regulation and Pancreatic Internal Secretion

Leucine-sensitive hypoglycemia of infancy: A disorder characterized by the development of hypoglycemia after high protein feedings. This form of hypoglycemia was also elicited by administration of oral or intravenous infusions of leucine. Leucine-sensitive hypoglycemia of infancy can be caused by either autosomal dominant or autosomal recessive mutations in the *ABCC8* gene which encodes the ATP-binding cassette (sub-family C member 8), a SUR1 subunit of the pancreatic beta cell inwardly rectifying potassium channel. Leucine-sensitive hypoglycemia of infancy is estimated to affect 1-5 in every 10,000 live births.

Persistent hyperinsulinemic hypoglycemia of infancy or Familial hyperinsulinemic hypoglycemia: A disorder characterized by inappropriate over-secretion of insulin by the endocrine pancreas. Familial hyperinsulinemic hypoglycemia is the most common genomic cause of hypoglycemia in early infancy. There are several underlying causes of this disorder with at least 7 autosomal genomic loci: Familial hyperinsulinemic hypoglycemia-1 is caused by mutations in the *ABCC8* gene which encodes the ATP-binding cassette (sub-family C member 8);; Familial hyperinsulinemic hypoglycemia-2 is caused by mutations in the *KCNJ11* gene which encodes the potassium inwardly-rectifying channel (subfamily J member 11), a Kir6.2 subunit of the pancreatic beta cell potassium channel;; Familial hyperinsulinemic hypoglycemia-3 is caused by mutations in the *GCK* gene which encodes the glucokinase;; Familial hyperinsulinemic hypoglycemia-4 is caused by mutations in the *HADH* gene which encodes the hydroxyacyl-CoA dehydrogenase;; Familial hyperinsulinemic hypoglycemia-5 is caused by mutations in the *INSR* gene which encodes the insulin receptor;; Familial hyperinsulinemic hypoglycemia-6 is caused by mutations in the *GLUD1* gene which encodes the glutamate dehydrogenase 1, a mitochondrial matrix enzyme that catalyzes the oxidative deamination of glutamate to alpha-ketoglutarate and ammonia;; and Familial hyperinsulinemic hypoglycemia-7 is caused by mutations in the *SLC16A1* gene which encodes the solute carrier (family 16 member A1), a monocarboxylic acid transporter. The affected children are often a product of either consanguineous parents or the strong founder effect. Familial hyperinsulinism is estimated to affect 1 in every 50,000 live births. There is a higher carrier frequency for Familial hyperinsulinemic hypoglycemia in Saudi Arabia, affecting 1 in 2,675 live births and in Central Finland, affecting 1 in 3,200.

Table 3.04.02.1 Recognizable Pancreatic Disorders

Disorder	OMIM	Gene	Biological Function	Product of Conception Alteration
Woodhouse-Sakati syndrome	241080	*DCAF17*	Nuclear transmembrane protein	AR
Wolfram syndrome-1 or Diabetes Insipidus, Diabetes Mellitus, Optic Atrophy, and Deafness	222300	*WFS1*	Endoplasmic reticulum protein	AR
Wolfram syndrome-2 or Diabetes Insipidus, Diabetes Mellitus, Optic Atrophy, and Deafness	604928	*CISD2*	Endoplasmic reticulum protein	AR

Table 3.04.02.1. Continued.

Disorder	OMIM	Gene	Biological Function	Product of Conception Alteration
Wolcott-Rallison syndrome	226980	EIF2AK3	Endoplasmic reticulum protein	AR
Diabetes and deafness or Ballinger and Wallace syndrome	520000	MT-TE	Mitochondrial protein	Mit
Diabetes and deafness or Ballinger and Wallace syndrome	520000	MT-TK	Mitochondrial protein	Mit
Diabetes and deafness or Ballinger and Wallace syndrome	520000	MT-TL1	Mitochondrial protein	Mit
Donohue syndrome or leprechaunism	246200	INSR	Cell surface receptor Enzyme-linked receptor	AD + AD = AR
Insulin-resistant diabetes mellitus with acanthosis nigricans	610549	INSR	Cell surface receptor Enzyme-linked receptor	AD or AR
Rabson-Mendenhall syndrome	262190	INSR	Cell surface receptor Enzyme-linked receptor	AR
Maturity-onset diabetes of the young-1	125850	HNF4A	Transcription factor Zinc finger	AD
Maturity-onset diabetes of the young-2	125851	GCK	Glucokinase	AD or AD + AD = AR
Maturity-onset diabetes of the young-3	600496	HNF1A	Transcription factor β-Scaffold factors	AD
Maturity-onset diabetes of the young-4	606392	PDX1	Transcription factor Helix-turn-helix domains	AD or AD + AD = AR
Maturity-onset diabetes of the young-5	137920	HNF1B	Transcription factor β-Scaffold factors	AD
Maturity-onset diabetes of the young-6	606394	NEUROD1	Transcription factor Basic helix-loop-helix	AD
Maturity-onset diabetes of the young-7	610508	KLF11	Transcription factor Zinc finger	AD
Maturity-onset diabetes of the young-8	609812	CEL	Carboxyl ester lipase	AD
Maturity-onset diabetes of the young-9	612225	PAX4	Transcription factor Helix-turn-helix domains	AD

Table 3.04.02.1. Continued.

Disorder	OMIM	Gene	Biological Function	Product of Conception Alteration
Maturity-onset diabetes of the young-10	613370	INS	Hormone	AD
Maturity-onset diabetes of the young-11	613375	BLK	Tyrosine kinase	AD
Leucine-sensitive hypoglycemia of infancy	240800	ABCC8	ABC-transporter	AR
Familial hyperinsulinemic hypoglycemia-1	256450	ABCC8	ABC-transporter	AR
Familial hyperinsulinemic hypoglycemia-2	601820	KCNJ11	Channelopathy Potassium	AR
Familial hyperinsulinemic hypoglycemia-3	602485	GCK	Glucokinase	AD
Familial hyperinsulinemic hypoglycemia-4	609975	HADH	Mitochondrial matrix protein	AR
Familial hyperinsulinemic hypoglycemia-5	609968	INSR	Cell surface receptor Enzyme-linked receptor	AD
Familial hyperinsulinemic hypoglycemia-6	606762	GLUD1	Dehydrogenase	AD
Familial hyperinsulinemic hypoglycemia-7	610021	SLC16A1	Solute carrier	AD
Chromosome 17q12 deletion syndrome or Renal cysts and diabetes syndrome	137920	HNF1B	Transcription factor β-Scaffold factors	AD

Table 3.04.02.2 Pancreatic Disorders in Different Ethnic Groups

Disease	Frequency of condition	People with higher prevalence (if available)
Maturity onset diabetes of the young	1-10 in 100,000	African-Americans and Romanians
Familial hyperinsulinemic hypoglycemia	2 in 100,000	Saudi Arabians and Finnish (Central)
Ballinger-Wallace syndrome	1-10 in 1,000,000	Worldwide
Wolfram syndrome	1-10 in 1,000,000	Unknown
Wolcott-Rallison syndrome	<1 in 1,000,000	Unknown
Woodhouse-Sakati syndrome	<1 in 1,000,000	Unknown
Donohue syndrome	<1 in 1,000,000	Unknown

3.04.03 Parathyroid Gland / PTH

The parathyroids are four small endocrine glands. These glands control the amount of calcium in the blood and within the bones. Several disorders affect these glands including: Di George syndrome which results in hypoparathyroidism. The only common condition is primary hyperparathyroidism. Table 3.04.03.1 displays the diseases, the genes, the biological functions, and the product of conception alterations. The biological function includes transcription factors, receptor G protein, imprinted antisense RNA, vesicle docking and fusion, parathyroid hormone, proto-oncogene, etc. All the conditions described are mapped on autosomal chromosome, so if there is a difference between males and females, it is not strongly supported by the contribution of the genomic factor. This subchapter illustrates how neonatal hyperparathyroidism could be inherited in a severe form (AD + AD = AR) when both parents are heterozygous and both parents provide the faulty copy to descendants.

Pseudohypoparathyroidism: A condition associated with end-organ resistance to the parathyroid hormone. Affected individuals have low serum calcium and high phosphate, but their parathyroid hormone level (PTH) is high. Pseudohypoparathyroidism type Ia is caused by maternal specific monoallelic mutations in the *GNAS* gene/locus which encodes the adenylate cyclase-stimulating G alpha protein, resulting in the loss-of-function of the stimulatory G-protein alpha subunit. The GNAS locus has a highly complex imprinted expression pattern. They give rise to maternally, paternally, and biallelically expressed transcripts that are derived from four alternative promoters and 5' exons; Pseudohypoparathyroidism type Ib is caused by imprinting/methylation defects at the GNAS locus, affecting the *GNAS* and the *STX16* genes, resulting in lack of expression of the maternal allele in renal tissue; Pseudohypoparathyroidism type Ic is characterized by PTH resistance, generalized hormone resistance, decreased cAMP response to PTH infusion, and normal erythrocyte Gs activity. Pseudohypoparathyroidism type Ic is caused by maternal specific monoallelic mutations in the *GNAS* gene; and Pseudohypoparathyroidism type II is characterized by a normal cAMP response to PTH infusion, but a deficient phosphaturic response, due to distal impairment to cAMP generation in renal cells.

Familial isolated hypoparathyroidism: A disorder characterized by abnormal calcium metabolism due to deficient secretion of PTH, without other endocrine disorders or developmental defects. Affected individuals in the first decade of life exhibit clinical signs of hypocalcemia, which is characterized by seizures, stridor, prolonged QTc, and tetany. Familial isolated hypoparathyroidism is the most common cause of genomic nonsyndromic hypoparathyroidism. There are several underlying causes of this disorder with at least 3 autosomal genomic loci: Familial isolated hypoparathyroidism can be caused by autosomal dominant mutations in either the *CASR* gene which encodes the calcium-sensing receptor, or the *PTH* gene which encodes the parathyroid hormone;; Familial isolated hypoparathyroidism can be caused by autosomal recessive mutations in the *GCM2* gene which encodes the glial cells missing homolog protein 2;; Hypocalcemia can be caused by an autosomal dominant activating gain-of-function mutations in the *CASR* gene. 42% of familial isolated hypoparathyroidism may be due to an activating mutation causing gain-of-function in the *CASR* gene. Familial isolated hypoparathyroidism is estimated to affect less than 1 in every 1,000,000 individuals.

Primary hyperparathyroidism: A disorder characterized by hypercalcemia through the excessive secretion of PTH, usually by an adenoma of the parathyroid glands. Primary hyperparathyroidism is a common cause of hypercalcemia. There are several underlying causes of this disorder with at least 3 autosomal genomic loci: Familial isolated hyperparathyroidism-1 can be caused by autosomal dominant mutations in the *CDC73* gene which encodes the cell division cycle 73, a tumor suppressor protein that is involved in transcriptional and post-transcriptional control pathways;; Familial isolated hyperparathyroidism-2 or hyperparathyroidism-jaw tumor syndrome can be caused by autosomal dominant mutations in the *CDC73* gene;; Familial isolated hyperparathyroidism-3 has been mapped but the gene has not been characterized; and Familial isolated hyperparathyroidism can be also caused by autosomal dominant mutations in the *MEN1* gene which encodes the multiple endocrine neoplasia I, a putative tumor suppressor. The prevalence in the United States has been estimated to be 3 in 1,000 individuals. Primary hyperparathyroidism is three times more common in women than in men.

Primary hyperparathyroidism of genomic origin: A rare cause of primary hyperparathyroidism. This condition is caused by hyperplasia rather than adenoma. Primary hyperparathyroidism of genomic origin can be caused familial endocrine disorders including: Multiple endocrine neoplasia type 1 and type 2A, and familial hyperparathyroidism. There are also severe inherited conditions when both parents are heterozygous and both parents provide the faulty copy to descendants (AD + AD = AR) as neonatal hyperparathyroidism. Neonatal severe primary hyperparathyroidism can be caused by biallelic loss-of-function mutations in the *CASR* gene. Neonatal severe primary hyperparathyroidism is estimated to affect less than 1 in every 1,000,000 individuals.

3.04.03.1 Other Forms of Hypoparathyroidism having a Genomic Basis include:

DiGeorge syndrome: This disease is discussed under Diseases of the blood and blood-forming organs and certain disorders involving the immune mechanism in the subchapter "Immunodeficiency Associated with other Major Defects" (3.03.6.3).

Barakat syndrome: This disease is discussed under congenital malformations, deformations and chromosomal abnormalities in the subchapter "Other Specified Congenital Malformation Syndromes (not elsewhere classified)" (3.17.09.6.7).

Kenny-Caffey syndrome: This disease is discussed under congenital malformations, deformations and chromosomal abnormalities in the subchapter "Craniofacial Dysostosis" (3.17.8.8.2).

Sanjad-Sakati syndrome: This disease is discussed under congenital malformations, deformations and chromosomal abnormalities in the subchapter "Congenital Malformation Syndromes predominantly Affecting Facial Appearance" (3.17.09.6.1).

Table 3.04.03 Recognizable Parathyroid Disorders

Disorder	OMIM	Gene	Biological Function	Product of Conception Alteration
Pseudohypoparathyroidism-1A	103580	GNAS	Receptor G protein	MAT
Pseudohypoparathyroidism-1B	603233	GNAS	Cell surface receptor G protein	MAT
Pseudohypoparathyroidism-1B	603233	GNAS-AS1	Imprinted antisense RNA	AD
Pseudohypoparathyroidism-1B	603233	STX16	Vesicle docking and fusion	AD
Pseudohypoparathyroidism-1C	612462	GNAS	Receptor G protein	AD
Hypocalcemia	146200	CASR	Receptor G protein	AD
Familial isolated hypoparathyroidism	146200	PTH	Parathyroid hormone	AR or AD
Familial isolated hypoparathyroidism	146200	GCM2	Transcription factor	AR
Hypoparathyroidism	146200	PTH	Parathyroid hormone	AR or AD
Familial primary hyperparathyroidism with and without jaw tumor syndrome	145000	CDC73	Enzyme-RNA polymerase 2	AD + de novo
Neonatal hyperparathyroidism	239200	CASR	Receptor G protein	AD + AD=AR
Multiple endocrine neoplasia-1	131100	MEN1	Nuclear signals	AD + 10%
Multiple endocrine neoplasia-2A	171400	RET	Proto-oncogene	AD + 10%

3.04.04 Pituitary Gland / ADH, Oxytocin, GH, ACTH, TSH, LH, FSH and Prolactin

The hypophysis is an endocrine gland; it produces and secretes nine hormones that regulate homeostasis. There are several rare syndromes included in this subchapter. There are conditions that the product of conception alteration is caused by a mechanism of dominant, recessive and X-linked inheritance. Also there are several examples of *de novo* alterations, such as central precocious puberty. Table 3.04.04.1 displays the diseases, the genes,

the biological functions, and the product of conception alterations. The biological function includes receptor G protein, hormones, transcription factor, enzyme-linked receptor, chromatin remodeling, etc. This section contains a clear example that sometimes is more useful to describe the diseases related to a single gene, such as AIP related conditions. And in some other instances, it is more useful to obtain the clinical information such as in hypogonadotropic hypogonadism with or without additional features.

3.04.04.1 Hyperfunction of Pituitary Gland

AIP related diseases: A few examples of pituitary tumors which are genomically passed from parents to offspring. It includes the overproduction of growth hormone and prolactin. There are condition dominant and/or *de novo* alterations.

Acromegaly: A condition caused by overproduction of growth hormone (GH) after epiphyseal plate closure at puberty. Acromegaly most commonly affects middle aged adults. Growth hormone-secreting pituitary adenomas or somatotropinomas can be caused by mutations in the *AIP* gene which encodes the aryl hydrocarbon receptor-interacting protein. Familial acromegaly can be caused by mutations in the *MEN1* gene, (associated with multiple endocrine neoplasia type I), or by mutations in the *PRKAR1A* gene which encodes the protein kinase A regulatory subunit-1-alpha (associated with Carney complex type 1). Familial acromegaly can also be caused by McCune-Albright syndrome, which is associated with early embryonic postzygotic somatic activating mutations in the *GNAS* gene.

Prolactinoma: A benign tumor of the pituitary gland caused by overproduction of prolactin. It is the most common type of pituitary tumor. Prolactin-secreting pituitary adenomas or prolactinoma can be caused by mutations in the *AIP* gene which encodes the aryl hydrocarbon receptor interacting protein. Prolactinoma can be caused by mutations in the *MEN1* gene, which is a feature of multiple endocrine neoplasia type I.

Syndrome of inappropriate antidiuretic hormone hypersecretion: A condition characterized by excessive production and release of antidiuretic hormone causing hyponatremia. Affected individuals exhibit inability to excrete a free water load with inappropriately concentrated urine, which results in hyponatremia, hypoosmolality and natriuresis. There are many non-genomics factors such as brain and lung tumors, head trauma, strokes, and others associated with this condition. There is also a genomic disorder with similar clinical features, the Nephrogenic syndrome of inappropriate antidiuresis. This nephrogenic condition is caused by X-linked gain-of-function mutations in the *AVPR2* gene which encodes the vasopressin V2 receptor. These gain-of-function mutations causes this receptor to become constitutive activated. On the contrary, X-linked nephrogenic diabetes insipidus is caused by loss-of-function mutations in the *AVPR2* gene. The *AVPR2* gene is located on the subtelomeric long arm of chromosome Xq28. Nephrogenic syndrome of inappropriate antidiuresis is estimated to affect less than 1 in every 1,000,000 individuals.

Precocious puberty: A condition defined as puberty occurring at a remarkably early age. Central precocious puberty can be traced back to a dysfunction in the hypothalamus or pituitary. Central precocious puberty can be caused by many non-genomic condition including trauma, infection, intracranial neoplasm, etc. Central precocious puberty can be caused by several underlying genomic causes including: McCune-Albright syndrome, Angelman syndrome and Central precocious puberty. Central precocious puberty can be caused by *de novo* mutations in *KISS1R* gene which encodes the KISS1 receptor, a galanin-like G protein-coupled receptor.

3.04.04.2 Hypofunction and other Disorders of Pituitary Gland

The Concise Encyclopedia of Genomic Diseases

Hypopituitarism: A disorder characterized by decreased production and secretion of all the hormones produced by the pituitary gland. The main causes are sellar tumor, transsphenoidal surgery, and radiotherapy.

Congenital hypopituitarism: A group of disorders that includegenomic diseases such as Prader-Willi syndrome, Bardet-Biedl syndrome and Kallmann syndrome, that are related to insufficient development or decreased function of the gland.

Combined pituitary hormone deficiency: A condition caused by deficiency of several hormones produced by the pituitary gland. It is associated with mutations of transcription factors involved in pituitary ontogenesis. These genes are almost exclusively expressed in the pituitary gland; they encode transcription factors, which function as activators and repressors of protein-coding genes. Combined pituitary hormone deficiency phenotype fluctuates depending on the transcription factor involved. There are several underlying causes of this disorder with at least 6 autosomal genomic loci and 1 X-linked genomic locus: Combined pituitary hormone deficiency-1 is caused by autosomal dominant or autosomal recessive mutations in the *POU1F1* gene which encodes the POU class 1 homeobox 1. Mutations in the *POU1F1* gene results in deficiencies of anterior pituitary hormones GH and TSH, pituitary hypoplasia;; Combined pituitary hormone deficiency-2 is associated with hypogonadism, and is caused by autosomal recessive mutations in the *PROP1* gene, which encodes the PROP paired-like homeobox 1. Mutations in the *PROP1* gene result in deficiencies of anterior pituitary hormones GH, TSH, PRL, LH, FSH, and ACTH. Mutation in the *PROP1* gene is the most commonly known cause of this condition, accounting for an esti-

mated 12-55% of all the cases;; Combined pituitary hormone deficiency-3 is associated with rigid cervical spine and variable sensorineural deafness, and is caused by mutations in the *LHX3* gene, which encodes the LIM homeobox 3. Mutations in the *LHX3* gene results in deficiencies of anterior pituitary hormones GH, TSH, PRL, LH, and FSH, and limited head and neck rotation;; Combined pituitary hormone deficiency-4 is caused by mutations in the *LHX4* gene which encodes the LIM homeobox 4. Mutations in the *LHX4* gene result in deficiencies of variable pituitary hormones, ectopic neurohypophysis, and cerebral abnormalities;; Combined pituitary hormone deficiency-5 is caused by mutations in the *HESX1* gene which encodes the HESX homeobox 1. Mutations in the HESX1 gene results in deficiencies of variable pituitary hormones and septo-optic dysplasia;; Combined pituitary hormone deficiency-6 is caused by mutations in the *OTX2* gene which encodes the orthodenticle homeobox 2. Mutations in the *OTX2* gene results in deficiencies of anterior pituitary hormones GH, ACTH, TSH, LH, and FSH;; and X-linked panhypopituitarism can be caused by X-linked mutations in the *SOX3* gene which encodes the SRY (sex determining region Y)-box 3. This form results in deficiencies of variable pituitary hormones associated with duplications in the *SOX3* gene. Replacement of hormone deficiencies is required to properly treat these patients. The incidence of congenital hypopituitarism is estimated to be 1 in 4,000-8,000 births.

Genomic disorder coursing with hypogonadotropic hypogonadism: There are several syndromes described below that overlap in clinical features and gene, but are commonly defined as a deficiency of the pituitary secretion of LH and FSH, which consequently impairs the pubertal maturation and reproductive functions.

Congenital idiopathic hypogonadotropic hypogonadism: A disorder characterized by lacking sexual maturation by the age of 18 years, in combined with almost absent circulating gonadotropins and testosterone. There are no other abnormalities of the hypothalamic-pituitary axis. Affected individuals can show isolated defects in the release or action of gonadotropin-releasing hormone (GNRH), or both. Other associated nonreproductive features may include anosmia, cleft palate and sensorineural hearing loss. If the patients are exhibiting different clinical features, such as anosmia, it could better fit the diagnosis of Kallman syndrome.

Kallmann syndrome or hypogonadotropic hypogonadism with or without anosmia: A disorder characterized by anosmia and hypogonadotropic hypogonadism. This condition was named after Franz Josef Kallmann in 1944. There are several underlying causes of this disorder with at least 1 X-linked genomic locus and 20 autosomal genomic loci: Kallmann syndrome-1, or hypogonadotropic hypogonadism-1, with or without anosmia, is caused by X-linked recessive mutations in the *KAL1* gene which encodes the Kallmann syndrome 1 sequence, an anosmin protein. The encoded protein functions in neural cell adhesion and axonal migration and plays a key role in the migration of GNRH neurons and olfactory nerves to the hypothalamus. Kallmann syndrome-1 mostly affects males;; Kallmann syndrome-2, or hypogonadotropic hypogonadism-2, with or without anosmia, can be caused by autosomal dominant mutations in the *FGFR1* gene which encodes the fibroblast growth factor receptor-1;; Hypogonadotropic hypogonadism-3, with or without anosmia, can be caused by mutations in the *PROKR2* gene which encodes the prokineticin receptor 2;; Hy-

pogonadotropic hypogonadism-4, with or without anosmia, can be caused by mutations in the *PROK2* gene which encodes the prokineticin 2;; Hypogonadotropic hypogonadism-5, with or without anosmia, can be caused by mutations in the *CHD7* gene which encodes the chromodomain helicase DNA binding protein 7;; Hypogonadotropic hypogonadism-6, with or without anosmia, can be caused by mutations in the *FGF8* gene which encodes the fibroblast growth factor 8 (androgen-induced);; Hypogonadotropic hypogonadism-7, with or without anosmia, can be caused by mutations in the *GNRHR* gene which encodes the gonadotropin-releasing hormone receptor;; Hypogonadotropic hypogonadism-8, with or without anosmia, can be caused by mutations in the *KISS1R* gene which encodes the KISS1 receptor;; Hypogonadotropic hypogonadism-9, with or without anosmia, can be caused by mutations in the *NSMF* gene which encodes the NMDA receptor synaptonuclear signaling and neuronal migration factor;; Hypogonadotropic hypogonadism-10, with or without anosmia, can be caused by mutations in the *TAC3* gene which encodes the tachykinin 3;; Hypogonadotropic hypogonadism-11, with or without anosmia, can be caused by mutations in the *TACR3* gene which encodes the tachykinin receptor 3;; Hypogonadotropic hypogonadism-12, with or without anosmia, can be caused by mutations in the *GNRH1* gene which encodes the gonadotropin-releasing hormone 1 (luteinizing-releasing hormone);; Hypogonadotropic hypogonadism-13, with or without anosmia, can be caused by mutations in the *KISS1* gene which encodes the KISS-1 metastasis-suppressor protein;; Hypogonadotropic hypogonadism-14, with or without anosmia, can be caused by mutations in the *WDR11* gene which encodes the WD repeat-containing protein domain 11;; Hypogonadotropic hypogonadism-15, with or without anosmia, can be caused by mutations in the *HS6ST1* gene which encodes the heparan sulfate 6-O-sulfotransferase 1;; Hypogonadotropic hypogonadism-16, with or without anosmia, can be caused by mutations in the *SEMA3A* gene which encodes the semaphorin 3A;; Hypogonadotropic hypogonadism-17, with or without anosmia, can be caused by mutations in the *SPRY4* gene which encodes the sprouty protein homolog 4;; Hypogonadotropic hypogonadism-18, with or without anosmia, can be caused by mutations in the *IL17RD* gene which encodes the interleukin 17 receptor D, a protein that is a component of the interleukin-17 receptor signaling complex;; Hypogonadotropic hypogonadism-19, with or without anosmia, can be caused by mutations in the *DUSP6* gene which encodes the dual specificity phosphatase 6;; Hypogonadotropic hypogonadism-20, with or without anosmia, can be caused by mutations in the *FGF17* gene which encodes the fibroblast growth factor 17;; and Hypogonadotropic hypogonadism-21 with or without anosmia, can be caused by mutations in the *FLRT3* gene which encodes the fibronectin leucine rich transmembrane protein 3. There is evidence that digenic mutations can contribute to produce GnRH-deficient conditions. There is also evidence that Kallmann Syndrome-1 can be caused by *de novo* alteration, since there are many cases of this disorder where there is no obvious family history. Kallmann Syndrome is estimated to affect 1 in 10,000 males and 1 in 50,000 females.

Neurohypophyseal diabetes insipidus: A disorder characterized by polyuria, polydipsia, and dehydration. Neurohypophyseal diabetes insipidus is caused by a lack of vasopressin production. Neurohypophyseal diabetes insipidus can be either acquired or familial. Neurohypophyseal diabetes insipidus is caused by either autosomal dominant or autosomal recessive mutations in the *AVP* gene which encodes the arginine vasopressin. The paraventricular and supraoptic nuclei exhibit a progressive loss of AVP-producing neurons relative to oxytocin-producing neurons. The acquired form occurs much more frequently than the familial form.

Isolated growth hormone deficiency: A disorder characterized by proportionate short stature, accompanied by a decreased growth velocity. There are several underlying causes of this disorder with at least 2 autosomal genomic loci and 1 X-linked genomic locus. There are 4 recognized clinical forms of this condition: Isolated growth hormone deficiency type IA is a severe form of dwarfism. Affected individuals often develop anti-GH antibodies when given exogenous growth hormone. Isolated growth hormone deficiency type IA is caused by autosomal recessive mutations in the *GH1* gene which encodes the growth hormone 1;; Isolated growth hormone deficiency type IB is a disorder characterized by a moderate form of dwarfism, which is less severe than in type IA. Patients respond to treatment with exogenous growth hormone. Isolated growth hormone deficiency type IB is caused by autosomal recessive mutations in the *GH1* gene which encodes the growth hormone 1 or the *GHRHR* gene which encodes the growth hormone receptor. Type IA and IB are prevalent in consanguineous families;; Isolated growth hormone deficiency type II is a disorder characterized by autosomal dominant mutations due to splice site or missense in the *GH1* gene. Isolated growth hormone deficiency type II is caused by monoallelic dominant-negative mutations in the *GH1* gene;; and Isolated growth hormone deficiency type III is a disorder often associated with hypogammaglobulinemia. Isolated growth hormone deficiency type III is caused by X-linked recessive mutations in the *BTK* gene which encodes the Bruton agammaglobulinemia tyrosine kinase. Isolated growth hormone deficiency is estimated to affect 1 in every 4,000-10,000 individuals.

Isolated adrenocorticotropic hormone (ACTH) deficiency: A disorder characterized by neonatal hypoglycemia, seizures, cholestatic jaundice, and very low plasmatic concentration in CRH, ACTH and cortisol.

Isolated follicle-stimulating hormone deficiency: A disorder affecting females characterized by primary amenorrhea. Affected individuals also exhibit high LH, undetectable serum FSH, and primordial follicles which had not matured. Isolated follicle-stimulating hormone deficiency is caused by mutations in the *FSHB* gene which encodes the beta chain of follicle-stimulating hormone.

Table 3.04.04.1 Recognizable Pituitary Disorders

Disorder	OMIM	Gene	Biological Function	Product of Conception Alteration
Growth hormone-secreting pituitary adenoma	102200	AIP	Receptor-interacting protein	AD + de novo
Prolactin-secreting pituitary adenoma	600634	AIP	Receptor-interacting protein	AD + de novo
Nephrogenic syndrome of inappropriate antidiuresis	300539	AVPR2	Cell surface receptor G protein	XR
Neurohypophyseal diabetes insipidus	304900	AVP	Arginine vasopressin hormone	AD
Central precocious puberty	176400	KISS1R	Cell surface receptor G protein	AD + 99%
Combined pituitary hormone deficiency-1	613038	POU1F1	Transcription factor diseases Helix-turn-helix domains	AD or AR
Combined pituitary hormone deficiency-2	262600	PROP1	Transcription factor	AR
Combined pituitary hormone deficiency-3	221750	LHX3	Transcription factor	AR
Combined pituitary hormone deficiency-4	262700	LHX4	Transcription factor	AR
Combined pituitary hormone deficiency-5	182230	HESX1	Transcriptional repressor	AR
Combined pituitary hormone deficiency-6	613986	OTX2	Transcription factor	AD + de novo
X-linked combined pituitary hormone deficiency	312000	SOX3	Transcription factor β-Scaffold factors	XR
Growth hormone deficiency with pituitary anomalies	182230	HESX1	Transcriptional repressor	AR

Table 3.04.04.1. Continued.

Disorder	OMIM	Gene	Biological Function	Product of Conception Alteration
Kallmann syndrome-1 with or without anosmia or hypogonadotropic hypogonadism-1 with or without anosmia	308700	KAL1	Neural cell adhesion and axonal migration	XR + de novo
Kallmann syndrome-2 with or without anosmia or hypogonadotropic hypogonadism-2 with or without anosmia	147950	FGFR1	Cell surface receptor Enzyme-linked receptor	AD
Hypogonadotropic hypogonadism-3 with or without anosmia	244200	PROKR2	Cell surface receptor G protein	
Kallmann syndrome-3	607123	PROKR2/ KAL	Genetic disease Cell surface receptor G protein	Dig
Hypogonadotropic hypogonadism-4 with or without anosmia	610628	PROK2	Circadian clock	AR
Hypogonadotropic hypogonadism-5 with or without anosmia	612370	CHD7	Chromatin remodeling	AD
Hypogonadotropic hypogonadism-6 with or without anosmia	612702	FGF8	Mitogenic and cell survival activities	AD
Hypogonadotropic hypogonadism-7 with or without anosmia	146110	GNRHR	Gonadotropin-releasing hormone receptor	AD
Hypogonadotropic hypogonadism-8 with or without anosmia	614837	KISS1R	Receptor G protein	AD
Hypogonadotropic hypogonadism-9 with or without anosmia	614838	NSMF	Signaling and neuronal migration factor	AD
Hypogonadotropic hypogonadism-10 with or without anosmia	614839	TAC3	Tachykinin peptide	AD
Hypogonadotropic hypogonadism-11 with or without anosmia	614840	TACR3	Receptor G protein	AD
Hypogonadotropic hypogonadism-12 with or without anosmia	614841	GNRH1	Gonadotropin-releasing hormone	AD
Hypogonadotropic hypogonadism-13 with or without anosmia	614842	KISS1	Metastasis-suppressor protein	AD
Hypogonadotropic hypogonadism-14 with or without anosmia	614858	WDR11	Tumor suppressor	AD
Hypogonadotropic hypogonadism-15 with or without anosmia	614880	HS6ST1	Heparan sulfate 6-O-sulfotransferase 1	AD

Table 3.04.04.1. Continued.

Disorder	OMIM	Gene	Biological Function	Product of Conception Alteration
Hypogonadotropic hypogonadism-16 with or without anosmia	614897	SEMA3A	Neuronal pattern development	AD
Hypogonadotropic hypogonadism-17 with or without anosmia	615266	SPRY4	Inhibitor of the receptor-transduced protein kinase	AD
Hypogonadotropic hypogonadism-18 with or without anosmia	615267	IL17RD	Interleukin 17 receptor	AD
Hypogonadotropic hypogonadism-19 with or without anosmia	615269	DUSP6	Phosphatase	AD
Hypogonadotropic hypogonadism-20 with or without anosmia	615270	FGF17	Fibroblast growth factor	AD
Hypogonadotropic hypogonadism-21 with or without anosmia	615271	FLRT3	Fibronectin	AD
Isolated growth hormone deficiency-1A	262400	GH1	Growth hormone	AR
Isolated growth hormone deficiency-1B	612781	GHRHR	Cell surface receptor G protein	AR
Isolated growth hormone deficiency-1B	612781	GH1	Growth hormone	AR
Isolated growth hormone deficiency-2	173100	GH1	Growth hormone	AD
Kowarski syndrome	262650	GH1	Growth hormone	AR
Neurohypophyseal diabetes insipidus	125700	AVP	Secreted	AD
Congenital isolated ACTH deficiency	201400	TBX19	Nuclear transcription factor	AR
Isolated follicle-stimulating hormone deficiency	229070	FSHB	Follicle-stimulating hormone	AR

Table 3.04.04.2 Pituitary Disorders in Different Ethnic Groups

Disease	Frequency of condition	People with higher prevalence (if available)
Congenital hypopituitarism	1 in 4,000-8,000	Worldwide
Isolated growth hormone deficiency	1 in 4,000-10,000	Worldwide
Kallmann syndrome or hypogonadotropic hypogonadism with or without anosmia	1 in 10,000 Males	Worldwide
Kallmann syndrome or hypogonadotropic hypogonadism with or without anosmia	1 in 50,000 Females	Worldwide

3.04.05 Adrenal Gland / Aldosterone, Cortisol, Epinephrine and Norepinephrine

The adrenal glands are endocrine organs that sit at the top of the kidneys. These glands produce and release hormones including aldosterone and cortisol, and androgens such as androstenedione and catecholamines. There are several rare syndromes included in this subchapter. There are conditions in which the product of conception alteration is caused by a mechanism of dominant, recessive, digenic and X-linked inheritance. Table 3.04.05.1 displays the diseases, the genes, the biological function and the product of conception alteration. The biological function includes receptor G protein, hormones, transcription factor, phosphodiesterase, proteins located in endoplasmic reticulum and mitochondria, etc.

Primary pigmented micronodular adrenocortical disease or familial Cushing syndrome: A disorder clinically characterized by suffering from Cushing syndrome. Affected individuals exhibit an ACTH-independent adrenal hyperplasia. It is usually seen as a manifestation of the Carney complex, a multiple neoplasia syndrome. These enlarged adrenal glands contain numerous yellow cortical nodules. Lipochromic pigment is a feature of this condition. There are several underlying causes of this disorder with at least 3 autosomal genomic loci: Primary pigmented nodular adrenocortical disease-1 is caused by autosomal dominant mutations in the *PRKAR1A* gene which encodes the protein kinase A regulatory subunit 1-alpha;; Primary pigmented nodular adrenocortical disease-2 is caused by mutations in the *PDE11A* gene which encodes the phosphodiesterase 11A. This enzyme catalyzes the hydrolysis of cAMP and cGMP to the corresponding 5'-monophosphates;; and Primary pigmented nodular adrenocortical disease-3 is caused by mutations in the *PDE8B* gene which encodes the phosphodiesterase 8B.

Aromatase deficiency: A disorder characterized by inappropriate virilization in XX fetuses. At birth, the female internal genitalia are normal, but the external genitalia could be ambiguous, such as clitoromegaly due to high androgen levels. At adolescence, affected individuals manifest primary amenorrhea and tall stature. In adulthood, affected patients suffer from osteoporosis. Aromatase deficiency is caused by autosomal recessive mutations in the *CYP19A1* gene which encodes the cytochrome P450 (family 19 subfamily A1). Aromatase deficiency is estimated to affect less than 1 in every 1,000,000 individuals.

Congenital adrenal hyperplasia: A disorder characterized by deficiency in any one of the enzymes of cortisol biosynthesis. This condition results from mutations of steroidogenesis genes, such as enzymes mediating the biochemical steps of production of cortisol from cholesterol by the adrenal glands. There are 4 recognized clinical forms of this condition: salt-wasting which is uncommon and life threatening; simple virilizing which is rare; nonclassic late-onset which is common, and is more a trait than a genomic disease; and cryptic form. The adrenal gland produces mineralocorticoid and androgen hormones, which depending on the condition could be decreased or increased. The decrease in mineralocorticoids is life threatening due to salt-wasting and the increase in androgens is virilizing. When Congenital adrenal hyperplasia is caused by 21-hydroxylase deficiency, which accounts for 90-95% of diagnosed cases, there are virilizations in affected females, who are born with ambiguous genitalia. Since the genomic material is located in a shared locus, there is a demonstrated close linkage between 21-hydroxylase deficiency and certain HLA haplotypes. Congenital adrenal hyperplasia is caused by autosomal recessive mutations in the *CYP21A2* gene which encodes the cytochrome P450 (subfamily 21 A polypeptide 2), a protein that localizes to the endoplasmic reticulum and hydroxylates steroids at the 21 position. Worldwide, the prevalence of congenital adrenal hyperplasia is around 1 in 15,000 newborns. In the canton of Zurich, Switzerland, the carrier frequency for Congenital adrenal hyperplasia trait is estimated to be 1 in 35 individuals, which means the frequency of this disorder is about 1 in 4,900 individuals or $1/(70)^2$. There is also Congenital adrenal hyperplasia caused by autosomal recessive mutations in the *HSD3B2* gene which encodes the hydroxy-delta-5-steroid dehydrogenase (3 beta- and steroid delta-isomerase 2), the *POR* gene which encodes the P450 (cytochrome) oxidoreductase, and the *CYP11B1* gene which encodes the cytochrome P450 (family 11 subfamily B1).

Cortisol resistance: A disorder characterized by mineralocorticoid excess resulting in hypertension, hypokalemia, metabolic alkalosis, and hypocortisolemia resistant to treatment. This condition is due to an affinity defect of the glucocorticoid receptor. Affected individuals exhibit a failure to regenerate the active glucocorticoid cortisol from cortisone. There are several underlying causes of this disorder with at least 2 autosomal genomic loci: Cortisone reductase deficiency-1 is caused by autosomal recessive mutations in the *H6PD* gene which encodes the hexose-6-phosphate dehydrogenase;; and Cortisone

reductase deficiency-2 is caused by mutations in the *HSD11B1* gene which encodes the enzyme 11-beta-hydroxysteroid dehydrogenase. There is also a digenic form of this condition.

Familial Hyperaldosteronism or Conn's syndrome: A condition characterized by excess in aldosterone, which can result in hypertension, hypokalemia, and metabolic alkalosis. There are several underlying causes of this disorder with at least 3 autosomal genomic loci: Familial hyperaldosteronism type I or glucocorticoid-remediable aldosteronism, is caused by autosomal dominant mutation that result in an anti-Lepore-type fusion of the *CYP11B2* gene which encodes the cytochrome P450 (family 11 subfamily B2) and the *CYP11B1* gene which encodes the cytochrome P450 (family 11 subfamily B1);; Familial hyperaldosteronism type II has been mapped, but the gene has not been characterized;; and Familial hyperaldosteronism type III is caused by monoallelic mutations in the *KCNJ5* gene which encodes the potassium inwardly-rectifying channel (subfamily J member 5).

Corticosteroid-binding globulin or CBG deficiency: A disorder characterized by chronic asthenia, depression, hypocortisolemia and hypotension. Corticosteroid-binding globulin deficiency can be caused by autosomal recessive or autosomal dominant mutations in the *SERPINA6* gene which encodes the serpin peptidase inhibitor (clade A member 6), an alpha-globulin protein with corticosteroid-binding properties.

Triple A syndrome or Allgrove syndrome: A disorder characterized by the triad of adrenocorticotropic hormone (ACTH) resistant adrenal insufficiency, achalasia, and alacrima. Triple A syndrome can be caused by autosomal recessive mutations in the *AAAS* gene which encodes the protein achalasia, adrenocortical insufficiency, alacrimia. The encoded protein is part of the nuclear pore complex and may be involved in normal development of the peripheral and central nervous system. In North Africans this pathogenic allele appears 1,000-1,175 years ago. Triple A syndrome is estimated to affect less than 1 in every 1,000,000 individuals.

X-linked congenital adrenal hypoplasia: A disorder characterized by defects in many endocrine tissues, especially the adrenal glands; it can also affect the testis resulting in hypogonadotropic hypogonadism. This genomic disorder mainly affects males. X-linked congenital adrenal hypoplasia is caused by X-linked recessive mutations in the *NR0B1* gene which encodes the nuclear receptor (subfamily 0 group B member 1). There is a contiguous gene syndrome involving a combination of congenital adrenal hypoplasia, glycerol kinase deficiency (infantile form) and Duchenne muscular dystrophy, and it is caused by deletion of multiple genes on chromosome Xp21.

Acute adrenal insufficiency in infancy: A disorder characterized by grossly elevated levels of ACTH and plasma renin activity with inappropriately low or absent adrenal steroids. Male infants exhibit female external genitalia, sometimes with clitoromegaly. Other features of these patients include: prematurity, complete underandrogenization, and severe early-onset adrenal failure. Acute adrenal insufficiency in infancy is caused by autosomal recessive mutations in the *CYP11A1* gene which encodes the cytochrome P450 cholesterol side-chain cleavage enzyme.

17-beta-hydroxysteroid dehydrogenase X deficiency: A disorder characterized by metabolic decompensation and lactic acidosis in infants. Affected individuals display a loss of mental and motor skills, with profound developmental regression, choreoathetosis, near blindness, and epilepsy. Patients showed a marked excretion of tiglyglycine and 2-methyl-3-hydroxybutyrate. Sometimes these patients responded to isoleucine restriction. 17-beta-hydroxysteroid dehydrogenase X deficiency is caused by X-linked mutations in the *HSD17B10* gene which encodes the 17-beta-hydroxysteroid dehydrogenase 10.

Familial glucocorticoid deficiency: An adrenocortical failure characterized by hypocortisolemia unresponsive to ACTH. There is not alteration in the renin-angiotensin-aldosterone pathway. There are several underlying causes of this disorder with at least 4 autosomal genomic loci: Familial glucocorticoid deficiency-1 is caused by autosomal recessive mutations in the *MC2R* gene which encodes the melanocortin-2 receptor;; Familial glucocorticoid deficiency-2 is caused by autosomal recessive mutations in the *MRAP* gene which encodes the melanocortin-2 receptor accessory protein;; Familial glucocorticoid deficiency-3 has been mapped;; and Familial glucocorticoid deficiency-4 is caused by autosomal recessive mutations in the *NNT* gene which encodes the nicotinamide nucleotide transhydrogenase. Familial glucocorticoid deficiency is estimated to affect less than 1 in every 1,000,000 individuals.

Familial hyperreninemic hypoaldosteronism: A disorder appearing in early infancy characterized by dehydration, failure to thrive, poor feeding, vomiting and intermittent fever. Affected individuals also showed hyponatremia and hyperkalemia, consistent with salt-wasting. Familial hyperreninemic hypoaldosteronism can be caused by autosomal recessive mutations in the *CYP11B2* gene which encodes the penultimate biochemical step of aldosterone biosynthesis, the changing of 18-hydroxylation of corticosterone into 18-hydroxycorticosterone. This disease is prevalent in consanguineous families.

Table 3.04.05.1 Recognizable Adrenal Disorders

Disorder	OMIM	Gene	Biological Function	Product of Conception Alteration
Primary pigmented nodular adrenocortical disease-1	610489	PRKAR1A	Protein kinase, cAMP-dependent	AD
Primary pigmented nodular adrenocortical disease-2	610475	PDE11A	Phosphodiesterase	AD
Primary pigmented nodular adrenocortical disease-3	614190	PDE8B	Phosphodiesterase	AD
Aromatase deficiency	613546	CYP19A1	Endoplasmic reticulum protein	AR
Congenital Adrenal hyperplasia, due to 21-hydroxylase deficiency	201910	CYP21A2	Endoplasmic reticulum protein	AR + KDN
Congenital adrenal hyperplasia, due to 11-beta-hydroxylase deficiency	202010	CYP11B1	Mitochondrial protein	AR
Congenital adrenal hyperplasia, due to 17-alpha-hydroxylase deficiency	202110	CYP17A1	Endoplasmic reticulum protein	AR
Congenital adrenal hyperplasia, due to 3-beta-hydroxysteroid dehydrogenase deficiency	201810	HSD3B2	Endoplasmic reticulum protein	AR
Disordered steroidogenesis, due to cytochrome P450 oxidoreductase deficiency	613571	POR	Endoplasmic reticulum protein	AR
Lipoid adrenal hyperplasia	201710	STAR	Mitochondrial protein	AR
Lipoid congenital adrenal hyperplasia	201710	CYP11A	Mitochondrial protein	AR
Cortisol resistance	138040	NR3C1	Nuclear receptor	AD + de novo
Cortisone reductase deficiency	604931	H6PD	Dehydrogenase	AR
Cortisone reductase deficiency	604931	HSD11B1/ H6PD	Endoplasmic reticulum protein	Dig
Familial hyperaldosteronism I or Conn syndrome	103900	CYP11B1	Mitochondrial protein	AD
Familial hyperaldosteronism I or Conn syndrome	103900	CYP11B2	Mitochondrial protein	AD
Familial hyperaldosteronism III or Conn syndrome		KCNJ5	Channelopathy Potassium	AD

Table 3.04.05.1. Continued.

Disorder	OMIM	Gene	Biological Function	Product of Conception Alteration
Corticosteroid-binding globulin deficiency	122500	*SERPINA6*	Serpin proteinase inhibitor	AR or AD
Achalasia-Addisonianism-Alacrimia syndrome or Allgrove syndrome	231550	*AAAS*	Nuclear pore complex	AR
Congenital adrenal hypoplasia with hypogonadotropic hypogonadism	300200	*NROB1*	Transcription factor Zinc finger	XR + de novo
Acute adrenal insufficiency	613743	*CYP11A1*	Cholesterol side-chain cleavage enzyme	AR
Familial Glucocorticoid deficiency-1	202200	*MC2R*	Cell surface receptor G protein	AR
Familial Glucocorticoid deficiency-2	607398	*MRAP*	Cell surface receptor G protein	AR
Familial Glucocorticoid deficiency-4	614736	*NNT*	Transhydrogenase	AR
Familial hyperreninemic hypoaldosteronism-1	203400	*CYP11B2*	Mitochondrial protein	AR
Familial hyperreninemic hypoaldosteronism-2	610600	*CYP11B2*	Mitochondrial protein	AR

3.04.06 Gonads / Estrogen, Androgens, Testosterone, etc.

This subchapter refers to the endocrine gland, not the gamete production. It describes 3 rare syndromes: Progesterone resistance, Aromatase excess syndrome, and Familial male-limited sexual precocity in endocrinal gonads. There is also a common condition, such as the premature ovarian failure, for which the main causes seem to be a combination of genomic and environmental factors. Most of the conditions in this subchapter are sex specific. There are conditions for which the product of conception alteration is caused by a mechanism of X dominant, autosomal recessive, etc. Also there is one example of *de novo* DNA alteration, such as Male precocious puberty. Table 3.04.06.1 dis-

plays the diseases, the genes, the biological function, and the product of conception alteration. The biological function includes receptor G protein, transcription factor, enzyme-linked receptor, etc.

Progesterone resistance: A female disorder characterized by infertility with a standard menstrual cycle, and concentrations of progesterone with histologically immature endometrium. Progesterone resistance is associated with a decrease in the *PGR* gene which encodes the progesterone receptor resulting in low concentration of endometrial progesterone receptors. Progesterone resistance syndrome may be due to mutations in the progesterone receptor gene.

Premature ovarian failure: A female disorder characterized by the loss-of-function of the ovaries before age 40. Affected individuals suffer from a frequently cited triad for the diagnosis, which is amenorrhea, hypoestrogenism, and hypergonadotropism. The main cause of this condition is produced by several non genomic factors, such as smoking, autoimmune diseases, genital tract infections, etc. There are several underlying causes of this disorder with at least 3 X-linked genomic loci and 5 autosomal genomic loci: Premature ovarian failure-1 is caused by X-linked premutations in the *FMR1* gene which encodes the fragile X mental retardation protein 1;; Premature ovarian failure-2A is caused by X-linked mutations in the *DIAPH2* gene which encodes the diaphanous homolog protein 2;; Premature ovarian failure-2B is caused by X-linked mutations in the *POF1B* gene which encodes the premature ovarian failure protein 1B;; Premature ovarian failure-3 is caused by mutations in the *FOXL2* gene which encodes the forkhead box L2;; Premature ovarian failure-4 is caused by mutations in the *BMP15* gene which encodes the bone morphogenetic protein 15;; Premature ovarian failure-5 is caused by monoallelic mutations in the *NOBOX* gene which encodes the NOBOX oogenesis homeobox;; Premature ovarian failure-6 is caused by mutations in the *FIGLA* gene which en-

codes the folliculogenesis specific basic helix-loop-helix;; and Premature ovarian failure-7 is caused by mutations in the *NR5A1* gene which encodes the nuclear receptor (subfamily 5 group A member 1). Premature ovarian failure is a common condition affecting 1% of the females.

Aromatase excess syndrome: A disorder characterized by increased extraglandular aromatization of steroids in which is present in both males and females with precocious puberty. In early childhood, males and females suffer from breast enlargement, growth, and bone age acceleration. Aromatase excess syndrome can be caused by autosomal dominant gain-of-function mutations in the *CYP19A1* gene which encodes the aromatase or cytochrome P450 (family 19 subfamily A1).

Male-limited precocious puberty: A disorder characterized by signs of puberty by 4 years of age. Male-limited precocious puberty can be caused by monoallelic either autosomal dominant or *de novo* gain-of-function mutations in the *LHCGR* gene which encodes the luteinizing hormone/choriogonadotropin receptor. Mutations in this gene cause constitutive receptor activation.

The Concise Encyclopedia of Genomic Diseases

Table 3.04.06.1 Recognizable Gonads Disorders

Disorder	OMIM	Gene	Biological Function	Product of Conception Alteration
Progesterone resistance	264080	PGR	Transcription factor Zinc finger	AR
Premature ovarian failure-1	311360	FMR1	mRNA trafficking from the nucleus to the cytoplasm	XD + Ant
Premature ovarian failure-2A	300511	DIAPH2	Development and normal function of the ovaries	XR
Premature ovarian failure-2B	300604	POF1B	Binds non-muscle actin filaments	XR
Premature ovarian failure-3	608996	FOXL2	Transcription factor Helix-turn-helix domains	AR
Premature ovarian failure-4	300510	BMP15	Cell surface receptor Enzyme-linked receptor	XR
Premature ovarian failure-5	611548	NOBOX	Transcription factor oogenesis	AD
Premature ovarian failure-6	612310	FIGLA	Transcription factor	AR
Premature ovarian failure-7	612964	NR5A1	Transcription factor β- Scaffold factors	AR
Aromatase excess syndrome	139300	CYP19A1	Endoplasmic reticulum protein	AD
Male precocious puberty	176410	LHCGR	Cell surface receptor G protein	AD + de novo

3.04.07 Other Endocrine Diseases

This subchapter only covers diseases that are related to endocrine dysfunction and are not classified anywhere else. There are several rare syndromes included in this subchapter. There are conditions for which the product of conception alteration is caused by a mechanism of dominant and *de novo*, such as Progeria. There are also recessive and X-linked inheritances. Table 3.04.07.1 emphasizes that Laron dwarfism, could be inherited in a severe form when both parents are heterozygous and both parents provide the faulty copy to descendants. The table displays the diseases, the genes, the biological function, and the product of conception alteration. The biological function includes Serine/threonine protein kinase, Insulin-like growth factor 1, transcription factor, enzyme-linked receptor, etc. This section clearly shows that sometimes it is more useful to describe the diseases related to a single gene; for example for growth hormone receptor or androgen receptor related conditions. But in some other occasions it is more useful to obtain the clinical information such as "short stature" with or without additional clinical features.

3.04.07.1 Polyglandular Dysfunction

Autoimmune polyendocrine syndromes: A heterogeneous group of disorders characterized by autoimmune activity against primarily endocrine organs, although non-endocrine organs can also be affected. There are 3 autoimmune polyendocrine syndromes; the clinical and genomic features shall be described below.

Autoimmune polyendocrine syndrome type 1 or APECED Syndrome (Auto-immune PolyEndocrinopathy, Candidiasis, and Ectodermal Dystrophy): A disorder characterized by onset of autoimmunity in childhood or early adulthood. Type 1 displays the combination of 2 of 3 major clinical symptoms, including: Addison disease, hypoparathyroidism and/or chronic mucocutaneous candidiasis. Autoimmune polyendocrinopathy syndrome type I is caused by mutations in the *AIRE* gene which encodes the autoimmune regulator protein, a transcriptional regulator.

Autoimmune polyendocrine syndrome type 2 or Schmidt syndrome: A disorder characterized by onset of autoimmunity in adulthood. Type 2 usually exhibits the combination of Addison disease and Hashimoto thyroiditis; less often it is the clinical combination of Addison disease, Graves disease, and type 1 diabetes mellitus. Autoimmune polyendocrine syndrome type 2 is more prevalent in middle-aged women. Autoimmune polyendocrine syndrome type 2 is associated with HLA-DR3 and/or HLA-DR4 haplotypes.

IPEX syndrome (Immune dysfunction, Polyendocrinopathy, and Enteropathy X-linked): A disorder characterized by onset in infancy of severe diarrhea due to enteropathy, type 1 diabetes mellitus, and dermatitis. IPEX syndrome is the most se-rious, but rarest form between the 3 auto-immune polyendocrine syndromes. IPEX syndrome is caused by X-linked recessive mutations in the *FOXP3* gene which encodes the forkhead/winged-helix family of transcriptional regulators. Most affected boys die due to the autoimmune activity which attacks many of their organs.

3.04.07.2 Other Endocrine Disorders

Endocrine-cerebro-osteodysplasia syndrome: A disorder characterized by various anomalies of the endocrine, cerebral and skeletal systems resulting in neonatal mortality. Endocrine-cerebro-osteodysplasia syndrome is caused by autosomal recessive mutations in the *ICK* gene which encodes the intestinal cell kinase, a serine/threonine protein kinase.

ANE syndrome (Alopecia, Neurologic defects, and Endocrinopathy): A disorder characterized by moderate to severe mental retardation and progressive motor deterioration during the second decade of life. Affected individuals also show central hypogonadotropic hypogonadism. Alopecia, neurologic defects, and endocrinopathy is caused by autosomal recessive mutations in the *RBM28* gene which encodes the RNA binding motif protein 28. ANE syndrome is estimated to affect less than 1 in every 1,000,000 individuals. This disease has been reported in consanguineous Arabs families.

3.04.07.3 Short Stature (not elsewhere classified)

Short stature: A disorder usually defined as a height below the third percentile for chronological age or minus 2 standard deviations (SD) from national height standards, in the absence of specific causative disorders. Short stature is associated with variations in the pseudoautosomal genes *SHOX* or *SHOXY*, which are X-linked and Y-linked respectively. Monoallelic mutations in the *SHOX* gene cause Léri-Weill dyschondrosteosis, and biallelic mutations cause Langer mesomelic dysplasia. In addition, haploinsufficiency of the *SHOX* gene is associated with short stature such as in Turner syndrome. Short stature can be caused by partial insensitivity to growth hormone due to mutations in either the *GHR* gene which encodes the growth hormone receptor, or the *GSHR* gene which encodes the growth hormone secretagogue.

Resistance to insulin-like growth factor I: A disorder characterized by variable intrauterine and postnatal growth retardation and elevated serum IGF-I levels. Resistance to insulin-like growth factor I in some cases may be caused by mutations in the *IGF1R* gene which encodes the insulin-like growth factor receptor.

Laron syndrome: A congenital disorder characterized by a severe form of growth hormone insensitivity; it is marked by short stature, growth impairment, dysfunctional growth hormone receptor, and failure to generate insulin-like growth factor I in response to growth hormone. Laron syndrome is caused by autosomal recessive mutations in the *GHR* gene which encodes the growth hormone receptor. Laron syndrome is caused when the product of conception has the two defective alleles and follows the al-teration in the form of AD + AD = AR and both parents provide the faulty copy to de-scendants. Males and females are equally affected. The disease has been reported in more than 250 cases and is more frequent among Semitic and Mediterranean individuals.

Laron syndrome with immunodeficiency: A disorder characterized by severe growth retardation associated with immunodeficiency. Affected individuals exhibit moderate lymphopenia which leads to recurrent infections. Laron syndrome with immunodeficiency is caused by mutations in the *STAT5B* gene which encodes the signal transducer and activator of transcription 5B.

3.04.07.4 Androgen Resistance Syndrome

Androgen insensitivity syndrome: A condition characterized by affecting SRY positive individuals (the *SRY* gene is mapped on the subtelomeric short arm on Y chromosome). Androgen insensitivity syndrome is the principal condition that leads to 46, XY undermasculinization. Affected individuals show a failure of the cell to respond to androgens, this effect prevents the masculinization of male genitalia in the developing fetus. In fact the external genitalia is that of a normal female. Androgen insensitivity syndrome can be divided according to the degree of genital masculinization: in complete androgen insensitivity syndrome, the external genitalia is that of a normal female; in partial androgen insensitivity syndrome, the external genitalia is partially, but not fully masculinized; and in mild androgen insensitivity syndrome, the external genitalia is that of a normal male. Affected males display female external genitalia, female breast development, absent uterus and female adnexa, blind vagina, and inguinal or abdominal testes. Androgen insensitivity is caused by X-linked recessive mutations in the *AR* gene which encodes the androgen receptor. Complete androgen insensitivity syndrome is estimated to affect about 1 in 20,000-64,000 male births.

3.04.07.5 Progeria (not elsewhere classified)

Hutchinson-Gilford progeria syndrome: A disorder characterized by premature aging of postnatal onset. Affected individuals exhibit average intelligence and a failure to thrive. Early death is caused by atherosclerosis or cerebrovascular disease. Hutchinson-Gilford progeria syndrome is caused by monoallelic *de novo* recurrent truncating mutations in the *LMNA* gene which encodes the lamin A/C, a protein involved in nuclear stability, chromatin structure and gene expression. Hutchinson-Gilford progeria syndrome is estimated to affect 1 in 8 million births. Apparently, patients with this condition do not survive long enough to reproduce.

Werner syndrome or adult progeria: A disorder which resembles accelerated aging. Affected individuals exhibit scleroderma-like skin changes, especially in the extremities, cataract, subcutaneous calcification, premature arteriosclerosis, diabetes mellitus, and wizened and prematurely aged facies. Werner syndrome is caused by autosomal recessive mutations in the *WRN* gene which encodes the Werner syndrome protein (RecQ helicase-like). This protein is involved in the repair of double strand DNA breaks. In Japan, Werner syndrome affects about 1 in 300,000 individuals.

Nestor-Guillermo progeria syndrome: A disorder characterized by normal development until 2 years of age. Then the patient develops a failure to thrive, dry skin, generalized lipoatrophy, severe osteoporosis and marked osteolysis. The condition was named Nestor-Guillermo' since this was the name of the first two patients in which this progeria syndrome was described. Nestor-Guillermo progeria syndrome is caused by autosomal recessive mutations in the *BANF1* gene which encodes the barrier to autointegration factor 1, a protein that is thought to facilitate nuclear reassembly. This condition has been reported in children of consanguineous Spanish parents.

Table 3.04.07.1 Other Recognizable Endocrine Disorders

Disorder	OMIM	Gene	Biological Function	Product of Conception Alteration
Autoimmune polyendocrinopathy type 1 or autoimmune polyendocrinopathy-candidiasis-ectodermal dystrophy or Whitaker syndrome	240300	*AIRE*	Transcription factor Zinc finger	AR
Immunodysregulation, polyendocrinopathy, and enteropathy	304790	*FOXP3*	Transcription factor Helix-turn-helix domains	XR
Endocrine-cerebroosteodysplasia	612651	*ICK*	Serine/threonine protein kinase	AR
ANE syndrome or Alopecia, neurologic defects and endocrinopathy	612079	*RBM28*	Nuclear hnRNP spliceosomes	AR
Idiopathic familial short stature	300582	*SHOX*	Nuclear protein	X
Idiopathic familial short stature	300582	*SHOXY*	Nuclear protein	Y

The Concise Encyclopedia of Genomic Diseases

Table 3.04.07.1. Continued.

Disorder	OMIM	Gene	Biological Function	Product of Conception Alteration
Short stature	604271	GHSR	Cell surface receptor G protein	AR or AD
Short stature	604271	GHR	Cell surface receptor JAK-STAT	AR or AD
Growth delay due to IGF-I resistance	270450	IGF1R	Cell surface receptor Enzyme-linked receptor	AD
Growth retardation with Deafness and Mental retardation	608747	IGF1	Growth factor	AR
Laron syndrome	262500	GHR	Cell surface receptor JAK-STAT	AD + AD = AR
Laron syndrome with immunodeficiency	245590	STAT5B	Transcription factor β-Scaffold factors	AR
Complete androgen insensitivity syndrome	300068	AR	Transcription factor Zinc finger	XR
Partial androgen insensitivity with or without breast cancer	312300	AR	Transcription factor Zinc finger	XR
Hutchinson-Gilford progeria	176670	LMNA	Nuclear Lamina Protein	AD + 90%
Werner syndrome	277700	WRN	DNA repair helicase	AR
Nestor-Guillermo progeria syndrome	614008	BANF1	Nuclear protein	AR

Table 3.04.07.2 Other Endocrine Disorders in Different Ethnic Groups

Disease	Frequency of condition	People with higher prevalence (if available)
Complete androgen insensitivity syndrome	1 in 20,000-64,000	Worldwide
Werner syndrome	<1 in 1,000,000	Japanese
ANE syndrome or Alopecia, neurologic defects and endocrinopathy	<1 in 1,000,000	Unknown
Hutchinson-Gilford progeria	1 in 8,000,000	Unknown

3.04.08 Nutritional Diseases

3.04.08.1 Nutritional Deficiencies

Autosomal dominant hypercarotenemia and vitamin A deficiency: A disorder characterized by intermittent yellow or orange discoloration of the skin, particularly on the palms, soles and face. Hypercarotenemia and vitamin A deficiency is caused by autosomal dominant mutations in the *BCMO1* gene which encodes the beta-carotene 15,15'-monooxygenase, a key enzyme in beta-carotene metabolism to vitamin A.

Biotinidase deficiency: A disorder characterized by behavioral features including: lack of coordination, learning disabilities and seizure. Some patients also suffer from hypotonia, alopecia, eczematoid rash, hearing loss, acidosis, fungal infections, etc... Biotin is vitamin B7 that is chemically bound to proteins. Biotinidase deficiency is associated with low biotin levels, and impairment in metabolism of fats, carbohydrates and proteins, particularly, essential branched-chain amino acids, such as leucine and isoleucine. Biotinidase deficiency is caused by autosomal recessive mutations in the *BTD* gene which encodes the biotinidase. This enzyme functions to recycle biotin. On average, the carrier frequency for biotinidase deficiency is estimated to be 1 in 120 individuals worldwide, which means the frequency of this condition is about 1 in 57,600 or $1/(240)^2$ individuals.

Holocarboxylase synthetase deficiency: A disorder characterized by similar symptoms as those of biotinidase deficiency. In holocarboxylase synthetase deficiency the individual is incapable of successfully using the vitamin biotin. Holocarboxylase synthetase deficiency is caused by auto-somal recessive mutations in the *HLCS* gene which encodes the holocarboxylase synthetase. This enzyme catalyzes the binding of biotin to carboxylases and histones. The exact incidence of this condition is unknown; the carrier frequency for holocarboxylase synthetase deficiency trait is estimated to be 1 in 147 individuals, which means the frequency of this condition is about 1 in 87,000 individuals or $1/(294)^2$.

Isolated ataxia with vitamin E deficiency: A disorder characterized by similar symptoms than those of Friedreich ataxia. Isolated ataxia with vitamin E deficiency is caused by autosomal recessive mutations in the *TTPA* gene which encodes the alpha-tocopherol transfer protein. This protein is involved in transporting vitamin E between membrane vesicles and facilitating the secretion of vitamin E from hepatocytes to circulating lipoproteins.

Trypsinogen deficiency: A disorder characterized by nutritional edema, failure to thrive and hypoproteinemia. Trypsinogen deficiency is caused by autosomal recessive mutations in the *PRSS1* gene which encodes the trypsinogen. This enzyme is secreted by the pancreas and cleaved to its active form in the small intestine. There is also a similar clinical condition caused by enterokinase deficiency.

3.04.08.2 Obesity and other Hyperalimentation Syndromes

Obesity: A disorder characterized by excess in body fat which could be detrimental to health, primarily reducing life expectancy. Obesity increases the probability of several diseases: mainly type 2 diabetes, heart disease, obstructive sleep apnea, cancer, etc. There are many inclusion criteria in obesity for it to be considered of genomic origin. There are syndromes associated with obesity which will be described in congenital malformations, deformations and chromosomal abnormalities. Several genomic syndromes cause obesity including Prader-Willi syndrome, Bardet-Biedl syndrome, Cohen syndrome, etc. Obesity is an example of polygenomic disease. There are also cases of obesity which typically follows Mendelian patterns of inheritance. Morbid obesity is caused by autosomal recessive mutations in the *LEP* gene which encodes the leptin. This protein is secreted by white adipocytes, and plays a major role in the regulation of body weight; Morbid obesity is caused by autosomal recessive mutations in the *LEPR* gene which encodes the leptin receptor;; Obesity with impaired prohormone processing is caused by autosomal recessive mutations in the *PCSK1* gene which encodes the proprotein convertase;; and "Obesity, adrenal insufficiency, and red hair" is caused by autosomal recessive mutations in the *POMC* gene which encodes the proopiomelanocortin. There is also abdominal obesity-metabolic syndrome mapped on 3q27. When patients with WAGR syndrome (Wilms tumor, Aniridia, Genitourinary anomalies, and mental Retardation syndrome) develop obesity it is referred to as WAGRO. The last condition is associated with haploinsufficiency on chromosome 11p14-p12 for the *BDNF* gene which encodes the brain-derived neurotrophic factor. The most common form of monogenic obesity is caused by mutations in *MC4R* gene which encodes the melanocortin 4 receptor. These *MC4R* gene mutations follow an autosomal dominant pattern of inheritance, and the most severe conditions occur when both parents are heterozygous for this gene alteration and both parents provide the faulty copy to descendants (AD + AD = AR). There are several genes that are associated with susceptibility to obesity including *ADIPOQ, AGRP, NTRK2, NR0B2, PYY, MC4R, PCSK1, POMC, APOE, SIM1, FTO, PPARG, UCP3, SDC3, POMC, ADRB2, ADRB3, UCP1, ENPP1, CARTPT, GHRL* and *PPARGC1B* genes. The counterpart in obesity is leanness; there are several SNPs in *PPARGC1B* and *AHSG* genes which are associated with a decrease in the frequency of obesity. There are large differences in the frequency of obesity among different ethnic groups. The contribution attributed to genomics in obesity ranges from 6-85%, depending on the group of people studied. Another clear example of genomics is the fact that less than 10% of the offspring of two parents who were of normal weight were obese; on the other hand 80% of the offspring of two obese parents were obese. There are few cases of obesity which are primarily due to genomics; still the main sources of obesity are caused by a mixture of physical inactivity and excessive food energy intake. The fact that obesity was rare before the 20th century is one of the stronger exclusion criteria for being of genomic origin. Nowadays, obesity rates are increasing worldwide and affecting both the developed and the developing world, particularly in urban settings.

Alström syndrome: A disorder characterized by multiorgan dysfunction with similar symptoms as Bardet-Biedl syndrome. Affected individuals display childhood obesity, sensorineural hearing loss, and blindness due to congenital retinal dystrophy. These patients also exhibit dilated cardiomyopathy renal failure, hepatic, pulmonary and urologic dysfunction. There are also manifestations of endocrinologic features such as hyperinsulinemia, early-onset type 2 diabetes and hypertriglyceridemia. In other words, these patients display a metabolic syndrome. This condition was named after Carl-Henry Alström in 1959. Alström syndrome is caused by autosomal recessive mutations in the *ALMS1* gene which encodes the Alstrom syndrome 1 protein. The encoded protein has been shown to function in ciliogenesis in inner medullary collecting duct cells.

Table 3.04.08.1 Other Recognizable Nutritional Disorders

Disorder	OMIM	Gene	Biological Function	Product of Conception Alteration
Hypercarotenemia and vitamin A deficiency	115300	*BCMO1*	Carotenoid oxygenases	AD
Biotinidase deficiency	253260	*BTD*	Biotinidase	AR
Holocarboxylase synthetase deficiency	253270	*HLCS*	Binding of biotin to carboxylases and histones	AR
Friedreich-like ataxia with isolated vitamin E deficiency	277460	*TTPA*	Transporting vitamin E between membrane vesicles	AR
Trypsinogen deficiency	614044	*PRSS1*	Trypsinogen	AR
Obesity, due to melanocortin-4 receptor deficiency	601665	*MC4R*	Cell surface receptor G protein	AD or AD + AD = AR
Morbid obesity, due to leptin deficiency	614962	*LEP*	Leptin	AR
Morbid obesity, due to leptin receptor deficiency	614963	*LEPR*	Leptin receptor	AR
Obesity with impaired prohormone processing	600955	*PCSK1*	Proprotein convertase	AR
Obesity, adrenal insufficiency, and red hair, due to POMC deficiency	609734	*POMC*	Proopiomelanocortin	AR
Alström syndrome	203800	*ALMS1*	Ciliary proteins Basal body	AR

3.04.09-20 Metabolic Diseases

3.04.09-13 Metabolic Disorders of Proteins, Fats, and Carbohydrates

3.04.09 Amino-acids Disorders

3.04.09.1 Disorders of Aromatic Amino-acid Metabolism

Phenylketonuria: A disorder characterized by accumulation of phenylalanine and phenylpyruvate in blood and urine. Postnatal hyperphenylalaninemia could result in impairment of the cognitive development due to a neurotoxic effect. Most patients with phenylketonuria display a reduced activity in phenylalanine hydroxylase, which is essential to metabolize the amino acid phenylalanine to the amino acid tyrosine. Phenylketonuria is caused by biallelic or autosomal recessive mutations in the *PAH* gene which encodes the phenylalanine hydroxylase. This enzyme is the rate-limiting step in phenylalanine catabolism. The product of conception alterations are defined as occasional *de novo* autosomal recessive syndrome, or autosomal recessive syndrome plus known *de novo* alterations (AR + KDN). Phenylketonuria is a rare disease with a 40 fold variation across different nationalities. In the United States, the carrier frequency for phenylketonuria trait is estimated to be 1 in 61 individuals, which means the frequency of this condition is about 1 in 15,000 individuals or $1/(122)^2$. There is a higher carrier frequency for phenylketonuria trait among Turkish people. The carrier frequency for phenylketonuria trait is estimated to be 1 in 25.5 individuals, which means the frequency of this condition is about 1 in 2,600 individuals or $1/(51)^2$.

On the contrary, in Finland the frequency is less than 1 in 100,000 individuals. There is also a moderately high carrier frequency for phenylketonuria trait among people from China, Italy, Ireland, and Yemen.

Tetrahydrobiopterin-deficient hyperphenylalaninemia: A disorder characterized by progressive cognitive and motor deficits. Affected individuals exhibit depletion of the neurotransmitters dopamine, serotonin and hyperphenylalaninemia. Tetrahydrobiopterin-deficient hyperphenylalaninemia is caused by deficiency in the synthesis or regeneration of Tetrahydrobiopterin, which is a cofactor for tyrosine hydroxylase, tryptophan hydroxylase, and phenylalanine hydroxylase. The first 2 enzymes are involved in neurotransmitter synthesis. There are several underlying causes of this disorder with at least 4 autosomal genomic loci: Tetrahydrobiopterin-deficient hyperphenylalaninemia type A is caused by autosomal recessive mutations in the *PTS* gene which encodes the 6-pyruvoyl-tetrahydropterin synthase. This enzyme catalyzes the elimination of inorganic triphosphate from dihydroneopterin triphosphate. Type A is the most common form;; Tetrahydrobiopterin-deficient hyperphenylalaninemia type B is caused by autosomal recessive mutations in the *GCH1* gene which encodes the GTP cyclohydrolase 1, a protein that is the first and rate-limiting enzyme in tetrahydrobiopterin biosynthesis;; Tetrahydrobiopterin-deficient hyperphenylalaninemia type C is caused by autosomal recessive mutations in the *QDPR* gene which encodes the quinoid dihydropteridine reductase;; and Tetrahydrobiopterin-deficient hyperphenylalaninemia type D is caused by autosomal recessive mutations in the *PCBD1* gene which encodes the pterin-4 alpha-carbinolamine dehydratase / dimerization cofactor of hepatocyte nuclear factor 1 alpha. There are also 2 conditions with similar neurologic symptoms but without hyperphenylalaninemia. These conditions are dopa-responsive dystonia and autosomal dominant dopa-responsive dystonia, caused by mutations in the *SPR* and *DYT5* genes respectively. Around 1-3% of patients with hyperphenylalaninemia have one of these Tetrahydrobiopterin-deficient forms.

Dopamine-responsive dystonia or hereditary progressive dystonia with diurnal fluctuation, or Segawa dystonia: This disease is discussed under neurological diseases in the subchapter "Dystonia" (3.06.03.2).

Tyrosinemia: A disorder characterized by the ineffective breakdown of the amino acid tyrosine. There are three forms: type I, type II and type III.

Tyrosinemia type I: A disorder characterized by progressive liver diseases, secondary kidney disturbances and hypophosphatemic rickets. Tyrosinemia type I is caused by autosomal recessive mutations in the *FAH* gene which encodes the fumarylacetoacetate hydrolase, the last enzyme in the tyrosine catabolism pathway. This condition is a rare syndrome worldwide, but is common in the Saguenay-Lac-Saint-Jean region (Canada). The carrier frequency for tyrosinemia type I trait is estimated to be 1 in 21 individuals, which means the frequency of this condition is about 1 in 1,764 individuals or $1/(42)^2$.

Tyrosinemia type II: A disorder characterized by hypertyrosinemia, palmoplantar hyperkeratosis, keratitis and mental retardation. Tyrosinemia type II is caused by autosomal recessive mutations in the *TAT* gene which encodes the tyrosine aminotransferase, a mitochondrial protein that catalyzes the conversion of L-tyrosine into p-hydroxyphenylpyruvate.

Tyrosinemia type III: A disorder characterized by mental retardation, and convulsions without liver damage. This is caused by elevated levels of blood tyrosine and enormous excretion of its derivatives into urine. Tyrosinemia type III is caused by autosomal recessive mutations in the *HPD* gene which encodes the 4-hydroxyphenylpyruvate dioxygenase, an enzyme in the catabolic pathway of tyrosine.

Hawkinsinuria or 4-Alpha-hydroxyphenylpyruvate hydroxylase deficiency: A disorder characterized by failure to thrive, fine and sparse hair, transient tyrosinemia, and metabolic acidosis. There is also excretion of the rare cyclic amino acid metabolite, hawkinsin, in the urine. This condition is caused by impairment in the metabolism of the sulfur amino acid hawkinsin. Hawkinsinuria is caused by autosomal dominant mutations in the *HPD* gene. Hawkinsinuria is an allelic disorder of tyrosinemia type III.

Alkaptonuria or black urine disease: A disorder characterized by excessive amounts of homogentisic acid which causes osteoarthritis, precipitate kidney stones and damage heart valves. This condition is due to the large excretion of homogentisic acid or alkapton, which is a toxic tyrosine by product. Alkaptonuria is caused by autosomal recessive mutations in the *HGD* gene which encodes the homogentisate 1,2-dioxygenase. This enzyme is involved in the catabolism of the amino acids tyrosine and phenylalanine. Historically, Alkaptonuria was one of the first conditions in which Mendelian recessive inheritance was proposed. Alkaptonuria is a rare condition, but is more prevalent in the Dominican Republic and Slovakia than in any other countries. Among Slovakian people, the carrier frequency for alkaptonuria trait is estimated to be 1 in 69 individuals, which means the frequency of this condition is about 1 in 19,000 individuals or $1/(138)^2$.

Albinism or Oculocutaneous albinism: A disorder characterized by the absence of melanin pigment in the skin, hair and eyes. There are several underlying causes of this disorder with at least 4 autosomal genomic loci: Oculocutaneous albinism type 1A is caused by autosomal recessive mutations in the *TYR* gene which encodes the tyrosinase. This enzyme catalyzes the first 2 steps in the conversion of tyrosine to melanin. This form is characterized by complete lack of tyrosinase activity;; Oculocutaneous albinism type 1B is caused by mutations in the *TYR* gene. This form is characterized by reduced activity of tyrosinase;; Oculocutaneous albinism type 2 is caused by autosomal recessive mutations in the *OCA2* gene which encodes the small molecule transport, specifically for tyrosine - a precursor of melanin;; Oculocutaneous albinism type 3 is caused by autosomal recessive mutations in the *TYRP1* gene which encodes the tyrosinase-related protein 1;; and Oculocutaneous albinism type 4 is caused by autosomal recessive mutations in the *SLC45A2* gene which encodes the solute carrier (family 45 member A2). In Eastern Asians and Caucasians this pathogenic allele appears 15,000-35,000 years ago. Types 2 to 4 are milder forms of the disorder. Affected individuals are more prone to sunburn and skin cancers. Besides the skin concerns, these patients also have vision defects. There are also other conditions which also courses with albinism, including Chédiak-Higashi syndrome, Griscelli syndrome, Hermansky-Pudlak syndrome, Waardenburg syndrome and Tietz syndrome. The syndrome with partial deficiency of the melanin is referred to as hypomelanism or hypomelanosis. On average, the carrier frequency for oculocutaneous albinism trait is estimated to be 1 in 65 individuals worldwide, which means the frequency of this condition is about 1 in 17,000 individuals or $1/(130)^2$. People of sub-Saharan African descent have the highest prevalence of oculocutaneous albinism.

Ocular albinism: A disorder characterized by primarily albinism in the eyes. Affected individuals exhibit impaired visual acuity, iris hypopigmentation with translucency, nystagmus, albinotic fundus and macular hypoplasia. Contrary to oculocutaneous albinism, patients with ocular albinism have normal pigmented skin and hair. There are several underlying causes of this disorder with at least 2 X-linked genomic loci: Ocular albinism type 1 or Nettleship-Falls syndrome is caused by X-linked mutations in the *GPR143* gene which encodes the G protein-coupled receptor 143, a protein that is thought to be involved in intracellular signal transduction mechanisms. This gene also has another allelic disorder called isolated X-linked congenital nystagmus-6;; and Ocular albinism type 2 or Åland Island eye disease or Forsius-Eriksson syndrome is caused by X-linked mutations in the *CACNA1F* gene which encodes the voltage-dependent calcium channel (L type alpha 1F subunit). Ocular albinism genes are on the X chromosome, which explains why males are most affected with this condition. Ocular albinism type 1, the most common form of this disorder, affects at least 1 in 60,000 males.

Chédiak–Higashi syndrome: A disorder characterized by partial albinism, recurrent pyogenic infections and peripheral neuropathy. This condition results of impairment in microtubule polymerization which leads to a decrease in phagocytosis.

Chédiak–Higashi syndrome is caused by autosomal recessive mutations in the *LYST* gene which encodes the lysosomal trafficking regulator, a protein that regulates intracellular protein trafficking in endosomes, and may be involved in pigmentation.

Griscelli syndrome: This disease is discussed under diseases of the skin and subcutaneous tissue in the subchapter "Other Disorders of Pigmentation" (3.12.7.1).

Hermansky–Pudlak syndrome: A disorder characterized by oculocutaneous albinism, bleeding, and lysosomal accumulation of ceroid lipofuscin. There are several underlying causes of this disorder with at least 9 autosomal genomic loci: Hermansky-Pudlak syndrome-1 is caused by autosomal recessive mutations in the *HPS1* gene which encodes the Hermansky-Pudlak syndrome 1 protein;; Hermansky-Pudlak syndrome-2, which includes immunodeficiency in its clinical features, is caused by autosomal recessive mutations in the *AP3B1* gene which encodes the adaptor-related protein complex 3 subunit beta 1;; Hermansky-Pudlak syndrome-3 is caused by autosomal recessive mutations in the *HPS3* gene which encodes the Hermansky-Pudlak syndrome 3 protein;; Hermansky-Pudlak syndrome-4 is caused by autosomal recessive mutations in the *HPS4* gene which encodes the Hermansky-Pudlak syndrome 4 protein;; Hermansky-Pudlak syndrome-5 is caused by autosomal recessive mutations in the *HPS5* gene which encodes the Hermansky-Pudlak syndrome 5 protein;; Hermansky-Pudlak syndrome-6 is caused by autosomal recessive mutations in the *HPS6* gene which encodes the Hermansky-Pudlak syndrome 6 protein;; Hermansky-Pudlak syndrome-7 is caused by autosomal recessive mutations in the *DTNBP1* gene which encodes the dystrobrevin binding protein 1, a protein that may play a role in organelle biogenesis associated with melanosomes, lysosomes and platelet dense granules;; Hermansky-Pudlak syndrome-8 is caused by autosomal recessive mutations in the *BLOC1S3* gene which encodes the biogenesis of lysosomal organelles complex-1 (subunit 3) , a protein that may play a role in intracellular vesicle trafficking;; and Hermansky-Pudlak syndrome-9 is caused by autosomal recessive mutations in the *BLOC1S6* gene which encodes the biogenesis of lysosomal organelles complex-1 (subunit 6), a protein that may play a role in intracellular vesicle trafficking. Hermansky-Pudlak syndrome is a rare syndrome, except in high, isolated mountain villages in the Swiss Alps and northwestern Puerto Rico. In this area of Puerto Rico, the carrier frequency for Hermansky-Pudlak syndrome type 3 trait is estimated to be 1 in 21 individuals, which means the frequency of this condition is about 1 in 1,800 individuals or $1/(42)^2$.

Waardenburg syndrome or Waardenburg-Shah Syndrome: A disorder characterized by pigmentation anomalies of the hair, including a white forelock and premature graying. Affected individuals also exhibit pigmentary changes of the iris and congenital sensorineural hearing loss. Waardenburg syndrome has been classified into 4 main clinical phenotypes. Type I patients have the additional feature of pigmentary changes of the iris, such as heterochromia iridis and brilliant blue eyes; congenital sensorineural hearing loss; and "dystopia canthorum." Type II patients are distinguished from type I by the absence of dystopia canthorum. Type III patients have dystopia canthorum and are distinguished by the presence of upper limb abnormalities. Type IV patients have the additional feature of Hirschsprung disease. There are several underlying causes of this disorder with at least 10 autosomal genomic loci: Waardenburg syndrome-1 is caused by monoallelic mutations in the *PAX3* gene which encodes the paired box 3 protein. There is evidence of advanced paternal age effect in de novo mutations;; Waardenburg syndrome-2A is caused by monoallelic mutations in the *MITF* gene which encodes the microphthalmia-associated transcription factor;; Waardenburg syndrome-2B has been mapped but the gene has not been characterized;; Waardenburg syndrome-2C has been mapped but the gene has not been characterized;; Waardenburg syndrome-2D is caused by autosomal recessive mutations in the *SNAI2* gene which encodes the snail family zinc finger 2 protein;; Waardenburg syndrome-2E, with or without neurologic involvement is caused by monoallelic mutations in the *SOX10* gene which encodes the SRY (sex determining region Y)-box 10;; Waardenburg syndrome-3 or Klein-Waardenburg syndrome is caused either by monoallelic or biallelic mutations in the *PAX3* gene;; Waardenburg-Shah syndrome or Waardenburg syndrome-4A is caused by autosomal recessive mutations in the *EDNRB* gene which encodes the endothelin receptor type B;; Waardenburg-Shah syndrome or Waardenburg syndrome-4B is caused by autosomal recessive or autosomal dominant mutations in the *EDN3* gene which encodes the endothelin 3;; and Waardenburg-Shah syndrome or Waardenburg syndrome-4C is caused by monoallelic mutations in the *SOX10* gene. There is also a variant which includes a digenic pattern of inheritance including *MITF/TYR* genes.

Tietz syndrome or Tietz albinism-deafness syndrome: A disorder characterized by deafness and leucism, or reduced pigmentation. Tietz syndrome is caused by autosomal dominant mutations in the *MITF* gene which encodes the microphthalmia-associated transcription factor.

Histidinemia: A benign condition characterized by increased levels of histidine in blood, urine and cerebrospinal fluid. Histidinemia is caused by autosomal recessive mutations in the *HAL* gene which encodes the histidine ammonia-lyase.

Brunner syndrome: A disorder characterized by mild mental retardation and impulsive aggressiveness. Brunner syndrome is caused by X-linked recessive mutations in the *MAOA* gene which encodes the monoamine oxidase A. This mutation results in an excess of monoamines in the brain, such as serotonin, dopamine, and epinephrine. Brunner syndrome primarily affects males.

Formiminoglutamic aciduria: A disorder characterized by mild developmental delay and elevated concentrations of formiminoglutamate in the urine. Formiminoglutamic aciduria is caused by autosomal recessive mutations in the *FTCD* gene which encodes the formiminotransferase-cyclodeaminase. This enzyme is involved in folic acid metabolism.

Urocanic aciduria or urocanase deficiency: A disorder characterized by severe mental retardation, ataxia, and periods of aggression and exaggerated affection-seeking. Urocanic aciduria is caused by autosomal recessive mutations in the *UROC1* gene which encodes the urocanate hydratase 1. This enzyme is involved in histidine catabolism.

Hypertryptophanemia: A disorder characterized by generalized joint laxity and pains, emotional lability including aggressive outbursts, hypertelorism, strabismus and myopia. Hypertryptophanemia is an autosomal recessive disorder but the gene/loci have not been characterized.

The Concise Encyclopedia of Genomic Diseases

Table 3.04.09.1.1 Recognizable Disorders of Aromatic Amino-acid Metabolism

Disorder	OMIM	Gene	Biological Function	Product of Conception Alteration
Phenylketonuria	261600	PAH	Phenylalanine hydroxylase	AR + KDN
Tetrahydrobiopterin-deficient hyperphenylalaninemia type A	261640	PTS	Pyruvoyl tetrahydro-biopterin synthase	AR
Tetrahydrobiopterin-deficient hyperphenylalaninemia type B	233910	GCH1	GTP cyclohydrolase	AR
Tetrahydrobiopterin-deficient hyperphenylalaninemia type C	261630	QDPR	Dihydropteridine reductase	AR
Tetrahydrobiopterin-deficient hyperphenylalaninemia type D	264070	PCBD1	Homodimerization of the transcription factor	AR
Tyrosinemia-1	276700	FAH	Hydrolase	AR
Tyrosinemia-2	276600	TAT	Cytosolic tyrosine aminotransferase	AR + KDN
Tyrosinemia-3	276710	HPD	Oxygenase	AR
Hawkinsinuria	140350	HPD	Oxygenase	AD
Alkaptonuria or ochronosis	203500	HGD	Oxidase	AR
Oculocutaneous albinism-1A	203100	TYR	Tyrosine hydroxylase and dopa oxidase	AR
Oculocutaneous albinism-1B	606952	TYR	Tyrosine hydroxylase and dopa oxidase	AR
Oculocutaneous albinism-2	203200	OCA2	Melanocyte-specific transporter protein	AR
Oculocutaneous albinism-3	203290	TYRP1	Melanosomal protein	AR
Oculocutaneous albinism-4	606574	SLC45A2	Solute carrier	AR
Ocular albinism-1 or Nettleship-Falls syndrome	300500	GPR143	Cell surface receptor G protein	XR
Ocular albinism-2 or Åland Island eye disease or Forsius-Eriksson syndrome	300600	CACNA1F	Channelopathy Calcium	XR
Chediak-Higashi syndrome	214500	LYST	Trafficking Vesicle formation	AR
Piebaldism	164920	KIT	Cell surface receptor Enzyme-linked receptor	AD
Hermansky-Pudlak syndrome-1	203300	HPS1	Trafficking Vesicle formation	AR
Hermansky-Pudlak syndrome-2	608233	AP3B1	Trafficking Vesicle formation	AR
Hermansky-Pudlak syndrome-3	606118	HPS3	Trafficking Vesicle formation	AR
Hermansky-Pudlak syndrome-4	606682	HPS4	Trafficking Vesicle formation	AR

Table 3.04.09.1.1. Continued.

Disorder	OMIM	Gene	Biological Function	Product of Conception Alteration
Hermansky-Pudlak syndrome-5	607521	HPS5	Trafficking Vesicle formation	AR
Hermansky-Pudlak syndrome-6	607522	HPS6	Trafficking Vesicle formation	AR
Hermansky-Pudlak syndrome-7	607145	DTNBP1	Trafficking Vesicle formation	AR
Hermansky-Pudlak syndrome-8	609762	BLOC1S3	Trafficking Vesicle formation	AR
Hermansky-Pudlak syndrome-9	614171	BLOC1S6	Trafficking Vesicle formation	AR
Waardenburg syndrome-1	193500	PAX3	Transcription factor Helix-turn-helix domains	AD + de novo
Waardenburg syndrome-2A	193510	MITF	Transcription factor Basic domains	AD
Waardenburg syndrome-2D	608890	SNAI2	Zinc finger protein	AR
Waardenburg syndrome-2E, with or without neurologic involvement	611584	SOX10	Transcription factor β-Scaffold factors	AD
Waardenburg syndrome-3	148820	PAX3	Transcription factor Helix-turn-helix domains	AR or AD + de novo
Waardenburg syndrome-4A	277580	EDNRB	Cell surface receptor G protein	AR
Waardenburg syndrome-4B	613265	EDN3	Endothelin	AR or AD
Waardenburg syndrome-4C	613266	SOX10	Transcription factor β-Scaffold factors	AD + de novo
Waardenburg syndrome/ocular albinism	103470	MITF/TYR	Transcription factor/Tyrosine hydroxylase	DigM
Tietz albinism-Deafness syndrome	103500	MITF	Transcription factor Basic domains	AD
Histidinemia or histidinuria	235800	HAL	Histidine ammonia lyase	AR
Brunner syndrome	300615	MAOA	Mitochondrial outer membrane protein	XR
Formiminoglutamic aciduria	229100	FTCD	Cyclodeaminase	AR
Urocanic aciduria or urocanase deficiency	276880	UROC1	Urocanate hydratase	AR

Table 3.04.09.1.2 Disorders of Aromatic Amino-acid Metabolism in Different Ethnic Groups

Disease	Frequency of condition	People with higher prevalence (if available)
Phenylketonuria	1-40 in 100,000	Turkish, Chinese, Italians, Irish and Yemenis
Oculocutaneous albinism	1 in 17,000	Sub-Saharan Africans
Ocular albinism type I	1 in 60,000 Males	Worldwide
Hermansky-Pudlak syndrome	1-9 in 1,000,000	Northwest Puerto Ricans and Swiss (Alps)
Alkaptonuria	1-9 in 1,000,000	Dominicans and Slovaks
Tyrosinemia type I	<1 in 1,000,000	French Canadians

3.04.09.2 Disorders of Branched-chain Amino-acid Metabolism and Fatty-acid Metabolism

Organic acidemia or organic aciduria: A group of disorders characterized by neurological symptoms caused by building up of acids in the blood and urine. These groups of disorders occur mainly due to alterations in amino acid metabolism, predominantly branched-chain amino acids and certain odd-chained fatty acids. There are four main types of organic acidemia: Maple syrup urine disease, Isovaleric acidemia, Methylmalonic acidemia and Propionic acidemia. Most of these conditions are inherited in autosomal recessive patterns.

Maple syrup urine disease or branched-chain ketoaciduria: A disorder characterized by mental and physical retardation, feeding problems, and urine with a characteristic maple syrup odor. Affected individuals exhibit an organic acidemia due to metabolic disorder affecting branched-chain amino acids. There are several underlying causes of this disorder with at least 4 autosomal genomic loci: This condition mainly follows autosomal recessive patterns of inheritance and few examples of *de novo* mutations. Maple syrup urine disease-1A is caused by autosomal recessive mutations in the *BCKDHA* gene which encodes the branched chain keto acid dehydrogenase E1 (alpha polypeptide);; Maple syrup urine disease-1B is caused by autosomal recessive mutations in the *BCKDHB* gene which encodes the branched chain keto acid dehydrogenase E1 (beta polypeptide);; Maple syrup urine disease-2 is caused by autosomal recessive mutations in the *DBT* gene which encodes the dihydrolipoamide branched chain transacylase E2;; and Maple syrup urine disease-3 is caused by autosomal recessive mutations in the *DLD* gene which encodes the dihydrolipoamide dehydrogenase. The product of conception alterations are defined as occasional *de novo* autosomal recessive syndrome, or autosomal recessive syndrome plus known *de novo* alterations (AR + KDN). These genes encode the catalytic components of the branched-chain alpha-keto acid dehydrogenase complex, which catalyzes the catabolism of the branched-chain amino acids, leucine, isoleucine and valine. On average, the carrier frequency for Maple syrup urine disease trait is estimated to be 1 in 212 individuals worldwide, which means the frequency of this condition is about 1 in 180,000 individuals or $1/(424)^2$. There is a higher carrier frequency for Maple syrup urine disease trait among the Old Order Amish, Mennonites, Ashkenazi Jews, and members of the Gypsy community in southern Portugal. In the Old Order Mennonites of Pennsylvania, the carrier frequency for Maple syrup urine disease trait is estimated to be 1 in 9.75 individuals, which means the frequency of this condition is about 1 in 380 individuals or $1/(19.5)^2$.

Isovaleric acidemia: A disorder of neonatal onset characterized by vomiting, dehydration, coma and abnormal movements. Affected neonates also manifest metabolic acidosis with ketosis, hyperammonemia, neutropenia, thrombopenia and hypocalcemia. Isovaleric acidemia is caused by autosomal recessive mutations in the *IVD* gene which encodes the isovaleryl CoA dehydrogenase. This enzyme is involved with leucine metabolism. On average in Europe, the carrier frequency for Isovaleric acidemia trait is estimated to be 1 in 158 individuals, which means the frequency of this condition is about 1 in 100,000 or $1/(316)^2$ individuals.

Methylmalonic acidemia or methylmalonic aciduria: A disorder characterized by progressive encephalopathy due to secondary hyperammonemia. There are different onset variants such as: infantile (with ketoacidotic coma, dehydration, hyperammonemia and leukothrombocytopenia); subacute (onset during early childhood with hypotonia, vomiting, growth and psychomotor retardation); and later onset variant (characterized by recurrent ketoacidotic comas). It is a classical type of organic acidemia. There are several underlying causes of this disorder with at least 8 autosomal genomic loci: Isolated methylmalonic acidurias are caused by autosomal recessive mutations in the *MUT* gene which encodes the methylmalonyl CoA mutase (mitochondrial enzyme that

catalyzes the isomerization of methylmalonyl-CoA to Succinyl-CoA), or *MMAA* gene which encodes the methylmalonic aciduria (cobalamin deficiency) cblA type, or *MMAB* gene which encodes the methylmalonic aciduria (cobalamin deficiency) cblB type, or *MCEE* gene which encodes the methylmalonyl CoA epimerase or the *CD320* gene which encodes the CD320 molecule. Mutations in the *MUT* gene account for 60% of all the cases of methylmalonic acidemia. The developing of methylmalonic acidemia is due to a defect in the metabolism or transport of vitamin B12. These disorders block the conversion of methylmalonyl-CoA to Succinyl-CoA. There are also patients who display combined signs of methylmalonic aciduria and homocystinuria. Patients with Combined methylmalonic aciduria and homocystinuria exhibit neurologic, metabolic, developmental, hematologic, ophthalmologic and dermatologic clinical features. There are at least 3 forms of combined conditions. "Combined methylmalonic aciduria and homocystinuria type C" is caused by autosomal recessive mutations in the *MMACHC* gene which encodes the methylmalonic aciduria cblC type, with homocystinuria (cobalamin deficiency), a protein that may have a role in the binding and intracellular trafficking of cobalamin;; "Combined methylmalonic aciduria and homocystinuria type D" is caused by autosomal recessive mutations in the *MMADHC* gene which encodes the methylmalonic aciduria cblD type, with homocystinuria (cobalamin deficiency), a mitochondrial protein that is involved in an early step of vitamin B12 metabolism;; and "Combined methylmalonic aciduria and homocystinuria type F" is caused by autosomal recessive mutations in the *LMBRD1* gene which encodes the LMBR1 domain containing protein 1. These mutations lead to a cobalamin metabolism defect. These mechanisms involve decreasing levels of the coenzymes adenosylcobalamin and methylcobalamin, which results in decreased activity in the enzymes methylmalonyl-CoA mutase and methyltetrahydrofolate: homocysteine methyltransferase. Methylmalonic acidemia affects an estimated 1 in every 50,000-100,000 individuals.

Propionic acidemia or propionyl-CoA carboxylase deficiency: A neonatal onset disorder characterized by lethargy, episodic vomiting, ketosis, neutropenia, developmental retardation, progressive encephalopathy and intolerance to protein. Propionic Acidemia is a branched-chain organic acidemia. Propionic acidemia is caused by autosomal recessive mutations in either the *PCCA* gene or the *PCCB* gene, which encode the mitochondrial enzyme propionyl CoA carboxylase (alpha and beta). This disorder is rare worldwide, except in certain parts of Saudi Arabia. Eighty percent of cases are reported in tribal areas, suggesting a founder effect. Propionic acidemia is also common among Amish, Mennonites, and the Inuit people of Greenland. In the United States, the carrier frequency for Propionic acidemia trait is estimated to be 1 in 93.5 individuals, which means the frequency of this condition is about 1 in 35,000 individuals or $1/(187)^2$.

3-Methylcrotonyl-CoA carboxylase deficiency or 3-Methylcrotonylglycinuria: A disorder characterized by inability to break down proteins containing the amino acid leucine. Affected infants appear normal at birth but frequently develop signs and symptoms in infancy or early childhood. Patients can exhibit a large range of features from mild to life-threatening, including lethargy, hypotonia, feeding difficulties, recurrent episodes of vomiting and diarrhea. If untreated, patients may manifest delayed development, seizures and coma. 3-Methylcrotonylglycinuria-1 is caused by autosomal recessive mutations in the *MCCC1* gene which encodes the methylcrotonoyl-CoA carboxylase 1; 3-Methylcrotonyl-glycinuria-2 is caused by autosomal recessive mutations in the *MCCC2* gene which encodes the methyl-crotonoyl-CoA carboxylase 2. These encoded proteins are subunits in the enzyme 3-methylcrotonyl-CoA carboxylase. This enzyme breaks down proteins from the diet; this multimeric enzyme is especially responsible for the fourth step in processing leucine. On average, the carrier frequency for 3-Methylcrotonylglycinuria trait is estimated to be 1 in 112 individuals worldwide, which means the frequency of this condition is about 1 in 50,000 individuals or $1/(224)^2$.

3-Methylglutaconic aciduria: A disorder characterized by either psychomotor retardation during childhood or progressive neurodegenerative disorder during adulthood. The latter form is characterized by white matter lesions in the brain, dementia, ataxia and spasticity. Affected individuals suffer from impairment in the ability to make energy in the mitochondria. Patients exhibit additional signs besides the 3-methylglutaconic aciduria. There are several underlying causes of this disorder with at least 4 autosomal genomic loci and 1 X-linked genomic locus: 3-methylglutaconic aciduria type I is caused by autosomal recessive mutations in the *AUH* gene which encodes the AU RNA binding protein/enoyl-CoA hydratase;; 3-methylglutaconic aciduria type II or Barth syndrome is caused by X-linked recessive mutations in the *TAZ* gene which encodes the tafazzin, a protein that is expressed at high levels in cardiac and skeletal muscle; 3-methylglutaconic aciduria type II exhibit male patients with mitochondrial cardiomyopathy, short stature, skeletal myopathy and recurrent infections; cognitive development is normal;; 3-methylglutaconic aciduria type III is caused by autosomal recessive mutations in the *OPA3* gene which encodes the optic atrophy 3 protein. 3-methylglutaconic aciduria type III is associated with optic atrophy, movement disorder and spastic paraplegia. This condition occurs in Iraqi Jews;; the 3-methylglutaconic aciduria type IV gene has not been characterized, but patients exhibit severe psychomotor retardation and cerebellar dysgenesis;; and 3-methylglutaconic aciduria type V is caused by autosomal recessive mutations in the *DNAJC19* gene which encodes the DnaJ homolog protein (subfamily C member 19). 3-methylglutaconic aciduria type V is associated with early-onset dilated cardiomyopathy with conduction defects, non-progressive cerebellar ataxia and testicular dysgenesis. 3-methylglutaconic aciduria affects the leucine catabolism. 3-Methylglutaconic aciduria is a rare syndrome worldwide except among the Jewish people of Iraq and in the Saguenay-Lac-Saint-Jean region of Canada.

2-methylbutyryl-CoA dehydrogenase deficiency or short/branched-chain acyl-CoA dehydrogenase deficiency: A disorder characterized by delayed neurological development and patients exhibiting neurological signs. This condition is almost always an asymptomatic disorder of impairment in isoleucine degradation. 2-methylbutyryl-CoA dehydrogenase deficiency is caused by autosomal recessive mutations in the *ACADSB* gene which encodes the short/branched-chain acyl-CoA dehydrogenase. This enzyme catalyzes the dehydrogenation of acyl-CoA derivatives in the metabolism of fatty acids or branch chained amino acids. On average, the carrier frequency for 2-methylbutyryl-CoA dehydrogenase deficiency trait is estimat-

ed to be 1 in 100-158 individuals worldwide, which means the frequency of this condition is about 1 in 40,000-100,000 individuals or $1/(200)^2 - 1/(316)^2$.

Alpha-methylacetoacetic aciduria: A disorder characterized by a progressive loss of mental and motor skills. Patients suffer from intermittent ketoacidotic episodes, due to dysfunction in the mitochondrial acetoacetyl-CoA thiolase. Alpha-methylacetoacetic aciduria is caused by autosomal recessive mutations in the *ACAT1* gene which encodes the mitochondrial acetyl-CoA acetyltransferase 1. This protein is involved in ketone body metabolism and isoleucine catabolism. Alpha-methylacetoacetic aciduria is estimated to affect less than 1 in every 1,000,000 individuals.

3-hydroxyisobutyryl-CoA hydrolase deficiency: A disorder characterized by progressive neurodegeneration. 3-hydroxyisobutyryl-CoA hydrolase deficiency is caused by autosomal recessive mutations in the *HIBCH* gene which encodes the 3-hydroxyisobutyryl-CoA hydrolase. This enzyme is involved in valine and propionate metabolism. 3-hydroxyisobutyryl-CoA hydrolase deficiency is estimated to affect less than 1 in every 1,000,000 individuals.

Isobutyryl-CoA dehydrogenase 8 deficiency: A disorder characterized by failure to thrive, dilated cardiomyopathy, anemia and carnitine deficiency. Isobutyryl-CoA dehydrogenase 8 deficiency is caused by autosomal recessive mutations in the *ACAD8* gene which encodes the acyl-CoA dehydrogenase (family member 8). This enzyme is involved in the degradation of the branched chain amino acid valine.

Table 3.04.09.2.1 Recognizable Disorders of Branched-chain Amino-acid Metabolism and Fatty-acid Metabolism

Disorder	OMIM	Gene	Biological Function	Product of Conception Alteration
Maple syrup urine disease-1A or Branched-chain ketoaciduria	248600	BCKDHA	Inner-mitochondrial protein	AR
Maple syrup urine disease-1B or Branched-chain ketoaciduria	248600	BCKDHB	Inner-mitochondrial protein	AR
Maple syrup urine disease-2 or Branched-chain ketoaciduria	248600	DBT	Inner-mitochondrial protein	AR + KDN
Maple syrup urine disease-3 or Branched-chain ketoaciduria	248600	DLD	Mitochondrial protein	AR
Isovaleric acidemia	243500	IVD	Mitochondrial protein	AR
Methylmalonic acidemia or methylmalonic aciduria	251000	MUT	Mitochondrial protein	AR
Methylmalonic acidemia or methylmalonic aciduria	251100	MMAA	Mitochondrial protein	AR
Methylmalonic acidemia or methylmalonic aciduria	251110	MMAB	Mitochondrial protein	AR
Methylmalonyl-CoA epimerase deficiency	251120	MCEE	Mitochondrial protein	AR
Methylmalonic acidemia or methylmalonic aciduria	613646	CD320	Cell surface receptor Other/ungrouped	AR
Combined methylmalonic aciduria and homocystinuria	277400	MMACHC	Mitochondrial protein	AR
Combined methylmalonic aciduria and homocystinuria	277410	MMADHC	Mitochondrial protein	AR
Combined methylmalonic aciduria and homocystinuria	277380	LMBRD1	Lysosomal protein	AR
Propionic acidemia	606054	PCCA	Mitochondrial protein	AR
Propionic acidemia	606054	PCCB	Mitochondrial protein	AR
3-methylcrotonylglycinuria or 3-Methylcrotonyl-CoA carboxylase deficiency	210200	MCCC1	Carboxylase	AR
3-methylcrotonylglycinuria or 3-Methylcrotonyl-CoA carboxylase deficiency	210210	MCCC2	Carboxylase	AR
3-methylglutaconic aciduria type 1	250950	AUH	Mitochondrial energy protein	AR

The Concise Encyclopedia of Genomic Diseases

Table 3.04.09.2.1. Continued.

Disorder	OMIM	Gene	Biological Function	Product of Conception Alteration
3-methylglutaconic aciduria type 2 or Barth syndrome	302060	TAZ	Mitochondrial energy protein	XR
3-methylglutaconic aciduria type 3 or Costeff syndrome	258501	OPA3	Mitochondrial protein	AR
3-methylglutaconic aciduria type 5	610198	DNAJC19	Mitochondrial energy protein	
2-methylbutyryl-CoA dehydrogenase deficiency	610006	ACADSB	Mitochondrial protein	AR
Alpha-methylacetoacetic aciduria	203750	ACAT1	Mitochondrial protein	AR
3-hydroxyisobutryl-CoA hydrolase deficiency	250620	HIBCH	Mitochondrial protein	AR
Isobutyryl-CoA dehydrogenase deficiency or valinuria	611283	ACAD8	Mitochondrial protein	AR

Table 3.04.09.2.2 Disorders of Branched-chain Amino-acid Metabolism and Fatty-acid Metabolism in Different Ethnic Groups

Disease	Frequency of condition	People with higher prevalence (if available)
Propionic acidemia	3 in 100,000	Amish, Mennonites, and Greenlandic Inuit
3-Methylcrotonylglycinuria	2 in 100,000	Worldwide
Methylmalonic acidemia	1-2 in 100,000	Worldwide
2-methylbutyryl-CoA dehydrogenase deficiency	1-2 in 100,000	Worldwide
Isovaleric acidemia	1 in 100,000	Unknown
Maple syrup urine disease or branched-chain ketoaciduria	6 in 1,000,000	Old Order Amish, Mennonites, Ashkenazi Jews, and Gypsies of Portugal (South)
Alpha-methylacetoacetic aciduria	<1 in 1,000,000	Unknown
3-hydroxyisobutyryl-CoA hydrolase deficiency	<1 in 1,000,000	Unknown

3.04.09.3 Disorders of Fatty-acid Metabolism

Very long-chain acyl-coenzyme A dehydrogenase deficiency: A disorder characterized by lethargy and hypoglycemia. Clinically this condition could be classified into 3 onset forms: early-onset, which is a severe form with high incidence of cardiomyopathy and high mortality; childhood onset, which manifests hypoketotic hypoglycemia and a more favorable outcome; and adult-onset, which is a myopathic form with isolated skeletal muscle involvement, rhabdomyolysis, and myoglobinuria after exercise or fasting. Affected individuals cannot convert long-chain fatty acids to energy, principally during periods without food. These partially degraded fatty acids could damage the heart, liver, retina and muscles. Very long-chain acyl-coenzyme A dehydrogenase deficiency is caused by autosomal recessive mutations in the *ACADVL* gene which encodes the enzyme very long-chain acyl-coenzyme A dehydrogenase. This enzyme catalyzes the first step of the mitochondrial fatty acid beta-oxidation pathway.

Long-chain 3-hydroxyacyl-coenzyme A dehydrogenase deficiency: A similar disorder as the previously described, but instead is due to long-chain 3-hydroxyacyl-coenzyme A dehydrogenase deficiency. Long-chain 3-hydroxyacyl-coenzyme A dehydrogenase deficiency is caused by autosomal recessive mutations in the *ACADL* gene which encodes the long-chain 3-hydroxyacyl-coenzyme A dehydrogenase, a mitochondrial flavoenzyme involved in fatty acid and branched chain amino-acid metabolism.

Medium-chain acyl-coenzyme A dehydrogenase deficiency: A disorder characterized by lethargy, hypotonia, and hypoglycemia during infancy or early childhood. This syndrome mimics sudden infant death syndrome, and sometimes is mistaken for Reye syndrome. Medium-chain acyl-coenzyme A dehydrogenase deficiency is caused by autosomal recessive mutations in the *ACADM* gene which encodes the acyl-CoA dehydrogenase C-4 to C-12 straight chain. This encoded protein oxidizes medium chain fatty acids having 4-12 carbons. Medium-chain acyl-coenzyme A dehydrogenase deficiency is a rare syndrome worldwide. There is a clear founder effect in northwestern Europe. In England, the carrier frequency for medium-chain acyl-coenzyme A dehydrogenase trait is estimated to be 1 in 40 individuals, which means the frequency of this condition is about 1 in 6,400 individuals or $1/(80)^2$. In the United States, the carrier frequency for medium-chain acyl-coenzyme A dehydrogenase trait is estimated to be 1 in 65 individuals, which means the frequency of this condition is about 1 in 17,000 individuals or $1/(130)^2$.

Short-chain/branched-chain acyl-coenzyme A dehydrogenase deficiency: This disease is discussed under endocrine, nutritional and metabolic diseases in the subchapter "Disorders of Branched-chain Amino-acid Metabolism and Fatty-acid Metabolism" (3.04.09.2).

3-hydroxyacyl-coenzyme A dehydrogenase deficiency: A disorder appearing during infancy or early childhood characterized by hyperinsulinism, hypoglycemia, poor appetite, vomiting, diarrhea, lethargy, hypotonia and liver problems. Affected individuals also suffer from heart and breathing problems, seizures, coma and sudden unexpected death. The deficiency of 3-hydroxyacyl-coenzyme A dehydrogenase displays a combined medium- and short-chain 3-hydroxyacyl-coenzyme A dehydrogenase deficiency in these patients. 3-hydroxyacyl-coenzyme A dehydrogenase deficiency is caused by autosomal recessive mutations in the *HADH* gene which encodes the mitochondrial hydroxyacyl-CoA dehydrogenase.

2-hydroxyglutaric aciduria: A disorder characterized by neurologic symptoms and high levels of urinary levels of hydroxyglutaric acid. Affected patients exhibit developmental delay, hypotonia, epilepsy and dysmorphic features. There are several underlying causes of this disorder with at least 3 autosomal genomic loci: L-2-hydroxy-glutaric aciduria, D-2-hydroxyglutaric aciduria-1 and D-2-hydroxyglutaric aciduria-2. L-2-hydroxyglutaric aciduria mainly disturbs the central nervous system particularly affecting the basal ganglia. Affected individuals also suffer from cystic cavitation in the white matter. Patients manifest symptoms of tremor, hypotonia and epilepsy. L-2-hydroxyglutaric aciduria is caused by autosomal recessive mutations in *L2HGDH* gene which encodes the L-2-hydroxyglutarate dehydrogenase, a FAD-dependent mitochondrial enzyme that oxidizes L-2-hydroxyglutarate to alpha-ketoglutarate;; The two forms of D-2-hydroxyglutaric aciduria are characterized by macrocephaly, cortical blindness, mental retardation, hypotonia and cardiomyopathy. D-2-hydroxyglutaric aciduria-1 is caused by autosomal recessive mutations in *D2HGDH* gene which encodes the D-2-hydroxyglutarate dehydrogenase, a mitochondrial enzyme that converts D-2-hydroxyglutarate to 2-ketoglutarate;; and D-2-hydroxyglutaric aciduria-2 is caused by either autosomal dominant or de novo gain-of-function mutations in *IDH2* gene which encodes the mitochondrial isocitrate dehydrogenase-2, a mitochondrial enzyme that plays a role in intermediary metabolism and energy production.

Acyl-CoA dehydrogenase deficiency 9: A disorder appearing after birth characterized by acute metabolic acidosis, hypertrophic cardiomyopathy and muscle weakness. This condition occurs due to deficiency of mitochondrial complex I activity in the muscle, liver and fibroblasts. Acyl-CoA dehydrogenase deficiency 9 is caused by autosomal recessive mutations in the *ACAD9* gene which encodes the acyl-CoA dehydrogenase (family member 9). The encoded protein is specifically active toward palmitoyl-CoA and long-chain unsaturated substrates.

Long-chain hydroxyacyl-CoA dehydrogenase deficiency: A disorder characterized by early-onset hypoglycemia, cardiomyopathy, neuropathy, pigmentary retinopathy and sudden death. Long-chain hydroxyacyl-CoA dehydrogenase deficiency is caused by autosomal recessive mutations in the *HADHA* gene which encodes the hydroxyacyl-CoA dehydrogenase/3-ketoacyl-CoA thiolase/ enoyl-CoA hydratase (alpha subunit).

Mitochondrial trifunctional protein deficiency: A disorder characterized by lethargy and hypoglycemia. Clinically this condition could be classified into 3 onset forms: early neonatal onset, which can result in sudden unexplained infant death; childhood onset, which can result in a hepatic Reye-like syndrome; and adult-onset, is usually a myopathic form. Affected individuals cannot convert long-chain fatty acids to energy, principally during periods without food. These partially degraded fatty acids could damage the heart, liver and muscles. Mitochondrial trifunctional protein deficiency is caused by autosomal recessive mutations in the *HADHA* gene, or the *HADHB* which encodes the hydroxyacyl-CoA dehydrogenase/3-ketoacyl-CoA thiolase/enoyl-CoA hydratase (beta subunit). The multimeric mitochondrial trifunctional protein catalyzes 3 steps in mitochondrial beta-oxidation of fatty acids: long-chain enoyl-CoA hydratase, long-chain 3-hydroxyacyl-CoA dehydrogenase and long-chain thiolase activities.

HMG-CoA synthase deficiency or 3-hydroxy 3-methylglutaryl-CoA synthase deficiency: A disorder appearing during childhood characterized by acute crisis ofhypoketotic hypoglycemic coma, with or without hepatomegaly. This condition appears during fasting or by a concomitant infectious illness. Affected individuals do not manifest symptoms between crises. Patients display usual profiles of acylcarnitine and urinary organic acid, but without elevation of ketone bodies. HMG-CoA synthase deficiency is caused by autosomal recessive mutations in the *HMGCS2* gene which encodes the mitochondrial 3-hydroxy-3-methylglutaryl-CoA synthase 2. HMG-CoA synthase deficiency is estimated to affect less than 1 in every 1,000,000 individuals.

Succinyl-CoA: 3-ketoacid CoA transferase deficiency: A disorder characterized by acute crisis of ketoacidosis, but without clinical symptoms between episodes. These ketoacidotic episodes usually start during the neonatal period. In these patients the extrahepatic tissues cannot use the ketone bodies produced by the liver. Contrary to HMG-CoA synthase deficiency, these patients do not show the characteristic organic acid and acylcarnitine profile. Succinyl-CoA: 3-ketoacid CoA transferase deficiency is caused by autosomal recessive mutations in the *OXCT1* gene which encodes the 3-oxoacid CoA transferase 1. Succinyl-CoA: 3-ketoacid CoA transferase deficiency is estimated to affect less than 1 in every 1,000,000 individuals.

2,4-Dienoyl-CoA reductase deficiency: A disorder characterized by persistent hypotonia since the neonatal period. This condition affects principally the metabolism of unsaturated fatty enoyl-CoA esters. Affected individuals also have shown hyperlysinemia, hypocarnitinemia and a normal organic acid profile. 2,4-Dienoyl-CoA reductase deficiency is caused by autosomal recessive mutations in the *DECR1* gene which encodes the mitochondrial 2,4-dienoyl CoA reductase 1, an accessory enzyme involved in the beta-oxidation and metabolism of unsaturated fatty enoyl-CoA esters.

Acute Fatty Liver Pregnancy: A disorder characterized by hyperemesis gravidum, acute fatty liver pregnancy syndrome; and HELLP syndrome (Hypertension or Hemolysis, Elevated Liver enzymes, and low Platelets). Patients exhibit symptoms of anorexia, nausea, vomiting, abdominal pain, jaundice and liver failure during the third trimester. These severe maternal illnesses arise during pregnancies due to affected fetuses with the common 1528G-C mutation on at least one allele in the *HADHA* gene. In other words, a primary genomic defect in the fetus causes symptoms in the mother.

Ethylmalonic encephalopathy: A disorder characterized by progressive brain dysfunction, and symptoms affecting the gastrointestinal tract and peripheral vessels. Affected individuals build up high levels of toxic compounds, such as ethylmalonic acid and lactic acid in the body. These patients' mitochondria are ineffective at producing energy. There is a failure to detoxify hydrogen sulfide from cysteine degradation. Ethylmalonic encephalopathy is caused by autosomal recessive mutations in the *ETHE1* gene which encodes the mitochondrial matrix ethylmalonic encephalopathy 1 protein. This enzyme functions in sulfide catabolism. Ethylmalonic encephalopathy is estimated to affect less than 1 in every 1,000,000 individuals. This condition mainly affects Mediterranean and Arab people.

Carnitine palmitoyltransferase I deficiency: A disorder characterized by episodes of metabolic crisis exhibiting hypoglycemia, hypoketonemia, hepatomegaly, muscle weakness and elevated levels of carnitine in the blood. Affected individuals cannot transport long-chain fatty acids into the mitochondria to be broken down, resulting in damaging of the liver, heart and brain. Carnitine palmitoyltransferase I deficiency is caused by autosomal recessive mutations in the *CPT1A* gene which encodes the hepatic carnitine palmitoyltransferase 1A. Carnitine palmitoyltransferase I deficiency is estimated to affect less than 1 in every 1,000,000 individuals.

Carnitine palmitoyltransferase II deficiency: A disorder characterized by similar symptoms as those of carnitine palmitoyltransferase I deficiency, but instead affecting the carnitine palmitoyltransferase II. This condition has different clinical presentations: the neonatal form, the infantile form and the adult form. With the neonatal form, affected patients usually experience hypoglycemia, seizures, hepatomegaly with liver failure, and cardiac and respiratory failure. The infantile form is an intermediate, between the neonatal and adult form. The adult form is the most frequent cause of adult hereditary myoglobinuria. The latter is the most common among the 3, where just the skeletal muscle is affected. Carnitine palmitoyltransferase II deficiency appears after prolonged exercise or fasting. Carnitine palmitoyltransferase II deficiency is caused by autosomal recessive mutations in the CPT2 gene which encodes the mitochondrial carnitine palmitoyltransferase 2.

Myopathic form of carnitine palmitoyltransferase II deficiency: A disorder characterized by muscle pain, recurrent attacks of rhabdomyolysis, and muscular weakness triggered by prolonged physical exercise, extremes in temperature and prolonged fasting. This disorder affects the mitochondrial oxidation of long chain fatty acids. Myopathic form of carnitine palmitoyltransferase II deficiency is caused by autosomal recessive mutations in *CPT2* gene which encodes the mitochondrial carnitine palmitoyltransferase 2. Myopathic form of carnitine palmitoyltransferase II deficiency accounts for 86% of all cases of *CPT2* gene mutations cases.

Primary systemic carnitine deficiency: A disorder characterized by acute episode of lethargy, hypoketotic hypoglycemia, somnolence, hepatomegaly and cardiomegaly. Affected individuals suffer from impaired ketogenesis, intermittent episodes of cerebral and hepatic dysfunction and underdeveloped muscles. Primary systemic carnitine deficiency display impaired fatty acid oxidation in skeletal and heart muscle. Primary systemic carnitine deficiency is caused by autosomal recessive mutations in the *SLC22A5* gene which encodes the solute carrier (family 22 member A5), an organic cation/carnitine transporter.

Neonatal adrenoleukodystrophy: A disorder characterized by hypotonia, liver dysfunction, developmental delay, retinal dystrophy, visual impairment and sensorineural hearing loss. Neonatal adrenoleukodystrophy belongs to the peroxisome biogenesis disorders. Neonatal adrenoleukodystrophy is a milder form of the Zellweger syndrome. There are several underlying causes of this disorder with at least 5 autosomal genomic loci: Neonatal adrenoleukodystrophy is caused by autosomal recessive mutations in the *PEX1*, or *PEX5*, or *PEX10*, or *PEX13*, or the *PEX26* genes. These genes encode the respective peroxisomal biogenesis factor. These proteins are cytoplasmic but are often anchored to a peroxisomal membrane where they form a heteromeric complex and play a role in the import of proteins into peroxisomes and peroxisome biogenesis. In the United States, Neonatal adrenoleukodystrophy is estimated to affect 1 in every 50,000 live births, whereas in Japan, Neonatal adreno-leukodystrophy is estimated to affect 1 in every 500,000 live births.

Peroxisomal acyl-CoA oxidase deficiency: A disorder characterized by seizures, hypotonia, facial dysmorphism, developmental delay, hepatomegaly and polydactyly. Hypotonia is an advanced feature of the disease and then is followed by hypertonia with hyperreflexia. Affected individuals also suffer from hearing loss, strabismus and nystagmus. Peroxisomal acyl-CoA oxidase deficiency is caused by autosomal recessive mutations in the *ACOX1* gene which encodes the palmitoyl acyl-CoA oxidase 1. Peroxisomal acyl-CoA oxidase deficiency is estimated to affect less than 1 in every 1,000,000 individuals.

Adrenoleukodystrophy or bronze Schilder disease: A disorder characterized by progressive brain damage and failure of the adrenal glands and the Leydig cells in the testes. Adrenoleukodystrophy gradually damages the myelin sheath which impairs the nerve conduction of the central and peripheral nervous systems, leading to increased disability. X-linked Adrenoleukodystrophy affects 5 times more males

than females. Adrenoleukodystrophy is caused by X-linked mutations in the *ABCD1* gene which encodes the ATP-binding cassette (sub-family D member 1). The product of conception alteration in adrenoleukodystrophy appears to either exhibit X-linked inheritance in most of the cases, or *de novo* X-linked alteration in a few cases. The ABCD1 protein functions as a transporter of very long-chain fatty acids from the diet. The defective adreno-leukodystrophy gene resides on the long arm subtelomeric band on the X chromosome (Xq28). Adrenoleukodystrophy is estimated to affect 1 in every 20,000 male births worldwide.

Adrenomyeloneuropathy: A disorder characterized by adrenal insufficiency, hypogonadism and progressive spastic paraparesis. This form does not include cerebral involvement. Affected individuals also exhibit other neurologic features including peripheral neuropathy, impotence and sphincter disturbances. Adrenomyeloneuropathy is caused by X-linked mutations in the *ABCD1* gene. This is an allelic disorder from adrenoleukodystrophy, but it is less severe and occurs later in life.

D-Bifunctional protein deficiency: A neonatal onset disorder characterized by hypotonia seizures, visual and hearing impairment, craniofacial disfiguration and psychomotor delay. This condition belongs to the severe peroxisomal disorders often resembling Zellweger syndrome. D-Bifunctional protein deficiency is caused by autosomal recessive mutations in the *HSD17B4* gene which encodes the 17-beta hydroxysteroid dehydrogenase 4.

The Concise Encyclopedia of Genomic Diseases

Table 3.04.09.3.1 Recognizable Disorders of Fatty-acid Metabolism

Disorder	OMIM	Gene	Biological Function	Product of Conception Alteration
Very long chain acyl-CoA dehydrogenase deficiency	201475	ACADVL	Mitochondrial protein	AR
Long chain acyl-CoA dehydrogenase deficiency	201460	ACADL	Mitochondrial protein	AR
Medium chain acyl-CoA dehydrogenase deficiency	201450	ACADM	Mitochondrial protein	AR
3-hydroxyacyl-CoA dehydrogenase deficiency	231530	HADH	Mitochondrial matrix protein	AR
L-2-hydroxyglutaric aciduria	236792	L2HGDH	Mitochondrial protein	AR
D-2-hydroxyglutaric aciduria	600721	D2HGDH	Mitochondrial protein	AR
D-2-hydrosyglutaric aciduria 2	613657	IDH2	Mitochondrial protein	AR
Acyl-CoA dehydrogenase 9 deficiency	611126	ACAD9	Mitochondrial protein	AR
Mitochondrial trifunctional protein deficiency	609015	HADHA	Mitochondrial trifunctional protein	AR
Mitochondrial trifunctional protein deficiency	609015	HADHB	Mitochondrial trifunctional protein	AR
HMG-CoA synthase-2 deficiency	605911	HMGCS2	Mitochondrial protein	AR
Succinyl CoA 3-oxoacid CoA transferase deficiency	245050	OXCT1	Mitochondrial matrix protein	AR
2,4-Dienoyl-CoA reductase deficiency	222745	DECR1	Mitochondrial protein	AR
Acute fatty liver of pregnancy	609016	HADHA	Mitochondrial trifunctional protein	AR
Ethylmalonic encephalopathy	602473	ETHE1	Mitochondrial protein	AR
Carnitine palmitoyltransferase I deficiency	255120	CPT1A	Mitochondrial protein	AR
Carnitine palmitoyltransferase II deficiency	600649	CPT2	Mitochondrial protein	AR
Myopathy due to CPT II deficiency	255110	CPT2	Mitochondrial protein	AR
Systemic primary Carnitine deficiency	212140	SLC22A5	Solute carrier	AR
Neonatal adrenoleukodystrophy	202370	PEX1	Peroxisomal and lysosomal structural protein	AR
Neonatal adrenoleukodystrophy	202370	PEX5	Peroxisomal and lysosomal structural protein	AR
Neonatal adrenoleukodystrophy	202370	PEX10	Peroxisomal and lysosomal structural protein	AR
Neonatal adrenoleukodystrophy	202370	PEX13	Peroxisomal and lysosomal structural protein	AR

Table 3.04.09.3.1. Continued.

Disorder	OMIM	Gene	Biological Function	Product of Conception Alteration
Neonatal adrenoleukodystrophy	202370	PEX26	Peroxisomal and lysosomal structural protein	AR
Peroxisomal acyl-CoA oxidase deficiency	264470	ACOX1	Peroxisomal and lysosomal structural protein	AR
Cerebral adrenoleukodystrophy	300100	ABCD1	ABC-transporter protein	XR + de novo
Adrenomyeloneuropathy	300100	ABCD1	ABC-transporter protein	XR + de novo
D-bifunctional protein deficiency	261515	HSD17B4	Peroxisomal and lysosomal structural protein	AR

Table 3.04.09.3.2 Disorders of Fatty-acid Metabolism in Different Ethnic Groups

Disease	Frequency of condition	People with higher prevalence (if available)
Medium chain acyl-CoA dehydrogenase deficiency	5-17 in 100,000	English
Cerebral adrenoleukodystrophy	1 in 20,000 Males	Worldwide
Neonatal adrenoleukodystrophy	2-20 in 1,000,000	North Americans
Peroxisomal acyl-CoA oxidase deficiency	<1 in 1,000,000	Unknown
Succinyl CoA 3-oxoacid CoA transferase deficiency	<1 in 1,000,000	Unknown
Ethylmalonic encephalopathy	<1 in 1,000,000	Unknown
Carnitine palmitoyltransferase I deficiency	<1 in 1,000,000	Unknown

3.04.09.4 Disorders of Amino-acid Transport

Cystinosis: A disorder characterized by intralysosomal accumulation of cystine due to a defect in the transportation mechanisms of this organelle. There are 3 clinical forms: infantile, juvenile and ocular. The most common is the infantile form which appears in patients over three months of age. Affected children exhibit marked height-weight growth delay and polyuro-polydipsic syndrome. There are also manifestations of intralysosomal accumulation of cysteine, which leads to in-sulin-dependent diabetes, hypothyroidism, and hepatosplenomegaly with portal hypertension. Cystinosis is the most common cause of Fanconi syndrome in children. The infantile form progresses to renal failure after 6 years of age. The juvenile cystinosis form usually appears around 8 years of age and displays an intermediate clinical picture with renal failure after 15 years of age. Finally, the ocular form, which only affects adults, is caused by accumulation of cystine in the cornea and conjunctiva, producing tearing and photophobia. Cystinosis is caused by autosomal recessive mutations in the CTNS gene which encodes the cystinosin, a lysosomal cystine transporter. In Brittany, France, the carrier frequency for Cystinosis trait is estimated to be 1 in 80.5 individuals, which means the frequency of this condition is about 1 in 26,000 individuals or $1/(161)^2$. There is a higher carrier frequency for Cystinosis trait among French Canadians. The frequency of this condition worldwide is estimated to be 1 in 100,000-200,000 individuals.

Cystinuria: A disorder characterized by the formation of cystine stones affecting the kidneys, ureter and bladder. Cystinuria is digenic disease caused by concomitant mutations in the *SLC3A1* gene and the *SLC7A9* gene. These genes encode protein transporter carrier which reabsorbs cystine and dibasic amino acids: lysine, ornithine and arginine. Digenic mutations in these genes interrupt the ability of this carrier to concentrate these amino acids in the urine, resulting in cystine crystals formation. Cystinuria affects 1 in 7,000 people worldwide, making it the most prevalent genomic error of amino acid transport. The range of prevalence of cystinuria varies from 1 in 2,500 individuals among Libyan Jews to 1 in 100,000 among Swedes.

Fanconi renotubular syndrome: A disorder characterized by polydipsia and polyuria with glycosuria, aminoaciduria and phosphaturia. Affected individuals also suffer from hypophosphatemic rickets or osteomalacia, acidosis and dehydration. Fanconi renotubular syndrome is caused by decreased solute and water reabsorption in the proximal tubule of the kidney. Patients also develop renal insufficiency. There are several underlying causes of this disorder with at least 2 autosomal genomic loci: Fanconi renotubular syndrome-1 has been mapped but the gene has not yet been characterized;; and Fanconi renotubular syndrome-2 is caused by autosomal recessive mutations in the *SLC34A1* gene which encodes the solute carrier (family 34 member A1), a sodium phosphate transporter. Fanconi syndrome can also be caused by Cystinosis, Wilson disease, Lowe syndrome and Dent disease. There are several non-genomic factors that cause Fanconi syndrome including the use of expired tetracyclines, lead poisoning, and the use of antiretroviral therapy.

Hartnup disease or Pellagra-like dermatosis: A disorder characterized by light-sensitive rash, cerebellar ataxia, emotional instability and amino aciduria. Affected individuals exhibit impairment in the absorption of nonpolar amino acids. Principally affecting the tryptophan; this is the precursor for serotonin, melatonin, nicotinamide and niacin. Hartnup disease is caused by autosomal recessive mutations in *SLC6A19* gene which encodes the solute carrier (family 6 member A19), a neutral amino acid transporter, a sodium-dependent and chloride-independent neutral amino acid transporter. In Massachusetts, Hartnup disease is estimated to affect 1 in every 14,219 live births.

Oculocerebrorenal syndrome or Lowe syndrome: A disorder characterized by cataracts, hydrophthalmia, mental retardation, reduced renal ammonia production, aminoaciduria and vitamin D-resistant rickets. Lowe syndrome resembles Fanconi syndrome, but is instead inherited in X-linked patterns. Lowe syndrome can be caused by X-linked mutations in the *OCRL1* gene which encodes the enzyme PIP2-5-phosphatase. This enzyme is involved in actin polymerization and is found in the trans-Golgi network. Mutations in the same gene also cause Dent disease. Lowe syndrome is estimated to affect 1 in every 500,000 individuals worldwide.

Dicarboxylicaminoaciduria: A disorder characterized by a defect in the transport of 2 acidic amino acids, glutamic and aspartic acids. Affected individuals are characterized by exhibiting hypoglycemia, hyperprolinemia and intellectual deficit of infantile-onset. Dicarboxylicaminoaciduria can be caused by autosomal recessive mutations in the *SLC1A1* gene which encodes the solute carrier (family 1 member A1), a neuronal and epithelial glutamate transporter.

Iminoglycinuria: A disorder characterized by inconsistent varieties of disease associations, including mental retardation, deafness and blindness. Affected individuals suffer from a defect of renal tubular reabsorption of certain amino acid resulting in glycinuria, prolinuria and hydroxyprolinuria, the last two of which are imino acids. Iminoglycinuria is a trigenic disease, caused by concomitant mutations in the *SLC6A19* gene which encodes the solute carrier (family 6 member A19), a neutral amino acid transporter, and/or *SLC6A20* gene which encodes the solute carrier (family 6 member A20), a proline IMINO transporter, and/or the *SLC36A2* gene which encodes the solute carrier (family 36 member A2), a proton/amino acid symporter. Among Caucasians, Iminoglycinuria is estimated to affect 1 in every 15,000 live births. Iminoglycinuria may be more prevalent among Ashkenazi Jewish than any others ethnic groups.

Table 3.04.09.4 Recognizable Disorders of Amino-acid Transport

Disorder	OMIM	Gene	Biological Function	Product of Conception Alteration
Nephropathic cystinosis	219800	*CTNS*	Lysosomal storage protein	AR
Ocular non nephropathic cystinosis	219750	*CTNS*	Lysosomal storage protein	AR
Cystinuria	604144	*SLC3A1/SLC7A9*	Solute carriers	DigB
Fanconi renotubular syndrome 2	613388	*SLC34A1*	Solute carrier	AR
Hartnup disorder	234500	*SLC6A19*	Solute carrier	AR
Oculocerebrorenal syndrome or Lowe syndrome	309000	*OCRL*	Trans-Golgi network	XR
Dicarboxylicaminoaciduria	222730	*SLC1A1*	Solute carrier	AR
Iminoglycinuria	242600	*SLC6A19/SLC6A20/SLC36A2*	Solute carriers	Tri

3.04.09.5 Disorders of Sulfur-bearing Amino-acid Metabolism

Cystathioninuria or Cystathionase deficiency: A benign biochemical anomaly characterized by accumulation of plasma cystathionine, which leads to increased urinary excretion. Cystathioninuria is caused by autosomal recessive mutations in the *CTH* gene which encodes the cystathionase or cystathionine gamma-lyase. Cystathioninuria is estimated to affect 1 in every 14,000 live births.

Homocystinuria or Cystathionine beta synthase deficiency: A disorder characterized by ectopia lentis, myopia, mental retardation, skeletal anomalies and thromboembolic events. Affected individuals also suffer from psychiatric disorders, hypopigmentation and hypermethioninemia. There are 2 main forms of homocystinuria: a more severe pyridoxine-nonresponsive form (already described), and a milder pyridoxine-responsive form. Homocystinuria is caused by either autosomal recessive or autosomal dominant mutations in the *CBS* gene which encodes the cystathionine beta-synthase. Heterozygous or monoallelic carriers have a 3-fold increase in cardiovascular risk events over wild type gene carriers. Severe homocystinuria is estimated to affect 1 in every 1 in 344,000 individuals worldwide. There is a higher incidence of severe homocystinuria affecting at least 1 in 20,500 individuals in Denmark, and 1 in 65,000 individuals in Ireland.

Homocystinuria, due to methylenetetrahydrofolate reductase deficiency: A disorder characterized by proximal muscle weakness, ataxia, incoordination, paresthesia and memory lapses. Affected individuals also display a high risk of coronary artery disease, neural tube defects, cleft lip/palate, cancer and spontaneous abortion. Homocystinuria, due to methylenetetrahydrofolate reductase deficiency is caused by autosomal recessive mutations in the *MTHFR* gene which encodes the methylenetetrahydrofolate reductase. This enzyme catalyzes the conversion of 5,10-methylenetetrahydro-folate to 5-methyltetrahydrofolate (co-substrate for homocysteine remethylation to methionine).

Homocystinuria-megaloblastic anemia, due to methionine synthase reductase deficiency: A disorder characterized by megaloblastic anemia and homocystinuria, but without methylmalonic aciduria. Homocystinuria-megaloblastic anemia, due to methionine synthase reductase deficiency is caused by autosomal recessive mutations in the *MTRR* gene which encodes the methionine synthase reductase.

Hypermethioninemia: A disorder characterized by developmental delay, mild dysmorphic features and early-onset seizures. This condition results from an excess of the amino acid methionine in the blood; due to failure in properly broke it down. There are several underlying causes of this disorder with at least 3 autosomal genomic loci: Hypermethioninemia due to adenosine kinase deficiency, Isolated hypermethioninemia and Hypermethioninemia with deficiency of S-adenosylhomocysteine hydrolase. Hypermethioninemia, due to adenosine kinase deficiency can be caused by autosomal recessive mutations in the *ADK* gene which encodes the adenosine kinase;; Isolated hypermethioninemia is caused by autosomal recessive mutations in the *MAT1A* gene which encodes the methionine adenosyltransferase, and there is also a persistent hypermethioninemia caused by autosomal dominant mutations in the *MAT1A* gene;; and Hypermethioninemia with deficiency of S-adenosylhomocysteine hydrolase can be caused by monoallelic mutations in the *AHCY* gene which encodes the adenosylhomocysteinase. The incidence of Hypermethioninemia is difficult to establish, since many individuals with hypermethioninemia have no symptoms.

Methylcobalamin deficiency, cblG type: A disorder characterized by mental retardation, megaloblastic anemia, and high folate activity in the serum and red cells. Methylcobalamin deficiency, cblG type is caused by autosomal recessive mutations in the *MTR* gene which encodes the methionine synthase.

Molybdenum cofactor deficiency: A disorder characterized by neurological damage due to sulphite accumulation. Affected children exhibit early onset seizures, high levels of sulphite and xanthine, and low blood levels of uric acid. Molybdenum cofactor deficiency results from the lack of active sulfite oxidase. Molybdenum cofactor deficiency type A is caused by autosomal recessive mutations in the *MOCS1* gene which encodes the molybdenum cofactor synthesis 1; Molybdenum cofactor deficiency type B is caused by autosomal recessive mutations in the *MOCS2* gene which encodes the molybdenum cofactor synthesis 2; and Molybdenum cofactor deficiency type C is caused by autosomal recessive mutations in the *GPHN* gene which encodes the gephyrin. Molybdenum cofactor deficiency is estimated to affect less than 1 in every 1,000,000 individuals.

Sulfite oxidase deficiency: A fatal disorder characterized by neurological damage due to sulphite accumulation. Affected children exhibit failure to thrive, developmental delay, regression and hypotonia. Sulfite oxidase deficiency is caused by autosomal recessive mutations in the *SUOX* gene which encodes the mitochondrial sulfite oxidase. This mitochondrial enzyme allows the generation of ATP, and belongs to the molybdenum oxotransferases.

Table 3.04.09.5.1 Recognizable Disorders of Sulfur-bearing Amino-acid Metabolism

Disorder	OMIM	Gene	Biological Function	Product of Conception Alteration
Cystathioninuria	219500	CTH	Cystathionine gamma-lyase	AR
Homocystinuria, B6-responsive and nonresponsive types	236200	CBS	Transsulfuration	AR
Homocystinuria	236250	MTHFR	Cytosol reductase	AR
Homocystinuria-megaloblastic anemia	236270	MTRR	Mitochondrial protein	AR
Hypermethioninemia, due to adenosine kinase deficiency	614300	ADK	Adenosine kinase	AR
Methionine adenosyltransferase deficiency	250850	MAT1A	Adenosyltransferase	AR
Hyperhomocysteinemic thrombosis	236200	CBS	Transsulfuration	AR
Hypermethioninemia with deficiency of S-adenosylhomocysteine hydrolase	613752	AHCY	Adenosylhomocysteinase	AD
Methylcobalamin deficiency cblG type	250940	MTR	Methyltransferase	AR
Molybdenum cofactor deficiency type A	252150	MOCS1	Molybdenum cofactor	AR
Molybdenum cofactor deficiency type B	252150	MOCS2	Molybdenum cofactor	AR
Molybdenum cofactor deficiency type C	252150	GPHN	Assembly protein	AR
Sulfite oxidase deficiency	272300	SUOX	Mitochondrial protein	AR

3.04.09.6 Disorders of Urea Cycle and Ornithine Metabolism

Disorders of urea cycle metabolism: Are a group of disorders characterized by the triad of encephalopathy, hyperammonemia and respiratory alkalosis. The urea cycle takes place in the liver, but the excess nitrogen generated is excreted by the kidneys.

Ornithine transcarbamylase deficiency: A disorder characterized by hyperammonemia, affecting various organs of the body. This condition primarily affects male neonates manifesting lethargic features, seizures or comas, who have poorly-controlled breathing rate or body temperature. Complications from this condition include progressive liver damage, skin lesions, developmental delay and mental retardation. Affected females may exhibit dislike for proteins, chronic vomiting, growth retardation, psychiatric disorders, hypotonia, psychomotor retardation, etc. Ornithine transcarbamylase deficiency is caused by X-linked mutations in the *OTC* gene which encodes the ornithine transcarbamylase, a mitochondrial matrix enzyme. Ornithine transcarbamylase deficiency is estimated to affect 1 in every 80,000 births.

Carbamylphosphate synthetase deficiency: A disorder characterized by chronic vomiting, a strong dislike of proteins, growth failure, hypotonia and intellectual deficit. Patients also exhibit hyperglutaminemia, hypocitrullinemia, and hyperammonemia with recurrent hyperammonemic coma during infancy. Carbamylphosphate synthetase deficiency is caused by autosomal recessive mutations in the *CPS1* gene which encodes the carbamoylphosphate synthetase, a mitochondrial enzyme that catalyzes the synthesis of carbamoyl phosphate from ammonia and bicarbonate. In Japan, Carbamyl phosphate synthetase deficiency is estimated to affect 1 in every 800,000 individuals.

Argininosuccinate synthetase deficiency or Citrullinemia type I: A disorder characterized by poor feeding, vomiting, lethargy, mental retardation, seizures, and loss of consciousness in infants after few days or months of life. Affected individuals also suffer from hyperammonemic coma, hypercitrullinemia, hypoargininemia and orotic aciduria. Citrullinemia type I is caused by autosomal recessive mutations in the *ASS1* gene which encodes the argininosuccinate synthetase. This enzyme catalyzes the penultimate step of the arginine biosynthetic pathway. Citrullinemia type I is estimated to affect 1 in every 100,000 births worldwide.

Citrullinemia type II: A disorder characterized by two presentations: intrahepatic cholestasis and no symptom of hyperammonemia in the neonatal form; and symptoms of hyperammonemia and intermediate citrulline levels in the adult form. The latter form affects mainly the nervous system, with features that include abnormal behaviors, confusion, seizures and coma. Citrullinemia type II is caused by autosomal recessive mutations in the *SLC25A13* gene which encodes the solute carrier (family 25 member A13), an aspartate/glutamate carrier. This mitochondrial protein usually shuttles certain molecules in and out. Citrullinemia type II affects mainly people from Japan, where it is estimated to affect 1 in every 100,000 to 230,000 individuals. In Chinese people this pathogenic allele appears 3,000-7,000 years ago. This condition also affects people from the Middle East.

Arginosuccinic aciduria or Argininosuccinate lyase deficiency: A disorder appearing soon after birth characterized by hyperammonemia and arginine deficiency. Affected individuals also suffer from severe hyperammonemic coma, anorexia, vomiting, hypotonia, growth failure and behavioral disorders. Arginosuccinic aciduria is caused by autosomal recessive mutations in the *ASL* gene which encodes the arginosuccinate lyase, a cytosolic enzyme that catalyzes the reversible hydrolytic cleavage of argininosuccinate into arginine and fumarate. Arginosuccinic aciduria is estimated to affect 1 in every 70,000 live births.

Argininemia or Arginase deficiency: A disorder characterized by spastic paraplegia, epileptic seizures, hyperargininemia and severe mental retardation. Affected individuals have impairment in the final step in the urea cycle; the hydrolysis of arginine to urea and ornithine. Argininemia is caused by autosomal recessive mutations in the *ARG1* gene which encodes the liver arginase 1, a cytosolic enzyme that catalyzes the hydrolysis of arginine to ornithine and urea. Argininemia is estimated to affect less than 1 in every 1,000,000 individuals.

N-Acetylglutamate synthase deficiency: A disorder characterized by neonatal hyperammonemia. Other features include vomiting, diarrhea, poor feeding, seizures, hypotonia, hyperactivity or lethargy, delayed psychomotor development and respiratory distress. N-Acetylglutamate synthase deficiency is caused by autosomal recessive mutations in the *NAGS* gene which encodes the mitochondrial N-Acetylglutamate synthase. This mitochondrial enzyme catalyzes the formation of N-acetylglutamate which is an allosteric activator of carbamylphosphate synthetase 1. N-Acetylglutamate synthase deficiency is estimated to affect less than 1 in every 1,000,000 individuals.

Hyperornithinemia, hyperammonemia and homocitrullinuria: A disorder characterized by 3 clinical presentations: neonatal, infantile and juvenile. The neonatal form exhibits hyperammonemia, convulsions and hypotonia, and results in anomalies in the white matter, mental deficiencies and spastic hemiplegia. There are also infantile and juvenile forms. "Hyperornithinemia, hyperammonemia and homocitrullinuria" is caused by defective transportation mechanisms of ornithine into the mitochondria. Affected individuals have plasmatic hyperornithinemia, and elevated excretions of orotic acid, hyperammonemia and homocitrullinuria (derived from lysine). "Hyperornithinemia, hyperammonemia and homocitrullinuria" is caused by autosomal recessive mutations in *SLC25A15* gene which encodes the solute carrier (family 25 member A15), a mitochondrial ornithine transporter. This disorder is rare worldwide, except in certain areas of Canada. In northern Saskatchewan, the carrier frequency for Hyperornithinemia, hyperammonemia and homocitrullinuria trait is estimated to be 1 in 19 individuals, which means the frequency of this condition is about 1 in 1,500 individuals or $1/(38)^2$.

Hyperornithinemia: A disorder characterized by hyperammonemic coma occurring during the neonatal period. Hyperornithinemia is caused by autosomal recessive mutations in *OAT* gene which encodes the ornithine aminotransferase. This condition also displays an allelic form referred to as Gyrate atrophy of the choroid and retina, with normal intelligence.

Gyrate atrophy of the choroid and retina: A disorder characterized by a triad of early cataract formation, type II muscle fiber atrophy, and progressive chorioretinal degeneration with constriction of the visual fields leading to blindness. Gyrate atrophy of the choroid and retina is caused by autosomal recessive mutations in *OAT*

gene. In Finland, the carrier frequency for Gyrate atrophy of the choroid and retina trait is estimated to be 1 in 110 individuals, which means the frequency of this condition is about 1 in 50,000 individuals or $1/(220)^2$.

Table 3.04.09.6.1 Recognizable Disorders of Urea Cycle and Ornithine Metabolism

Disorder	OMIM	Gene	Biological Function	Product of Conception Alteration
Ornithine transcarbamylase deficiency	311250	*OTC*	Mitochondrial protein	X + de novo
Carbamoylphosphate synthetase I deficiency	237300	*CPS1*	Mitochondrial protein	AR
Citrullinemia	215700	*ASS1*	Argininosuccinate synthase	AR
Citrullinemia 2	603471	*SLC25A13*	Solute carrier	AR
Citrullinemia 2, neonatal-onset type	605814	*SLC25A13*	Solute carrier	AR
Argininosuccinic aciduria or argininosuccinic acidemia	207900	*ASL*	Argininosuccinate lyase	AR
Argininemia	207800	*ARG1*	Arginase	AR
N-acetylglutamate synthase deficiency	237310	*NAGS*	Mitochondrial protein	AR
Hyperornithinemia-hyperammonemia-homocitrullinemia syndrome	238970	*SLC25A15*	Solute carrier	AR
Hyperornithinemia	258870	*OAT*	Mitochondrial protein	AR
Gyrate atrophy of choroid and retina with or without ornithinemia	258870	*OAT*	Mitochondrial protein	AR

3.04.09.7 Disorders of Lysine and Hydroxylysine Metabolism

Glutaric acidemia type I or Glutaric aciduria type I: A disorder characterized by patients exhibiting macrocephaly from birth. This metabolic disorder produce gliosis and neuronal loss in the basal ganglia creating a progressive movement disorder that usually begins during the first year of life. Affected patients also suffer from rigidity, hypotonia, spasms and mental retardation; sometimes this condition is referred to as cerebral palsy of genetic origins. Affected individuals are unable to completely break down the amino acids lysine, hydroxylysine and tryptophan. There are excessive levels of brain-toxic intermediate breakdown products, including glutaric acid, glutaryl-CoA, 3-hydroxyglutaric acid and glutaconic acid. Glutaric acidemia type I is caused by autosomal recessive mutations in the *GCDH* gene which encodes the glutaryl-CoA dehydrogenase. This enzyme catalyzes the oxidative decarboxylation of glutaryl-CoA to crotonyl-CoA and CO_2 in the degradative pathway of L-lysine, L-hydroxylysine and L-tryptophan metabolism. Glutaric acidemia type I is estimated to affect 1 in every 100,000 infants worldwide. Amongst the Lancaster County Old Order Amish, the carrier frequency for Glutaric acidemia type I trait is estimated to be 1 in 10 individuals, which means the frequency of this condition is about 1 in 400 individuals or $1/(20)^2$. Among people of Germanic origin, there is a higher carrier frequency than the worldwide average for Glutaric acidemia type I trait.

Glutaric acidemia type II or glutaric aciduria type II: A disorder characterized by two clinical presentations: the neonatal form, which exhibits severe nonketotic hypoglycemia, metabolic acidosis, and excretion of large amounts of fatty acid- and amino acid-derived metabolites; and the adult form, manifesting symptoms of hypoglycemia, metabolic acidosis, lethargy, vomiting and hepatomegaly. Glutaric aciduria type II is an inherited disorder of fatty acid, amino acid and choline metabolism. There are several underlying causes of this disorder with at least 3 autosomal genomic loci: Glutaric aciduria type IIA is caused by autosomal recessive mutations in the *ETFA* gene which encodes the electron-transfer-flavoprotein alpha polypeptide;; Glutaric aciduria type IIB is caused by autosomal recessive mutations in the *ETFB* gene which encodes the electron-transfer-flavoprotein beta polypeptide;; and Glutaric aciduria type IIC is caused by autosomal recessive mutations in the *ETFDH* gene which encodes the electron-transferring-flavoprotein dehydrogenase. These genes encode the subunits of electron transfer flavoproteins. Glutaric aciduria type II impairs a multiple acyl-CoA dehydrogenase. This condition affects not only glutaric acid, but also lactic, ethylmalonic, butyric, isobutyric, 2-methylbutyric, and isovaleric acids.

Hydroxylysinuria: A disorder characterized by mental retardation, myoclonic and other major motor seizures. The gene/loci causing this condition have not been identified. This condition is presumed to be autosomal recessive since the parents of the affected children were related in the reported cases.

Saccharopinemia and/or hyperlysinemia: A disorder characterized by moderate mental retardation with EEG abnormalities. Affected individuals' urine contains large amounts of saccharopine, lysine, citrulline and histidine. Saccharopinemia and/or hyperlysinemia is caused by autosomal recessive mutations in the *AASS* gene which encodes the aminoadipate-semialdehyde synthase, a bifunctional enzyme that catalyzes the first two steps in the lysine degradation pathway.

Lysinuric protein intolerance or dibasic amino aciduria: A disorder characterized by protein intolerance. Signs and symptoms of this condition are relieved by protein restriction. Affected individuals manifest hepatomegaly, low blood urea, hyperammonemia, vomiting, diarrhea and failure to thrive. Lysinuric protein intolerance is caused by autosomal recessive mutations in the *SLC7A7* gene which encodes the solute carrier (family 7 member A7), an amino acid transporter light chain (y+L system). In Finland, Lysinuric protein intolerance is estimated to affect 1 in every 60,000 individuals. There is a higher carrier frequency for Lysinuric protein intolerance trait in Northern Japan. Lysinuric protein intolerance is estimated to affect 1 in every 57,000 individuals in Japan. There is an allelic disorder form referred to as Dibasic amino aciduria that causes large excretions of lysine, ornithine, and arginine. The inheritance described in this condition was autosomal recessive in most families, and autosomal dominant in a French-Canadian kindred.

Table 3.04.09.7 Recognizable Disorders of Lysine and Hydroxylysine Metabolism

Disorder	OMIM	Gene	Biological Function	Product of Conception Alteration
Glutaricaciduria-1	231670	*GCDH*	Mitochondrial protein	AR
Glutaricaciduria-2A	231680	*ETFA*	Mitochondrial protein	AR
Glutaricaciduria-2B	231680	*ETFB*	Mitochondrial protein	AR
Glutaricaciduria-2C	231680	*ETFDH*	Mitochondrial protein	AR
Saccharopinuria or Hyperlysinemia	268700	*AASS*	Bifunctional enzyme	AR
Lysinuric protein intolerance	222700	*SLC7A7*	Solute carrier	AR or AD

3.04.09.8 Disorders of Glycine Metabolism

D-glyceric aciduria: A disorder characterized by patients exhibiting microcephaly, seizures, delayed psychomotor development, mental retardation and neonatal hypotonia. Affected individuals suffer from nonketotic hyperglycinemia with the excretion of D-glyceric acid in the urine. D-glyceric aciduria is caused by autosomal recessive mutations in the *GLYCTK* gene which encodes the glycerate kinase. This enzyme catalyzes the phosphorylation of (R)-glycerate and may be involved in serine degradation and fructose metabolism.

Hyperhydroxyprolinemia: A disorder characterized by hyperhydroxyprolinemia without a clear metabolic phenotype. The gene responsible for this condition has not been identified. This condition is presumed to follow an autosomal recessive pattern of inheritance.

Hyperprolinemia or prolinuria: A disorder characterized by mental retardation, seizures or other neurological problems, but most of the people with this condition are asymptomatic. Affected individuals cannot break down the amino acid proline, causing a buildup of proline in the body. Affected individuals also have hydroxyprolinuria and glycinuria. There are several underlying causes of this disorder with at least 2 autosomal genomic loci: Hyperprolinemia-1 is caused by autosomal recessive mutations in the *PRODH* gene which encodes the proline dehydrogenase. This protein starts the process of degrading proline. The *PRODH* gene falls within the region deleted in the 22q11 deletion syndrome, including DiGeorge syndrome and velocardiofacial syndrome;; and Hyperprolinemia-2 is caused by autosomal recessive mutations in the *ALDH4A1* gene which encodes the aldehyde dehydrogenase (family4 member A1), a mitochondrial matrix dehydrogenase. This enzyme catalyzes the second step of the proline degradation pathway, breaking down the pyrroline-5-carboxylate.

Prolidase deficiency: A disorder characterized by skin lesions predominantly on the face, palms, lower legs and soles. Affected individuals excrete considerable amounts of imidodipeptides in urine. Prolidase deficiency is caused by autosomal recessive mutations in the *PEPD* gene which encodes the exopeptidase prolidase. Prolidase deficiency is estimated to affect less than 1 in every 1,000,000 individuals.

Sarcosinemia or Hypersarcosinemia: A relatively benign condition characterized by an increased concentration of sarcosine in blood plasma and urine. Sarcosinemia is caused by autosomal recessive mutations in the *SARDH* gene which encodes the sarcosine dehydrogenase, a mitochondrial matrix enzyme that catalyzes the oxidative demethylation of sarcosine. Sarcosinemia is estimated to affect 1 in every 28,000-350,000 live births.

Dimethylglycine dehydrogenase deficiency: A disorder characterized by an abnormal fish-like body odor. The odor increases under stress and effort, and leads to professional and psychological problems. Dimethylglycine dehydrogenase deficiency is caused by autosomal recessive mutations in the *DMGDH* gene which encodes the dimethylglycine dehydrogenase. This enzyme is involved in the catabolism of choline, catalyzing the oxidative demethylation of dimethylglycine to form sarcosine. Dimethylglycine dehydrogenase deficiency is estimated to affect less than 1 in every 1,000,000 individuals.

Glycine encephalopathy or non-ketotic hyperglycinemia: A disorder characterized by encephalopathy. During the first few days of life, affected neonates manifest hypotonia, lethargy, and myoclonic jerks that progress into apnea and often to death. Other features include intractable seizures and profound mental retardation. Glycine encephalopathy is caused by a defect in the glycine cleavage system. This enzyme is responsible for glycine catabolism. Glycine encephalopathy is caused by autosomal recessive mutations in the *AMT* gene which encodes the aminomethyltransferase, or *GCSH* gene which encodes the glycine cleavage system protein H, or the *GLDC* gene which encodes the pyridoxal phosphate-dependent glycine decarboxylase. Most patients with Glycine encephalopathy have a defect in the *GLDC* gene. On average, the carrier frequency for Glycine encephalopathy trait is estimated to be 1 in 122.5 individuals worldwide, which means the frequency of this condition is about 1 in 60,000 individuals or $1/(245)^2$. After phenylketonuria, this condition is the second most prevalent disorder of amino acid metabolism.

Glycine N-methyltransferase deficiency: A disorder characterized by mild hepatomegaly, persistent isolated hypermethioninemia, and chronic elevation of serum transaminases. Glycine N-methyltransferase deficiency is caused by autosomal recessive mutations in the *GNMT* gene which encodes the Glycine N-methyltransferase.

Glutathione synthetase deficiency: A disorder characterized by three presentations: mild, moderate and severe forms. Affected patients cannot produce enough glutathione, which helps to prevent damage to cells. In severe cases, patients with this condition manifest neurological symptoms; including ataxia, seizures, psychomotor, metabolic acidosis, mental retardation and recurrent bacterial infections. The moderate form of glutathione synthetase deficiency courses with hemolytic anemia and metabolic acidosis of onset after birth. The mild form of glutathione synthetase deficiency courses with hemolytic anemia. Glutathione synthetase deficiency is caused by autosomal recessive mutations in the *GSS* gene which encodes the glutathione synthetase. This enzyme catalyzes the conversion of gamma-L-glutamyl-L-cysteine to glutathione. Glutathione syn-

thetase deficiency is estimated to affect less than 1 in every 1,000,000 individuals.

Table 3.04.09.8 Recognizable Disorders of Glycine Metabolism

Disorder	OMIM	Gene	Biological Function	Product of Conception Alteration
D-glyceric aciduria	220120	GLYCTK	Glycerate kinase	AR
Hyperprolinemia-1	239500	PRODH	Mitochondrial protein	AR
Hyperprolinemia-2	239510	ALDH4A1	Oxidoreductase	AR
Prolidase deficiency	170100	PEPD	Recycling of proline	AR
Sarcosinemia or hypersarcosinemia	268900	SARDH	Mitochondrial matrix protein	AR
Dimethylglycine dehydrogenase deficiency	605850	DMGDH	Mitochondrial protein	AR
Glycine encephalopathy or non-ketotic hyperglycinemia	605899	AMT	Mitochondrial protein	AR
Glycine encephalopathy or non-ketotic hyperglycinemia	605899	GCSH	Mitochondrial protein	AR
Glycine encephalopathy or non-ketotic hyperglycinemia	605899	GLDC	Mitochondrial protein	AR
Glycine N-methyltransferase deficiency	606664	GNMT	Methyltransferase	AR
Glutathione synthetase deficiency	231900	GSS	Glutathione synthetase	AR

3.04.09.9 Other Specified Disorders of Amino-acid Metabolism

Succinic semialdehyde dehydrogenase deficiency or 4-hydroxybutyricaciduria deficiency: A disorder characterized by neurological symptoms. Affected individuals manifest psychomotor retardation, delayed speech development, hypotonia and ataxia. Patients display the accumulation of gamma-hydroxybutyrate in urine, plasma and cerebro-spinal fluid. This condition is an inborn error in the metabolism of the neurotransmitter GABA or 4-aminobutyric acid. Succinic semialdehyde dehydrogenase deficiency is caused by

autosomal recessive mutations in the *ALDH5A1* gene which encodes the aldehyde dehydrogenase (family 5 member A1), a mitochondrial NAD(+)-dependent succinic semialdehyde dehydrogenase. Succinic semialdehyde dehydrogenase deficiency is estimated to affect less than 1 in every 1,000,000 individuals.

Congenital glutamine synthetase deficiency: A disorder characterized by encephalopathy after birth. Affected individuals suffer from hyperreflexia, seizures, generalized muscular hypotonia and variable brain abnormalities. Congenital glutamine synthetase deficiency is caused by autosomal recessive mutations in the

GLUL gene which encodes the glutamate-ammonia ligase. This enzyme catalyzes the synthesis of glutamine from glutamate and ammonia. Most of the affected patients are children of consanguineous parents. Congenital glutamine deficiency is estimated to affect less than 1 in every 1,000,000 individuals.

Gamma aminobutyric acid transaminase deficiency: A disorder characterized by hypotonia, hyperreflexia, lethargy, severe psychomotor retardation and refractory seizures. Gamma aminobutyric acid transaminase deficiency is caused by autosomal recessive mutations in the *ABAT* gene which encodes the gamma aminobutyric acid transaminase. This enzyme metabolizes the GABA neurotransmitter. Gamma aminobutyric acid transaminase deficiency is estimated to affect less than 1 in every 1,000,000 individuals.

Arginine:glycine amidinotransferase deficiency: A disorder characterized by psychomotor retardation, due to impairment in energetic metabolism which affects the level of creatine synthesis. These patients displays low plasmatic levels for both guanidinoacetate and creatine concentrations. Arginine:glycine amidinotransferase deficiency is caused by autosomal recessive mutations in the *GATM* gene which encodes the glycine amidinotransferase (L-arginine:glycine amidinotransferase). Arginine:glycine amidinotransferase deficiency is estimated to affect less than 1 in every 1,000,000 individuals.

Phosphoglycerate dehydrogenase deficiency: A disorder characterized by microcephaly, psychomotor retardation, seizures, hypogonadism and hypertonia. This condition is a disorder of L-serine biosynthesis. Phosphoglycerate dehydrogenase deficiency is caused by autosomal reces-

sive mutations in the *PHGDH* gene which encodes the phosphoglycerate dehydrogenase.

Phosphoserine aminotransferase deficiency: A disorder characterized by microcephaly, hypertonia, intractable seizures and psychomotor retardation. Affected individuals show low concentrations of serine and glycine in both the plasma and CSF. Phosphoserine aminotransferase deficiency is caused by autosomal recessive mutations in the *PSAT1* gene which encodes the phosphoserine aminotransferase 1. Phosphoserine aminotransferase deficiency is estimated to affect less than 1 in every 1,000,000 individuals.

Creatine deficiency syndrome: A disorder characterized by failure to thrive, seizures, hypotonia, severe speech delay, mental retardation and behavioral abnormalities. Creatine deficiency syndrome is caused by X-linked recessive mutations in the *SLC6A8* gene which encodes the solute carrier (family 6 member A8), a creatine transporter. Most of the patients are males, sometimes female carriers show mild neuropsychological impairment.

Guanidinoacetate methyltransferase deficiency: A disorder characterized by intractable seizures, movement disturbances, developmental delay, mental retardation, and severe disturbance of expressive and cognitive speech. The effects of deficiency in this enzyme are more prominent among organs and tissues that require large amounts of energy, such as the brain and muscles. Guanidinoacetate methyltransferase deficiency is caused by autosomal recessive mutations in the *GAMT* gene which encodes the guanidinoacetate methyltransferase. This enzyme is involved in creatine synthesis. Guanidinoacetate methyltransferase deficiency is estimated to affect less than 1 in every 1,000,000 individuals.

Hyperinsulinism-hyperammonemia syndrome: A disorder characterized by hypoglycemia, hyperinsulinism and hyperammonemia. Hyperinsulinism-hyperammonemia syndrome is caused by monoallelic mutations in the *GLUD1* gene which encodes the glutamate dehydrogenase 1, a mitochondrial matrix enzyme that catalyzes the oxidative deamination of glutamate to alpha-ketoglutarate and ammonia. The product of conception in hyperinsulinism-hyperammonemia syndrome appears to either exhibit *de novo* alteration in a few cases or autosomal dominant inheritance in most cases.

Malonyl-CoA decarboxylase deficiency or Malonic aciduria: A disorder characterized by hypoglycemia, metabolic acidosis, ketosis, lactic acidemia, diarrhea, vomiting, cardiomyopathy, hypotonia, seizures, and developmental delay in early childhood. Many tissues cannot use fatty acids as a source of energy. Malonyl-CoA decarboxylase deficiency is caused by autosomal recessive mutations in the *MLYCD* gene which encodes the malonyl-CoA decarboxylase. This enzyme breaks down malonyl-CoA into acetyl-CoA and carbon dioxide. Malonyl-CoA decarboxylase deficiency is estimated to affect less than 1 in every 1,000,000 individuals.

Table 3.04.09.9.1 Other Recognizable Disorders of Amino-acid Metabolism

Disorder	OMIM	Gene	Biological Function	Product of Conception Alteration
Succinic semialdehyde dehydrogenase deficiency or 4-hydroxybutyricaciduria deficiency	271980	ALDH5A1	Oxidoreductase	AR
Congenital glutamine synthetase deficiency	610015	GLUL	Glutamine synthetase	AR
Gamma aminobutyric acid transaminase deficiency	613163	ABAT	Mitochondrial protein	AR
Arginine:glycine amidinotransferase deficiency	612718	GATM	Mitochondrial protein	AR
Phosphoglycerate dehydrogenase deficiency	601815	PHGDH	Dehydrogenase	AR
Phosphoserine aminotransferase deficiency	610992	PSAT1	Aminotransferase	AR
Creatine deficiency syndrome	300352	SLC6A8	Solute carrier	XR + de novo
Guanidinoacetate methyltransferase deficiency	612736	GAMT	Methyltransferase	AR
Hyperinsulinism-hyperammonemia syndrome	606762	GLUD1	Mitochondrial protein	AD
Malonyl-CoA decarboxylase deficiency	248360	MLYCD	Mitochondrial protein	AR

3.04.10 Carbohydrates Disorders

Adult-type hypolactasia or Lactose intolerance: A common disorder characterized by flatulence, diarrhea, nausea, abdominal pain, bloating, and acid reflux after consumption of milk and other dairy products. These symptoms starts during adulthood and are caused by the inability to digest and metabolize lactose, by the lactase enzyme. The activities of lactase and most of the other digestive hydrolases are maximal at birth. The decline in lactase occurs at approximately 3 to 5 years of age. This condition is more common among African, Asian, and Southern European people than people from northern European ancestry. Lactose intolerance is caused by either genomic or environmentally factors with insufficient levels of lactase enzyme in the lining of the duodenum. Lactose intolerance is caused by autosomal recessive mutations in the *MCM6* gene which encodes the minichromosome maintenance complex (component 6), an essential protein for the initiation of eukaryotic genome replication. This gene is also associated with lactase persistence/nonpersistence.

Lactase persistence: A disorder with the opposite phenotype than lactose intolerance. Lactase persistence is associated with prolonging the activity of lactase beyond infancy. This condition is caused by autosomal dominant mutations in the *MCM6* gene, which is also associated with lactase persistence/nonpersistence. This condition affects mainly northern European people.

Congenital lactase deficiency: A disorder characterized by severe gastrointestinal watery diarrhea. This neonatal condition usually occurs on the third day of life in either breast fed or other lactose-containing formulas infants. Congenital lactase deficiency is caused by autosomal recessive mutations in the *LCT* gene which encodes the lactase. This enzyme has both phlorizin hydrolase activity and lactase activity. The worldwide incidence of this condition is unknown. In Finland, the carrier frequency for Congenital lactase deficiency trait is estimated to be 1 in 122.5 individuals, which means the frequency of this condition is about 1 in 60,000 individuals or $1/(245)^2$.

3.04.10.1 Glycogen Storage Disease

Glycogen storage disease: A group of disorder characterized by an alteration in the processing of glycogen, either affecting the synthesis or breakdown within muscles, liver and other cell types. Glycogen storage disease could be caused by genomic and environmental factors such as intoxication with the alkaloid castanospermine. Glycogen storage disease is mainly caused by defective enzymes and transporters. In the United States, the carrier frequency for Glycogen storage disease trait is estimated to be 1 in 70-80 individuals, which means the frequency of this condition is about 1 in 20,000-25,000 individuals or $1/(140)^2$ - $1/(160)^2$. There are 15 distinct forms of diseases that are commonly considered to be glycogen storage diseases. These conditions are caused by mutations in at least 24 genes. Some of these conditions have been reclassified. Glycogen storage diseases type 0 does not result in storage of extra glycogen in the liver. Glycogen storage diseases type VIII has been removed. Glycogen storage diseases type XI has two distinct conditions.

Muscle glycogen storage disease type 0 or muscle glycogen synthase deficiency: A disorder characterized by muscle glycogen storage. Muscle glycogen storage disease type 0 is caused by autosomal recessive mutations in the *GYS1* gene which encodes the muscle glycogen synthase. This condition has been reported in the offspring of consanguineous parents of Syrian origin.

Liver glycogen storage disease type 0 or liver glycogen synthase deficiency: A disorder characterized by fasting hypoglycemia and postprandial hyperglycemia. This hepatic glycogen storage is caused by autosomal recessive mutations in the *GYS2* gene which encodes the liver glycogen synthase.

Lethal congenital Glycogen storage disease of heart: A disorder characterized by extreme cardiomegaly and death during infancy. Lethal congenital Glycogen storage disease of heart can be caused by monoallelic mutations in the *PRKAG2* gene which encodes the noncatalytic gamma-2 subunit of AMP-activated protein kinase. Mutations in the *PRKAG2* gene are also associated with Wolff-Parkinson-White syndrome and with familial hypertrophic cardiomyopathy. Lethal congenital Glycogen storage disease of heart is estimated to affect less than 1 in every 1,000,000 individuals.

Glycogen storage disease type I or von Gierke disease: A disorder characterized by severe hypoglycemia, which can lead to seizures, and hepatomegaly appearing during the first 3 or 4 months of life. These infants also exhibit hyperlipidemia and xanthomas in the skin, hyperuricemia, lactic acidemia, growth retardation, delayed puberty, diarrhea, hepatic adenomas and pulmonary hypertension. Glycogen storage disease type I results from the accumulation of glycogen in certain organs and tissues, particularly the liver, kidneys and small intestines, impairing their function.

Glycogen storage disease Ia is caused by autosomal recessive mutations in the *G6PC* gene which encodes the glucose-6-phosphatase. Among Caucasians, the carrier frequency for Glycogen storage disease type I trait is estimated to be 1 in 158-274 individuals, which means the frequency of this condition is about 1 in 100,000-300,000 individuals or $1/(316)^2$ - $1/(548)^2$. There is a higher carrier frequency for Glycogen storage disease type I trait among Ashkenazi Jews, affecting 1 in 20,000 individuals.

Glycogen storage disease type Ib: A similar disorder as von Gierke disease, but instead is characterized by inability to liberate glucose. Affected individuals show a normal activity in glucose-6-phosphatase. Glycogen storage disease Ib is caused by autosomal recessive mutations in the *SLC37A4* gene which encodes the solute carrier (family 37 member A4), a glucose-6-phosphate transporter.

Glycogen storage disease type Ic: A similar disorder as von Gierke disease, but instead is characterized by inability to liberate glucose. Affected individuals show a normal activity in glucose-6-phosphatase. Glycogen storage disease Ic is caused by autosomal recessive mutations in the *SLC17A3* gene which encodes the solute carrier (family 17 member A3), a glucose-6-phosphate transporter.

Glycogen storage disease type II or Pompe disease: A disorder that is the prototype of lysosomal storage disease. This disorder is characterized by three clinical presentations: classic infantile, non-classic form of infantile-onset, and late-onset forms. Patients with the classic infantile presentation of Pompe disease exhibit muscular hypotonia and cardiomyopathy within a few months of birth. They also have a failure to thrive, hepatomegaly and breathing problems. In the non-classic infantile-onset presentation of Pompe disease, patients are characterized by delayed motor skills and progressive muscle weakness by the 1st year of age. Affected individuals also exhibit cardiomegaly but usually without experiencing heart failure. Patients with the late-onset presentation of Pompe disease are apparently unaware until later in childhood. It is characterized by muscle weakness leading to respiratory failure. This form is less likely to involve the heart. Glycogen storage disease type II is caused by autosomal recessive mutations in the *GAA* gene which encodes the acid alpha-1,4-glucosidase or is also known as acid maltase. This is an essential enzyme for the degradation of glycogen to glucose in lysosomes In the United States, the carrier frequency for Pompe disease trait is estimated to be 1 in 100 individuals, which means the frequency of this condition is about 1 in 40,000 individuals or $1/(200)^2$. The incidence of Pompe disease varies among different ethnic groups, occuring most frequently among Palestinian Arabs.

Glycogen storage disease type IIb or Danon disease: A disorder characterized by heart muscle abnormalities resembling severe hypertrophic cardiomyopathy. This condition mainly affects male patients who also suffer from skeletal myopathy and intellectual disability. Patients with Danon disease also have palpitations and arrhythmia, including a clear genomic cause of Wolff-Parkinson-White disease. This condition was named after Moris Danon in 1981. Most male patients with this condition have intellectual disability, which is less common among females. Males with Danon disease habitually develop more severe symptoms than affected females, and their condition more often progresses to earlier onset form. The average lifespan is 19 years for males and 34 years for fe-

males. Danon disease is caused by X-linked dominant mutations in the *LAMP2* gene which encodes the lysosomal-associated membrane protein 2. This protein plays a role in autophagosome-lysosome fusion. Danon disease is estimated to affect less than 1 in every 1,000,000 individuals.

Glycogen storage disease type III or Cori Disease or glycogen debranching enzyme deficiency: A disorder characterized by hypoglycemia, hyperlipidemia, and hepatomegaly with elevated blood levels of liver enzymes. Affected individuals suffer from a slowly progressing muscle weakness and muscle wasting. There are patients who display signs of congestive heart failure. Glycogen storage disease type III is caused by autosomal recessive mutations in the *AGL* gene which encodes the glycogen debrancher enzyme. In the United States, the carrier frequency for Cori disease trait is estimated to be 1 in 158 individuals, which means the frequency of this condition is about 1 in 100,000 individuals or $1/(316)^2$. There is a higher carrier frequency for Cori disease trait among North African Jews. This trait is estimated to be 1 in 36 individuals, which means the frequency of this condition is about 1 in 5,400 individuals or $1/(72)^2$.

Glycogen storage disease type IV or Andersen disease or Glycogen branching enzyme deficiency: A disorder in which some patients exhibit a severe phenotype with underdeveloped lungs and hydrops fetalis before birth. There are other signs and symptoms within the first few months of life including hepatomegaly leading to liver failure, arthrogryposis, hypotonia and muscle atrophy, cardiomyopathy and splenomegaly. Glycogen storage disease type IV is caused by autosomal recessive mutations in the *GBE1* gene which encodes the glycogen branching enzyme. Mutations in the *GBE1* gene are also associated with adult polyglucosan body disease. Glycogen storage disease type IV is estimated to affect 1 in every 600,000-800,000 individuals.

Glycogen storage disease type V or McArdle disease: A disorder characterized by intolerance of exercise past the first few minutes. Affected patients experience fatigue, muscle pain and cramps when performing intense exercises, and can later experience rhabdomyolysis. Affected individuals, in severe episodes, experience myoglobinuria with life-threatening kidney failure. Glycogen storage disease type V is caused by autosomal recessive mutations in the *PYGM* gene which encodes the muscle glycogen phosphorylase. In the Dallas-Fort Worth area, the carrier frequency for Glycogen storage disease type V trait is estimated to be 1 in 158 individuals, which means the frequency of this condition is about 1 in 100,000 individuals or $1/(316)^2$.

Glycogen storage disease type VI or Hers disease: A disorder characterized by prominent hepatomegaly, mild to moderate hypoglycemia, mild ketosis and growth retardation. Hers disease does not affect the heart and/or the skeletal muscle. Glycogen storage disease type VI is caused by autosomal recessive mutations in the *PYGL* gene which encodes the liver glycogen phosphorylase. This disorder is rare worldwide except among Mennonites. The carrier frequency for Glycogen storage disease type VI trait is estimated to be 1 in 15.5 individuals, which means the frequency of this condition is about 1 in 1,000 individuals or $1/(33)^2$.

Glycogen storage disease type VII or Tarui disease or muscle phosphofructokinase deficiency: A disorder characterized by exercise intolerance, associated

nausea and vomiting. Patients experience severe muscle cramps and myoglobinuria after vigorous exercise. Glycogen storage disease type VII is caused by autosomal recessive mutations in the *PFKM* gene which encodes the muscle phosphofructokinase. Glycogen storage disease type VII is estimated to affect less than 1 in every 1,000,000 individuals. This disorder is particularly prevalent among people of Ashkenazi Jewish ancestry.

Glycogen storage disease type IXa1-IXa2 or liver glycogenosis: A disorder characterized by hepatomegaly, hypoglycemia and ketosis. Affected individuals also suffer from growth retardation, delayed puberty, liver fibrosis and cirrhosis. Glycogen storage disease type IXa1- IXa2 is caused by X-linked recessive mutations in the *PHKA2* gene which encodes the hepatic phosphorylase kinase. On average, Glycogen storage disease type IX trait is estimated to affect 1 in 100,000 individuals worldwide.

Glycogen storage disease type IXb or Phosphorylase kinase deficiency of liver and muscle: A disorder characterized by liver and muscle symptoms. Glycogen storage disease type IXb is caused by autosomal recessive mutations in the *PHKB* gene which encodes the liver and muscle phosphorylase kinase.

Glycogen storage disease type IXc: A disorder characterized by liver symptoms. Glycogen storage disease type IXc is caused by autosomal recessive mutations in the *PHKG2* gene which encodes the liver phosphorylase kinase.

Glycogen storage disease type IXd or muscle glycogenosis: A disorder characterized by muscle symptoms. Glycogen storage disease type IXd is caused by autosomal recessive mutations in the *PHKA1* gene which encodes the muscle phosphorylase kinase.

Glycogen storage disease type X or muscle phosphoglycerate mutase deficiency: A disorder characterized by exercise-induced cramps, occasional myoglobinuria and intolerance for strenuous exercise. Glycogen storage disease type X is caused by autosomal recessive mutations in the *PGAM2* gene which encodes the muscle phosphoglycerate mutase. Glycogen storage disease type X is estimated to affect less than 1 in every 1,000,000 individuals.

Fanconi-Bickel syndrome or Glycogen storage disease type XI: A disorder characterized by hepatorenal glycogen accumulation, proximal renal tubular dysfunction, and impaired utilization of glucose and galactose. Fanconi-Bickel syndrome is caused by autosomal recessive or autosomal dominant mutations in the *SLC2A2* gene which encodes the solute carrier (family 2 member A2), a facilitated glucose transporter. Fanconi-Bickel syndrome is estimated to affect less than 1 in every 1,000,000 individuals.

Glycogen storage disease type XI or lactate dehydrogenase A deficiency: A disorder characterized by painful muscle stiffness, exertional myoglobinuria, easy fatigue, and renal failure. Lactate dehydrogenase A deficiency is caused by autosomal recessive mutations in the *LDHA* gene which encodes the A subunit of lactate dehydrogenase. In the Shizuoka Prefecture, Japan, the carrier frequency for Lactate dehydrogenase A deficiency trait is estimated to be 1 in 540 individuals, which means the frequency of this condition is about 1 in 1,167,000 individuals or $1/(1,080)^2$.

Glycogen storage disease type XII or Aldolase A deficiency: A disorder characterized by hemolytic anemia leading to myopathy with exercise intolerance and rhabdomyolysis. Glycogen storage disease type XII is caused by autosomal recessive mutations in the *ALDOA* gene which encodes the aldolase A. This enzyme is primarily located in red blood cells and skeletal muscle.

Glycogen storage disease type XIII or Enolase 3 deficiency: A disorder characterized by generalized muscle weakness, exercise-induced myalgia and fatigability. Glycogen storage disease type XIII is caused by autosomal recessive mutations in the *ENO3* gene which encodes the enolase 3.

Glycogen storage disease type XIV: A disorder characterized by myopathy, exercise-induced intolerance and rhabdomyolysis episodes. Glycogen storage disease type XIV is caused by autosomal recessive mutations in the *PGM1* gene which encodes the phosphoglucomutases. This enzyme catalyzes the transfer of phosphate between the 1 and 6 positions of glucose.

Glycogen storage disease type XV or glycogenin deficiency: A disorder characterized by glycogen depletion in skeletal and cardiac muscle. Affected individuals exhibits muscle weakness and cardiac arrhythmia. Glycogen storage disease type XV is caused by autosomal recessive mutations in the *GYG1* gene which encodes the glycogenin. This enzyme is essential to the initiation reactions of glycogen synthesis. Glycogen storage disease type XV is estimated to affect less than 1 in every 1,000,000 individuals.

3.04.10.2 Disorders of Fructose Metabolism

Essential fructosuria or hepatic fructokinase deficiency: A benign disorder characterized by asymptomatic fructosuria. The degree of fructosuria is contingent on the dietary fructose intake. Essential fructosuria does not have any clinical manifestations. Essential fructosuria is caused by autosomal recessive mutations in the *KHK* gene which encodes the ketohexokinase (fructokinase). This enzyme catalyzes the conversion of fructose to fructose-1-phosphate. Essential fructosuria is estimated to affect 1 in every 130,000 individuals. The actual prevalence is probably higher.

Fructose intolerance: A disorder characterized by severe hypoglycemia at the time of weaning during infancy, when fruits and juices are incorporated into the diet. Patients also exhibit recurrent nausea, vomiting, bloating, diarrhea and abdominal pain. Affected individuals also develop metabolic acidosis, fructosemia, hypophosphatemia and hyperuricemia. In the cases of long-term exposure to fructose, patients can develop liver failure, renal tubulopathy and growth retardation. Fructose intolerance is caused by autosomal recessive mutations in the *ALDOB* gene which encodes the aldolase B. This enzyme converts fructose-1-phosphate to DHAP and glyceraldehyde. Among Caucasians, the carrier frequency for Fructose intolerance trait is estimated to be 1 in 71 individuals, which means the frequency of this condition is about 1 in 20,000 individuals or $1/(142)^2$.

Fructose-1,6-bisphosphatase deficiency: A disorder characterized by recurrent attacks of metabolic acidosis and hypoglycemia. Patients are asymptomatic between attacks or after ingestion of fructose. Fructose-1,6-bisphosphatase deficiency is caused by autosomal recessive mutations in the *FBP1* gene which encodes the fructose-1,6-bisphos-phatase 1. Fructose-1,6-bisphosphatase deficiency is estimated to affect 1 in every 20,000 live births worldwide.

Fructose malabsorption: A disorder characterized by stomach pain, bloating, diarrhea, and flatulence when fructose is ingested. This condition is caused by inability of the intestine to absorb fructose normally. This condition is common among people suffering from other digestive disorders, such as irritable bowel syndrome, lactose intolerance and celiac disease. In the Western hemisphere, fructose malabsorption is thought to affect 30 to 40 percent of individuals.

3.04.10.3 Disorders of Galactose Metabolism

Galactosemia type I or Classic Galactosemia: A disorder characterized by food intolerance, hypoglycemia, jaundice, hepatomegaly, hepatocellular and renal insufficiency. This condition if untreated could cause cirrhosis, cataracts and brain damage. It is caused by inability to metabolize galactose properly which leads to toxic levels of galactose 1-phosphate. Galactosemia type I is caused by autosomal recessive mutations in the *GALT* gene which encodes the galactose-1-phosphate uridyltransferase. This enzyme catalyzes the conversion of UDP-glucose + galactose-1-phosphate to glucose-1-phosphate + UDP-galactose. Among Caucasians, the carrier frequency for Galactosemia type I trait is estimated to be 1 in 122.5 individuals, which means the frequency of this condition is about 1 in 60,000 individuals or $1/(245)^2$. There is a higher carrier frequency for Galactosemia trait among Irish traveler people.

Galactosemia type II or Galactokinase deficiency with cataracts: A disorder characterized by cataract formation in children. This condition results from the accumulation of galactitol in the lens, due to an osmotic phenomenon. Galactosemia type II is caused by autosomal recessive mutations in the *GALK1* gene which encodes the galactokinase.

Galactosemia type III or Epimerase-deficiency galactosemia: A benign disorder characterized by complete absence of galactose epimerase activity. Galactosemia type III is caused by autosomal recessive mutations in the *GALE* gene which encodes the UDP-galactose-4-epimerase gene. Among African Americans in Maryland, the carrier frequency for Galactosemia type III trait is estimated to be 1 in 39.5 individuals, which means the frequency of this condition is about 1 in 6,200 individuals or $1/(79)^2$. This condition seems to be even less prevalent among other non-African Americans from Maryland whereas 1 in 64,800 individuals were affected. This benign condition is less prevalent among the Japanese where the complete absence of galactose epimerase activity affects 1 in 23,000 individuals.

3.04.10.4 Other Disorders of Intestinal Carbohydrate Absorption

Congenital sucrase-isomaltase deficiency: A disorder characterized by osmotic diarrhea when the disaccharide sucrase-isomaltase is ingested. This condition results from inability to metabolize sucrose in the small intestine brush-border. Congenital sucrase-isomaltase deficiency is caused by autosomal recessive mutations in the *SI* gene which encodes the sucrase-isomaltase or alpha-glucosidase. This enzyme is expressed in the intestinal brush border. In adults, intestinal sucrase deficiency is a common cause of diarrhea affecting almost 0.2% of people. This condition is even more common among Greenland Eskimos and Alaskan Eskimos affecting at least 10 % of these among these groups.

Glucose-galactose malabsorption: A disorder characterized by severe diarrhea, dehydration and intermittent glucosuria, if glucose and galactose are not eliminated from the diet. Glucose-galactose malabsorption is caused by autosomal recessive mutations in the *SLC5A1* gene which encodes the solute carrier (family 5 member A1), a sodium/glucose cotransporter protein. Glucose-galactose malabsorption is estimated to affect less than 1 in every 1,000,000 individuals.

Trehalase deficiency: A disorder characterized by diarrhea and vomiting after ingestion of trehalose, a disaccharide found primarily in mushrooms. Trehalase deficiency is caused by autosomal recessive mutations in the *TREH* gene which encodes the trehalase, an enzyme that hydrolyses trehalose. This membrane glycoprotein is expressed in the intestinal brush border. Trehalase deficiency is common among people from Greenland, affecting an estimated 1 in 13 individuals. In most other groups of people, this condition is rare.

3.04.10.5 Disorders of Pyruvate Metabolism and Gluconeogenesis

Pyruvate carboxylase deficiency: A disorder characterized by two clinical presentations: type A and type B. Patients with the Pyruvate carboxylase deficiency type

A manifest moderately severe symptoms of infancy onset. This form has been identified mostly in people from North America. Affected individuals exhibit developmental delay and lactic acidosis. Usually, these infants display vomiting, abdominal pain, muscle weakness, fatigue and difficulty breathing. These children normally survive only into early childhood. Patients with the Pyruvate carboxylase deficiency type B manifest a life-threatening condition, starting shortly after birth. This form has mostly been identified in people from Europe, principally from France. Affected individuals develop severe lactic acidosis, hyperammonemia, liver failure, hypotonia, seizures and coma. These infants normally survive for less than 3 months after birth. Pyruvate carboxylase deficiency is caused by autosomal recessive mutations in the *PC* gene which encodes the mitochondrial pyruvate carboxylase. This enzyme requires biotin and ATP to catalyse the carboxylation of pyruvate to oxaloacetate. This gene is involved in gluconeogenesis, lipogenesis, insulin secretion and synthesis of the neurotransmitter glutamate. This condition affects the nervous system by impairing myelin formation and neurotransmitter production. On average, the carrier frequency for Pyruvate carboxylase deficiency trait is estimated to be 1 in 250 individuals worldwide, which means the frequency of this condition is about 1 in 250,000 individuals or $1/(500)^2$. There is a higher carrier frequency for Pyruvate carboxylase deficiency trait among Algonkian Indian tribes in eastern Canada.

Pyruvate dehydrogenase deficiency: A disorder characterized by lactic acidemia. Affected individuals exhibit developmental defects, particularly of the brain and nervous system, displaying muscular spasticity and early death. The genes involved in pyruvate dehydrogenase deficiency are essential to life, and are involved in chem-ical reaction between glycolysis and the citric acid cycle. There are several underlying causes of this disorder with at least 1 X-linked genomic locus and 6 autosomal genomic loci and: Pyruvate dehydrogenase E1-alpha deficiency is caused by X-linked dominant mutations in the *PDHA1* gene which encodes the alpha 1 pyruvate dehydrogenase;; Pyruvate dehydrogenase E1-beta deficiency is caused by autosomal recessive mutations in the *PDHB* gene which encodes the beta 1 pyruvate dehydrogenase;; Pyruvate dehydrogenase E2 deficiency is caused by autosomal recessive mutations in the *DLAT* gene which encodes the dihydrolipoamide S-acetyltransferase;; Pyruvate dehydrogenase E3-binding protein deficiency or Lacticacidemia is caused by autosomal recessive mutations in the *PDX1* gene which encodes the pyruvate dehydrogenase complex;; Pyruvate dehydrogenase phosphatase deficiency is caused by autosomal recessive mutations in the *PDP1* gene which encodes the pyruvate dehydrogenase phosphatase (catalytic subunit 1);; and Pyruvate dehydrogenase lipoic acid synthetase deficiency is caused by autosomal recessive mutations in the *LIAS* gene which encodes the lipoic acid synthetase. Pyruvate dehydrogenase deficiency is estimated to affect 1 in every 1,000,000 live births.

Glycerol kinase deficiency: A disorder characterized by 3 clinical presentations: infantile, juvenile and adult. The infantile form of glycerol kinase deficiency is associated with hyperglycerolemia and severe developmental delay, and those with the adult form are almost asymptomatic. Glycerol kinase deficiency is caused by X-linked recessive mutations in the *GK* gene which encodes the glycerol kinase. This enzyme catalyzes the phosphorylation of glycerol to glycerol-3-phosphate. There is a contiguous gene syndrome involving a combination of glycerol kinase deficiency (infantile form), Duchenne muscular dystrophy and congenital adrenal hypoplasia, and it is caused by deletion of multiple genes on chromosome Xp21.

3.04.10.6 Other Specified Disorders of Carbohydrate Metabolism

Pentosuria: A benign disorder characterized by daily excretion in the urine of 1 to 4 grams of the pentose L-xylulose. This condition may give false diagnosis of diabetes. Pentosuria is caused by autosomal recessive mutations in the *DCXR* gene which encodes the dicarbonyl/L-xylulose reductase. The encoded protein displays diacetyl reductase and L-xylulose reductase activities. This benign condition occurs principally in Jews and Lebanese people. In American Jews, the carrier frequency for Pentosuria trait is estimated to be 1 in 22-25 individuals, which means the frequency of this condition is about 1 in 2,000-2,500 individuals or $1/(44)^2$ - $1/(50)^2$.

Primary hyperoxaluria: A disorder characterized by increased excretion of oxalate, with oxalate stones formation. There are several underlying causes of this disorder with at least 3 autosomal genomic loci: Primary hyperoxaluria type I is caused by autosomal recessive mutations in the *AGXT* gene which encodes the alanine-glyoxylate aminotransferase, a peroxisomal enzyme involved in glyoxylate detoxification. On average, the carrier frequency for Primary hyperoxaluria type I trait is estimated to be 1 in 287-500 individuals worldwide, which means the frequency of this condition is about 1 in 330,000-1,000,000 individuals or $1/(574)^2$ - $1/(1,000)^2$. There is a higher carrier fre-

quency for Primary hyperoxaluria type I trait among Mediterranean countries, particularly Tunisia;; Primary hyperoxaluria type II is caused by autosomal recessive mutations in the *GRHPR* gene which encodes the glyoxylate reductase/hydroxypyruvate reductase. The prevalence of this condition is unknown;; and Primary hyperoxaluria type III is caused by autosomal recessive mutations in the *HOGA1* gene which encodes the dihydrodipicolinate synthase.

Renal glucosuria: A disorder characterized by abnormal high levels of glucose in urine. This condition results of improper tubular reabsorption of glucose of the renal tubules. Renal glucosuria is caused by autosomal recessive mutations in the *SLC5A2* gene which encodes the solute carrier (family 5 member A2), a sodium glucose cotransporter. In the United States, primary renal glucosuria is estimated to affect 1 in every 20,000 live births.

Transaldolase deficiency: A disorder characterized by hepatosplenomegaly, telangiectasia of the skin and persistent thrombocytopenia. These patients also exhibit elevated concentrations of ribitol, D-arabitol, and erythritol in both plasma and urine. Transaldolase deficiency is caused by autosomal recessive mutations in the *TALDO1* gene which encodes the transaldolase 1. This nonoxidative pentose phosphate pathway enzyme provides the ribose-5-phosphate for nucleic acid synthesis and NADPH for lipid biosynthesis. Transaldolase deficiency is estimated to affect less than 1 in every 1,000,000 individuals. This disorder is rare worldwide, and has been described in children born to consanguineous parents of Turkish and Arabic origin.

Table 3.04.10.1 Recognizable Disorders of Carbohydrate Metabolism

Disorder	OMIM	Gene	Biological Function	Product of Conception Alteration
Lactase persistance/nonpersistance	223100	*MCM6*	DNA unwinding enzyme	AR
Congenital lactase deficiency	223000	*LCT*	Integral membrane protein	AR
Muscle glycogen storage disease type 0 or muscle glycogen synthase deficiency	611556	*GYS1*	Glycogen synthase	AR
Liver glycogen storage disease type 0 or liver glycogen synthase deficiency	240600	*GYS2*	Glycogen synthase	AR
Lethal congenital Glycogen storage disease of heart	261740	*PRKAG2*	Protein kinase, AMP-activated	AD + de novo
Glycogen storage disease type Ia or von Gierke disease	232200	*G6PC*	Endoplasmic reticulum protein	AR
Glycogen storage disease type Ib	232220	*SLC37A4*	Solute carrier	AR
Glycogen storage disease type Ic	232240	*SLC17A3*	Solute carrier	AR
Glycogen storage disease type II or Pompe disease	232300	*GAA*	Lysosomal storage protein	AR
Glycogen storage disease type IIb or Danon disease	300257	*LAMP2*	Lysosomal storage protein	XD + de novo
Glycogen storage disease type III or Cori Disease or glycogen debranching enzyme deficiency	232400	*AGL*	Debrancher enzyme	AR
Glycogen storage disease type IV or Andersen disease or Glycogen branching enzyme deficiency	232500	*GBE1*	Glucan branching enzyme	AR
Glycogen storage disease type V or McArdle disease	232600	*PYGM*	Phosphorylase	AR
Glycogen storage disease type VI or Hers disease	232700	*PYGL*	Phosphorylase	AR
Glycogen storage disease type VII or Tarui disease or muscle phosphofructokinase deficiency	232800	*PFKM*	Phosphofructokinase	AR
Glycogen storage disease type IXa1-IXa2 or liver glycogenosis	306000	*PHKA2*	Phosphorylase kinase	XR
Glycogen storage disease type IXb or Phosphorylase kinase deficiency of liver and muscle	261750	*PHKB*	Phosphorylase kinase	AR
Glycogen storage disease type IXc	613027	*PHKG2*	Phosphorylase kinase	AR
Glycogen storage disease type IXd or muscle glycogenosis	300559	*PHKA1*	Phosphorylase kinase	XR

Table 3.04.10.1. Continued.

Disorder	OMIM	Gene	Biological Function	Product of Conception Alteration
Glycogen storage disease type X or muscle phosphoglycerate mutase deficiency	261670	PGAM2	Dimeric enzyme	AR
Fanconi-Bickel syndrome or Glycogen storage disease type XI	227810	SLC2A2	Solute carrier	AR or AD
Glycogen storage disease type XI or lactate dehydrogenase A deficiency	612933	LDHA	Dehydrogenase	AR
Glycogen storage disease type XII or Aldolase A deficiency	611881	ALDOA	Glycolytic enzyme	AR
Glycogen storage disease type XIII or Enolase 3 deficiency	612932	ENO3	Enolase	AR
Glycogen storage disease type XIV	612934	PGM1	Cytosol phosphoglu-comutase	AR
Glycogen storage disease type XV or glycogenin deficiency	613507	GYG1	Glycosyltransferase	AR
Essential fructosuria or hepatic fructokinase deficiency	229800	KHK	Hepatic fructokinase	AR
Fructose intolerance	229600	ALDOB	Glycolytic enzyme	AR
Fructose-1,6-biphosphatase deficiency	229700	FBP1	Cytosol phosphatase	AR
Galactosemia type I or Classic Galactosemia	230400	GALT	Cytosol uridylyltransferase	AR
Galactosemia type II or Galactokinase deficiency with cataracts	230200	GALK1	Galactokinase	AR
Galactosemia type III or Epimerase-deficiency galactosemia	230350	GALE	Epimerase	AR
Congenital sucrase-isomaltase deficiency	222900	SI	Sucrase-isomaltase	AR
Glucose/galactose malabsorption	606824	SLC5A1	Solute carrier	AR
Trehalase deficiency	612119	TREH	Brush-border protein	AR
Fructose malabsorption	138230	SLC2A5	Solute carrier	AR
Pyruvate carboxylase deficiency	266150	PC	Mitochondrial protein	AR
Pyruvate dehydrogenase E1-alpha deficiency	312170	PDHA1	Mitochondrial protein	XD
Pyruvate dehydrogenase E1-beta deficiency	614111	PDHB	Mitochondrial protein	AR
Pyruvate dehydrogenase E2 deficiency	245348	DLAT	Mitochondrial protein	AR
Pyruvate dehydrogenase E3-binding protein deficiency or Lacticacidemia	245349	PDX1	Transcription factor diseases Helix-turn-helix domains	AR

Table 3.04.10.1. Continued.

Disorder	OMIM	Gene	Biological Function	Product of Conception Alteration
Pyruvate dehydrogenase phosphatase deficiency	608782	PDP1	Mitochondrial protein	AR
Pyruvate dehydrogenase lipoic acid synthetase deficiency	614462	LIAS	Mitochondrial protein	AR
Glycerol kinase deficiency	307030	GK	Glycerol kinase	XR
Pentosuria	260800	DCXR	Reductase, L-xylulose	AD
Primary hyperoxaluria type I	259900	AGXT	Peroxisomal and lysosomal structural protein	AR
Primary hyperoxaluria type II	260000	GRHPR	Hydroxypyruvate reductase	AR
Primary hyperoxaluria type III	613616	HOGA1	Mitochondrial protein	AR
Renal glucosuria	233100	SLC5A2	Solute carrier	AR
Transaldolase deficiency	606003	TALDO1	Dihydroxyacetone transferase	AR

Table 3.04.10.2 Disorders of Carbohydrate Metabolism in Different Ethnic Groups

Disease	Frequency of condition	People with higher prevalence (if available)
Fructose intolerance	5 in 100,000	Unknown
Fructose-1,6-biphosphatase deficiency	5 in 100,000	Unknown
Renal glucosuria	5 in 100,000	Unknown
Pompe disease	2-3 in 100,000	Palestinian Arabs
Congenital sucrase-isomaltase deficiency	1-5 in 10,000	Greenlander Eskimos and Alaskan Eskimos
Galactosemia type I or Classic Galactosemia	2 in 100,000	Irish traveler people
Glycogen storage disease type III or Cori Disease or glycogen debranching enzyme deficiency	1 in 100,000	North African Jews
Glycogen storage disease type V or McArdle disease	1 in 100,000	Unknown
Glycogen storage disease type IXa1- IXa2 or liver glycogenosis	1 in 100,000	Worldwide
Essential fructosuria or hepatic fructokinase deficiency	1 in 100,000	Unknown
Pyruvate carboxylase deficiency	4 in 1,000,000	Algonkian Indians (East Canada)
Glycogen storage disease type IV or Andersen disease or Glycogen branching enzyme deficiency	1-2 in 1,000,000	Unknown

Table 3.04.10.2 Disorders of Carbohydrate Metabolism in Different Ethnic Groups

Disease	Frequency of condition	People with higher prevalence (if available)
Glycogen storage disease type Ia or von Gierke disease	1-3 in 1,000,000	Ashkenazi Jews
Primary hyperoxaluria type I	1-3 in 1,000,000	Tunisians and Mediterraneans
Congenital lactase deficiency	<1 in 1,000,000	Finnish
Glycogen storage disease type IIb or Danon disease	<1 in 1,000,000	Unknown
Glycogen storage disease type VI or Hers disease	<1 in 1,000,000	Mennonites
Glycogen storage disease type VII or Tarui disease or muscle phosphofructokinase deficiency	<1 in 1,000,000	Ashkenazi Jews
Glycogen storage disease type X or muscle phosphoglycerate mutase deficiency	<1 in 1,000,000	Unknown
Fanconi-Bickel syndrome or Glycogen storage disease type XI	<1 in 1,000,000	Unknown
Glycogen storage disease type XV or glycogenin deficiency	<1 in 1,000,000	Unknown
Glucose/galactose malabsorption	<1 in 1,000,000	Unknown
Pyruvate dehydrogenase deficiency	<1 in 1,000,000	Worldwide

3.04.11 Lipids Disorders

3.04.11.1 Disorders of Sphingolipid Metabolism and other Lipid Storage Disorders

Tay–Sachs disease or Hexosaminidase A deficiency: A disorder characterized by developmental regression, deterioration of mental and physical abilities. This condition appears at around 6 months of age resulting in death by age 2 or 3 years. Affected individuals exhibit a failure to thrive, developmental retardation, paralysis, dementia and blindness. Patients also display a central 'cherry-red' spot in funduscopy. Tay–Sachs disease is caused by accumulation of harmful quantities of gangliosides in the brain. The disease is named after Warren Tay and Bernard Sachs. Tay–Sachs disease is caused by autosomal recessive mutations in the *HEXA* gene which encodes the alpha-subunit of beta-N-acetylhexosaminidase A, a lysosomal enzyme. This enzyme together with the cofactor GM2 activator protein catalyzes the degradation of the ganglioside GM2, and other molecules containing terminal N-acetyl hexosamines. In Ashkenazi Jews, the carrier frequency for Tay–Sachs disease trait is estimated to be 1 in 27 individuals, which means the frequency of this condition is about 1 in 3,000 individuals or $1/(54)^2$. The current frequency among Ashkenazi Jews is 1 in 30,000 since they have implemented effective programs of disassortative mating. There is also a similar prevalence for Tay–Sachs disease trait among French Canadians of southeastern Quebec and Cajuns of southern Louisiana. The carrier frequency in Irish-Americans is 1 in 50 individuals and Sephardi Jews is 1 in 110 individuals. On average in the United States, the carrier frequency for Tay–Sachs disease trait is estimated to be 1 in 287 individuals, which means the frequency of this condition is about 1 in 330,000 individuals or $1/(574)^2$.

Sandhoff disease: A disorder with similar clinically features as those of Tay-Sachs disease. These two conditions are clinically indistinguishable from each other. These patients exhibit cherry red spots, early blindness, progressive mental and motor deterioration, doll-like face, and macrocephaly. Sandhoff disease is caused by autosomal recessive mutations in the *HEXB* gene which encodes the beta-subunit of beta-hexosaminidase, a lysosomal enzyme. On average, the carrier frequency for Sandhoff disease trait is estimated to be 1 in 310 individuals worldwide, which means the frequency of this condition is about 1 in 384,000 individuals or $1/(620)^2$. There is a higher carrier frequency for Sandhoff disease trait among Creoles of Northern Argentina, Christian Maronites of Cyprus, and Indians in Saskatchewan (Canada).

3.04.11.2 Other Gangliosidosis

GM1 gangliosidoses: A disorder characterized by 3 clinical presentations: classic infantile, juvenile and adult late onset. Classic infantile is the most common and severe form. Affected individuals begin digressing in development before 6 months of age. Generally, these patients lose the ability to sit up by themselves or crawl. Affected patients exhibit seizures, cherry-red spots in the retina, blindness, deafness, paralysis, organomegaly and

mental retardation. The juvenile and adult forms are less severe than classic infantile, and also include red spots in retina, cognitive impairment, ataxia, muscle weakness and impaired walking. Affected individuals display variable degrees of neurodegeneration and skeletal abnormalities. This condition results from the abnormal storage of acidic lipid materials in the nerve cells. GM1 gangliosidoses is caused by autosomal recessive mutations in the *GLB1* gene which encodes the beta-galactosidase-1. This lysosomal enzyme hydrolyzes the terminal beta-galactose from ganglioside substrates and other glycoconjugates. Mutations in the *GLB1* gene also cause Morquio B disease which is an allelic disorder with skeletal anomalies and no neurologic involvement. GM1 gangliosidoses is estimated to affect 1 to 2 in every 200,000 live births. There is a higher carrier frequency for GM1 gangliosidoses trait among individuals from Malta, Cyprus and Brazil.

GM2-gangliosidosis AB variant: A disorder with similar clinically features as those of Tay-Sachs disease. Affected individuals are characterized by usual hexosaminidase A and B activity, but the inability to form a functional GM2 activator complex. GM2-gangliosidoses AB variant is caused by autosomal recessive mutations in the *GM2A* gene which encodes the GM2 activator protein. This small glycolipid transport protein acts as a substrate specific co-factor for the lysosomal enzyme beta-hexosaminidase A.

Mucolipidosis type IV: A disorder now classified as a gangliosidosis. This condition is characterized by developmental regression, psychomotor retardation and ophthalmologic abnormalities. Mucolipidosis type IV is caused by autosomal recessive mutations in the *MCOLN1* gene which encodes the mucolipin 1. This protein affects membrane sorting and endocytosis functions. In Ashkenazi Jews, the carrier frequency for Mucolipidosis type IV trait is estimated to be 1 in 110 individuals, which means the frequency of this condition is about 1 in 48,000 individuals or $1/(220)^2$. The current frequency among Ashkenazi Jews is probably lower since they have implemented programs of disassortative mating. Over 80% of the patients with Mucolipidosis type IV are Ashkenazi Jews, particularly from northern Poland and Lithuania.

3.04.11.3 Other sphingolipidoses

Fabry disease or alpha-galactosidase A deficiency: A multisystemic disorder manifesting cardiac, cerebrovascular and renal symptoms. Patients with this condition exhibit progressive renal failure, neuropathy and skin lesions. This is lysosomal storage disorder of glycolipid that is known as globotriaosylceramide. Most of the patients are males, there are females that also display a milder form of Fabry disease and even experience significant life-threatening conditions. There are also a cardiac variant of Fabry Disease with normal coronary arteries and symptoms of angina, exercise intolerance and signs of electrocardiographic changes. The disease is named after Johannes Fabry. Fabry disease is caused by X-linked mutations in the *GLA* gene which encodes the alpha-galactosidase. Fabry disease is estimated to affect 1 in every 55,000 male births. The real prevalence is probable higher.

Gaucher's disease: A disorder characterized by 3 clinical presentations: Type I or non-neuropathic type, Type II or acute infantile, and Type III or chronic form. Type I or non-neuropathic type occurs most often among persons of Ashkenazi Jewish background. Affected individuals display hepatosplenomegaly, skeletal weakness and pancytopenia from childhood. Most of these patients have a mild form of the disease and live well into adulthood. Patients with types II and III have central nervous system involvement and neurologic manifestations. Type II or acute infantile, generally starts within 6 months of birth. Affected individuals show progressive brain damage, seizures, spasticity and limb rigidity. Most of these children generally die by age 2. Type III or chronic form is a milder version of type II. The disease is named after Philippe Gaucher. Gaucher disease does not exhibit a good correlation between the genotypes and the clinical phenotypes, due to a great variability in predicting the disease course. Gaucher's disease is caused by autosomal recessive mutations in the *GBA* gene which encodes the acid beta-glucosidase or glucocerebrosidase, a lysosomal membrane protein. This enzyme metabolizes glucocerebroside. On average, the carrier frequency for Gaucher's disease trait in the United States is estimated to be 1 in 100 individuals, which means the frequency of this condition is about 1 in 40,000 individuals or $1/(200)^2$. Type I form accounts for 80 % of all cases of Gaucher's disease. There is a higher carrier frequency for Gaucher's disease trait among Ashkenazi Jews. This trait is estimated to be 1 in 11 individuals, which means the frequency of this condition is about 1 in 450 individuals or $1/(22)^2$. This condition is optionally included in a program of disassortative mating, due to large clinical variability and poor genotype-phenotype correlation.

Krabbe disease or globoid cell leukodystrophy: A disorder characterized by extreme irritability, spasticity and developmental delay starting at 6 month of age. Most of these children generally die by age 2. This lysosomal disorder affects the white matter of both the central and peripheral nervous systems. The disease is named after Knud Haraldsen Krabbe. Krabbe disease is caused by autosomal recessive mutations in the *GALC* gene which encodes the galactosylceramidase. This lysosomal protein hydrolyzes the galactose ester bonds of galactosylsphingosine, galactosylceramide, lactosylceramide and monogalactosyldiglyceride galactocerebrosidase. In Druze communities in Israel, the carrier frequency for Krabbe disease trait is estimated to be 1 in 39 individuals, which means the frequency of this condition is about 1 in 6,000 individuals or $1/(78)^2$. Arab communities in Israel also have a large prevalence of this condition. On average, Krabbe disease is estimated to affect 1 in every 100,000 births worldwide, with a slightly higher prevalence in the city of Catania in Sicily, the Netherlands, Portugal, and Scandinavian countries.

Niemann–Pick disease: A group of fatal lysosomal storage diseases. There are three classifications of this disease: Neimann-Pick disease type A and B; Neimann-Pick disease type C1 and D; and Niemann-Pick disease type C2. They will be discussed below:

Niemann–Pick disease type A and B: A disorder characterized by severe neurologic disturbances, including hypotonia, rigidity and mental retardation. These patients also exhibit failure to thrive and hepatosplenomegaly. Affected individuals usually developed symptoms within 6 months after birth and died by 3 years of age. This condition is due to accumulation of sphingomyelin in reticuloendothelial and other cell types throughout the body. Niemann-Pick disease type B is a milder form, characterized by visceral involvement only and survival into adulthood. Niemann–Pick disease type A and B is caused by autosomal recessive mutations

in the *SMPD1* gene which encodes the acid lysosomal sphingomyelin phosphodiesterase 1. This enzyme converts sphingomyelin to ceramide. On average, the carrier frequency for Niemann–Pick disease type A and B trait is estimated to be 1 in 250 individuals worldwide, which means the frequency of this condition is about 1 in 250,000 individuals or $1/(500)^2$. There is a higher carrier frequency for Niemann–Pick disease type A and B trait among Ashkenazi Jews. This trait is estimated to be 1 in 100 individuals, which means the frequency of this condition is about 1 in 40,000 individuals or $1/(200)^2$.

Niemann–Pick disease type C1 and D: A disorder appearing between 2 to 4 years of age characterized by neurologic symptoms including ataxia, grand mal seizures and regression. The hepatosplenomegaly is less striking than those of types A and B. Affected individuals usually die at age 5 to 15. Niemann–Pick disease type C1 and D is caused by autosomal recessive mutations in the *NPC1* gene which encodes the Niemann-Pick C1 protein. This protein shows sequence similarity to the morphogen receptor 'Patched'. On average, the carrier frequency for Niemann–Pick disease type C1 and D trait is estimated to be 1 in 194 individuals worldwide, which means the frequency of this condition is about 1 in 150,000 individuals or $1/(388)^2$. There is a higher carrier frequency for Niemann–Pick disease type C1 and D trait among the French-Acadians in Nova Scotia, particularly in Yarmouth County. This trait is estimated to be 1 in 4-10 individuals, which means the frequency of this condition is about 1 in 64-400 individuals or $1/(8)^2$ - $1/(20)^2$.

Niemann–Pick disease type C2: A condition caused by autosomal recessive mutations in the *NPC2* gene. This condition accounts for only 5 % of all the Niemann–Pick disease type C cases.

Canavan disease or aminoacylase 2 deficiency: A neurodegenerative disorder characterized by leukodystrophy. Affected individuals have myelin degeneration leading to progressive damage to the nerve cells in the brain. Canavan disease patients exhibit hyperextension of legs and flexion of arms, atonia of neck muscles, hypotonia, blindness, severe mental defect, megalocephaly with demyelination and leukodystrophy. On average, these patients usually die within 18 months of birth. Canavan disease is caused by autosomal recessive mutations in the *ASPA* gene which encodes the aspartoacylase. This enzyme catalyzes the conversion of N-acetyl_L-aspartic acid to aspartate and acetate. In Ashkenazi Jews, the carrier frequency for Canavan disease trait is estimated to be 1 in 40 individuals, which means the frequency of this condition is about 1 in 6,400 individuals or $1/(80)^2$. In the United States, this condition is particularly common among infants of Jewish extraction whose ancestors lived in Vilna Ghetto. The current frequency among Ashkenazi Jews is probably lower since they have implemented programs of disassortative mating. This condition has also been reported in Saudi Arabia, Iran and Australia.

Alexander disease: A slow progressing neurodegenerative disorder, with similar features as those of Canavan disease. There are different presentations; in the infantile form, patients exhibit megalencephaly, developmental delay, seizures and spasticity; in the adult form the symptoms resemble multiple sclerosis. Alexander disease is caused by *de novo* monoallelic mutations in the *GFAP* gene which encodes the glial fibrillary acidic protein, an intermediate filament protein. Alexan-

der disease is estimated to affect less than 1 in every 1,000,000 individuals.

Metachromatic leukodystrophy: A disorder characterized by at least 3 clinical presentations: late infantile, juvenile and adult forms. In the late infantile form, the children exhibit motor symptoms, rigidity, seizure and mental deterioration. Affected individuals begin with symptoms in the second year of life and death occurs in most cases before 5 years of age. The age of demarcation of juvenile forms from adult forms is between 16 to 21 years of age. In the adult form of metachromatic leukodystrophy, patients display symptoms similar to those of schizophrenia. Metachromatic leukodystrophy is caused by autosomal recessive mutations in the *ARSA* gene which encodes the arylsulfatase A. This enzyme hydrolyzes cerebroside sulfate to cerebroside and sulfate. In Navajo Indians, the carrier frequency for Metachromatic leukodystrophy trait is estimated to be 1 in 25 individuals, which means the frequency of this condition is about 1 in 2,520 individuals or $1/(50)^2$. There is a higher carrier frequency for Metachromatic leukodystrophy trait amongst Arabs living in Israel. On average, the carrier frequency for Metachromatic leukodystrophy trait is estimated to be 1 in 100 individuals worldwide, which means the frequency of this condition is about 1 in 40,000 individuals or $1/(200)^2$. There is also a 'pseudo'-deficiency that affects 7% of the people worldwide.

Farber disease or Farber lipogranulomatosis: A disorder characterized by hoarse cry, irritability and erythematous swellings starting during the first few weeks of life. Affected individuals also exhibit severe motor and mental retardation. Most of these children generally die by age 2. Farber disease is caused by autosomal recessive mutations in the *ASAH1* gene which encodes the N-acylsphingosine amidohydrolase (acid ceramidase) 1 or acid ceramidase. Mutations in the same gene, causes spinal muscular atrophy with progressive myoclonic epilepsy which is an allelic disorder. Farber disease is estimated to affect less than 1 in every 1,000,000 individuals.

Encephalopathy due to prosaposin deficiency: A neurovisceral disorder characterized by a progressive fatal course. Affected individuals exhibit massive myoclonic bursts, seizures, hypotonia, abnormal ocular movements, dystonia and hepatosplenomegaly. This sphingolipidoses start manifesting immediately after birth. Most of these children generally die within 4 month of birth, from respiratory failures following repeated pulmonary infections. Encephalopathy due to prosaposin deficiency is caused by autosomal recessive mutations in the *PSAP* gene which encodes the prosaposin. This protein is the common precursor for the lysosomal degradation of several sphingolipids. Mutations in the *PSAP* gene also produce other allelic disorders including Metachromatic leukodystrophy, Atypical Gaucher and Atypical Krabbe disease. Encephalopathy due to prosaposin deficiency is estimated to affect less than 1 in every 1,000,000 individuals.

Cystic leukoencephalopathy without megalencephaly: A disorder characterized by bilateral cysts in the anterior part of the temporal lobe, cerebral white matter anomalies, severe motor and intellectual retardation, spasticity, and non-progressive leukoencephalopathy. Cystic leukoencephalopathy without megalencephaly is caused by autosomal recessive mutations in the *RNASET2* gene which encodes the ribonuclease T2. Cystic leukoencephalopathy without megalencephaly is esti-

mated to affect less than 1 in every 1,000,000 individuals.

Leukoencephalopathy and peripheral neuropathy: A disorder characterized by autosomal recessive mutations in the *RPIA* gene which encodes the ribose 5-phosphate isomerase A. This enzyme catalyzes the reversible conversion between ribose-5-phosphate and ribulose-5-phosphate in the pentose-phosphate pathway.

Leukoencephalopathy with brain stem and spinal cord involvement - lactate elevation: A disorder characterized by progressive cerebellar ataxia, epilepsy, cognitive decline, pyramidal and spinal cord dysfunction with spasticity, and dorsal column dysfunction. This leukoencephalopathy with high lactate appears during early childhood. Leukoencephalopathy with brain stem and spinal cord involvement - lactate elevation is caused by autosomal recessive mutations in the *DARS2* gene which encodes the mitochondrial aspartyl-tRNA synthetase. This enzyme aminoacylates the aspartyl-tRNA. Leukoencephalopathy with brain stem and spinal cord involvement - lactate elevation is estimated to affect less than 1 in every 1,000,000 individuals.

Diffuse leukoencephalopathy with spheroids or familial progressive subcortical gliosis: A disorder characterized by variable behavioral, cognitive and motor changes during the fourth decade of life. Patients usually exhibit bradykinesia, hyperreflexia, apraxia, rigidity and spasticity. They often die of dementia within 6 years of the symptoms appearing. Diffuse leukoencephalopathy with spheroids is caused by autosomal dominant mutations in the *CSF1R* gene which encodes the colony stimulating factor 1 receptor. The encoded protein is a member of tyrosine kinase transmembrane receptor. Diffuse leukoencephalopathy with spheroids is estimated to affect less than 1 in every 1,000,000 individuals.

Megalencephalic leukoencephalopathy with subcortical cysts: A disorder characterized by ataxia, epilepsy, progressive pyramidal signs and mild mental deterioration. This condition appears in the first year of life. Megalencephalic leukoencephalopathy with subcortical cysts can be caused by autosomal recessive mutations in either the *MLC1* gene which encodes the megalencephalic leukoencephalopathy with subcortical cysts 1, or the *HEPACAM* gene which encodes the hepatic and glial cell adhesion molecule. This condition occurs more frequently in people with high degree of consanguinity, although the prevalence for this condition is unknown. Mutations in the *HEPACAM* gene also cause another allelic disorder: the remitting megalencephalic leukoencephalopathy with subcortical cysts 2B with or without mental retardation.

Multiple sulfatase deficiency: A disorder that combines features as those of mucopolysaccharidosis and metachromatic leukodystrophy. Multiple sulfatase deficiency is characterized by skeletal anomalies, organomegaly, ichthyosis and mental retardation. Multiple sulfatase deficiency is caused by autosomal recessive mutations in the *SUMF1* gene which encodes the sulfatase-modifying factor-1. Multiple sulfatase deficiency is estimated to affect 1 in every 1.4 million live births.

Hypomyelinating leukodystrophy-1 or Pelizaeus-Merzbacher disease: A disorder characterized by developmental delay, hypotonia, spastic quadriplegia, ataxia, nystagmus and variable intellectual deficit. In this condition myelin is improperly formed in the central nervous system.

Pelizaeus-Merzbacher disease affects males in its most severe form since it is X-linked leukodystrophy. Females with this condition exhibit a milder phenotype. Pelizaeus-Merzbacher disease is caused by X-linked mutations in the *PLP1* gene which encodes the proteolipid protein-1. Mutations in the same gene also have another allelic disorder - the Spastic paraplegia-2. Pelizaeus-Merzbacher disease is estimated to affect 1 in every 400,000 individuals.

Hypomyelinating leukodystrophy: A disorder with similar clinical features as those of Pelizaeus-Merzbacher disease. There are several underlying causes of this disorder with at least 7 autosomal genomic loci and 2 X-linked genomic loci: Hypomyelinating leukodystrophy-1 or Pelizaeus-Merzbacher disease is caused by X-linked mutations in the *PLP1* gene;; Hypomyelinating leukodystrophy-2 is caused by autosomal recessive mutations in the *GJC2* gene which encodes the gap junction protein gamma 2;; Hypomyelinating leukodystrophy-3 is caused by autosomal recessive mutations in the *AIMP1* gene which encodes the aminoacyl tRNA synthetase complex-interacting multifunctional protein 1;; Hypomyelinating leukodystrophy-4 is caused by autosomal recessive mutations in the *HSPD1* gene which encodes the heat shock 60kDa protein 1 (chaperonin);; Hypomyelinating leukodystrophy-5 is caused by autosomal recessive mutations in the *FAM126A* gene which encodes the family with sequence similarity 126 (member A);; Hypomyelinating leukodystrophy-6 with atrophy of the basal ganglia and cerebellum but the gene/loci have not been characterized;; Hypomyelinating leukodystrophy-7 is caused by autosomal recessive mutations in the *POLR3A* gene which encodes the DNA-directed RNA polymerase III (polypeptide A);; Hypomyelinating leukodystrophy-8 is caused by autosomal recessive mutations in the *POLR3B* gene which encodes the DNA-directed RNA polymerase III (polypeptide B);; and Allan-Herndon-Dudley syndrome is caused by X-linked recessive mutations in the *SLC16A2* gene which encodes the solute carrier (family 16 member A2), a thyroid hormone transporter.

Adult-onset demyelinating leukodystrophy: A chronic progressive neurologic disorder characterized by cerebellar, pyramidal and autonomic abnormalities. Neurological symptoms first appear during the 4th and 5th decades of life, with patients often living on for 20 years after that. Adult-onset demyelinating leukodystrophy is caused by autosomal dominant or monoallelic microduplication of the *LMNB1* gene which encodes the lamin B1. This condition results in an extra copy of the *LMNB1* gene. These lamins proteins are nucleoskeletal intermediate filaments, which are the major components of the nuclear lamina.

Leukoencephalopathy with vanishing white matter: A progressive disorder that affects the brain and spinal cord. This condition usually has an onset age ranging from 3-19 years of age. This disorder causes deterioration of the central nervous system's white matter, delaying the development of motor skills. Patients also exhibit spasticity, ataxia and deterioration of mental functions. Leukoencephalopathy with vanishing white matter is caused by autosomal recessive mutations in the *EIF2B1, EIF2B2, EIF2B3, EIF2B4* and *EIF2B5* genes. These genes encode subunits of the translation initiation factor. Leukoencephalopathy with vanishing white matter is estimated to affect less than 1 in every 1,000,000 individuals. There is a severe and early-onset form of this condition affecting the Cree and Chippewayan people of northern Quebec and Manitoba.

Ovarioleukodystrophy: A disorder characterized by ovarian failure with vanishing white matter leukodystrophy. Affected individuals also display poor fine motor performance, behavior problems and subsequent neurologic deterioration. Ovarioleukodystrophy is caused by autosomal recessive mutations in the *EIF2B4* gene, the *EIF2B5* gene or the *EIF2B2* gene.

Neurologic variant of Waardenburg-Shah syndrome or PCWH syndrome: A disorder characterized by the combination of the 4 distinct syndromes: Peripheral demyelinating neuropathy, Central dysmyelination, Waardenburg syndrome and Hirschsprung disease. PCWH syndrome is caused by *de novo* mutations in the *SOX10* gene, either by a large deletion or point mutation. The *SOX10* gene belongs to the SOX gene family member; these genes are related to the SRY. Neurologic variant of Waardenburg-Shah syndrome is estimated to affect less than 1 in every 1,000,000 individuals.

Polycystic lipomembranous osteodysplasia with sclerosing leukoencephalopathy or Nasu-Hakola disease: A progressive disorder characterized by affecting the bones and brain. Affected individuals lose the ability to walk, speak or care for themselves. Patients usually live only into their thirties or forties. Nasu-Hakola disease is caused by autosomal recessive mutations in either the *TYROBP* gene which encodes the TYRO protein (tyrosine kinase binding protein), or the *TREM2* gene which encodes the triggering receptor expressed on myeloid cells 2. Polycystic lipomembranous osteodysplasia with sclerosing leukoencephalopathy have mostly been diagnosed in Finland and Japan. In Finland, Polycystic lipomembranous osteodysplasia with sclerosing leukoencepha- lopathy is estimated to affect 1 in every 500,000 individuals.

Retinal vasculopathy with cerebral leukodystrophy: A disorder characterized by vasculopathy involving the microvessels of the brain. This adult-onset condition results in central nervous system degeneration with progressive loss of vision, stroke, motor impairment and cognitive decline. Death occurs in most patients 5-10 years after the neurodegenerative symptoms begin. Retinal vasculopathy with cerebral leukodystrophy is caused by autosomal dominant mutations in the *TREX1* gene which encodes the three prime repair exonuclease 1 protein. Retinal vasculopathy with cerebral leukodystrophy is estimated to affect less than 1 in every 1,000,000 individuals.

3.04.11.4 Neuronal Ceroid Lipofuscinosis

Neuronal Ceroid Lipofuscinosis: A family of neurodegenerative disorders characterized by excessive buildup of lipofuscin in the neuronal cells, myocardium, liver and other organs. Affected individuals exhibit visual loss, progressive intellectual and motor deterioration, seizures, and ataxia. There are different ages of onset: infantile, late-infantile, juvenile and adult. There are several underlying causes of this disorder with at least 12 autosomal genomic loci: Neuronal ceroid lipofuscinosis-1 is caused by autosomal recessive mutations in the *PPT1* gene which encodes the palmitoyl-protein thioesterase-1;; Neuronal ceroid lipofuscinosis-2 is caused by autosomal recessive mutations in the *TPP1* gene which encodes the tripeptidyl peptidase I;; Neuronal ceroid lipofuscinosis-3 is caused by autosomal recessive mutations in the *CLN3* gene which encodes the neuronal ceroid-lipofuscinosis protein 3;; Neuronal ceroid lipofuscinosis-4A or Kufs disease is caused by autosomal recessive mutations in the *CLN6* gene which encodes the neuronal ceroid-lipofuscinosis protein 6. Mutation in the *CLN6* gene can also cause earlier onset of neuronal ceroid lipofuscinosis with ocular involvement;; Neuronal ceroid lipofuscinosis-4B is caused by autosomal dominant mutations in the *DNAJC5* gene which encodes the DnaJ homolog protein (subfamily C member 5);; Neuronal ceroid lipofuscinosis-5 is caused by autosomal recessive mutations in the *CLN5* gene which encodes the neuronal ceroid-lipofuscinosis protein 5. In Finnish people this pathogenic allele appears 400-750 years ago;; Neuronal ceroid lipofuscinosis-6 is caused by autosomal recessive mutations in the *CLN6* gene;; Neuronal ceroid lipofuscinosis-7 is caused by autosomal recessive mutations in the *MFSD8* gene which encodes the major facilitator superfamily domain containing protein 8;; Neuronal ceroid lipofuscinosis-8 and the Northern epilepsy variant of CLN8 are caused by autosomal recessive mutations in the *CLN8* gene which encodes the neuronal ceroid-lipofuscinosis protein 8;; Neuronal ceroid lipofuscinosis-9 the gene/loci have not been characterized;; Neuronal ceroid lipofuscinosis-10 is caused by autosomal recessive mutations in the *CTSD* gene which encodes the cathepsin D;; Neuronal ceroid lipofuscinosis-11 is caused by autosomal recessive mutations in the *GRN* gene which encodes the granulin;; and Neuronal ceroid lipofuscinosis-14 is caused by autosomal recessive mutations in the *KCTD7* gene which encodes the potassium channel tetramerization domain containing protein 7. Neuronal ceroid lipofuscinosis is estimated to affect 2-4 in every 100,000 individuals worldwide.

3.04.11.5 Other Lipid Storage Disorders

Cerebrotendinous xanthomatosis or cerebral cholesterosis: A disorder characterized by progressive cerebellar ataxia beginning after puberty. Affected individuals also suffer from juvenile cataracts and tendinous or tuberous xanthomas. Patients exhibit deposition of a cholestanol in the brain and other tissues. Cerebrotendinous xanthomatosis is caused by autosomal recessive mutations in the *CYP27A1* gene which encodes the sterol 27-hydroxylase, a mitochondrial cytochrome P450. In Caucasians, Cerebrotendinous xanthomatosis is estimated to affect 1 in every 50,000 live births.

Neutral lipid storage disease or Chanarin-Dorfman syndrome: A disorder characterized by nonbullous congenital ichthyosiform erythroderma. Affected individuals also display hepatosplenomegaly, myopathy and vacuolated granulocytes. Neutral lipid storage disease is caused by autosomal recessive mutations in the *ABHD5* gene which encodes the acyltransferase. This protein is involved in the synthesis of phosphatidic acid. Neutral lipid storage disease is estimated to affect less than 1 in every 1,000,000 individuals.

Neutral lipid storage disease with myopathy: A disorder characterized by lipid accumulation in liver, muscle and other visceral tissue. Affected individuals suffer from a mild myopathy but without ichthyosis. Patients exhibit multisystem triglyceride storage with cardiac abnormalities and hepatomegaly. Neutral lipid storage disease with myopathy is caused by autosomal recessive mutations in the *PNPLA2* gene which encodes the adipose triglyceride lipase. Neutral lipid storage disease with myopathy is estimated to affect less than 1 in every 1,000,000 individuals.

Wolman Disease or Lysosomal acid lipase deficiency: A disorder characterized by xanthomatous changes in the liver, small intestine, lungs, adrenal, spleen, lymph nodes, bone marrow and thymus, and to a smaller degree in the retina, skin and central nervous system. This condition starts manifesting during infancy. Wolman disease is caused by the inability to break down triglycerides and cholesterol esters, leading to their progressive accumulation in lysosomes in the aforementioned organs and tissues. Wolman disease is caused by autosomal recessive mutations in the *LIPA* gene which encodes the lysosomal acid lipase A. This lysosomal enzyme catalyzes the hydrolysis of cholesteryl esters and triglycerides. In the United States, the carrier frequency for Wolman disease trait is estimated to be 1 in 354 individualss which means the frequency of this condition is about 1 in 500,000 individuals or $1/(708)^2$. Wolman disease seems to be more prevalent among people of Iranian-Jewish ancestry.

Cholesteryl ester storage disease: A milder clinical phenotype of Wolman Disease. Cholesteryl ester storage disease is characterized by later-onset with primarily hepatic involvement. In this condition the macrophages are engorged with cholesteryl esters. Cholesteryl ester storage disease is caused by autosomal recessive mutations in the *LIPA* gene. Cholesteryl ester storage disease is estimated to affect less than 1 in every 1,000,000 individuals.

Table 3.04.11.1 Recognizable Lipid Storage Disorders

Disorder	OMIM	Gene	Biological Function	Product of Conception Alteration
Tay–Sachs disease or Hexosaminidase A deficiency	272800	HEXA	Hexosaminidase A	AR
Sandhoff disease	268800	HEXB	Hexosaminidase B	AR
GM1-gangliosidosis type I	230500	GLB1	Galactosidase	AR
GM1-gangliosidosis type II	230600	GLB1	Galactosidase	AR
GM1-gangliosidosis type III	230650	GLB1	Galactosidase	AR
GM2-gangliosidosis AB variant	272750	GM2A	GM2 ganglioside activator	AR
Mucolipidosis type IV	252650	MCOLN1	Mucolipin	AR
Fabry disease or alpha-galactosidase A deficiency	301500	GLA	Galactosidase	XR
Gaucher disease type I	230800	GBA	Glucosidase	AR + KDN
Gaucher disease type II	230900	GBA	Glucosidase	AR
Gaucher disease type III	231000	GBA	Glucosidase	AR
Gaucher disease type IIIC	231005	GBA	Glucosidase	AR
Gaucher disease perinatal lethal	608013	GBA	Glucosidase	AR
Krabbe disease or globoid cell leukodystrophy	245200	GALC	Galactosylceramidase	AR
Niemann-Pick disease type A	257200	SMPD1	Sphingomyelin phosphodiesterase	AR
Niemann-Pick disease type B	607616	SMPD1	Sphingomyelin phosphodiesterase	AR
Niemann-Pick disease type C1	257220	NPC1	Lysosomal protein	AR
Niemann-pick disease type C2	607625	NPC2	Lysosomal protein	AR
Niemann-Pick disease type D	257220	NPC1	Lysosomal protein	AR
Canavan disease or aminoacylase 2 deficiency	271900	ASPA	Aspartoacylase	AR
Alexander disease	203450	GFAP	Cytoskeletal Intermediate filaments	AD + 99%
Metachromatic leukodystrophy	250100	ARSA	Arylsulfatase	AR
Farber disease or Farber lipogranulomatosis	228000	ASAH1	N-acylsphingosine amidohydrolase	AR
Encephalopathy, due to prosaposin deficiency	611721	PSAP	Prosaposin	AR
Atypical Gaucher disease	610539	PSAP	Prosaposin	AR
Atypical Krabbe disease	611722	PSAP	Prosaposin	AR
Metachromatic leukodystrophy due to SAP-b deficiency	249900	PSAP	Prosaposin	AR
Cystic leukoencephalopathy without megalencephaly	612951	RNASET2	Ribonuclease T2	AR

Table 3.04.11.1. Continued.

Disorder	OMIM	Gene	Biological Function	Product of Conception Alteration
Leukoencephalopathy and peripheral neuropathy	608611	RPIA	Isomerase	AR
Leukoencephalopathy with brain stem and spinal cord involvement and lactate elevation	611105	DARS2	Mitochondrial protein	AR
Diffuse leukoencephalopathy with spheroids or familial progressive subcortical gliosis	221820	CSF1R	Colony stimulating factor 1 receptor	AD
Megalencephalic leukoencephalopathy with subcortical cysts	604004	MLC1	Integral membrane transporter	AR
Megalencephalic leukoencephalopathy with subcortical cysts 2A	613925	HEPACAM	Hepatic and glial cell adhesion molecule	AR
Multiple sulfatase deficiency	272200	SUMF1	Sulfatase modifying factor	AR
Hypomyelinating leukodystrophy-1 or Pelizaeus-Merzbacher disease	312080	PLP1	Transmembrane proteolipid protein	XR
Hypomyelinating leukodystrophy-2	608804	GJC2	Channelopathy Connexin	AR
Hypomyelinating leukodystrophy-3	260600	AIMP1	Aminoacyl tRNA synthetase complex-interacting protein	AR
Hypomyelinating leukodystrophy-4	612233	HSPD1	Heat shock protein	AR
Hypomyelinating leukodystrophy-5	610532	FAM126A	Signal transducer activity	AR
Hypomyelinating leukodystrophy-7, with or without oligodontia and/or hypogonadotropic hypogonadism	607694	POLR3A	RNA polymerases	AR
Hypomyelinating leukodystrophy-8, with or without oligodontia and/or hypogonadotropic hypogonadism	614381	POLR3B	RNA polymerases	AR
Allan-Herndon-Dudley syndrome	300523	SLC16A2	Solute carrier	XR
Adult-onset demyelinating leukodystrophy	169500	LMNB1	Lamin B1	AD
Ovarioleukodystrophy and Leukoencephalopathy with vanishing white matter	603896	EIF2B1	Translation initiation factor	AR
Ovarioleukodystrophy and Leukoencephalopathy with vanishing white matter	603896	EIF2B2	Translation initiation factor	AR

Table 3.04.11.1. Continued.

Disorder	OMIM	Gene	Biological Function	Product of Conception Alteration
Ovarioleukodystrophy and leukoencephalopathy with vanishing white matter	603896	*EIF2B3*	Translation initiation factor	AR
Ovarioleukodystrophy and Leukoencephalopathy with vanishing white matter	603896	*EIF2B4*	Translation initiation factor	AR
Ovarioleukodystrophy and Leukoencephalopathy with vanishing white matter	603896	*EIF2B5*	Translation initiation factor	AR
Neurologic variant of Waardenburg-Shah syndrome or PCWH syndrome	609136	*SOX10*	Transcription factor β-Scaffold factors	AD + 99%
Polycystic lipomembranous osteodysplasia with sclerosing leukoencephalopathy or Nasu-Hakola disease	221770	*TYROBP*	Tyrosine kinase binding protein	AR
Polycystic lipomembranous osteodysplasia with sclerosing leukoencephalopathy or Nasu-Hakola disease	221770	*TREM2*	Triggering receptor	AR
Retinal vasculopathy with cerebral leukodystrophy	192315	*TREX1*	Exonuclease	AD
Neuronal ceroid lipofuscinosis-1 or Santavuori-Haltia disease	256730	*PPT1*	Palmitoyl-protein thioesterase	AR
Neuronal ceroid lipofuscinosis-2 or Jansky-Bielschowsky disease	204500	*TPP1*	Tripeptidyl peptidase	AR
Neuronal ceroid lipofuscinosis-3 or Batten disease	204200	*CLN3*	Lysosomal protein	AR
Neuronal ceroid lipofuscinosis-4A or Kufs disease	204300	*CLN6*	Lysosomal protein	AR
Neuronal ceroid lipofuscinosis-4B or Kufs disease	162350	*DNAJC5*	Membrane trafficking and protein folding	AD
Neuronal ceroid lipofuscinosis-5 or Finnish Late Infantile Variant	256731	*CLN5*	Lysosomal protein	AR
Neuronal ceroid lipofuscinosis-6 or Late Infantile Variant	601780	*CLN6*	Lysosomal protein	AR
Neuronal ceroid lipofuscinosis-7	610951	*MFSD8*	Lysosomal protein	AR
Neuronal ceroid lipofuscinosis-8 or Northern epilepsy variant	610003	*CLN8*	Lysosomal protein	AR
Neuronal ceroid lipofuscinosis-8 or Turkish Late Infantile Variant	600143	*CLN8*	Lysosomal protein	AR
Neuronal ceroid lipofuscinosis-10	610127	*CTSD*	Cathepsin D	AR

Table 3.04.11.1. Continued.

Disorder	OMIM	Gene	Biological Function	Product of Conception Alteration
Neuronal ceroid lipofuscinosis-11	614706	GRN	Granulin	AR
Neuronal ceroid lipofuscinosis-14 or progressive myoclonic epilepsy-3 with or without intracellular inclusions	611726	KCTD7	Potassium channel tetramerisation domain	AR
Cerebrotendinous xanthomatosis or cerebral cholesterosis	213700	CYP27A1	Mitochondrial disease	AR
Neutral lipid storage disease or Chanarin-Dorfman syndrome	275630	ABHD5	Abhydrolase domain	AR
Neutral lipid storage disease with myopathy	610717	PNPLA2	Phospholipase	AR
Wolman Disease or Lysosomal acid lipase deficiency	278000	LIPA	Lysosomal acid lipase	AR
Cholesteryl ester storage disease	278000	LIPA	Lysosomal acid lipase	AR

Table 3.04.11.2 Lipid Storage Disorders in Different Ethnic Groups

Disease	Frequency of condition	People with higher prevalence (if available)
Neuronal ceroid lipofuscinosis	2-4 in 100,000	Worldwide
Cerebrotendinous xanthomatosis or cerebral cholesterosis	1-9 in 100,000	Unknown
Hypomyelinating leukodystrophy-1 or Pelizaeus-Merzbacher disease	2-3 in 1,000,000	Unknown
Gaucher disease	2-3 in 100,000	Ashkenazi Jews
Metachromatic leukodystrophy	2-3 in 100,000	Navajo Indians and Arab communities in Israel
Fabry disease or alpha-galactosidase A deficiency	1 in 100,000	Worldwide
Krabbe disease or globoid cell leukodystrophy	1 in 100,000	Arab communities in Israel, Catanesi (Sicily), Dutch, Portuguese and Scandinavians
Tay–Sachs disease or Hexosaminidase A deficiency	3 in 1,000,000	Ashkenazi Jews, French Canadians, Cajuns (Louisiana), Irish-Americans, and Sephardi Jews
Sandhoff disease	3 in 1,000,000	Creoles of Northern Argentina, Christian Maronites (Cyprus), and Indians in Saskatchewan (Canada)
Mucolipidosis type IV	4 in 1,000,000	Ashkenazi Jews

The Concise Encyclopedia of Genomic Diseases

Table 3.04.11.2. Continued.

Disease	Frequency of condition	People with higher prevalence (if available)
Niemann-Pick disease	4 in 1,000,000	Ashkenazi Jews and French-Acadians
GM1-gangliosidosis	5-10 in 1,000,000	Maltese, Cypriots and Brazilians
Canavan disease or aminoacylase 2 deficiency	<1 in 1,000,000	Ashkenazi Jews, Saudi Arabians, Iranians, and Australians
Alexander disease	<1 in 1,000,000	Unknown
Farber disease or Farber lipogranulomatosis	<1 in 1,000,000	Unknown
Encephalopathy, due to prosaposin deficiency	<1 in 1,000,000	Unknown
Cystic leukoencephalopathy without megalencephaly	<1 in 1,000,000	Unknown
Leukoencephalopathy with brain stem and spinal cord involvement and lactate elevation	<1 in 1,000,000	Unknown
Diffuse leukoencephalopathy with spheroids or familial progressive subcortical gliosis	<1 in 1,000,000	Unknown
Multiple sulfatase deficiency	<1 in 1,000,000	Unknown
Leukoencephalopathy with vanishing white matter	<1 in 1,000,000	Cree and Chippewayan people (Quebec and Manitoba)
Neurologic variant of Waardenburg-Shah syndrome or PCWH syndrome	<1 in 1,000,000	Unknown
Polycystic lipomembranous osteodysplasia with sclerosing leukoencephalopathy or Nasu-Hakola disease	<1 in 1,000,000	Finnish
Retinal vasculopathy with cerebral leukodystrophy	<1 in 1,000,000	Unknown
Neutral lipid storage disease or Chanarin-Dorfman syndrome	<1 in 1,000,000	Unknown
Neutral lipid storage disease with myopathy	<1 in 1,000,000	Unknown
Wolman Disease or Lysosomal acid lipase deficiency	<1 in 1,000,000	Iranian-Jews

3.04.12 Combinations Disorders

3.04.12.1 Disorders of Glycosaminoglycan Metabolism

Disorders of glycosaminoglycan metabolism: Are a group of inherited disorders of glycosaminoglycan metabolism caused by the absence or malfunctioning of lysosomal enzymes.

Mucopolysaccharidosis type I: A disorder characterized by 3 clinical presentations: Hurler, Hurler-Scheie and Scheie syndromes. Hurler syndrome is the most severe, Hurler-Scheie is intermediate, and Scheie syndrome is a mild form of the disease. Patients with Hurler syndrome exhibit coarse facies, dwarfism, micrognathia, macroglossia corneal clouding, retinal degeneration, mental retardation, hernias, dysostosis multiplex, cardiomyopathy and hepatosplenomegaly. Affected individuals start showing symptoms at 1 to 3 years of age, and most of these kids die by 10 years of age. Mucopolysaccharidosis type I is caused by autosomal recessive mutations in the *IDUA* gene which encodes the alpha-L-iduronidase. This enzyme hydrolyzes the terminal alpha-L-iduronic acid residues of 2 glycosaminoglycans, heparan sulfate and dermatan sulfate. This hydrolysis is prerequisite for the lysosomal degradation of these glycosaminoglycans. On average, the carrier frequency for Mucopolysaccharidosis type I trait is estimated to be 1 in 158 individuals worldwide, which means the frequency of this condition is about 1 in 100,000 individuals or $1/(316)^2$.

Mucopolysaccharidosis type II or Hunter syndrome: A multisystem disorder characterized by skeletal deformities, cardiomyopathy, airway obstruction and neurologic regression. Affected individuals excrete excessive amounts of dermatan and heparan sulfate in the urine. Mucopolysaccharidosis type II predominantly affects males, but some female cases have also been reported. Mucopolysaccharidosis type II is similar, but with milder symptoms than Hurler syndrome. Patients usually die in the second decade of life but sometimes survive into adulthood. Mucopolysaccharidosis type II is caused by X-linked mutations in *IDS* gene which encodes the iduronate 2-sulfatase. This enzyme is required for the lysosomal degradation of heparan sulfate and dermatan sulfate. Mucopolysaccharidosis type II is estimated to affect 1 in every 100,000 male live births.

3.04.12.2 Other mucopolysaccharidosis

Mucopolysaccharidosis type III or Sanfilippo syndrome: A disorder characterized mainly by severe central nervous system degeneration. Affected individuals exhibit motor dysfunction, developmental delay, severe hyperactivity and spasticity. Most of these patients die by the second or third decade of life. The condition is named after Sylvester Sanfilippo. There are several underlying causes of this disorder with at least 4 autosomal genomic loci: Mucopolysaccharidosis type IIIA or Sanfilippo syndrome A is caused by autosomal recessive mutations in the *SGSH* gene which encodes the N-sulfoglucosamine sulfohydrolase;; Mucopolysaccharidosis type IIIB or Sanfilippo syndrome B is caused by autosomal recessive mutations in the *NAGLU* gene which encodes the alpha-N-acetylglucosaminidase;; Mucopolysaccharidosis type IIIC or Sanfilippo syndrome C is caused by autosomal recessive mutations in the *HGSNAT* gene which encodes the acetyl CoA:alpha-glucosaminide acetyltransferase;; and Mucopolysaccharidosis type IIID or Sanfilippo syndrome D is caused by autosomal recessive mutations in the *GNS* gene which encodes the N-acetylglucosamine 6-sulfatase. Type A appears to be the most severe, with earlier onset and rapid progression of symptoms and shorter survival. In the Netherlands and Australia, Mucopolysaccharidosis type III is the most prevalent form of mucopolysaccharidosis. In the Netherlands and Australia, Mucopolysaccharidosis type III is estimated to affect 1 in 53,000 and 1 in 67,000 individuals respectively.

Mucopolysaccharidosis IV or Morquio's syndrome: A disorder characterized by patients exhibiting short stature, severe skeletal dysplasia, dental anomalies, corneal clouding and motor dysfunction. This condition is named after Luis Morquio. There are several underlying causes of this disorder with at least 2 autosomal genomic loci: Mucopolysaccharidosis type IVA is caused by autosomal recessive mutations in the *GALNS* gene which encodes the galactosamine-6-sulfate sulfatase;; and Mucopolysaccharidosis type IVB is caused by autosomal recessive mutations in the *GLB1* gene which encodes the beta-galactosidase. Mucopolysaccharidosis type IV is estimated to affect 1 in every 200,000-300,000 individuals.

Mucopolysaccharidosis type VI or Maroteaux–Lamy syndrome: A disorder characterized by severe skeletal dysplasia, short stature, kyphosis, hepatosplenomegaly, corneal clouding, cardiac abnormalities and facial dysmorphism. Mucopoly-

saccharidosis type VI is caused by autosomal recessive mutations in the *ARSB* gene which encodes the arylsulfatase B. On average, the carrier frequency for mucopolysaccharidosis type VI trait is estimated to be 1 in 250 individuals worldwide, which means the frequency of this condition is about 1 in 250,000 individuals or $1/(500)^2$.

Mucopolysaccharidosis type VII or Sly syndrome: A disorder characterized by a highly variable phenotype. This condition ranges from severe lethal hydrops fetalis to intermediate and mild forms. In the mild form, patients can survive into adulthood. The most common mucopolysaccharidosis type VII form is the intermediate phenotype which is characterized by skeletal anomalies, coarse facies, corneal clouding, hepatomegaly and mental retardation. Mucopolysaccharidosis type VII is caused by autosomal recessive mutations in the *GUSB* gene which encodes the beta-glucuronidase. This condition affects less than 1 in 250,000 individuals.

Mucopolysaccharidosis type IX or Natowicz syndrome: A disorder characterized by short stature, mental retardation, excessive coarse hair, hepatomegaly, hypoplasia of the odontoid, mild dysostosis multiplex and facial changes. Mucopolysaccharidosis type IX is caused by autosomal recessive mutations in the *HYAL1* gene which encodes the hyaluronoglucosaminidase 1. Mucopolysaccharidosis type IX is estimated to affect less than 1 in every 1,000,000 individuals.

Table 3.04.12.2 Recognizable Disorders of Glycosaminoglycan Metabolism

Disorder	OMIM	Gene	Biological Function	Product of Conception Alteration
Mucopolysaccharidosis-1 or Hurler syndrome	607014	IDUA	Iduronidase	AR
Mucopolysaccharidosis-1 or Hurler-Scheie syndrome	607015	IDUA	Iduronidase	AR
Mucopolysaccharidosis-1 or Scheie syndrome	607016	IDUA	Iduronidase	AR
Mucopolysaccharidosis-2 or Hunter syndrome	309900	IDS	Iduronate 2-sulfatase	XR
Mucopolysaccharidosis-3A or Sanfilippo syndrome A	252900	SGSH	Sulfohydrolase	AR
Mucopolysaccharidosis-3B or Sanfilippo syndrome B	252920	NAGLU	N-acetylglucosaminidase	AR
Mucopolysaccharidosis-3C or Sanfilippo syndrome C	252930	HGSNAT	Heparan-alpha-glucosaminide N-acetyltransferase	AR
Mucopolysaccharidosis-3D or Sanfilippo syndrome D	252940	GNS	Glucosamine sulfatase	AR
Mucopolysaccharidosis-4A or Morquio syndrome A	253000	GALNS	Galactosamine sulfatase	AR
Mucopolysaccharidosis-4B or Morquio syndrome B	253010	GLB1	Galactosidase	AR
Mucopolysaccharidosis-6 or Maroteaux-Lamy syndrome	253200	ARSB	Arylsulfatase	AR
Mucopolysaccharidosis-7 or Sly syndrome	253220	GUSB	Glucuronidase	AR
Mucopolysaccharidosis-9	601492	HYAL1	Hyaluronoglucosaminidase	AR

3.04.12.3-7 Disorders of Glycoprotein Metabolism

3.04.12.4 Defects in Post-translational Modification of Lysosomal Enzymes

Mucolipidosis type II alpha/beta or I-cell or Inclusion cells disease: A disorder characterized by cardiomegaly, skeletal abnormalities, short stature and developmental delay. Affected individuals display clinical and radiographic features similar to those of Hurler syndrome but with an earlier onset. Most of these patients die before their seventh year of life. I-cell disease is caused by autosomal recessive mutations in the *GNPTAB* gene which encodes the N-acetylglucosamine-1-phosphotransferase. I-cell disease is a lysosomal storage disease, resulting from a defect in phosphotransferase. This enzyme localizes in the Golgi apparatus. In Mediterraneans this pathogenic allele appears 2,063 years ago. The prevalence of this condition ranges from 1 in 100,000 to 1 in 2,500,000 births worldwide. The carrier frequency estimated for I-cell disease varies several folds among different ethnic groups. In the Irish travelers community, Republic of Ireland, the carrier frequency for I-cell disease trait is estimated to be 1 in 15 individuals, which means the frequency of this condition is about 1 in 880 individuals or $1/(30)^2$. In the province of Quebec, Canada, the carrier frequency for I-cell disease trait is estimated to be 1 in 39 individuals, which means the frequency of this condition is about 1 in 6,184 individuals or $1/(78)^2$. I-cell disease is estimated to affect 1 in every 123,500 individuals in Portugal, 1 in every 252,500 individuals in Japan, and 1 in every 625,500 individuals in the Netherlands.

Mucolipidosis type III alpha/beta or Pseudo-Hurler polydystrophy: A disorder phenotypically less severe than the allelic disorder I-cell disease. This disorder is characterized by cardiomegaly, short stature, skeletal abnormalities and developmental delay. Mucolipidosis type III alpha/beta is caused by autosomal recessive mutations in the *GNPTAB* gene. On average, the carrier frequency for Mucolipidosis type III alpha/beta trait is estimated to be 1 in 283 individuals worldwide, which means the frequency of this condition is about 1 in 320,000 individuals or $1/(566)^2$.

Mucolipidosis type III gamma: A disorder characterized by cardiomegaly, skeletal abnormalities, short stature and developmental delay. Mucolipidosis type III gamma is caused by autosomal recessive mutations in the *GNPTG* gene which encodes the N-acetylglucosamine-1-phosphotransferase (gamma subunit). Mucolipidosis type III gamma is estimated to affect 1 in every 100,000 to 400,000 individuals worldwide.

3.04.12.5 Defects in Glycoprotein Degradation

Sialidosis type I and II or Mucolipidosis type I: An oligosaccharidosis or glycoproteinosis disorder characterized by 2 clinical presentations: neonatal and infantile forms. Patients with the neonatal onset form manifest edema, hepatosplenomegaly with ascites, and hydrops fetalis. Patients with the infantile onset form show dysmorphic and neurological abnormalities. The dysmorphic features include coarse facies, cherry red spot, inner ear hearing loss, short trunk, barrel chest, spinal deformity and mental retardation. Affected individuals also manifest many neurologic symptoms including ataxia, myoclonus, muscular hypotonia and hypotrophy, and seizures. Sialidosis type I is a disorder characterized by a later onset with minor dysmorphic features. Affected individuals exhibit progressive reduction of visual acuity, red-green blindness, bilateral cherry red spots, punctate opacities of the lens and minimal neurologic symptoms. Sialidosis type I and II are caused by autosomal recessive mutations in the *NEU1* gene which encodes the N-acetyl-alpha-neuraminidase. Sialidosis, in all its forms, is estimated to affect 1 in every 4,200,000 live births worldwide. Sialidosis type I seems to be particularly frequent in Italians, and Sialidosis type II is more prevalent among Japanese.

Fucosidosis: A disorder characterized by 2 clinical presentations: type I and type II. Patients with the type I form manifest the most severe phenotype, with symptoms typically appearing in the first 3 to 18 months of life. Affected patients exhibit neurologic signs, progressive psychomotor retardation, coarse facial features and dysostosis multiplex. Other features include elevated sodium and chloride excretion in the sweat, but vascular lesion is not associated with this form. Patients with the type II form first manifest symptoms during the first 12 to 24 months of life. Affected individuals have a mild form with similar symptoms, angiokeratomas are frequent, and a standard salinity in the sweat. Fucosidosis is caused by autosomal recessive mutations in the *FUCA1* gene which encodes the alpha-fucosidase. This enzyme is involved in breaking down fucose. Fucosidosis is estimated to affect less than 1 in every 1,000,000 individuals. Most cases of fucosidosis were reported among Italians from Calabria. This condition has been also reported among Cubans, and Hispanic-American people of New Mexico and Colorado.

Alpha-mannosidosis: A disorder characterized by immune deficiency with recurrent infections, facial and skeletal abnormalities, moderate-to-severe sensorineural hearing loss and mental retardation. Alpha-mannosidosis is caused by autosomal recessive mutations in the *MAN2B1* gene which encodes the alpha-D-mannosidase. This enzyme hydrolyzes terminal, non-reducing alpha-D-mannose residues in alpha-D-mannosides. This enzyme activity is required for the catabolism of N-linked carbohydrates released during glycoprotein turnover. On average, the carrier frequency for alpha-mannosidosis trait is estimated to be 1 in 353.5 individuals worldwide, which means the frequency of this condition is about 1 in 500,000 individuals or $1/(707)^2$.

Beta-mannosidosis: A disorder with similar features as those of Alpha-mannosidosis. This form includes dysmorphic features, hearing loss, mental retardation, repeated respiratory infections, and sometimes neurological symptoms such as hypotonia and convulsions. Beta-mannosidosis is caused by autosomal recessive mutations in the *MANBA* gene which encodes the lysosomal beta-mannosidase. This enzyme is the final exoglycosidase in the pathway for N-linked glycoprotein oligosaccharide catabolism.

Aspartylglucosaminuria: A developmental disorder characterized by the slow development of mental retardation and skeletal abnormalities such as mild facial dysmorphism and slight kyphoscoliosis. Aspartylglucosaminuria is caused by autosomal recessive mutations in the *AGA* gene which encodes the N-aspartyl-beta-glucosaminidase. This enzyme is involved in the catabolism of N-linked oligosaccharides of glycoproteins. In Finland, the carrier frequency for Aspartylglucosaminuria trait is estimated to be 1 in 65 individuals, which means the frequency of this condition is about 1 in 17,000 individuals or $1/(130)^2$. In Finland, Aspartylglycosaminuria is the third most frequent multiple congenital anomaly/mental retardation syndrome after Down syndrome and Fragile X syndrome

Galactosialidosis: A disorder characterized by 3 clinical presentations: early infantile, late infantile and juvenile/adult. Patients with the early infantile form display the most severe phenotype including hydrops fetalis, skeletal dysplasia and early death. Patients with the late infantile form display a normal or mildly affected mental state, hepatosplenomegaly, growth retardation and cardiac involvement. Patients with the juvenile/adult form display angiokeratoma, neurologic deterioration, myoclonus, ataxia, mental retardation and absence of visceromegaly. Galactosialidosis is caused by autosomal recessive mutations in the *CTSA* gene which encodes the cathepsin A. The prevalence of this condition is unknown. The juvenile/adult form has mostly been reported among Japanese.

Schindler disease: A disorder characterized by 3 clinical presentations: type I or infantile-onset neuroaxonal dystrophy, type II or Kanzaki disease and type III. Patients with the type I undergo standard development until about a year old and then display rapid psychomotor regression with progressive hypotonia, extrapyramidal signs, blindness and seizures. Most of these patients die within 4 years of birth. Patients with the type II or adult form exhibit milder symptoms such as angiokeratomas, but without neurological degeneration. Patients with type III manifest an intermediate disorder exhibiting mental retardation, seizures, delayed speech, behavioral problems and mild autistic features. In this last form, patients might start exhibiting signs and symptoms during infancy. Patients manifest autistic or behavioral problems, seizures and mental retardation. Schindler disease is named after Detlev Schindler. Schindler disease is caused by autosomal recessive mutations in the *NAGA* gene which encodes the alpha-N-acetylgalactosaminidase, a lysosomal enzyme that cleaves alpha-N-acetylgalac-tosaminyl moieties from glycoconjugates. Schindler disease is estimated to affect less than 1 in every 1,000,000 individuals.

3.04.12.6 Other Disorders of Glycoprotein Metabolism

Free sialic acid storage disease or Salla disease: A disorder characterized by severe developmental delay, failure to thrive, coarse facies, seizures, hypotonia, bone malformations, cardiomegaly and hepatosplenomegaly with ascites. There are different forms of free sialic acid storage disease; the moderate type is referred to as Salla disease. Free sialic acid storage disease is caused by autosomal recessive mutations in the *SLC17A5* gene which en-

codes the solute carrier (family 17 member A5), an acidic sugar transporter. On average, the carrier frequency for Free sialic acid storage disease trait is estimated to be 1 in 363 individuals worldwide, which means the frequency of this condition is about 1 in 528,000 individuals or $1/(726)^2$. There is a higher carrier frequency for Salla disease trait among people of Northern Finland. This trait is estimated to be 1 in 40 individuals, which means the frequency of this condition is about 1 in 6,400 individuals or $1/(80)^2$.

Sialuria, French type: A disorder characterized by hepatosplenomegaly, coarse facies, and massive urinary excretion of free N-acetylneuraminic acid, with near normal growth and development. Sialuria, French type is caused by autosomal recessive mutations in the *GNE* gene which encodes the glucosamine (UDP-N-acetyl)-2-epimerase /N-acetylmannosamine kinase.

Congenital disorder of glycosylation: A family of disorders characterized by dysfunction of several different organ systems, especially the nervous system, muscles and intestines. These conditions usually begin in infancy. Affected individuals manifest a large and diverse set of symptoms: from severe developmental delay to hypoglycemia to protein-losing enteropathy. Most of the causes of Congenital disorders of glycosylation are due to impairment in genes involved in N-linked glycosylation. These conditions have been classified in two groups: Type I disorders involve disrupted synthesis and transfer of the lipid-linked oligosaccharide precursor. Type II disorders involve impairment in the modification processing of the protein-bound oligosaccharide chain. Most Congenital disorders of glycosylation forms have been described in only a few individuals. There are several underlying causes of this disorder with at least 26 autosomal genomic loci: Congenital disorders of glycosylation type IA is the only form that has been described throughout the world. Congenital disorder of glycosylation type IA is caused by autosomal recessive mutations in the *PMM2* gene which encodes the phosphomannomutase-2. This enzyme catalyzes the isomerization of mannose 6-phosphate to mannose 1-phosphate, which is a precursor to GDP-mannose necessary for the synthesis of dolichol-P-oligosaccharides;; Congenital disorder of glycosylation type IB is caused by autosomal recessive mutations in the *MPI* gene which encodes the mannose phosphate isomerase;; Congenital disorder of glycosylation type IC is caused by autosomal recessive mutations in the *ALG6* gene which encodes the alpha-1,3-glucosyltransferase;; Congenital disorder of glycosylation type ID is caused by autosomal recessive mutations in the *ALG3* gene which encodes the alpha-1,3- mannosyltransferase;; Congenital disorder of glycosylation type IE is caused by autosomal recessive mutations in the *DPM1* gene which encodes the dolichyl-phosphate mannosyltransferase polypeptide 1 (catalytic subunit);; Congenital disorder of glycosylation type IF is caused by autosomal recessive mutations in the *MPDU1* gene which encodes the mannose-P-dolichol utilization defect 1;; Congenital disorder of glycosylation type IG is caused by autosomal recessive mutations in the *ALG12* gene which encodes the alpha-1,6-mannosyltransferase;; Congenital disorder of glycosylation type IH is caused by autosomal recessive mutations in the *ALG8* gene which encodes the alpha-1,3-glucosyltransferase;; Congenital disorder of glycosylation type IJ is caused by autosomal recessive mutations in the *DPAGT1* gene which encodes the dolichyl-phosphate (UDP-N-acetylglucosamine) N-acetylgluco-saminephosphotransferase 1;; Congenital disorder of glycosylation type

IK is caused by autosomal recessive mutations in the *ALG1* gene which encodes the chitobiosyldiphosphodolichol beta-mannosyl-transferase;; Congenital disorder of glycosylation type IL is caused by autosomal recessive mutations in the *ALG9* gene which encodes the alpha-1,2-mannosyltransferase;; Congenital disorder of glycosylation type IM is caused by autosomal recessive mutations in the *DOLK* gene which encodes the dolichol kinase;; Congenital disorder of glycosylation type IN is caused by autosomal recessive mutations in the *RFT1* gene which encodes the RFT1 homolog, a putative endoplasmic reticulum multispan transmembrane protein;; Congenital disorder of glycosylation type IO is caused by autosomal recessive mutations in the *DPM3* gene which encodes the dolichyl-phosphate mannosyltransferase (polypeptide 3);; Congenital disorder of glycosylation type IQ is caused by autosomal recessive mutations in the *SRD5A3* gene which encodes the steroid 5 alpha-reductase 3;; Congenital disorder of glycosylation type II is caused by autosomal recessive mutations in the *ALG2* gene which encodes the alpha-1,3/1,6-mannosyl-transferase;; Congenital disorder of glycosylation type IIA is caused by autosomal recessive mutations in the *MGAT2* gene which encodes the mannosyl (alpha-1,6-)-glycoprotein beta-1,2-N-acetylgluco-saminyltransferase;; Congenital disorder of glycosylation type IIB is caused by autosomal recessive mutations in the *MOGS* gene which encodes the mannosyl-oligosaccharide glucosidase;; Congenital disorder of glycosylation type IIC is caused by autosomal recessive mutations in the *SLC35C1* gene which encodes the solute carrier (family 35 member C1), a GDP-fucose transporter;; Congenital disorder of glycosylation type IID is caused by autosomal recessive mutations in the *B4GALT1* gene which encodes the UDP-Gal:betaGlcNAc beta 1,4- galactosyltransferase (polypeptide 1);; Congenital disorder of glycosylation type IIE is caused by autosomal recessive mutations in the *COG7* gene which encodes the component of oligomeric golgi complex 7;; Congenital disorder of glycosylation type IIF is caused by autosomal recessive mutations in the *SLC35A1* gene which encodes the solute carrier (family 35 member A1), a solute carrier-sialic acid transporter;; Congenital disorder of glycosylation type IIG is caused by autosomal recessive mutations in the *COG1* gene which encodes the component of oligomeric golgi complex 1;; Congenital disorder of glycosylation type IIH is caused by autosomal recessive mutations in the *COG8* gene which encodes the component of oligomeric golgi complex 8;; and Congenital disorder of glycosylation type IIJ is caused by autosomal recessive mutations in the *COG4* gene which encodes the component of oligomeric golgi complex 4. Half of the known worldwide cases have been described among Scandinavian individuals. In Danish people, the carrier frequency for Congenital disorder of glycosylation trait is estimated to be 1 in 60-79 individuals, which means the frequency of this condition is about 1 in 14,400-25,000 individuals or $1/(120)^2 - 1/(158)^2$.

3.04.12.7 Unspecified Disorder of Glycoprotein Metabolism

Familial hypercholanemia: A disorder characterized by elevated serum bile acid concentrations, fat malabsorption and itching. There are several underlying causes of this disorder with at least 3 autosomal genomic loci: Familial hypercholanemia is caused by autosomal recessive mutations in the *TJP2* gene which encodes the tight junction protein 2, or *EPHX1* gene which encodes the microsomal epoxide hydrolase 1, or the *BAAT* gene which encodes the bile acid CoA: amino acid N-acyltransferase (glycine N-choloyltransferase). Familial hypercholanemia is estimated to affect less than 1 in every 1,000,000 individuals. This condition has been described mostly in families of Lancaster County in Old Order Amish descent. This condition responds to treatment with ursodeoxycholic acid.

Table 3.04.12.3-7 Recognizable Disorders of Glycoprotein Metabolism

Disorder	OMIM	Gene	Biological Function	Product of Conception Alteration
Mucolipidosis type II alpha/beta or I-cells or Inclusions cells disease	252500	*GNPTAB*	N-acetylglucosamine-1-phosphate transferase	AR
Mucolipidosis type III alpha/beta or pseudo-Hurler polydystrophy	252600	*GNPTAB*	N-acetylglucosamine-1-phosphate transferase	AR
Mucolipidosis type III gamma	252605	*GNPTG*	N-acetylglucosamine-1-phosphate transferase	AR
Sialidosis type I and II or Mucolipidosis type I	256550	*NEU1*	Lysosomal sialidase	AR
Fucosidosis	230000	*FUCA1*	Fucosidase	AR
Alpha-mannosidosis	248500	*MAN2B1*	Mannosidase	AR

Table 3.04.12.3-7. Continued.

Disorder	OMIM	Gene	Biological Function	Product of Conception Alteration
Beta-mannosidosis	248510	MANBA	Mannosidase	AR
Aspartylglucosaminuria	208400	AGA	Aspartylglucosaminidase	AR
Galactosialidosis	256540	CTSA	Cathepsin	AR
Schindler disease	609241	NAGA	N-acetylgalactosaminidase	AR
Kanzaki disease	609242	NAGA	N-acetylgalactosaminidase	AR
Free sialic acid storage disease or Salla disease	269920	SLC17A5	Solute carrier	AR
Sialuria, French type	269921	GNE	Bifunctional enzyme	AR
Congenital disorder of glycosylation-1A	212065	PMM2	Isomerization of mannose 6-phosphate to mannose 1-phosphate	AR
Congenital disorder of glycosylation-1B	602579	MPI	Isomerase	AR
Congenital disorder of glycosylation-1C	603147	ALG6	Mannosyltransferase	AR
Congenital disorder of glycosylation-1D	601110	ALG3	Mannosyltransferase	AR
Congenital disorder of glycosylation-1E	608799	DPM1	Endoplasmic reticulum protein	AR
Congenital disorder of glycosylation-1F	609180	MPDU1	Endoplasmic reticulum protein	AR
Congenital disorder of glycosylation-1G	607143	ALG12	Mannosyltransferase	AR
Congenital disorder of glycosylation-1H	608104	ALG8	Mannosyltransferase	AR
Congenital disorder of glycosylation-1I	607906	ALG2	Mannosyltransferase	AR
Congenital disorder of glycosylation-1J	608093	DPAGT1	Endoplasmic reticulum protein	AR
Congenital disorder of glycosylation-1K	608540	ALG1	Mannosyltransferase	AR
Congenital disorder of glycosylation-1L	608776	ALG9	Mannosyltransferase	AR
Congenital disorder of glycosylation-1M	610768	DOLK	Endoplasmic reticulum protein	AR
Congenital disorder of glycosylation-1N	612015	RFT1	Endoplasmic reticulum protein	AR
Congenital disorder of glycosylation-1O	612937	DPM3	Endoplasmic reticulum protein	AR
Congenital disorder of glycosylation-1Q	612379	SRD5A3	Oxidoreductase	AR

Table 3.04.12.3-7. Continued.

Disorder	OMIM	Gene	Biological Function	Product of Conception Alteration
Congenital disorder of glycosylation-2A	212066	MGAT2	Golgi complex	AR
Congenital disorder of glycosylation-2B	606056	MOGS	Endoplasmic reticulum protein	AR
Congenital disorder of glycosylation-2C	266265	SLC35C1	Solute carrier	AR
Congenital disorder of glycosylation 2D	607091	B4GALT1	Galactosyltransferase	AR
Congenital disorder of glycosylation-2E	608779	COG7	Golgi complex	AR
Congenital disorder of glycosylation-2F	603585	SLC35A1	Solute carrier	AR
Congenital disorder of glycosylation-2G	611209	COG1	Golgi complex	AR
Congenital disorder of glycosylation-2H	611182	COG8	Golgi complex	AR
Congenital disorder of glycosylation-2J	613489	COG4	Golgi complex	AR
Familial hypercholanemia	607748	BAAT	Acyltransferase	AR
Familial hypercholanemia	607748	TJP2	Assembly of tight junctions	AR
Familial hypercholanemia	607748	EPHX1	Endoplasmic reticulum hydrolase	AR

3.04.13 Disorders of Lipoprotein Metabolism and other Lipidemias

3.04.13.1 Pure hypercholesterolemia

Hyperlipidemia and/or hyperlipoproteinemia: Are a group of disorders characterized by abnormally elevated levels of lipids and/or lipoproteins in the blood. In this condition, it is difficult to separate the effects of heritability and environment factors. Nowadays, the main factors that cause this condition are environmental. There are primary causes of hyperlipidemia, which are of genomic origin. There are several exclusion criteria in hyperlipidemia for it to be considered of genomic or familial origin. There are clear environmental risk factors that contribute to this condition, including lifestyle choices such as diet, exercise and tobacco smoking. There are also additional factors that contribute to hyperlipidemia such as the individual's gender and age. Other factors include medical problems such as diabetes and obesity which impact cholesterol levels. There are 4 genes which are related to Familial hypercholesterolemia *LDLR, APOB, PCSK9* and *LDLRAP1*. There are also genes which cause xanthomatosis susceptibility such as the *APOA2, GHR* and *EPHX2*. These genes contribute to modification of the phenotype of Familial hypercholesterolemia. There are also others genes such as *PPP1R17* and *ITIH4* which play a role in susceptibility to hypercholesterolemia

Familial hypercholesterolemia or Hyperlipoproteinemia type II: A disorder characterized by elevation of serum cholesterol, ranging from 250 to 450 mg per dl of LDL-cholesterol. Affected individuals develop tendinous xanthomas, corneal arcus and coronary artery disease. Familial hypercholesterolemia has a least 3 patterns of inheritance. First, the most common condition is caused by autosomal dominant or monoallelic mutations in the *LDLR* gene, the *APOB* gene, and/or the *PCSK9* gene. The *LDLR* gene encodes the low density lipoprotein receptor which in cell surface proteins is involved in receptor-mediated endocytosis of low density lipoprotein. The *APOB* gene encodes the main apolipoprotein of chylomicrons (apoB-48) and low density lipoproteins (apoB-100). The apoB-48 is synthesized exclusively in the intestine due to a unique process of RNA editing. The apoB-100 is synthesized in the liver. The *PCSK9* gene encodes the proprotein convertase subtilisin/kexin type 9, which is a serine protease that plays a role in cholesterol homeostasis. Familial hypercholesterolemia due to mutations in the *LDLR* gene is estimated to affect 1 in every 500 individuals worldwide. Familial hypercholesterolemia due to mutations in the *APOB* gene is estimated to affect 1 in every 1,000 individuals worldwide. Familial hypercholesterolemia due to mutations in the *PCSK9* gene is estimated to affect 1 in every 2,500 individuals worldwide. The condition caused by *LDLR* is more prevalent among Ashkenazi Jews, Lebanese Christians, Afrikaners, French Canadians, and Finnish. Second, the same *LDLR* gene, when the product of conception alteration has the two defective alleles and follows the form of AD + AD = AR, since both

parents provide the faulty copy to descendants, exhibits an extremely severe phenotype. On average, the carrier frequency for Familial hypercholesterolemia (due to monoallelic mutations in the *LDLR* gene) is estimated to be 1 in 500 individuals worldwide, which means the frequency of this condition in biallelic form is about 1 in 1,000,000 individuals or $1/(1,000)^2$. In this homozygous Familial hypercholesterolemia, the aortic root develops atherosclerotic plaque at an early age. The LDL-cholesterol in these patients is usually greater than 500 mg per dl. The condition is lethal between 10 and 30 years of age. There is a higher prevalence for this condition among Ashkenazi Jews, Lebanese Christians, Afrikaners, French Canadians and Finnish. In Ashkenazi Jews, the carrier frequency for Familial hypercholesterolemia (due to monoallelic mutations in the *LDLR* gene) is estimated to be 1 in 69 individuals, which means the frequency of this condition in biallelic form is about 1 in 19,000 individuals or $1/(138)^2$. In Lebanese Christians, the carrier frequency for Familial hypercholesterolemia (due to monoallelic mutations in the *LDLR* gene) is estimated to be 1 in 71 individuals, which means the frequency of this condition in biallelic form is about 1 in 20,000 individuals or $1/(142)^2$. In Afrikaner from Transvaal Province, the carrier frequency for Familial hypercholesterolemia (due to monoallelic mutations in the *LDLR* gene) is estimated to be 1 in 100 individuals, which means the frequency of this condition in biallelic form is about 1 in 40,000 individuals or $1/(200)^2$. In French Canadians from Saguenay-Lac-Saint-Jean region, Quebec, the carrier frequency for Familial hypercholesterolemia (due to monoallelic mutations in the *LDLR* gene) is estimated to be 1 in 122 individuals, which means the frequency of this condition in biallelic form is about 1 in 60,000 individuals or $1/(244)^2$. The same form of AD + AD = AR could be due to mutations in the *APOB* gene where both parents provide the faulty copy to descendants. On average, the carrier frequency for Familial hypercholesterolemia (due to monoallelic mutations in the *APOB* gene) is estimated to be 1 in 1,000 individuals worldwide, which means the frequency of this condition in biallelic form is about 1 in 4,000,000 individuals or $1/(2,000)^2$. There is a higher carrier frequency for Familial hypercholesterolemia due to mutations in the *APOB* gene among central European people; this allele appears 6,000 years ago. Third, Familial autosomal recessive hypercholesterolemia is caused by biallelic mutations in the *LDLRAP1* gene which encodes the low density lipoprotein receptor adaptor protein 1. This protein interacts with low density lipoprotein receptor (LDLR) and the endocytic machinery.

Familial dysbetalipoproteinemia or Hyperlipoproteinemia type III: A disorder characterized by xanthomatosis and premature coronary and/or peripheral vascular disease. This condition results from a defect in apolipoprotein E which increases plasma cholesterol and triglycerides, due to an impaired clearance of chylomicron and VLDL remnants. Most cases of Familial dysbetalipoproteinemia involve the 2 allele APOE-E2 isoform, or a combination of APOE-E2 and APOE-E4 isoform. Only 1-4% of people with the two allele APOE-E2 isoform develop Familial dysbetalipoproteinemia. Apolipoprotein E is essential for the normal catabolism of triglyceride-rich lipoprotein.

Hyperlipoproteinemia type V or Familial hyperchylomicronemia: A disorder characterized by abdominal pain and eruptive xanthoma. Hyperlipoproteinemia type V is caused by autosomal recessive mutations in the *APOA5* gene which encodes the Apolipoprotein A-V. This protein is a component of high density lipoprotein and is an important determinant of plasma triglyceride levels.

Familial hypertriglyceridemia: A disorder associated with mutations in either the *RP1* gene which encodes the retinitis pigmentosa protein 1, or the *LIP1* gene which encodes the lipase member I.

Sitosterolemia or Phytosterolemia: A disorder characterized by hyperabsorption and decreased biliary excretion of dietary sterols including cholesterol, plant and shellfish sterols. Affected individuals exhibit hypercholesterolemia, premature development of atherosclerosis, tendon and tuberous xanthomas, and abnormal hematologic and liver function test results. Sitosterolemia is caused by autosomal recessive mutations in either the *ABCG5* gene, or the *ABCG8* gene which encodes the ATP-binding cassette transporter (family 5 and 8, respectively). Sitosterolemia is estimated to affect less than 1 in every 1,000,000 individuals. There is a higher carrier frequency for Sitosterolemia trait among Amish-Mennonite people of Lancaster County, Pennsylvania, as well as other persons of Swiss-German ancestry. In this group, Sitosterolemia is caused by a common SNP in the *ABCG8* gene. In Amish-Mennonites this pathogenic allele appears 250 years ago. Sitosterolemia is significantly underdiagnosed or probably misdiagnosed with hyperlipidemia.

Congenital deficiency of Lipoprotein (a): A trait in which persons with this congenital deficiency appeared to be clinically healthy. There are several SNPs that have a significant correlation with coronary artery disease in genome wide association study.

Familial lipoprotein lipase deficiency or Hyperlipoproteinemia type I: A disorder characterized by very severe hypertriglyceridemia (>2000 mg/dL) with episodes of abdominal pain of childhood-onset. Affected infants also exhibit hyperchylomicronemia, recurrent acute pancreatitis, eruptive cutaneous xanthomata and hepatosplenomegaly. Familial lipoprotein lipase deficiency is caused when the product of conception has the two defective alleles and follows the alteration in the form of AD + AD = AR and both parents provide the faulty copy to descendants. Severe familial lipoprotein lipase deficiency is caused by two defective alleles in the *LPL* gene which encodes the lipoprotein lipase. On average, the carrier frequency for monoallelic familial lipoprotein lipase deficiency is estimated to be 1 in 500 individuals worldwide, which means the frequency of this condition in biallelic form is about 1 in 1,000,000 individuals or $1/(1,000)^2$.

Familial combined hyperlipidemia or familial apoprotein C2 deficiency: A disorder characterized by similar symptoms as those of Familial lipoprotein lipase deficiency, since the mechanistic pathway is shared. Familial apoprotein C2 deficiency is caused by autosomal recessive mutations in the *APOC2* gene which encodes the Apolipoprotein C-II. This protein is a cofactor for the activation of lipoprotein lipase in tissues.

Combined lipase deficiency: A disorder characterized by similar symptoms as those of Familial lipoprotein lipase deficiency. Patients with Combined lipase deficiency display severe hypertriglyceridemia. Combined lipase deficiency is caused by autosomal recessive mutations in the *LMF1* gene which encodes the lipase maturation factor-1. This is an endoplasmic reticulum protein involved in the maturation and transport of lipoprotein lipase through the secretory pathway.

Combined lipase deficiency is estimated to affect 1 in every 1,000,000 individuals.

Familial combined hyperlipidemia or Hyperalphalipoproteinemia type I: A trait that is generally asymptomatic without signs of atherosclerosis and/or cardiovascular disease. This condition is characterized by elevated levels of alpha-lipoprotein resulting in high levels of HDL-cholesterol (more than 150 mg/dl). Hyperalphalipoproteinemia type I is caused by loss-of-function mutations in the *CETP* gene which encodes the cholesteryl ester transfer protein.

Familial combined hyperlipidemia or Hyperalphalipoproteinemia type II: A trait associated with a cardioprotective effect. Individuals with this trait are characterized by lower fasting and postprandial serum triglycerides, lower levels of LDL cholesterol, and higher levels of HDL cholesterol than non-carriers. Hyperalphalipoproteinemia type II is caused by loss-of-function mutations in the *APOC3* gene which encodes the Apolipoprotein C-III. In the Old Order Amish communities of Lancaster Pennsylvania, the carrier frequency for heterozygous mutations in the *APOC3* gene is estimated to be 1 in 20 individuals, which means the frequency of homozygous is 1 in 1,600 individuals or $1/(40)^2$. Heterozygous carriers individuals showed a cardioprotective effect due to this DNA mutation.

Abetalipoproteinemia or Bassen-Kornzweig syndrome: A disorder appearing during the first year of life or in young childhood characterized by malabsorption with features of celiac syndrome. Affected individuals exhibit diarrhea with steatorrhea and fat malabsorption. Patients also exhibit growth delay, hepatomegaly, and neurological and neuromuscular manifestations. Patients also have defective intestinal absorption of lipids, low levels of apolipoprotein B and low LDL cholesterol. Other features of this condition include progressive ataxic neuropathy, atypical retinitis pigmentosa, acanthocytosis and cirrhosis. Abetalipoproteinemia is caused by autosomal recessive mutations in the *MTTP* gene which encodes the microsomal triglyceride transfer protein. This protein is essential for creating beta-lipoproteins. On average, the carrier frequency for the Abetalipoproteinemia trait is estimated to be 1 in 500 individuals worldwide, which means the frequency of this biallelic condition is about 1 in 1,000,000 individuals or $1/(1,000)^2$.

Chylomicron retention disease or Anderson disease: A disorder appearing during early infancy (1-6 months) characterized by malabsorption of fat. Affected individuals exhibit malnutrition, failure to thrive, growth delay, hepatomegaly, cardiomyopathy, neurolopathy and ophthalmologic complications. Chylomicron retention disease is caused by autosomal recessive mutations in the *SAR1B* gene which encodes the SAR1 homolog GTP-binding protein B. This protein is involved in protein transport from the endoplasmic reticulum to the Golgi. This condition results in accumulation of pre-chylomicron transport vesicles in the cytoplasm of enterocytes. Chylomicron retention disease is estimated to affect less than 1 in every 1,000,000 individuals.

Tangier disease or Familial alpha-lipoprotein deficiency: A disorder characterized by corneal opacities, tonsil enlargement, hepatosplenomegaly and neurologic symptoms. This condition is caused by cholesterol deposition in tissue and organs. Tangier disease is caused by autosomal recessive mutations in the *ABCA1* gene which encodes the ATP-binding cassette (sub-family A member 1).

This is a protein involved in cholesterol efflux, acting as a pump in the cellular lipid removal pathway. On average, the carrier frequency for Tangier disease trait is estimated to be 1 in 500 individuals worldwide, which means the frequency of this condition is about 1 in 1,000,000 individuals or $1/(1,000)^2$. There is a higher carrier frequency for this condition in the Tangier Island in the Chesapeake Bay, Virginia. This gene also has 2 other allelic disorders: the Familial hypoalphalipoproteinemia or HDL deficiency type 2, and the Protection against coronary artery disease in familial hypercholesterolemia.

Familial hypoalphalipoproteinemia: A similar condition as those of Tangier disease. Familial hypoalphalipoproteinemia is caused by autosomal recessive mutations in the *APOA1* gene which encodes the Apolipoprotein A-I. This protein is a major apoprotein of high density lipoprotein (HDL).

Fish-eye disease: A disorder characterized by corneal opacities, in which patients' eyes resemble those of a boiled fish. This form of dyslipoproteinemia is characterized by standard levels of cholesterolemia with hypertriglyceridemia. This condition displays elevated levels of VLDL, strikingly high LDL, and reduced levels of HDL cholesterol. Fish-eye disease is caused by autosomal recessive mutations in the *LCAT* gene which encodes the lecithin cholesterol acyltransferase. This protein catalyzes the formation of cholesterol esters in lipoproteins. This gene also has another allelic disorder the Norum disease. The latter condition display normochromic anemia, proteinuria and corneal deposits of lipid. Fish-eye disease is estimated to affect less than 1 in every 1,000,000 individuals.

3.04.13.2 Other Disorders of Lipoprotein Metabolism

Urbach–Wiethe disease or Lipoid proteinosis: A disorder characterized by a progressive hoarseness since birth and beaded papules on the eyelids. Affected individuals exhibit poor wound healing, and generalized thickening of skin, mucosa and certain viscera. Some patients also have lesion in brain tissue that leads to epilepsy and neuropsychiatric abnormalities. The condition in these patients does not cause a decreased life span. Urbach–Wiethe disease is caused by autosomal recessive mutations in the *ECM1* gene which encodes the extracellular matrix protein 1. Urbach–Wiethe disease is estimated to affect less than 1 in every 1,000,000 individuals. This condition has been mostly described in South Africa among people of Dutch, German, and Khoisan heritage.

Table 3.04.13.1 Recognizable Disorders of Lipoprotein Metabolism and other Lipidemias

Disorder	OMIM	Gene	Biological Function	Product of Conception Alteration
Familial hypercholesterolemia or Hyperlipoproteinemia type II	143890	LDLR	Cell surface receptor Lipid receptor	AD or AD + AD = AR
Familial hypercholesterolemia or Hyperlipoproteinemia type II	144010	APOB	Apolipoprotein	AD or AD + AD = AR
Familial hypercholesterolemia or Hyperlipoproteinemia type II	603776	PCSK9	Proprotein convertase	AD
Familial hypercholesterolemia or Hyperlipoproteinemia type II or Familial autosomal recessive hypercholesterolemia	603813	LDLRAP1	Receptor adaptor protein	AR
Familial hypercholesterolemia	143890	APOA2	Apolipoprotein	AD
Familial modification of hypercholesterolemia	143890	GHR	Cell surface receptor JAK-STAT	AD
Familial modification of hypercholesterolemia, due to LDLR defect	143890	EPHX2	Endoplasmic reticulum hydrolase	AD
Susceptibility to hypercholesterolemia	143890	PPP1R17	Phosphatase	AD
Susceptibility to hypercholesterolemia	143890	ITIH4	Trypsin inhibitor	AD
Familial dysbetalipoproteinemia or hyperlipoproteinemia type III	107741	APOE	Apolipoprotein	Acod
Hyperlipoproteinemia type V or Familial hyperchylomicronemia	144650	APOA5	Apolipoprotein	AR
Familial hypertriglyceridemia	145750	RP1	Ciliary proteins, connecting cilia	
Familial hypertriglyceridemia	145750	LIPI	Phospholipase	
Familial hypertriglyceridemia	144650	APOA5	Apolipoprotein	AR
Sitosterolemia or Phytosterolemia	210250	ABCG5	ABC-transporter protein	AR

Table 3.04.13.1. Continued.

Disorder	OMIM	Gene	Biological Function	Product of Conception Alteration
Sitosterolemia or Phytosterolemia	210250	ABCG8	ABC-transporter protein	AR
Congenital LPA deficiency	152200	LPA	Lipoprotein	
Familial lipoprotein lipase deficiency or Hyperlipoproteinemia type I	144250	LPL	Lipoprotein lipase	AD + de novo
Familial lipoprotein lipase deficiency or Hyperlipoproteinemia type I	238600	LPL	Lipoprotein lipase	AD + AD = AR
Familial combined hyperlipidemia or Familial apoprotein C2 deficiency	207750	APOC2	Apolipoprotein	AR
Combined lipase deficiency	246650	LMF1	Endoplasmic reticulum protein	AR
Familial combined hyperlipidemia or Hyperalphalipoproteinemia type I	143470	CETP	Cholesteryl ester transfer protein	AD
Familial combined hyperlipidemia or Hyperalphalipoproteinemia type II	614028	APOC3	Apolipoprotein	AD or AD + AD = AR
Abetalipoproteinemia or Bassen-Kornzweig syndrome	200100	MTTP	Endoplasmic reticulum protein	AR
Chylomicron retention disease or Anderson disease	246700	SAR1B	Protein transport from the endoplasmic reticulum to the Golgi	AR
Tangier disease or Familial alpha-lipoprotein deficiency	205400	ABCA1	ABC-transporter protein	AR
Protection against coronary artery disease in familial hypercholesterolemia	143890	ABCA1	ABC-transporter protein	AD
HDL deficiency 2	604091	ABCA1	ABC-transporter protein	AR
Familial hypoalphalipoproteinemia	604091	APOA1	Apolipoprotein	AR
Fish-eye disease	136120	LCAT	Lecithin-cholesterol acyltransferase	AR
Norum disease	245900	LCAT	Lecithin-cholesterol acyltransferase	AR
Urbach–Wiethe disease or Lipoid proteinosis	247100	ECM1	Extracellular matrix disease	AR

3.04.14 Other Metabolic Disorders

3.04.14.1 Disorders of Purine and Pyrimidine Metabolism

Kelley-Seegmiller syndrome: A disorder characterized by the development of gouty arthritis and the formation of uric acid stones in the urinary tract. Kelley-Seegmiller syndrome is mild form of hypoxanthine-guanine phosphoribosyltransferase deficiency. Kelley-Seegmiller syndrome is caused by X-linked recessive mutations in the *HPRT1* gene which encodes the hypoxanthine-guanine phosphoribosyltransferase. This condition predominantly affects males, since females are usually asymptomatic carriers.

Phosphoribosylpyrophosphate synthetase superactivity: A disorder characterized by hyperuricemia and hyperuricosuria, which results in gout and urolithiasis. This condition appears during late adolescence or early adulthood in males. There is a more severe form phosphoribosylpyrophosphate synthetase superactivity with crystalluria, and neurodevelopmental impairment, with hypotonia, locomotor delay, mental retardation, and high frequency hearing loss since childhood. Phosphoribosylpyrophosphate synthetase superactivity is caused by gain-of-function mutation with acceleration of *PRPS1* gene transcription, resulting in increased enzyme activity. There are 3 additional conditions caused by mutations in the *PRPS1* gene including: Arts syndrome, X-linked Charcot-Marie-Tooth disease-5 and isolated X-linked sensorineural deafness.

Lesch-Nyhan syndrome: A disorder characterized by gouty arthritis, kidney stones, neurologic and behavioral problems with psychomotor delay. This condi-tion becomes evident within 3-6 months of birth. Affected individuals manifest severe dystonia with hypotonia and involuntary movements such as choreoathetosis and ballismus. Patients exhibit obsessive-compulsive self-mutilation and aggressive behaviors. Females are heterozygous asymptomatic carriers and males are usually affected. Affected young boys are wheelchair-bound. Lesch-Nyhan syndrome is caused by X-linked recessive mutations in the *HPRT1* gene which encodes the hypoxanthine phosphoribosyltransferase 1. This enzyme catalyzes the conversion of hypoxanthine to inosine monophosphate and guanine to guanosine monophosphate via transfer of the 5-phosphoribosyl group from 5-phosphoribosyl 1-pyrophosphate. This protein is responsible for recycling purines. Lesch-Nyhan syndrome is estimated to affect 1 in every 235,000-380,000 live births. This condition occurs with a similar frequency in all ethnic groups.

3.04.14.2 Other Disorders of Purine and Pyrimidine Metabolism

Xanthinuria type I: A disorder characterized by tendency to form xanthine stones. During the first decade of life, affected patients can exhibit urolithiasis. Other features include nephropathy, myopathy and arthropathy. The last complication of this condition occurs more often in older patients with xanthinuria. In these individuals, uric acid is unusually diminished in serum and urine. Xanthinuria also occurs in molybdenum cofactor deficiency. Xanthinuria type I is caused by autosomal recessive mutations in the *XDH* gene which encodes the xanthine dehydrogenase. Xanthinuria type I is an isolated deficiency of xanthine dehydrogenase, whereas xanthinuria type II is a dual deficiency of xanthine dehydrogenase and aldehyde oxidase. The prevalence of Xanthinuria is unknown since it is a condition rarely reported. The estimated prevalence varies from 1 in every 6,000-69,000 individuals. This condition is particularly high in Mediterranean and Middle Eastern countries. The main reason for clinical xanthinuria in those countries includes genomic factors such as consanguinity and environmental factors such as climate aridity.

Renal hypouricemia: A generally benign medical condition characterized by impaired uric acid reabsorption at the apical membrane of proximal renal tubule cells. Most patients with this condition are asymptomatic, but 10% of patients could suffer from nephrolithiasis and exercise-induced acute renal failure. There are several underlying causes of this disorder with at least 2 autosomal genomic loci: Renal hypouricemia-1 can be caused by autosomal recessive mutations in the *SLC22A12* gene which encodes the solute carrier (family 22 member A12), an organic anion/urate transporter;; and Renal hypouricemia-2 can be caused by autosomal recessive mutations in the *SLC2A9* gene which encodes the solute carrier (family 22 member A12). These genes encode a high-affinity urate transporter. Among Japanese children, Renal hypouricemia-1 is estimated to affect 1 in every 1,730.

Adenine phosphoribosyltransferase deficiency or Urolithiasis, 2,8-dihydroxyadenine: A disorder characterized by crystalluria and the formation of urinary stones. This condition is caused by accumulation of the insoluble purine 2,8-dihydroxyadenine in the kidney. Affected individuals exhibit renal colic, dysuria, hematuria, urinary tract infection and renal failure. Adenine phosphoribosyltransferase deficiency is caused by autosomal reces-

sive mutations in the *APRT* gene which encodes the adenine phosphoribosyltransferase. There are 2 forms of APRT deficiency. Type I deficiency is characterized by complete enzyme deficiency, whereas type II deficiency is characterized by partial deficiency in cell lysates. Type II deficiency is most common among Japanese. On average in Japan, the carrier frequency for Adenine phosphoribosyltransferase deficiency type II trait is estimated to be 1 in 82 individuals, which means the frequency of this condition is about 1 in 27,000 individuals or $1/(164)^2$. There is a lower carrier frequency for this condition among Europeans. The disease is estimated to affect 1 in every 50,000-100,000 individuals.

Myopathy due to myoadenylate deaminase deficiency: A disorder characterized by myopathy usually beginning at the fourth decade of life. Patients with this condition exhibit muscle weakness, myalgia, cramps and muscle atrophy. Myopathy due to myoadenylate deaminase deficiency is caused by autosomal recessive mutations in the *AMPD1* gene which encodes the AMP deaminase. This enzyme catalyzes the deamination of AMP to IMP in skeletal muscle. This protein also plays a role in the purine nucleotide cycle. Genomic mutations in the *AMPD1* gene are common among Caucasians, but this condition rarely causes significant symptoms. In Caucasians, myopathy due to myoadenylate deaminase deficiency is one of the most prevalent genomic disorders and is estimated to affect 1 in every 50-100 individuals. This condition is less common among Japanese people.

Adenylosuccinate lyase deficiency or Adenylosuccinase deficiency: A disorder characterized by severe psychomotor delay, epilepsy and autism. Occasionally, affected patients also show growth retardation and muscle wasting. Affected individuals exhibit the appearance of succinylaminoimidazolecarboxamide riboside and succinyladenosine in the body fluids. Adenylosuccinate lyase deficiency is caused by autosomal recessive mutations in *ADSL* gene which encodes the adenylosuccinate lyase. This enzyme catalyzes an important reaction in the *de novo* pathway of purine biosynthesis. Adenylosuccinate lyase deficiency is estimated to affect less than 1 in every 1,000,000 individuals.

AICA-ribosiduria due to ATIC deficiency: A disorder characterized by devastating neurological symptoms. Affected individuals exhibit profound mental retardation, epilepsy, dysmorphic features and congenital blindness. AICA-ribosiduria is caused by autosomal recessive mutations in the *ATIC* gene which encodes the bifunctional enzyme AICAR transformylase/IMP cyclohydrolase. AICA-riboside or 5-amino-4-imidazole-carboxamide riboside is an intermediate of the *de novo* purine biosynthetic pathway. AICA-ribosiduria is estimated to affect less than 1 in every 1,000,000 individuals.

Beta-ureidopropionase deficiency: A disorder characterized by severe neurological symptoms. Affected individuals exhibit muscular hypotonia, dystonic movements, mental retardation, seizures, microcephaly and severe developmental delay. Patients showed an increased N-carbamyl-beta-alanine and N-carbamyl-beta-aminoisobutyric acid in cerebrospinal fluid (CSF), plasma, and urine. Beta-ureidopropionase deficiency is caused by autosomal recessive mutations in the *UPB1* gene which encodes the beta-ureidopropionase. This enzyme catalyzes the last step in the pyrimidine degradation pathway. Beta-ureidopropionase deficiency is estimated to affect less than 1 in every 1,000,000 individuals.

Dihydropyrimidine dehydrogenase deficiency: A disorder characterized by convulsive disorder, mental retardation, microcephaly, hypertonia and autistic behavior. This condition starts manifesting during infancy. There are individuals with total Dihydropyrimidine dehydrogenase deficiency who are asymptomatic. There are also individuals with partial deficiency of dihydropyrimidine dehydrogenase who may be vulnerable to fluoropyrimidine drugs. Dihydropyrimidine dehydrogenase deficiency is caused by autosomal recessive mutations in the *DPYD* gene which encodes the dihydropyrimidine dehydrogenase. This enzyme is involved in the breakdown of uracil and thymine. The partial deficiency of dihydropyrimidine dehydrogenase is a common condition affecting 2-8% of people. On average, the carrier frequency for partial Dihydropyrimidine dehydrogenase deficiency in Japan is estimated to be 1 in 50 individuals, which means the frequency of total Dihydropyrimidine dehydrogenase deficiency is about 1 in 10,000 individuals or $1/(100)^2$.

Dihydropyrimidinuria: A disorder characterized by neurologic abnormalities. Affected individuals showed severe retardation, convulsions, metabolic acidosis, extrapyramidal dyskinesia and pyramidal signs. There are individuals with dihydropyrimidinase deficiency who are asymptomatic. Patients exhibit increased uracil, dihydrouracil, thymine and dihydrothymine in bodily fluids. Dihydropyrimidinuria is caused by autosomal recessive mutations in the *DPYS* gene which encodes the dihydropyrimidinase. This is the second enzyme in the 3-step degradation pathway of uracil and thymine. Dihydropyrimidinuria is estimated to affect less than 1 in every 1,000,000 individuals.

Table 3.04.14.1 Recognizable Disorders of Purine and Pyrimidine Metabolism

Disorder	OMIM	Gene	Biological Function	Product of Conception Alteration
Kelley-Seegmiller syndrome	300323	HPRT1	Phosphoribosyltransferase	XR + de novo
Phosphoribosylpyrophosphate synthetase superactivity	300661	PRPS1	Phosphoribosylation	XR
Lesch-Nyhan syndrome	300322	HPRT1	Phosphoribosyltransferase	XR + 50%
Xanthinuria type I	278300	XDH	Xanthine dehydrogenase	AR
Renal hypouricemia-1	220150	SLC22A12	Solute carrier	AR
Renal hypouricemia-2	612076	SLC2A9	Solute carrier	AR
Adenine phosphoribosyltransferase deficiency or Urolithiasis, 2,8-dihydroxyadenine	102600	APRT	Adenine phosphoribosyltransferase	AR
Myopathy due to myoadenylate deaminase deficiency	102770	AMPD1	Adenosine monophosphate deaminase	AR
Adenylosuccinate lyase deficiency or adenylosuccinase deficiency	103050	ADSL	Adenylosuccinate lyase	AR
AICA-ribosiduria	608688	ATIC	Formyltransferase/IMP cyclohydrolase	AR
Beta-ureidopropionase deficiency	613161	UPB1	Hydrolase	AR
Dihydropyrimidine dehydrogenase deficiency	274270	DPYD	Dehydrogenase	AR
Dihydropyrimidinuria	222748	DPYS	Dihydropyrimidinase	AR

3.04.15.1 Disorders of Porphyrin and Bilirubin Metabolism

Porphyrias: Are a group of disorders in which the heme group is not made properly. These conditions result in the accumulation and increased excretion of porphyrins or porphyrin precursors.

X-linked erythropoietic protoporphyria: A disorder characterized by acute photosensitivity. Affected individuals did not show either anemia or iron overload.

X-linked erythropoietic protoporphyria is caused by X-linked dominant gain-of-function mutations in the *ALAS2* gene which encodes the delta-aminolevulinate synthase 2. This mitochondrial enzyme catalyzes the first step in the heme biosynthetic pathway. In this X-linked condition both sexes are equally affected. On the contrary, loss-of-function mutations in the *ALAS2* gene cause X-linked sideroblastic anemia. X-linked erythropoietic protoporphyria is estimated to affect 1 in 75,000 to 200,000 individuals worldwide.

Erythropoietic protoporphyria: A disorder beginning in childhood characterized by light-sensitive dermatitis. Most patients with this condition experience itching, burning and erythema after exposure to bright light. Besides this painful photosensitivity, there are also patients who develop liver complications, due to the accumulation of protoporphyrin in this organ. Affected individuals also exhibit elevated red cell protoporphyrin levels. Erythropoietic protoporphyria is caused by autosomal recessive mutations in the *FECH* gene which encodes the ferrochelatase. This is the last enzyme of the heme biosynthetic pathway. This enzyme catalyzes the insertion of iron into protoporphyrin to form heme. Erythropoietic protoporphyria is estimated to affect 1 in 75,000-200,000 individuals worldwide.

Congenital erythropoietic porphyria or Gunther disease: A disorder that is considered the most severe and dramatic form of porphyria. This childhood onset condition is characterized by severe damage to skin exhibiting hypertrichosis, blistering and scarring of exposed areas. This condition leads to mutilating deformity. Uroporphyrin I and coproporphyrin I are found in many organs and tissue, staining the urine and the teeth red. Patients also manifest hemolytic anemia. Congenital erythropoietic porphyria is caused by autosomal recessive mutations in the *UROS* gene which encodes the uroporphyrinogen III synthase. This is the fourth enzyme in the 8-enzyme pathway in the conversion of glycine and succinyl-CoA to heme. Congenital erythropoietic porphyria is estimated to affect less than 1 in every 1,000,000 individuals.

Porphyria cutanea tarda: A disorder appearing during adulthood characterized by bullous photodermatitis. Affected individuals suffer from extremely fragile skin followed by bullous cutaneous lesions on the surface of skin exposed to the sun. Patients have scarring with hyper and hypopigmentation, and sclerodermic signs. Affected patients also have hepatic lesions such as steatosis, siderosis and chronic inflammatory disorders. There are two forms: Type I is the most common form, and is more common in men than in women. The cause of this condition is probably multifactorial. There are several risk factors including excessive alcohol consumption, hepatitis C, hemochromatosis, estrogen use and other factors. Porphyria cutanea tarda type II is caused by autosomal dominant mutations in the *UROD* gene which encodes the uroporphyrinogen decarboxylase. This is the fifth enzyme in the heme biosynthesis pathway. In the United States, Porphyria cutanea tarda is estimated to affect 1 in every 25,000 individuals. There is a higher frequency for Porphyria cutanea in the Czech Republic and Slovakia, where 1 in every 5,000 individuals are affected.

3.04.15.2 Other Porphyria

Acute intermittent porphyria: A disorder starting after puberty characterized by episodes of acute gastrointestinal and neuropathic symptoms. The most common symptom associated with this condition is abdominal discomfort, sometimes accompanied by constipation and urinary retention. Affected individuals also exhibit paraesthesias, paralysis, psychotic episodes, seizures and hypertension. This condition may occur in acute attacks; between episodes, the patients are healthy. Only 10-20% of people who are gene mutations carriers experience the aforementioned symptoms because the Acute intermittent porphyria is a low penetrance condition. Acute intermittent porphyria is caused by autosomal dominant mutations in the *HMBS* gene which encodes the porphobilinogen deaminase. This is the third enzyme of the biosynthetic pathway leading to the production of heme. Acute intermittent porphyria is estimated to affect in 1 in 50,000 individuals worldwide, probably across all ethnic groups. In Lapland, northern Sweden, Acute intermittent porphyria is estimated to affect 1 in 1,500 individuals. In Europeans this pathogenic allele appears 1,000 years ago.

Hereditary coproporphyria: A disorder characterized by acute episodes of neurologic dysfunction. This condition is often provoked by drugs, fasting, menstrual cycle and other precipitants. Affected individuals could exhibit skin photosensitivity, constipation and abdominal colic. Patients display a large excretion of coproporphyrin III, mostly in feces and urine. Hereditary coproporphyria is caused by autosomal dominant mutations in the *CPOX* gene which encodes the coproporphyrinogen oxidase. This is the sixth enzyme of the heme biosynthetic pathway. Hereditary coproporphyria is estimated to affect in 1 in 500,000 individuals worldwide.

Harderoporphyria: A severe disorder appearing at birth characterized by hemolytic anemia with intense jaundice and hepatosplenomegaly. Affected children also exhibit skin lesions and massive excretion of harderoporphyrin in feces. Patients with harderoporphyria exhibit iron overload secondary to dyserythropoiesis. Harderoporphyria is caused by autosomal recessive mutations in the CPOX gene. In Europe, Harderoporphyria is estimated to affect 1 in every 1,000,000 individuals.

Variegate porphyria or South African porphyria: A disorder with similar features as those of Acute intermittent porphyria. Affected individuals exhibit cutaneous manifestations, hyperpigmentation, hypertrichosis, photosensitivity, blistering and skin fragility. Patients also manifest acute episodes of abdominal pain, constipation, muscular paralysis, sensory disturbances, tachycardia and hypertension. Episodes are often related to iron overload and/or are drug-induced. Most people who are gene mutation carriers never experience the aforementioned symptoms because the Variegate porphyria is a low penetrance condition. Variegate porphyria is caused by autosomal dominant or monoallelic mutations in the *PPOX* gene which encodes the protoporphyrinogen oxidase. This protein is located in the inner membrane of mitochondria. This enzyme is responsible for the seventh step in heme biosynthetic pathway. In Afrikaners, variegate porphyria is estimated to affect 1 in every 300 individuals. The high frequency of porphyria variegata in Afrikaners is a clear example of founder effect. In Finland, variegate porphyria is estimated to affect 1 in every 77,000 individuals.

Homozygous variant of Variegate Porphyria: A severe disorder appearing during early childhood characterized by photosensitization, growth retardation, skeletal abnormalities of the hand, short stature, mental retardation, nystagmus and convulsions. Homozygous variant of Variegate Porphyria is caused by biallelic mutations in the *PPOX* gene. This variant is characterized by severe PPOX deficiency, caused by biallelic inherited defective alleles.

Acute hepatic porphyria: A disorder characterized by recurrent attacks of pain, vomiting and hyponatremia. Affected individuals excreted large amounts of 5-aminolevulinic acid and coproporphyrin. Patients with hepatic porphyria are sensitive to alcohol ingestion or lead exposure. Acute hepatic porphyria is caused by autosomal recessive mutations in the *ALAD* gene which encodes the delta-aminolevulinate dehydratase. This is a cytosolic enzyme that catalyzes the second step in the porphyrin and heme biosynthetic pathway. Acute hepatic porphyria is estimated to affect less than 1 in every 1,000,000 individuals.

Hepatoerythropoietic porphyria: A severe form of porphyria starting from infancy manifesting variable degrees of liver damage or fibrosis. Affected individuals also exhibit skin photosensitive of early onset, hypertrichosis, and severe scleroderma-like lesions of the hands. Patients display excessive excretion of acetate-substituted porphyrins and accumulation of protoporphyrin in erythrocytes. Hepatoerythropoietic porphyria is caused by autosomal recessive mutations in the *UROD* gene which encodes the uroporphyrinogen decarboxylase. This enzyme catalyzes the conversion of uroporphyrinogen to coproporphyrinogen through the removal of 4 carboxymethyl side chains Hepatoerythropoietic porphyria is estimated to affect less than 1 in every 1,000,000 individuals.

3.04.15.3 Defects of Catalase and Peroxidase

Acatalasemia or Takahara disease: A disorder characterized by progressive oral gangrene but most individuals are usually asymptomatic. Other features include oral ulcerations, diabetes mellitus and atherosclerosis. Acatalasemia is caused by autosomal recessive mutations in the *CAT* gene which encodes the catalase. This enzyme catalyzes the decomposition of hydrogen peroxide to oxygen and water. Mutations in the *CAT* gene have also been associated with aniridia and hypertension. Acatalasemia is estimated to affect 1 in every 31,250 individuals.

Myeloperoxidase deficiency: A disorder characterized by impairment in the myeloperoxidase function of phagocytic cells, particularly affecting the polymorphonuclear leukocytes. This condition appears to show similarities with Chronic granulomatous disease. Some patient exhibit candidiasis, but most of the individuals with myeloperoxidase deficiency show no signs of immunodeficiency. Myeloperoxidase deficiency is caused by autosomal recessive mutations in the *MPO* gene which encodes the myeloperoxidase. In the United States, myeloperoxidase deficiency is estimated to affect 1 in every 2,000 individuals.

3.04.15.4 Disorder of Bilirubin Metabolism

Gilbert syndrome: A common benign condition characterized by delayed clearance of bilirubin from the blood and mild jaundice. Affected individuals usually have total serum bilirubin levels from 1-6 mg/dL. Gilbert syndrome is caused by either autosomal recessive or autosomal dominant SNPs in the promoter and exon 1 or more commonly due to variation in TATAA element of the 5-prime promoter region in the *UGT1A1* gene. This gene encodes the uridinediphosphoglucuronate glucuronosyltransferase. This enzyme catalyzes the glucuronidation of many lipophilic xenobiotics, and endobiotics such as bilirubin. Gilbert syndrome affects at least 5% of all people worldwide. The autosomal dominant form is more common among Asians. Mutations in the *UGT1A1* gene are also associated with Crigler-Najjar syndrome type I, Crigler-Najjar syndrome type II and Transient familial neonatal hyperbilirubinemia.

Crigler-Najjar Syndrome: A disorder characterized by non-hemolytic jaundice with unconjugated hyperbilirubinemia leading to brain damage in infants. There are two forms: Type I is the more severe form, with total serum bilirubin levels ranging from 20-45 mg/dL, and is frequently accompanied by kernicterus. Type II is a moderate form, with total serum bilirubin levels ranging from 6.2-18.8 mg/dL. Crigler-Najjar syndromes are caused by autosomal recessive mutations in the *UGT1A1* gene. Type I associated with the absence of uridinediphosphoglucuronate glucuronosyltransferase activity. Most patients with type I have mutations in one of the common exons (2 to 5), particularly nonsense and frameshift mutations. Type II is associated with incomplete deficiency of hepatic uridinediphosphoglucuronate glucuronosyltransferase activity. On average, the carrier frequency for Crigler-Najjar syndrome type I trait is estimated to be 1 in 500 individuals worldwide, which means the frequency of this condition is about 1 in 1,000,000 individuals or $1/(1,000)^2$.

Dubin–Johnson syndrome: A benign and asymptomatic condition in most cases.

Dubin–Johnson syndrome is characterized by an increase in conjugated bilirubin in the serum with abnormal hepatic pigmentation and without liver damage. This condition results from inability of hepatocytes to secrete conjugated bilirubin into the bile. Dubin–Johnson syndrome is caused by autosomal recessive mutations in the *ABCC2* gene which encodes the multispecific anion transporter. In Iranian Jews, the carrier frequency for Dubin–Johnson syndrome trait is estimated to be 1 in 18 individuals, which means the frequency of this condition is about 1 in 1,300 individuals or $1/(36)^2$.

Rotor syndrome: A similar condition as Dubin–Johnson syndrome, but without abnormal hepatic pigmentation. Rotor syndrome is caused by digenic biallelic mutations in the *SLCO1B1* and *SLCO1B3* genes which encode the solute carrier organic anion transporter (member 1B1 and 1B3). Rotor syndrome is estimated to affect less than 1 in every 1,000,000 individuals.

Table 3.04.15.1 Recognizable Disorders of Porphyrin and Bilirubin Metabolism

Disorder	OMIM	Gene	Biological Function	Product of Conception Alteration
X-linked Erythropoietic protoporphyria	300752	*ALAS2*	Mitochondrial protein	XD
Erythropoietic protoporphyria	177000	*FECH*	Mitochondrial protein	AR
Congenital erythropoietic porphyria or Gunther disease	263700	*UROS*	Uroporphyrinogen synthase	AR
Porphyria cutanea tarda	176100	*UROD*	Decarboxylase	AD
Porphyria cutanea tarda	176100	*UROD/ HFE*	Decarboxylase/Receptor	Dig
Acute intermittent porphyria	176000	*HMBS*	Hydroxymethylbilane synthase	AD + de novo
Hereditary coproporphyria	612732	*CPOX*	Mitochondrial protein	Dig

The Concise Encyclopedia of Genomic Diseases

Table 3.04.15.1. Continued.

Disorder	OMIM	Gene	Biological Function	Product of Conception Alteration
Hereditary coproporphyria	612732	CPOX/ ALAD	Mitochondrial protein/Dehydratase	Dig
Harderoporphyria	121300	CPOX	Mitochondrial protein	AD + AD = AR
Variegate porphyria or South African porphyria	176200	PPOX	Inner mitochondrial membrane	AD
Homozygous Variant of Variegate Porphyria	176200	PPOX	Inner mitochondrial membrane	AD + AD = AR
Acute hepatic porphyria	612740	ALAD	Dehydratase	AR
Hepatoerythropoietic porphyria	176100	UROD	Decarboxylase	AD + AD = AR
Acatalasia or acatalasemia, or Takahara disease	614097	CAT	Peroxisomal and lysosomal structural protein	AR
Myeloperoxidase deficiency	254600	MPO	Microbicidal activity of netrophils	AR
Gilbert syndrome	143500	UGT1A1	Glucuronidation	AR or AD
Crigler-Najjar syndrome type I	218800	UGT1A1	Glucuronidation	AR
Crigler-Najjar syndrome type II	606785	UGT1A1	Glucuronidation	AR
Dubin-Johnson syndrome	237500	ABCC2	ABC-transporter protein	AR
Rotor syndrome	237450	SLCO1B1/ SLCO1B3	Solute carriers	DigB

3.04.16 Disorders of Mineral Metabolism

3.04.16.1 Disorders of Copper Metabolism

Wilson disease or Hepatolenticular degeneration: A disorder characterized by hepatic, neurological and psychiatric signs and symptoms. This condition typically appears in people under 40 years of age. Affected individuals exhibit confusion, loss of memory, tremors, and lack of coordination and muscle control. Patients display the Kayser-Fleischer ring in the cornea, hypercalciuria and nephrocalcinosis. This condition is named after Samuel Alexander Kinnier Wilson,, and is caused by copper accumulation in tissues. Wilson disease causes copper overload in the liver, leading to cirrhosis at a young age. It is caused by autosomal recessive mutations in the *ATP7B* gene which encodes the cation transport ATPase (Cu2+ transporting beta polypeptide). This protein's functions include effluxing copper in the cells into bile and incorporating copper into ceruloplasmin. On average, the carrier frequency for Wilson disease trait is estimated to be 1 in 90 individuals worldwide, which means the frequency of this condition is about 1 in 32,400 individuals or $1/(180)^2$. There is a higher carrier frequency for this condition among Sardinians (Italy), and also in the Canarians (Canary Islands-Spain). In the latter group, the estimated prevalence for Wilson disease is 1 in 2,600 individuals. This condition is also prevalent in Korea and China. Among Han Chinese from Hong Kong, the estimated prevalence for Wilson disease is 1 in every 5,400 individuals.

Menkes disease: A disorder characterized by generalized copper deficiency. This condition results in clinical features from the impairment of copper-dependent enzymes such as tyrosinase, lysyl oxidase, monoamine oxidase, cytochrome c oxidase and ascorbate oxidase. These enzymes have major functions in the nervous system, skin, hair, bone and blood vessels. Affected individuals exhibit neurologic impairment within a month or two of birth. Patients are characterized by spastic dementia, seizures, and focal cerebral and cerebellar degeneration. This condition is named after John Hans Menkes. These patients also display kinky hair and growth retardation. Menkes disease is caused by X-linked recessive mutations in the *ATP7A* gene which encodes the cation transport ATPase (Cu2+ transporting alpha polypeptide). This protein is localized in the trans-Golgi network. Mutations in the same gene cause another allele disorder called the Occipital horn syndrome. On average, Menkes disease is estimated to affect 3-4 in every 1,000,000 live-born babies worldwide. There is a higher prevalence of Menkes disease in Melbourne, Australia.

3.04.16.2 Disorders of Iron Metabolism

Hemochromatosis: A group of disorders characterized by pathological increase in total body iron stores. The excessive iron accumulation disrupts, the normal function of organs including the liver, adrenal glands, pancreas, heart, skin, joints, gonads and others. Affected individuals exhibit signs or symptoms of liver disease leading to cirrhosis, adrenal insufficiency, heart failure, diabetes, polyarthropathy and hypogonadotropic hypogonadism. There are 4 forms of Hemochromatosis. Type 1 is caused by autosomal recessive gain-of-function SNPs in the *HFE* gene. This gene encodes the hemochromatosis protein, a membrane protein that functions to regulate iron absorption. The 2 most common SNPs or alleles which cause Hemochromatosis type 1 are Cys282Tyr and His63Asp. The *HFE* gene is closely linked to the HLA-A3 locus. Among people of Northern European ancestry, particularly those of Celtic descent, the carrier frequency for Hemochromatosis type 1 trait is estimated to be 1 in 10 individuals, which means the frequency of this condition is about 1 in 400 individuals or $1/(20)^2$. This condition seems to be even more common among Australians, of whom 1 in 200 individuals have this condition. The disease hasvariable penetration, affecting 1 out of every 2 males with this biallelic form and only 1 out of every 4 females. Hemochromatosis type 2 or Juvenile hemochromatosis is a form of hemochromatosis that appears during youth. There are two forms: Hemochromatosis type 2A is caused by biallelic mutations in the *HFE2* gene which encodes the hemochromatosis type 2; Hemochromatosis type 2B is caused by biallelic mutations in the *HAMP* gene which encodes the hepcidin antimicrobial peptide; Hemochromatosis type 3 is caused by biallelic mutations in the *TFR2* gene which encodes the transferrin receptor 2; Hemochromatosis type 4 or African iron overload is caused by monoallelic mutations in the *SLC40A1* gene which encodes the solute carrier (family 40 member A1), an iron-regulated transporter or ferroportin. This condition is common in rural areas of central and southern Africa, where up to 10% of the people may be affected.

Aceruloplasminemia (Neurodegeneration with brain iron accumulation): A disorder characterized by progressive neurodegeneration of the retina and basal ganglia, and diabetes mellitus. This condition appears during the fourth decade of life. Affected individuals also exhibit torticollis, cerebellar ataxia chorea and other extrapyramidal disorders. Aceruloplasminemia is caused by autosomal recessive mutations in the *CP* gene which encodes the ceruloplasmin (ferroxidase). This metalloprotein binds most of the copper in plasma and is involved in the peroxidation of Fe(II) transferrin to Fe(III) transferrin This protein is involved in iron transport and processing. The worldwide prevalence of aceruloplasminemia is unknown. In Japan, Aceruloplasminemia is estimated to affect 1 in every 100,000 individuals.

Atransferrinemia: A disorder characterized by anemia and hemosiderosis in the heart and liver. This condition leads to heart failure. Atransferrinemia is caused by autosomal recessive mutations in the *TF* gene which encodes the transferrin. This protein transports iron through the blood. Atransferrinemia is estimated to affect less than 1 in every 1,000,000 individuals.

3.04.16.3 Disorders of Zinc Metabolism

Acrodermatitis enteropathica: A disorder characterized by diarrhea, alopecia, and periorificial and acral dermatitis. Affected individuals start manifesting this condition as infants when weaned off breast milk. This condition is caused by impairment in the uptake of zinc. Acrodermatitis enteropathica is caused by autosomal recessive mutations in the *SLC39A4* gene which encodes the solute carrier (family 39 member A4), a zinc-specific transporter. On average, the carrier frequency for Acrodermatitis enteropathica trait is estimated to be 1 in 353.5 individuals worldwide, which means the frequency of this condition is about 1 in 500,000 individuals or $1/(707)^2$.

3.04.16.4 Disorders of Phosphorus Metabolism

Hypophosphatasia: A disorder characterized by impairment in the development of bones and teeth. There is a large range of clinical symptoms associated with this condition from the rapidly fatal perinatal variant to a milder form with progressive osteomalacia later in life. Patients with Infantile hypophosphatasia have soft and weak bones. This condition results in skeletal abnormalities such as soft skull bones, short limbs and abnormally shaped chest. Affected individuals also exhibit hypercalcemia, and respiratory and kidney problems. Infantile hypophosphatasia is caused by autosomal recessive mutations in the *ALPL* gene which encodes the alkaline phosphatase (liver/bone/kidney). This enzyme is involved in matrix mineralization. Defective enzyme impairs bone mineralization, leading to rickets. Mutations in this gene are also associated with Odontohypophosphatasia. On average, the carrier frequency for Infantile hypophosphatasia trait is estimated to be 1 in 158 individuals worldwide, which means the frequency of this condition is about 1 in 100,000 individuals or $1/(316)^2$. There is a higher carrier frequency for Infantile hypophosphatasia trait among Mennonite people in Manitoba, Canada. This trait is estimated to be present in 1 in every 25 individuals, which means the frequency of this condition is about 1 in 2,500 individuals or $1/(50)^2$.

Hypophosphatemic rickets: A disorder characterized by low levels of phosphate, which is a mineral that is essential for the normal formation of bones and teeth. Hypophosphatemic rickets can have several patterns of inheritance: X-linked dominant, autosomal recessive, autosomal dominant and X-linked recessive. There are several underlying causes of this disorder with at least 4 autosomal genomic loci and 2 X-linked genomic loci.

X-linked dominant hypophosphatemia or hypophosphatemic vitamin D-resistant rickets: A disorder characterized by bone deformity including short stature and genu varum. The bone disease is more severe in males. This condition is caused by X-linked dominant mutations in the *PHEX* gene which encodes the phosphate-regulating endopeptidase. This protein inactivates phosphatonins that promote phosphate excretion. X-linked dominant hypophosphatemia is estimated to affect 1 in every 20,000 individuals. This condition is the most common genomic cause of hypophosphatemic rickets.

Autosomal recessive hypophosphatemia: A disorder caused by biallelic mutations in either the *DMP1* gene which encodes the dentin matrix acidic phosphoprotein 1, or the *ENPP1* gene which encodes the ectonucleotide pyrophosphatase/phosphodiesterase 1.

Autosomal dominant hypophosphatemia: A disorder caused by monoallelic mutations in the *FGF23* gene which encodes the fibroblast growth factor. Mutations in this gene also cause Tumor-induced osteomalacia and Familial hyperphosphatemic tumoral calcinosis. Autosomal dominant hypophosphatemia is estimated to affect less than 1 in every 1,000,000 individuals.

X-linked recessive hypophosphatemia: A form of hypercalciuric nephrolithiasis characterized by proximal renal tubular reabsorptive failure, nephrocalcinosis, hypercalciuria and renal insufficiency. X-linked recessive hypophosphatemia is caused by mutations in the *CLCN5* gene which encodes the voltage-gated chloride ion channel 5. Mutations in this gene also cause Dent disease, type I Nephrolithiasis, and low molecular weight proteinuria with hypercalciuric nephrocalcinosis.

Hypophosphatemic rickets with hypercalciuria: A disorder caused by autosomal recessive mutations in the *SLC34A3* gene which encodes the solute carrier (family 34 member A3), a type II sodium/phosphate contransporter.

Lysosomal acid phosphatase deficiency: A disorder characterized by hypotonia, lethargy, intermittent vomiting, opisthotonos, terminal bleeding and death in early infancy. This condition is caused by autosomal recessive mutations in the *ACP2* gene which encodes the lysosomal acid phosphatase-2. This enzyme hydrolyzes orthophosphoric monoesters to alcohol and phosphate. Lysosomal acid phosphatase deficiency is estimated to affect less than 1 in every 1,000,000 individuals.

Vitamin D resistant rickets: A disorder characterized by defective mineralization mechanism on cartilage growth plates during the first months of life. Affected individuals exhibit bone deformities, mainly on the lower limbs. There are several underlying causes of this disorder with at least 4 autosomal genomic loci: Vitamin D-dependent rickets type 1A is caused by autosomal recessive mutations in the *CYP27B1* gene which encodes the cytochrome P450 (family 27 B, polypeptide 1). This enzyme catalyzes the hydroxylation and metabolic activation of 25-

hydroxyvitamin D3 into 1,25-dihydroxy-vitamin D3. This reaction occurs in the inner mitochondrial membrane of renal proximal tubule. Vitamin D-dependent rickets type 1A is prevalent among French Canadians in the Saguenay region, Quebec. The carrier frequency for vitamin D-dependent rickets type 1A is estimated to be 1 in 26 individuals, which means the frequency of this condition is about 1 in 2,700 individuals or $1/(52)^2$;; Vitamin D-dependent rickets type 1B is caused by autosomal recessive mutations in the *CYP2R1* gene which encodesthe cytochrome P450 (family 2 R polypeptide 1). This enzyme catalyzes hydroxylation of vitamin D at carbon 25 in the liver;; Vitamin D-dependent rickets type 2A is caused by autosomal recessive mutations in the *VDR* gene which encodes the vitamin D (1,25- dihydroxyvitamin D3) receptor. This gene shares function and sequence similarity to the steroid and thyroid hormone receptors;; and Vitamin D-dependent rickets type 2B is a disorder which has been described in a rural areas of the Cauca department in the southwest part of Colombia. The gene has not been characterized. Most of the patients are of African descent, and have lower limb deformities.

3.04.16.5 Disorders of Magnesium Metabolism

Hypomagnesemia with secondary hypocalcemia or Intestinal hypomagnesemia-1: A disorder appearing during the first 6 months of life characterized by convulsions and spasms. This condition is caused by decreased intestinal magnesium reabsorption, leading to hypomagnesemia, hypocalcemia and lowered parathyroid hormone. Hypomagnesemia with secondary hypocalcemia is caused by autosomal recessive mutations in the *TRPM6* gene which encodes the melastatin-related transient receptor channel 6.

Renal hypomagnesemia-2 or Renal magnesium wasting: A disorder characterized by hypomagnesemia without other plasma electrolyte abnormalities. This condition is caused by autosomal dominant mutations in the *FXYD2* gene which encodes the sodium-potassium-ATPase (gamma-subunit). This Na+,K+-ATPase maintains the transmembrane potential and provides the driving force for active transport processes in the kidney.

Renal hypomagnesemia-3: A disorder associated with hypercalciuria and nephrocalcinosis. Renal hypomagnesemia-3 is caused by autosomal recessive mutations in the *CLDN16* gene which encodes the tight-junction claudin 16.

Renal hypomagnesemia-4: A disorder associated with normocalciuria. This condition is caused by autosomal recessive mutations in the *EGF* gene which encodes the epidermal growth factor.

Renal hypomagnesemia-5 or renal hypomagnesemia with ocular involvement: A disorder associated with severe ocular involvement, hypercalciuria and nephrocalcinosis. The visual impairment includes bilateral chorioretinal scars in the macula, nystagmus and malignant myopia. Renal hypomagnesemia-5 is caused by autosomal recessive mutations in the *CLDN19* gene which encodes the tight-junction claudin 19.

Renal hypomagnesemia-6: A disorder caused by autosomal recessive mutations in the *CNNM2* gene which encodes the cyclin M2, a protein similar to the cyclins. This protein plays an important role in magnesium homeostasis by mediating the epithelial transport and renal reabsorption of Mg2+. There are other genomic causes of hypomagnesemia including Gitelman syndrome and Bartter syndrome.

3.04.16.6 Disorders of Calcium Metabolism

Familial hypocalciuric hypercalcemia: A disorder characterized by symptoms of hypercalcemia. Familial hypocalciuric hypercalcemia is a benign stable condition that does not lower life expectancy. There are several underlying causes of this disorder with at least 3 autosomal genomic loci: Familial hypocalciuric hypercalcemia is caused by autosomal dominant loss-of-function mutations in the *CASR* gene which encodes the calcium-sensing receptor. This protein is expressed in parathy-roid and kidney tissue. On the contrary, gain-of-function mutations in the *CASR* gene are also associated with Neonatal severe hyperparathyroidism. There are two additional forms of Familial hypocalciuric hypercalcemia in which the genes have not yet been identified.

Absorptive hypercalciuria: A disorder characterized by absorptive hypercalciuria with recurrent calcium oxalate stones. Absorptive hypercalciuria is caused by autosomal dominant mutations in the *ADCY10* gene which encodes the soluble adenylyl cyclase 10. There is also another gene/locus for a condition with similar clinical features but the gene has not been identified.

Table 3.04.16.1 Recognizable Disorders of Mineral Metabolism

Disorder	OMIM	Gene	Biological Function	Product of Conception Alteration
Wilson disease or hepatolenticular de-generation	277900	*ATP7B*	ATPase disorders	AR
Menkes disease	309400	*ATP7A*	ATPase disorders	XR
Occipital horn syn-drome	304150	*ATP7A*	ATPase disorders	XR
Hemochromatosis-1	235200	*HFE*	Cell surface receptor Other/ungrouped	AR + Env
Hemochromatosis-2A	602390	HFE2	Iron overload	AR
Hemochromatosis-2B	613313	HAMP	Iron homeostasis	AR
Hemochromatosis	606464	HFE/ HAMP	Cell surface receptor Other/ungrouped	Dig
Hemochromatosis-3	604250	TFR2	Membrane protein	AR
Hemochromatosis-4	606069	*SLC40A 1*	Solute carrier	AD
Modifier of HFE hemo-chromatosis	235200	*BMP2*	Cell surface receptor Enzyme-linked receptor	AR
Aceruloplasminemia	604290	*CP*	Iron transport and processing	AR
Atransferrinemia	209300	*TF*	Transferrin	AR
Acrodermatitis entero-pathica	201100	*SLC39A 4*	Solute carrier	AR

Table 3.04.16.1. Continued.

Disorder	OMIM	Gene	Biological Function	Product of Conception Alteration
Childhood-onset hypophosphatasia	241510	ALPL	Alkaline phosphatase	AR
Adult hypophosphatasia	146300	ALPL	Alkaline phosphatase	AR
Odontohypophosphatasia	146300	ALPL	Alkaline phosphatase	AR
X-linked dominant hypophosphatemia or hypophosphatemic vitamin D-resistant rickets	307800	PHEX	Transmembrane endopeptidase	XD
Autosomal recessive hypophosphatemia	241520	DMP1	Osteoblast maturated extracellular matrix	AR
Autosomal recessive hypophosphatemia	613312	ENPP1	Pyrophosphatase/phosphodiesterase	AR
Autosomal dominant hypophosphatemia	193100	FGF23	Mitogenic and cell survival activities	AD
X-linked recessive hypophosphatemia	300554	CLCN5	Channelopathy Chloride	XR
Hypophosphatemic rickets with hypercalciuria	241530	SLC34A3	Solute carrier	AR
Lysosomal acid phosphatase deficiency	200950	ACP2	Lysosomal protein	AR
Vitamin D-dependent rickets type 1A	264700	CYP27B1	Mitochondrial inner membrane	AR
Vitamin D-dependent rickets type 1B	600081	CYP2R1	Endoplasmic reticulum protein	AR
Vitamin D-dependent rickets type 2A	277440	VDR	Nuclear receptor	AR
Hypomagnesemia with secondary hypocalcemia or intestinal hypomagnesemia-1	602014	TRPM6	Channelopathy transient receptor	AR
Renal hypomagnesemia-2 or renal magnesium wasting	154020	FXYD2	Transmembrane proteins	AD
Renal hypomagnesemia-4	611718	EGF	Mitogenic factor	AR
Renal hypomagnesemia-5 or renal hypomagnesemia with ocular involvement	248190	CLDN19	Channelopathy tight junctions	AR
Renal hypomagnesemia-6	613882	CNNM2	Simil to cyclins	AR

Table 3.04.16.1. Continued.

Disorder	OMIM	Gene	Biological Function	Product of Conception Alteration
Familial hypocalciuric hypercalcemia	145980	*CASR*	Cell surface receptor G protein	AD
Absorptive hypercalciuria	143870	*ADCY10*	Adenylyl cyclase	AD

3.04.17 Cystic Fibrosis

Cystic fibrosis or Mucoviscidosis: A disorder characterized by the production of mucus secretions with abnormal viscosity, and sweat with high salt content due to abnormal transport of chloride. This condition mainly affects the epithelium in the lungs, pancreas, liver and intestine. Affected patients can suffer from chronic bronchitis, pancreatic insufficiency, adolescent diabetes, pancreatitis, intestinal obstruction and liver cirrhosis. This condition is also associated with infertility and carcinoma. This disease usually starts at birth with meconium ileus or during early childhood, but there are also late onset forms. Patients with Cystic Fibrosis often live on to only 30 to 50 years of age. The pulmonary disease is the most serious complication, resulting from recurrent lung infections. Cystic fibrosis is caused by autosomal recessive loss-of-function mutations in the *CFTR* gene which encodes the cystic fibrosis transmembrane conductance regulator (ATP-binding cassette sub-family C, member 7). This is a halide anion channel essential to regulate the components of mucus, sweat and digestive juices. This condition is an example of loss-of-function mutation during early migration trait. These pathogenic alleles started to appear 10,000-52,000 years ago most likely in Europe, the Middle East and Western Asia. Cystic fibrosis traits are present mainly due to founder effect and endogamy. The product of conception alterations are defined as occasional *de novo* autosomal recessive syndrome (AR + KDN) with unknown prevalence of *de novo* polymorphism. Cystic fibrosis is the most common life-limiting autosomal recessive disease among people of European inheritance, except among the Finnish. Table 3.04.17 displays the trait frequency, allele frequency and disease frequency of Cystic Fibrosis by country. There is also similar phenotype consisting of bronchiectasis with or without elevated sweat chloride caused by autosomal recessive mutations in the *SCNN1B, SCNN1A,* and *SCNN1G* genes, encoding the 3 subunits of the epithelial sodium channel.

The Concise Encyclopedia of Genomic Diseases

Table 3.04.17 Cystic Fibrosis Worldwide Prevalence

Region	Trait Frequency	Allele Frequency	Disease Frequency
EUROPE			
Ireland	1 in 19	1 in 38	1 in 1,400
Scotland (United Kingdom)	1 in 22	1 in 44	1 in 1,984
Switzerland	1 in 22	1 in 44	1 in 2,000
France	1 in 24	1 in 48	1 in 2,350
Italy	1 in 24.5	1 in 49	1 in 2,438
United Kingdom	1 in 25.5	1 in 51	1 in 2,600
Czech Republic	1 in 26.5	1 in 53	1 in 2,833
Germany	1 in 29	1 in 58	1 in 3,300
Greece	1 in 29.5	1 in 59	1 in 3,500
Spain	1 in 29.5	1 in 59	1 in 3,500
Netherlands	1 in 30	1 in 60	1 in 3,650
Estonia	1 in 33	1 in 66	1 in 4,500
Norway	1 in 33	1 in 66	1 in 4,500
Denmark	1 in 34	1 in 68	1 in 4,700
Russian Federation	1 in 35	1 in 70	1 in 4,900
Poland	1 in 39	1 in 78	1 in 6,000
Sweden	1 in 42	1 in 84	1 in 7,300
Turkey	1 in 50	1 in 60	1 in 10,000
Finland	1 in 79	1 in 158	1 in 25,000
NORTH AMERICA			
Canada	1 in 30	1 in 60	1 in 3,500
United States	1 in 30	1 in 60	1 in 3,500
(By Ethnic Group):			
Caucasian-Americans	1 in 25	1 in 50	1 in 2,500
Hispanic-Americans	1 in 46	1 in 92	1 in 8,500
African-Americans	1 in 65	1 in 130	1 in 17,000
Asian-Americans	1 in 90	1 in 180	1 in 32,000
LATIN AMERICA			
Cuba	1 in 31	1 in 62	1 in 3,900
Chile	1 in 31.5	1 in 63	1 in 4,000
Brazil	1 in 41.5	1 in 83	1 in 6,900
Mexico	1 in 46	1 in 92	1 in 8,500
MIDDLE EAST			
Bahrain	1 in 38	1 in 76	1 in 5,800
United Arab Emirates	1 in 63	1 in 126	1 in 15,800
ASIA			
India	1 in 50	1 in 100	1 in 10,000
Japan	1 in 158	1 in 316	1 in 100,000
AFRICA			
South Africa (non-Afrikaners)	1 in 42	1 in 81	1 in 7,056
OCEANIA			
Australia	1 in 25	1 in 50	1 in 2,500

3.04.18 Amyloidosis

3.04.18.1 Heredofamilial Amyloidosis

Amyloidosis: A group of conditions in which amyloid proteins are abnormally deposited in organs and/or tissues. These conditions may be inherited or acquired. Symptoms of this condition depend on the site of amyloid deposition.

Familial Mediterranean fever: A disorder characterized by recurrent attacks of fever, rashes and painful inflammation of the lungs, abdomen and joints. These acute episodes generally last 24-48 hours. Affected individuals display symptoms of peritonitis, skin rashes, arthritis and chest pain. This condition starts manifesting between the ages of 5-15. Familial Mediterranean fever may leads to amyloidosis with renal failure. In most patients, this condition is caused by autosomal recessive polymorphism in the *MEFV* gene which encodes the pyrin also known as marenostrin protein, an important modulator of innate immunity. This protein is expressed almost exclusively in mature granulocytes. There are also few cases of autosomal dominant inheritance. Familial Mediterranean fever is common among people of the Mediterranean region, especially Armenians, Arabs, Turks, Greeks and Sephardi Jews. In this area of the world, the carrier frequency for Familial Mediterranean fever trait is estimated to be 1 in 6-8 individuals, which means the frequency of this condition is about 1 in 150-250 individuals or $1/(12)^2$ to $1/(16)^2$. In this condition, there is a reduced penetrance which means the homozygous genotype is more common than the condition. This condition can be modifyed by other genes; for instance, mutations in the *SAA1* gene, which makes the alpha version of the serum amyloid A1 protein, increase the risk of amyloidosis.

Hereditary Amyloidosis: A disorder characterized by polyneuropathy, carpal tunnel syndrome, autonomic insufficiency, cardiomyopathy, vitreous opacities, renal insufficiency and gastrointestinal symptoms. Patients with this condition develop severe diarrhea with malabsorption, cachexia, orthostatic hypotension, incapacitating neuropathy, etc. Most of these patient die after 5-15 years after the onset of symptoms. Hereditary amyloidosis is caused by autosomal dominant mutations in the *TTR* gene which encodes the transthyretin, a prealbumin that transports thyroid hormones and vitamin A. Hereditary amyloidosis is estimated to affect 1 in every 100,000 individuals worldwide. This condition particularly affects people from Sweden and northern Portugal. Hereditary amyloidosis is estimated to affect 1 in every 538 people from northern Portugal; this allele appears 375-750 years ago.

Amyloidosis, Finnish type: A disorder characterized by corneal lattice dystrophy, skin changes, cranial neuropathy and bulbar signs. Some patients may develop peripheral neuropathy and renal failure. Amyloidosis, Finnish type is caused by autosomal dominant mutations in the *GSN* gene which encodes the gelsolin. This protein functions in both assembly and disassembly of actin filaments. There are also homozygous patients, for whom both parents are affected. Patients with the homozygous form are more severely affected than those with the monoallelic form. In the biallelic form, patients develop nephrotic syndrome, renal failure and cardiac symptoms.

Familial primary localized cutaneous amyloidosis: A disorder characterized by pruritus and skin-scratching. Affected individuals show deposits of amyloid staining on keratinous debris in the papillary dermis. There are several underlying causes of this disorder with at least 2 autosomal genomic loci: Familial primary localized cutaneous amyloidosis-1 can be caused by autosomal dominant mutations in the *OSMR* gene which encodes the oncostatin M receptor-beta;; and Familial primary localized cutaneous amyloidosis-2 can be caused mutations in the *IL31RA* gene which encodes the interleukin 31 receptor A. This protein is part of an IL31RA/OSMR receptor complex. Most cases of primary localized cutaneous amyloidosis are sporadic, although 10% of cases have been reported to be familial origin. Familial primary cutaneous amyloidosis occurs more frequently among people from China, Southeast Asia and South America.

Familial visceral amyloidosis or familial renal amyloidosis: A systemic nonneuro-pathic amyloidosis disorder primarily characterized by affecting the kidney. Affected individuals display signs of chronic nephropathy with albuminuria, hematuria, edema, arterial hypertension and hepatosplenomegaly. There are several underlying causes of this disorder with at least 4 autosomal genomic loci: Familial visceral amyloidosis can be caused by autosomal dominant mutations in the *APOA1* gene which encodes the apolipoprotein A1, or *FGA* gene which encodes the fibrinogen alpha-chain, or the *LYZ* gene which encodes the lysozyme, or the *B2M* gene which encodes the beta-2-microglobulin.

Muckle–Wells syndrome or Urticaria - deafness - amyloidosis syndrome: A disorder characterized by sensorineural deafness, and recurrent hives which lead to amyloidosis. Affected individuals display episodes of fever, chills and joint pain. Muckle–Wells syndrome is caused by autosomal dominant mutations in the *NLRP3* gene which encodes the NLR family pyrin domain containing protein 3. This protein is expressed predominantly in peripheral blood leukocytes.

3.04.17-18.1 Cystic Fibrosis and Amyloidosis

Disorder	OMIM	Gene	Biological Function	Product of Conception Alteration
Cystic fibrosis	219700	*CFTR*	Channelopathy Chloride	AR + KDN
Familial Mediterranean fever	249100	*MEFV*	Modulator of innate immunity	AR
Familial Mediterranean fever	134610	*MEFV*	Modulator of innate immunity	AD + de novo
Hereditary Amyloidosis	105210	*TTR*	Secreted carrier protein	AD
Amyloidosis, Finnish type	105120	*GSN*	Actin filaments	AD or AD + AD = AR
Familial primary localized cutaneous amyloidosis-1	105250	*OSMR*	Cell surface receptor JAK-STAT	AD

3.04.17-18.1. Continued.

Disorder	OMIM	Gene	Biological Function	Product of Conception Alteration
Familial primary localized cutaneous amyloidosis-2	613955	IL31RA	Cell surface receptor JAK-STAT	AD
Familial visceral amyloidosis or familial renal amyloidosis	105200	LYZ	Anti-microbial agents	AD
Familial visceral amyloidosis or familial renal amyloidosis	105200	FGA	Fibrinogen	AD
Familial visceral amyloidosis or familial renal amyloidosis	105200	APOA1	Apolipoprotein	AD
Familial visceral amyloidosis or familial renal amyloidosis	105200	B2M	Microglobulin	AD
Muckle–Wells syndrome or Urticaria - deafness - amyloidosis syndrome	191900	NLRP3	Nucleotide Oligomerization Domain receptors	AD

3.04.19 Disorders of Fluid, Electrolyte and Acid-base Balance

3.04.19.1 Combined Oxidative Phosphorylation Deficiency

Combined oxidative phosphorylation deficiency: A group of disorders characterized by neonatal lactic acidosis, rapidly progressive encephalopathy, microcephaly, hypertonicity, axial hypotonia, cardiomyopathy, liver dysfunction and growth retardation. Death generally occurs in the first weeks or years of life. This condition is caused by a defect in the mitochondrial oxidative phosphorylation system. There are several underlying causes of this disorder with at least 14 autosomal genomic loci and 1 X-linked locus: Combined oxidative phosphorylation deficiency-1 is caused by autosomal recessive mutations in the *GFM1* gene which encodes the mitochondrial elongation factor G1;; Combined oxidative phosphorylation deficiency-2 is caused by autosomal recessive autosomal recessive mutations in the *MRPS16* gene which encodes the mitochondrial ribosomal protein S16;; Combined oxidative phosphorylation deficiency-3 is caused by autosomal recessive mutations in the *TSFM* gene which encodes the mitochondrial Ts translation elongation factor;; Combined oxidative phosphorylation deficiency-4 is caused by autosomal recessive mutations in the *TUFM* gene which encodes the mitochondrial Tu translation elongation factor;; Combined oxidative phosphorylation deficiency-5 is caused by autosomal recessive mutations in the *MRPS22* gene which encodes the mitochondrial ribosomal protein S22;; Combined oxidative phosphorylation deficiency-6 is caused by X-linked recessive mutations in the *AIFM1* gene which encodes the mitochondrion-associated apoptosis-inducing factor 1;; Combined oxidative phosphorylation deficiency-7 is caused by autosomal recessive mutations in the *HOGA1* gene which encodes the 4-hydroxy-2-oxoglutarate aldolase 1;; Combined oxidative phosphorylation deficiency-8 is caused by autosomal recessive mutations in the *AARS2* gene which encodes the mitochondrial alanyl-tRNA synthetase 2;; Combined oxidative phosphorylation deficiency-9 is caused by autosomal recessive mutations in the *MRPL3* gene which encodes the mitochondrial ribosomal protein L3;; Combined oxidative phosphorylation deficiency-10 is caused by autosomal recessive mutations in the *MTO1* gene which encodes the mitochondrial translation optimization 1 homolog;; Combined oxidative phosphorylation deficiency-11 is caused by autosomal recessive mutations in the *RMND1* gene which encodes the required for meiotic nuclear division 1 homolog;; Combined oxidative phosphorylation deficiency-12 is caused by autosomal recessive mutations in the *EARS2* gene which encodes the mitochondrial glutamyl-tRNA synthetase 2;; Combined oxidative phosphorylation deficiency-13 is caused by autosomal recessive mutations in the *PNPT1* gene which encodes the polyribonucleotide nucleotidyltransferase 1;; Combined oxidative phosphorylation deficiency-14 is caused by autosomal recessive mutations in the *FARS2* gene which encodes the mitochondrial phenylalanyl-tRNA synthetase 2;; and Combined oxidative phosphorylation deficiency-15 is caused by autosomal recessive mutations in the *MTFMT* gene which encodes the mitochondrial methionyl-tRNA formyltransferase.

Mitochondrial DNA depletion syndrome: A group of infancy onset disorders characterized by totally reduced cellular levels of mitochondrial DNA. Affected individuals do not display any problems until birth, but develop symptoms in the early neonatal period. Mitochondrial DNA depletion syndrome is caused by a defect affecting the control of mitochondrial DNA copy number after birth. There are several underlying causes of this disorder with at least 9 autosomal genomic loci. The prevalence of the disease is unknown.

In one report, 11% of young children (less than 2 years old) referred for weakness, hypotonia, and developmental delay had Mitochondrial DNA depletion syndrome.

Mitochondrial DNA depletion syndrome-1 or MNGIE syndrome (Mitochondrial NeuroGastroIntestinal Encephalomyopathy syndrome): A disorder characterized by the association of gastrointestinal dysmotility, peripheral neuropathy, chronic progressive external ophthalmoplegia and leukoencephalopathy. Affected individuals display gastrointestinal symptoms during the second decade of life. Mitochondrial DNA depletion syndrome-1 is caused by autosomal recessive mutations in the *TYMP* gene which encodes the thymidine phosphorylase. This enzyme catalyzes the phosphorylation of thymidine or deoxyuridine to thymine or uracil, and is consequently essential for the nucleotide salvage pathway.

Mitochondrial DNA depletion syndrome-2: A disorder characterized by severe lactic acidosis, progressive external ophthalmoplegia, and fatal mitochondrial myopathy with progressive generalized hypotonia. Mitochondrial DNA depletion syndrome-2 is caused by autosomal recessive mutations in the *TK2* gene which encodes the thymidine kinase.

Mitochondrial DNA depletion syndrome-3 or Hepatocerebral type: A disorder of infancy onset characterized by severe lactic acidosis, hypoglycemia, progressive liver failure and neurologic abnormalities. Mitochondrial DNA depletion syndrome-3 is caused by autosomal recessive mutations in the *DGUOK* gene which encodes the deoxyguanosine kinase.

Mitochondrial DNA depletion syndrome-4A or Alpers type: A disorder appearing in infants and young children characterized by triad of intractable epilepsy, psychomotor retardation and liver failure. Mitochondrial DNA depletion syndrome-4A is caused by autosomal recessive mutations in the *POLG* gene which encodes the DNA polymerase gamma.

Mitochondrial DNA depletion syndrome-4B or MNGIE syndrome (Mitochondrial NeuroGastroIntestinal Encephalomyopathy syndrome): A disorder manifesting as a neurogastrointestinal encephalopathy. Affected individuals also display gastrointestinal malabsorption, peripheral neuropathy, ataxia, ophthalmoplegia and diffuse muscle weakness. Mitochondrial DNA depletion syndrome-4B is caused by autosomal recessive mutations in the *POLG* gene which encodes the DNA polymerase gamma.

Mitochondrial DNA depletion syndrome-5: A disorder characterized by progressive neurologic deterioration, hypotonia, lactic acidosis, severe psychomotor retardation and methylmalonic aciduria of infantile onset. Mitochondrial DNA depletion syndrome-5 is caused by autosomal recessive mutations in the *SUCLA2* gene which encodes the succinate-CoA ligase (beta subunit).

Mitochondrial DNA depletion syndrome-6 or Navajo neurohepatopathy or Hepatocerebral type: A disorder of infantile onset characterized by progressive damage to the liver and neurologic symptoms including ataxia, hypotonia, dystonia and psychomotor regression. Affected Navajo children display a mutilating neuropathy with severe motor involvement, painless fractures, severe anesthesia leading to corneal ulceration and progressive white matter lesions. Other features of this condition included poor weight gain, macronodular cirrhosis, hepa-

tomegaly, persistent neonatal jaundice and a Reye-like syndrome. Death often occurs in the first decade due to liver disease. Mitochondrial DNA depletion syndrome-6 is caused by autosomal recessive mutations in the *MPV17* gene which encodes the mitochondrial inner membrane protein 17. In Navajo people, the carrier frequency for Mitochondrial DNA depletion syndrome-6 trait is estimated to be 1 in 20 individuals, which means the frequency of this condition is about 1 in 1,600 individuals or $1/(40)^2$.

Mitochondrial DNA depletion syndrome-7 or hepatocerebral type: A disorder characterized by mtDNA depletion in the brain and liver. Affected individuals exhibit hypotonia, ataxia, ophthalmoplegia, hearing loss, seizures and sensory axonal neuropathy. Mitochondrial DNA depletion syndrome-7 was originally classified as a form of spinocerebellar ataxia. Mitochondrial DNA depletion syndrome-7 is caused by autosomal recessive mutations in the *C10orf2* gene which encodes the chromosome 10 open reading frame 2, a hexameric DNA helicase which unwinds short stretches of double-stranded DNA.

Mitochondrial DNA depletion syndrome-8A or Encephalomyopathic type with renal tubulopathy: A disorder of neonatal onset characterized lactic acidosis, hypotonia, neurologic deterioration, and renal tubular impairment. Mitochondrial DNA depletion syndrome-8A is caused by autosomal recessive mutations in the *RRM2B* gene which encodes the ribonucleotide reductase M2B.

Mitochondrial DNA depletion syndrome-8B or MNGIE syndrome (Mitochondrial NeuroGastroIntestinal Encephalomyopathy syndrome): A disorder characterized by brain MRI changes, ophthalmoplegia, ptosis, peripheral neuropathy, gastrointestinal dysmotility and ca-chexia. Mitochondrial DNA depletion syndrome-8B is caused by autosomal recessive mutations in the *RRM2B* gene which encodes the ribonucleotide reductase M2B.

Mitochondrial DNA depletion syndrome-9: A disorder of infantile onset characterized by progressive neurologic deterioration, hypotonia, lactic acidosis, severe psychomotor retardation and methylmalonic aciduria. Mitochondrial DNA depletion syndrome-9 is caused by autosomal recessive mutations in the *SUCLG1* gene which encodes the the succinate-CoA ligase (alpha subunit).

Table 3.04.19.1 Recognizable Disorders of Fluid, Electrolyte and Acid-base Balance

Disorder	OMIM	Gene	Biological Function	Product of Conception Alteration
Combined oxidative phosphorylation deficiency-1	609060	GFM1	Mitochondrial protein	AR
Combined oxidative phosphorylation deficiency-2	610498	MRPS16	Mitochondrial protein	AR
Combined oxidative phosphorylation deficiency-3	610505	TSFM	Mitochondrial protein	AR
Combined oxidative phosphorylation deficiency-4	610678	TUFM	Mitochondrial protein	AR
Combined oxidative phosphorylation deficiency-5	611719	MRPS22	Mitochondrial protein	AR
Combined oxidative phosphorylation deficiency-6	300816	AIFM1	Mitochondrial protein	XR
Combined oxidative phosphorylation deficiency-7	613559	HOGA1	Mitochondrial protein	AR
Combined oxidative phosphorylation deficiency-8	614096	AARS2	Mitochondrial protein	AR
Combined oxidative phosphorylation deficiency-9	614582	MRPL3	Mitochondrial protein	AR
Combined oxidative phosphorylation deficiency-10	614702	MTO1	Mitochondrial protein	AR
Combined oxidative phosphorylation deficiency-11	614922	RMND1	Mitochondrial protein	AR
Combined oxidative phosphorylation deficiency-12	614924	EARS2	Mitochondrial protein	AR
Combined oxidative phosphorylation deficiency-13	614932	PNPT1	Mitochondrial protein	AR
Combined oxidative phosphorylation deficiency-14	614946	FARS2	Mitochondrial protein	AR
Combined oxidative phosphorylation deficiency-15	614947	MTFMT	Mitochondrial protein	AR
Mitochondrial DNA depletion syndrome-1 or Mitochondrial Neuro-GastroIntestinal Encephalomyopathy syndrome or MNGIE syndrome	603041	TYMP	Thymidine phosphorylase	AR

Table 3.04.19.1. Continued.

Disorder	OMIM	Gene	Biological Function	Product of Conception Alteration
Mitochondrial DNA depletion syndrome-2	609560	TK2	Thymidine kinase	AR
Mitochondrial DNA depletion syndrome-3 or hepatocerebral type	251880	DGUOK	Deoxyguanosine kinase	AR
Mitochondrial DNA depletion syndrome-4A or Alpers type	203700	POLG	DNA polymerase gamma	AR
Mitochondrial DNA depletion syndrome-4B or MNGIE syndrome	613662	POLG	DNA polymerase gamma	AR
Mitochondrial DNA depletion syndrome-5	612073	SUCLA2	Mitochondrial protein	AR
Mitochondrial DNA depletion syndrome-6 or Navajo neuro-hepatopathy or hepatocerebral type	256810	MPV17	Mitochondrial protein	AR
Mitochondrial DNA depletion syndrome-7 or hepatocerebral type	271245	C10orf2	Mitochondrial protein	AR
Mitochondrial DNA depletion syndrome-8A or encephalomyopathic type with renal tubulopathy	612075	RRM2B	Mitochondrial protein	AR
Mitochondrial DNA depletion syndrome-8B or MNGIE syndrome	612075	RRM2B	Mitochondrial protein	AR
Mitochondrial DNA depletion syndrome-9	245400	SUCLG1	Mitochondrial protein	AR

3.04.20 Other Metabolic Disorders (not elsewhere classified)

3.04.20.1 Disorders of Plasma-protein Metabolism (not elsewhere classified)

Alpha-1-antitrypsin deficiency: A disorder characterized by emphysema, liver disease leading to cirrhosis, hemorrhagic diathesis and panniculitis. The most common manifestation of thess biallelic alterations in this gene is emphysema. The earliest symptoms include shortness of breath following mild activity, reduced ability to exercise and wheezing in the third to fourth decade of life. There are also environmental factors that contribute to worsening this condition; cigarette smoking or exposure to tobacco smoke for instance, accelerates the appearance of emphysema symptoms and damage to the lungs. Approximately 15% of adults with Alpha-1 antitrypsin deficiency also develop liver disease. Alpha-1-antitrypsin deficiency is caused by autosomal recessive mutations in the *SERPINA1* gene which encodes the serpin peptidase inhibitor (clade A). Among people of European ancestry, the carrier frequency for Alpha-1-antitrypsin deficiency trait is estimated to be 1 in 19.5-29.5 individuals, which means the frequency of this condition is about 1 in 1,500-3,500 individuals or $1/ (39)^2$ - $1/$

$(59)^2$. In Northern Europeans at least one pathogenic allele appears 6,000 years ago.

3.04.20.2 Lipodystrophy (not elsewhere classified)

Acquired partial lipodystrophy or Barraquer–Simons syndrome: A disorder characterized by a gradual onset of bilateral lipodystrophy in cephalocaudal sequence. This condition starts manifesting from symmetrical loss of subcutaneous fat from the face, neck, upper extremities, thorax and abdomen, sparing the lower extremities. Other patient's abnormalities include glomerulonephritis, diabetes mellitus, hyperlipidemia, mental retardation and complement deficiency. Females are 4 times more frequently affected than males. This condition is named after Luis Barraquer Roviralta and Arthur Simons. This disorder is not inherited in a classic Mendelian pattern. Susceptibility to the development of acquired partial lipodystrophy can be conferred in some cases by monoallelic mutations in the *LMNB2* gene which encodes the lamin B2, a nuclear lamina protein.

Congenital generalized lipodystrophy or Berardinelli–Seip syndrome: A disorder appearing at birth or early infancy characterized by a combination of severe insulin resistance and the absence of subcutaneous fat, and muscular hypertrophy. Affected individuals exhibit hypertriglyceridemia, diabetes mellitus, hepatic steatosis, hypertension, metabolic syndrome and hypertrophic cardiomyopathy. This condition leads to heart failure and sudden death. There are several underlying causes of this disorder with at least 4 autosomal genomic loci: Congenital generalized lipodystrophy-1 is caused by autosomal recessive mutations in the *AGPAT2* gene which encodes the 1-acylglycerol-3-phosphate O-acyltransferase-2;; Congenital generalized lipodystrophy-2 is caused by autosomal recessive mutations in the *BSCL2* gene which encodes the Berardinelli-Seip congenital lipodystrophy protein 2 (seipin). In Norwegians this pathogenic allele appears 400 years ago;; Congenital generalized lipodystrophy-3 is caused by autosomal recessive mutations in the *CAV1* gene which encodes the caveolin 1;; and Congenital generalized lipodystrophy-4 is caused by autosomal recessive mutations in the *PTRF* gene which encodes the polymerase I and transcript release factor. This protein is also thought to modify lipid metabolism and insulin-regulated gene expression. Berardinelli-Seip congenital lipodystrophy is estimated to affect 1 in every 10 million people worldwide. This condition may be more common in Lebanon, Brazil and Portugal.

Familial partial lipodystrophy or Köbberling–Dunnigan syndrome: A disorder characterized by loss of subcutaneous adipose tissue, primarily affecting the lower limbs. Other features include metabolic syndrome with insulin-resistant diabetes mellitus, hypertriglyceridemia, and hypertension of childhood or young adult onset. There are several underlying causes of this disorder with at least 4 autosomal genomic loci: Familial partial lipodystrophy-1 is characterized by loss of subcutaneous fat confined to the limbs. The gene causing type 1 has not been characterized;; Familial partial lipodystrophy-2 is characterized by loss of subcutaneous fat from the limbs and trunk. Familial partial lipodystrophy-2 is caused by autosomal dominant mutations in the *LMNA* gene which encodes the lamin A/C;; Familial partial lipodystrophy-3 is caused by autosomal dominant mutations in the *PPARG* gene which encodes the peroxisome proliferator-activated receptor gamma;; and Familial partial lipodystrophy-4 is caused by autosomal dominant mutations in the *PLIN1* gene which encodes the perilipin 1. There is also a severe insulin resistance associated with digenic mutations in the *PPARG* gene and the *PPP1R3A* gene which encodes the protein phosphatase 1 regulatory subunit 3A. Familial partial lipodystrophy is estimated to affect less than 1 in every 1,000,000 individuals.

3.04.20.3 Mitochondrial Metabolism Disorders

MELAS syndrome (Myopathy, Encephalopathy, Lactic Acidosis and Stroke-like episodes): A disorder characterized by muscle weakness and pain, recurrent headaches, loss of appetite, vomiting and seizures. MELAS syndrome can be caused by mutations in several mitochondrial encoded genes, including *MT-TL1, MT-TQ, MT-TH, MT-TK, MT-TC, MT-TS1, MT-ND1, MT-ND5, MT-ND6* and *MT-TS2*. In Europeans, MELAS syndrome is estimated to affect 1 in every 6,250 individuals.

MERRF syndrome (Myoclonic Epilepsy associated with Ragged-Red Fibers): A disorder characterized by exhibiting myoclonus, myopathy, spasticity, recurrent seizures, ataxia, peripheral neuropathy, dementia and cardiomyopathy. MERRF syndrome can be caused by mutations in several mitochondrial encoded genes, including *MT-TK, MT-TL1, MT-TH, MT-TS1, MT-TS2, MT-TF* and *MT-ND5*. In Europeans, MELAS syndrome is estimated to affect 1 in every 110,000 individuals, but the disorder seems to be more prevalent in the United States.

3.04.20.4 Other Specified Metabolic Disorders

Trimethylaminuria or Fish malodor syndrome: A disorder characterized by inability to break down trimethylamine. Affected individuals released large amounts of volatile and malodorous trimethylamine in the sweat, urine and breath. Trimethylaminuria is caused by autosomal recessive mutations in the *FMO3* gene which encodes the flavin-containing monooxygenase 3, an endoplasmic reticulum protein. Affected patients many times suffer rejection due to the malodor. There are also patients with fish-like body odor, which results from deficiency of dimethylglycine dehydrogenase. In a British report, the carrier frequency for Trimethylaminuria trait was estimated to be 1 in 100 individuals, which means the frequency of this condition was about 1 in 40,000 individuals or $1/(200)^2$.

Mevalonic aciduria or Mevalonate kinase deficiency: A disorder of infancy onset characterized by recurrent episodes of fever, lasting about 3-6 days. These episodes include lymphadenopathy, joint pain, skin rashes, hepatosplenomegaly, abdominal pain, diarrhea and headache. Affected children also exhibit failure to thrive, developmental delay, ataxia, and progressive problems with vision. Mevalonic aciduria is caused by autosomal recessive mutations in the *MVK* gene which encodes the mevalonate kinase. This enzyme is involved in the biosynthesis of cholesterol, which is later converted into steroid hormones and bile acids. There are two types of mevalonate kinase deficiency: a severe type called Mevalonic aciduria, and a less severe type called Hyperimmunoglobulinemia D syndrome. Mevalonic aciduria is estimated to affect less than 1 in every 1,000,000 individuals.

Polygamist Down or Fumarase deficiency: A disorder with similar features as those of Down syndrome. Polygamist

Down is characterized by profound psychomotor retardation with IQ of around 25, hypotonia, and brain abnormalities, such as ventriculomegaly, agenesis of the corpus callosum and gyral defects. Affected individuals also display metabolic acidosis, distress and encephalopathy. Polygamist Down is caused by autosomal recessive mutations in the *FH* gene which encodes the fumarate hydratase. This mitochondrial enzyme catalyzes the formation of L-malate from fumarate. Deficiency of this enzyme builds up fumaric acid leading to high levels of fumarate excretion in the urine. Polygamist Down is estimated to affect less than 1 in every 1,000,000 individuals. Polygamist Down has only been reported in the southwestern United States, specifically in Colorado City, Arizona, and Hildale, Utah. This condition is ethnic specific affecting members of the Fundamentalist Church of Jesus Christ of Latter Day Saints.

Glycosylphosphatidylinositol deficiency: A disorder characterized by propensity for portal vein thrombosis, portal hypertension and seizures. Glycosylphosphatidylinositol deficiency is caused by autosomal recessive mutations in the promoter region of the *PIGM* gene which encodes the phosphatidylinositol glycan anchor biosynthesis (class M) protein. Mutations in this gene reduce the transcription of PIGM and block mannosylation of glycosylphosphatidylinositol. Hypercoagulability syndrome due to glycosylphosphatidylinositol deficiency is estimated to affect less than 1 in every 1,000,000 individuals.

Alpha-ketoglutarate dehydrogenase deficiency or Oxoglutaric aciduria: A disorder appearing immediately after birth characterized by psychomotor retardation, hypotonia, ataxia, convulsions, metabolic acidosis and hyperlactacidemia. Affected individuals also exhibit hepatic disorders, myocardiopathy and sudden death. Oxoglutaric aciduria is caused by autosomal recessive mutations in the *OGDH* gene which encodes the alpha-ketoglutarate dehydrogenase complex. This enzyme catalyzes the transformation of alpha ketoglutarate into succinyl-CoA in the Krebs tricarboxylic acid cycle. Most cases have been reported in children of consanguineous parents.

Mitochondrial respiratory chain complex I deficiency: A group of disorders which are the most common enzymatic defect of the oxidative phosphorylation disorders. These conditions are characterized by a wide range of clinical disorders, from lethal neonatal disease to adult-onset neurodegenerative disorders. Affected individuals exhibit also nonspecific encephalopathy, macrocephaly, progressive leukodystrophy, hypertrophic cardiomyopathy, myopathy, liver disease, Leigh syndrome, Leber hereditary optic neuropathy and some forms of Parkinson disease. There are several underlying causes of this disorder with at least 21 nuclear-encoded genes and 7 mitochondrial-encoded genes. This impairment is caused by alteration in the respiratory chain complex I or NADH-ubiquinone reductase. The majority of cases are caused by mutations in nuclear-encoded genes; there are two types of inheritance: X-linked dominant and autosomal recessive. X-linked dominant inheritance results from mutations in the *NDUFA1* gene which encodes the NADH dehydrogenase (ubiquinone) 1 alpha subcomplex. Autosomal recessive inheritance results from mutations in the nuclear-encoded subunit genes, including the genes *NDUFV1* gene which encodes the NADH dehydrogenase (ubiquinone) flavoprotein 1, *NDUFV2* gene which encodes the NADH dehydrogenase (ubiquinone) flavoprotein 2, *NDUFS1* gene which encodes the NADH dehydrogenase (ubiquinone) Fe-S protein 1, (NADH-coenzyme Q reductase), *NDUFS2* gene which encodes the NADH dehydrogenase (ubiquinone) Fe-S protein 2, (NADH-coenzyme Q reductase), *NDUFS3* gene which encodes the NADH dehydrogenase (ubiquinone) Fe-S protein 3, (NADH-coenzyme Q reductase), *NDUFS4* gene which encodes the NADH dehydrogenase (ubiquinone) Fe-S protein 4, (NADH-coenzyme Q reductase), *NDUFS6* gene which encodes the NADH dehydrogenase (ubiquinone) Fe-S protein 6, (NADH-coenzyme Q reductase), *NDUFS7* gene which encodes the NADH dehydrogenase (ubiquinone) Fe-S protein 7, (NADH-coenzyme Q reductase), *NDUFS8* gene which encodes the NADH dehydrogenase (ubiquinone) Fe-S protein 8, (NADH-coenzyme Q reductase), *NDUFA2* gene which encodes the NADH dehydrogenase (ubiquinone) 1 alpha subcomplex 2, *NDUFA11* gene which encodes the NADH dehydrogenase (ubiquinone) 1 alpha subcomplex 11, *NDUFAF3* gene which encodes the NADH dehydrogenase (ubiquinone) complex I assembly factor 3, *NDUFA10* gene which encodes the NADH dehydrogenase (ubiquinone) 1 alpha subcomplex 10, and primarily the complex I assembly genes *NDUFAF2* gene which encodes the NADH dehydrogenase (ubiquinone) complex I assembly factor 2, *NDUFAF4* gene which encodes the NADH dehydrogenase (ubiquinone) complex I assembly factor 4, *NDUFAF5* gene which encodes the NADH dehydrogenase (ubiquinone) complex I assembly factor 5, *NUBPL* gene which encodes the nucleotide binding protein-like, *NDUFAF1* gene which encodes the NADH dehydrogenase (ubiquinone) complex I assembly factor 1, *ACAD9* gene which encodes the acyl-CoA dehydrogenase family member 9, and the *FOXRED1* gene which encodes the FAD-dependent oxidoreductase domain containing protein 1. Mitochondrial inheritance has been as-sociated with mutations in the 7 mito-chondrial-encoded components of com-plex I including: *MT-ND1, MT-ND2, MT-ND3, MT-ND4, MT-ND5, MT-ND6,* and *MT-TS2* genes. Most of these patients have a phenotype of Leigh syndrome and/or Leber hereditary optic neuropathy. Collec-tively, Mitochondrial respiratory chain complex deficiency is estimated to affect 1 in every 5,000-8,000 live births. Mito-chondrial respiratory chain complex I de-ficiency is probably the most common en-zyme defect.

Mitochondrial respiratory chain complex II deficiency: A disorder character-ized by progressive encephalomyopathy with dementia, ataxia, myoclonic seizures, external ophthalmoplegia, pigmentary ret-inopathy, cardiac conduction defects, hy-pertrophic cardiopathy, skeletal muscle myopathy and short stature. Mitochondrial respiratory chain complex II deficiency can be caused by autosomal recessive mu-tations either in the *SDHA* gene which en-codes the succinate dehydrogenase com-plex (subunit A flavoprotein), or the *SDHAF1* gene which encodes the succin-ate dehydrogenase complex assembly fac-tor 1.

Mitochondrial respiratory chain complex III deficiency: A disorder of neona-tal onset characterized by lactic acidosis, hypotonia, hypoglycemia, hepatopathy, renal tubulopathy, failure to thrive, en-cephalopathy and delayed psychomotor development. There are several underlying causes of this disorder with at least 4 auto-somal genomic loci: Mitochondrial respir-atory chain complex III deficiency can be caused by autosomal recessive mutations in the nuclear-encoded *BCS1L* gene which encodes the BC1 synthesis-like protein (ubiquinol-cytochrome c reductase), the *TTC19* gene which encodes the tetratrico-peptide repeat domain 19, the *UQCRB* gene which encodes the ubiquinol-cytochrome c reductase (binding protein), or the *UQCRQ* gene which encodes the ubiquinol-cytochrome c reductase (complex III subunit VII).

GRACILE syndrome (fetal Growth Retardation, Aminoaciduria, Cholestasis, Iron overload, Lactacidosis and Early death): A disorder caused by autosomal recessive mutations in the *BCS1L* gene. Mutations in the same gene also causes Leigh syndrome and Björnstad syndrome. The last form is characterized by sensorineural hearing loss and pili torti. In Finland, the carrier frequency for GRACILE syndrome trait is estimated to be 1 in 110 individuals, which means the frequency of this condition is about 1 in 47,000 individuals or $1/(220)^2$.

Mitochondrial respiratory chain complex IV deficiency: A disorder characterized by clinical heterogeneous features, ranging from isolated myopathy to severe multisystem disease. This condition starts manifesting between infancy and adulthood, with the adult onset form manifesting in mildly affected individuals. There are several underlying causes of this disorder with at least 7 nuclear-encoded genes and 3 mitochondrial-encoded genes. The majority of cases are caused by mutations in nuclear-encoded genes. This impairment is caused by alteration in the respiratory chain complex IV or cytochrome c oxidase. Mitochondrial respiratory chain complex IV is the terminal enzyme of the respiratory chain. Autosomal recessive inheritance results from mutations in the nuclear-encoded subunit genes, including the *COA5* gene which encodes the cytochrome c oxidase (assembly factor 5), the *COX10* gene which encodes the cytochrome c oxidase (assembly homolog 10), the *FASTKD2* gene which encodes the FAST kinase domains 2, the *COX14* gene which encodes the COX14 cytochrome c oxidase (assembly homolog), the *SCO1* gene which encodes the SCO1 cytochrome c oxidase (assembly protein), the *SCO2* gene which encodes the SCO2 cytochrome c oxidase (assembly protein), and the *COX6B1* gene which encodes the cytochrome c oxidase (subunit VIb polypeptide 1). Mitochondrial inheritance has been associated with mutations in 3 mitochondrial-encoded components of complex I: *MT-COI, MT-CO2*, and *MT-CO3* genes. Patients with cytochrome c oxidase deficiency often have features of Leigh syndrome. This condition includes movement problems, loss of mental function, hypertrophic cardiomyopathy, eating difficulties and brain abnormalities. Cytochrome c oxidase deficiency is one of the many causes of Leigh syndrome. Cytochrome c oxidase deficiency associated with the French-Canadian type of Leigh syndrome is caused by mutations in the *LRPPRC* gene which encodes the leucine-rich pentatricopeptide repeat containing protein. There are patients that have a combination of Leigh syndrome with Cytochrome c oxidase deficiency. There are several underlying causes of the latter condition with at least 3 autosomal genomic loci: Cytochrome c oxidase deficiency associated with Leigh syndrome can be caused by autosomal recessive mutations in the *SURF1* gene which encodes the surfeit 1, or *COX15* gene which encodes the cytochrome c oxidase (assembly homolog 15), or the *TACO1* gene which encodes the translational activator of mitochondrially encoded cytochrome c oxidase I. In French-Canadians, the carrier frequency for Leigh syndrome trait is estimated to be 1 in 25 individuals, which means the frequency of this condition is about 1 in 2,473 individuals or $1/(50)^2$. In Eastern Europe, Cytochrome c oxidase deficiency is estimated to occur in 1 in every 35,000 individuals.

Table 3.04.20.1 Other Recognizable Metabolic Disorders (not elsewhere classified)

Disorder	OMIM	Gene	Biological Function	Product of Conception Alteration
Alpha1-Antitrypsin Deficiency	613490	*SERPINA1*	Serine protease inhibitor	AR
Acquired partial lipodystrophy or Barraquer–Simons syndrome	608709	*LMNB2*	Nuclear Lamina Protein	AD
Congenital generalized lipodystrophy-1 or Berardinelli-Seip congenital lipodystrophy	608594	*AGPAT2*	Endoplasmic reticulum membrane	AR
Congenital generalized lipodystrophy-2 or Berardinelli-Seip congenital lipodystrophy	269700	*BSCL2*	Intracellular signaling peptides and proteins	AR
Congenital generalized lipodystrophy-3 or Berardinelli-Seip congenital lipodystrophy	612526	*CAV1*	Trafficking Vesicle fusion	AR
Congenital generalized lipodystrophy-4 or Berardinelli-Seip congenital lipodystrophy	613327	*PTRF*	RNA polymerase	AR
Familial partial lipodystrophy-2 or Köbberling–Dunnigan syndrome	151660	*LMNA*	Nuclear Lamina Protein	AD
Familial partial lipodystrophy-3 or Köbberling–Dunnigan syndrome	604367	*PPARG*	Peroxisome proliferator-activated receptor	AD + de novo
Familial partial lipodystrophy-4 or Köbberling–Dunnigan syndrome	604368	*PLIN1*	Perilipin	AD
Severe insulin resistance	604367	*PPARG/ PPP1R3A*	Peroxisome proliferator-activated receptor/phosphatases	Dig
MELAS syndrome or myopathy, encephalopathy, lactic acidosis, and stroke-like episodes	516000	*MT-ND1*	Mitochondrial protein	Mit
MELAS syndrome or myopathy, encephalopathy, lactic acidosis, and stroke-like episodes	516005	*MT-ND5*	Mitochondrial protein	Mit
MELAS syndrome or myopathy, encephalopathy, lactic acidosis, and stroke-like episodes	516006	*MT-ND6*	Mitochondrial protein	Mit

Table 3.04.20.1. Continued.

Disorder	OMIM	Gene	Biological Function	Product of Conception Alteration
MELAS syndrome or myopathy, encephalopathy, lactic acidosis, and stroke-like episodes	590020	*MT-TC*	Mitochondrial protein	Mit
MELAS syndrome or myopathy, encephalopathy, lactic acidosis, and stroke-like episodes	590040	*MT-TH*	Mitochondrial protein	Mit
MELAS syndrome or myopathy, encephalopathy, lactic acidosis, and stroke-like episodes	590060	*MT-TK*	Mitochondrial protein	Mit
MELAS syndrome or myopathy, encephalopathy, lactic acidosis, and stroke-like episodes	590050	*MT-TL1*	Mitochondrial protein	Mit
MELAS syndrome or myopathy, encephalopathy, lactic acidosis, and stroke-like episodes	590030	*MT-TQ*	Mitochondrial protein	Mit
MELAS syndrome or myopathy, encephalopathy, lactic acidosis, and stroke-like episodes	590080	*MT-TS1*	Mitochondrial protein	Mit
MELAS syndrome or myopathy, encephalopathy, lactic acidosis, and stroke-like episodes	590085	*MT-TS2*	Mitochondrial protein	Mit
MERRF syndrome or myoclonic epilepsy associated with ragged-red fibers	516005	*MT-ND5*	Mitochondrial protein	Mit
MERRF syndrome or myoclonic epilepsy associated with ragged-red fibers	590070	*MT-TF*	Mitochondrial protein	Mit
MERRF syndrome or myoclonic epilepsy associated with ragged-red fibers	590040	*MT-TH*	Mitochondrial protein	Mit
MERRF syndrome or myoclonic epilepsy associated with ragged-red fibers	590060	*MT-TK*	Mitochondrial protein	Mit
MERRF syndrome or myoclonic epilepsy associated with ragged-red fibers	590050	*MT-TL1*	Mitochondrial protein	Mit

Table 3.04.20.1. Continued.

Disorder	OMIM	Gene	Biological Function	Product of Conception Alteration
MERRF syndrome or myo-clonic epilepsy associated with ragged-red fibers	590080	*MT-TS1*	Mitochondrial protein	Mit
MERRF syndrome or myo-clonic epilepsy associated with ragged-red fibers	590085	*MT-TS2*	Mitochondrial protein	Mit
Trimethylaminuria or fish malodor syndrome	602079	*FMO3*	Endoplasmic reticulum protein	AR
Mevalonic aciduria or Meva-lonate kinase deficiency	610377	*MVK*	Peroxisomal and lyso-somal structural pro-tein	AR
Polygamist Down or Fumarase deficiency	606812	*FH*	Mitochondrial protein	AR
Glycosylphosphatidylinositol deficiency	610293	*PIGM*	Endoplasmic reticulum protein	AR
Alpha-ketoglutarate dehy-drogenase deficiency or Ox-oglutaric aciduria	203740	*OGDH*	Mitochondrial protein	AR
Mitochondrial complex I de-ficiency	252010	*NDUFA1*	Mitochondrial protein	XD
Mitochondrial complex I de-ficiency	252010	*NDUFS2*	Mitochondrial protein	AR
Mitochondrial complex I de-ficiency	252010	*NDUFB3*	Mitochondrial protein	AR
Mitochondrial complex I de-ficiency	252010	*NDUFS1*	Mitochondrial protein	AR
Mitochondrial complex I de-ficiency	252010	*NDUFAF3*	Mitochondrial protein	AR
Mitochondrial complex I de-ficiency	252010	*NDUFS6*	Mitochondrial protein	AR
Mitochondrial complex I de-ficiency	252010	*NDUFS4*	Mitochondrial protein	AR
Mitochondrial complex I de-ficiency	252010	*NDUFAF2*	Mitochondrial protein	AR
Mitochondrial complex I de-ficiency	252010	*NDUFAF4*	Mitochondrial protein	AR
Mitochondrial complex I de-ficiency	252010	*NDUFS3*	Mitochondrial protein	AR
Mitochondrial complex I de-ficiency	252010	*NDUFV1*	Mitochondrial protein	AR
Mitochondrial complex I de-ficiency	252010	*FOXRED1*	Mitochondrial protein	AR
Mitochondrial complex I de-ficiency	252010	*NUBPL*	Mitochondrial protein	AR

The Concise Encyclopedia of Genomic Diseases

Table 3.04.20.1. Continued.

Disorder	OMIM	Gene	Biological Function	Product of Conception Alteration
Mitochondrial complex I deficiency	252010	NDUFAF1	Mitochondrial protein	AR
Mitochondrial complex I deficiency	252010	NDUFV2	Mitochondrial protein	AR
Mitochondrial complex I deficiency	252010	NDUFA11	Mitochondrial protein	AR
Mitochondrial complex I deficiency	252010	NDUFAF5	Mitochondrial protein	AR
Mitochondrial complex I deficiency	516000	MT-ND1	Mitochondrial protein	Mit
Mitochondrial complex I deficiency	516001	MT-ND2	Mitochondrial protein	Mit
Mitochondrial complex I deficiency	516002	MT-ND3	Mitochondrial protein	Mit
Mitochondrial complex I deficiency	516003	MT-ND4	Mitochondrial protein	Mit
Mitochondrial complex I deficiency	516005	MT-ND5	Mitochondrial protein	Mit
Mitochondrial complex I deficiency	516006	MT-ND6	Mitochondrial protein	Mit
Mitochondrial respiratory chain complex II deficiency	252011	SDHA	Mitochondrial protein	AR
Mitochondrial respiratory chain complex II deficiency	252011	SDHAF1	Mitochondrial protein	AR
Mitochondrial respiratory chain complex III deficiency	124000	BCS1L	ATPase, AAA-type	AR
Mitochondrial respiratory chain complex III deficiency	124000	UQCRB	Mitochondrial protein	AR
Mitochondrial respiratory chain complex III deficiency	124000	UQCRQ	Mitochondrial protein	AR
GRACILE syndrome	603358	BCS1L	ATPase, AAA-type	AR
Mitochondrial respiratory chain complex IV deficiency	220110	COX6B1	Mitochondrial protein	AR
Mitochondrial respiratory chain complex IV deficiency	220110	FASTKD2	Mitochondrial protein	AR
Mitochondrial respiratory chain complex IV deficiency	220110	COA5	Mitochondrial protein	AR
Cytochrome c oxidase deficiency	220110	COX6B1	Mitochondrial protein	AR

3.04.21 Disease Prevention for Disassortative Mating in Ethnic Groups

Table 3.04.21 displays a few relevant examples of autosomal recessive diseases that require the implementation of some methodology to decrease the prevalence of their associated conditions. Besides the disease, the table also includes the condition associated with the gene, the incidence of expected disassortative mating frequency, and the ethnic group with a high prevalence of the condition. For example, 1 in every 60 families of Caucasian ancestry is at risk of having a child with Hemochromatosis-1 because both parents are carriers of a polymorphism in the *HFE* gene. In Ireland, 1 in every 350 families is at risk of having a child with Cystic fibrosis because both parents are carriers of a polymorphism in the *CFTR* gene. In the United States, 1 in every 625 families of Caucasian ancestry is at risk of having a child with Cystic fibrosis because both parents are carriers of a polymorphism in the *CFTR* gene. In the United States, 1 in every 4,250 families of African American ancestry is at risk of having a child with Cystic fibrosis because both parents are carriers of a polymorphism in the *CFTR* gene. In the United States, 1 in every 1,625 families of interethnic marriages between one of the parents is Caucasian and the other is African-American African American ancestry are at risk of having a child with Cystic fibrosis becuase both parents are carriers of a polymorphism in the *CFTR* gene. In Ashkenazi Jews, 1 in every 730 families is at risk of having a child with Tay-Sachs diseasebecause both parents are carriers of a polymorphism in the *HEXA* gene. In Irish-Americans, 1 in every 2,500 families is at risk of having a child with Tay-Sachs disease because both parents are carriers of mutations in the *HEXA* gene. In Saguenay-Lac-Saint-Jean individuals (Quebec), 1 in every 440 families is at risk of having a child with Tyrosinemia type 1 because both parents are carriers of a polymorphism in the *FAH* gene. In Old Order Mennonites, 1 in every 75 families is at risk of having a child with Maple syrup urine disease 1A because both parents are carriers of a polymorphism in the *BCKDHA* gene. In Western Europeans, 1 in every 375 families is at risk of having a child with Emphysema and COPD because both parents are carriers of mutations in the *SERPINA1* gene. In Turkey, 1 in every 650 families is at risk of having a child with Phenylketonuria because both parents are carriers of a polymorphism in the *PAH* gene. In Saudi Arabia, 1 in every 500 families is at risk of having a child with Propionic acidemia because both parents are carriers of a polymorphism in the *PCCA* gene. In Saguenay-Lac-Saint-Jean individuals (Quebec), 1 in every 1,550 families is at risk of having a child with I-cell disease because both parents are carriers of a polymorphism in the *GNPTAB* gene. In Slovakia, 1 in every 4,750 families is at risk of having a child with Alkaptonuria because both parents are carriers of a polymorphism in the *HGD* gene. Each of these families has a 1 in 4 risk of having a child with any of these conditions. Each of these families should receive counseling to prevent these autosomal recessive diseases.

The Concise Encyclopedia of Genomic Diseases

Table 3.04.21 Frequency of Disassortative Mating in Multiple Ethnic Groups

Disease	Gene	Disassortative mating frequency	People with higher prevalence (if available)
Hemochromatosis-1	HFE	1 in 60	Northern Europeans
Cystic fibrosis	CFTR	1 in 350	Irish
Cystic fibrosis	CFTR	1 in 625	Caucasians (United States)
Cystic fibrosis	CFTR	1 in 4,250	African Americans (United States)
Cystic fibrosis	CFTR	1 in 1,625	Caucasians/African Americans (United States)
Tay-Sachs disease	HEXA	1 in 730	Ashkenazi Jews
Tay-Sachs disease	HEXA	1 in 2,500	Irish-Americans
Tyrosinemia	FAH	1 in 440	Fench Canadians (Quebec)
Gaucher disease	GBA	1 in 125	Ashkenazi Jews
Gaucher disease	GBA	1 in 12,500	Americans
Maple syrup urine disease-1A or branched-chain ketoaciduria	BCKDHA	1 in 75	Old Order Amish and Mennonites
Emphysema and COPD, due to alpha-1 antitrypsin Pittsburgh mutation	SERPINA1	1 in 375	Western Europeans
Congenital adrenal hyperplasia	CYP21A2	1 in 1,250	Swiss
Phenylketonuria	PAH	1 in 650	Turkish
Phenylketonuria	PAH	1 in 3,750	Americans
Propionic acidemia	PCCA or PCCB	1 in 500	Saudi Arabians, Amish and Mennonites
Familial hypercholesterolemia	LDLR	1 in 4,800	Ashkenazi Jews
Familial hypercholesterolemia	LDLR	1 in 250,000	Americans
Glycogen Storage Disease type 1A	G6PC	1 in 5,000	Ashkenazi Jews
Hyperinsulinemic hypoglycemia	ABCC8	1 in 700	Saudi Arabians and Finnish
Hyperinsulinemic hypoglycemia	ABCC8	1 in 2,500	Ashkenazi Jews
Mucolipidosis or I-cell disease	GNPTAB	1 in 1,550	French Canadians (Quebec)
Alkaptonuria or ochronosis	HGD	1 in 4,750	Slovaks

3.05.0 Subchapter V

Mental and Behavioral Disorders

Table 3.05.0 Mental and Behavioral Disorders

3.05.00-10	Mental and behavioral disorders	
	3.05.1	Organic Mental Disorders (including symptomatic)
	3.05.2	Mental and Behavioral Disorders due to Psychoactive Substance use
	3.05.3	Schizophrenia, Schizotypal and Delusional Disorders
	3.05.4	Mood (affective) Disorders
	3.05.5	Neurotic, Stress-related and Somatoform Disorders
	3.05.6	Behavioral Syndromes Associated with Physiological Disturbances and Physical Factors
	3.05.7	Disorders of Adult Personality and Behavior
	3.05.8	Mental Retardation
	3.05.9	Disorders of Psychological Development
	3.05.10	Behavioral and Emotional Disorders with onset usually occurring in Childhood and Adolescence

3.05.1 Organic Mental Disorders (including symptomatic)

CADASIL syndrome (Cerebral Autosomal Dominant Arteriopathy with Subcortical Infarcts and Leukoencephalopathy): A disorder characterized by a progressive arteriopathy of the small vessels of the brain. Affected individuals manifest symptoms such as strokes, migraine and neuropsychiatric symptoms. Other features of this condition include white matter lesions with resultant cognitive impairment, leading to dementia. This hereditary multi-infarct dementia results from relapsing strokes affecting relatively young adults of both sexes. CADASIL syndrome is caused by autosomal dominant mutations in the *NOTCH3* gene which encodes the notch 3 protein. This protein establishes an intercellular signalling pathway that plays a key role in neural development. The worldwide prevalence of CADASIL syndrome is unknown. In Scotland, CADASIL syndrome is estimated to affect 1 in 50,000 adults.

Pick's disease or frontotemporal dementia: A disorder characterized by progressive destruction of nerve cells in the brain. Patients with this condition exhibit aphasia affecting behavior, language and movement. In the fourth or fifth decade of life these patients start to exhibit dementia. A defining characteristic of the disease is buildup of argyrophilic, intraneuronal inclusions, and 'Pick cells,' which are enlarged neurons. There are several underlying causes of this disorder with at least 2 autosomal genomic loci: Pick's disease is caused by autosomal dominant mutations in either the *MAPT* gene which encodes the microtubule-associated protein tau, or the *PSEN1* gene which encodes the presenilin 1. The worldwide prevalence of genomic Pick's disease is unknown. In the Netherlands, genomic Pick's disease is estimated to affect 1 in 1,000,000 adults. However, the disorder is likely underdiagnosed.

Creutzfeldt–Jakob disease: A disorder of adulthood onset characterized by rapid progressive neurologic deterioration. Patients with this condition have impairment in brain function, causing personality changes, memory changes and dementia. Creutzfeldt–Jakob disease has several causative factors including those of sporadic, transmissible, and genomic or familial origins. The sporadic form seems to be the most common, but homozygosis for met129 in the *PRNP* gene confers susceptibility for the development of sporadic Creutzfeldt–Jakob disease. The familial form is milder than the sporadic and accounts for 5-10% of all the cases. Familial Creutzfeldt–Jakob disease is caused by autosomal dominant mutations in *PRNP* gene which encodes the prion protein. This protein is water-insoluble and can be misfolded. This gene has been associated with transmissible neurodegenerative spongiform encephalopathies, including Gerstmann–Sträussler disease, kuru, Huntington disease-like 1 and Familial fatal insomnia. Creutzfeldt–Jakob disease affects about 1 in 1,000,000 individuals worldwide annually. This condition is more frequent among people from Chile and among Libyan Jews. In the latter group, this condition is 43 times more common than average prevalence worldwide.

Gerstmann-Sträussler disease: A disorder of adulthood onset characterized by cognitive decline, memory loss, dementia, dysarthria, and progressive limb and truncal ataxia. Gerstmann-Sträussler disease can be caused by mutations in the *PRNP* gene.

Prion disease with protracted course: A disorder characterized by presenile dementia with a rapidly progressive and protracted clinical course. Prion disease with protracted course can be caused by mutations in the *PRNP* gene.

Huntington's disease and Huntington like disease: This disease is discussed under diseases of the nervous system in the subchapter "Systemic Atrophies primarily affecting the Central Nervous System" (3.06.02).

Table 3.05.1 Recognizable Organic Mental Disorders (including symptomatic)

Disorder	OMIM	Gene	Biological Function	Product of Conception Alteration
CADASIL syndrome	125310	*NOTCH3*	Notch signaling pathway	AD
Pick disease	172700	*MAPT*	Cytoskeletal Microtubules	AD
Pick disease	172700	*PSEN1*	Regulate APP processing	AD
Creutzfeldt-Jakob disease	123400	*PRNP*	Prion protein	AD + de novo or AR
Gerstmann-Straussler-Scheinker syndrome	137440	*PRNP*	Prion protein	AD
Prion disease with protracted course	606688	*PRNP*	Prion protein	AD
Huntington disease	143100	*HTT*	Huntingtin	AD + Ant

3.05.2 Mental and Behavioral Disorders due to Psychoactive Substance use

Susceptibility for alcoholism: A disorder characterized by compulsive and uncontrolled consumption of alcoholic beverages. This condition habitually causes a decline in the drinker's health, and social and personal relationships. The cause of alcoholism is probably multifactorial. There are several genomics factors associated with this condition including family and twin studies. Plato and Aristotle recognized the tendency of children's alcohol consumption patterns to resemble that of their parents. The concordance rate for alcoholism in monozygotic twins is 55% while in dizygotic twins it is only 28%. The lifetime risk for alcoholism in sons and brothers of severely alcoholic men is 25-50%. A genomic basis of organ specificity for complication of this condition is evident in pancreatitis from type V hyperlipidemia and Wernicke-Korsakoff syndrome. There are 2 separate heritable forms of alcoholism. Patients with type 1 exhibit a pattern of alcohol abuse with habitual onset after 25 years of age, which is characterized by affecting males and females with severe psychological dependence and guilt. Patients with type 2 exhibit a pattern of alcohol abuse with onset before the age of 25 and almost exclusively affecting males; persons with type 2 form of alcoholism are frequently aggressive with antisocial behavior and inability to abstain from alcohol. There are several genes associated with susceptibility to alcohol dependence including: *GABRA2, ADH1B, ADH1C, TAS2R16, HTR2A* and *RCBTB1* genes.

3.05.3 Schizophrenia, Schizotypal and Delusional Disorders

Schizophrenia: A disorder characterized by distortions of thinking and perception, and blunted or inappropriate affects. The most common manifestation includes mild impairment of cognitive function, severe inappropriate emotional responses, auditory hallucinations, paranoid or bizarre delusions, or disorganized thinking and concentration. Affected individuals exhibit erratic behavior leading to social and occupational deterioration. The onset of symptoms usually occurs in young adulthood. This condition has a 1% worldwide prevalence. Schizophrenic patients have a higher suicide rate (5-10%) and a lower average life expectancy (12-15 years) than non-schizophrenic people. The main causes of schizophrenia include a combination of genomic and environmental factors. In this condition, it is difficult to separate the effects of heritability and environment factors. The environmental factors include the living environment, social isolation, unsupportive parenting style, urban environment and drug use. This condition is slightly more common in males than females and usually appears earlier in men. In schizophrenia, there is a large evidence for a genomic component which comes from first degree relatives and twin studies. The risk for developing schizophrenia is 6-9% when a first-degree relative has this condition. The concordance rate for schizophrenia in monozygotic twins is over 40%. There are several genes with unknown patterns of transmission which are associated with susceptibility to schizophrenia including *APOL2, APOL4, DISC2, DAOA, DAO, DTNBP1, DISC1, CHI3L1, MTHFR, DRD3, HTR2A, LGR4, COMT, DISC1, RTN4R, AKT1, SYN2, SHANK3, VIPR2* and *PRODH* genes. The genomics factors that cause schizophrenia and bipolar disorder overlap.

3.05.4 Mood (affective) Disorders

Major depressive disorder: A disorder characterized by at least 2 weeks of low mood accompanied by low self-esteem, and by reduction of energy and decrease in normally enjoyable activities. These patients manifest a marked tiredness, and reduced enjoyment, interest and concentration. Other features include feelings of worthlessness or guilt, disturbed sleep patterns, and change in eating habits. During these episodes, there are recurrent thoughts of death or suicidal ideation, plans or attempts. The time of onset is between the ages of 20-40 years and is twice as common in women as in men. The lifetime risk of major depressive disorder varies from 5-12% for men to 10-25% for women. The risk of major depression is increased with disabling diseases including stroke, Parkinson's disease, cardiovascular illnesses, multiple sclerosis and others. Major depressive disorder is more common in urban than in rural environments. This condition harmfully affects a person's family, work or school life and general health. In high-income countries, 1 in 30 individuals with major depression commit suicide, and up to 60% of people who commit suicide had depression or another mood disorder. There is a 4-fold increase in the death rate of individuals with major depressive disorder over 55 years of age. The course of the condition varies extensively, from one episode lasting weeks to a lifelong disorder with recurrent major depressive episodes. Major depressive disorder patients have shorter life expectancies than those without depression, in part because they have a larger susceptibility to suicide and medical illnesses. Depression affects over 120 million persons worldwide, and is a major cause of morbidity. Panic disorder is the most common comorbidity affecting females, whereas substance use is more common in affected males. In Major depressive disorder cases is difficult to separate the effects of heritability and environment factors. In Major depressive disorder, there is evidence for a genomic component which comes from first-degree relatives and twin studies. The risk of major depressive disorder in first-degree relatives is 2-8 times higher than average. Twin studies estimate the heritability of major depressive disorder at 36-70%. There are several genes with unknown patterns of transmission which are associated with susceptibility to major depressive disorder including *TOR1A, MTHFR, CHRM2, DRD4, FKBP5, BCR, SLC6A4, CREB1, TPH1* and *TPH2* genes.

Perry syndrome: A disorder appearing in the fifth decade of life characterized by unresponsive mental depression with sleep disturbances, exhaustion and marked weight loss. Years later, patients develop respiratory failure and Parkinsonism features. Perry syndrome patients have diminished taurine in plasma and cerebrospinal fluid. Taurine is a putative inhibitory synaptic transmitter. Perry syndrome is caused by autosomal dominant mutations in the *DCTN1* gene which encodes the dynactin 1. This protein is a microtubule-based biologic motor protein. Mutations in this gene also cause distal hereditary motor neuropathy type VIIB and susceptibility to amyotrophic lateral sclerosis. Perry syndrome is estimated to affect less than 1 in every 1,000,000 individuals.

3.05.5 Neurotic, Stress-related and Somatoform Disorders

Obsessive–compulsive disorder: A disorder characterized by recurring obsessions and/or compulsions. Patients with this condition exhibit anxiety, and intrusive thoughts which lead to repetitive behaviors aimed at reducing the associated anxiety. Symptoms of the disorder include nervous rituals, excessive washing or cleaning, preoccupation with sexual, violent or religious thoughts. These time-consuming symptoms alienate these individuals and produce a severe emotional and financial distress. Obsessive–compulsive disorder individuals have above-average intelligence, high attention to detail, avoidance of risk, careful planning and other features. The prevalence of the disorder has been estimated in 1-3% of people. The lifetime prevalence is higher among highly educated people, than any other groups. This condition has a similar prevalence among genders. In Obsessive–compulsive disorder, there is a large evidence for a genomic component which comes from twin studies, family genetics studies and segregation analyses. Susceptibility to obsessive-compulsive disorder has been associated with SNPs in the *SLC6A4* gene which encodes the solute carrier (family 6 member A4-neurotransmitter transporter), and the promoter region in the *HTR2A* gene which encodes the serotonin 5-HT2A receptor (G protein-coupled). There is a protective allele associated with variation in the *BDNF* gene which encodes the brain-derived neurotrophic factor.

Paroxysmal extreme pain disorder: A disorder characterized by pain and flushing affecting the submandibular, ocular and rectal areas. Paroxysmal extreme pain disorder and Primary erythromelalgia shared similarities in flushing and episodic pain, although pain is typically present in the extremities for Primary erythromelalgia patients. Paroxysmal extreme pain disorder is caused by autosomal dominant mutations in the *SCN9A* gene which encodes the voltage-gated sodium channel (type 9 alpha subunit). Mutations in the same gene also cause Generalized epilepsy with febrile seizures plus type 7, Familial febrile seizures 3B, Small fiber neuropathy and Channelopathy-associated with insensitivity to pain. Paroxysmal extreme pain disorder is estimated to affect less than 1 in every 1,000,000 individuals.

3.05.6 Behavioral Syndromes Associated with Physiological Disturbances and Physical Factors

Fatal familial insomnia: A disorder characterized by insomnia of progressive course, which leads to hallucinations, delirium, and confused states similar to those of dementia. Patients sometimes exhibit diurnal dreaming state and dysautonomia preceding motor and cognitive deterioration. The age at onset varies between 37 and 61 years of age. The average survival period after the onset of symptoms is 18 months. Fatal familial insomnia is caused by autosomal recessive mutations in the *PRNP* gene. Homozygosity for the met129 affects the anterior and dorsomedial thalamus. This protein plays a key role integrating and expressing sleep, autonomic functions and neuroendocrine circadian rhythm. Mutations in the same gene are associated with Creutzfeldt–Jakob disease. Fatal familial insomnia is estimated to affect less than 1 in every 1,000,000 individuals.

Brunner syndrome: A disorder characterized by impulsive aggressiveness, mild mental retardation, sleep disorders and mood swings. Brunner syndrome patients manifest uncharacteristic and sometimes violent behaviors including arson, attempted rape and exhibitionism. Brunner syndrome is caused by X linked recessive mutations in the *MAOA* gene which encodes the monoamine oxidase A. This protein breaks down monoamines.

3.05.7 Disorders of Adult Personality and Behavior

Anxiety-related personality traits: Are a group of disorders characterized by moodiness, low self-esteem, anxiety, depression and lack of self-confidence. There are genomic and environmental factors which shape the personality. In Anxiety-related personality traits are difficult to separate the effects of heritability and environment factors. These genomic factors contribute 40-60% of this trait variance. Anxiety-related personality traits are associated with variation in the *SLC6A4* gene which encodes the solute carrier (family 6 member A4), a serotonin neurotransmitter transporter.

Novelty seeking personality: A disorder characterized by risk attitudes such as substance abuse, unprotected sex, gambling, theft and aggression. This condition varies with age, but there is a considerable heritable component in personality traits. Novelty seeking personality is associated with autosomal recessive promoter polymorphism in the *DRD4* gene which encodes the dopamine receptor D4, a G-protein coupled receptor which inhibits adenylyl cyclase.

Trichotillomania: A disorder characterized by chronic, repetitive compulsivity to pull out one's own hair. This condition leads to visible hair loss, distress, and social or functional impairment. The peak age of onset of trichotillomania is 9-13 years, usually elicited by stress or depression. Trichotillomania is a common condition; the lifetime prevalence is estimated to be between 0.6-0.8%. This condition is slightly more common in females than in males. There are overlapping psychological conditions, such as Tourette syndrome or obsessive-compulsive disorder. Heritability trichotillomanic behavior has been determined using twin studies; the genomic factors have been estimated to account for a high proportion - around 76-78% - of the total factors. Trichotillomania has been associated with mutations in the *SLITRK1* gene which encodes the SLIT and NTRK-like family member protein 1. This protein is involved in neurite outgrowth.

3.05.8 Mental Retardation

Mental retardation: A disorder characterized by significantly impaired cognitive functioning. This condition has been traditionally defined as a triad of Intelligence quotient score under 70, significant limitations in adaptive behavior, such as daily-living skills, social skills and communication, and evidence that the limitations became apparent before the age of 18. Mental retardation is a common disorder that affects 1–3% of all people worldwide. There are different levels of disability with mild mental retardation (IQ 50–69), moderate mental retardation (IQ 35–49), severe mental retardation (IQ 20-34), and profound mental retardation (IQ below 20). Mild retardation is the most common affecting 75–90% of all cases. This heterogeneous condition has genomic and environmental factors. The 3 more common causes of mental retardation in the United States and Europe include Fetal alcohol syndrome affecting 0.5-2.0 cases out of 1,000 live births, Down syndrome affecting 1.5 cases out of 1,000 live births, and DiGeorge syndrome affecting 2-3 cases out of 10,000 live births. There are other causes of mental retardation in underdeveloped countries. In those countries the main cause of mental retardation is attributed to iodine deficiency and malnutrition. There are other causes of mental retardation including: problems during pregnancy such as rubella infection, problems during labor and birth such as developmental disability due to oxygen deprivation, and exposure to infection or toxins. There are other genomic causes of mental retardation described elsewhere in this book, including Klinefelter's syndrome, Down syndrome, DiGeorge syndrome, Fragile X syndrome, FG syndrome, Neurofibromatosis, Congenital hypothyroidism, Williams-Beuren syndrome, Phenylketonuria, Rett syndrome, Wolf-Hirschhorn syndrome, Mowat-Wilson syndrome, Prader-Willi syndrome, Angelman syndrome, Cornelia de Lange syndrome, Rubinstein-Taybi syndrome, CHARGE syndrome, Koolen-De Vries syndrome, Pitt-Hopkins syndrome, Bohring-Opitz syndrome, Gillespie syndrome, WAGR syndrome, Temtamy syndrome, Silver-Russell syndrome, Kabuki syndrome and Branchiooculofacial syndrome. There are several underlying causes of this disorder with at least 56 autosomal genomic loci and 120 X-linked genomic loci. Mental retardation can be further classified as syndromic or nonsyndromic. In several forms of mental retardation the genes have not been characterized. The various types of mental retardation will be listed on Table 3.05.8.1, along with their OMIM, their associated genes, their biological functions and the product of conception alterations.

Lujan–Fryns syndrome or X-linked mental retardation with Marfanoid habitus: A disorder characterized by mild to moderate mental retardation and marfanoid habitus with long, narrow face, small mandible, high-arched palate and hypernasal voice. Affected individuals al-

so exhibit features of malformations in the brain and heart. Lujan-Fryns syndrome is caused by X-linked dominant mutations in the *MED12* gene which encodes the mediator complex subunit 12, a mediator in transcriptional activation or repression. Mutations in the same gene are also associated with Opitz-Kaveggia syndrome.

Woodhouse–Sakati syndrome: This disease is discussed under endocrine, nutritional and metabolic diseases in the subchapter "Diabetes Mellitus" (3.04.02.1).

SeSAME syndrome (SEizures, Sensorineural deafness, Ataxia, Mental retardation, and Electrolyte imbalance): A disorder which is caused by autosomal recessive mutations in the *KCNJ10* gene which encodes the inwardly rectifying potassium channel (subfamily J member 10). This protein is expressed in the brain, inner ear and kidney. SeSAME syndrome is estimated to affect less than 1 in every 1,000,000 individuals.

N syndrome: A disorder characterized by mental retardation, deafness, ocular anomalies, spasticity, cryptorchidism, hypospadias and susceptibility to T-cell leukemia. N syndrome is caused by X-linked recessive mutations in the *POLA1* gene which encodes the DNA polymerase alpha. This enzyme leads to increased chromosome breakage.

Coffin-Lowry syndrome: A disorder characterized by cognitive impairment, skeletal and cardiac malformations, growth retardation, hearing and visual abnormalities, and paroxysmal movement disorders. Coffin-Lowry syndrome is caused by X-linked dominant mutations in the *RPS6KA3* gene which encodes the ribosomal protein S6 kinase (90kDa, polypeptide 3), a serine/threonine kinase. This condition is more common among males than females. Mutations in the *RPS6KA3* gene are associated with nonsyndromic X-linked mental retardation-19. Coffin-Lowry syndrome is estimated to affect 1 in every 50,000-100,000 individuals.

Birk-Barel mental retardation dysmorphism syndrome: A disorder characterized by mental retardation, hypotonia and a characteristic dysmorphism. Birk-Barel mental retardation dysmorphism syndrome can be caused by monoallelic mutations in the maternally transmitted (imprinted with paternal silencing) *KCNK9* gene which encodes the potassium channel (subfamily K member 9).

Mental retardation, stereotypic movements, epilepsy, and/or cerebral malformations or autosomal dominant mental retardation syndrome-20: A disorder characterized by moderate intellectual disability, attention deficit-hyperactivity disorder, bilateral iris coloboma, dental anomaly, hearing loss and dysmorphic facial features. "Mental retardation, stereotypic movements, epilepsy, and/or cerebral malformations" is caused by either monoallelic mutation in the *MEF2C* gene, or heterozygous deletion of the chromosome 5q14.3 region.

Renpenning syndrome: A disorder characterized by mental retardation, microcephaly, short stature, small testes and typical facies. Renpenning syndrome can be caused by X-linked mutation in the *PQBP1* gene which encodes the polyglutamine binding protein 1.

Table 3.05.8.1 Recognizable Mental Retardation Syndrome

Disorder	OMIM	Gene	Biological Function	Product of Conception Alteration
Mental retardation, FRA12A type	136630	*DIP2B*	Transcriptional regulator	AD
Aarskog-Scott syndrome or X-linked syndromic mental retardation-16	305400	*FGD1*	Zinc finger protein	XR
Alpha-thalassemia/Mental retardation syndrome	301040	*ATRX*	Chromatin remodeling	XR + de novo
Arts syndrome or X-linked syndromic mental retardation-18	301835	*PRPS1*	Pyrophosphate synthetase	XR
Birk-Barel Mental retardation dysmorphism syndrome	612292	*KCNK9*	Channelopathy Potassium	MAT
Borjeson-Forssman-Lehmann syndrome	301900	*PHF6*	Zinc finger protein	XR + de novo
Coffin-Lowry syndrome	303600	*RPS6KA3*	Chromatin structure	XD
Corpus callosum, agenesis of, with mental retardation, ocular coloboma and micrognathia or X-linked syndromic mental retardation-28	300472	*IGBP1*	Immunoglobulin binding protein	XR
Lujan–Fryns syndrome or X-linked mental retardation with Marfanoid habitus	309520	*MED12*	Mediator of RNA polymerase 2	XD
Mental retardation	300495	*NLGN4X*	Central nervous system synapses	XR

Table 3.05.8.1. Continued.

Disorder	OMIM	Gene	Biological Function	Product of Conception Alteration
Mental retardation with epilepsy	300423	ATP6AP2	ATPase protein	XR
Mental retardation with language impairment and autistic features	613670	FOXP1	Transcription factor	AD + de novo
Mental retardation, Lubs type	300260	MECP2	Chromatin structure	XD
Mental retardation, Siderius type	300263	PHF8	Zinc finger protein	XR + de novo
Mental retardation, Snyder-Robinson type	309583	SMS	Spermine synthase	XR
Mental retardation, truncal obesity, retinal dystrophy, and micropenis	610156	INPP5E	Ciliary proteins, primary cilia	AR
Mental retardation-hypotonic facies syndrome	309580	ATRX	Chromatin remodeling	XR + de novo
N syndrome	310465	POLA1	DNA polymerase	XR
Neurocutaneous syndrome, Bicknell type	612652	ALDH18A1	Oxidoreductase	AR
Partington syndrome or X-linked mental retardation-36	309510	ARX	Transcription factor Helix-turn-helix domains	XR
Renpenning syndrome	309500	PQBP1	Nuclear polyglutamine-binding protein transcription activation	XR
SeSAME syndrome	612780	KCNJ10	Channelopathy Potassium	AR
Skeletal defects, genital hypoplasia, and mental retardation	612447	ZBTB16	Nuclear zinc finger protein	AD + 99%
Wilson-Turner syndrome or X-linked syndromic mental retardation-6	309585	HDAC8	Histone deacetylase	XD
Woodhouse-Sakati syndrome	241080	DCAF17	Nuclear transmembrane protein	AR
Autosomal dominant mental retardation-1 or Chromosome 2q23.1 deletion syndrome	156200	MBD5	Methylation of DNA	AD + de novo
X-linked mental retardation and microcephaly with pontine and cerebellar hypoplasia	300749	CASK	Serine protein kinase	XR + de novo
X-linked mental retardation syndrome of Stocco dos Santos	300434	SHROOM4	Cytoskeletal architecture	XR
X-linked mental retardation with cerebellar hypoplasia and distinctive facial appearance	300486	OPHN1	Rho-dependent signals	XR

Table 3.05.8.1. Continued.

Disorder	OMIM	Gene	Biological Function	Product of Conception Alteration
X-linked mental retardation with isolated growth hormone deficiency	300123	SOX3	Transcription factor β-Scaffold factors	XR
X-linked mental retardation, FRAXE type	309548	AFF2	Transcriptional activator	XR
X-linked mental retardation-hypotonic facies syndrome	301040	ATRX	Chromatin re-modeling	XR + de novo
X-linked syndromic mental retardation, Christianson type	300243	SLC9A6	Solute carrier	XD
X-linked syndromic mental retardation, Fried type	300630	AP1S2	Trafficking Vesicle formation	XR
X-linked syndromic mental retardation, JARID1C-related	300534	KDM5C	Regulation of transcription and chromatin re-modeling	XR
X-linked syndromic mental retardation, Nascimento-type or X-linked syndromic mental retardation-30	300860	UBE2A	Ubiquitin-conjugating enzyme	XR
X-linked syndromic mental retardation, Raymond type	300799	ZDHHC9	Nuclear zinc finger protein	XR
X-linked syndromic mental retardation, Turner type	300706	HUWE1	E3 ubiquitin protein ligase	XD
X-linked syndromic mental retardation, Wu type	300699	GRIA3	Ligand-gated ion channels	XR
X-linked syndromic mental retardation-1	309510	ARX	Transcription factor Helix-turn-helix domains	XR
X-linked syndromic mental retardation-10	300220	HSD17B10	Mitochondrial protein	XR
X-linked syndromic mental retardation-13	300055	MECP2	Chromatin structure	XR + de novo
X-linked syndromic mental retardation-14	300676	UPF3B	mRNA nuclear export and mRNA surveillance	XR
X-linked syndromic mental retardation-15 or Cabezas type	300354	CUL4B	E3 ubiquitin protein ligase	XR + de novo
X-linked syndromic mental retardation-21	300630	AP1S2	Trafficking Vesicle formation	XR
X-linked syndromic mental retardation-29	300699	GRIA3	Ligand-gated ion channels	XR

Table 3.05.8.1. Continued.

Disorder	OMIM	Gene	Biological Function	Product of Conception Alteration
X-linked syndromic mental retardation-3/8	309500	PQBP1	Nuclear polyglutamine-binding protein transcription activation	XR
X-linked syndromic mental retardation-32	300886	CLIC2	Chloride intracellular channel	XR
Autosomal dominant mental retardation-3	612580	CDH15	Cytoskeleton	AD
Autosomal dominant mental retardation-4	612581	KIRREL3	Transmembrane protein	AD
Autosomal dominant mental retardation-5	612621	SYNGAP1	Ras GTPase activating protein	AD + de novo
Autosomal dominant mental retardation-6	613970	GRIN2B	Channelopathy	AD + de novo
Autosomal dominant mental retardation-7	614104	DYRK1A	Tyrosine phosphorylation regulated kinase	AD + de novo
Autosomal dominant mental retardation-8	614254	GRIN1	Channelopathy	AD + de novo
Autosomal dominant mental retardation-9	614255	KIF1A	Axonal transport of synaptic vesicles	AD + de novo
Autosomal dominant mental retardation-10	614256	CACNG2	Channelopathy calcium	AD + de novo
Autosomal dominant mental retardation-11	614257	EPB41L1	Erythrocyte membrane protein	AD + de novo
Autosomal dominant mental retardation-12	614562	ARID1B	AT rich interactive domain	AD + de novo
Autosomal dominant mental retardation-13	614563	DYNC1H1	Cytoplasmic dynein	AD + de novo
Autosomal dominant mental retardation-14	614607	ARID1A	AT rich interactive domain	AD
Autosomal dominant mental retardation-15	614608	SMARCB1	Regulator of chromatin	AD
Autosomal dominant mental retardation-16	614609	SMARCA4	Regulator of chromatin	AD + de novo
Autosomal dominant mental retardation-17	615009	PACS1	Sorting protein	AD + de novo
Autosomal dominant mental retardation-18	615074	GATAD2B	GATA zinc finger domain	AD + de novo
Autosomal dominant mental retardation-19	615075	CTNNB1	Cytoskeleton	AD + de novo

Table 3.05.8.1. Continued.

Disorder	OMIM	Gene	Biological Function	Product of Conception Alteration
Autosomal dominant mental retardation-2	614113	DOCK8	Dedicator of cytokinesis	AD + de novo
Autosomal dominant mental retardation-20	613443	MEF2C	Transcription enhancer factor 2	AD + de novo
Autosomal recessive nonsyndromic mental retardation-1	249500	PRSS12	Serine proteases	AR
Autosomal recessive nonsyndromic mental retardation-2	607417	CRBN	Proteases	AR
Autosomal recessive nonsyndromic mental retardation-3	608443	CC2D1A	DNA binding and transcriptional repressor	AR
Autosomal recessive nonsyndromic mental retardation-5	611091	NSUN2	RNA methyltransferase	AR
Autosomal recessive nonsyndromic mental retardation-6	611092	GRIK2	Ligand-gated ion channels	AR
Autosomal recessive nonsyndromic mental retardation-7/22	611093	TUSC3	Tumor suppressor	AR
Autosomal recessive nonsyndromic mental retardation-12	611090	ST3GAL3	Galactoside sialyltransferase	AR
Autosomal recessive nonsyndromic mental retardation-13	613192	TRAPPC9	Trafficking protein	AR
Autosomal recessive nonsyndromic mental retardation-14	614020	TECR	Trans-2,3-enoyl-CoA reductase	AR
Autosomal recessive nonsyndromic mental retardation-15	614202	MAN1B1	Mannosidase	AR
Autosomal recessive nonsyndromic mental retardation-17/21	614207	PGAP2	Attachment to proteins	AR
Autosomal recessive nonsyndromic mental retardation-18	614249	MED23	Mediator of RNA polymerase 2	AR
Autosomal recessive nonsyndromic mental retardation-34	614499	CRADD	Signal transduction complex	AR
X-linked mental retardation-1/18	309530	IQSEC2	Cytoskeletal and synaptic organization	XR
X-linked mental retardation-3	309541	HCFC1	Host cell factor	XR
X-linked mental retardation-9/44	309549	FTSJ1	Nucleolar protein	XR
X-linked mental retardation-16/79	300055	MECP2	Chromatin structure	XR
X-linked mental retardation-17/31	300705	HSD17B10	Mitochondrial protein	XR
X-linked mental retardation-19	300844	RPS6KA3	Chromatin structure	XR

Table 3.05.8.1. Continued.

Disorder	OMIM	Gene	Biological Function	Product of Conception Alteration
X-linked mental retardation-21/34	300143	IL1RAPL1	Interleukin receptor accessory protein	XR
X-linked mental retardation-29/32/33/38/43/54/76/87	300419	ARX	Transcription factor Helix-turn-helix domains	XR
X-linked mental retardation-30/47	300558	PAK3	Serine/threonine protein kinase	XR
X-linked mental retardation-41/48	300849	GDI1	Vesicle trafficking	XsD
X-linked mental retardation-45	300498	ZNF81	Transcription factor Zinc finger	XR + de novo
X-linked mental retardation-46	300436	ARHGEF6	Rho-dependent signals	XR
X-linked mental retardation-55	309500	PQBP1	Nuclear polyglutamine-binding protein transcription activation	XR
X-linked mental retardation-58	300210	TSPAN7	Cell surface glycoprotein	XR
X-linked mental retardation-59	300630	AP1S2	Trafficking Vesicle formation	XR
X-linked mental retardation-63/68	300387	ACSL4	Ligase, long chain fatty acid-CoA	XR
X-linked mental retardation-72	300271	RAB39B	Ras oncogene	XR
X-linked mental retardation-88	300852	AGTR2	Angiotensin II receptor	XR + de novo
X-linked mental retardation-89	300848	ZNF41	Transcription factor Zinc finger	XR + de novo
X-linked mental retardation-90	300850	DLG3	Clustering of NMDA receptors	XR
X-linked mental retardation-91	300577	ZDHHC15	Nuclear zinc finger protein	XR + de novo
X-linked mental retardation-92	300851	ZNF674	Transcription factor Zinc finger	XR
X-linked mental retardation-93	300659	BRWD3	Chromatin-modifying function	XR
X-linked mental retardation-94	300699	GRIA3	Ligand-gated ion channels	XR
X-linked mental retardation-95	300716	MAGT1	Cell membrane magnesium cation transporter	XR

Table 3.05.8.1. Continued.

Disorder	OMIM	Gene	Biological Function	Product of Conception Alteration
X-linked mental retardation-96	300802	*SYP*	Integral membrane protein of small synaptic vesicles	XR
X-linked mental retardation-97	300803	*ZNF711*	Transcription factor Zinc finger	XR

3.05.9 Disorders of Psychological Development

Autism: A neurodevelopmental disorder that usually becomes apparent by 3 years of age. This condition is characterized by reduced social interaction and communication, and by restricted and repetitive behavior. By the time the child is 18 months old, most parents of autistic children suspect difficulties in: social interactions, pretend play, and verbal and nonverbal communication. Autism can be associated with another comorbid condition including: Fragile X syndrome, mental retardation, tuberous sclerosis, seizures, and sleep disorders such as insomnia. Around 25-70% of autistic individuals also have criteria for mental retardation. Autism is a common disorder with its prevalence estimated to be from 1-6 cases in every 1,000 children. The true prevalence may be underestimated. In the last 20 years, there has been a large increase in the numbers of autism cases diagnoses. This increase in number of cases is attributable to parent awareness, changes in diagnostic practices and other factors. Autism is 4 times more common among males than females. The causes of autism have been associated with advanced parental age at the time of reproduction, particularly paternal age.

Autism does not correlate well with ethnicity and socioeconomic background. Around 10-15% of cases of autism are associated with a Mendelian condition, such as Fragile X syndrome or phenylketonuria. In autism, there is a large evidence for a genomic component which comes from twin studies, since there are 97% concordances in autism among the monozygotic twins and 23% concordance among the dizygotic twins. Affected individuals exhibit a high concordance for cognitive abnormalities; in fact, 82% for monozygotic pairs are concordant and is only 10% for dizygotic pairs. There are several loci and genes that have been associated with autism. There are at least 17 loci for autism in autosomes, and several genes including mutations in the *CNTNAP2* gene which encodes the contactin associated protein-like 2, or *SLC9A9* gene which encodes the solute carrier (family 9 member A9), a Na(+)/H(+) cation proton antiporter, or the *SHANK2* gene which encodes the SH3 and multiple ankyrin repeat domains 2. There are at least 6 loci for autism which are X-linked forms, and mutations in several genes have been associated with autism including *NLGN3* gene which encodes the neuroligin 3, *NLGN4X* gene which encodes the neuroligin 4, *MECP2* gene which encodes the methyl CpG binding protein 2, *PTCHD1* gene which encodes the patched domain containing protein 1, *RPL10* gene which encodes the ribosomal protein L10, and *TMLHE* gene which encodes the trimethyllysine hydroxylase epsilon.

Rett syndrome: A neurodevelopmental disorder characterized by arrested development between 6 and 18 months of age. This condition occurs almost exclusively in females. Affected individuals manifest regressions of acquired skills, loss of speech, repetitive hand movements, deceleration of the rate of head growth, seizures, mental retardation, and gastrointestinal disorders such as severe constipation. This condition is named after Andreas Rett. Rett syndrome is caused by X-linked dominant *de novo* mutations in the *MECP2* gene which encodes the methyl-CpG-binding protein-2. This nuclear protein binds specifically to methylated DNA. These *de novo* mutations in the *MECP2* gene are usually derived from the paternal copy of the X chromosome. This condition has not been associated with either parental age or consanguineous marriages. The product of conception alterations are defined as classic X-linked dominant syndrome plus 99 % (XD + 99%). There is also a Congenital variant of Rett syndrome caused by autosomal monoallelic mutations in the *FOXG1* gene which encodes the forkhead box G1. There is also an Atypical Rett syndrome caused by X-linked mutations in the *CDKL5* gene which encodes the cyclin-dependent kinase-like 5. During pregnancies with a male fetus, Rett syndrome is lethal, so males are rarely affected except in cases of extra X chromosome or somatic mosaicism. Rett syndrome is estimated to affect 1 in every 10,000-23,000 girls worldwide.

Asperger disorder: A form of high-functioning autism characterized by impairment in social interaction, together with restricted and repetitive patterns of behavior such as, eye-to-eye gaze, facial expression, body postures and gestures. Affected individuals exhibit narrow interests, physical clumsiness and atypical use of language. Asperger disorder differs from other autism spectrum disorders by its relative preservation of cognitive development and linguistic. This condition is named after Hans Asperger. There are at least 3 loci associated with susceptibility to Asperger syndrome in autosomes on chromosome 1q21-q22, 3q and 17p. There are also 2 loci associated with X-linked susceptibility to Asperger syndrome. Asperger disorder is associated with mutations in the *NLGN3* gene or the *NLGN4* gene. Asperger syndrome is estimated to affect 2.7-3.6 in every 1,000 children. This condition affects males at least twice more often than females.

3.05.10 Behavioral and Emotional Disorders with onset usually occurring in Childhood and Adolescence

Gilles de la Tourette syndrome: A disorder characterized by involuntary or semi-voluntary movements and phonic or vocal tics associated with behavioral abnormalities. Affected individuals manifest sudden tics consisting of simple, coordinated, repetitive movements, gestures or utterances that mimic fragments of normal behavior. This condition is named after Georges Albert Édouard Brutus Gilles de la Tourette. In Gilles de la Tourette, there is a large evidence for a genomic component which comes from twin studies. The concordance rate in monozygotic twins is 89-94% for the disorder. There are several loci associated with susceptibility to Gilles de la Tourette. This condition has been associated with mutations in the *HDC* gene which encodes the histidine decarboxylase, and the *SLITRK1* gene which encodes the SLIT and NTRK-like family (member 1). Gilles de la Tourette is a common disorder affecting 1% of people worldwide.

3.06.0 Subchapter VI

Diseases of the Nervous System

Table 3.06.0 Diseases of the Nervous System

3.06.00-11	Diseases of the Nervous System	
	3.06.01	Inflammatory Diseases of the Central Nervous System
	3.06.02	Systemic Atrophies Primarily Affecting the Central Nervous System
	3.06.03	Extrapyramidal and Movement Disorders
	3.06.04	Other Degenerative Diseases of the Nervous System
	3.06.05	Demyelinating Diseases of the Central Nervous System
	3.06.06	Episodic and Paroxysmal Disorders
	3.06.07	Nerve, Nerve Root and Plexus Disorders
	3.06.08	Polyneuropathies and other Disorders of the Peripheral Nervous System
	3.06.09	Diseases of Myoneural Junction and Muscle
	3.06.10	Cerebral Palsy and other Paralytic Syndromes
	3.06.11	Other Disorders of the Nervous System
3.06.12	Disease Prevention for Disassortative Mating in Ethnic Groups	

3.06.00-11 Diseases of the Nervous System

3.06.01 Inflammatory Diseases of the Central Nervous System

NOMID disease (Neonatal Onset Multisystem Inflammatory Disease) or CINCA syndrome (Chronic Infantile Neurologic Cutaneous and Articular): A disorder usually appearing during the neonatal period characterized by periodic fever syndrome which causes uncontrolled inflammation in multiple parts of the body. NOMID disease patients manifest severe arthritis, skin rashes, and chronic meningitis leading to neurologic damage. This condition is one of the cryopyrin-associated periodic syndromes. NOMID disease is associated with either inherited or *de novo* autosomal dominant mutations

in the *NLRP3* gene. This gene encodes a pyrin-like protein which helps control inflammation. Mutations in this gene also cause Muckle-Wells syndrome and Familial cold urticaria. NOMID disease is estimated to affect less than 1 in 1,000,000 live births.

Familial advanced sleep-phase disorder: A disorder characterized by very early sleep onset and offset. Most of these patients feel very sleepy and go to bed early in the evening and wake up very early in the morning. This genomic condition affects both men and women equally. This condition is linked to the body's circadian rhythms. This condition does not necessarily affect the health, but severely impacts social life. Familial advanced sleep-phase syndrome can be caused by autosomal dominant mutations in either the *PER2* gene which encodes the period circadian clock protein 2, or the *CSNK1D* gene which encodes the delta casein kinase 1, a serine/threonine-specific protein kinase. The real prevalence for this condition is unknown; most people suffering this condition don't seek medical help except when it starts to severely impact their social life.

3.06.02 Systemic Atrophies Primarily Affecting the Central Nervous System

Huntington's disease or Huntington's chorea: A disorder characterized by chorea, cognitive decline, incoordination, dystonia and behavioral difficulties. This progressive neurodegenerative genomic disorder is caused by selective neural cell loss and atrophy in the caudate and putamen. Adult-onset Huntington disease usually appears in the third or fourth decade of life. Affected individuals often live 10-15 years after the onset of signs and symptoms. Early symptoms could appear anywhere from 5-80 years of age. Huntington disease is caused by autosomal dominant CAG expansion in the *HTT* gene which encodes the huntingtin. This protein appears to play an important role in caudate and putamen neurons. The length of this CAG amplification reaches a certain threshold; it produces an altered form of the protein, called mutant Huntingtin protein. The expansion of CAG repeats creates a polyglutamine tract which produces an adverse gain-of-function mutation, due to increased aggregation. This condition is an almost fully penetrant disorder when it involves alleles with 41 or more CAG repeats, but there is a reduced penetrance form that involves only 36-39 CAG repeats. The *HTT* gene is located on the subtelomeric short arm of chromosome 4 at 4p16.3. Huntington disease can be inherited from the mother or the father, but the main phenomenon of anticipation in this case is due to CAG expansion that occurs during spermatogenesis in the paternal allele. The product of conception alteration in this condition is AD + Ant. The *de novo* anticipation event can also be classified as being reversible or not, or a combination of both; the most clear example is Huntington disease in which there are trinucleotide expansion in paternal allele, and contraction in the maternal. At least one allele that matches with the trinucleotide expansion in Huntington disease has an origin in Europe from 4,700-10,000 years ago. The pathogenic allele in Huntington disease is an example of DNA alteration during early migration trait, founder effect, parental advancing age, subtelomeric condition, and gain-of-function mutation. Huntington disease is estimated to affect 30-70 in every 1,000,000 individuals of European ancestry. The disorder seems to be less common in some other groups, affecting 0.6 in every 1,000,000 non Afrikaner South Africans, 3.8 in every 1,000,000 Japanese, 5 in every 1,000,000 Finnish, and 15 in every 1,000,000 African Americans. The disorder seems to be more common in several regions of the world including Scotland, Wales, Sweden, Tasmania and in the Lake Maracaibo region of Venezuela. In this region of Venezuela, particularly in the State of Zulia, Huntington disease is estimated to affect 7,000 in every 1,000,000 individuals. There is a similar disorder refered to as Huntington like disease and there are several underlying causes of this disorder with at least 4 autosomal genomic loci. Huntington like disease accounts for 1% of people with Huntington symptomatology. Huntington like disease-1 is caused by autosomal dominant mutations in the *PRNP* gene;; the same gene also is responsible for Creutzfeldt–Jakob disease; Huntington like disease-2 is caused by autosomal dominant mutations in the *JPH3* gene which encodes the junctophilin 3. Type 2 is the second most common and occurs in high prevalence among non-Afrikaner South Africans;; Huntington like disease-3 is inherited in an autosomal recessive pattern, but the gene involved has not been characterized;; and Huntington like disease-4 is caused by autosomal dominant mutations in the *TBP* gene which encodes the TATA box binding protein. Type 4 is the most common cause of Huntington like disease. Mutations in the *TBP* gene also are responsible for Spinocerebellar ataxia type 17. Huntington like disease type 2 and 4 are also often associated with anticipation events.

3.06.02.1 Hereditary Ataxia

Ataxias: Are a group of disorders characterized by signs and symptoms of muscle movement incoordination. Ataxia implies dysfunction of the nervous system that coordinates movement, such as cerebellum or spinal cord degeneration. Ataxia can be caused by non-genomic factors such as: severe head injury due to trauma or infection, or due to disruption in the supply of blood to the brain including stroke, brain hemorrhage and transient ischemic attack. Other medical conditions that also are associated with ataxia include prolonged long-term alcohol abuse, cerebral palsy, epilepsy, autoimmune conditions such as multiple sclerosis and lupus, cancer, medication and others conditions. There are two main forms of ataxia described in other subchapters: ataxia caused by underlying inborn errors of metabolism such as Wilson Disease, Niemann Pick disease and Abetalipoproteinemia, and ataxia caused by progressive degenerative conditions such as chronic ethanol abuse and genomics factors described subsequently. Ataxias can also be classified according to the mode of inheritance or *de novo* occurrence.

Dentatorubral-pallidoluysian atrophy or Haw River Syndrome or Naito-Oyanagi disease: A disorder characterized by myoclonic epilepsy, dementia, ataxia and choreoathetosis. Patients with this condition start exhibiting symptoms in their twenties, and death frequently occurs in their forties. Dentatorubral-pallidoluysian atrophy is a spinocerebellar degeneration caused by an autosomal dominant expansion of a CAG repeat encoding a polyglutamine tract in the *ATN1* gene, which encodes the atrophin 1 protein. This condition is associated with anticipation due to CAG expansion. This expansion is usually associated with paternal transmission. Dentatorubral-pallidoluysian atrophy occurs almost exclusively among Japanese due to founder effect. Dentatorubral-pallidoluysian atrophy is estimated to affect 1 in every 208,000 individuals in Japan.

Infantile-onset ascending hereditary spastic paralysis: An infantile onset disorder characterized by ascending spastic paralysis. Patients with this condition start to exhibit symptoms during the first 2 years of life manifesting spastic paraplegia. This condition leads to upper limb paraplegia within the next few years. This disease progresses to tetraplegia, anarthria, dysphagia and slow eye movements. Infantile-onset ascending hereditary spastic paralysis is caused by autosomal recessive mutations in the *ALS2* gene which encodes the alsin, a guanine nucleotide exchange factor for the small GTPase RAB5. Mutations in the *ALS2* gene cause Juvenile primary lateral sclerosis or juvenile amyotrophic lateral sclerosis. Infantile-onset ascending hereditary spastic paralysis is estimated to affect less than 1 in every 1,000,000 individuals.

Mitochondrial DNA depletion syndrome-1 or MNGIE syndrome (Mitochondrial NeuroGastroIntestinal Encephalomyopathy syndrome): This disease is discussed under endocrine, nutritional and metabolic diseases in the subchapter "Combined Oxidative Phosphorylation Deficiency" (3.04.19.1).

Friedreich's ataxia: A disorder characterized by a large variation of symptoms from gait disturbance to speech problems, and also heart disease and diabetes. This condition is caused by progressive damage to the nervous system, particularly in the sensory neurons of the spinal cord. These neurons connect with the cerebellum, and direct the muscle movement of the arms and legs. This condition is named after Nikolaus Friedreich. Friedreich's ataxia is caused by autosomal recessive mutations in the *FXN* gene which encodes the frataxin. This protein is essential for appropriate functioning of mitochondria. The absence of frataxin in these nerve and muscle cells makes them sensitive to iron build-up, resulting in free radical toxicity. Friedreich's ataxia is mostly caused by expansion of GAA triplet repeats in the first intron in the *FXN* gene. This trinucleotide repeats expansion leads to anticipation. The product of conception alterations are defined as autosomal recessive syndrome plus anticipation (AR + Ant). This expanded allele originated between 9,000 and 24,000 years ago. This expansion is a form of loss-of-function mutation, which leads to gene silencing due to decrease in the *FXN* gene transcription. This expansion changes the heterochromatin structure. This mutation or allele expansion does not produce an abnormal frataxin protein. In Caucasian Americans, the carrier frequency for Friedreich's ataxia trait is estimated to be 1 in 110 which means the frequency of this condition is about 1 in 50,000 individuals or $1/(220)^2$, making it the most prevalent inherited ataxia in this group. There are 2 regions with a higher prevalence worldwide for Friedreich's ataxia: the neighboring villages of the Paphos district (Cyprus) and the Rimouski area (Province of Quebec-Canada). In Italians, the frequency of this condition is about 1 in 22,000-25,000 individuals. This condition seems to be less prevalent among sub-Saharan Africans, Finnish and southeastern Asians.

Marinesco-Sjögren syndrome: A disorder characterized by a combination of cerebellar ataxia with dysarthria, mild to severe mental retardation, and congenital cataracts evident since early childhood. Affected individuals also exhibit skeletal deformities due to muscle weakness, thin brittle fingernails, sparse hair, strabismus, nystagmus and hypergonadotropic hypogonadism. This condition is not associated with a reduction in life expectancy. This condition is named after Gheorghe Marinescu and Torsten Sjögren. Marinesco-Sjögren syndrome is caused by autosomal recessive mutations in the *SIL1* gene which encodes the SIL1 homolog, an endoplasmic reticulum chaperone. This protein interacts with ATPase domains. Marinesco-Sjögren syndrome is estimated to affect 1-9 in every 1,000,000 individuals. There is a higher carrier frequency for Marinesco-Sjögren syndrome trait in Romani families of north of Belgrade (Serbia).

Gillespie syndrome: A disorder characterized by cerebellar ataxia, aniridia and mental retardation. Gillespie syndrome is caused by autosomal recessive mutations in the *PAX6* gene which encodes the paired box 6, a transcriptional regulator. This protein is involved in oculogenesis and other developmental processes. Gillespie syndrome is estimated to affect less than 1 in every 1,000,000 individuals.

NARP syndrome (Neuropathy, Ataxia, and Retinitis Pigmentosa): A disorder characterized by early childhood manifestations of learning difficulties, developmental delay and ataxia, and young adult onset sensory-motor neuropathy, muscle weakness and vision loss. Affected individuals also exhibit a variable combination of dementia, seizures, hearing loss and cardiac conduction defects. This maternal transmitted condition is caused by mutation in the mitochondrial genome affecting the *MT-ATP6* gene. This gene encodes the subunit 6 of H (+)-ATPase, which is a subunit of ATP synthase. Most individuals with NARP syndrome have an 8993T-G heteroplasmic mutation in the MT-ATP6 gene. This condition appears when the heteroplasmic mutation affects 70-90% of their mitochondrial DNA. When the same heteroplasmic mutation affects more than 90% of their mitochondrial DNA, this causes Leigh syndrome. NARP syndrome is estimated to affect 1 in every 12,000 individuals.

Fragile X-associated tremor/ataxia syndrome: A disorder appearing during the sixth decade of life characterized by progressive multiple system atrophy with intentional tremor, Parkinsonism, dysautonomia, peripheral neuropathy, cognitive decline and dementia. Fragile X tremor/ataxia syndrome is caused by trinucleotide repeats expansion ranging from 55-200 repeats in the *FMR1* gene which encodes the fragile X mental retardation protein 1. This form of trinucleotide repeats expansion is also referred to as premutation. The same gene alteration results in fragile X mental retardation syndrome when the repeat expansions are greater than 200 repeats. This X-linked dominant condition has at least 33% penetrance in male carriers aged 50 years and over, whereas in female carriers the penetrance is approximately 5-10%. Fragile X-

associated tremor/ataxia syndrome is estimated to affect 1 in every 4,000-6,000 individuals.

Ataxia telangiectasia or Boder-Sedgwick syndrome or Louis–Bar syndrome: A disorder characterized by cerebellar ataxia, telangiectasia of small dilated blood vessels, and a predisposition to infection and malignancy, particularly leukemia and Hodgkin lymphoma. Most of the neurological symptoms in these patients appear at the toddler stage with a "drunken walk". Ataxia telangiectasia is caused by autosomal recessive mutations in the *ATM* gene which encodes the ataxia telangiectasia mutated protein, a cell cycle checkpoint kinase. This protein is responsible for DNA repair and/or cell cycle control. This condition is defined as AD + AD = AR, since both parents provide the faulty copy to descendants. Patients that are heterozygous for this condition have a higher prevalence of certain cancers, such as breast cancer and stomach cancer. In Europeans, Middle Easterns, and Western Asians at least one pathogenic allele appears 50,000 years ago. On average, the carrier frequency for Ataxia telangiectasia trait is estimated to be 1 in 100-158 individuals worldwide, which means the frequency of this condition is about 1 in 40,000-100,000 individuals or $1/(200)^2$ to $1/(316)^2$. There are also similar conditions named Ataxia-telangiectasia like disorder which has similar features in the progressive cerebellar ataxia, hypersensitivity to ionizing radiation and genomic instability, but differs in that there is an absence of telangiectasia and immune deficiency. Ataxia-telangiectasia like disorder is caused by autosomal recessive mutations in the *MRE11A* gene which encodes the MRE11 meiotic recombination protein 11 (homolog A). This protein has a hyperrecombinational phenotype and confers profound sensitivity to double-strand break damage.

Ataxia and oculomotor apraxia: A disorder characterized by exhibiting resemblance to the neurologic features of ataxia-telangiectasia with oculomotor apraxia, ataxia and choreoathetosis, but without the extraneurologic symptoms. Affected individuals also display peripheral axonal neuropathy and hypoalbuminemia. There are several underlying causes of this disorder with at least 2 autosomal genomic loci: Ataxia-oculomotor apraxia-1 is caused by autosomal recessive mutations in the *APTX* gene which encodes the aprataxin. This protein may play a role in single-stranded DNA repair through its nucleotide-binding activity and its diadenosine polyphosphate hydrolase activity;; and Ataxia-oculomotor apraxia-2 is caused by autosomal recessive mutations in the *SETX* gene which encodes the senataxin. In Portugal, the carrier frequency for Ataxia with oculomotor apraxia-1 trait is estimated to be 1 in 67 individuals, which means the frequency of this condition is about 1 in 18,000 individuals or $1/(134)^2$. Type 2 is more prevalent among French-Canadians than among any other groups of people worldwide.

Hereditary spastic paraplegias or Strumpell-Lorrain disease: A group of disorders characterized by progressive stiffness and spasticity in the lower extremity with onset in the 2nd to the 3rd decade of life. Sometimes, patients with this condition also exhibit other associations such as optic neuropathy, retinopathy, deafness, ataxia, extrapyramidal disturbance, dementia, ichthyosis and mental retardation. Familial spastic paraplegias are caused by defects of cellular transport, which affects mainly long nerves due to degeneration of the corticospinal tracts. Hereditary spastic paraplegia is named af-

ter Adolph Strümpell and Maurice Lorrain. There are at least 53 different spastic paraplegics' loci. These conditions are caused mainly by autosomal recessive inheritance. There are also 10-30% of cases which follow a dominant pattern of inheritance, and there are also examples of X-linked conditions. Combined hereditary spastic paraplegias is estimated to affect 2-8 in every 100,000 people worldwide. Table 3.06.02.1 lists the different types of Hereditary spastic paraplegias, along with their OMIM, biological functions, associated genes, and product of conception alterations.

MASA syndrome (Mental retardation, Aphasia, Shuffling gait, and Adducted thumbs) or spastic paraplegia-1: A disorder caused by X-linked recessive mutations in the *L1CAM* gene which encodes the L1 cell adhesion molecule. Mutations in the same gene are also associated with X-linked aqueductal stenosis or hydrocephalus. X-linked spastic paraplegia-2 is caused by X-linked mutations in the *PLP1* gene which encodes the myelin proteolipid protein. This transmembrane protein plays a role in the compaction, stabilization, and maintenance of myelin sheaths, as well as in oligodendrocyte development and axonal survival. There are also two other forms of X-linked spastic paraplegia type 16 and 34, for which the gene has not been characterized. These symptomatic conditions affect males, since females are usually asymptomatic carriers. MASA syndrome is estimated to affect 1 in every 25,000-60,000 males.

Spinocerebellar ataxia: A group of disorders characterized by progressive cerebellar ataxias with variable involvement of the brainstem and spinal cord. There are at least 31 different loci associated with these conditions. These disorders can be autosomal dominant, autosomal recessive, or

X-linked. A large subgroup of spinocerebellar ataxia forms part of a group of polyglutamine diseases which are caused when a disease-associated protein contains a polyglutamine tract beyond a certain threshold. In most cases these polyglutamine disorders are autosomal dominant with gender specific CAG amplification origin including spinocerebellar ataxia type 1, type 2, type 3, type 5, type 6, type 7, type 10 and type 17. Type 8 is caused by CTG expansion, which is the reverse strand of the trinucleotide CAG. The product of conception alteration in this condition is AD + Ant. Spinocerebellar ataxias can be inherited from the mother or the father, but the main phenomenon of anticipation is either of maternal or paternal origin. For example, Spinocerebellar ataxia type 8 is caused by CTG amplification in the *ATXN8OS* gene, which is of maternal origin appearing during ovogenesis, Spinocerebellar ataxia type 2 is either maternal or paternal, Spinocerebellar ataxia type 1 and type 7 are mostly paternal, and Spinocerebellar ataxia type 3, type 10 and type 17 are of paternal origin and appear during spermatogenesis. Some of these conditions have allele expansion that is region specific. For example, Spinocerebellar ataxia type 2 mainly affects individuals from Cuba, Spinocerebellar ataxia type 3 or Machado-Joseph disease affects individuals from the Azores Islands, Portugal, and Spinocerebellar ataxia type 10 affects individuals from Mexico. Spinocerebellar ataxia type 5, type 8 and type 9, Spinocerebellar ataxia with axonal neuropathy, Spinocerebellar ataxia with epilepsy and "Sensory ataxic neuropathy, dysarthria, and ophthalmoparesis" are caused by an autosomal recessive pattern of inheritance. Autosomal dominant Spinocerebellar ataxias is estimated to affect 0.8-3.5 in every 100,000 individuals.

Machado–Joseph disease or Spinocerebellar ataxia type 3: A disorder characterized by ataxia, ophthalmoparesis, pyramidal signs such as spasticity and hyperreflexia, and extrapyramidal signs including dystonia and other movement disorders. This condition is named after William Machado and Antone Joseph. Machado–Joseph disease is caused by autosomal dominant mutations in the *ATXN3* gene which encodes the ataxin-3. These mutations lead to CAG expansion and consequently, the creation of polyglutamine tracts. This condition is associated with anticipation due to unidirectional CAG expansion. This condition is associated more with paternal than maternal transmission. Machado–Joseph disease can be inherited from the mother or the father, but the main phenomenon of anticipation in this case is CAG expansion that occurs during spermatogenesis in the paternal allele. The product of conception alteration in this condition is AD + Ant. This condition results in degeneration of cells in the rhombencephalon. This degeneration leads to symptoms, such as clumsiness and rigidity. Machado–Joseph disease is the most common cause of autosomal-dominant ataxia. Machado–Joseph disease is estimated to affect 1-2 in every 100,000 individuals worldwide, with significant ethnic and geographical variations. The highest prevalence has been found in the Azores Islands (Flores and Sao Miguel Island with a prevalence of 1 in 140 and 1 in 239 individuals, respectively), and the central Honshu island of Japan. Among immigrants of Portuguese ancestry in New England (the northeastern corner of the United States), it is estimated to affect 1 in every 4,000 individuals. There are intermediate prevalence rates in other parts of Portugal, Japan, Germany, the Netherlands and China. There is a lower prevalence of this condition in Australia, India and the rest of North America.

Episodic ataxia: A disorder appearing in infancy characterized by sporadic attacks of incoordination and imbalance, often associated with progressive ataxia. Ataxia can be triggered by stress, startle or heavy exertion. Episodic ataxia patients also experience migraine, familial hemiplegic migraine, progressive cerebellar degenerative disorders and seizures. These conditions are caused by misfiring of Purkinje cells in the cerebellum. There are several underlying causes of this disorder with at least 7 loci associated with 4 known genes including the *KCNA1* gene, the *CACNA1A* gene, the *CACNB4* gene and the *SLC1A3* gene. Episodic ataxia follows an autosomal dominant pattern of inheritance. Episodic ataxia-1 is caused by monoallelic or autosomal dominant mutations in the *KCNA1* gene which encodes the potassium channel;; Episodic ataxia-2 is caused by mutations in the *CACNA1A* gene which encodes the voltage-gated calcium channel (subunit alpha 1A). Type 2 is the most common form;; Episodic ataxia-3 has been mapped, but the gene has not been characterized;; Episodic ataxia-4 has been mapped, but the gene has not been characterized;; Episodic ataxia-5 is caused by mutations in the *CACNB4* gene which encodes the calcium channel voltage-dependent (subunit beta 4);; Episodic ataxia-6 is caused by mutations in the *SLC1A3* gene which encodes the solute carrier (family 1 member A3), a glial high affinity glutamate transporter;; and Episodic ataxia-7 has been mapped, but the gene has not been characterized. A similar condition referred to as Isolated myokymia-2 is associated with mutations in the *KCNQ2* gene which encodes the voltage-gated potassium channel (KQT-like subfamily member 2). Episodic ataxia is estimated to affect less than 1 in every 100,000 individuals.

3.06.02.2 Spinal Muscular Atrophy and Related Syndromes

Spinal muscular atrophy: A disorder characterized by incurable general muscle wasting and mobility impairment of variable onset. The first muscles affected in these patients are the lower extremities, then the upper extremities, spine, neck and, in more severe cases, pulmonary and mastication muscles. These conditions affect earlier and to a greater degree the proximal muscles than distal. There are 4 onset forms: infantile, intermediate, juvenile and Adult. Infantile-onset or Werdnig–Hoffmann disease is the most severe form. This condition starts manifesting in the first 6 months of life, with most of these "floppy babies" dying by 2 years of age; the intermediate-onset or Dubowitz disease starts to manifest within the first 6–18 months of life; juvenile-onset or Kugelberg–Welander disease starts to manifest after 18 months of age; and adult-onset manifests after the 3rd decade of life. Spinal muscular atrophy is caused by biallelic mutations in pericentromeric genes. There are two nearly identical genes at location 5q13: a telomeric copy *SMN1* and a centromeric copy *SMN2*. Spinal muscular atrophy is caused by autosomal recessive mutations in the either *SMN1* gene or the *SMN2* gene. These genes encode SMN proteins which are involved in survival of motor neurons. Patients with this condition do not express this protein, and with the absence of this protein, neuronal cells die of the anterior horn of spinal cord leading to system-wide muscle atrophy. The severity of this condition depends of how many *SMN1* or *SMN2* genes are remaining to comply with the survival of motor neurons functions. In 96-98% of Spinal muscular atrophy cases the inheritance is autosomal recessive. The remaining cases appear to be caused by *de novo* mutation events. The product of conception alterations are defined as occasional *de novo* autosomal recessive syndrome, or autosomal recessive syndrome plus known *de novo* alterations (AR + KDN). Spinal muscular atrophy is the most common genomic cause of infant death, particularly affecting Caucasian children. In the United States, the estimated carrier prevalence varies among all the ethnic groups. In Caucasian Americans, the carrier frequency for Spinal muscular atrophy trait is estimated to be 1 in 46 individuals, which means the frequency of this condition is about 1 in 8,400 individuals or $1/ (92)^2$. In East Asians Americans, the carrier frequency for Spinal muscular atrophy trait is estimated to be 1 in 56 individuals, which means the frequency of this condition is about 1 in 12,500 individuals or $1/ (112)^2$. In African Americans, the carrier frequency for Spinal muscular atrophy trait is estimated to be 1 in 91 individuals, which means the frequency of this condition is about 1 in 33,100 individuals or $1/ (182)^2$. In Hispanic Americans, the carrier frequency for Spinal muscular atrophy trait is estimated to be 1 in 125 individuals, which means the frequency of this condition is about 1 in 62,500 individuals or $1/ (250)^2$.

Spinal and bulbar muscular atrophy of Kennedy: A disorder appearing during the third to fifth decade of life characterized by a slow-progressing limb and bulbar muscle weakness with cramps and fasciculation, muscle atrophy, erectile dysfunction and gynecomastia. Males have symptoms of this condition, while females are asymptomatic carriers. Spinal and bulbar muscular atrophy of Kennedy is caused by X-linked recessive mutations in the *AR* gene which encodes the androgen receptor. These mutations lead to CAG expansion and consequently polyglutamine tract. The CAG expansion particularly occurs more commonly during male meiosis (spermatogenesis) than in female meiosis. This condition is named after W. R. Kennedy. Spinal and bulbar muscular atrophy of Kennedy is estimated to affect 1 in every 150,000 live male births worldwide. This condition is more common among Scandinavian people due to founder effect. There is a higher frequency for this condition in the Vasa region of western Finland. The prevalence in this area is 13 cases in 85,000 live male births.

Scapuloperoneal syndrome, myopathic type: A disorder characterized by proximal myopathy, with clinical features of foot drop and proximal arm weakness with normal motor neurons and peripheral nerves. There are several underlying causes of this disorder with at least 1 autosomal genomic locus and 1 X-linked genomic locus: Scapuloperoneal syndrome, myopathic type is caused by autosomal dominant mutations in the *MYH7* gene which encodes the beta-myosin heavy chain. Scapuloperoneal syndrome, myopathic type is caused by X-linked dominant mutations in the *FHL1* gene which encodes the four and a half LIM domains 1. Males are affected at an earlier age of onset and are more severely affected than females. Overall lifespan is not affected significantly. Scapuloperoneal syndrome, myopathic type is estimated to affect 1 in every 25,000 individuals.

Amyotrophic lateral sclerosis or Lou Gehrig's disease: A disorder appearing during the fourth to sixth decade of life characterized by the degeneration leading to the death of motor neurons in the ventral horn of the spinal cord, brain and brainstem. Amyotrophic lateral sclerosis patients typically begin with asymmetric involvement of the muscles. This condition exhibits rapidly progressing weakness, muscle atrophy and fasciculation, dysarthria and spasticity. This progressive condition leads to respiratory compromise and fatal pneumonia after 2 to 5 years from the onset of signs and symptoms. Amyotrophic lateral sclerosis is estimated to affect 6 cases in every 100,000 individuals worldwide. The highest prevalence of this condition has been described in the West Pacific; particularly in Guam which is inhabited by the Chamorro people. In the latter group, Amyotrophic lateral sclerosis is estimated to affect 143 in every 100,000 individuals. This condition is also largely prevalent in West Papua New Guinea and the Kii Peninsula of Japan. There is a slightly higher prevalence in males than in females. Genomics factors account for approximately 5-20% of all the Amyotrophic lateral sclerosis cases. There are several examples of this condition that were believed to be apparently sporadic, but in some patients from Sweden and Finland were caused by autosomal recessive inheritance. There are several underlying causes of this disorder with at least 17 autosomal genomic loci and 1 X-linked genomic locus: Amyotrophic lateral sclerosis-1 is associated with mutations in the *SOD1* gene which encodes the copper/zinc superoxide dismutase 1. This enzyme is responsible for scavenging free radicals. Sometimes, Amyotrophic lateral sclerosis is caused by autosomal recessive inheritance, primarily caused by mutations in the *SOD1* gene, but is more frequently caused due to autosomal dominant mutations in the *SOD1* gene. In Amerindians this pathogenic allele appears 400-600 years ago;; Amyotrophic lateral sclerosis-6 is caused by mutations in the *FUS* gene which encodes the fused in sarcoma, a RNA-binding protein;; Amyotrophic lateral sclerosis-8 is caused by mutations in the *VAPB* gene which encodes the VAMP (vesicle-associated membrane protein)-associated protein B and C;; Amyotrophic lateral sclerosis-9 is caused by mutations in the *ANG* gene which encodes the angi-

ogenin ribonuclease;; Amyotrophic lateral sclerosis-10 is caused by mutations in the *TARDBP* gene which encodes the TAR DNA binding protein;; Amyotrophic lateral sclerosis-11 is caused by mutations in the *FIG4* gene which encodes the FIG4 homolog protein (SAC1 lipid phosphatase domain containing protein);; Amyotrophic lateral sclerosis-12 is caused by mutations in the *OPTN* gene which encodes the optineurin;; Amyotrophic lateral sclerosis-14 is caused by mutations in the *VCP* gene which encodes the valosin containing protein;; Amyotrophic lateral sclerosis-15 is caused by X-linked mutations in the *UBQLN2* gene which encodes the ubiquilin 2;; Amyotrophic lateral sclerosis-17 is caused by mutations in the *CHMP2B* gene which encodes the charged multivesicular body protein 2B. This protein is involved in the sorting of cell-surface receptors into multivesicular endosomes;; and Amyotrophic lateral sclerosis-18 is caused by mutations in the *PFN1* gene which encodes the profilin 1. "Amyotrophic lateral sclerosis and/or frontotemporal dementia" is caused by mutations in the *C9orf72* gene which encode the chromosome 9 open reading frame 72. Amyotrophic lateral sclerosis-3 has been mapped, but the gene has not been characterized;; and Amyotrophic lateral sclerosis-7 has been mapped, but the gene has not been characterized. The phenotype for Amyotrophic lateral sclerosis-13 is contributed by intermediate-length polyglutamine repeat expansions in the *ATXN2* gene which encodes the ataxin 2. Susceptibility to Amyotrophic lateral sclerosis has been associated with mutations in other genes, including deletions or insertions in the *NEFH* gene which encodes the heavy neurofilament subunit; deletions in the *PRPH* gene which encodes the peripherin; and mutations in the *DCTN1* gene which encodes the dynactin. There are some forms of Amyotrophic lateral sclerosis which show juvenile onset including: Amyotrophic lateral sclerosis-2, caused by mutations in the *ALS2* gene which encodes the alsin;; Amyotrophic lateral sclerosis-4, caused by mutations in the *SETX* gene which encodes the senataxin;; and Amyotrophic lateral sclerosis-16, caused by mutations in the *SIGMAR1* gene which encodes the sigma non-opioid intracellular receptor 1. Amyotrophic lateral sclerosis-5 is a juvenile-onset form which has been mapped, but the gene has not been characterized.

Brown-Vialetto-Van Laere syndrome: A disorder characterized by a variety of cranial nerve palsies. In the second decade of life, affected individuals manifest sensorineural hearing loss and motor palsies of the 7th and 9th-12th cranial nerves. Brown-Vialetto-Van Laere syndrome resembles amyotrophic lateral sclerosis. There are several underlying causes of this disorder with at least 2 autosomal genomic loci: Brown-Vialetto-Van Laere syndrome-1 can be caused by biallelic mutations in the *SLC52A3* gene which encodes the solute carrier (family 52 member A3), a riboflavin transporter;; and Brown-Vialetto-Van Laere syndrome-2 can be caused by biallelic mutations in the *SLC52A2* gene which encodes the solute carrier (family 52 member A2), a riboflavin transporter.

Distal hereditary motor neuronopathies: Are a group of disorders starting to manifest symptoms between 20-40 years of age characterized by gradual progression of symmetrical distal atrophy and weakness of the extremities associated with hyporeflexia. These neuronopathies are caused by anterior horn cell degeneration. The main pathologic process resides in the neuron cell body and not in the axons. There is a large genomic heterogeneity of distal hereditary motor neuronopathies with 3 inheritance forms including autosomal dominant, autosomal recessive

and X-linked. According to age at onset, mode of inheritance, and presence of additional features these conditions can be classified into 7 phenotypic subtypes. Those that show autosomal dominant inheritance include Distal hereditary motor neuronopathy distal type I and II, and are characterized by juvenile and adult onset, respectively; Distal hereditary motor neuronopathy type V is characterized by upper limb involvement; and Distal hereditary motor neuronopathy is characterized by vocal cord paralysis. There are several forms of these condition in which the gene has not yet been characterized; Distal hereditary motor neuronopathy-2A can be caused by monoallelic mutations in the *HSPB8* gene which encodes the heat shock 22kDa protein 8;; Distal hereditary motor neuronopathy-2B can be caused by monoallelic mutations in the *HSPB1* gene which encodes the heat shock 27kDa protein 1;; Distal hereditary motor neuronopathy-2C can be caused by monoallelic mutations in the *HSPB3* gene which encodes the heat shock 27kDa protein 3;; Distal hereditary motor neuronopathy-5A can be caused by monoallelic mutations in the *GARS* gene which encodes the glycyl-tRNA synthetase;; Distal hereditary motor neuronopathy-5B can be caused by monoallelic mutations in the *BSCL2* gene which encodes the Berardinelli-Seip congenital lipodystrophy protein 2 (seipin);; Distal hereditary motor neuronopathy-7A can be caused by monoallelic mutations in the *SLC5A7* gene which encodes the solute carrier (family 5 member A7), a sodium/choline cotransporter;; and Distal hereditary motor neuronopathy-7B can be caused by monoallelic mutations in the *DCTN1* gene which encodes the dynactin 1. Those that show autosomal recessive inheritance include Distal hereditary motor neuronopathy type III, IV, and VI also known as Autosomal recessive distal spinal muscular atrophy. Type III and IV have juvenile onset and differ only by a less severe involvement in type III. Autosomal recessive distal spinal muscular atrophy-1 or Distal hereditary motor neuronopathy-6 or Spinal muscular atrophy with respiratory distress-1 can be caused by biallelic mutations in the *IGHMBP2* gene which encodes the immunoglobulin mu (binding protein 2);; Autosomal recessive distal spinal muscular atrophy-2 has been mapped, but the gene has not been characterized;; Autosomal recessive distal spinal muscular atrophy-3 has been mapped, but the gene has not been characterized;; Autosomal recessive distal spinal

muscular atrophy-4 can be caused by biallelic mutations in the *PLEKHG5* gene which encodes the pleckstrin homology domain containing protein (family G member 5);; and Autosomal recessive distal spinal muscular atrophy-5 can be caused by biallelic mutations in the *DNAJB2* gene which encodes the DnaJ homolog (subfamily B member 2). X-linked distal spinal muscular atrophy-3 is caused by X-linked mutations in the *ATP7A* gene which encodes the copper transporting ATPase (alpha polypeptide). The prevalence of Distal hereditary motor neuronopathy is unknown.

Table 3.06.02.1 Recognizable Systemic Atrophies Primarily Affecting the Central Nervous System

Disorder	OMIM	Gene	Biological Function	Product of Conception Alteration
Neonatal onset multisystem inflammatory disease or NOMID or CINCA syndrome	607115	*NLRP3*	Nucleotide Oligomerization Domain receptors	AD + de novo
Familial Advanced sleep phase syndrome	604348	*PER2*	Circadian pattern	AD
Huntington disease	143100	*HTT*	Huntingtin	AD + Ant
Huntington disease-like-1	603218	*PRNP*	Prion protein	AD + de novo
Huntington disease-like-2	606438	*JPH3*	Junctional complexes	AD + Ant
Huntington disease-like-4 or Spinocerebellar ataxia-17	607136	*TBP*	TATA box binding protein	AD + Ant
Dentatorubral-pallidoluysian atrophy or Haw River Syndrome or Naito-Oyanagi disease	125370	*ATN1*	Atrophin	AD + Ant
Infantile onset ascending spastic paralysis	607225	*ALS2*	GTP-binding protein regulators	AR
Friedreich ataxia	229300	*FXN*	Mitochondrial iron transport	AR + Ant
Marinesco-Sjögren syndrome	248800	*SIL1*	Endoplasmic reticulum protein	AR
Gillespie syndrome	206700	*PAX6*	Transcription factor Helix-turn-helix domains	AR
Neuropathy, ataxia and retinitis pigmentosa or NARP syndrome	516060	*MT-ATP6*	Mitochondrial diseases	Mit

Table 3.06.02.1. Continued.

Disorder	OMIM	Gene	Biological Function	Product of Conception Alteration
Fragile X-associated tremor/ataxia syndrome	300623	FMR1	mRNA trafficking from the nucleus to the cytoplasm	XD
Ataxia-telangiectasia or Louis–Bar syndrome	208900	ATM	DNA repair MRN complex	AD + AD = AR
Ataxia-telangiectasia-like disorder	604391	MRE11A	DNA recombination and repair protein	AR
Ataxia-oculomotor apraxia-1	208920	APTX	Single-stranded DNA repair	AR
Ataxia-oculomotor apraxia-2	606002	SETX	DNA/RNA helicase	AR
Primary coenzyme Q10 deficiency-1	607426	COQ2	Mitochondrial protein	AR
Primary coenzyme Q10 deficiency-2	614651	PDSS1	Inner mitochondrial membrane protein	AR
Primary coenzyme Q10 deficiency-3	614652	PDSS2	Inner mitochondrial membrane protein	AR
Primary coenzyme Q10 deficiency-4	612016	ADCK3	Mitochondrial protein	AR
Primary coenzyme Q10 deficiency-5	614654	COQ9	Mitochondrial protein	AR
Primary coenzyme Q10 deficiency-6	614650	COQ6	Mitochondrial protein	AR
Early-onset ataxia with oculomotor apraxia and hypoalbuminemia	208920	APTX	Single-stranded DNA repair	
Cerebellar ataxia, Cayman type	601238	ATCAY	Neuron-restricted protein	AR
Spastic ataxia, Charlevoix-Saguenay type	270550	SACS	Protein folding	AR
X-linked spastic paraplegia-1 or MASA syndrome	303350	L1CAM	Axonal glycoprotein	XR
X-linked spastic paraplegia-2	312920	PLP1	Transmembrane proteolipid protein	XR
Autosomal dominant spastic paraplegia-3A	182600	ATL1	Golgi body transmembrane protein	AD + de novo
Autosomal dominant spastic paraplegia-4	182601	SPAST	ATPase, AAA-type	AD
Autosomal recessive spastic paraplegia-5A	270800	CYP7B1	Endoplasmic reticulum protein	AR

The Concise Encyclopedia of Genomic Diseases

Table 3.06.02.1. Continued.

Disorder	OMIM	Gene	Biological Function	Product of Conception Alteration
Autosomal dominant spastic paraplegia-6	600363	NIPA1	Magnesium transporter	AD
Autosomal recessive spastic paraplegia-7	607259	SPG7	ATPase, AAA-type	AR
Autosomal dominant spastic paraplegia-8	603563	KIAA0196	Transmembrane and spectrin	AD
Autosomal dominant spastic paraplegia-10	604187	KIF5A	Cytoskeletal protein	AD + de novo
Autosomal recessive spastic paraplegia-11	604360	SPG11	Transmembrane protein that is phosphorylated upon DNA damage	AR
Autosomal dominant spastic paraplegia-13	605280	HSPD1	Lysosomal storage diseases	AD
Autosomal recessive spastic paraplegia-15	270700	ZFYVE26	Nuclear zinc finger centrosomal protein	AR
Autosomal dominant spastic paraplegia-17	270685	BSCL2	Intracellular signaling peptides and proteins	AD
Autosomal recessive spastic paraplegia 20 or Troyer syndrome	275900	SPG20	Endosomal trafficking and mitochondria function	AR
Autosomal recessive spastic paraplegia-30	610357	KIF1A	Axonal transport of synaptic vesicles	AR
Autosomal dominant spastic paraplegia-31	610250	REEP1	Mitochondrial disease	AD
Autosomal dominant spastic paraplegia-33	610244	ZFYVE27	Nuclear zinc finger protein	AD
Autosomal recessive spastic paraplegia-35	612319	FA2H	Endoplasmic reticulum protein	AR
Autosomal recessive spastic paraplegia-39	612020	PNPLA6	Phospholipase	AR
Autosomal dominant spastic paraplegia-42	612539	SLC33A1	Solute carrier	AD
Autosomal recessive spastic paraplegia-44	613206	GJC2	Channelopathy Connexin	AR
Silver spastic paraplegia syndrome	270685	BSCL2	Intracellular signaling peptides and proteins	AD
Infantile-onset spinocerebellar ataxia	271245	C10orf2	Mitochondrial protein	AR

Table 3.06.02.1. Continued.

Disorder	OMIM	Gene	Biological Function	Product of Conception Alteration
Spinocerebellar ataxia-1	164400	*ATXN1*	Ataxin	AD + Ant
Spinocerebellar ataxia-2	183090	*ATXN2*	Ataxin	AD + Ant
Spinocerebellar ataxia-3 or Machado-Joseph disease	109150	*ATXN3*	Ataxin	AD + Ant
Spinocerebellar ataxia-5	600224	*SPTBN2*	Cytoskeletal protein	AD + Ant
Spinocerebellar ataxia-6	183086	*CACNA1A*	Channelopathy Calcium	AD + Ant
Spinocerebellar ataxia-7	164500	*ATXN7*	Nucleolus protein	AD + Ant
Spinocerebellar ataxia-8	608768	*ATXN8OS*	Long non-coding RNAs disease	AD + Ant
Spinocerebellar ataxia-10	603516	*ATXN10*	Ataxin	AD + Ant
Spinocerebellar ataxia-11	604432	*TTBK2*	Cytoskeletal protein	AD
Spinocerebellar ataxia-12	604326	*PPP2R2B*	Ser/Thr phosphatases	AD
Spinocerebellar ataxia-13	605259	*KCNC3*	Channelopathy Potassium	AD
Spinocerebellar ataxia-14	605361	*PRKCG*	Protein kinase C	AD
Spinocerebellar ataxia-15	606658	*ITPR1*	Intracellular receptor for inositol	AD
Spinocerebellar ataxia-17	607136	*TBP*	Nuclear transcription initiation factor	AD + Ant
Spinocerebellar ataxia-23	610245	*PDYN*	Prodynorphin	AD
Spinocerebellar ataxia-27	609307	*FGF14*	Mitogenic and cell survival activities	AD
Spinocerebellar ataxia-28	610246	*AFG3L2*	ATPase, AAA-type	AD
Spinocerebellar ataxia-31	117210	*BEAN1*	Brain-expressed protein	AD
Autosomal recessive spinocerebellar ataxia-5	606937	*ZNF592*	Transcription factor Zinc finger	AR
Autosomal recessive spinocerebellar ataxia-8	610743	*SYNE1*	Nuclear envelope protein	AR
Autosomal recessive spinocerebellar ataxia-9	612016	*ADCK3*	Mitochondrial diseases	AR
Autosomal recessive spinocerebellar ataxia-11	614229	*SYT14*	membrane trafficking	AR
Autosomal recessive spinocerebellar ataxia with axonal neuropathy	607250	*TDP1*	Repairing DNA complexes	AR

Table 3.06.02.1. Continued.

Disorder	OMIM	Gene	Biological Function	Product of Conception Alteration
Spinocerebellar ataxia with epilepsy and Sensory ataxic neuropathy, dysarthria, and ophthalmoparesis	607459	POLG	Mitochondrial diseases	AR
X-linked spinocerebellar ataxia-1	302500	ATP2B3	Calcium-transporting ATPase	XR
Episodic ataxia-1 or Episodic ataxia/myokymia syndrome	160120	KCNA1	Channelopathy Potassium	AD + de novo
Episodic ataxia-2	108500	CACNA1A	Channelopathy Calcium	AD + de novo
Episodic ataxia-5	613855	CACNB4	Channelopathy Calcium	AD
Episodic ataxia-6	612656	SLC1A3	Solute carrier	AD
Spinal muscular atrophy-1 or Werdnig-Hoffmann disease	253300	SMN1	Nucleolus protein	AR + KDN
Spinal muscular atrophy-2	253550	SMN1	Nucleolus protein	AR + KDN
Spinal muscular atrophy-3 or Kugelberg-Welander disease	253400	SMN1	Nucleolus protein	AR + KDN
Spinal muscular atrophy-4 or Adult form	271150	SMN1	Nucleolus protein	AR + KDN
Modifier of Spinal muscular atrophy	253400	SMN2	Nucleolus protein	AR + KDN
Late-onset spinal muscular atrophy, Finkel type	182980	VAPB	Vesicle-associated membrane protein	AD
Spinal and bulbar muscular atrophy of Kennedy	313200	AR	Transcription factor Zinc finger	XR + Ant
Scapuloperoneal syndrome, myopathic type	181430	MYH7	Cytoskeletal protein	AD
Scapuloperoneal syndrome, myopathic type	300695	FHL1	Zinc finger protein	XD
Scapuloperoneal spinal muscular atrophy	181405	TRPV4	Channelopathy transient receptor potential channels	AD
Neurogenic scapuloperoneal syndrome, Kaeser type	181400	DES	Cytoskeletal Intermediate filaments	AD
Amyotrophic lateral sclerosis-1	105400	SOD1	Superoxide dismutase	AD or AR
Amyotrophic lateral sclerosis-6	137070	FUS	Nucleus protein	AR
Amyotrophic lateral sclerosis-6, with or without frontotemporal dementia	608030	FUS	Nucleus protein	AD

The Concise Encyclopedia of Genomic Diseases

Table 3.06.02.1. Continued.

Disorder	OMIM	Gene	Biological Function	Product of Conception Alteration
Amyotrophic lateral sclerosis-8	608627	VAPB	Vesicle-associated membrane protein	AD
Amyotrophic lateral sclerosis-9	611895	ANG	Esterases	AD
Amyotrophic lateral sclerosis-10 and Frontotemporal lobar degeneration	612069	TARDBP	DNA-binding protein	AD
Amyotrophic lateral sclerosis-11	612577	FIG4	Phosphatase	AD
Amyotrophic lateral sclerosis-12	613435	OPTN	Cellular morphogenesis	AR
Amyotrophic lateral sclerosis-14, with or without frontotemporal dementia	613954	VCP	ATPase, AAA-type	AD
Amyotrophic lateral sclerosis-15	300857	UBQLN2	Ubiquilin	XR
Amyotrophic lateral sclerosis-17	614696	CHMP2B	Charged multivesicular body protein	AD
Amyotrophic lateral sclerosis-18	614808	PFN1	Profilin	AD
Juvenile amyotrophic lateral sclerosis-2	205100	ALS2	GTP-binding protein regulators	AR
Juvenile amyotrophic lateral sclerosis-4	602433	SETX	DNA/RNA helicase	AD
Juvenile amyotrophic lateral sclerosis-16	614373	SIGMAR1	Sigma non-opioid intracellular receptor	AD
Susceptibility to amyotrophic lateral sclerosis	105400	NEFH	Cytoskeletal Intermediate filaments	AD
Susceptibility to amyotrophic lateral sclerosis	105400	PRPH	Cytoskeletal Intermediate filaments	AD
Susceptibility to amyotrophic lateral sclerosis	105400	DCTN1	Dynactin is involved in a diverse array of cellular functions	AD
Brown-Vialetto-Van Laere syndrome-1	211530	SLC52A3	Solute carrier	AR
Brown-Vialetto-Van Laere syndrome-2	614707	SLC52A2	Solute carrier	AR
Distal hereditary motor neuronopathy-2A	158590	HSPB8	Folding and unfolding of other proteins	AD

The Concise Encyclopedia of Genomic Diseases

Table 3.06.02.1. Continued.

Disorder	OMIM	Gene	Biological Function	Product of Conception Alteration
Distal hereditary motor neuronopathy-2B	608634	HSPB1	Translocates from the cytoplasm to the nucleus upon stress	AD
Distal hereditary motor neuronopathy-2C	613376	HSPB3	Folding and unfolding of other proteins	AD
Distal hereditary motor neuronopathy-5A	600794	GARS	Glycyl-tRNA synthetase	AD
Distal hereditary motor neuronopathy-5B	600794	BSCL2	Endoplasmic reticulum protein	AD
Autosomal recessive distal spinal muscular atrophy-1 or Distal hereditary motor neuronopathy-6 or spinal muscular atrophy with respiratory distress-1	604320	IGHMBP2	Transcription factor	AR
Distal hereditary motor neuronopathy-7A	158580	SLC5A7	Choline transporter	AD
Distal hereditary motor neuronopathy-7B	607641	DCTN1	Dynactin is involved in a diverse array of cellular functions	AD
Autosomal recessive distal spinal muscular atrophy-3	611067	PLEKHG5	Activates the nuclear factor kappa B	AR
Autosomal recessive distal spinal muscular atrophy-5	614881	DNAJB2	DnaJ homolog	AR
X-linked distal spinal muscular atrophy-3	300489	ATP7A	ATPase protein	XR

Table 3.06.02.2 Systemic Atrophies Primarily Affecting the Central Nervous System in Different Ethnic Groups

Disease	Frequency of condition	People with higher prevalence (if available)
Spinal muscular atrophy	2-12 in 100,000	Caucasians
Huntington disease	3-7 in 100,000	Venezuelans (Lake Maracaibo region), Tasmanians and Mauritians
Fragile X-associated tremor/ataxia syndrome	1-2 in 10,000	Worldwide
Neuropathy, ataxia and retinitis pigmentosa or NARP syndrome	1 in 12,000	Worldwide
Hereditary spastic paraplegias	2-8 in 100,000	Worldwide
Amyotrophic lateral sclerosis	6 in 100,000	People of the West Pacific
Spinocerebellar ataxia-3	1-3 in 100,000	People of Azores Islands
Friedreich ataxia	1-3 in 100,000	Inhabitants of Paphos district of Cyprus and in Rimouski area (Quebec)
Ataxia-telangiectasia or Louis–Bar syndrome	1-2 in 100,000	Sephardic Jews
X-linked spastic paraplegia-1 or MASA syndrome	2-4 in 100,000 Males	Worldwide
Scapuloperoneal syndrome, myopathic type	4 in 100,000	Worldwide
Episodic ataxia	1 in 100,000	Worldwide
Marinesco-Sjögren syndrome	1-9 in 1,000,000	Romani families (Serbia)
Spinal and bulbar muscular atrophy of Kennedy	6 in 1,000,000 Males	Western Finnish
Dentatorubral-pallidoluysian atrophy or Haw River Syndrome or Naito-Oyanagi disease	1 in 1,000,000	African-American of North Carolina and Japanese
Huntington disease-like 2	<1 in 1,000,000	Non-Afrikaner South Africans
Gillespie syndrome	<1 in 1,000,000	Unknown
Ataxia-telangiectasia-like disorder	<1 in 1,000,000	Unknown

3.06.03 Extrapyramidal and Movement Disorders

Parkinson's disease: A progressive disorder characterized by motor symptoms caused by the death of dopamine-generating cells in the substantia nigra, a region of the midbrain. Often the first symptoms include shaking of a limb, especially when the body is at rest, and also rigidity, slowness of movement and difficulty with walking and gait. Later symptoms include alteration in the affection, emotions and thinking ability leading to depression, visual hallucinations and dementia. This condition occurs more commonly in late-onset form, i.e. after the age of 50. Early-onset disease occurs when signs and symptoms begin before age 50, and the juvenile-onset occurs when signs and symptoms begin before age 20. The disease was named after James Parkinson in 1817. Parkinson's disease is the second most common neurodegenerative disorder after Alzheimer's disease. Parkinson's disease is more common in the elderly and prevalence ranges from 1% in those over 60 years of age to 4% of the people over 80. Early-onset Parkinson's disease accounts for 5-10% of all cases. The most important non-genomics risk factors include the patient's exposure to neurotoxin pesticides and living in rural environments. Protective factors include caffeine consumption and the harmful habit of cigarette smoking. There are several genomic factors that contribute to Parkinson's disease. Susceptibility to late-onset or sporadic Parkinson's disease has been associated with polymorphisms and/or mutations in several genes, including *GBA, ADH1C, TBP,* and *MAPT* and genes at the HLA locus. The late-onset condition exhibits a poor correlation in twin studies, which makes genomics a small contributing factor in this age group. On the contrary, Early-onset Parkinson disease has large genomic or familial factors that contributed to this condition. There are at least 18 loci associated with Familial Parkinson disease with autosomal dominant, autosomal recessive or X-linked pattern of inheritance. There are several forms of this condition for which the gene has not yet been characterized. There are several gene loci implicated in autosomal dominant forms of Parkinson's disease, including: Parkinson disease-1 is caused by monoallelic mutations in the *SNCA* gene which encodes the alpha-synuclein. This protein integrates presynaptic signaling and membrane trafficking.;; Parkinson disease-4 is caused by triplication of the *SNCA* gene;; Parkinson disease-5 is caused by monoallelic mutations in the *UCHL1* gene which encodes the ubiquitin carboxyl-terminal esterase L1 (ubiquitin thiolesterase);; Parkinson disease-8 is caused by monoallelic mutations in the *LRRK2* gene which encodes the leucine-rich repeat kinase 2. In Chi-

nese people this pathogenic allele appears 2,500 years ago;; Parkinson disease-11 is caused by monoallelic mutations in the *GIGYF2* gene which encodes the GRB10 interacting GYF protein 2;; and Parkinson disease-13 is caused by monoallelic mutations in the *HTRA2* gene which encodes the HtrA serine peptidase 2;; Parkinson disease-17 is caused by monoallelic mutations in the *VPS35* gene which encodes the vacuolar protein sorting 35 homolog;; and Parkinson disease-18 is caused by monoallelic mutations in the *EIF4G1* gene which encodes the eukaryotic translation initiation factor 4 gamma 1. There are several gene loci implicated in autosomal recessive early-onset Parkinson's disease including: Parkinson disease-2, caused by biallelic mutations in the *PARK2* gene which encodes the parkin 2;; Parkinson disease-6, caused by biallelic mutations in the *PINK1* gene which encodes the PTEN induced putative kinase 1;; Parkinson disease-7, caused by biallelic mutations in the *PARK7* gene which encodes the parkinson protein 7;; Parkinson disease-9, caused by biallelic mutations in the *ATP13A2* gene which encodes the ATPase type 13A2;; Parkinson disease-14, caused by biallelic mutations in the *PLA2G6* gene which encodes the calcium-independent phospholipase A2;; and Parkinson disease-15, caused by biallelic mutations in the *FBXO7* gene which encodes the F-box protein 7. There are also forms of Parkinson's disease that have been mapped, but whose genes have not been characterized including type 3, type 10 and type 16. Type 12 is a form of Parkinson's disease X-linked. There is also evidence that mitochondrial mutations may cause or contribute to Parkinson's disease. Susceptibility to late-onset form of Parkinson's disease has been associated with polymorphisms or mutations in several genes, including *MC1R, ADH1C, GBA, MAPT,* and genes at the HLA locus. Each of these genes functions as independent risk factors to the development of the disease. Susceptibility to Parkinson's disease may also be conferred by conditions with expanded trinucleotide repeats. This susceptibility is particularly caused by the spinocerebellar ataxia syndromes type 2, 3, 8 and 17. These conditions are caused by trinucleotide expansions in the *ATXN2, ATXN3, ATXN8OS*, and *TBP* genes, respectively.

3.06.03.1 Other Degenerative Diseases of Basal Ganglia

Progressive supranuclear palsy or Richardson-Steele-Olszewski syndrome: A disorder that is the second most common cause of degenerative Parkinsonism. Progressive supranuclear palsy accounts for 4-6% of patients with Parkinsonism. Progressive supranuclear palsy is a disorder characterized by Parkinsonism usually with onset in the sixth decade of life. Affected individuals also exhibit supranuclear palsy of vertical gaze, followed by proximal rigidity, postural instability, language problems, mild-to-moderate dementia, and visual disturbances such as diplopia. There are several underlying causes of this disorder with at least 3 autosomal genomic loci: Progressive supranuclear palsy-1 is caused by autosomal dominant mutations in the *MAPT* gene which encodes the microtubule-associated protein tau. There are two additional loci associated with progressive supranuclear palsy but the genes have not been characterized yet. There is also an Atypical Supranuclear palsy form which is caused by autosomal recessive mutations in the *MAPT* gene. This form also follows the inheritance type (AD + AD = AR). This atypical form is more severe with onset in the third decade of life, and has high prevalence among consanguineous marriages.

Thiamine metabolism dysfunction syndrome or biotin-responsive basal ganglia disease or thiamine-responsive encephalopathy: A disorder characterized by periodic encephalopathy, often elicited by febrile illness. Affected individuals manifest megaloblastic anemia, diabetes mellitus and sensorineural deafness. Other features may include optic atrophy, congenital heart defects, short stature and stroke. This condition starts manifesting as confusion, external ophthalmoplegia, seizures, dysphagia, and sometimes coma and death. There are several underlying causes of this disorder with at least 5 autosomal genomic loci: Thiamine metabolism dysfunction syndrome-1, or Thiamine-responsive megaloblastic anemia syndrome, is caused by autosomal recessive mutations in the *SLC19A2* gene which encodes the solute carrier (family 19 member A2), a thiamine transporter;; Thiamine metabolism dysfunction syndrome-2, or Biotin-responsive basal ganglia disease, is caused by autosomal recessive mutations in the *SLC19A3* gene which encodes the solute carrier (family 19 member A3), a thiamine transporter;; Thiamine metabolism dysfunction syndrome-3, or Amish lethal microcephaly, is caused by autosomal recessive mutations in the *SLC25A19* gene which encodes the solute carrier (family 25 member A19), a mitochondrial thiamine pyrophosphate carrier;; Thiamine metabolism dysfunction syndrome-4, or Bilateral striatal necrosis and progressive polyneuropathy, is caused by autosomal recessive mutations in the *SLC25A19* gene;; and Thiamine metabolism dysfunction syndrome-5 is caused by autosomal recessive mutations in the *TPK1* gene which encodes the thiamine pyrophosphokinase 1.

Adult-onset bilateral striopallidodentate calcinosis or idiopathic basal ganglia calcification: A neurodegenerative disorder appearing between 30 and 50 years of age characterized by the accumulation of calcium deposits in different brain regions, predominantly in the basal ganglia. Patients may exhibit a progressive movement disorders, including Parkinsonism, chorea, tremor, dystonia, ataxia, athetosis and orofacial dyskinesia. Other features include difficultly with concentration and memory, personality and/or behavior changes and dementia. There are several underlying causes of this disorder with at least 4 autosomal genomic loci: Idiopathic basal ganglia calcification-1 has been mapped but the gene has not been characterized;; Idiopathic basal ganglia calcification-2 has been mapped but the gene has not been characterized; Idiopathic basal ganglia calcification-3;; Idiopathic basal ganglia calcification-3 is caused by autosomal dominant mutations in the *SLC20A2* gene which encodes the solute carrier (family 20 member A2), a phosphate transporter;; and Idiopathic basal ganglia calcification-4 is caused by autosomal dominant mutations in the *PDGFRB* gene which encodes the platelet-derived growth factor receptor (beta polypeptide). Idiopathic basal ganglia calcification is estimated to affect less than 1 in every 1,000,000 individuals.

Choreoacanthocytosis: A progressive disorder appearing in the third to fifth decade of life, characterized by neurodegeneration and red cell acanthocytosis, with normal serum lipoproteins. Affected individuals exhibit epilepsy, behavioral changes, generalized weakness, and involuntary movements including grimacing, dystonia and chorea. This condition shares similarities of symptoms with those of Huntington's disease. Choreoacanthocytosis is caused by autosomal recessive mutations and in some cases by autosomal dominant mutations in the *VPS13A* gene which encodes the vacuolar protein sorting 13 homolog A (chorein). This protein is involved in protein sorting. There is also a similar phenotype referring as McLeod syndrome with or without chronic granulomatous disease which is caused by X-linked recessive mutations in the *XK* gene which encodes the X-linked Kx blood group (McLeod syndrome), an antigen of the Kell blood group system. Choreoacanthocytosis is estimated to affect less than 1 in every 1,000,000 individuals.

Autosomal dominant striatal degeneration: A disorder characterized by mild slurring of speech and stiffness in tongue of insidious onset in the third to fifth decade of life. Patients develop a progressive dysarthria, gait disturbance, mild bradykinesia, dysdiadochokinesia, and brisk lower extremity reflexes. Autosomal dominant striatal degeneration is caused by monoallelic mutations in the *PDE8B* gene which encodes the phosphodiesterase 8B. Autosomal dominant striatal degeneration is estimated to affect less than 1 in every 1,000,000 individuals.

Infantile striatonigral degeneration: A more severe form of striatonigral degeneration than the previously described condition. Infantile striatonigral degeneration is caused by autosomal recessive mutations in the *NUP62* gene which encodes the nucleoporin 62kDa. Infantile striatonigral degeneration has been reported among Israeli Bedouin families. There is also a mitochondrial form of Infantile striatonigral degeneration and Infantile bilateral striatal necrosis which is caused by mutations in the *MT-ATP6* gene. This gene encodes the mitochondrial ATP synthase (subunit 6). The prevalence of the sporadic form is estimated to affect 1-9 in every 1,000,000 and the familial form is estimated to affect less than 1 in every 1,000,000 individuals.

HARP syndrome (Hypoprebetalipoproteinemia, Acanthocytosis, Retinitis pig-

mentosa, and P̲allidal degeneration): An early childhood onset disorder characterized by severe spasticity and dystonia. HARP syndrome is caused by autosomal recessive mutations in the *PANK2* gene which encodes the pantothenate kinase-2. There is also a most severe disorder - the Pantothenate kinase-associated neurodegeneration or Hallervorden-Spatz disease. HARP syndrome is estimated to affect less than 1 in every 1,000,000 individuals.

Aceruloplasminemia: This disease is discussed under endocrine, nutritional and metabolic diseases in the subchapter "Disorders of Iron Metabolism" (3.04.16.2).

Neurodegeneration with brain iron accumulation-1 or Pantothenate kinase-associated neurodegeneration: A severe disorder characterized by progressive iron accumulation in the basal ganglia and other regions of the brain. Affected individuals manifest extrapyramidal movements, such as Parkinsonism and dystonia. Neurodegeneration with brain iron accumulation-1 is also caused by autosomal recessive mutations in the *PANK2* gene. On average, the carrier frequency for Neurodegeneration with brain iron accumulation trait is estimated to be 1 in 289-500 individuals worldwide, which means the frequency of this condition is about 1 in 334,000-1,000,000 individuals or 1 in 1/$(578)^2$ - 1/ $(1,000)^2$. There is a higher carrier frequency for Neurodegeneration with brain iron accumulation trait among people from Friesland, a northern province of the Netherlands, and the Agrawal community in northern India.

Neurodegeneration with brain iron accumulation-2B: A disorder characterized by excessive cerebral iron accumulation. Affected individuals exhibit ataxia, speech delay, optic atrophy, progressive dystonia and dysarthria, spastic or areflexic tetraparesis, nystagmus and seizures. Patients also exhibit neurobehavioral disturbances with hyperactivity, impulsivity, emotional lability, distractibility and poor attention span;; Neurodegeneration with brain iron accumulation-2B is caused by autosomal recessive mutations in the *PLA2G6* gene which encodes the phospholipase A2 (cytosolic calcium-independent);; Neurodegeneration with brain iron accumulation-3 caused by autosomal recessive mutations in the *FTL* gene which encodes the ferritin light polypeptide;; Neurodegeneration with brain iron accumulation-4 caused by autosomal recessive mutations in the *C19orf12* gene which encodes the chromosome 19 open reading frame 12 protein;; and Neurodegeneration with brain iron accumulation-5 is caused by X-linked dominant de novo mutations in the C19orf12 gene which encodes the WD repeat-containing protein 45.

Karak syndrome: A disorder characterized by early-onset progressive cerebellar ataxia, dystonia, spasticity and intellectual decline. Affected individuals exhibit cerebellar atrophy, and feature iron deposition in the substantia nigra and putamen with the "eye of the tiger" sign. This condition has been reported in a consanguineous family who lived in Karak, a town in southern Jordan. Karak syndrome is also caused by autosomal recessive mutations in the *PLA2G6* gene.

Leukodystrophy - spastic paraplegia - dystonia: A disorder characterized by progressive cognitive dysfunction, spasticity, dystonia and periventricular white matter disease of childhood onset. Leukodystrophy - spastic paraplegia - dystonia is caused by autosomal recessive mutations in the *FA2H* gene which encodes the fatty acid 2-hydroxylase. This enzyme is involved in myelinization. The same gene has an allelic disorder the spastic paraplegia-35. Leukodystrophy - spastic paraplegia - dystonia has been described mainly among consanguineous families. Leukodystrophy - spastic paraplegia - dystonia is estimated to affect less than 1 in every 1,000,000 individuals.

3.06.03.2 Dystonia

Dystonia: Are a group of disorders characterized by sustained muscle contractions, leading to twisting and repetitive movements or abnormal postures. These disorders can be caused by genomics and other non-genomics factors including birth-related conditions, physical trauma, and infection, and reactions to pharmaceutical drugs, primarily neuroleptics. There are many forms of classification of these conditions such as: on the basis primary versus associated with other features; by age of onset; by muscle groups affected and by pattern of inheritance. There are at least 21 loci associated with different forms of dystonia with autosomal dominant, autosomal recessive and X-linked recessive pattern of inheritance. There are different forms of dystonia, including early-onset torsion dystonia, dopamine-responsive, primary cervical, paroxysmal and kinesigenic nonkinesigenic dyskinesia, episodic choreoathetosis/spasticity and myoclonic dystonia. There are several forms of this condition for which that the genes have not been characterized yet. The autosomal dominant Dystonia-12 is caused by monoallelic mutations in the *ATP1A3* gene which encodes the Na+/K+ transporting ATPase (alpha 3 polypeptide);; Dystonia-16 is caused by monoallelic mutations in the *PRKRA* gene which encodes the interferon-inducible double stranded RNA-dependent protein kinase activator A;; Myoclonic dystonia-11 is caused by monoallelic paternal mutations in the *SGCE* gene which encodes the epsilon-sarcoglycan;; Torsion dystonia-1 is caused by monoallelic mutations in the *TOR1A* gene which encodes the torsin A;; and Torsion dystonia-6 is caused by monoallelic mutations in the *THAP1* gene which encodes the THAP domain containing apoptosis associated protein 1. The autosomal recessive is associated with mutations in the *SPR* gene which encodes the sepiapterin reductase, *GCH1* gene which encodes the GTP cyclohydrolase 1, Dystonia-parkinsonism, Paisan-Ruiz type is caused by biallelic mutations in the *PLA2G6* gene; and Dystonia-parkinsonism is caused by X-linked recessive mutations in *TAF1* gene which encodes the TAF1 RNA polymerase II (TATA box binding protein-associated factor). There is a higher prevalence for X-linked dystonia-parkinsonism syndrome in the Philippines, mainly in the Panay Islands. The prevalence is 5.24 in every 100,000 in these islands, with the highest rate being 18.9 in every 100,000 in the province of Capiz.

Torsion dystonia: A disorder appearing during childhood or adolescence characterized by painful muscle contractions of the trunk, neck or limbs. These contractions results in uncontrollable distortions. Torsion dystonia is caused by autosomal dominant mutations in the *TOR1A* gene which encodes the torsin-A, an ATP-binding protein. More than 90% of Torsion dystonia cases that occur among Ashkenazi people are due to GAG base pairs deletion, resulting in the loss of 1 of glutamic acid residues. The penetrance for Torsion dystonia due to the GAG deletion in the *TOR1A* gene is around 30%. This deletion is an example of founder mutation, which happened approximately 350 years ago among Jews from Lithuania and Byelorussia. This gene is located in the subtelomeric region on chromosome 9. There is evidence of advanced paternal age effect in de novo mutations. Among Ashkenazi Jews, Torsion dystonia is estimated to affect 1 in every 15,000-23,000 individuals.

Juvenile-onset Dystonia: A disorder characterized by rapidly progressive and generalized levodopa-unresponsive dystonia. Juvenile-onset dystonia can be caused by autosomal recessive mutations in the *ACTB* gene which encodes the beta-actin.

Dopa-responsive dystonia with or without hyperphenylalaninemia: A disorder appearing during early childhood characterized by dystonia and Parkinsonism features. This condition has a circadian variation features with diurnal fluctuation. These features are absent in the morning or after rest but worsen during the day and with exertion. Children with Dopa-responsive dystonia are often misdiagnosed as having cerebral palsy. Dopa-responsive dystonia can be caused by autosomal dominant mutations in the *GCH1* gene which encodes the GTP cyclohydrolase I. This is the rate-limiting enzyme in the production of a cofactor for tyrosine hydroxylase. Penetrance seems to be higher (almost twice) in females than in males.

Segawa syndrome or Dopa-responsive dystonia: A disorder appearing during childhood characterized by walking problems due to dystonia of the lower limbs. Children with Segawa syndrome are often misdiagnosed as having cerebral palsy. Segawa syndrome is caused by autosomal recessive mutations in the *TH* gene which encodes the tyrosine hydroxylase. This enzyme is involved in the conversion of phenylalanine to dopamine. Among Europeans, the carrier frequency for the Segawa syndrome trait is estimated to be 1 in 224-500 individuals, which means the frequency of this condition is about 1 in

200,000-1,000,000 individuals or $1/(448)^2$ - $1/(1,000)^2$.

Dopa-responsive dystonia due to sepiapterin reductase deficiency: A severe disorder characterized by neurologic deterioration with progressive psychomotor retardation and dystonia. Affected individuals exhibit also hypersomnolence, myoclonus, choreoathetosis, abnormal gait, oculomotor apraxia and dysarthria. This condition is caused by severe dopamine and serotonin deficiencies in the central nervous system. "Dopa-responsive dystonia due to sepiapterin reductase deficiency" is caused by autosomal recessive mutations in the *SPR* gene which encodes the sepiapterin reductase. This enzyme is a component of the tetrahydrobiopterin synthetic pathway. Dopa-responsive dystonia due to sepiapterin reductase deficiency is estimated to affect less than 1 in every 1,000,000 individuals.

Aromatic L-amino acid decarboxylase deficiency: A disorder usually appearing during infancy or childhood and is characterized by severe hypotonia, oculogyric crises, dystonia, neurologic dysfunction, developmental retardation, and vegetative symptoms such as temperature dysregulation and postural hypotension. This neurotransmitter metabolism disorder leads to combined serotonin and catecholamine deficiency. Aromatic L-amino acid decarboxylase deficiency is caused by autosomal recessive mutations in the *DDC* gene which encodes the dopa decarboxylase (aromatic L-amino acid decarboxylase). This enzyme catalyzes the decarboxylation of L-3,4-dihydroxy-phenylalanine (DOPA) to dopamine, L-5-hydroxytryptophan to serotonin and L-tryptophan to tryptamine. Aromatic L-amino acid decarboxylase deficiency is estimated to affect less than 1 in every 1,000,000 individuals.

Mohr–Tranebjærg syndrome or Deafness-dystonia syndrome: A disorder characterized by severe deafness and dystonia, and occasionally, features of cortical deterioration with vision and mental deterioration. Mohr–Tranebjærg syndrome is caused by X-linked recessive mutations in the *TIMM8A* gene which encodes the translocase of inner mitochondrial membrane 8 (homolog A). Mohr–Tranebjærg syndrome is estimated to affect less than 1 in every 1,000,000 individuals.

3.06.03.3 Other Extrapyramidal and Movement Disorders

Myoclonic dystonia or Myoclonus-dystonia syndrome: A disorder characterized by bilateral myoclonic jerks affecting mostly proximal muscles including arms, neck, axial muscles and mild dystonia, usually torticollis or writer's cramp. Affected individuals can also exhibit psychiatric abnormalities such as obsessive-compulsive behavior, panic attacks and alcohol-abuse. Onset of the disorder is usually during the first or second decade of life. Myoclonus-dystonia is also referred to as alcohol-responsive dystonia since these patients can often ameliorate the symptoms with alcohol use. Myoclonus-dystonia-11 is caused by autosomal monoallelic paternal mutations in the *SGCE* gene which encodes the epsilon-sarcoglycan. The maternal allele is null probably due to imprinting mechanism presumably by CpG methylation.

Episodic kinesigenic dyskinesia-1 or paroxysmal kinesigenic dyskinesia: A disorder appearing during childhood or early adulthood characterized by recurrent and brief attacks of involuntary movement; triggered by sudden voluntary movement. Affected individuals also exhibit dystonic postures, chorea or athetosis. This condition is the most common type of paroxysmal movement disorder. There are several underlying causes of this disorder with at least 2 autosomal genomic loci: Episodic kinesigenic dyskinesia-1 is caused by autosomal dominant mutations in the *PRRT2* gene which encodes the proline-rich transmembrane protein 2;; and Episodic kinesigenic dyskinesia-2 or "Paroxysmal kinesigenic choreoathetosis" has been mapped, but the gene has not yet been characterized; and "Rolandic epilepsy with paroxysmal exercise-induced dystonia and writer's cramp" has been mapped, but the gene has not been characterized yet.

Paroxysmal nonkinesigenic dyskinesia: A disorder characterized by paroxysmal choreoathetosis with similar features as those of Huntington chorea. These attacks occur several times a day and last only a few minutes. There are several precipitating factors associated with these episodes including the use of alcohol, coffee and tobacco. Other precipitating factors are hunger, fatigue and stress. There are several underlying causes of this disorder with at least 2 autosomal genomic loci: Paroxysmal nonkinesigenic dyskinesia-1 is caused by autosomal dominant mutations in the *MR1* gene which encodes the myofibrillogenesis regulator 1;; and Paroxysmal nonkinesigenic dyskinesia-2 has been mapped, but the gene has not yet been characterized.

Paroxysmal exercise-induced dyskinesia with or without epilepsy and/or hemolytic anemia or Dystonia-18: A disorder appearing during childhood characterized by paroxysmal exercise-induced dyskinesia affecting the exercised limbs. Affected individuals also exhibit epilepsy, most commonly childhood absence epilepsy, hemolytic anemia and mild mental retardation. Paroxysmal exercise-induced dyskinesia is caused by autosomal dominant mutations in the *SLC2A1* gene which encodes the solute carrier (family 2 member A1), a facilitated glucose transporter. Mutations in the *SLC2A1* gene manifest other allelic disorders such as GLUT1 deficiency syndrome-1, dystonia-9, and idiopathic generalized epilepsy-12.

Progressive myoclonic epilepsy-4 or Action myoclonus-renal failure syndrome: A disorder characterized by severe progressive action myoclonus, ataxia, dysarthria, infrequent generalized seizures, and renal failure requiring dialysis of adolescence or young adulthood onset. Action myoclonus-renal failure syndrome is caused by autosomal recessive mutations in the *SCARB2* gene which encodes the scavenger receptor class B. This is a type III glycoprotein that is located principally in limiting membranes of lysosomes and endosomes.

Hereditary benign chorea: A nonprogressive form of chorea of early-onset without intellectual deterioration. Hereditary benign chorea is a socially embarrassing condition. Hereditary benign chorea is caused by autosomal dominant mutations in the *NKX2-1* gene which encodes the NK2 homeobox 1. The same gene is also associated with Choreoathetosis, congenital hypothyroidism and neonatal respiratory distress.

Hyperekplexia: An early-onset neurologic disorder characterized by exaggerated startle responses to sudden acoustic or tactile stimuli. Years after onset, patients manifest generalized stiffness. Individuals with this condition also exhibit brief episodes of intense truncal hypertonia in response to stimulation. Neonates may have prolonged periods of rigidity and are at risk for sudden death from apnea or aspiration. Hyperekplexia has been mainly linked to glycine neurotransmission. Glycine is an inhibitory neurotransmitter mediating postsynaptic inhibition in the spinal cord and other regions of the central nervous system. Hyperekplexia is in general a genomic disorder, but some brain traumas can produce similar features. There are several underlying causes of this disorder with at least 4 autosomal genomic loci: Hereditary hyperekplexia-1 is caused by either autosomal recessive or autosomal dominant mutations in the *GLRA1* gene which encodes the glycine receptor (subunit alpha 1). The alpha-1 subunit of the glycine receptor is a ligand-gated chloride channel;; Hereditary hyperekplexia-2 is caused by autosomal recessive mutations in the *GLRB* gene which encodes the glycine receptor (subunit beta);; Hereditary hyperekplexia-3 is caused by autosomal recessive mutations in the *SLC6A5* gene which encodes the solute carrier (family 6 member A5), a sodium and chloride-dependent glycine neurotransmitter transporter;; and Hyperekplexia can also occur in early infantile epileptic encephalopathy-8, caused by X-linked mutations in *ARHGEF9* gene which encodes the Cdc42 guanine nucleotide exchange (factor 9). Hereditary hyperekplexia is estimated to affect less than 1 in every 1,000,000 individuals.

Elevated serum Creatine phosphokinase: A disorder characterized by muscle cramps with exertion, but without evidence of neuromuscular disease. Elevated serum Creatine phosphokinase is caused by either inherited or *de novo* autosomal monoallelic mutations in the *CAV3* gene which encodes the caveolin 3.

The Concise Encyclopedia of Genomic Diseases

Table 3.06.03.1 Recognizable Extrapyramidal and Movement Disorders

Disorder	OMIM	Gene	Biological Function	Product of Conception Alteration
Parkinson disease-1	168601	SNCA	Inhibit phospholipase	AD + de novo
Autosomal recessive parkinson disease-2	600116	PARK2	E3 ubiquitin protein ligase	AR
Parkinson disease-4	605543	SNCA	Inhibit phospholipase	AD
Parkinson disease-5	613643	UCHL1	Thiol protease	AD
Autosomal recessive parkinson disease-6	605909	PINK1	Mitochondrial protein	AR
Autosomal recessive parkinson disease-7	606324	PARK7	Cytoskeletal Microtubules	AR
Parkinson disease di-genic	602533	PARK7/PINK1	Cytoskele-tal/Mitochondrial protein	Dig
Parkinson disease-8	607060	LRRK2	Serine/threonine-protein kinase	AD + de novo
Parkinson disease-9 or Kufor-Rakeb syndrome	606693	ATP13A2	ATPase protein	AR
Parkinson disease-11	607688	GIGYF2	Regulation of tyrosine kinase	AD
Parkinson disease-13	610297	HTRA2	Serine proteases	AD
Autosomal recessive parkinson disease-14	612953	PLA2G6	Phospholipase	AR
Autosomal recessive parkinson disease-15 or parkinsonian-pyramidal syndrome	260300	FBXO7	Phosphorylation-dependent ubiquitination	AR
Parkinson disease-17	614203	VPS35	Vacuolar protein sorting	AD
Parkinson disease-18	614251	EIF4G1	Translation initiation fac-tor 4	AD
Infantile Parkinsonism-dystonia	613135	SLC6A3	Solute carrier	AR
Resistance to Parkin-son disease	613643	UCHL1	Thiol protease	AD
Susceptibility to Par-kinson disease	168600	ADH1C	Alcohol dehydrogenase	
Susceptibility to Par-kinson disease	168600	MAPT	Cytoskeletal Microtubules	
Susceptibility to Par-kinson disease	168600	TBP	Nuclear transcription initiation factor	
Progressive supranu-clear palsy-1	601104	MAPT	Cytoskeletal Microtubules	AD
Progressive atypical supranuclear palsy	260540	MAPT	Cytoskeletal Microtubules	AD + AD = AR

Table 3.06.03.1. Continued.

Disorder	OMIM	Gene	Biological Function	Product of Conception Alteration
Thiamine metabolism dysfunction syndrome-2 or Biotin-responsive basal ganglia disease	607483	SLC19A3	Solute carrier	AR
Thiamine metabolism dysfunction syndrome-1 or Thiamine-responsive megalo-blastic anemia syndrome	249270	SLC19A2	Solute carrier	AR
Thiamine metabolism dysfunction syndrome-3 or Amish lethal microcephaly	607196	SLC25A19	Mitochondrial thiamine pyrophosphate carrier	AR
Thiamine metabolism dysfunction syndrome-4 or Bilateral striatal necrosis and progressive polyneuropathy	613710	SLC25A19	Mitochondrial thiamine pyrophosphate carrier	AR
Thiamine metabolism dysfunction syndrome-5	614458	TPK1	Thiamin pyrophosphokinase 1	AR
Idiopathic basal ganglia calcification-3	614540	SLC20A2	Phosphate transporter	AD
Idiopathic basal ganglia calcification-4	615007	PDGFRB	Growth factor receptor	AD
Choreoacanthocytosis	200150	VPS13A	Vacuolar protein sorting	AR or AD
McLeod neuroacanthocytosis syndrome with or without chronic granulomatous disease	300842	XK	Membrane transport protein	XR
Autosomal dominant striatal degeneration	609161	PDE8B	Phosphodiesterase	AD
Infantile striatonigral degeneration	271930	NUP62	Nuclear pore complex	AR
Infantile striatonigral degeneration (mitochondrial)	516060	MT-ATP6	Mitochondrial protein	Mit
HARP syndrome	607236	PANK2	Mitochondrial protein	AR
Hereditary hypoceruloplasminemia	604290	CP	Metal metabolism	AR

Table 3.06.03.1. Continued.

Disorder	OMIM	Gene	Biological Function	Product of Conception Alteration
Neurodegeneration with brain iron accumulation-1 or pantothenate kinase-associated neurodegeneration	234200	*PANK2*	Mitochondrial protein	AR
Neurodegeneration with brain iron accumulation-2A	256600	*PLA2G6*	Phospholipase	AR
Neurodegeneration with brain iron accumulation-2B	610217	*PLA2G6*	Phospholipase	AR
Neurodegeneration with brain iron accumulation-3	606159	*FTL*	Intracellular iron storage	AD
Neurodegeneration with brain iron accumulation-4	614298	*C19orf12*	Transmembrane protein	AR
Neurodegeneration with brain iron accumulation-5	300894	*WDR45*	WD repeat-containing protein 45	XD + de novo
Karak syndrome	610217	*PLA2G6*	Phospholipase	AR
Infantile neuroaxonal dystrophy 1	256600	*PLA2G6*	Phospholipase	AR
Cerebellar ataxia	604290	*CP*	Metal metabolism	AR
Leukodystrophy - spastic paraplegia - dystonia	612319	*FA2H*	Endoplasmic reticulum protein	AR
Dystonia-12	128235	*ATP1A3*	ATPase protein	AD + de novo
Dystonia-16	612067	*PRKRA*	Protein kinase activated by double-stranded RNA	AR
Myoclonic dystonia-11	159900	*SGCE*	Linkage of the actin cytoskeleton to the extracellular matrix	Pat
Torsion dystonia-1	128100	*TOR1A*	ATPase, AAA-type	AD + de novo
Torsion dystonia-6	602629	*THAP1*	Nuclear proapoptotic factor	AD
Dopa-responsive dystonia	612716	*SPR*	Sepiapterin reductase	AR
Dopa-responsive dystonia with or without hyperphenylalainemia	233910	*GCH1*	GTP cyclohydrolase	AR
Dystonia-parkinsonism, Paisan-Ruiz type	612953	*PLA2G6*	Phospholipase	AR
Dystonia-parkinsonism	314250	*TAF1*	RNA polymerase 2	XR
Juvenile-onset dystonia	607371	*ACTB*	Cytoskeletal Microfilaments	AD
Dopa-responsive dystonia with or without hyperphenylalaninemia	233910	*GCH1*	GTP cyclohydrolase	AR

Table 3.06.03.1. Continued.

Disorder	OMIM	Gene	Biological Function	Product of Conception Alteration
Segawa syndrome or Dopa-responsive dystonia	605407	TH	Tyrosine hydroxylase	AR
Dopa-responsive dystonia, due to sepiapterin reductase deficiency	612716	SPR	Sepiapterin reductase	AR
Aromatic L-amino acid de-carboxylase deficiency	608643	DDC	Dopa decarboxylase	AR
Mohr-Tranebjaerg syndrome or Deafness-dystonia syn-drome	304700	TIMM8A	Mitochondrial protein	XR + de novo
Episodic kinesigenic dyskine-sia-1 or paroxysmal kinesi-genic dyskinesia	128200	PRRT2	Transmembrane pro-tein	AD + de novo
Paroxysmal nonkinesigenic dyskinesia	118800	MR1	MHC 1	AD
Paroxysmal exercise-induced dyskinesia with or without epilepsy and/or hemolytic anemia or dystonia 18	612126	SLC2A1	Solute carrier	AD + de novo
Progressive myoclonic epi-lepsy-4 or Action myoclonus-renal failure syndrome	254900	SCARB2	Lysosomal membrane sialoglycoprotein	AR
Hereditary benign chorea	118700	NKX2-1	Transcription factor	AD + de novo
Hyperekplexia-1	149400	GLRA1	Neurotransmitter-gated ion channel	AD or AR
Hyperekplexia-2	614619	GLRB	Neurotransmitter-gated ion channel	AR
Hyperekplexia-3	614618	SLC6A5	Solute carrier	AR
Hyperekplexia and epilepsy	300607	ARHGEF9	Rho-dependent signals	X
Elevated serum creatine phosphokinase	123320	CAV3	Trafficking Vesicle fu-sion	AD + de novo

Table 3.06.03.2 Extrapyramidal and Movement Disorders in Different Ethnic Groups

Disease	Frequency	People with higher prevalence (if available)
Early-onset Parkinson disease	1-5 in 10,000	Worldwide
Progressive supranuclear palsy or Richardson-Steele-Olszewski syndrome	1-9 in 100,000	Worldwide
Neurodegeneration with brain iron accumulation-1 or Pantothenate kinase-associated neurodegeneration	1-3 in 1,000,000	People in Friesland (Netherlands) and in the Agrawal community (India)
Parkinson disease-8	15 in 1,000,000	Ashkenazi Jews and North African Arabs
Segawa syndrome or Dopa-responsive dystonia	1-5 in 1,000,000	Europeans
Torsion dystonia	1-9 in 1,000,000	Ashkenazi Jews
Idiopathic basal ganglia calcification	<1 in 1,000,000	Unknown
Choreoacanthocytosis	<1 in 1,000,000	Unknown
Autosomal dominant striatal degeneration	<1 in 1,000,000	Unknown
Infantile striatonigral degeneration	<1 in 1,000,000	Israeli Bedouins
HARP syndrome	<1 in 1,000,000	Unknown
Leukodystrophy - spastic paraplegia - dystonia	<1 in 1,000,000	Unknown
X-linked dystonia-parkinsonism syndrome	<1 in 1,000,000	People of Panay Island (Philippines)
Dopa-responsive dystonia due to sepiapterin reductase deficiency	<1 in 1,000,000	Unknown
Aromatic L-amino acid decarboxylase deficiency	<1 in 1,000,000	Unknown
Mohr–Tranebjærg syndrome or Deafness-dystonia syndrome	<1 in 1,000,000	Unknown
Hereditary hyperekplexia	<1 in 1,000,000	Unknown

3.06.04 Other Degenerative Diseases of the Nervous System

Alzheimer's disease: A disorder that is the most common form of progressive dementia accounting for 60-80% of all the dementia cases. 1 in 8 Americans aged 65 or older has mild cognitive impairment leading to Alzheimer's disease; 42% of those 85 or older have the neurodegenerative disease. Alzheimer's disease is a disorder that in the early stages is characterized by difficulty in remembering recent events, leading to inability to generate coherent speech, recognize or identify objects, execute motor activities and think abstractly. When Alzheimer's disease advances, the patients also exhibit confusion, irritability, aggression, mood swings, trouble with language, long-term memory loss, and withdrawal from family or society. This condition is named after Alois Alzheimer. Life expectancy following diagnosis is approximately 7 years. There are two main forms: early-onset familial and sporadic Alzheimer's disease. The early onset is when this condition is diagnosis under 65 years of age. The early onset form affects less than 1% of all the cases. There are at least 16 loci associated with this familial or early-onset form of Alzheimer's disease. The early onset form can be caused in 69% of all the cases by mutations in the *PSEN1* gene which encodes the presenilin 1. The early onset form can be caused in 13% of all the cases by mutations in the *APP* gene which encodes the amyloid beta (A4) precursor protein. The early onset form can be caused in 7.5% of all the cases by duplication in the *APP* gene (due to Down syndrome). The early onset form can be caused in 2% of all the cases by mutations in the *PSEN2* gene which encodes the presenilin 2. The early onset form can also be caused by the E4 allele of the apolipoprotein E (APOE*4 allele). The *APOE* gene encodes the apolipoprotein E. Alzheimer's disease can be associated with cerebral amyloid angiopathy in individuals with Down syndrome. The latter condition is caused by the APP locus duplication. The APP locus is located on chromosome 21q21. Familial Alzheimer disease-1 is caused by mutation in the *APP* gene;; Familial Alzheimer disease-2 is associated with the APOE*4 allele;; Familial Alzheimer disease-3 is caused by mutations in the *PSEN1* gene;; and Familial Alzheimer disease-4 is caused by mutation in the *PSEN2* gene. There are 12 autosomal loci causing this condition in which the loci have been mapped, but the genes have not been characterized. There is also autosomal recessive form mapped to chromosome 10q24. This biallelic form has been described in people of Arab community located in Wadi Ara, northern Israel. This community has a high prevalence for Alzheimer's disease. For example, 20% of those aged 65 or older are affected with Alzheimer's disease; 60.5% of those aged 85 or older are also affected. There are also forms of Alzheimer disease caused by autosomal recessive mutations in the APP gene. This biallelic form leads to early-onset progressive

Alzheimer's Disease in the third decade of life. Using twin studies the heritability for Alzheimer's Disease was estimated to be 58%. Most cases of Alzheimer's disease do not exhibit either autosomal dominant or recessive inheritance, and they are referred to as sporadic Alzheimer's disease. Nevertheless, this sporadic condition is rooted in our genes. These genes may act as risk factors - for example, the inheritance of the E4 allele of the apolipoprotein E increases the risk of the disease 3-fold in monoallelic individuals and 15-fold in biallelic individuals. 40-80% of people with this condition have at least one apo E4 allele. Besides the gene described, there is also susceptibility to late-onset or sporadic Alzheimer disease which has been associated with polymorphisms and/or mutations in several genes, including *APBB2, BLMH, SORL1, HFE, ACE, MPO, PAXIP1, A2M, PLAU* and *NOS3*. There are also mitochondrial DNA polymorphisms that may be risk factors in Alzheimer's disease.

3.06.04.1 Other Degenerative Diseases of Nervous System (not elsewhere classified)

Dementia with Lewy bodies: A disorder characterized by Parkinsonism and dementia. Affected individuals exhibit progressive cognitive decline, visual hallucinations, variations in attention and alertness, falls, syncopal episodes and extrapyramidal symptoms. The pathologic findings in this condition include Lewy bodies, which are intracytoplasmic eosinophilic neuronal inclusions found in subcortical and cortical regions. This condition is the second most common form of dementia among the elderly after Alzheimer's disease. There are overlapping forms between Alzheimer's disease and Dementia with Lewy bodies. Dementia with Lewy bodies can be caused by autosomal dominant mutations in the either *SNCA* gene which encodes the alpha-synuclein, or the *SNCB* gene which encodes the beta-synuclein. There are also another genes/alleles associated with Dementia with Lewy bodies including the epsilon-4 allele of the *APOE* gene and the B allele of the *CYP2D6* gene, a cytochrome P-450 monooxygenase. There are also other mutations in other genes which are associated with Dementia with Lewy bodies and other diseases, such as: mutations in the *LRRK2* gene which is associated with Parkinson disease-8, and mutations in the *PRNP* gene which is associated Creutzfeldt-Jakob disease. In individuals over the age of 65 years, Dementia with Lewy Bodies is estimated to affect 3-26% of all Dementia cases.

Leigh disease or Subacute Necrotizing Encephalomyelopathy: An early-onset progressive neurometabolic disorder which often appears after a viral infection. Affected individuals are characterized by exhibiting focal and bilateral lesions in one or more areas of the central nervous system. These lesions include demyelination, gliosis, necrosis, spongiosis or capillary proliferation. The symptoms associated with this condition depend on the area of the central nervous system affected. The specific affected areas include the brainstem, thalamus, basal ganglia, cerebellum and the spinal cord. These infants exhibit symptoms of brain stem and/or basal ganglia involvement, and motor and intellectual developmental regression or delay. Affected patients often start with loss of motor milestones, hypotonia with poor head control, recurrent vomiting and a movement disorder. Later, these patients also exhibit pyramidal and extrapyramidal signs such as nystagmus, breathing disorders, ophthalmoplegia and peripheral neuropathy. This condition is named after Denis Archibald Leigh. There are at least

24 genes associated with this condition including both nuclear and mitochondrial-encoded genes. Most of these genes are involved in energy metabolism, including mitochondrial respiratory chain enzyme defects complexes I, II, III, IV, and V, which are involved in oxidative phosphorylation and the generation of ATP, and components of the pyruvate dehydrogenase complex. This condition could either be inherited or occur as a *de novo* event. The patterns of inheritance for this condition include autosomal recessive, X-linked recessive, and mitochondrial or maternal. The autosomal recessive is caused by mutations in *BCS1L, DLD, NDUFS3, NDUFS4, NDUFS7, NDUFS8, NDUFV1, SDHA, FOXRED1, NDUFAF2, COX15, C8orf38, NDUFA2, SURF1* and *LRPPRC* genes. Mutations in complex I include both mitochondrial and nuclear encoded genes, including: the mitochondrial-encoded *MT-ND2, MT-ND3, MT-ND5,* and *MT-ND6* genes, and the nuclear-encoded NDUFS1, *NDUFS3, NDUFS4, NDUFS7, NDUFS8, NDUFA2, NDUFA9, NDUFA10, NDUFA12* and *C8orf38* genes, and the complex I assembly factor *NDUFAF2* and *NDUFAF5* genes. This condition is also associated with mutations in the *MTFMT* gene, which is involved in mitochondrial translation. Mutations in complex II include mutations in the *SDHA* gene which encodes the flavoprotein subunit A of complex II, which is a major catalytic subunit of succinate-ubiquinone oxidoreductase. Mutations in complex III include mutations in the *BCS1L* gene, which is involved in the assembly of complex III. Mutations in the complex IV include genes that are mitochondrial-encoded *MT-CO3* and nuclear-encoded *COX10, COX15, SCO2, SURF1* and *TACO1*. Mutations in complex V include the mitochondrial-encoded *MT-ATP6* gene. There is also Leigh's Disease caused by mutations in components of the pyruvate

dehydrogenase complex, which include mutations in the *DLD* and *PDHA1* genes. The latter is associated with the X-linked form of Leigh syndrome. There is also a Saguenay-Lac-Saint-Jean or French Canadian type of Leigh syndrome. The French Canadian type is associated with COX deficiency caused by mutations in the *LRP-PRC* gene. There are also other uncommon mitochondrial variants of Leigh syndrome caused by maternal mutations in the *MT-TV, MT-TK, MT-TW* and *MT-TL1* genes. On average, Leigh disease is estimated to affect 1 in every 36,000-77,000 births worldwide.

Aminoacylase 1 deficiency: A disorder characterized by acute encephalopathy with seizures and transient hemiplegia. This condition starts manifesting during the third day of life for duration of about 2 weeks. Affected neonates also exhibit apnea, vomiting, hypotonia and sensorineural hearing loss. Aminoacylase 1 deficiency is caused by autosomal recessive mutations in the *ACY1* gene which encodes the aminoacylase 1. This enzyme catalyzes the formation of free amino acids from N-acetylated amino acids, including methionine, glutamine, alanine, leucine, glycine, valine and isoleucine. Most of the ACY1 deficiencies are detected through newborn screening. Neurological conditions associated with aminoacylase 1 deficiency are estimated to affect less than 1 in every 1,000,000 individuals.

Familial British dementia: A disorder appearing between 40 and 60 years of age characterized by spasticity with increased deep tendon reflexes and tone. Affected individuals also exhibit muscular rigidity, ataxia without tremor and impairment of the optic nerves. Mental decline is progressive and begins with apathy and impaired memory for recent events, and proceeds to complete disorientation in around

one decade after the onset. Familial British dementia is caused by autosomal dominant mutations in the *ITM2B* gene which encodes the integral membrane protein 2B. This transmembrane protein is processed at the C-terminus by furin or furin-like proteases to produce a small secreted peptide which inhibits the deposition of beta-amyloid. This gene is also associated with a similar condition called Familial Danish dementia.

Pick's disease or frontotemporal dementia: This disease is discussed under mental and behavioral disorders in the subchapter "Organic Mental Disorders (including symptomatic)" (3.05.1).

Aicardi–Goutières syndrome or Cree encephalitis and pseudo-TORCH syndrome: A disorder appearing within the first few weeks of life characterized by the association of leukodystrophy, basal ganglia calcification and cerebrospinal fluid lymphocytosis. Affected individuals begin exhibiting symptoms of subacute encephalopathy associated with epilepsy, chilblain skin lesions on the extremities and episodes of aseptic febrile illness. Several months later, patients develop microcephaly and pyramidal signs. These conditions are result in death within the first decade of life. Pseudo-TORCH syndrome shared clinical features with congenital TORCH syndrome, which is caused by infections due to toxoplasmosis, rubella, cytomegalovirus and herpes simplex virus. There are several underlying causes of this disorder with at least 6 autosomal genomic loci: Aicardi–Goutières syndrome-1 is caused by either autosomal dominant and recessive mutations in the *TREX1* gene which encodes the 3'->5' exonuclease. There are also a form caused by *de novo* mutations in the *TREX1* gene;; Aicardi–Goutières syndrome-2 is caused by autosomal recessive mutations in the *RNASEH2B* gene which encodes the subunit of the RNase H2 endonuclease complex;; Aicardi–Goutières syndrome-3 is caused by autosomal recessive mutations in the *RNASEH2C* gene which encodes the subunit of the RNase H2 endonuclease complex;; Aicardi–Goutières syndrome-4 is caused by autosomal recessive mutations in the *RNASEH2A* gene which encodes the subunit of the RNase H2 endonuclease complex;; Aicardi–Goutières syndrome-5 is caused by autosomal recessive mutations in the *SAMHD1* gene which encodes the SAM domain and HD domain 1;; and Aicardi–Goutières syndrome-6 is caused by autosomal recessive or autosomal dominant mutations in the *ADAR* gene which encodes the adenosine deaminase RNA-specific. Most of the cases follow autosomal recessive patterns of inheritance. Consequently, these conditions are more prevalent in consanguineous families. Aicardi–Goutières syndrome is estimated to affect less than 1 in every 1,000,000 individuals, but it may be under-diagnosed due to the TORCH syndrome.

Familial encephalopathy with neuroserpin inclusion bodies: A disorder appearing within the second to the fifth decade of life characterized by cognitive decline. Affected individuals exhibit attention and concentration deficit, response regulation difficulties, and impaired visuospatial skills with memorial impairment. Affected patients manifest a smaller degree of symptoms than those usually seen in patients with Alzheimer's disease. This form of dementia is unique in that it doesn't affect recall memory. Familial encephalopathy with neuroserpin inclusion bodies is caused by autosomal dominant mutations in the *SERPINI1* gene which encodes the neuroserpin. Familial encephalopathy with neuroserpin inclusion bodies is estimated to affect less than 1 in every 1,000,000 individuals. This condition has been reported in 2 unrelated Caucasian families living in the United States, and a few more families around the world.

Neurodegeneration due to cerebral folate transport deficiency: A disorder characterized by severe developmental regression, movement disturbances, epilepsy, and leukodystrophy of late infancy onset. Affected individuals are severely handicapped and wheelchair-bound, and suffer from epileptic seizures resistant to treatment. This condition starts manifesting neurodegeneration signs in the second or third year of life. Neurodegeneration due to cerebral folate transport deficiency can be caused by autosomal recessive mutations in the *FOLR1* gene which encodes the folate receptor 1. This protein is transporter brain-specific of folate. This condition is responsive to folinic acid therapy which can reverse the clinical symptoms. Neurodegeneration due to cerebral folate transport deficiency is estimated to affect less than 1 in every 1,000,000 individuals.

The Concise Encyclopedia of Genomic Diseases

Table 3.06.04.1 Recognizable Degenerative Diseases of the Nervous System

Disorder	OMIM	Gene	Biological Function	Product of Conception Alteration
Alzheimer disease-1	104300	APP	Integral membrane protein	AD or AR
Alzheimer disease-2	104310	APOE	Triglyceride-rich lipoprotein	AD or AD + AD = AR
Alzheimer disease-3, with spastic paraparesis and apraxia and unusual plaques	607822	PSEN1	Regulate APP processing	AD
Alzheimer disease-4	606889	PSEN2	Regulate APP processing	AD
Late-onset Alzheimer disease	602710	APBB2	Cytoplasmic domain	
Susceptibility to Alzheimer disease	602403	BLMH	Cytoplasmic peptidase	
Susceptibility to Alzheimer disease	602005	SORL1	Endocytosis and sorting	
Susceptibility to Alzheimer disease	613609	HFE	Cell surface receptor Other/ungrouped	
Susceptibility to Alzheimer disease	106180	ACE	Integral membrane protein	
Susceptibility to Alzheimer disease	606989	MPO	Microbicidal activity of netrophils	
Susceptibility to Alzheimer disease	608254	PAXIP1	Nuclear protein transcription-activation	
Susceptibility to Alzheimer disease	103950	A2M	Plasma and extracellular spaces protein	
Susceptibility to late-onset Alzheimer disease	191840	PLAU	Serine protease	
Susceptibility to late-onset Alzheimer disease	163729	NOS3	Nitric oxide synthase	
Alzheimer disease	614263	BACE1-AS	Long non-coding RNAs disease	
Dementia with Lewy bodies	127750	SNCA	Inhibit phospholipase	AD + de novo
Dementia with Lewy bodies	127750	SNCB	Inhibit phospholipase	AD + de novo
Leigh syndrome	603647	BCS1L	ATPase, AAA-type	AR
Leigh syndrome	256000	DLD	Mitochondrial protein	AR
Leigh syndrome	603846	NDUFS3	Mitochondrial protein	AR
Leigh syndrome	602694	NDUFS4	Mitochondrial protein	AR
Leigh syndrome	601825	NDUFS7	Mitochondrial protein	AR

Table 3.06.04.1. Continued.

Disorder	OMIM	Gene	Biological Function	Product of Conception Alteration
Leigh syndrome	602141	NDUFS8	Mitochondrial protein	AR
Leigh syndrome	161015	NDUFV1	Mitochondrial protein	AR
Leigh syndrome	308930	PDHA1	Mitochondrial protein	XR
Leigh syndrome	600857	SDHA	Mitochondrial protein	AR
Leigh syndrome	613622	FOXRED1	Mitochondrial protein	AR
Leigh syndrome	609653	NDUFAF2	Mitochondrial protein	AR
Leigh syndrome due to cytochrome c oxidase deficiency	603646	COX15	Mitochondrial protein	AR
Leigh syndrome due to mitochondrial complex I deficiency	612392	C8orf38	Mitochondrial protein	AR
Leigh syndrome due to mitochondrial complex I deficiency	602137	NDUFA2	Mitochondrial protein	AR
Leigh syndrome, due to COX deficiency	185620	SURF1	Inner mitochondrial membrane protein	AR
Leigh syndrome, French-Canadian type	220111	LRPPRC	Mitochondrial protein	AR
Mitochondrial Leigh syndrome	590060	MT-TK	Mitochondrial protein	Mit
Mitochondrial Leigh syndrome	590050	MT-TL1	Mitochondrial protein	Mit
Mitochondrial Leigh syndrome	590105	MT-TV	Mitochondrial protein	Mit
Mitochondrial Leigh syndrome	590095	MT-TW	Mitochondrial protein	Mit
Pyruvate carboxylase deficiency	266150	PC	Mitochondrial protein	AR
Aminoacylase 1 deficiency	609924	ACY1	Recycling of N-acetylated amino acids	AR
Familial British dementia	176500	ITM2B	Integral membrane protein	AD
Familial Danish dementia	117300	ITM2B	Integral membrane protein	AD
Familial nonspecific dementia	600795	CHMP2B	Endosomal sorting complex required for transport	AD

Table 3.06.04.1. Continued.

Disorder	OMIM	Gene	Biological Function	Product of Conception Alteration
Frontotemporal dementia	600274	PSEN1	Regulate APP processing	AD
Frontotemporal dementia with or without parkinsonism	600274	MAPT	Cytoskeletal Microtubules	AD
Frontotemporal dementia, with or without parkinsonism	600274	MAPT	Cytoskeletal Microtubules	AD
Frontotemporal lobar degeneration with ubiquitin-positive inclusions	607485	GRN	Granulin	AD
Frontotemporal lobar degeneration, TARDBP-related	612069	TARDBP	DNA-binding protein	AD
Pick disease	172700	MAPT	Cytoskeletal Microtubules	AD
Pick disease	172700	PSEN1	Regulate APP processing	AD
Aicardi-Goutieres syndrome-1	225750	TREX1	Exonuclease	AR or AD + de novo
Aicardi-Goutieres syndrome-2	610181	RNASEH2B	DNA replication	AR
Aicardi-Goutieres syndrome-3	610329	RNASEH2C	Nucleus ribonuclease	AR
Aicardi-Goutieres syndrome-4	610333	RNASEH2A	DNA replication Separation/initiation	AR
Aicardi-Goutieres syndrome-5	612952	SAMHD1	Innate immune response	AR
Aicardi-Goutieres syndrome-6	615010	ADAR	Adenosine deaminase RNA-specific	AR or AD
Familial encephalopathy with neuroserpin inclusion bodies	604218	SERPINI1	Serine protease inhibitor	AD
Neurodegeneration due to cerebral folate transport deficiency	613068	FOLR1	Folate receptor	AR

3.06.05 Demyelinating Diseases of the Central Nervous System

Multiple sclerosis: An autoimmune and inflammatory demyelinating disorder characterized by relapsing episodes of neurologic impairment followed by remissions. Genomic and environmental factors influence susceptibility to the disease, but multiple sclerosis is not a clearly genomically inherited condition. Environmental factors include geographic latitude, as Multiple sclerosis is more common in people who live farther from the equator. Other non-genomic factors include stress, smoking, occupational exposures, toxins and vaccinations. There are several underlying causes of this disorder with at least 5 autosomal genomic loci: Susceptibility to multiple sclerosis-1 is associated with variation in certain HLA genes on chromosome 6p21, including HLA-A, HLA-DRB1, HLA-DQB1 and HLA-DRA. An HLA-DRB1*1501-DQB1*0602 haplotype (HLA-DR15) has been repetitively demonstrated in high-risk people of Northern European ancestry;; Susceptibility to multiple sclerosis-2 has been mapped, but the gene has not yet been characterized;; Susceptibility to multiple sclerosis-3 has been mapped, but the gene has not yet been characterized;; Susceptibility to multiple sclerosis-4 has been mapped, but the gene has not yet been characterized;; and Susceptibility to multiple sclerosis-5 is influenced by variation in the *TNFRSF1A* gene which encodes the tumor necrosis factor receptor superfamily (member 1A). Another inclusion criterion is that monozygotic twin's studies showed a 25.9% of concordance rate for this condition. First degree relatives have a 15-20-fold larger than average risk for this inflammatory condition. Multiple sclerosis is estimated to affect 2-150 in every 100,000 individuals, and is more common in women.

Global cerebral hypomyelination: A disorder characterized by severe psychomotor retardation, hypotonia and hypomyelination of the central nervous system. This condition starts manifesting in the first months of life. Years later, patients develop seizures, episodic apnea and severe spasticity with hyperreflexia. Global cerebral hypomyelination can be caused by autosomal recessive mutations in the *SLC25A12* gene which encodes the solute carrier (family 25 member A12), an aspartate/glutamate carrier. This protein is a transporter of aspartate from mitochondria to cytosol in exchange for glutamate. Aralar protein also plays a role in the malate-aspartate NADH shuttle. This condition has been reported in a child born of distantly related Swedish parents.

3.06.06 Episodic and Paroxysmal Disorders

3.06.06.1 Epilepsy

Autosomal dominant nocturnal frontal lobe epilepsy: A disorder appearing during childhood characterized by clusters of violent seizures during sleep, shortly before awakening, that last from a few seconds to a few minutes. These seizures regularly involve both vocalizations and complex motor movements such as hand clenching and knee bending. The seizures associated with this condition begin in the frontal lobes, leading to inability to perform critical functions including reasoning, planning, judgment and problem-solving. The prevalence of this condition is unknown. Symptoms of Autosomal dominant nocturnal frontal lobe epilepsy are frequently mistaken for nightmares. There are several underlying causes of this disorder with at least 4 autosomal genomic loci. Three of these loci have been characterized at the gene level including nicotinic acetylcholine receptor alpha and beta subunits *CHRNA4, CHRNB2,* and *CHRNA2* genes. Autosomal dominant nocturnal frontal lobe epilepsy-1 can be caused by monoallelic mutations in the *CHRNA4* gene which encodes the neuronal nicotinic acetylcholine receptor (subunit alpha-4);; Autosomal dominant nocturnal frontal lobe epilepsy-3 can be caused by monoallelic mutations in the *CHRNB2* gene which encodes the neuronal nicotinic acetylcholine receptor (subunit beta-2);; and Autosomal dominant nocturnal frontal lobe epilepsy-4 can be caused by monoallelic mutations in the *CHRNA2* gene which encodes the neuronal nicotinic acetylcholine receptor (subunit alpha-2). Type 2 has been mapped, but the gene has not yet been characterized. Autosomal dominant nocturnal frontal lobe epilepsy has been reported in more than 100 families worldwide.

Autosomal dominant familial temporal lobe epilepsy: A disorder characterized by partial seizures originating from the temporal lobe with auditory features such as buzzing, humming or ringing. There are several underlying causes of this disorder with at least 5 autosomal genomic loci: Autosomal dominant familial temporal lobe epilepsy-1 is caused by monoallelic mutations in the *LGI1* gene which encodes the leucine-rich glioma-inactivated protein 1;; Autosomal dominant familial temporal lobe epilepsy-5 is caused by monoallelic mutations in the *CPA6* gene which encodes the carboxypeptidase A6; and Autosomal dominant familial temporal lobe epilepsy 2, 3, and 4 are familial forms which are believed to be caused by autosomal dominant or monoallelic mutations; the loci responsible have been mapped, but the genes have not yet been characterized. The prevalence of this condition is unknown.

Childhood absence epilepsy or pyknolepsy: A disorder appearing during 4-10 years of age characterized by absence seizures. Most of the time these absence seizures are brief (~4–20 seconds), but they occur hundreds of times per day. The prognosis is favorable with most patients "growing out" of their epilepsy. Childhood absence epilepsy is estimated to affect 1 in every 1,000 children. In 5-15% of these cases the susceptibility to the development of childhood absence epilepsy may be conferred by genomic factors. There are

several underlying causes of this disorder with at least 5 autosomal genomic loci: Childhood absence epilepsy-1 has been mapped but the gene has not yet been characterized;; Childhood absence epilepsy-2 is conferred by variations in the *GABRG2* gene which encodes the gamma-aminobutyric acid (GABA) A receptor (subunit gamma 2);; Childhood absence epilepsy-4 is conferred by variations in the *GABRA1* gene which encodes the gamma-aminobutyric acid (GABA) A receptor (subunit alpha 1);; Childhood absence epilepsy-5 is conferred by variations in the *GABRB3* gene which encodes the gamma-aminobutyric acid (GABA) A receptor (subunit beta 3);; and Childhood absence epilepsy-6 is conferred by variations in the *CACNA1H* gene which encodes the voltage-dependent calcium channel (type T subunit alpha 1H). Childhood absence epilepsy-3 has been renamed.

Juvenile myoclonic epilepsy or Janz syndrome: A disorder appearing during 12-18 years of age characterized by myoclonic jerks. Myoclonic epilepsy patients usually manifest symptoms in the morning. The condition is prevalent in between 4-10% of all patients with epilepsy. There are several underlying causes of this disorder with at least 9 autosomal genomic loci (5 of them with known causative genes); Most of these genes are either ion channels or affecting ion channel currents: Juvenile myoclonic epilepsy-1 is caused by mutations in the *EFHC1* gene which encodes the EF-hand domain (C-terminal) containing protein 1;; Juvenile myoclonic epilepsy-2, 3, 4, and 9 have been mapped, but the genes have not yet been characterized;; Juvenile myoclonic epilepsy-5 is caused by mutations in the *GABRA1* gene which encodes the gamma-aminobutyric acid (GABA) A receptor (alpha 1);; Juvenile myoclonic epilepsy-6 is caused by mutations in the *CACNB4* gene which encodes the calcium channel voltage-dependent (subunit beta 4);; Juvenile myoclonic epilepsy-7 is caused by mutations in the *GABRD* gene which encodes the gamma-aminobutyric acid (GABA) A receptor (subunit delta);; and Juvenile myoclonic epilepsy-8 is caused by mutations in the *CLCN2* gene which encodes the chloride channel voltage-sensitive 2.

Progressive myoclonus epilepsies: Are a group of disorders characterized by progressive neurodegeneration and cerebellar ataxia. These conditions start manifesting symptoms between 6 and 13 years of age. Patients with this condition exhibit myoclonic and tonic-clonic seizures associated with progressive neurological decline. There are several underlying causes of this disorder with at least 8 autosomal genomic loci: Progressive myoclonic epilepsy-1 or Myoclonic epilepsy of Unverricht and Lundborg is caused by autosomal recessive mutations in the *CSTB* gene which encodes the cystatin B (stefin B). This protein functions as an intracellular thiol protease inhibitor. The most typical mutation includes a dodecamer repeat expansion upstream of the transcription start. This expansion causes in the variant alleles (containng more than 60 repeats) a reduced transcription of CSTB mRNA;; Progressive myoclonic epilepsy-1B is caused by autosomal recessive and autosomal dominant mutations in the *PRICKLE1* gene which encodes the prickle homolog protein 1;; Myoclonic epilepsy of Lafora or Progressive myoclonic epilepsy-2A is caused by autosomal recessive mutations in the *EPM2A* gene which encodes the epilepsy progressive myoclonus type 2;; Myoclonic epilepsy of Lafora or Progressive myoclonic epilepsy-2B is caused by autosomal recessive mutations in the *NHLRC1* gene which encodes the NHL repeat containing protein 1;; Progressive myoclonic epilepsy-3 is caused by auto-somal recessive mutations in the *KCTD7* gene which encodes the potassium channel tetramerization domain containing protein 7;; Progressive myoclonic epilepsy-4 is caused by autosomal recessive mutations in the *SCARB2* gene which encodes the scavenger receptor (class B member 2);; Progressive myoclonic epilepsy-5 is caused by autosomal recessive mutations in the *PRICKLE2* gene which encodes the prickle homolog protein 2;; and Progressive myoclonic epilepsy-6 is caused by autosomal recessive mutations in the *GOSR2* gene which encodes the golgi SNAP receptor complex (member 2). There are also other disorders that course with progressive myoclonic epilepsy features including MERFF syndrome, Neuronal ceroid lipofuscinoses; Sialidosis, Dentatorubral-pallidoluysian atrophy and others. In Finland, the carrier frequency for myoclonic epilepsy of Unverricht and Lundborg trait is estimated to be 1 in 71 individuals, which means the prevalence of this condition is 1 in 20,000 or $1/(142)^2$ individuals. This condition is also prevalent among people from North Africa. Progressive myoclonic epilepsy accounts for less than 1% of all forms of epilepsy.

Amish infantile epilepsy syndrome: A disorder characterized by epilepsy syndrome associated with developmental stagnation and blindness of infantile onset. Affected children start manifesting irritability, poor feeding, vomiting and failure to thrive noted between the ages of 2 weeks and 3 months. Amish infantile epilepsy syndrome is caused by autosomal recessive mutations in the *ST3GAL5* gene which encodes the ST3 beta-galactoside alpha-2,3-sialyltransferase,5 a glycosyltransferase that may be localized to the Golgi apparatus. Among Old Order Amish of Geauga County, Ohio, the carrier frequency for Amish infantile epilepsy syndrome trait is estimated to be 1 in 11 individuals, which means the frequency of this condition is about 1 in 500 individuals or $1/(22)^2$.

Epilepsy with neurodevelopmental defects: A neurodevelopmental disorder characterized by mental retardation or learning difficulties and epilepsy of onset in the first decade of life. Epilepsy with neurodevelopmental defects can be caused by either inherited or *de novo* autosomal monoallelic mutations in the *GRIN2A* gene which encodes the regulatory subunits of ionotropic glutamate receptor N-methyl D-aspartate 2A (NMDA receptor).

Polyhydramnios, megalencephaly and symptomatic epilepsy: A disorder characterized by a combination of complicated pregnancies by polyhydramnios, spontaneous onset of labor occurring between 25 and 36 weeks' of gestation, macrocephaly, complex partial seizures and severe psychomotor retardation. Most of these patients die during the first decade of life. "Polyhydramnios, megalencephaly and symptomatic epilepsy" is caused by autosomal recessive mutations in the *STRADA* gene which encodes the STE20-related kinase adaptor alpha. This condition is caused by a biallelic or homozygous 7-kb deletion. In Old Order Mennonites, the carrier frequency for Polyhydramnios, megalencephaly and symptomatic epilepsy trait is estimated to be 1 in 25 individuals, which means the frequency of this condition is about 1 in 2,500 individuals or $1/(50)^2$.

Rolandic epilepsy with speech dyspraxia and mental retardation: A disorder appearing during childhood characterized by oral and speech dyspraxia, rolandic seizures and mental retardation. Rolandic epilepsy with speech dyspraxia and mental retardation can be caused by X-linked dominant mutations in the *SRPX2* gene

which encodes the X-linked sushi-repeat containing protein 2. There is also a form of this disorder that shows autosomal dominant inheritance, but the gene has not been characterized yet. The prevalence of this condition is unknown.

Early infantile epileptic encephalopathy or Ohtahara syndrome: A severe form of epilepsy of infancy onset, characterized by recurrent tonic seizures or spasms. Affected children exhibit developmental regression, such as sitting, rolling over and babbling. There are several underlying causes of Early infantile epileptic encephalopathy with at least 13 autosomal genomic loci and 2 X-linked genomic loci: Early infantile epileptic encephalopathy-1 can be caused by *de novo* X-linked recessive mutations in the *ARX* gene which encodes the aristaless-related homeobox. Males are more severely affected for this condition than females. Mutations in the *ARX* gene has been associated with developmental disorders including syndromic and non-syndromic mental retardation, lissencephaly, Proud syndrome and Infantile spasms without brain malformations;; Early infantile epileptic encephalopathy-2 is caused by X-linked dominant mutations in the *CDKL5* gene which encodes the cyclin-dependent kinase-like 5;; Early infantile epileptic encephalopathy-3 is caused by autosomal recessive mutations in the *SLC25A22* gene which encodes the solute carrier (family 25 member A22), a mitochondrial glutamate carrier 1;; Early infantile epileptic encephalopathy-4 is caused by monoallelic mutations in the *STXBP1* gene which encodes the syntaxin binding protein 1;; Early infantile epileptic encephalopathy-5 is caused by monoallelic mutations in the *SPTAN1* gene which encodes the spectrin alpha chain (non-erythrocytic 1);; Early infantile epileptic encephalopathy-6 or Dravet syndrome is caused by *de novo* monoallelic mutations in the *SCN1A* gene which encodes the sodium channel protein (subunit alpha-1);; Early infantile epileptic encephalopathy-7 is caused by autosomal recessive mutations in the *KCNQ2* gene which encodes the potassium voltage-gated channel (subfamily KQT member 2);; Early infantile epileptic encephalopathy-8 is caused by X-linked mutations in the *ARHGEF9* gene which encodes the Cdc42 guanine nucleotide exchange factor 9;; Early infantile epileptic encephalopathy-9 is caused by autosomal recessive mutations in the *PCDH19* gene which encodes the protocadherin 19;; Early infantile epileptic encephalopathy-10 is caused by autosomal recessive mutations in the *PNKP* gene which encodes the polynucleotide kinase 3'-phosphatase;; Early infantile epileptic encephalopathy-11 is caused by autosomal dominant mutations in the *SCN2A* gene which encodes the sodium channel protein (subunit alpha-2);; Early infantile epileptic encephalopathy-12 is caused by autosomal recessive mutations in the *PLCB1* gene which encodes the phospholipase C, beta 1 (phosphoinositide-specific);; Early infantile epileptic encephalopathy-13 is caused by autosomal dominant mutations in the *SCN8A* gene which encodes the sodium channel protein (subunit alpha-8);; Early infantile epileptic encephalopathy-14 is caused by autosomal dominant mutations in the *KCNT1* gene which encodes the potassium channel (subfamily T member 1);; and Early infantile epileptic encephalopathy-15 is caused by autosomal recessive mutations in the *ST3GAL3* gene which encodes the sialyltransferase 3. Early infantile epileptic encephalopathy, with the exception of Dravet syndrome, is estimated to affect 1-1.6 in every 100,000 live births, and around 75% of these cases progress to West syndrome.

Dravet syndrome: A disorder characterized by refractory epileptic encephalopathy occurring in otherwise healthy infants. This form of epileptic encephalopathy appears during the first year of life and is characterized by generalized tonic, clonic and tonic-clonic seizures. These episodes of seizure have been associated with fever and occur every 1-2 months. Other seizure types affecting these patients include myoclonus, atypical absences and complex partial seizures starting during the 2nd or 3rd year of life. Affected individuals manifest ataxia, behavioral disorders, photosensitivity and slowing the psychomotor development. This condition is also associated with an increased risk of sudden death in childhood, especially between 2 and 4 years of age. This condition is named after Charlotte Dravet. Dravet syndrome can be caused mainly by *de novo* monoallelic mutations in the *SCN1A* gene which encodes the voltage-gated sodium channel. The product of conception alterations are defined as autosomal dominant syndrome plus 90% (AD + 90%). Most of the *de novo* mutations in Dravet syndrome seem to originate in the paternal allele. Dravet syndrome belongs to the group of channelopathies. Dravet syndrome is estimated to affect 1 in every 20,000-40,000 individuals. Dravet syndrome is two times more frequent in males than in females.

Benign familial neonatal epilepsy: A disorder characterized by tonic-clonic seizures in an otherwise healthy neonate, without precipitating factors. The seizures start in the first week of life and remit spontaneously by 4-12 months of age. Seizures episodes are brief, usually lasting 1-2 min, and may be as frequent as 20-30 per day. Around 10-15% of patients have febrile or afebrile seizures later in childhood. There are several underlying causes of this disorder with at least 3 autosomal genomic loci: Benign familial neonatal epilepsy type 1 can be caused by autosomal dominant mutations in the *KCNQ2* gene which encodes the potassium voltage-gated channel (subfamily KQT member 2);; Benign familial neonatal epilepsy type 2 can be caused by autosomal dominant mutations in the *KCNQ3* gene which encodes the potassium voltage-gated channel (subfamily KQT member 3);; and Benign familial neonatal epilepsy type 3 can be caused by pericentric inversion on chromosome 5;; and there is also an autosomal recessive form reported in an Iranian Jewish kindred, the loci has not been characterized yet. Benign familial neonatal epilepsy is estimated to affect 1-14.4 in every 100,000 births.

Febrile seizure: A disorder characterized by seizures in children between 6 months to 6 years of age, associated with fever but without any evidence of intracranial infection or defined pathological or traumatic cause. Patients with febrile seizures have 5-7 times higher risk of developing seizures later in life than those without a history of febrile seizures. This condition affects 3% of children and is twice more common in males than in females. There is a higher prevalence of Febrile seizures among people from Guam, where 14% of children are affected and Japan, where 7% of children are affected. Sometimes these seizures are precipitated by a recent upper respiratory infection or gastroenteritis. The mode of inheritance follows mainly autosomal dominance pattern or polygenic inheritance. There seems to be no contribution in X-linked or mitochondrial inheritance. There are several underlying causes of this disorder with at least 11 autosomal genomic loci: There are at least 11 genomic forms of febrile seizures: types 1, 2, 5, 7, 9 and 10 have been mapped, but the genes have not yet been characterized;; Febrile seizures-3A is caused by mutations in the *SCN1A* gene which encodes the sodium channel (subunit alpha 1);; Febrile seizures-3B is caused by mutations in the

SCN9A gene which encodes the sodium channel (subunit alpha 9);; Febrile seizures-4 is caused by mutations in the *GPR98* gene which encodes the G protein-coupled receptor 98;; Febrile seizures-8 is caused by mutations in the *GABRG2* gene which encodes the gamma-aminobutyric acid (GABA) A receptor (subunit gamma 2);; and Febrile seizures-11 is caused by mutations in the *CPA6* gene which encodes the carboxypeptidase A6. Many febrile seizures genes and proteins belong to the group of channelopathies, including mutations in the *SCN1A* gene, *SCN9A* gene, and the *GABRG2* gene.

Generalized epilepsy with febrile seizures plus: A disorder characterized by features beyond the classic febrile seizures, which persist after the age of 6 years, and/or a variety of afebrile seizure types. There are several underlying causes of this disorder with at least 8 autosomal genomic loci: Generalized epilepsy with febrile seizures plus-1 is caused by autosomal dominant mutations in the *SCN1B* gene which encodes the sodium channel (subunit beta 1);; Generalized epilepsy with febrile seizures plus-2 is caused by autosomal dominant mutations in the *SCN1A* gene;; Generalized epilepsy with febrile seizures plus-3 is caused by autosomal dominant mutations in the *GABRG2* gene;; Generalized epilepsy with febrile seizures plus-5 is associated with variations in the *GABRD* gene which encodes the gamma-aminobutyric acid (GABA) A receptor (subunit delta);; Generalized epilepsy with febrile seizures plus-7 is caused by autosomal dominant mutations in the *SCN9A* gene;; and Generalized epilepsy with febrile seizures plus type 4, 6, and 8 have been mapped, but the genes have not yet been characterized. The known genes which cause these conditions include voltage-gated sodium channels; alpha subunit genes encoded by *SCN1A* and *SCN9A*, beta subunit genes encoded by *SCN1B*, and a *GABAA* receptor gamma subunit gene, encoded by the *GABRG2* gene and the *GABRD* gene. Generalized epilepsy with febrile seizures plus belongs to the group of channelopathies, and its prevalence is unknown.

West syndrome: A disorder appearing between 3 and 12 months of age characterized by a triad of: clusters of axial spasms, hypsarrhythmic interictal EEG pattern and developmental regression. This condition is named after William James West. West syndrome is the most frequent type of epileptic encephalopathy and is estimated to affect 1 in every 3,200-3,500 of live births. Boys are slightly more affected than girls. The etiologies of the syndrome include environmental and genomic factors. In this condition, it is difficult to separate the effects of heritability and environmental factors. The environmental factors include cerebral malformations, sequelae of meningoencephalitis or ischemia organic disorder of the brain, congenital infections, post-vaccination reactions, etc. The genomic factors that cause West syndrome include: Down syndrome, Patau syndrome, 1p36 deletion syndrome, Tuberous sclerosis, and X-linked form of Early infantile epileptic encephalopathy, caused by mutations in either the *ARX* gene which encodes the aristaless related homeobox, or the *CDKL5* gene which encodes the cyclin-dependent kinase-like 5.

Lennox–Gastaut syndrome: A form of pharmacoresistant epileptic encephalopathy characterized by frequent seizures and different seizure types including: axial tonic seizures, atypical absences, and sudden atonic or myoclonic falls. This early onset condition in general appears between 2-7 years of age. This condition is also associated with developmental delay, and psychological and behavioral problems. The causes of Lennox-Gastaut syndrome include environmental and genomic factors. The environmental factors include: cortical dysplasia, cranial trauma, perinatal asphyxia, sequelae of meningoencephalitis and other forms of epilepsy such as West syndrome. The genomic factors can be caused by either *de novo* or autosomal dominant monoallelic mutations in the *MAPK10* gene which encodes the mitogen-activated protein kinase 10. Other genomic conditions associated with Lennox-Gastaut syndrome include Tuberous sclerosis and genomics forms of West syndrome. Lennox-Gastaut syndrome is estimated to affect 15 in every 100,000 births.

Pyridoxamine 5'-phosphate oxidase deficiency or Pyridoxal phosphate-responsive seizures: A form of epileptic encephalopathy characterized by severe seizures within hours of birth. These episodes of seizure are responsive to treatment with pyridoxal phosphate, but they are not responsive to anticonvulsants drug therapy. Pyridoxamine 5'-phosphate oxidase deficiency can be caused by autosomal recessive mutations in the *PNPO* gene which encodes the pyridoxamine 5'-phosphate oxidase. This enzyme catalyzes the terminal, rate-limiting step in the synthesis of pyridoxal 5'-phosphate, also known as vitamin B6. This condition has been reported among children of consanguineous parents.

Pyridoxine dependent seizures: A disorder characterized by a combination of various seizure types appearing during the first hours of life that is unresponsive to any anticonvulsant therapy except pyridoxine. The dependence is permanent, and the interruption of daily pyridoxine supplementation leads to the recurrence of seizures. The seizure types include myoclonic, atonic, infantile spasms, and partial or generalized seizures. Pyridoxine dependent seizures can be caused by autosomal recessive mutations in the *ALDH7A1* gene which encodes the aldehyde dehydrogenase (family 7 member A1). This enzyme is an alpha-aminoadipic semialdehyde dehydrogenase acting in the pipecolic acid pathway of lysine catabolism. "Pyridoxine dependent seizures" is estimated to affect 1 in every 400.000-700.000 live births.

3.06.06.2 Headaches

Migraine: A chronic disorder characterized by recurrent moderate to severe headaches often in association with a number of autonomic nervous system symptoms. There are several underlying causes of this disorder with at least 12 autosomal genomic loci and 1 X-linked genomic locus: Migraine with or without aura-1 to 12 have been mapped, but the genes have not yet been characterized; and Migraine with or without aura-13 is caused by monoallelic mutations in the *KCNK18* gene which encodes the potassium channel (subfamily K member 18). Susceptibility to migraine can be conferred by a polymorphism in the *ESR1* gene which encodes the estrogen receptor and a polymorphism in the *TNF* gene which encodes the tumor necrosis factor. A polymorphism in the *EDNRA* gene which encodes the endothelin receptor type A may confer resistance to migraine. Genomics factors include a 2-fold increase in the risk of migraine among first-degree relatives of patients with this condition. Migraine is the most common type of chronic episodic headache.

Familial hemiplegic migraine: A form of classical migraine that usually includes hemiparesis during the aura phase. Affected individuals can also manifest other symptoms such as ataxia, coma and epileptic seizures. There are several underly-

ing causes of this disorder with at least 4 autosomal genomic loci: Familial hemiplegic migraine-1 is caused by monoallelic mutations in the *CACNA1A* gene which encodes the P/Q-type calcium channel (alpha subunit);; Familial hemiplegic migraine-2 is caused by monoallelic mutations in the *ATP1A2* gene which encodes the Na+/K+-ATPase (alpha 2 polypeptide);; Familial hemiplegic migraine-3 is caused by autosomal dominant mutations in the *SCNA1* gene which encodes the sodium channel (subunit alpha 1);; and Familial hemiplegic migraine-4 has been mapped, but the gene has not yet been characterized. Type 1 is the most commonly occurring type in 50% of patients with this condition, type 2 is the second most commonly occurring type in 25%. Type 1 and 2 can be either inherited or *de novo* in 30-45% of cases. The product of conception alterations are defined as autosomal dominant syndrome plus 30% (AD + 30%). The prevalence of all the familial and *de novo* hemiplegic migraine is unknown. In Denmark, Familial hemiplegic migraine is estimated to affect 1 in every 10,000 individuals.

Familial cluster headache: A disorder occurring once every other day to 8 times per day and lasting 15-180 minutes. This condition is characterized by excruciating unilateral periorbital pain associated with ipsilateral autonomic signs such as lacrimation, nasal congestion, ptosis, miosis, lid edema and eye redness. There are 2 forms of cluster headache: episodic and chronic. Around 85% of cluster headache patients have the episodic subtype, in which the headaches occur in cluster periods lasting from 7 days to 1 year and separated by attack-free intervals of 1 month or more. The remainder of patients has the chronic subtype, in which attacks reappear for greater than 1 year without remission or with remissions lasting less than 1 month. This condition has a familial origin in about 10% of cases. Genomics factors include a 39-fold increase in the risk of cluster headache among first-degree relatives of patients with this condition. Cluster headache patients seem to follow an autosomal dominant pattern with a penetrance of 0.3-0.34 in males and 0.17-0.21 in females. Cluster headache is associated with the G1246A polymorphism in the *HCRTR2* gene which encodes the hypocretin receptor 2. Homozygosity for the G allele confers a 2-5-fold increased risk for the disorder, compared to G/A and A/A. Cluster headache is estimated to affect 0.5-1 in every 1,000 individuals, predominantly males.

3.06.06.3 Sleep Disorders

Congenital central hypoventilation syndrome or Ondine's Curse: A disorder characterized by abnormal control of respiration leading to respiratory arrest during sleep. Affected newborns usually develop cyanosis and increased carbon dioxide during sleep during the first hours of life. Affected neonates have no identifiable brainstem lesions, or neuromuscular, lung or cardiac disease. Affected individuals have a deficiency in the arousal responses to hypercapnia and hypoxemia, which involves an inborn failure of autonomic control of breathing. Affected individuals require a lifelong dependence to mechanical ventilation. Congenital central hypoventilation syndrome can be caused by *de novo* monoallelic mutations of *PHOX2B* gene which encodes the homeobox protein 2b. The protein functions as a transcription factor involved in the development of the autonomic nervous system, several major noradrenergic neurons and the determination of neurotransmitter phenotype. Mutations of *PHOX2B* gene account for 90% of all cases. At least 60% of all cases are caused by *de novo* mutations in *PHOX2B* gene, due to alanine expansions. Most mutations consist of 5 to 9 alanine expansions within a 20-residue polyalanine tract. In the remaining 10% of all the cases, this condition could also be caused by mutations in other genes, including *RET, GDNF, EDN3, BDNF* and *ASCL1*. Haddad syndrome is a combination of Congenital central hypoventilation syndrome associated with Hirschsprung disease. Haddad syndrome is caused by autosomal dominant mutations in the *ASCL1* gene which encodes the achaete-scute complex homolog protein 1. Congenital central hypoventilation syndrome can be associated with Hirschsprung disease in 16% of the cases. Congenital central hypoventilation syndrome is estimated to affect 1 in every 200,000 live births.

Narcolepsy: A chronic sleep disorder characterized by excessive day-time sleepiness associated with sleep attacks at inappropriate times and sometimes paralysis at sleep, hypnagogic hallucinations and automatic behavior. Affected individuals experience the REM stage of sleep within 5 minutes; while most people do not experience REM sleep until at least until one hour. This condition usually starts between the ages of 10-30 years and is a lifetime disease. This condition is also associated with cataplexy which is a sudden muscular weakness usually brought on by strong emotions. The disabling features of this condition include educational, occupational, cognitive and psychosocial problems associated with the excessive daytime sleepiness symptom of narcolepsy. There are several underlying causes of this disorder with at least 7 autosomal genomic loci: Narcolepsy-1 is caused by either inherited or *de novo* monoallelic mutations in the *HCRT* gene which encodes the hypocretin (orexin), a neuropeptide precursor;; Narcolepsy 2, 3, 4, 5, and 6 have been mapped, but the genes have not yet been characterized;; and Narcolepsy-7 is caused by autosomal dominant mutations in the *MOG* gene which encodes the myelin oligodendrocyte glycoprotein. There are also alleles that confer resistance to narcolepsy such as a marker in the proximity of *LINC00163* gene which maps on the long intergenic non-protein coding RNA 163. Narcolepsy is associated with the HLA Region on Chromosome 6p21. In fact, almost all the affected individuals of European descent with narcolepsy carry the HLA haplotype DRB5*0101-DRB1*1501-DQA1*0102 -DQB1*0602. This is a common allele affecting 15-25% of European descent individuals. Other genomic factor includes the monozygotic twin's studies showing a 25-31% of concordance rate for this condition. Narcolepsy is a common condition and is estimated to affect 1 in every 1,500 persons.

Table 3.06.06.1 Recognizable Episodic and Paroxysmal Disorders

Disorder	OMIM	Gene	Biological Function	Product of Conception Alteration
Autosomal dominant nocturnal frontal lobe epilepsy-1	600513	*CHRNA4*	Cell surface receptor Ligand-gated ion channels	AD
Autosomal dominant nocturnal frontal lobe epilepsy-3	605375	*CHRNB2*	Cell surface receptor Ligand-gated ion channels	AD
Autosomal dominant nocturnal frontal lobe epilepsy-4	610353	*CHRNA2*	Cell surface receptor Ligand-gated ion channels	AD
Autosomal dominant familial temporal lobe epilepsy-1	600512	*LGI1*	Transmembrane protein	AD
Autosomal dominant familial temporal lobe epilepsy-5	614417	*CPA6*	Carboxypeptidase	AD
Childhood absence epilepsy-2	607681	*GABRG2*	Cell surface receptor Ligand-gated chloride channels	AD
Childhood absence epilepsy-4	611136	*GABRA1*	Cell surface receptor Ligand-gated chloride channels	AD
Childhood absence epilepsy-5	612269	*GABRB3*	Cell surface receptor Ligand-gated chloride channels	AD
Childhood absence epilepsy-6	611942	*CAC-NA1H*	Channelopathy Calcium	AD
Juvenile myoclonic epilepsy-1	254770	*EFHC1*	Calcium homeostasis	AR or AD
Juvenile myoclonic epilepsy-5	611136	*GABRA1*	Cell surface receptor Ligand-gated chloride channels	AD
Juvenile myoclonic epilepsy-6	607682	*CACNB4*	Channelopathy Calcium	AD

Table 3.06.06.1. Continued.

Disorder	OMIM	Gene	Biological Function	Product of Conception Alteration
Juvenile myoclonic epilepsy-7	613060	GABRD	Cell surface receptor Ligand-gated chloride channels	AD
Juvenile myoclonic epilepsy-8	607628	CLCN2	Channelopathy Chloride	AD
Neonatal myoclonic epilepsy with suppression-burst pattern	609304	SLC25A22	Solute carrier	AR
Progressive myoclonic epilepsy-1 or Myoclonic epilepsy of Unverricht and Lundborg	254800	CSTB	intracellular thiol protease inhibitor	AR
Progressive myoclonic epilepsy-1B	612437	PRICKLE1	Nuclear membrane protein	AR or AD
Myoclonic epilepsy of Lafora or Progressive myoclonic epilepsy-2A	254780	EPM2A	Phosphatase associates with polyribosomes	AR
Myoclonic epilepsy of Lafora or Progressive myoclonic epilepsy-2B	254780	NHLRC1	E3 ubiquitin protein ligase	AR
Progressive myoclonic epilepsy-3	611726	KCTD7	Channelopathy Potassium	AR
Progressive myoclonic epilepsy-4, with or without renal failure	254900	SCARB2	Scavenger receptor	AR
Progressive myoclonic epilepsy-5	613832	PRICKLE2	Nuclear membrane protein	AR
Progressive myoclonic epilepsy-6	614018	GOSR2	Golgi SNAP receptor complex	AR
Myoclonic epilepsy with mental retardation and spasticity	300382	ARX	Transcription factor Helix-turn-helix domains	XR
Susceptibility to generalized and juvenile myoclonic Epilepsy	613060	GABRD	Cell surface receptor Ligand-gated chloride channels	AD
Amish infantile epilepsy syndrome	609056	ST3GAL5	Golgi apparatus protein	AR
Epilepsy with neurodevelopmental defects	613971	GRIN2A	Channelopathy	AD + de novo
Polyhydramnios, megalencephaly, and symptomatic epilepsy	611087	STRADA	Pseudokinase	AR
Rolandic epilepsy with speech dyspraxia and mental retardation	300643	SRPX2	Development of speech and language centers	XD

Table 3.06.06.1. Continued.

Disorder	OMIM	Gene	Biological Function	Product of Conception Alteration
Early infantile epileptic encephalopathy-1	308350	*ARX*	Transcription factor Helix-turn-helix domains	XR + de novo
Early infantile epileptic encephalopathy-2	300672	*CDKL5*	Serine/threonine kinase	XD + de novo
Early infantile epileptic encephalopathy-3	609304	*SLC25A22*	Mitochondrial glutamate carrier	AR
Early infantile epileptic encephalopathy-4	612164	*STXBP1*	Release of neurotransmitters	AD + de novo
Early infantile epileptic encephalopathy-5	613477	*SPTAN1*	Cytoskeletal-Membrane	AD + de novo
Dravet syndrome or Early infantile epileptic encephalopathy-6	607208	*SCN1A*	Channelopathy Sodium	AD + 90%
Early infantile epileptic encephalopathy-7	613720	*KCNQ2*	Channelopathy Potassium	AR
Early infantile epileptic encephalopathy-8	300607	*ARHGEF9*	Rho-dependent signals	X
Early infantile epileptic encephalopathy-9	300088	*PCDH19*	Calcium-dependent adhesion	AR
Early infantile epileptic encephalopathy-10	613402	*PNKP*	Polynucleotide kinase	AR
Early infantile epileptic encephalopathy-11	613721	*SCN2A*	Channelopathy Sodium	AD
Early infantile epileptic encephalopathy-12	613722	*PLCB1*	Phosphodiesterase	AR
Early infantile epileptic encephalopathy-13	614558	*SCN8A*	Channelopathy Sodium	AD + de novo
Early infantile epileptic encephalopathy-14	614959	*KCNT1*	Channelopathy Potassium	AD + 50%
Early infantile epileptic encephalopathy-15	615006	*ST3GAL3*	Galactoside sialyltransferase	AR
Benign familial infantile seizures-3	607745	*SCN2A*	Channelopathy Sodium	AD
Benign familial infantile seizures-1	121200	*KCNQ2*	Channelopathy Potassium	AD
Benign familial infantile seizures-2	121201	*KCNQ3*	Channelopathy Potassium	AD
Familial febrile convulsions-3A	182389	*SCN1A*	Channelopathy Sodium	AD
Familial febrile convulsions-3B	613863	*SCN9A*	Channelopathy Sodium	AD
Familial febrile convulsions-4	604352	*GPR98*	Cell surface receptor G protein	AD

Table 3.06.06.1. Continued.

Disorder	OMIM	Gene	Biological Function	Product of Conception Alteration
Familial febrile convulsions-8	611277	GABRG2	Cell surface receptor Ligand-gated chloride channels	AD
Familial febrile convulsions-11	614418	CPA6	Carboxypeptidase	AD + de novo
Generalized epilepsy with febrile seizures plus-1	604233	SCN1B	Channelopathy Sodium	AD
Generalized epilepsy with febrile seizures plus-2	604403	SCN1A	Channelopathy Sodium	AD
Generalized epilepsy with febrile seizures plus-3	611277	GABRG2	Cell surface receptor Ligand-gated chloride channels	AD
Generalized epilepsy with febrile seizures plus-5	613060	GABRD	Cell surface receptor Ligand-gated chloride channels	AD
Generalized epilepsy with febrile seizures plus-7	613863	SCN9A	Channelopathy Sodium	AD
West syndrome	308350	ARX	Transcription factor Helix-turn-helix domains	XR + de novo
Epileptic encephalopathy, Lennox-Gastaut type	606369	MAPK10	Serine/threonine protein kinase	AD + de novo
Pyridoxamine 5'-phosphate oxidase deficiency	610090	PNPO	Oxidase	AR
Pyridoxine-dependent epilepsy	266100	ALDH7A1	Oxidoreductase	AR
Migraine with or without aura-13	613656	KCNK18	Channelopathy Potassium	AD
Familial hemiplegic migraine-1	141500	CACNA1A	Channelopathy Calcium	AD + 30%
Familial hemiplegic migraine-2	602481	ATP1A2	ATPase protein	AD + 30%
Familial hemiplegic migraine-3	609634	SCN1A	Channelopathy Sodium	AD
Familial hemiplegic migraine, with progressive cerebellar ataxia	141500	CACNA1A	Channelopathy Calcium	AD
Congenital central hypoventilation syndrome	603851	PHOX2B	Transcription factor	AD + 99%
Congenital central hypoventilation syndrome	600837	BDNF	Growth factors	AD
Congenital central hypoventilation syndrome	164761	RET	Proto-oncogene	AD
Congenital central hypoventilation syndrome	131242	EDN3	Endothelin	AD

Table 3.06.06.1. Continued.

Disorder	OMIM	Gene	Biological Function	Product of Conception Alteration
Congenital central hypoventilation syndrome	113505	GDNF	Growth factors	AD
Congenital central hypoventilation syndrome	100790	ASCL1	Transcription factor	AD
Congenital Haddad syndrome	209880	ASCL1	Transcription factor	AD
Narcolepsy-1	161400	HCRT	Hypocretin	AD + de no-vo
Narcolepsy-7	614250	MOG	Myelin oligodendro-cyte glycoprotein	AD

Table 3.06.06.2 Episodic and Paroxysmal Disorders in Different Ethnic Groups

Disease	Frequency of condition	People with higher prevalence (if available)
Febrile seizures	3-7 in 100	People of Guam and Japan
Cluster headache	1-2 in 1,000	Worldwide
Epilepsy	7 in 1,000	Worldwide
Childhood absence epilepsy or pyknolepsy	1-9 in 10,000	Worldwide
Narcolepsy	1-9 in 10,000	Worldwide
West syndrome	3 in 10,000	Worldwide
Lennox-Gastaut syndrome	15 in 100,000	Worldwide
Dravet syndrome	2-5 in 100,000	Worldwide
Early infantile epileptic encephalopathy	1-2 in 100,000	Worldwide
Benign familial neonatal epilepsy	1-9 in 100,000	Worldwide
Familial hemiplegic migraine	1-9 in 100,000	Worldwide
Congenital central hypoventilation syndrome or Ondine's Curse	5 in 1,000,000	Worldwide
Pyridoxine dependent seizures	1-2 in 1,000,000	Unknown
Amish infantile epilepsy syndrome	<1 in 1,000,000	Old Order Amish (Ohio)
Polyhydramnios, megalencephaly, and symptomatic epilepsy	<1 in 1,000,000	Old Order Mennonites

3.06.07 Nerve, Nerve Root and Plexus Disorders

Congenital mirror movements: A disorder characterized by contralateral involuntary movements that mirror voluntary ones. Mirror movements are sometimes found in healthy young children, and persistence beyond the age of 10 years is considered pathological. There are several underlying causes of this disorder with at least 2 autosomal genomic loci: Congenital mirror movements-1 can be caused by autosomal dominant loss-of-function mutations in the *DCC* gene which encodes the deleted in colorectal carcinoma protein, a tumor suppressor;; and Congenital mirror movements-2 can be caused by autosomal dominant mutations in the *RAD51* gene which encodes the RAD51 protein homolog. The prevalence of this condition is unknown. "Familial congenital mirror movements" is estimated to affect less than 1 in every 1,000,000 individuals.

Hereditary neuralgic amyotrophy: A disorder characterized by acute recurring attacks of brachial plexus neuropathy. These attacks are incapacitating due to pain, weakness, wasting, depression of reflexes and sensory loss. Hereditary neuralgic amyotrophy can be caused by autosomal dominant mutations in the *SEPT9* gene which encodes the septin 9. Hereditary neuralgic amyotrophy is estimated to affect 1 in every 30,000-50,000 individuals, but under-recognition and initial misdiagnosis is common.

Familial carpal tunnel syndrome: A disorder characterized by signs and symptoms of entrapment of the median nerve within the carpal tunnel. Carpal tunnel syndrome is a common condition, and is estimated to affect 1 in every 100 individuals, but the familial form prevalence is unknown. Familial carpal tunnel syndrome can be caused by autosomal dominant mutations in the *TTR* gene which encodes the transthyretin. There is also susceptibility to the development of carpal tunnel syndrome by autosomal dominant mutations in the *SH3TC2* gene which encodes the SH3 domain and tetratricopeptide repeat-containing protein 2.

3.06.08 Polyneuropathies and other Disorders of the Peripheral Nervous System

3.06.08.1 Hereditary and Idiopathic Neuropathy

Charcot–Marie–Tooth disease: A group of disorders often appearing during the first or second decade of life characterized by sensorineural peripheral polyneuropathy. These conditions have an insidious start in the peroneal compartment, causing these patients to frequently trip while walking. Other manifestations include foot drop, steppage gait, impaired sensation and absent or hypoactive deep tendon reflexes. The disease is named after Jean-Martin Charcot, Pierre Marie and Howard Henry Tooth. There are at least 50 genomic forms of this condition with different patterns of inheritance including autosomal dominant, autosomal recessive and X-linked. Charcot-Marie-Tooth disease is estimated to affect 1 in every 2,500 individuals worldwide. These conditions can be inherited or occur as *de novo* events. There are several forms of Charcot–Marie–Tooth disease that have been mapped, for which the genes have not yet been identified. These conditions can be classified in at least 8 major groups including: type 1, type 2, type 3, type 4, type 5, type 6, type intermediate and X-linked type. Charcot-Marie-Tooth disease type 1 is a primary peripheral demyelinating disease and is characterized by severely reduced motor nerve conduction velocity and segmental demyelination and remyelination; Charcot-Marie-Tooth disease type 2 is a primary peripheral axonal neuropathies characterized by chronic axonal degeneration and regeneration on nerve biopsy; Charcot-Marie-Tooth disease type 3 or Dejerine-Sottas disease is a severe early childhood form of neuropathy; Charcot-Marie-Tooth disease type 4 affects either the axon or myelin and follows an autosomal recessive pattern of inheritance; Charcot–Marie–Tooth disease type 5 or Charcot–Marie–Tooth disease with pyramidal features, appears in the second decade of life with distal muscle wasting, particularly in the legs; Charcot–Marie–Tooth disease type 6 shows optic atrophy; Charcot–Marie–Tooth disease intermediate type is characterized by a broader range of nerve conduction velocities values; and Charcot-Marie-Tooth disease type X is caused by mutations in a gene on the X chromosome. Charcot-Marie-Tooth disease type 1 and 2 follows mainly autosomal dominant patterns of inheritance. Each group has several types of Charcot-Marie-Tooth disease due to specific gene that are altered. Charcot-Marie-Tooth disease-1A is caused by monoallelic either by duplication or mutations in the *PMP22* gene which encodes the peripheral myelin protein 22. This integral membrane protein is a major component of myelin in the peripheral nervous system. Type 1A produces a defect in neuronal proteins. These duplications are inherited in 90% of cases and *de novo* in the remaining 10% of cases. *De novo* cases usually originate in the paternal allele. The product of conception alterations are defined as autosomal dominant syndrome plus 10% (AD + 10%). On the contrary, the autosomal dominant deletion of the *PMP22* gene results in hereditary neuropathy with liability to pressure palsies. Other point mutations in the *PMP22* gene are associated with hypertrophic neuropathy of Dejerine-Sottas. The most common cause of type 1 Charcot-Marie-Tooth disease accounting for 70-80% of the cases is the microduplication of 1.5 Mb in chromosome 17p12;; Charcot-Marie-Tooth disease-1B is caused by monoallelic mutations in the *MPZ* gene which encodes the myelin protein zero;; Charcot–Marie–Tooth disease-1C is caused by monoallelic mutations in the *LITAF* gene which encodes the lipopolysaccharide-induced TNF factor;; Charcot–Marie–Tooth disease-1D is caused by monoallelic mutations in the *EGR2* gene which encodes the early growth response 2;; Charcot–Marie–Tooth disease-1E is caused by monoallelic mutations in the *PMP22* gene;; Charcot–Marie–Tooth disease-1F is caused by monoallelic mutations in the *NEFL* gene which encodes the neurofilament light polypeptide;; Charcot–Marie–Tooth disease type 2 accounts for 20–40% of all the cases. Charcot–Marie–Tooth disease-2A1 is caused by monoallelic mutations in the *KIF1B* gene which encodes the kinesin family member 1B;; Charcot–Marie–Tooth disease-2A2 is caused by monoallelic mutations in the *MFN2* gene which encodes the mitofusin 2;; Charcot–Marie–Tooth disease-2B is caused by monoallelic mutations in the *RAB7A* gene which encodes the RAB7A member RAS oncogene family;; Charcot–Marie–Tooth disease-2B1 is caused by biallelic mutations in the *LMNA* gene which encodes the lamin A/C;; Charcot–Marie–Tooth disease-2B2 is caused by monoallelic mutations in the *MED25* gene which encodes the mediator complex (subunit 25);; Charcot–Marie–Tooth disease-2C is caused by monoallelic mutations in the *TRPV4* gene which encodes the transient receptor potential cation channel (subfamily V member 4);; Charcot–Marie–Tooth disease-2D is caused by monoallelic mutations in the *GARS* gene which encodes the glycyl-tRNA synthetase;; Charcot–Marie–Tooth disease-2E is caused by monoallelic mutations in the *NEFL* gene which encodes the neurofilament (light polypeptide);; Charcot–Marie–Tooth disease-2F is caused by monoallelic mutations in the *HSPB1* gene which encodes the heat shock 27kDa protein 1;; Charcot–Marie–Tooth disease-2G has been mapped, but the gene has not been characterized;; Charcot–Marie–Tooth disease-2H is caused by monoallelic mutations in the *GDAP1* gene which encodes the ganglioside induced differentiation associated protein 1;; Charcot–Marie–Tooth disease-2I is caused by monoallelic mutations in the *MPZ* gene;; Charcot–Marie–Tooth disease-2J is caused by monoallelic mutations in the *MPZ* gene;; Charcot–Marie–Tooth disease-2K is caused by monoallelic mutations in the *GDAP1* gene which encodes the ganglioside induced differentiation associated protein 1;; Charcot–Marie–Tooth disease-2L is caused by monoallelic mutations in the *HSPB8* gene which encodes the heat shock 22kDa protein 8;; Charcot–Marie–Tooth disease-2M is caused by monoallelic mutations in the *DNM2* gene which encodes the dynamin 2;; Charcot–Marie–Tooth disease-2N is caused by monoallelic mutations in the *AARS* gene which encodes the alanyl-tRNA synthetase;; Charcot–Marie–Tooth disease type 3 is caused by monoallelic or biallelic mutations in the *MPZ, EGR2, PMP22*, or the *PRX* genes. Charcot–Marie–Tooth disease-4A is caused by biallelic mutations in the *GDAP1* gene;; Charcot–Marie–Tooth disease-4B1 is caused by biallelic mutations in the *MTMR2* gene which encodes the myotubularin related protein 2;; Charcot–Marie–Tooth disease-4B2 is caused by biallelic mutations in the *SBF2* gene which encodes the SET binding factor 2;; Charcot–Marie–Tooth disease-4C is caused by biallelic mutations in the *SH3TC2* gene which encodes the SH3

domain and tetratricopeptide repeats 2. In Spanish Gypsies this pathogenic allele appears 225 years ago;; Charcot–Marie–Tooth disease-4D is caused by biallelic mutations in the *NDRG1* gene which encodes the N-myc downstream regulated 1;; Charcot–Marie–Tooth disease-4E is caused by concomitants mutations in the *MPZ* gene and the *EGR2* gene;; Charcot–Marie–Tooth disease-4F is caused by biallelic mutations in the *PRX* gene which encodes the periaxin;; Charcot–Marie–Tooth disease-4G is caused by biallelic mutations in the NMSR locus;; Charcot–Marie–Tooth disease-44-H is caused by biallelic mutations in the *FGD4* gene which encodes the FYVE, RhoGEF and PH domain containing protein 4;; Charcot–Marie–Tooth disease-44-J is caused by biallelic mutations in the *FIG4* gene which encodes the FIG4 protein homolog;; Charcot–Marie–Tooth disease type 5, or Charcot–Marie–Tooth disease with pyramidal features, follows an autosomal dominant pattern of inheritance, but the gene has not been characterized. Charcot–Marie–Tooth disease type 6, or Charcot–Marie–Tooth disease with optic atrophy, is caused by monoallelic mutations in the *MFN2* gene which encodes the mitofusin 2. Charcot–Marie–Tooth disease intermediate type can be caused by either autosomal dominant or recessive inheritance. Autosomal dominant intermediate Charcot–Marie–Tooth disease A follows an autosomal dominant pattern of inheritance, but the gene has not been characterized;; Autosomal dominant intermediate Charcot–Marie–Tooth disease B is caused by monoallelic mutations in the *DNM2* gene which encodes the dynamin 2;; Autosomal dominant intermediate Charcot–Marie–Tooth disease C is caused by monoallelic mutations in the *YARS* gene which encodes the tyrosyl-tRNA synthetase;; Autosomal dominant intermediate Charcot–Marie–Tooth disease D is caused by monoallelic mutations in the *MPZ* gene;; Autosomal dominant intermediate Charcot–Marie–Tooth disease E is caused by monoallelic mutations in the *INF2* gene which encodes the inverted formin-2;; Autosomal recessive intermediate Charcot–Marie–Tooth disease A is caused by biallelic mutations in the *GDAP1* gene;; Autosomal recessive intermediate Charcot–Marie–Tooth disease B is caused by biallelic mutations in the *KARS* gene which encodes the lysyl-tRNA synthetase. Charcot–Marie–Tooth disease type X accounts for 10–20% of all cases. Charcot–Marie–Tooth disease-X1 is caused by X-linked dominant mutations in the *GJB1* gene which encodes the gap junction protein (beta 1). The last form accounts for 90% of all the type X conditions;; Charcot–Marie–Tooth disease-X 2 is caused by X-linked recessive mutations in the CMTX2 locus;; Charcot–Marie–Tooth disease-X 3 is caused by X-linked recessive mutations in the CMTX3 locus;; Charcot–Marie–Tooth disease-X 4 or Cowchock syndrome is caused by X-linked recessive mutations in the *AIFM1* gene which encodes the apoptosis-inducing factor (mitochondrion-associated 1);; and Charcot–Marie–Tooth disease-X 5 is caused by X-linked recessive mutations in the *PRPS1* gene which encodes the phosphoribosyl pyrophosphate synthetase 1. This enzyme catalyzes the phosphoribosylation of ribose 5-phosphate to 5-phosphoribosyl-1-pyrophos-phate, which is necessary for nucleotide biosynthesis and purine metabolism.

Dejerine–Sottas disease or Charcot–Marie–Tooth disease type 3: A demyelinating peripheral neuropathy characterized by delayed motor development, difficulties in gait and generalized hypotonia. This condition is a form of Charcot–Marie–Tooth disease with early onset. Sometimes this condition also includes other features such as pes cavus, scoliosis and sensory ataxia. This condition is named after Joseph Jules Dejerine and Jules Sottas. Dejerine-Sottas syndrome can be caused by either autosomal dominant or autosomal recessive mutations in the *MPZ* gene, the *PMP22* gene, the P*RX* gene and the *EGR2* gene. Mutations in the *GJB1* gene may contribute to this condition.

Hereditary motor and sensory neuropathies: Are a group of conditions characterized by affecting both afferent and efferent neural communications. These conditions are other forms of classification for neuropathies. It includes Charcot-Marie-Tooth disease, and a few other conditions. These group of neuropathic disorders display features of hereditary motor and sensory neuropathies. These conditions include at least 6 groups of disorders Charcot–Marie–Tooth disease type 1, 2, 3, 5 and 6 correspond to hereditary motor and sensory neuropathy type 1, type 2, type 3, type 5 and type 6 respectively. Hereditary motor and sensory neuropathy type 4 is due to Refsum disease, which is caused by mutations in the *PHYH* gene which encodes the phytanoyl-CoA 2-hydroxylase.

Autosomal dominant slowed nerve conduction velocity: A disorder characterized by bilateral pes cavus, distal areflexia in the lower limbs, mild weakness of the peroneal muscles, and decreased motor and sensory nerve conduction velocity without other phenotypic features. Autosomal dominant slowed nerve conduction velocity is caused by mutations in the *ARHGEF10* gene which encodes the Rho guanine-nucleotide exchange factor-10.

Ribose 5-phosphate isomerase deficiency: A disorder characterized by peripheral neuropathy and leukoencephalopathy. Ribose 5-phosphate isomerase deficiency is caused by biallelic mutations in the *RPIA* gene which encodes the ribose 5-phosphate isomerase A.

3.06.08.2 Refsum's Disease

Refsum disease or phytanic acid storage disease: A disorder characterized by peripheral neuropathy, cerebellar ataxia, ichthyosis and eye problems. This condition leads to vision loss due to retinitis pigmentosa. Affected individuals also exhibit other variable features such as cardiac dysfunction, difficulty hearing, anosmia and multiple epiphyseal dysplasia. Refsum disease is a progressive condition of childhood or adolescence onset. This condition is named after Sigvald Bernhard Refsum. In 90% of cases, Refsum disease is caused by autosomal recessive mutations in the *PHYH* gene which encodes the phytanoyl-CoA hydroxylase. This peroxisomal protein catalyzes the first step in the alpha-oxidation of phytanic acid; a branched-chain fatty acid. This neurological disease leads to the over-accumulation of phytanic acid in cells and tissues. There is also a less common form of Refsum disease which is caused by autosomal recessive mutations in the *PEX7* gene which encodes the peroxisomal biogenesis factor 7. On average, the carrier frequency for Refsum disease trait is estimated to be 1 in 250 individuals worldwide, which means the frequency of this condition is 1 in 250,000 individuals or $1/(500)^2$.

3.06.08.3 Other Hereditary and Idiopathic Neuropathies

Hereditary sensory and autonomic neuropathies: Are a group of neuropathic disorders characterized by progressive dysfunction affecting primarily the peripheral sensory nerves. There are at least 6 different clinical forms with a combined incidence estimated to be about 1 in 25,000 individuals.

Hereditary sensory and autonomic neuropathy type 1: A disorder characterized by a sensory deficit in the distal portion of the lower limbs causing chronic perforating ulcerations of the feet and progressive destruction of underlying bones. This condition starts manifesting during late childhood or early adolescence. There are patients who also exhibit deafness and atrophy of the peroneal muscles. There are 5 loci causing this condition, but only 4 genes have been characterized. Hereditary sensory and autonomic neuropathy-1A is caused by autosomal dominant mutations in the *SPTLC1* gene which encodes the serine palmitoyltransferase long chain base (subunit 1);; Hereditary sensory and autonomic neuropathy-1B with cough and gastroesophageal reflux has been mapped;; Hereditary sensory and autonomic neuropathy-1C is caused by mutations in the *SPTLC2* gene which encodes the serine palmitoyltransferase long chain base (subunit 2);; Hereditary sensory and autonomic neuropathy-1D is caused by mutations in the *ATL1* gene which encodes the atlastin GTPase 1;; and Hereditary sensory and autonomic neuropathy-1E is caused by mutations in the *DNMT1* gene which encodes the DNA (cytosine-5-)-methyltransferase 1.

Hereditary sensory and autonomic neuropathy type 2 or Congenital sensory neuropathy: A disorder characterized by chronic ulcerations in the upper and lower limbs. These lesions cause autoamputation of the distal phalanges, neuropathic joint degeneration and reduction in the deep tendon reflexes. This condition starts manifesting during early infancy or childhood. This condition follows an autosomal recessive pattern of inheritance. Hereditary sensory and autonomic neuropathy-2A is caused by mutations in the *WNK1* gene which encodes the WNK lysine deficient protein kinase 1; Hereditary sensory and autonomic neuropathy-2B is caused by mutations in the *FAM134B* gene which encodes the family with sequence similarity 134 (member B); and Hereditary sensory and autonomic neuropathy-2C is caused by mutations in the *KIF1A* gene which encodes the kinesin family member 1A.

Hereditary sensory and autonomic neuropathy type 3 or Familial dysautonomia or Riley-Day syndrome: A disorder characterized by sensory and autonomic impairment of neonatal onset. These newborns have weak suck reflex, hypotonia, hypothermia, and retarded physical development with average intelligence. Affected individuals also exhibit indifference to pain, postural hypotension, reduced or absent tears and depressed deep tendon reflexes. Familial dysautonomia is caused by autosomal recessive mutations in the *IKBKAP* gene which encodes the inhibitor of kappa light polypeptide gene enhancer in B-cells (kinase complex-associated protein). This condition is ethnic specific among Ashkenazim from Poland. In Ashkenazi Jews, the carrier frequency for Familial dysautonomia trait is estimated to be 1 in 30 individuals, which means the frequency of this condition is about 1 in 3,600 individuals or $1/(60)^2$. The current frequency among Ashkenazi Jews is lower since they have implemented programs of disassortative mating.

Hereditary sensory and autonomic neuropathy type 4 or Congenital insensitivity to pain with anhidrosis: A disorder characterized by anhidrosis, insensitivity to pain, self-mutilating behavior and episodes of fever with extreme hyperpyrexia. Affected individuals also exhibit hypotonia, delayed developmental, palmar skin thickening, and susceptibility to repeated trauma leading to Charcot joints. Most of these patients are from consanguineous parents and die in the first decade of life. Congenital insensitivity to pain with anhidrosis is caused by autosomal recessive mutations in the *NTRK1* gene which encodes the neurotrophic tyrosine kinase receptor. The same gene is also associated with familial medullary thyroid carcinoma. Congenital insensitivity to pain with anhidrosis is estimated to affect less than 1 in every 1,000,000 individuals. This condition is extremely rare worldwide, and has mainly been reported in Japan. There are also a few cases among Bedouins in northern Israel, the United States, New Zealand and Morocco.

Hereditary sensory and autonomic neuropathy type 5 or Congenital insensitivity to pain with partial anhidrosis: A condition with similar features as those of Congenital insensitivity to pain with anhidrosis. Hereditary sensory and autonomic neuropathy-5 is characterized by a selective absence of small myelinated fibers. Congenital insensitivity to pain with partial anhidrosis is caused by autosomal recessive mutations in the *NGF* gene which encodes the nerve growth factor. This protein is involved in the regulation of growth and differentiation of sympathetic and certain sensory neurons. Congenital insensitivity to pain with partial anhidrosis is estimated to affect less than 1 in every 1,000,000 individuals.

Hereditary sensory and autonomic neuropathy type 6: A severe disorder characterized by neonatal hypotonia, lack of psychomotor development, areflexia, respiratory and feeding difficulties, and autonomic abnormalities including lack of corneal reflexes and labile cardiovascular function. Hereditary sensory and autonomic neuropathy-6 is caused by autosomal recessive mutations in the *DST* gene which encodes the dystonin. This gene encodes a member of the plakin protein family of adhesion junction plaque proteins. Hereditary sensory and autonomic neuropathy type 6 has been reported among a large consanguineous family of Ashkenazi Jewish ancestry. There is an adult-onset Hereditary sensory and autonomic neuropathy with anosmia that has been mapped.

Congenital insensitivity to pain or Congenital analgesia: A disorder appearing during infancy characterized by inability to feel pain causing them to not respond to problems such as biting off their tongue, infections and fractures of bones. Affected individuals do not display abnormal sensory symptoms and have conserved reflexes and autonomic responses. Congenital insensitivity to pain is caused by autosomal recessive mutations in the *SCN9A* gene which encodes the voltage-gated sodium channel (type IX alpha subunit). This condition belongs to the channelopathy group. Congenital insensitivity to pain is caused by loss-of-function mutations in the *SCN9A* gene. On the contrary, Primary erythermalgia is caused by gain-of-function mutations in the same gene. Congenital insensitivity to pain is estimated to affect less than 1 in every 1,000,000 individuals.

Giant axonal neuropathy: A progressive disorder appearing during early childhood characterized by polyneuropathy. This condition affects both the peripheral and central nervous systems. This condition is associated with kinky hair and unique pos-

ture of legs. Affected individuals also exhibit epilepsy, mental retardation, and cerebellar and pyramidal tract signs. Most patients become bedridden and often die by late adolescence. Giant axonal neuropathy is caused by autosomal recessive mutations in the *GAN* gene which encodes the gigaxonin. This is a ubiquitously expressed cytoskeletal protein. The prevalence of this condition is unknown.

Cold-induced sweating syndrome: A disorder characterized by profuse sweating mainly of the upper body, in response to cold temperatures, and sweating very little in heat. Additional abnormalities may include progressive kyphoscoliosis, characteristic facial anomalies such as round face, micrognathia, high-arched palate, chubby cheeks, low-set ears, depressed nasal bridge, dental decay, camptodactyly, and impaired peripheral sensitivity to pain and temperature. There are several underlying causes of this disorder with at least 2 autosomal genomic loci: Cold-induced sweating syndrome-1 is caused by autosomal recessive mutations in the *CRLF1* gene which encodes the cytokine receptor-like factor 1. This protein acts on cells expressing ciliary neurotrophic factor receptors. Mutations in this gene result in Crisponi syndrome;; and Cold-induced sweating syndrome-2 is caused by autosomal recessive mutations in the *CLCF1* gene which encodes the cardiotrophin-like cytokine factor 1. Cold-induced sweating syndrome is estimated to affect less than 1 in every 1,000,000 individuals.

The Concise Encyclopedia of Genomic Diseases

Table 3.06.08.1 Recognizable Polyneuropathies and other Disorders of the Peripheral Nervous System

Disorder	OMIM	Gene	Biological Function	Product of Conception Alteration
Charcot-Marie-Tooth disease-1A	118220	PMP22	Integral membrane protein	AD + 10%
Charcot-Marie-Tooth disease-1B	118200	MPZ	Membrane myelin protein	AD + de novo
Charcot-Marie-Tooth disease-1C	601098	LITAF	Stimulator of secretion of tumor necrosis factor-alpha	AD + de novo
Charcot-Marie-Tooth disease-1D	607678	EGR2	Transcription factor	AD + de novo
Charcot-Marie-Tooth disease-1E	118300	PMP22	Integral membrane protein	AD
Charcot-Marie-Tooth disease-1F	607734	NEFL	Cytoskeletal Intermediate filaments	AD
Charcot-Marie-Tooth disease-2A1	118210	KIF1B	Cytoskeletal Microtubules	AD
Charcot-Marie-Tooth disease-2A2	609260	MFN2	Mitochondrial assembly protein	AD + de novo
Charcot-Marie-Tooth disease-2B	600882	RAB7A	Ras oncogene	AD + de novo
Charcot-Marie-Tooth disease-2B1	605588	LMNA	Nuclear Lamina Protein	AR
Charcot-Marie-Tooth disease-2B2	605589	MED25	RNA polymerase 2 transcription	AD
Charcot-Marie-Tooth disease-2C	606071	TRPV4	Channelopathy, transient receptor potential channels	AD + de novo
Charcot-Marie-Tooth disease-2D	601472	GARS	Glycyl-tRNA synthetase	AD
Charcot-Marie-Tooth disease-2E	607684	NEFL	Cytoskeletal Intermediate filaments	AD
Charcot-Marie-Tooth disease-2F	606595	HSPB1	Translocates from the cytoplasm to the nucleus upon stress	AD
Charcot-Marie-Tooth disease-2H	607731	GDAP1	Signal transducer activity	AD
Charcot-Marie-Tooth disease-2I	607677	MPZ	Membrane myelin protein	AD
Charcot-Marie-Tooth disease-2J	607736	MPZ	Membrane myelin protein	AD

Table 3.06.08.1. Continued.

Disorder	OMIM	Gene	Biological Function	Product of Conception Alteration
Charcot-Marie-Tooth disease-2K	607831	GDAP1	Signal transducer activity	AD
Charcot-Marie-Tooth disease-2L	608673	HSPB8	Folding and unfolding of other proteins	AD
Charcot-Marie-Tooth disease-2M	606482	DNM2	Cytoskeletal Microtubules	AD + de novo
Charcot-Marie-Tooth disease-2N	613287	AARS	Alanyl-tRNA synthetase	AD
Dejerine-Sottas disease or Charcot–Marie–Tooth disease type 3	145900	PMP22	Integral membrane protein	AD + de novo
Dejerine-Sottas disease or Charcot–Marie–Tooth disease type 3	145900	EGR2	Transcription factor	AD
Autosomal recessive Dejerine-Sottas disease or Charcot–Marie–Tooth disease type 3	145900	PRX	Peripheral nerve myelin upkeep	AR
Dejerine-Sottas disease or Charcot–Marie–Tooth disease type 3	145900	MPZ	Membrane myelin protein	AD + de novo
Charcot-Marie-Tooth disease-4A	214400	GDAP1	Signal transducer activity	AR
Charcot-Marie-Tooth disease-4B1	601382	MTMR2	Tyrosine phosphatase	AR
Charcot-Marie-Tooth disease-4B2	604563	SBF2	Myotubularin-related protein	AR
Charcot-Marie-Tooth disease-4C	601596	SH3TC2	Adapter or docking molecule	AR
Charcot-Marie-Tooth disease-4D	601455	NDRG1	N-myc downstream regulated	AR
Charcot-Marie-Tooth disease-4E	605253	EGR2	Transcription factor	AR
Charcot-Marie-Tooth disease-4F	145900	PRX	Peripheral nerve myelin upkeep	AR
Charcot-Marie-Tooth disease-4H	609311	FGD4	Regulation of the actin cytoskeleton and cell shape	AR
Charcot-Marie-Tooth disease-4J	611228	FIG4	Phosphatase	AR
Charcot-Marie-Tooth disease with hoarseness	607706	GDAP1	Signal transducer activity	AR
Charcot-Marie-Tooth disease-6 or Charcot–Marie–Tooth disease with optic atrophy	601152	MFN2	Mitochondrial assembly protein	AD
Autosomal dominant intermediate Charcot–Marie–Tooth disease B	606482	DNM2	Cytoskeletal Microtubules	AD

Table 3.06.08.1. Continued.

Disorder	OMIM	Gene	Biological Function	Product of Conception Alteration
Autosomal dominant intermediate Charcot–Marie–Tooth disease C	608323	YARS	Tyrosyl-tRNA ligase	AD
Autosomal dominant intermediate Charcot–Marie–Tooth disease D	607791	MPZ	Membrane myelin protein	AD
Autosomal dominant intermediate Charcot–Marie–Tooth disease E	614455	INF2	Inverted formin	AD
Autosomal recessive intermediate Charcot–Marie–Tooth disease A	608340	GDAP1	Signal transducer activity	AR
Autosomal recessive intermediate Charcot–Marie–Tooth disease B	613641	KARS	Lysyl-tRNA synthetase	AR
Charcot-Marie-Tooth disease-X1	302800	GJB1	Channelopathy Connexin	XD + de novo
Charcot-Marie-Tooth disease-X 4 or or Cowchock syndrome	310490	AIFM1	Apoptosis-inducing factor	XR
Charcot-Marie-Tooth disease-X 5	311070	PRPS1	Phosphoribosylation	XR
Recurrent neuropathy with pressure palsies	162500	PMP22	Integral membrane protein	AD
Roussy-Levy syndrome	180800	PMP22	Integral membrane protein	AD
Roussy-Levy syndrome	180800	MPZ	Membrane myelin protein	AD
Congenital hypomyelinating neuropathy	605253	MPZ	Membrane myelin protein	AR
Congenital hypomyelinating neuropathy 1	605253	EGR2	Transcription factor	AR
Slowed nerve conduction velocity	608236	ARHGEF10	Rho-dependent signals	AD
Ribose 5-phosphate isomerase deficiency	608611	RPIA	Isomerase	AR
Refsum disease or Hereditary motor and sensory neuropathy-4	266500	PHYH	Peroxisomal and lysosomal structural protein	AR
Refsum disease	266500	PEX7	Peroxisomal and lysosomal structural protein	AR
Infantile Refsum disease	266510	PEX1	Peroxisomal and lysosomal structural protein	AR
Infantile Refsum disease	266510	PEX26	Peroxisomal and lysosomal structural protein	AR

Table 3.06.08.1. Continued.

Disorder	OMIM	Gene	Biological Function	Product of Conception Alteration
Infantile Refsum disease	266510	PEX2	Peroxisomal and lysosomal structural protein	AR
Hereditary sensory and autonomic neuropathy-1A	162400	SPTLC1	Serine palmito-yltransferase	AD + de novo
Hereditary sensory and autonomic neuropathy-1C	613640	SPTLC2	Serine palmito-yltransferase	AD + de novo
Hereditary sensory and autonomic neuropathy-1D	613708	ATL1	Atlastin GTPase	AD
Hereditary sensory and autonomic neuropathy-1E	614116	DNMT1	DNA methyltrans-ferase	AD
Hereditary sensory and autonomic neuropathy-2A	201300	WNK1	Serine/threonine-protein kinase	AR
Hereditary sensory and autonomic neuropathy-2B	613115	FAM134B	Golgi transmem-brane protein	AR
Hereditary sensory and autonomic neuropathy-2C	614213	KIF1A	Axonal transport of synaptic vesicles	AR
Hereditary sensory and autonomic neuropathy-3 or familial dysauto-nomia or Riley-Day syndrome	223900	IKBKAP	Scaffold protein and a regulator for 3 different kinases	AR
Hereditary sensory and autonomic neuropathy-4 or congenital insen-sitivity to pain with anhidrosis	256800	NTRK1	Cell surface recep-tor Enzyme-linked receptor	AR
Hereditary sensory and autonomic neuropathy-5 or congenital insen-sitivity to pain with partial anhi-drosis	608654	NGF	Growth factors	AR
Hereditary sensory and autonomic neuropathy-6	614653	DST	Dystonin	AR
Hereditary sensory neuropathy with spastic paraplegia	256840	CCT5	Folding, ATP-dependent	AR
Congenital insensitivity to pain or congenital analgesia	243000	SCN9A	Channelopathy So-dium	AR
Giant axonal neuropathy	256850	GAN	Cytoskeletal Caten-in	AR
Cold-induced sweating syndrome-1	272430	CRLF1	Cell surface recep-tor JAK-STAT	AR
Cold-induced sweating syndrome-2	610313	CLCF1	Channelopathy Chloride	AR

Table 3.06.08.2 Polyneuropathies and other Disorders of the Peripheral Nervous System in Different Ethnic Groups

Disease	Frequency of condition	People with higher prevalence (if available)
Charcot-Marie-Tooth disease	4 in 10,000	Worldwide
Hereditary sensory and autonomic neuropathies	4 in 100,000	Worldwide
Hereditary sensory and autonomic neuropathy-3 or Familial dysautonomia or Riley-Day syndrome	<1 in 1,000,000	Ashkenazi Jews
Hereditary sensory and autonomic neuropathy-4 or Congenital insensitivity to pain with anhidrosis	<1 in 1,000,000	Japanese
Hereditary sensory and autonomic neuropathy-5 or Congenital insensitivity to pain with partial anhidrosis	<1 in 1,000,000	Unknown
Refsum disease	4 in 1,000,000	Unknown
Congenital insensitivity to pain or congenital analgesia	<1 in 1,000,000	Unknown
Cold-induced sweating syndrome	<1 in 1,000,000	Unknown

3.06.09 Diseases of Myoneural Junction and Muscle

3.06.09.1 Myasthenia Gravis and other Myoneural Disorders

Congenital myasthenic syndrome: Are a group of disorders of neonatal onset affecting neuromuscular transmission. Sometimes, patients start manifesting symptoms of this condition later during childhood, adolescence or even adulthood. Affected children exhibit hypotonia that worsens with physical effort. The affected musculatures include: the axial and limb muscles, the facial and bulbar muscle (impairing sucking and swallowing, and causing dysphonia), and the ocular muscles (causing ptosis and ophthalmoplegia). These non-immune conditions can be classified by the site of the transmission defect: presynaptic, synaptic and postsynaptic. Presynaptic accounts for 10% of cases, Synaptic accounts for 15% of cases, and Postsynaptic accounts for around 75% of them. This last form is mainly caused by

Acetylcholine receptor deficiency. Presynaptic congenital myasthenic syndrome with episodic ataxia is caused by either autosomal recessive or dominant mutations in the *CHAT* gene which encodes the choline O-acetyltransferase. This enzyme catalyzes the biosynthesis of the neurotransmitter acetylcholine;; Synaptic congenital myasthenic syndrome is caused by autosomal recessive mutations in the *COLQ* gene which encodes the collagen-like tail subunit (single strand of homotrimer) of asymmetric acetylcholinesterase. This form leads to endplate acetylcholinesterase deficiency. Postsynaptic disorders can be divided in: Acetylcholine receptor deficiency which is the most prevalent of all of these conditions; and into 2 kinetic defects, fast-channel and slow-channel congenital myasthenic syndrome. There are 4 forms of Postsynaptic congenital myasthenic syndrome associated with acetylcholine receptor deficiency. They can be caused by autosomal recessive mutations in either the *CHRNE* gene which encodes the epsilon subunit of the acetylcholine receptor or the *CHRNB1* gene which encodes the beta subunit of the acetylcholine receptor;; Postsynaptic congenital myasthenic syndrome associated with acetylcholine receptor deficiency can be caused by autosomal recessive mutations in the *RAPSN* gene which encodes the receptor-associated protein of the synapse. This encoded protein plays an essential role in the clustering of Acetylcholine receptor at the endplate;; and Postsynaptic congenital myasthenic syndrome associated with acetylcholine receptor deficiency can be caused by autosomal recessive mutations in the *MUSK* gene which encodes the muscle, skeletal receptor tyrosine kinase. This encoded protein is critical for synaptic differentiation. There are 4 forms of Postsynaptic slow-channel congenital myasthenic syndrome caused by either autosomal recessive or dominant mutations in the *CHRNA1*, the *CHRNB1*, the *CHRND*, and the *CHRNE* genes, encoded for the subunits of the acetylcholine receptor. Gain-of-function mutations are the most common cause of slow-channel Congenital myasthenic syndrome. There are 2 forms of Postsynaptic fast-channel congenital myasthenic syndrome caused by autosomal recessive mutations in either the *CHRNA1* or the *CHRND* genes. These genes encode subunits of the acetylcholine receptor. Besides these conditions, there are also overlapping phenotypes associated with other conditions such as: Acetazolamide-responsive congenital myasthenic syndrome is caused by mutations in the *SCN4A* gene which encodes the sodium channel gene; Familial limb-girdle myasthenic syndrome is caused by autosomal recessive mutations in either the *DOK7* gene which encodes the docking protein 7, or the *AGRN* gene which encodes the agrin; and Limb-girdle myasthenia with tubular aggregates is caused by autosomal recessive mutations in the *GFPT1* gene which encodes the glutamine--fructose-6-phosphate transaminase 1. The combined prevalence for all the forms of these non-immune Congenital myasthenic syndromes is estimated to be 1-2 in every 500,000 individuals. There is a higher prevalence of this condition among the Romani Gypsies, in whom it is caused by a single base pairs deletion in the *CHRNE* gene. This mutation is also present in people from India and Pakistan. In North African families, the *CHRNE* gene also has another mutated haplotype. This mutation is an example of founder effect that occurred about 700 years ago.

3.06.09.2 Primary Disorders of Muscles

Myosclerosis: A disorder characterized by symmetrical congenital contractures of the joints leading to sclerosis of both muscle and skin. Affected individuals exhibit difficulty in running and climbing stairs during early childhood. Affected individuals start manifesting Achilles tendon contractures, which progress to contractures of all joints. Myosclerosis can be caused by autosomal recessive mutations in the *COL6A2* gene which encodes the alpha-2 subunit of collagen VI. Most of the cases occurred in children born to consanguineous parents.

Early-onset myopathy with fatal cardiomyopathy: A disorder appearing during ages 5-12 characterized by delayed skeletal motor development and a progressive dilated cardiomyopathy with rhythmic disturbances of later onset. Death from cardiomyopathy occurs during their teenage years. Early-onset myopathy with fatal cardiomyopathy is caused by autosomal recessive partial deletion of the *TTN* gene which encodes the titin. This condition has been described among 2 consanguineous families of Moroccan and Sudanese origin.

Hereditary myopathy with lactic acidosis or Myopathy with exercise intolerance, Swedish type: A disorder appearing during childhood characterized by exercise intolerance with muscle tenderness, cramping, dyspnea and palpitations. This is a chronic disorder with exacerbation and remission of the muscular symptoms. Hereditary myopathy with lactic acidosis can be caused by autosomal recessive mutations in the *ISCU* gene which encodes the iron-sulfur cluster scaffold protein. The prevalence of this condition is unknown worldwide. Myopathy with lactic acidosis has been reported among families originating in northern Sweden, due to the founder effect phenomenon; this pathogenic allele appears 300 years ago.

3.06.09.3 Muscular dystrophy

Duchenne muscular dystrophy: A disorder characterized by muscle degeneration, difficulties in walking and breathing, and death. Symptoms of this condition frequently start to manifest in children before age 5; with progressive proximal muscle weakness with a loss of muscle mass of the legs and pelvis. The classic early signs associated with this condition include pseudohypertrophy of calf and deltoid muscles. Affected children manifest muscle weakness that spreads to the arms, neck and other areas. Patients with this condition progresses, causing muscle tissue wasting and replacement with fibrotic tissue. Most patients become wheelchair dependent by 12 years of age. Affected individuals also exhibit skeletal deformities, intellectual impairment and dilated cardiomyopathy. Affected individuals generally die between 20 and 30 years of age. Duchenne muscular dystrophy can be caused by X-linked recessive mutations in the *DMD* gene which encodes the dystrophin. This rod-like cytoskeletal protein is found at the inner surface of muscle fibers. Dystrophin is an important structural component within muscle tissue, providing structural stability to the dystroglycan complex. There is a contiguous gene syndrome involving a combination of Duchenne muscular dystrophy, glycerol kinase deficiency (infantile form) and congenital adrenal hypoplasia, and it is caused by deletion of multiple genes on chromosome Xp21. Mutations within the dystrophin gene are inherited in 76% of cases and it occur *de novo* in 24% of cases. The product of conception alterations are defined as X-linked recessive syndrome plus 30% (XR + 30). These alterations in the DNA mainly include large deletions in 79% of cases. Other less common alterations in the DNA include small rearrangements and large duplications. Duchenne muscular dystrophy is estimated to affect 1 in every 3,600 newborn males worldwide. Females are infrequently affected and are more often carriers.

Becker muscular dystrophy or Benign pseudohypertrophic muscular dystrophy: A similar condition to Duchenne muscular dystrophy but is less severe. Symptoms frequently start to manifest in children after 12 years of age. Affected individuals generally die between 40 and 50 years of age. Becker muscular dystrophy can be caused by either inherited or *de novo* X-linked recessive mutations in the *DMD* gene. These alterations in the DNA mainly include large deletions. In Becker muscular dystrophy these large deletions do not cause frameshifts while those in Duchenne muscular dystrophy cause frameshifts. Becker muscular dystrophy is estimated to affect 3-6 in every 100,000 newborn males worldwide. Females are infrequently affected and are more often carriers.

Congenital muscular dystrophy-dystroglycanopathy with brain and eye anomalies (Type A): A disorder characterized by congenital muscular dystrophy, profound mental retardation, distinctive brain and eye malformations, and early death. Affected neonates display cerebellar malformations, cobblestone lissencephaly and retinal malformations. There are several underlying causes of this disorder with at least 12 autosomal genomic loci: Congenital muscular dystrophy-dystroglycanopathy with brain and eye anomalies-A1 is caused by autosomal recessive mutations in the *POMT1* gene which encodes the protein-O-mannosyltransferase 1;; Congenital muscular dystrophy-dystroglycanopathy with brain and eye anomalies-A2 is caused by autosomal recessive mutations in the *POMT2* gene which encodes the protein-O-mannosyltransferase 2;; Congenital muscular dystrophy-dystroglycanopathy with brain and eye anomalies-A3 is caused by autosomal recessive mutations in the *POMGNT1* gene which encodes the protein O-linked mannose beta 1,2-N-acetylglucosaminyltransferase;; Congenital muscular dystrophy-dystroglycanopathy with brain and eye anomalies-A4 is caused by autosomal recessive mutations in the *FKTN* gene which encodes the fukutin;; Congenital muscular dystrophy-dystroglycanopathy with brain and eye anomalies-A5 is caused by autosomal recessive mutations in the *FKRP* gene which encodes the fukutin related protein;; Congenital muscular dystrophy-dystroglycanopathy with brain and eye anomalies-A6 is caused by autosomal recessive mutations in the *LARGE* gene which encodes the like-glycosyltransferase;; Congenital muscular dystrophy-dystroglycanopathy with brain and eye anomalies-A7 is caused by autosomal recessive mutations in the *ISPD* gene which encodes the isoprenoid syn-

thase domain containing protein;; Congenital muscular dystrophy-dystroglycanopathy with brain and eye anomalies-A8 is caused by autosomal recessive mutations in the *GTDC2* gene which encodes the glycosyltransferase-like domain containing protein 2;; Congenital muscular dystrophy-dystroglycanopathy with brain and eye anomalies-A10 is caused by autosomal recessive mutations in the *TMEM5* gene which encodes the transmembrane protein 5;; Congenital muscular dystrophy-dystroglycanopathy with brain and eye anomalies-A11 is caused by autosomal recessive mutations in the *B3GALNT2* gene which encodes the beta-1,3-N-acetylgalactosaminyltransferase 2;; and Congenital muscular dystrophy-dystroglyca-nopathy with brain and eye anomalies-A12 is caused by autosomal recessive mutations in the *SGK196* gene which encodes the protein kinase-like protein SgK196.

Congenital muscular dystrophy-dystroglycanopathy with mental retardation (Type B): A disorder characterized by early onset of muscle weakness, usually before ambulation is achieved, mental retardation and mild brain anomalies. There are several underlying causes of this disorder with at least 6 autosomal genomic loci: Congenital muscular dystrophy-dystrogly-canopathy with mental retardation-B1 is caused by autosomal recessive mutations in the *POMT1* gene;; Congenital muscular dystrophy-dystroglycanopathy with mental retardation-B2 is caused by autosomal recessive mutations in the *POMT2* gene;; Congenital muscular dystrophy-dystrogly-canopathy with mental retardation-B3 is caused by autosomal recessive mutations in the *POMGNT1* gene;; Congenital muscular dystrophy-dystroglycanopathy with mental retardation-B4 is caused by autosomal recessive mutations in the *FKTN* gene;; Congenital muscular dystrophy-dystroglyca-nopathy with mental retardation-B5 is caused by autosomal recessive mutations in the *FKRP* gene;; and Congenital muscular dystrophy-dystroglycanopathy with mental retardation-B6 is caused by autosomal recessive mutations in the *LARGE* gene. Type B is an intermediate form; is more severe than limb-girdle muscular dystrophy-dystroglycanopathy (Type C), and is less severe than muscular dystrophy-dystroglycanopathy with brain and eye anomalies (Type A).

Limb-girdle muscular dystrophy-dystroglycanopathy (Type C): A disorder characterized byless severe features than Type B. There are several underlying causes of this disorder with at least 6 autosomal genomic loci: Limb-girdle muscular dystrophy-dystroglycanopathy-C1 is caused by autosomal recessive mutations in the *POMT1* gene;; Limb-girdle muscular dystrophy-dystroglycanopathy-C2 is caused by autosomal recessive mutations in the *POMT2* gene;; Limb-girdle muscular dystrophy-dystroglycanopathy-C3 is caused by autosomal recessive mutations in the *POMGNT1* gene;; Limb-girdle muscular dystrophy-dystroglycanopathy-C4 is caused by autosomal recessive mutations in the *FKTN* gene;; Limb-girdle muscular dystrophy-dystroglycanopathy-C5 is caused by autosomal recessive mutations in the *FKRP* gene;; and Limb-girdle muscular dystrophy-dystroglycanopathy-C9 is caused by autosomal recessive mutations in the *DAG1* gene which encodes the dystroglycan 1 (dystrophin-associated glycoprotein 1).

Facioscapulohumeral muscular dystrophy: A disorder appearing during ages 3 to 50 characterized by progressive muscle weakness with focal involvement of the facial, shoulder and arm muscles. There are several underlying causes of this disorder with at least 3 autosomal genomic loci: Facioscapulohumeral muscular dystrophy-1 is caused by monoallelic mutations in the *FSHMD1A* gene which encodes the facioscapulohumeral muscular dystrophy 1A. This condition is associated with contraction of the D4Z4 macrosatellite repeat in the subtelomeric region of chromosome 4q35. In unaffected individuals, the D4Z4 array consists of 11-150 repeats (each polymorphic repeat structure consisting of 3.3 kb repeat units), whereas affected patients have 1-10 repeats;; and Facioscapulohumeral muscular dystrophy-2 is caused by concomitant digenic inheritance due to monoallelic mutations in the *SMCHD1* gene and the *DUX4* gene. Facioscapulohumeral muscular dystrophy is the third most frequent form of myopathy, and is estimated to affect 1 in every 20,000 individuals.

Bethlem myopathy: A slow progressive form of muscular dystrophy characterized by weakness of childhood onset. This condition leads to walking difficulties after age 50 and a normal life span. Affected individuals exhibit also symptoms of contractures of the fingers and Gower's sign. There are several underlying causes of this disorder with at least 3 autosomal genomic loci. Bethlem myopathy can be caused by autosomal dominant in any genes including *COL6A1* or *COL6A2* or the *COL6A3* genes. These genes encode alpha-1, alpha-2, and alpha-3 subunits of collagen VI. Bethlem myopathy is estimated to affect 1 in every 130,000 individuals.

Ullrich congenital muscular dystrophy: A similar condition to Bethlem myopathy but is more severe and progressive. Affected individuals exhibit muscle weakness and multiple contractures usually noted at birth or in early infancy. Affected individuals also display marked hypermobility of the distal joints, failure to thrive, delayed motor milestones, spinal rigidity, scoliosis, normal intelligence and respiratory failure. Most patients with this condition are unable to walk. There are several underlying causes of this disorder with at least 3 autosomal genomic loci. Ullrich congenital muscular dystrophy can be caused by autosomal recessive mutations in any genes including *COL6A1* or *COL6A2* or *COL6A3*. These genes encode the subunits of collagen type VI. Ullrich congenital muscular dystrophy is estimated to affect 1 in every 770,000 individuals.

Emery–Dreifuss muscular dystrophy: A disorder characterized by a triad of: joint contractures, weakness and atrophy of skeletal muscles, and cardiac conduction defect. The last defect is the most serious and life-threatening clinical manifestation of the disease. Most of these patients are males and the cardiac involvement usually occurs after the second decade of life. This condition is named after Alan Eglin H. Emery and Fritz E. Dreifuss. The different types of Emery–Dreifuss muscular dystrophy can be classified according to either their pattern of inheritance, as in X-linked recessive, autosomal dominant or autosomal recessive; or by the gene/loci affected. There are at least 6 gene/loci that have been characterized for this condition including: Emery-Dreifuss muscular dystrophy-1 is caused by X-linked recessive mutation in the subtelomeric *EMD* gene which encodes the emerin. This serine-rich nuclear membrane protein mediates membrane anchorage to the cytoskeleton. This is another example of the importance of subtelomeric regions;; Emery-Dreifuss muscular dystrophy-2 is caused by autosomal dominant mutations in the *LMNA* gene which encodes the lamin A/C gene. *EMD* and *LMNA* genes encode components of the nuclear envelope. These proteins are involved regulating the move-

ment of molecules into and in the nucleus;; Emery-Dreifuss muscular dystrophy-3 is caused by autosomal recessive mutations in the *LMNA* gene. Type 3 does not manifest cardiac features;; Emery-Dreifuss muscular dystrophy-4 is caused by autosomal dominant mutations in the *SYNE1* gene which encodes the nuclear envelope spectrin repeat containing protein 1;; Emery-Dreifuss muscular dystrophy-5 is caused by autosomal dominant mutations in the *SYNE2* gene which encodes the nuclear envelope spectrin repeat containing protein 2;; and Emery-Dreifuss muscular dystrophy-6 is caused by X-linked recessive mutations in the *FHL1* gene which encodes the four and a half LIM domains protein 1. Emery-Dreifuss muscular dystrophy is estimated to affect 1 in every 100,000 individuals; the forms 2 to 6 are the rarest forms.

Distal muscular dystrophies: A group of slow progressing disorders appearing in the 4th-5th decade of life. These groups of disorders affect the muscles of the extremities, the hands, feet, lower arms or lower legs. There are at least 9 known types of Distal muscular dystrophies including: Tibial muscular dystrophy, Laing distal myopathy, Distal myopathy with vocal cord and pharyngeal weakness, ZASP-related myofibrillar myopathy, Nonaka myopathy, Miyoshi muscular dystrophy-1, Miyoshi muscular dystrophy-2, Miyoshi muscular dystrophy-3 and Welander distal myopathy. Tibial muscular dystrophy is caused by autosomal dominant mutations in the *TTN* gene which encodes the titin, and autosomal recessive or biallelic mutations in the *TTN* gene is also associated with a more severe phenotype the Limb-girdle muscular dystrophy-2J;; Laing distal myopathy or distal myopathy-1 can be caused by autosomal dominant mutations in the *MYH7* gene which encodes the myosin heavy chain of type 1 fibers of skeletal muscle and cardiac ventricles. Mutations in the *MYH7* gene is associated with myosin storage myopathy, and with both hypertrophic and dilated cardiomyopathy;; Distal myopathy with vocal cord and pharyngeal weakness or Distal myopathy-2 is caused by autosomal dominant mutations in the *MATR3* gene which encodes the matrin-3;; ZASP-related myofibrillar myopathy is caused by autosomal dominant mutations in the *LDB3* gene which encodes the LIM domain binding 3;; Nonaka myopathy is caused by autosomal recessive mutations in the *GNE* gene which encodes the UDP-N-acetylglucosami- ne-2-epimerase/N-acetylmannosamine kinase. Nonaka myopathy is reported among Japanese people; there is an allelic disorder with a similar phenotype described as autosomal recessive inclusion body myopathy type 2 in Middle Eastern Jews, Karaites, and Arab Muslims of Palestinian and Bedouin origin. This mutation in the *GNE* gene is an example of the founder effect; it happened around 1,300 years ago;; Miyoshi muscular dystrophy-1 can be caused by autosomal recessive mutations in the *DYSF* gene which encodes the dysferlin. Mutations in the *DYSF* gene is also associated with either distal myopathy with anterior tibial onset; or a form of limb-girdle muscular dystrophy;; Miyoshi muscular dystrophy-2 has been mapped, but the gene has not yet been characterized;; and Miyoshi muscular dystrophy-3 is caused by mutations in the *ANO5* gene which encodes the anoctamin 5. Miyoshi muscular dystrophy has been described among Japanese people;; and Welander distal myopathy has been mapped, but the gene has not yet been characterized. The last condition is a late-onset distal myopathy that follows an autosomal dominant pattern of inheritance, affecting people from Sweden and Finland.

Limb-girdle muscular dystrophy: A group of disorders characterized by similar clinical phenotype than Duchenne muscular dystrophy and Becker's muscular dystrophy, but with a different pattern of inheritance. Limb-girdle muscular dystrophy follows autosomal inheritance: type 1 is dominant and type 2 is recessive. Limb-girdle muscular dystrophy affects the proximal muscles, especially those involving the pelvic or shoulder girdle musculature. The onset of these conditions is generally between the ages of 8-15. Limb-girdle muscular dystrophy-1A is caused by autosomal dominant mutations in the *MYOT* gene which encodes the myotilin;; Limb-girdle muscular dystrophy-1B is caused by autosomal dominant mutations in the *LMNA* gene which encodes the lamin A/C;; Limb-girdle muscular dystrophy-1C is caused by autosomal dominant or autosomal recessive mutations in the *CAV3* gene which encodes the caveolin 3;; Limb-girdle muscular dystrophy type 1D, type 1E, type 1F, and type 1G are caused by autosomal dominant patterns of inheritance, the loci have been identified, but the genes have not yet been characterized;; Limb-girdle muscular dystrophy-2A is caused by autosomal recessive mutations in the *CAPN3* gene which encodes the proteolytic enzyme calpain-3;; Limb-girdle muscular dystrophy-2B is caused by autosomal recessive mutations in the *DYSF* gene which encodes the dysferlin;; Limb-girdle muscular dystrophy-2C is caused by autosomal recessive mutations in the *SGCG* gene which encodes the gamma-sarcoglycan;; Limb-girdle muscular dystrophy type 2D is caused by autosomal recessive mutations in the *SGCA* gene which encodes the alpha-sarcoglycan;; Limb-girdle muscular dystrophy-2E is caused by autosomal recessive mutations in the *SGCB* gene which encodes the beta-sarcoglycan;; Limb-girdle muscular dystrophy-2F is caused by autosomal recessive mutations in the *SGCD* gene which encodes the delta-sarcoglycan;; Limb-girdle muscular dystrophy-2G is caused by autosomal recessive mutations in the *TCAP* gene which encodes the titin-cap;; Limb-girdle muscular dystrophy-2H is caused by autosomal recessive mutations in the *TRIM32* gene which encodes the tripartite motif containing protein 32;; Limb-girdle muscular dystrophy-2I is caused by autosomal recessive mutations in the *FKRP* gene which encodes the fukutin related protein;; Limb-girdle muscular dystrophy-2J is caused by autosomal recessive mutations in the *TTN* gene which encodes the titin;; Limb-girdle muscular dystrophy-2K is caused by autosomal recessive mutations in the *POMT1* gene which encodes the protein-O-mannosyltransferase 1;; Limb-girdle muscular dystrophy-2L is caused by autosomal recessive mutations in the *ANO5* gene which encodes the anoctamin 5;; Limb-girdle muscular dystrophy-2M is caused by autosomal recessive mutations in the *FKTN* gene which encodes the fukutin;; Limb-girdle muscular dystrophy-2N is caused by autosomal recessive mutations in the *POMT2* gene which encodes the protein-O-mannosyltransferase 2;; Limb-girdle muscular dystrophy-2O is caused by autosomal recessive mutations in the POMGNT1 gene which encodes the protein O-linked mannose beta1,2-N-acetylglucosaminyltrans-ferase;; and Limb-girdle muscular dystrophy-2Q is caused by autosomal recessive mutations in the *PLEC* gene which encodes the plectin. In the United Kingdom, Limb-girdle muscular dystrophy is estimated to affect 1 in every 44,000 individuals.

3.06.09.4 Myotonic disorders

Myotonia Congenita or Thomsen disease: A disorder characterized by muscle stiffness and an inability of the muscle to relax after voluntary contraction. Myotonia congenita is caused by autosomal dominant mutations in the *CLCN1* gene which encodes the skeletal muscle chloride channel-1. There is also an autosomal recessive form called Myotonia congenita or Becker disease caused by biallelic mutations in the *CLCN1* gene. Becker disease is more common and more severe than Thomsen disease. There are indistinguishable overlapping phenotypes caused by mutations in the *SCN4A* gene which encodes the sodium channel protein skeletal muscle subunit alpha; including paramyotonia congenita and potassium-aggravated myotonia. On average, Myotonia congenita is estimated to affect 1 in every 100,000 individuals worldwide. There is a higher prevalence among the northern Norwegian and Finnish people. It is estimated to affect 9 in every 100,000 individuals.

Schwartz-Jampel syndrome: A disorder characterized by myotonia and osteoarticular abnormalities. Affected individuals exhibit short stature, myotonic myopathy, dystrophy of epiphyseal cartilages, joint contractures, blepharophimosis, external ear anomalies and myopia. There are several underlying causes of this disorder with at least 2 autosomal genomic loci: Schwartz-Jampel syndrome-1 is caused by autosomal recessive mutations in the *HSPG2* gene which encodes the perlecan. This protein is a major component of the cellular matrix. There is also an allelic disorder with a more severe phenotype; the Silverman-Handmaker type of dyssegmental dysplasia. The last condition is a lethal form of neonatal short-limbed dwarfism;; and there is also a more severe phenotype referred to as Neonatal Schwartz-Jampel syndrome-2 or Stüve-Wiedemann syndrome, caused by autosomal recessive mutations in the *LIFR* gene which encodes the leukemia inhibitory factor receptor alpha. Schwartz-Jampel syndrome is estimated to affect less than 1 in every 1,000,000 individuals.

3.06.09.5 Congenital myopathies

Myotonic dystrophy: A multisystem disorder mainly characterized by myotonia, muscular dystrophy, cataracts, endocrine damage, baldness, cataract, sleep disorders, arrhythmia and/or cardiac conduction disorders. There are several underlying causes of this disorder with at least 2 autosomal genomic loci: Myotonic dystrophy-1 is caused by monoallelic mutations in the *DMPK* gene which encodes the dystrophia myotonica-protein kinase. This condition results from a trinucleotide repeat expansion in the 3-prime untranslated region of a dystrophia myotonica-protein kinase. Most individuals have a repeat length of 5-37. A repeat length exceeding 50 CTG repeats is pathogenic, but most individuals with 50-99 repeats were asymptomatic except for cataracts. Mildly affected individuals have 50-150 repeats, patients with classic form have 100-1,000 repeats, and those with congenital onset can have more than 2,000 repeats. The disorder displays genomic anticipation, with expansion of the repeat number dependent on the sex of the transmitting parent. Alleles of 40-80 repeats are typically extended when transmitted by males, whereas only alleles longer than 80 repeats tend to increase in length during maternal transmissions. Repeat contraction events also occur;; and Myotonic dystrophy-2 is caused by monoallelic mutations in the *CNBP* gene which encodes the CCHC-type zinc finger (nucleic acid binding protein). This condition results from a tetranucleotide repeat expansion (CCTG) in intron 1 of the *CNBP* gene. Most individuals have a repeat length of less than 30. A repeat length exceeding 75 CCTG repeats is pathogenic. Myotonic dystrophy is estimated to affect 1 in every 8,000-20,000 individuals. There is a higher prevalence among French-Canadians. The prevalence of myotonic dystrophy in the Saguenay-Lac-Saint-Jean region of Quebec province is 30-60 times higher than the worldwide average.

Centronuclear myopathies: Are a group of disorders characterized by slowly progressive muscular weakness and wasting. Affected individuals exhibit severe hypotonia, hypoxia-requiring breathing assistance and scaphocephaly. In these forms of congenital myopathies the typically histopathologic features include cell nuclei which are uncharacteristically located in the center of the skeletal muscle cells. There are several underlying causes of this disorder with at least 1 X-linked genomic locus and 4 autosomal genomic loci: X-linked myotubular myopathy-1 or X-linked centronuclear myopathy is caused by X-linked recessive mutations in the *MTM1* gene which encodes the myotubularin;; Centronuclear myopathy-1 is caused by autosomal dominant mutations in the *DNM2* gene which encodes the dynamin-2;; Centronuclear myopathy-2 is caused by autosomal recessive mutations in the *BIN1* gene which encodes the bridging integrator 1;; Centronuclear myopathy-3 is caused by autosomal dominant mutations in the *MYF6* gene which encodes the myogenic factor 6 (herculin);; and Centronuclear myopathy-4 is caused by autosomal dominant mutations in the *CCDC78* gene which encodes the coiled-coil domain containing protein 78. Furthermore, there is a clinical overlapping with Central core disease, caused by mutations in the *RYR1* gene which encodes the ryanodine receptor 1 (skeletal). X-linked myotubular myopathy is estimated to affect 1 in every 50,000 male births. The X-linked form is more severe and typically start manifesting at birth whereas the autosomal forms start later in life. Many patients with X-linked centronuclear myopathy die in infancy prior to receiving a formal diagnosis. The combined incidences of all the autosomal forms of Centronuclear myopathy are even less common than the X-linked form.

Nemaline myopathy: A group of disorders characterized by delayed motor development and slowly progressing muscle weakness throughout the body. This condition is characteristically most severe in the muscles of the face, arm, throat, neck and limbs. In Nemaline myopathy patients, the typically histopathologic features include abnormal thread-like rods called nemaline bodies in the muscle cells. Nemaline myopathy is often clinically categorized into several groups, including: severe congenital, Amish, intermediate congenital, typical congenital, childhood-onset and adult-onset. However, these different clinical forms are fairly ambiguous, as the categories frequently overlap. Muscle weakness typically involves proximal muscles, but respiratory problems are a primary concern for people with all forms of Nemaline myopathy. There are at least 7 genomic forms of Nemaline myopathy; 6 of them encode a component of skeletal muscle sarcomeric thin filaments. Most cases follow an autosomal recessive pattern of inheritance; few cases follow autosomal dominant pattern. Nemaline myopathy-1 can be caused by either autosomal dominant or autosomal recessive mutations in the *TPM3* gene which encodes the alpha-tropomyosin-3;; Nemaline myopathy-2 can be caused by autosomal recessive mutations in the *NEB* gene which encodes the nebulin;; Nemaline myopathy-3 is caused by autosomal dominant mutations in the *ACTA1* gene which encodes the alpha-actin-1;; Nemaline myopathy-4 caused by autosomal dominant mutations in the *TPM2* gene which encodes the beta-tropomyosin;; Nemaline myopathy-5 or Amish nemaline myopathy can be caused by autosomal recessive mutations in the *TNNT1* gene which encodes the troponin T1;; Nemaline myopathy-6 can be caused by autosomal dominant mutations in the *KBTBD13* gene which encodes the kelch repeat and BTB (POZ) domain containing protein 13;; and Nemaline myopathy-7 can be caused by autosomal recessive mutations in the *CFL2* gene which encodes the cofilin-2. The most common form of Nemaline myopathy worldwide is type 2 due to mutations in the *NEB* gene. Half of all cases occur due to mutations in the *NEB* gene. In Amish people the most common form is type 5. Nemaline myopathy is estimated to affect 1 in every 50,000 individuals worldwide. In people from the Old Order Amish of Lancaster County, Pennsylvania, the carrier frequency for Nemaline myopathy-5 or Amish nemaline myopathy trait is estimated to be 1 in 11 individuals, which means the frequency of this condition is about 1 in 500 individuals or $1/(22)^2$.

Myopathy with lactic acidosis and sideroblastic anemia: A disorder appearing during childhood characterized by progressive exercise intolerance. Affected individuals during adolescence also develop sideroblastic anemia, lactic acidemia and mitochondrial myopathy. There are several underlying causes of this disorder with at least 2 autosomal genomic loci: Myopathy with lactic acidosis and sideroblastic anemia-1 is caused by autosomal recessive mutations in the *PUS1* gene which encodes the nuclear pseudouridine synthase 1. This enzyme leads to defective pseudouridylation of mitochondrial tRNAs which may be responsible for the oxidative phosphorylation disorder specifically affecting the skeletal muscle and bone marrow;; and Myopathy with lactic acidosis and sideroblastic anemia-2 is caused by autosomal recessive mutations in the *YARS2* gene which encodes the mitochondrial tyrosyl-tRNA synthetase 2. Myopathy with lactic acidosis and sideroblastic anemia is estimated to affect less than 1 in every 1,000,000 individuals.

Mitochondrial phosphate carrier deficiency: A disorder appearing at birth characterized by hypertrophic cardiomyopathy, muscular hypotonia and the presence of lactic acidosis. Mitochondrial phosphate carrier deficiency is caused by autosomal recessive mutations in the *SLC25A3* gene which encodes the solute carrier (family 25 member A3), a mitochondrial membrane transporter (phosphate carrier). Mitochondrial phosphate carrier deficiency is estimated to affect less than 1 in every 1,000,000 individuals.

Mitochondrial respiratory chain complex II deficiency: This disease is discussed under endocrine, nutritional and metabolic diseases in the subchapter "Other Specified Metabolic Disorders" (3.04.20.4).

Fatal infantile cardioencephalomyopathy due to cytochrome c oxidase deficiency: A disorder characterized by hypertrophic cardiomyopathy, lactic acidosis, hypotonia, respiratory difficulties and central nervous system gliosis. This condition results in death during infancy. There are several underlying causes of this disorder with at least 2 autosomal genomic loci: Fatal infantile cardioencephalomyopathy due to cytochrome c oxidase deficiency-1 is caused by autosomal recessive mutations in the *SCO2* gene which encodes the SCO2 cytochrome c oxidase assembly protein. This enzyme catalyzes the transfer of reducing equivalents from cytochrome c to molecular oxygen and pumps protons across the inner mitochondrial membrane;; and Fatal infantile cardioencephalomyopa-

thy due to cytochrome c oxidase deficiency-1 is caused by autosomal recessive mutations in the *COX15* gene which encodes the cytochrome c oxidase assembly homolog 15.

Brody myopathy: A disorder appearing during childhood characterized by painless muscle cramping and exercise-induced impairment of muscle relaxation. Affected individuals sometimes exhibit stiffening, cramps, and muscle pain. Symptoms in these patients are exacerbated by cold temperatures. The most common affected muscles are from the arms, legs and face, particularly the eyelids. Brody myopathy can be caused by autosomal recessive mutations in the *ATP2A1* gene which encodes the fast-twitch muscle Ca++ transporting ATPase. The ATP2A1 encoded protein pumps back calcium ions from the cell into the sarcoplasmic reticulum, triggering muscle relaxation.

Hereditary inclusion body myopathies: Are a group of disorders appearing in young adults characterized by muscle weakness and the development of wasting. In these forms of myopathies, the typically histopathologic features include abundant lined vacuoles and characteristic cytoplasmic inclusions with immunoreactivity to neural cell adhesion molecule. There are at least 4 forms of hereditary inclusion body myopathies, encompassing both autosomal recessive and autosomal dominant muscle disorders, including: Myofibrillar myopathy-1 or formerly known as inclusion body myopathy-1 is caused by autosomal dominant mutations in the *DES* gene which encodes the desmin;; Inclusion body myopathy-2 is caused by autosomal recessive mutations in the *GNE* gene which encodes the glucosamine (UDP-N-acetyl)-2-epimerase/N-acetyl-mannosamine kinase. Nonaka myopathy is an allelic disorder of Inclusion body myopathy-2 with a similar phenotype among Japanese and other East Asian people, this disorder is known in South East Asia as distal myopathy with rimmed vacuoles. Inclusion body myopathy-2 and Nonaka myopathy has been described in distal muscular dystrophy, since it mainly affects leg muscles while sparing the quadriceps;; Inclusion body myopathy-3 is caused by autosomal dominant mutations in the *MYH2* gene which encodes the myosin heavy chain IIa;; "Inclusion body myopathy with early-onset Paget disease and frontotemporal dementia" is caused by autosomal dominant mutations in the *VCP* gene which encodes the valosin-containing protein. The prevalence of these conditions is unknown, except among people of Iranian Jewish descent. In this group, the carrier frequency for Inclusion body myopathy-2 trait is estimated to be 1 in 19 individuals, which means the frequency of this condition is about 1 in 1,500 individuals or $1/(38)^2$. In Japan, Nonaka myopathy is estimated to affect 1 in every 1,000,000 individuals.

Myofibrillar myopathy: A group of skeletal muscle disorders characterized by a pathologic morphologic pattern of myofibrillar degradation and abnormal accumulation of proteins including desmin, alpha-B-crystallin, dystrophin and myotilin. This condition results from disintegration of the sarcomeric Z disc. The clinical manifestations are variable and the dominant clinical feature is usually a slowly progressive muscular weakness. A subset of patients manifests cardiomyopathy and peripheral neuropathy. There are several underlying causes of this disorder with at least 6 autosomal genomic loci: Myofibrillar myopathy-1 is caused mainly by autosomal dominant and less often due to autosomal recessive mutations in the *DES* gene which encodes the desmin;; Myofibrillar myopathy-2 or Alpha-B crystallin-related myofi-

brillar myopathy can be caused by autosomal dominant mutations in the *CRYAB* gene which encodes the alpha-B-crystallin;; Severe fatal infantile hypertonic form of myofibrillar myopathy can be caused by autosomal recessive patterns of inheritance, due to founder mutations in the *CRYAB* gene. The last condition has been reported among Canadian aboriginal infants of Cree ancestry; Mutations in the *CRYAB* gene can also cause cataracts without causing myofibrillar myopathy;; Myofibrillar myopathy-3 or Myotilinopathy can be caused by autosomal dominant mutations in the *MYOT* gene which encodes the myotilin, this protein is also known as titin immunoglobulin domain protein. Limb-girdle muscular dystrophy type 1A is an allelic disorder with some overlapping features; Myotilinopathy and Spheroid body myopathy can both be caused by mutations in the *MYOT* gene;; Myofibrillar myopathy-4 or ZASP-related myofibrillar myopathy can be caused by mutations in the *LDB3* gene which encodes the LIM domain-binding protein 3;; Myofibrillar myopathy-5 or FLNC-related myofibrillar myopathy can be caused by mutations in the *FLNC* gene which encodes the filamin-C;; and Myofibrillar myopathy-6 or BAG3-related myofibrillar myopathy can be caused by mutations in the *BAG3* gene which encodes the BAG family molecular chaperone regulator 3. There is also a Desmin-related form of myopathy which results in Myofibrillar myopathy due to intrasarcoplasmic aggregates of desmin, often in addition to other sarcomeric proteins. Examples of Desmin-related myopathy include Rigid spine syndrome which can be caused by mutation in the *SEPN1* gene which encodes the selenoprotein N 1. The prevalence of myofibrillar myopathy is unknown.

Rippling muscle disease: A benign disorder appearing during childhood or adolescence characterized by mechanically triggered contractions of skeletal muscle. In rippling muscle disease, local muscle compression evokes muscle contractions that spread to neighboring fibers which cause visible ripples to move over the muscle. The patients also exhibit muscle cramps, pain and stiffness, particularly during or following exercise. There are several underlying causes of this disorder with at least 2 autosomal genomic loci: Rippling muscle disease-1 has been mapped, but the gene has yet not been characterized;; and Rippling muscle disease-2 is mostly caused by autosomal dominant mutations in the *CAV3* gene which encodes the caveolin-3. There are also few examples of autosomal recessive Rippling muscle disease-2. Limb-girdle muscular dystrophy-1C is an allelic disorder with an overlapping phenotype.

3.06.09.6 Other myopathies

Hyperkalemic periodic paralysis: A disorder characterized by episodes of flaccid muscle weakness which can be exacerbated by an increase in serum potassium concentration or cold temperatures. These episodes can lead to uncontrolled shaking followed by paralysis. This condition appears in the 1st or 2nd decade of life. These episodes vary in frequency, in duration, ranging from a few minutes to hours, and in severity, ranging from focal paresis to total paralysis. This condition affects the limb muscles, and spares the facial and respiratory musculature. Hyperkalemic periodic paralysis is caused by autosomal dominant mutations in the *SCN4A* gene which encodes the voltage-dependent sodium channel type IV (alpha subunit), found at the neuromuscular junction. Mutations of this sodium channel disrupt regulation of muscle contraction, leading to episodes of severe muscle weakness or paralysis. Hy-perkalemic periodic paralysis is a sodium muscle channelopathy. Hyperkalemic periodic paralysis is caused by gain-of-function mutations in sodium channels. Hyperkalemic periodic paralysis is estimated to affect 1 in every 200,000 individuals. There are also allelic disorders with overlapping features including Paramyotonia congenita and the Potassium-aggravated myotonia. Mutations in the *SCN4A* gene could also cause additional conditions: Normokalemic potassium-sensitive periodic paralysis and Hypokalemic periodic paralysis.

Hypokalemic periodic paralysis: A disorder with symptoms similar to those of Hyperkalemic periodic paralysis, except that the attacks are due to low potassium levels. These episodes can be triggered by: strenuous exercise followed by rest, sudden changes in temperature, hyperthyroidism, high sodium or carbohydrate meals, and other factors. There are several underlying causes of this disorder with at least 2 autosomal genomic loci: Hypokalemic periodic paralysis-1 is caused by autosomal dominant mutations in the *CACNA1S* gene which encodes the voltage-gated calcium channel (L type alpha 1S subunit). This protein is found in the transverse tubules of skeletal muscle cells;; and Hypokalemic periodic paralysis-2 is caused by autosomal dominant mutations in the *SCN4A* gene which encodes the voltage-dependent sodium channel type IV (alpha subunit). Hypokalemic periodic paralysis is estimated to affect 1 in every 100,000 individuals.

Thyrotoxic periodic paralysis: A disorder appearing during the 2nd-4th decade of life characterized by episodes of muscle weakness in the presence of hyperthyroidism. The condition may be life-threatening due to cardiac arrhythmias or respiratory failure. There are several underlying causes of this disorder with at least 3 autosomal genomic loci: Thyrotoxic periodic paralysis-1 can be conferred by polymorphisms in the *CACNA1S* gene which encodes the calcium channel (L-type alpha1-subunit) in skeletal muscle. The *CACNA1S* gene is also associated with Hypokalemic periodic paralysis and Malignant hyperthermia;; Susceptibility to thyrotoxic periodic paralysis-2 can be conferred by polymorphisms in the *KCNJ18* gene which encodes the inwardly rectifying potassium channels;; and Susceptibility to thyrotoxic periodic paralysis-3 has been mapped, but the gene has not yet been characterized.

Thyrotoxic periodic paralysis in Chinese males has also been associated with the HLA types BW22 and BW17. This condition predominantly affects males of Chinese, Japanese, Vietnamese, Filipino, Thais, Korean and Native American ancestry. The overall incidence of Thyrotoxic periodic paralysis in Chinese and Japanese thyrotoxic patients had been estimated at 1.8% and 1.9%, respectively. In North America, the incidence is much lower, at about 0.1-0.2% in thyrotoxic patients. This condition is 17-70 times more prevalent in males than in females.

Table 3.06.09.1 Recognizable Diseases of Myoneural Junction and Muscle

Disorder	OMIM	Gene	Biological Function	Product of Conception Alteration
Presynaptic Congenital myasthenic syndrome with episodic ataxia	254210	CHAT	Biosynthesis of the neuro-transmitter acetylcholine	AR or AD
Synaptic Congenital myasthenic syndrome	603034	COLQ	Extracellular matrix protein	AR
Postsynaptic Congenital myasthenic syndrome	608931	CHRNE	Cell surface receptor Ligand-gated ion channels	AR
Postsynaptic Congenital myasthenic syndrome	608931	CHRNB1	Cell surface receptor Ligand-gated ion channels	AR
Postsynaptic Congenital myasthenic syndrome	608931	RAPSN	Receptor associated proteins of the synapse	AR
Postsynaptic Congenital myasthenic syndrome	608931	MUSK	Tyrosine-protein kinase receptor	AR
Postsynaptic slow-channel Congenital myasthenic syndrome	601462	CHRNA1	Cell surface receptor Ligand-gated ion channels	AR or AD
Postsynaptic slow-channel Congenital myasthenic syndrome	601462	CHRNB1	Cell surface receptor Ligand-gated ion channels	AR or AD
Postsynaptic slow-channel Congenital myasthenic syndrome	601462	CHRND	Cell surface receptor Ligand-gated ion channels	AR or AD
Postsynaptic slow-channel Congenital myasthenic syndrome	601462	CHRNE	Cell surface receptor Ligand-gated ion channels	AR or AD
Postsynaptic fast-channel Congenital myasthenic syndrome	608930	CHRNA1	Cell surface receptor Ligand-gated ion channels	AR

Table 3.06.09.1. Continued.

Disorder	OMIM	Gene	Biological Function	Product of Conception Alteration
Postsynaptic slow-channel Congenital myasthenic syndrome	601462	CHRNB1	Cell surface receptor Ligand-gated ion channels	AR or AD
Postsynaptic slow-channel Congenital myasthenic syndrome	601462	CHRND	Cell surface receptor Ligand-gated ion channels	AR or AD
Postsynaptic slow-channel Congenital myasthenic syndrome	601462	CHRNE	Cell surface receptor Ligand-gated ion channels	AR or AD
Postsynaptic fast-channel Congenital myasthenic syndrome	608930	CHRNA1	Cell surface receptor Ligand-gated ion channels	AR
Postsynaptic fast-channel Congenital myasthenic syndrome	608930	CHRND	Cell surface receptor Ligand-gated ion channels	AR
Lethal form of multiple pterygium syndrome	253290	CHRNA1	Cell surface receptor Ligand-gated ion channels	AD + AD = AR
Lethal form of multiple pterygium syndrome	253290	CHRND	Cell surface receptor Ligand-gated ion channels	AD + AD = AR
Lethal form of multiple pterygium syndrome	253290	CHRNG	Cell surface receptor Ligand-gated ion channels	AD + AD = AR
Acetazolamide-responsive congenital myasthenic syndrome	608390	SCN4A	Channelopathy Sodium	AD
Familial limb-girdle myasthenia	254300	DOK7	Receptor kinase, muscle-specific	AR
Familial limb-girdle myasthenia	254300	AGRN	Extracellular matrix protein	AR
Myosclerosis	255600	COL6A2	Extracellular matrix protein	AR
Early-onset myopathy with fatal cardiomyopathy	611705	TTN	Cystoskeletal protein	AR
Hereditary myopathy with lactic acidosis or myopathy with exercise intolerance, Swedish type	255125	ISCU	Mitochondrial protein	AR
Duchenne muscular dystrophy	310200	DMD	Cytoskeletal-Membrane	XR + 30%
Becker muscular dystrophy or Benign pseudohypertrophic muscular dystrophy	300376	DMD	Cytoskeletal-Membrane	XR + 30%

The Concise Encyclopedia of Genomic Diseases

Table 3.06.09.1. Continued.

Disorder	OMIM	Gene	Biological Function	Product of Conception Alteration
Muscular dystrophy, megaconial type	602541	*CHKB*	Choline kinase	AR
Congenital muscular dystrophy	607855	*LAMA2*	Extracellular matrix protein	AR
Congenital muscular dystrophy	613204	*ITGA7*	Cell surface receptor Other/ungrouped	AR
Muscular dystrophy-dystroglycanopathy with brain and eye anomalies-A1	236670	*POMT1*	Endoplasmic reticulum protein	AR
Muscular dystrophy-dystroglycanopathy with brain and eye anomalies-A2	613150	*POMT2*	Endoplasmic reticulum protein	AR
Muscular dystrophy-dystroglycanopathy with brain and eye anomalies-A3	253280	*POMGNT1*	Golgi transmembrane protein	AR
Muscular dystrophy-dystroglycanopathy with brain and eye anomalies-A4	253800	*FKTN*	Golgi-cis protein	AR
Muscular dystrophy-dystroglycanopathy with brain and eye anomalies-A5	613153	*FKRP*	Golgi apparatus protein	AR
Muscular dystrophy-dystroglycanopathy with brain and eye anomalies-A6	613154	*LARGE*	Glycosyltransferase	AR
Muscular dystrophy-dystroglycanopathy with brain and eye anomalies-A7	614643	*ISPD*	Isoprenoid synthase	AR
Muscular dystrophy-dystroglycanopathy with brain and eye anomalies-A8	614830	*GTDC2*	Glycosyltransferase	AR
Muscular dystrophy-dystroglycanopathy with brain and eye anomalies-A10	615041	*TMEM5*	Transmembrane protein	AR

Table 3.06.09.1. Continued.

Disorder	OMIM	Gene	Biological Function	Product of Conception Alteration
Muscular dystrophy-dystroglycanopathy with brain and eye anomalies-A11	615181	B3GALNT2	Acetylgalactosaminyltransferase	AR
Muscular dystrophy-dystroglycanopathy with brain and eye anomalies-A12	615249	SGK196	Protein kinase	AR
Congenital muscular dystrophy-dystroglycanopathy with mental retardation-B1	236670	POMT1	Endoplasmic reticulum protein	AR
Congenital muscular dystrophy-dystroglycanopathy with mental retardation-B2	613150	POMT2	Endoplasmic reticulum protein	AR
Congenital muscular dystrophy-dystroglycanopathy with mental retardation-B3	253280	POMGNT1	Golgi transmembrane protein	AR
Congenital muscular dystrophy-dystroglycanopathy with mental retardation-B4	253800	FKTN	Golgi-cis protein	AR
Congenital muscular dystrophy-dystroglycanopathy with mental retardation-B5	613153	FKRP	Golgi apparatus protein	AR
Congenital muscular dystrophy-dystroglycanopathy with mental retardation-B6	613154	LARGE	Glycosyltransferase	AR
Limb-girdle muscular dystrophy-dystroglycanopathy-C1	236670	POMT1	Endoplasmic reticulum protein	AR
Limb-girdle muscular dystrophy-dystroglycanopathy-C2	613150	POMT2	Endoplasmic reticulum protein	AR
Limb-girdle muscular dystrophy-dystroglycanopathy-C3	253280	POMGNT1	Golgi transmembrane protein	AR
Limb-girdle muscular dystrophy-dystroglycanopathy-C4	253800	FKTN	Golgi-cis protein	AR

Table 3.06.09.1. Continued.

Disorder	OMIM	Gene	Biological Function	Product of Conception Alteration
Limb-girdle muscular dystrophy-dystroglycanopathy-C5	613153	*FKRP*	Golgi apparatus protein	AR
Limb-girdle muscular dystrophy-dystroglycanopathy-C9	613154	*DAG1*	Cytoskeletal-Membrane	AR
Rigid spine muscular dystrophy-1	602771	*SEPN1*	Selenoprotein	AR
Facioscapulohumeral muscular dystrophy-1	158900	*FSHMD1A*	Facioscapulohumeral muscular dystrophy	AD + Ant
Facioscapulohumeral muscular dystrophy-2	158901	*SMCHD1-DUX4*	Structural maintenance of chromosomes	Dig
Bethlem myopathy	158810	*COL6A1*	Extracellular matrix protein	AD + de novo or AR
Bethlem myopathy	158810	*COL6A2*	Extracellular matrix protein	AD + de novo
Bethlem myopathy	158810	*COL6A3*	Extracellular matrix protein	AD + de novo
Ullrich congenital muscular dystrophy	254090	*COL6A1*	Extracellular matrix protein	AD + AD = AR
Ullrich congenital muscular dystrophy	254090	*COL6A2*	Extracellular matrix protein	AD + AD = AR
Ullrich congenital muscular dystrophy	254090	*COL6A3*	Extracellular matrix protein	AD + AD = AR
Ullrich congenital muscular dystrophy	120240	*COL6A1/ COL6A2*	Extracellular matrix protein	Dig
Emery-Dreifuss muscular dystrophy-1	310300	*EMD*	Nuclear envelope protein	XR
Emery-Dreifuss muscular dystrophy-2	181350	*LMNA*	Nuclear Lamina Protein	AD + de novo
Emery-Dreifuss muscular dystrophy-3	181350	*LMNA*	Nuclear Lamina Protein	AR
Emery-Dreifuss muscular dystrophy-4	612998	*SYNE1*	Nuclear envelope protein	AD
Emery-Dreifuss muscular dystrophy-5	612999	*SYNE2*	Nuclear envelope protein	AD
Emery-Dreifuss muscular dystrophy-6	300696	*FHL1*	Zinc finger protein	XR
Emery-Dreifuss muscular dystrophy	310300	*EMD/ LMNA*	Nuclear envelope/lamina proteins	Dig
Tardive tibial muscular dystrophy	600334	*TTN*	Cystoskeletal protein	AD

Table 3.06.09.1. Continued.

Disorder	OMIM	Gene	Biological Function	Product of Conception Alteration
Proximal myopathy with early respiratory muscle involvement or Edström Myopathy	603689	TTN	Cystoskeletal protein	AD
Laing distal myopathy	160500	MYH7	Cystoskeletal protein	AD
Distal myopathy 2	606070	MATR3	Nuclear matrix protein	AD
Myofibrillar myopathy	609452	LDB3	Protein-protein interaction	AD
Nonaka distal myopathy	605820	GNE	Bifunctional enzyme	AR
Miyoshi muscular dystrophy-1	254130	DYSF	Trafficking Vesicle fusion	AR
Miyoshi muscular dystrophy-3	613319	ANO5	Putative calcium-activated chloride channel	AR
Limb-girdle muscular dystrophy-1A	159000	MYOT	Cystoskeletal protein	AD
Limb-girdle Muscular dystrophy-1B	159001	LMNA	Nuclear Lamina Protein	AD + de novo
Limb-girdle muscular dystrophy-1C	607801	CAV3	Trafficking Vesicle fusion	AD or AR
Limb-girdle muscular dystrophy-2A	253600	CAPN3	Proteases	AR
Limb-girdle muscular dystrophy-2B	253601	DYSF	Trafficking Vesicle fusion	AR
Limb-girdle muscular dystrophy-2C	253700	SGCG	Linking of the subsarcolemmal cytoskeleton and the extracellular matrix	AR
Limb-girdle muscular dystrophy-2D	608099	SGCA	Linkage of the actin cytoskeleton to the extracellular matrix	AR
Limb-girdle muscular dystrophy-2E	604286	SGCB	Linkage of the actin cytoskeleton to the extracellular matrix	AR
Limb-girdle muscular dystrophy-2E/F	601411	SGCB/ SGCD	Linkage of the actin cytoskeleton to the extracellular matrix	Dig
Limb-girdle muscular dystrophy-2F	601287	SGCD	Linkage of the actin cytoskeleton to the extracellular matrix	AR
Limb-girdle muscular dystrophy-2G	601954	TCAP	Sarcomere assembly	AR
Limb-girdle muscular dystrophy-2H	254110	TRIM32	Ciliary proteins Basal body	AR
Limb-girdle muscular dystrophy-2I	607155	FKRP	Golgi apparatus protein	AR
Limb-girdle muscular dystrophy-2J	608807	TTN	Cystoskeletal protein	AR
Limb-girdle muscular dystrophy-2K	609308	POMT1	Endoplasmic reticulum protein	AR

Table 3.06.09.1. Continued.

Disorder	OMIM	Gene	Biological Function	Product of Conception Alteration
Limb-girdle muscular dystrophy-2L or Erb's muscular dystrophy	611307	ANO5	Putative calcium-activated chloride channel	AR
Limb-girdle muscular dystrophy-2M	611588	FKTN	Golgi-cis protein	AR
Limb-girdle muscular dystrophy-2N	607439	POMT2	Endoplasmic reticulum protein	AR
Limb-girdle muscular dystrophy-2O	606822	POMGNT1	Golgi transmembrane protein	AR
Limb-girdle muscular dystrophy-2Q	613723	PLEC1	Cystoskeletal protein	AR
Muscular dystrophy with epidermolysis bullosa simplex	226670	PLEC1	Cystoskeletal protein	AR
Myotonia congenita or Thomsen disease	160800	CLCN1	Channelopathy Chloride	AD
Myotonia congenita or Becker myotonia	255700	CLCN1	Channelopathy Chloride	AD + AD = AR
Schwartz-Jampel syndrome-1	255800	HSPG2	Extracellular matrix protein	AR
Neonatal Schwartz-Jampel syndrome-2 or Stüve-Wiedemann syndrome	601559	LIFR	Cell surface receptor JAK-STAT	AR
Myotonic dystrophy-1 or Steinert disease	160900	DMPK	Protein kinase	AD + Ant
Myotonic dystrophy-2 or proximal myotonic myopathy	602668	CNBP	Transcription factor Zinc finger	AD + Ant
X-linked myotubular myopathy-1 or X-linked centronuclear myopathy	310400	MTM1	Tyrosine phosphatase	XR
Centronuclear myopathy-1	160150	DNM2	Cystoskeletal protein	AD + de novo
Centronuclear myopathy-2	255200	BIN1	Nucleocytoplasmic adaptor protein	AR
Centronuclear myopathy-3	614408	MYF6	Basic helix-loop-helix DNA binding	AD
Centronuclear myopathy-4	614807	CCDC78	Coiled-coil domain-containing protein	AD
Distal myopathy with anterior tibial onset	606768	DYSF	Trafficking Vesicle fusion	AR

Table 3.06.09.1. Continued.

Disorder	OMIM	Gene	Biological Function	Product of Conception Alteration
Distal myopathy with decreased caveolin 3 digenic phenotype KCNH2/CAV3	601253	KCNH2	Channelopathy Potassium	Dig
Distal myopathy with decreased caveolin 3 digenic phenotype KCNH2/CAV3	601253	CAV3	Trafficking Vesicle fusion	Dig
Distal myopathy, Tateyama type	614321	CAV3	Trafficking Vesicle fusion	AD
Myopathy with postural muscle atrophy	300696	FHL1	Zinc finger protein	XR
Central core disease	117000	RYR1	Channelopathy Calcium	AD
Minicore myopathy with external ophthalmoplegia	255320	RYR1	Channelopathy Calcium	AR
Modifiers of the centronuclear myopathy	160150	MTMR14	Tyrosine phosphatase	AD
Nemaline myopathy-1	609284	TPM3	Cystoskeletal protein	AD
Nemaline myopathy-2	256030	NEB	Cystoskeletal protein	AR
Nemaline myopathy-3	161800	ACTA1	Cystoskeletal protein	AD + de novo
Nemaline myopathy-4	609285	TPM2	Cystoskeletal protein	AD
Nemaline myopathy-5 or Amish nemaline myopathy	605355	TNNT1	Cystoskeletal protein	AR
Nemaline myopathy-6	609273	KBTBD13	Cytoskeleton regulation, ion channel tetramerization and gating	AD
Nemaline myopathy-7	610687	CFL2	Intranuclear and cytoplasmic actin	AR
Myopathy with lactic acidosis and sideroblastic anemia-1	600462	PUS1	Mitochondrial protein	AR
Myopathy with lactic acidosis and sideroblastic anemia-2	613561	YARS2	Tyrosyl-tRNA synthetase 2	AR
Mitochondrial phosphate carrier deficiency	610773	SLC25A3	Solute carrier	AR
Fatal infantile cardioencephalomyopathy, due to cytochrome c oxidase deficiency-1	604377	SCO2	Mitochondrial protein	AR
Fatal infantile cardioencephalomyopathy, due to cytochrome c oxidase deficiency-2	615119	COX15	Mitochondrial protein	AR

The Concise Encyclopedia of Genomic Diseases

Table 3.06.09.1. Continued.

Disorder	OMIM	Gene	Biological Function	Product of Conception Alteration
Fatal infantile cardioencephalomyopathy, due to cytochrome c oxidase deficiency-2	615119	COX15	Mitochondrial protein	AR
Cytochrome c oxidase deficiency	220110	COX6B1	Mitochondrial protein	AR
Erythrocyte lactate transporter defect	245340	SLC16A1	Solute carrier	AD
Myopathy with congenital cataract, hearing loss, and developmental delay	613076	GFER	Hepatic regenerative stimulation substance	AR
Neonatal encephalocardiomyopathy mitochondrial, due to ATP synthase deficiency	604273	TMEM70	Mitochondrial protein	AR
Nuclear-encoded ATP synthase deficiency	604273	ATPAF2	ATPase protein	AR
Brody myopathy	601003	ATP2A1	ATPase protein	AR
Myofibrillar myopathy-1 or Inclusion body myopathy-1	601419	DES	Cystoskeletal protein	AD
Inclusion body myopathy-2	600737	GNE	Bifunctional enzyme	AR
Inclusion body myopathy-3	605637	MYH2	Cystoskeletal protein	AD
Inclusion body myopathy with early-onset Paget disease and frontotemporal dementia	167320	VCP	ATPase, AAA-type	AD
Myosin storage myopathy	608358	MYH7	Cystoskeletal protein	AD
Myotubular myopathy	160150	MTMR14	Tyrosine phosphatase	AD
Myotubular myopathy	160150	MTMR14 / DNM2	Tyrosine phosphatase/Cytoskeletal Microtubules	Dig
Reducing body myopathy	300717	FHL1	Zinc finger protein	XD + de novo
Reducing body myopathy	300718	FHL1	Zinc finger protein	XR
Spheroid body myopathy	182920	MYOT	Cystoskeletal protein	AD
Paramyotonia congenita	168300	SCN4A	Channelopathy Sodium	AD + de novo
Myofibrillar myopathy-1	601419	DES	Cystoskeletal protein	AD

Table 3.06.09.1. Continued.

Disorder	OMIM	Gene	Biological Function	Product of Conception Alteration
Myofibrillar myopathy-2 or alpha-B crystallin-related myofibrillar myopathy	608810	CRYAB	Crystallin	AD
Severe fatal infantile hypertonic form of myofibrillar myopathy	613869	CRYAB	Crystallin	AD + AD = AR
Myofibrillar myopathy-3	182920	MYOT	Cystoskeletal protein	AD
Congenital fiber-type disproportion myopathy	606210	SEPN1	Selenoprotein	AR
Congenital fiber-type disproportion myopathy	191030	TPM3	Cystoskeletal protein	AD
Congenital fiber-type disproportion myopathy	255310	ACTA1	Cystoskeletal protein	AR
Myofibrillar myopathy-5 or FLNC-related myofibrillar myopathy	609524	FLNC	Cystoskeletal protein	AD
Oculopharyngeal muscular dystrophy	164300	PABPN1	Nuclear protein	AD
Myofibrillar myopathy-6 or BAG3-related myofibrillar myopathy	612954	BAG3	Cytoprotective protein	AD
Rigid spine muscular dystrophy 1	606210	SEPN1	Selenoprotein	AR
Myotilinopathy	609200	MYOT	Cystoskeletal protein	AD
Congenital myopathy Compton-North	612540	CNTN1	Cell adhesion molecule	AR
Rippling muscle disease-2	606072	CAV3	Trafficking Vesicle fusion	AD
Hyperkalemic periodic paralysis or Gamstorp disease	170500	SCN4A	Channelopathy Sodium	AD + de novo
Hypokalemic periodic paralysis-1 or Westphall disease	170400	CACNA1S	Channelopathy Calcium	AD
Hypokalemic periodic paralysis-2 or Westphall disease	613345	SCN4A	Channelopathy Sodium	AD
Thyrotoxic periodic paralysis-1	188580	CACNA1S	Channelopathy Calcium	
Thyrotoxic periodic paralysis-2	613239	KCNJ18	Channelopathy Potassium	

Table 3.06.09.2 Diseases of Myoneural Junction and Muscle in Different Ethnic Groups

Disease	Frequency of condition	People with higher prevalence (if available)
Duchenne muscular dystrophy	3 in 10,000 Males	Worldwide
Myotonic dystrophy	5-12 in 100,000	French-Canadians
Becker muscular dystrophy	3-6 in 100,000 Males	Worldwide
Facioscapulohumeral muscular dystrophy	5/100,000	Worldwide
X-linked myotubular myopathy	5 in 100,000 Males	Unknown
Nemaline myopathy	2 in 100,000	Worldwide
Nemaline myopathy-5 or Amish nemaline myopathy	2 in 1,000	Old Order Amish (Pennsylvania)
Emery-Dreifuss muscular dystrophy	1 in 100,000	Worldwide
Myotonia Congenita or Thomsen disease	1 in 100,000	Norwegians and Finns
Hypokalemic periodic paralysis	1 in 100,000	Worldwide
Congenital myasthenic syndrome	1-2 in 1,000,000	Romani Gypsies and North Africans
Ullrich congenital muscular dystrophy	1-2 in 1,000,000	Unknown
Bethlem myopathy	8 in 1,000,000	Unknown
Limb-girdle muscular dystrophy	1-9 in 1,000,000	Worldwide
Inclusion body myopathy	1-9 in 1,000,000	Iranian Jews
Hyperkalemic periodic paralysis	1-9 in 1,000,000	Worldwide
Schwartz-Jampel syndrome	<1 in 1,000,000	Unknown
Myopathy with lactic acidosis and sideroblastic anemia	<1 in 1,000,000	Unknown
Mitochondrial phosphate carrier deficiency	<1 in 1,000,000	Unknown
Thyrotoxic periodic paralysis	<1 in 1,000,000	Chinese, Japanese, Vietnamese, Filipinos, Thais, Koreans, and Native Americans

3.06.10 Cerebral Palsy and other Paralytic Syndromes

Spastic quadriplegic cerebral palsy: A disorder characterized by spasticity affecting all four extremities. There are several underlying causes of this disorder with at least 2 autosomal genomic loci: Spastic quadriplegic cerebral palsy-1 is caused by autosomal recessive mutations in the *GAD1* gene which encodes the glutamate decarboxylate 1 (brain, 67kDa). This enzyme catalyzes the production of gamma-aminobutyric acid from L-glutamic acid;; and Spastic quadriplegic cerebral palsy-2 is caused by deletion in the paternal allele of the *KANK1* gene which encodes the KN motif and ankyrin repeat domain-containing protein 1. This gene contains nuclear localization and exporting signals. This gene seems to be maternally imprinted, and is expressed only from the paternal allele. There are also similar conditions that were formerly classified in the Spastic quadriplegic cerebral palsy group such as Spastic paraplegia-47, Spastic paraplegia-50, Spastic paraplegia-51 and Spastic paraplegia-52. Cerebral palsy is a common condition with an incidence of 1 in every 250-1,000 births. In English and Swedish children, around 2% of cases of cerebral palsy are caused by genomic factors.

3.06.11 Other Disorders of the Nervous System

Dopamine beta-hydroxylase deficiency: A disorder of perinatal onset characterized by primary autonomic failure. Affected individuals also exhibit cardiovascular disorders, severe orthostatic hypotension, hypotonia, hypothermia, hypoglycemia and inability to exercise. Congenital dopamine beta-hydroxylase deficiency is caused by autosomal recessive mutations in the *DBH* gene which encodes the dopamine beta-hydroxylase. This enzyme catalyzes the oxidative hydroxylation of dopamine to norepinephrine. Congenital dopamine beta-hydroxylase deficiency is estimated to affect less than 1 in every 1,000,000 individuals.

Proliferative vasculopathy and hydranencephaly-hydrocephaly syndrome or Encephaloclastic proliferative vasculopathy: A prenatally lethal disorder characterized by hydranencephaly. Affected fetuses have a distinctive glomerular vasculopathy in the central nervous system and retina. These fetuses display diffuse ischemic lesions of the brain stem, basal ganglia and spinal cord with calcifications.

"Proliferative vasculopathy and hydranencephaly-hydrocephaly syndrome" is caused by autosomal recessive mutations in the *FLVCR2* gene which encodes the feline leukemia virus subgroup C cellular receptor family (member 2). The encoded protein belongs to the major facilitator superfamily of secondary carriers.

Neonatal severe encephalopathy: A disorder characterized by developmental delay, microcephaly, respiratory insufficiency with apnea, central hypoventilation and poor feeding. Most patients suffering this condition are males. This disorder usually results in death before the age of 2 years. Neonatal severe encephalopathy is caused by either *de novo* or inherited X-linked dominant mutations in the *MECP2* gene which encodes the methyl CpG binding protein 2. This nuclear protein binds specifically to methylated DNA. Mutations in the *MECP2* gene are also associated with Rett syndrome, particularly in females. Mutations in the *MECP2* gene are also associated with Nonspecific X-linked mental retardation, X-linked mental retardation due to increased dosage of the *MECP2* gene and X-linked mental retardation with spasticity. Neonatal severe encephalopathy is estimated to affect 1 in every 50,000-100,000 live males.

Glucose transporter type 1 deficiency syndrome: A neurologic disorder characterized by marked encephalopathy. Affected children manifest spasticity of onset between 1 and 4 months of age, childhood epilepsy and acquired microcephaly. Glucose transporter type 1 deficiency syndrome-1 is caused by *de novo* mutations in 90% of cases and in 10%, by autosomal dominant mutations in the *SLC2A1* gene which encodes the solute carrier (family 2 member A1), a facilitated glucose transporter. This protein is a major glucose transporter in the brain, placenta and the erythrocytes. There is also an autosomal recessive form of glucose transporter type 1 deficiency syndrome reported in a child of consanguineous Arab parents from Bedouin kindred in Qatar. Mutations in the *SLC2A1* gene are also associated with GLUT1 deficiency syndrome-2, Dystonia-9 and Idiopathic generalized epilepsy-12. Glucose transporter type 1 deficiency syndrome is estimated to affect less than 1 in every 1,000,000 individuals.

Lethal encephalopathy due to defective mitochondrial and peroxisomal fission: A disorder appearing during the 1st week of life characterized by poor feeding and neurologic impairment. This condition appears after an uncomplicated delivery. Affected children exhibit hypotonia, little spontaneous movement, no tendon reflexes, no response to light stimulation and poor visual fixation. Lethal encephalopathy due to defective mitochondrial and peroxisomal fission is caused by *de novo* autosomal monoallelic mutations in the *DNM1L* gene which encodes the dynamin-1-like protein. This protein mediates mitochondrial and peroxisomal division.

Alternating hemiplegia of childhood: A disorder characterized by recurrent episodes of hemi- or quadriplegia lasting minutes to days. This condition appears before 18 months of age. Most of these children also exhibit tonic-clonic seizures, choreoathetoid movements, dystonic posturing, nystagmus, autonomic disturbances and progressive developmental retardation. There are several underlying causes of this disorder with at least 2 autosomal genomic loci: Alternating hemiplegia of childhood-1 is caused by monoallelic mutations in the *ATP1A2* gene which encodes the alpha-2 subunit of the sodium/potassium ATPase pump. Mutations in the *ATP1A2* gene are also associated with migraine or hemiplegic migraine;; and Al-

ternating hemiplegia of childhood-2 is caused by monoallelic mutations in the *ATP1A3* gene which encodes the alpha-3 subunit of the sodium/potassium ATPase pump. Most of the cases are due to *de novo* mutations, and there are also a few familial cases. Alternating hemiplegia of childhood is estimated to affect 1 in every 110,000 newborns.

3.06.12 Disease Prevention for Disassortative Mating in Ethnic Groups

Table 3.06.12 displays a few relevant examples of autosomal recessive diseases that require the implementation of some methodology to decrease the prevalence of their associated conditions. Besides the disease, the table also includes the condition associated with the gene, the incidence of expected disassortative mating frequency, and the ethnic group/s with a high prevalence of the condition. For example, 1 in every 12,100 families of the United States is at risk of having a child with Friedreich ataxia because both parents are carriers of mutations in the *FXN* gene. In the United States, 1 in every 2,100 Caucasian American families are at risk of having a child with Spinal muscular atrophy 1 because both parents are carriers of mutations in the *SMN1* gene. In Ashkenazi Jews, 1 in every 900 families is at risk of having a child with Familial dysautonomia because both parents are carriers of a polymorphism in the *IKBKAP* gene. In Guam (Chamorro people), 1 in every 700 families is at risk of having a child with Amyotrophic lateral sclerosis 6 because both parents are carriers of a polymorphism in the *FUS* gene. In Old Order Amish of Geauga County (Ohio), 1 in every 120 families is at risk of having a child with Amish infantile epilepsy syndrome because both parents are carriers of a polymorphism in the *ST3GAL5* gene. In Finland (North/Eastern region), 1 in every 6,250 families of Caucasians American ancestry are at risk of having a child with Lethal congenital contracture syndrome because both parents are carriers of muta-

tions in the *GLE1* gene. Each of these families has a 1 in 4 risk of having a child with any of these conditions. Each of these families should receive counseling to prevent these autosomal recessive diseases.

Table 3.06.12 Frequency of Disassortative Mating in Multiple Ethnic Groups

Disease	Gene	Disassortative mating frequency	People with higher prevalence (if available)
Spinal muscular atrophy-1 or Werdnig-Hoffmann disease	SMN1	1 in 2,100	Caucasian Americans
Spinal muscular atrophy-1 or Werdnig-Hoffmann disease	SMN1	1 in 15,625	Mexican Americans
Spinal muscular atrophy-1 or Werdnig-Hoffmann disease	SMN1	1 in 5,750	Caucasian/Mexican Americans
Friedreich ataxia	FXN	1 in 12,100	Americans
Friedreich ataxia	FXN	1 in 6,000	Italians
Familial dysautonomia	IKBKAP	1 in 900	Ashkenazi Jews
Ataxia telangiectasia	ATM	1 in 10,000	Worldwide
Amyotrophic lateral sclerosis-6	FUS	1 in 700	Chamorro people (Guam)
Lethal congenital contracture syndrome	GLE1	1 in 6,250	Northern and Eastern Finnish
Amish infantile epilepsy syndrome	ST3GAL5	1 in 120	Old Order Amish (Ohio)
Polyhydramnios, megalencephaly, and symptomatic epilepsy	STRADA	1 in 625	Old Order Mennonites
Myoclonic epilepsy of Unverricht and Lundborg	CSTB	1 in 5,000	Finnish

3.07.0 Subchapter VII

Diseases of the Eye and Adnexa

Table 3.07.00-10 Diseases of the Eye and Adnexa

3.07.00-10	Diseases of the Eye and Adnexa	
	3.07.01	Disorders of Eyelid, Lacrimal System and Orbit
	3.07.02	Disorders of Conjunctiva
	3.07.03	Disorders of Sclera and Cornea
	3.07.04	Disorders of Lens
	3.07.05	Disorders of Choroid and Retina
	3.07.06	Glaucoma
	3.07.07	Disorders of Optic Nerve and Visual Pathways
	3.07.08	Disorders of Ocular Muscles, Binocular Movement, Accommodation and Refraction
	3.07.09	Visual Disturbances and Blindness
	3.07.10	Other Disorders of Eye and Adnexa

3.07.01 Disorders of Eyelid, Lacrimal System and Orbit

LADD syndrome (LacrimoAuriculoDentoDigital syndrome): A disorder characterized by different features including: Lacrimal with aplasia or hypoplasia of the puncta with obstruction of the nasal lacrimal ducts leading to epiphora, and chronic conjunctivitis due to alacrymia; Auricular with cup-shaped pinnas with sensorineural or mixed hearing deficit; Dental with small and peg-shaped lateral maxillary incisors and mild enamel dysplasia, other features include poor saliva production, hypoplasia of the salivary glands leading to dry mouth and early onset of severe dental caries; and Digital with fifth finger clinodactyly, duplication of the distal phalanx of the thumb, triphalangeal thumb and syndactyly. LADD syndrome patients also display other features including: radial aplasia, or radial-ulnar synostosis, congenital renal disease, cleft lip and palate, hypospadias, aneurysm of the interventricular septum and others. This condition could be inherited or occur as a *de novo* event. There are several underlying causes of this disorder with at least 3 autosomal genomic loci: LADD syndrome can be caused by monoallelic mutations in the tyrosine kinase domains of either the *FGFR2* gene which encodes the fibroblast growth factor receptors 2, or the *FGFR3* gene which encodes the fibroblast growth factor receptors 3;; and LADD syndrome can also be caused by autosomal dominant mutations in the *FGF10* gene which encodes the FGFR ligand. Mutations in the *FGF10* gene produce an allelic disorder resulting in aplasia of the lacrimal and salivary glands. LADD syndrome is estimated to affect less than 1 in every 1,000,000 individuals.

3.07.02 Disorders of Conjunctiva

Ligneous conjunctivitis: A form of chronic conjunctivitis characterized by the recurrent formation of pseudomembranous lesions; most frequently occurring on the palpebral surfaces. The initial symptoms of ligneous conjunctivitis include chronic tearing and redness of the conjunctiva with the consequent formation of fibrin-rich pseudomembranes. These lesions progress by replacing the normal mucosa. The disease is bilateral of infant onset. Ligneous conjunctivitis seems to be the ocular manifestation of a systemic disease. Patients with Ligneous conjunctivitis display other symptoms including: occlusive hydrocephalus, and lesions on the mucosa of the mouth, nasopharynx, ear, tracheobronchial tree, intestines, kidneys and female genital tract. This condition could be triggered by non genomic factors including local injuries, local and systemic infections, and various types of eye surgery. Ligneous conjunctivitis is caused by autosomal recessive mutations in the *PLG* gene which encodes the plasminogen, a circulating zymogen. Mutations in the *PLG* gene are also associated with Types I and II plasminogen deficiency, Plasminogen Tochigi disease and Dysplasminogenemic thrombophilia. Ligneous conjunctivitis is estimated to affect 1-9 in every 1,000,000 individuals.

3.07.03 Disorders of Sclera and Cornea

3.07.03.1 Keratitis

Hereditary keratitis: A disorder characterized by opacification and vascularization of the cornea. This condition is frequently associated with macula hypoplasia of childhood onset. Hereditary keratitis can be caused by autosomal dominant mutations in the *PAX6* gene which encodes the paired box 6, a transcriptional regulator. This protein is involved in oculogenesis and other developmental processes.

Mutations in the *PAX6* gene cause several other eye anomalies, including Aniridia, Coloboma of optic nerve, Gillespie syndrome and others. The prevalence of hereditary keratitis is unknown.

3.07.03.2 Other Disorders of Cornea

Adult onset foveomacular dystrophy with choroidal neovascularization: A disorder appearing during the 4th or 5th decade of life characterized by decreased visual acuity. This condition belongs to the patterned dystrophy of the retinal pigment epithelium. Affected individuals exhibit bilateral subfoveal vitelliform lesions, associated with gradually progressing vision loss. There are several underlying causes of this disorder with at least 2 autosomal genomic loci: Adult-onset vitelliform macular dystrophy is caused by autosomal dominant mutations in the *PRPH2* gene which encodes the peripherin 2 (slow retinal degeneration). Mutations in the *PRPH2* gene also cause a form of Retinitis pigmentosa, Patterned dystrophies of the retinal pigment epithelium, and other Macular dystrophies forms;; and Adult-onset vitelliform macular dystrophy is caused by autosomal dominant mutations in the *BEST1* gene which encodes the bestrophin 1. Mutations in the *BEST1* gene also cause Juvenile-onset vitelliform macular dystrophy or Best disease.

3.07.03.3 Corneal Dystrophies

Corneal dystrophy: A group of disorders characterized by bilateral corneal opacity leading to vision loss. These conditions can be classified: According to the location affecting the epithelial and subepithelial area including: Epithelial basement membrane corneal dystrophy, Gelatinous drop-like corneal dystrophy, Meesmann epithelial corneal dystrophy, Lisch epithelial corneal dystrophy and Subepithelial mucinous corneal dystrophy; According to the location affecting the Bowman layer area including: Reis-Bücklers corneal dystrophy and Thiel-Behnke corneal dystrophy; According to the location affecting the stroma area including: Avellino type of corneal dystrophy, Congenital stromal corneal dystrophy, Schnyder crystalline corneal dystrophy, Groenouw type I granular corneal dystrophy, Lattice corneal dystrophy type I, Lattice corneal dystrophy type II, Lattice corneal dystrophy type IIIA, Macular corneal dystrophy, Fleck corneal dystrophy and Posterior amorphous corneal dystrophy; Affecting the Descemet/endothelial area including: Posterior polymorphous corneal dystrophy 1, Posterior polymorphous corneal dystrophy 2, Posterior polymorphous corneal dystrophy 3, Corneal endothelial dystrophy and perceptive deafness, Fuchs endothelial corneal dystrophy 1, Fuchs endothelial corneal dystrophy 4, Fuchs endothelial corneal dystrophy 6, Corneal endothelial dystrophy 1, Corneal endothelial dystrophy 2 and X-linked endothelial corneal dystrophy. In the United States, Corneal dystrophy is estimated to affect 1 in every 1,150 individuals. Endothelial and anterior corneal dystrophies account for most of the cases.

3.07.03.4 Epithelial and Subepithelial Corneal dystrophy

Epithelial basement membrane corneal dystrophy: A bilateral anterior corneal dystrophy characterized by grayish epithelial fingerprint lines, geographic map-like lines and dots on slit-lamp examination. In most cases this disorder is not considered to be inherited, but there are cases of familial form due to autosomal dominant

point mutations in the *TGFBI* gene which encodes the transforming growth factor-beta-induced.

Gelatinous drop-like corneal dystrophy: A disorder appearing during the 1st decade of life characterized by severe corneal amyloidosis. Affected individuals start exhibiting photophobia, foreign body sensation in the cornea, and severe loss of vision. Usually in the 3rd decade of life, patients display mulberry-shaped gelatinous masses beneath the corneal epithelium, leading to blindness. Gelatinous drop-like corneal dystrophy can be caused by autosomal recessive mutations in the *TACSTD2* gene which encodes the tumor-associated calcium signal transducer 2. In Japan, the carrier frequency for Gelatinous drop-like corneal dystrophy trait is estimated to be 1 in 90 individuals, which means the frequency of this condition is about 1 in 32,500 individuals or $1/(180)^2$.

Meesmann epithelial corneal dystrophy: A disorder characterized by fragility of the anterior corneal epithelium. This condition starts to manifest with signs of irritation during the 1st or 2nd year of life. In most cases vision is rarely impaired. This condition is named after Alois Meesmann. There are several underlying causes of this disorder with at least 2 autosomal genomic loci: Meesmann corneal dystrophy is caused by autosomal dominant mutation in the *KRT3* gene which encodes the cornea-specific keratins K3, an intermediate filament protein;; and Meesmann corneal dystrophy is caused by autosomal dominant mutation in the *KRT12* gene which encodes the cornea-specific keratins K12, an intermediate filament protein. Meesmann corneal dystrophy due to monoallelic mutations in the *KRT12* gene has been reported among people from Germany. This mutated allele was traced back to the year 1620. Meesmann corneal dystrophy due to monoallelic in the *KRT3* gene has been reported among people from Northern Ireland.

Lisch epithelial corneal dystrophy: A disorder appearing during childhood characterized by gray band-shaped and feathery opacities. Lisch epithelial corneal dystrophy has been mapped to chromosomal region Xp22.3, but the gene has not yet been characterized. Males and females seem to be equally affected.

Subepithelial mucinous corneal dystrophy: A disorder characterized by frequent, recurrent corneal erosions in the first decade of life. The gene/loci have not been mapped.

3.07.03.5 Bowman Layer Corneal Dystrophy

Reis-Bücklers corneal dystrophy: A disorder characterized by disintegration of the Bowman's layer. The corneas of affected children exhibit symmetrical dusty opacity and a rough map-like surface. This condition leads to corneal erosions and attacks of ocular hyperemia, pain and photophobia. Most of the affected persons with this condition also have strabismus. Reis-Bücklers corneal dystrophy is caused by autosomal dominant mutations in the *TGFBI* gene which encodes the transforming growth factor-beta-induced. This protein binds to type I, II and IV collagens playing a role in cell-collagen interactions.

Thiel-Behnke corneal dystrophy: A disorder characterized by progressive honeycomb-like, subepithelial corneal opacities with recurrent erosions. There are several underlying causes of this disorder with at least 2 autosomal genomic loci: Thiel-Behnke corneal dystrophy is caused by autosomal dominant mutations in *TGFBI* gene;; and the other locus has been mapped, but the gene has not yet been characterized.

3.07.03.6 Stromal Corneal Dystrophy

Avellino type of corneal dystrophy: A disorder characterized by clinical granular and pathologic examination displaying both granular and lattice corneal dystrophies. Avellino type of corneal dystrophy is caused by autosomal dominant mutations in the *TGFBI* gene which encodes the transforming growth factor-beta-induced. This condition has been reported in patients whose family origin was traced back to the Italian province of Avellino. Autosomal dominant mutations in the same gene has also been associated with Epithelial basement membrane corneal dystrophy, Groenouw type I granular corneal dystrophy, Type I lattice corneal dystrophy, Type IIIA lattice corneal dystrophy, Reis-Bücklers corneal dystrophy and Thiel-Behnke corneal dystrophy. Most of the pathogenic mutations in the *TGFBI* gene that result in corneal dystrophies occur due to DNA changes at the CpG dinucleotide level, causing codon substitution.

Congenital stromal corneal dystrophy: A disorder characterized by opacities caused by large numbers of flakes and spots throughout all layers of the stroma. Congenital stromal corneal dystrophy is caused by autosomal dominant mutation in the *DCN* gene which encodes the decorin. This protein is a proteoglycan found in many connective tissues.

Schnyder crystalline corneal dystrophy: A disorder appearing early in life characterized by oval or annular clouding of the central part of the cornea with the periphery remaining clear. Schnyder crystalline corneal dystrophy is caused by autosomal dominant mutation in the *UBIAD1* gene which encodes the UbiA prenyltransferase domain containing protein 1. This protein seems to be involved in cholesterol and phospholipid metabolism.

Groenouw type I granular corneal dystrophy: A disorder characterized by changes in the corneal stroma rather than the epithelium. Groenouw type I granular corneal dystrophy is caused by autosomal dominant mutations in the *TGFBI* gene.

Lattice corneal dystrophy type I: A disorder characterized by changes involving the central portion of the cornea. This condition starts to manifest in adolescence resulting in a form of reticular pattern in the corneal stroma. Affected individuals manifest recurrent corneal ulceration, leading to progressively severe visual impairment by the 5th or 6th decade of life. Lattice corneal dystrophy type I is caused by autosomal dominant mutations in the *TGFBI* gene.

Lattice corneal dystrophy type II: A disorder appearing during the 2nd decade of life characterized by local amyloid deposition affecting the eyes. Years later, these patients' nerves and skin are also affected. Lattice corneal dystrophy type II is caused by autosomal dominant mutations in the *GSN* gene which encodes the gelsolin. This condition seems to be more common among Finnish people.

Lattice corneal dystrophy type IIIA: A disorder characterized by late onset, appearing during the 7th-9th decade of life. Lattice corneal dystrophy type IIIA is caused by autosomal dominant mutations in the *TGFBI* gene.

Macular corneal dystrophy: A progressive disorder with onset during the first decade of life. This condition is characterized by painful attacks with photophobia, foreign body sensations and recurrent erosions. Macular corneal dystrophy is caused by autosomal recessive mutations in the *CHST6* gene which encodes the corneal N-acetylglu-cosamine-6-sulfotransferase. This enzyme catalyzes the transfer of a sulfate group to the GlcNAc residues of keratan. Macular corneal dystrophy is common among people from Iceland.

Fleck corneal dystrophy: A condition characterized by small corneal opacities, some of which resemble "flecks". These opacities or flecks are non-progressive and in most cases asymptomatic. Fleck corneal dystrophy is caused by autosomal dominant mutations in *PIKFYVE* gene which encodes the phosphoinositide kinase FYVE finger containing protein. This enzyme phosphorylates the D-5 position in phosphatidylinositol-3-phosphate.

Posterior amorphous corneal dystrophy: A disorder characterized by irregular sheet-like areas of opacification with involvement of the Descemet membrane. The gene/loci have not been mapped but seem to follows an autosomal dominant pattern of inheritance.

3.07.03.7 Descemet/Endothelial Corneal Dystrophy

Posterior polymorphous corneal dystrophy: A disorder characterized by metaplasia and overgrowth of corneal endothelial cells. There are several underlying causes of this disorder with at least 3 autosomal genomic loci: Posterior polymorphous corneal dystrophy-1 is caused by autosomal dominant mutations in the *VSX1* gene which encodes the visual system homeobox 1; Posterior polymorphous corneal dystrophy-2 is caused by autosomal dominant mutations in the *COL8A2* gene which encodes the alpha 2 type VIII collagen; and Posterior polymorphous corneal dystrophy-3 is caused by autosomal dominant mutations in the *ZEB1* gene which encodes the zinc finger E-box binding homeobox 1.

Fuchs endothelial corneal dystrophy: A progressive disorder characterized by bilateral dysfunction of the corneal epithelium. This condition appears during 3rd-4th decade of life affecting vision two decades later. This condition is slightly more common in women than in men. In this condition, it is difficult to separate the effects of heritability and environmental factors. This condition is named after Ernst Fuchs. There are several underlying causes of this disorder with at least 7 autosomal genomic loci: Early-onset Fuchs endothelial corneal dystrophy or Fuchs endothelial corneal dystrophy-1 can be caused by autosomal dominant mutations in the *COL8A2* gene which encodes the alpha 2 type VIII collagen;; Fuchs endothelial corneal dystrophy-4 can be caused by mutations in the *SLC4A11* gene which encodes the solute carrier (family 4 member A11), a sodium borate transporter;; and Fuchs endothelial corneal dystrophy-6 can be caused by mutations in the *ZEB1* gene which encodes the zinc finger E-box (binding homeobox 1). Type 4 and 6 forms are most common, and they are late-onset forms. Other late-onset loci include Fuchs endothelial corneal dystrophy 2, 3, 5 and 7. These forms have been mapped, but the genes have not yet been characterized. In the United States, the prevalence of this condition is estimated to be about 4% among persons over the age of 40 years. The early-onset variant of Fuchs endothelial dystrophy is rare, although the exact prevalence is unknown. Fuchs endothelial corneal dystrophy seems to be un-common in Saudi Arabia, China and Singapore, and very rare in Japan.

Corneal endothelial dystrophy: A disorder characterized by a thickening and opacification of the cornea, altered morphology of the endothelium, and secretion of an abnormal collagenous layer at the Descemet membrane. This condition can be either autosomal dominant or recessive. The latter form is more common and more severe. There are several underlying causes of this disorder with at least 2 autosomal genomic loci: Autosomal dominant Corneal endothelial dystrophy or Type 1 has been mapped, but the gene has not yet been characterized;; and Autosomal recessive Corneal endothelial dystrophy or Type 2 is caused by biallelic mutations in the *SLC4A11* gene which encodes the solute carrier (family 4 member A11), a sodium borate transporter.

X-linked endothelial corneal dystrophy: A disorder characterized by severe corneal opacifications with corneal clouding and milky appearance. X-linked endothelial corneal dystrophy has been mapped, but the gene has not yet been characterized. X-linked endothelial corneal dystrophy is estimated to affect less than 1 in every 1,000,000 individuals.

3.07.03.8 Keratoconus

Keratoconus: A degenerative disorder characterized by bilateral protrusion of the cornea, displacing the apex downward and nasally. Patients with this condition display substantial distortion of vision, with multiple images, streaking and sensitivity to light. Keratoconus is the most common corneal dystrophy appearing during the adolescent years. Environmental and genomic factors are possible causes for this condition. The genomics factors include the fact that this condition is over 4 times more common among people from South Asian heritage than Caucasians. Patients with Down syndrome have both a higher prevalence and a more rapidly progressing form of keratoconus. There are several underlying causes of this disorder with at least 8 autosomal genomic loci: Kerato-

conus-1 can be caused by autosomal dominant mutations in the *VSX1* gene which encodes the visual system homeobox 1;; and in the 7 remaining loci, the genes have not yet been characterized. Keratoconus is estimated to affect around 1 in every 1,000-2,000 individuals of Caucasian origin.

Table 3.07.03.1 Recognizable Disorders of Sclera and Cornea

Disorder	OMIM	Gene	Biological Function	Product of Conception Alteration
Hereditary keratitis	148190	PAX6	Transcription factor Helix-turn-helix domains	AD
Adult onset foveomacular dystrophy with choroidal neovascularization	608161	BEST1	Channelopathy chloride	AD
Adult onset foveomacular dystrophy with choroidal neovascularization	608161	PRPH2	Cytoskeletal Intermediate filaments	AD
Epithelial basement membrane corneal dystrophy	121820	TGFBI	Cell-collagen interactions	AD
Gelatinous drop-like corneal dystrophy	204870	TACSTD2	Calcium signal transducer	AR
Meesmann corneal dystrophy	122100	KRT12	Cytoskeletal Intermediate filaments	AD
Meesmann corneal dystrophy	122100	KRT3	Cytoskeletal Intermediate filaments	AD
Reis-Bücklers corneal dystrophy	608470	TGFBI	Cell-collagen interactions	AD
Thiel-Behnke corneal dystrophy	602082	TGFBI	Cell-collagen interactions	AD
Avellino type of corneal dystrophy	607541	TGFBI	Cell-collagen interactions	AD
Congenital stromal corneal dystrophy	610048	DCN	Extracellular matrix protein	AD
Schnyder crystalline corneal dystrophy	121800	UBIAD1	Transferase activity	AD
Groenouw type I granular corneal dystrophy	121900	TGFBI	Cell-collagen interactions	AD
Lattice corneal dystrophy type I	122200	TGFBI	Cell-collagen interactions	AD
Lattice corneal dystrophy type II	122000	DCN	Extracellular matrix protein	AD
Lattice corneal dystrophy type IIIA	608471	TGFBI	Cell-collagen interactions	AD
Macular corneal dystrophy	217800	CHST6	Sulfotransferases	AR

Table 3.07.03.1. Continued.

Disorder	OMIM	Gene	Biological Function	Product of Conception Alteration
Fleck corneal dystrophy	121850	PIKFYVE	Endomembrane homeostasis	AD
Polymorphous posterior corneal dystrophy-1	122000	VSX1	Nuclear transcription factor	AD
Polymorphous posterior corneal dystrophy-2	609140	COL8A2	Extracellular matrix protein	AD
Polymorphous posterior corneal dystrophy-3	609141	ZEB1	Transcription factor Helix-turn-helix domains	AD
Fuchs endothelial corneal dystrophy-1	136800	COL8A2	Extracellular matrix protein	AD
Fuchs endothelial corneal dystrophy-4	613268	SLC4A11	Solute carrier	AD
Fuchs endothelial corneal dystrophy-6	613270	ZEB1	Transcription factor Helix-turn-helix domains	AD
Corneal endothelial dystrophy-2	217700	SLC4A11	Solute carrier	AR
Corneal endothelial dystrophy and perceptive Deafness	217400	SLC4A11	Solute carrier	AR
Patterned macular dystrophy	169150	PRPH2	Cytoskeletal Intermediate filaments	AD
Keratoconus-1	148300	VSX1	Nuclear transcription factor	AD

3.07.04 Disorders of Lens

EDICT syndrome (Endothelial Dystrophy, Iris hypoplasia, congenital Cataract and Thinning of the corneal stroma): An anterior segment dysgenesis disorder caused by autosomal dominant mutation in the *MIR184* gene (microRNA 184). This gene codes for single-stranded noncoding RNAs. EDICT syndrome is estimated to affect less than 1 in every 1,000,000 individuals.

Juvenile-onset cortical cataract: A disorder characterized by cotton-like cortical opacities with occasional grape-like cysts in the anterior cortex. Juvenile-onset cortical cataract is caused by autosomal recessive mutations in the *BFSP1* gene which encodes the beaded filament structural protein 1, a cytoskeletal protein. This condition has been reported among a large consanguineous Indian family.

Hyperferritinemia-cataract syndrome: A disorder characterized by the association of early onset cataract with an excess storage of iron in the plasma and tissues in the absence of iron overload. Hyperferritinemia-cataract syndrome is caused by monoallelic mutations in the *FTL* gene which encodes the ferritin light chain. Specifically, mutations occur in the iron-responsive element in the 5-prime noncoding region of this gene. Hyperferritinemia-cataract syndrome is estimated to affect at least 1 in 200,000 individuals.

Foveal hypoplasia: A poorly defined syndrome with several features including: foveal hypoplasia, congenital nystagmus, corneal epithelial changes, peripheral corneal pannus and presenile cataract. The cataract appears before the age of 40. Sometimes, the foveal hypoplasia may occur without other anomalies, although the fundus is usually lightly pigmented. Foveal hypoplasia is caused by autosomal dominant mutation in the *PAX6* gene which encodes the paired box 6, a transcriptional regulator, involved in oculogenesis and other developmental processes. Mutations in the same gene are also associated with Foveal hypoplasia and presenile cataract syndrome, Isolated foveal hypoplasia, Foveal hyperplasia, Foveal hypoplasia with anterior segment anomalies, Aniridia, Cataract with late-onset corneal dystrophy, Coloboma of optic nerve, Ocular coloboma, Gillespie syndrome, Keratitis, Morning glory disc anomaly, Optic nerve hypoplasia and Peters anomaly.

3.07.05 Disorders of Choroid and Retina

3.07.05.1 Disorders of Choroid

Bietti crystalline corneoretinal dystrophy: A progressive disorder characterized by choroidal vessels sclerosis and multiple sparkling crystals scattered in the posterior retina and corneal limbus. Most of these patients exhibit in the third decade of life, night blindness, decreased vision, and paracentral scotoma, eventually leading to legal blindness. Bietti crystalline corneoretinal dystrophy is caused by autosomal recessive mutations in the *CYP4V2* gene which encodes the CYP450 (family

4V polypeptide 2). This protein is involved in fatty acid and steroid metabolism. Bietti crystalline corneoretinal dystrophy appears to be most common in East Asians, especially in people from China and Japan.

Pigmented paravenous chorioretinal atrophy: A stationary disease of the ocular fundus that is almost always asymptomatic. In symptomatic cases, patients exhibit strabismus, hyperopia, esotropia and vitreoretinal degeneration. Pigmented paravenous chorioretinal atrophy can be caused by autosomal dominant mutations in the *CRB1* gene which encodes the crumbs homolog 1. Specifically, mutations occur in the EGF-like domain. This condition could also be X-linked, but the locus has not been identified.

Central areolar choroidal dystrophy: A retinal disorder appearing during the 3rd-6th decade of life characterized by a large area of atrophy in the center of the macula. Affected individuals exhibit loss or absence of photoreceptors, retinal pigment epithelium and choriocapillaris in this area, resulting in progressive decrease in visual acuity. There are several underlying causes of this disorder with at least 3 autosomal genomic loci: Central areolar choroidal dystrophy-1 has been mapped, but the gene has not been characterized;; Central areolar choroidal dystrophy-2 can be caused by autosomal dominant mutations in the *PRPH2* gene which encodes the peripherin 2;; and Central areolar choroidal dystrophy-3 gene/loci have not been characterized.

Choroideremia: A progressive disorder characterized by degeneration of the choriocapillaris, the retinal pigment epithelium and the photoreceptor of the eye. Male patients exhibit nyctalopia in the 1st or 2nd decade of life. Years later, these young adults show peripheral visual field constriction with progression from annular scotomas to concentric visual field loss. By mid-adulthood, affected individuals display severe visual acuity impairment. Choroideremia is caused by X-linked recessive mutations in the *CHM* gene which encodes the Ras-related GTPase Rab escort protein. This protein is involved in controlling vesicle trafficking in secretory and endocytic pathways. Choroideremia is estimated to affect 1-2 in every 100,000 individuals. Female carriers usually do not show serious visual impairment.

3.07.05.2 Retinal Detachments and Breaks

Retinoschisis: A disorder characterized by the abnormal splitting of the retina's neurosensory layers, resulting in a loss of vision in the corresponding visual field. Retinoschisis is caused by X-linked recessive mutations in the *RS1* gene which encodes the retinoschisin. This protein is implicated in cell-cell interactions and cell adhesion. Retinoschisis is estimated to affect 1 in every 5,000-25,000 individuals, primarily young males.

Knobloch syndrome: A disorder appearing in the first year of life characterized by myopia vitreoretinal degeneration with retinal detachment, macular abnormalities, hydrocephalus and occipital encephalocele. Knobloch syndrome features in the eyes are severe and progressive, leading to bilateral blindness at a young age. Patients also exhibit both ocular and extraocular abnormalities: the ocular include congenital cataract, iris abnormalities and lens subluxation; the extraocular include pyloric stenosis, cardiac dextroversion, abnormal lymphatic vessels in the lung and patent ductus arteriosus. There are several underlying causes of this disorder with at

least 3 autosomal genomic loci: Knobloch syndrome-1 is caused by autosomal recessive mutations in the *COL18A1* gene which encodes the type XVIII alpha 1 collagen;; Knobloch syndrome-2 can be caused by mutations in the *ADAMTS18* gene which encodes the ADAM metallopeptidase with thrombospondin type 1 motif 18;; and Knobloch syndrome-3 has been mapped, but the gene has not been characterized. Knobloch syndrome is estimated to affect less than 1 in every 1,000,000 individuals.

3.07.05.3 Retinal Disorders

Basal laminar drusen: A disorder characterized by slightly raised yellow subretinal nodules that are randomly scattered in the macula. The basal laminar drusen may lead to a serous pigment epithelial detachment of the macula, resulting in vision loss. Basal laminar drusen can be caused by mutations in the *CFH* gene which encodes the complement factor H. This gene influences the development of basal laminar drusen. Bilateral large drusen is a significant risk factor for the advancement of early age-related macular dystrophy with loss of central vision.

Bradyopsia: A disorder appearing during childhood characterized by prolonged electroretinal response suppression. Affected individuals show difficulties adjusting to changes in luminance, normal to subnormal acuity and photophobia. There are several underlying causes of this disorder with at least 2 autosomal genomic loci: Bradyopsia can be caused by autosomal recessive mutations in the *RGS9* gene which encodes the regulator of G-protein signaling 9;; and Bradyopsia can be caused by autosomal recessive mutations in the *R9AP* gene which encodes the regulator of G-protein signaling 9 binding protein. Bradyopsia is estimated to affect less than 1 in every 1,000,000 individuals.

Familial exudative vitreoretinopathy: A disorder characterized by retinal traction, peripheral vitreous opacities, and subretinal and intraretinal exudates. This condition frequently leads to retinal folds, tears and detachments. There are several underlying causes of this disorder with at least 4 autosomal genomic loci and 1 X-linked genomic locus: Familial exudative vitreoretinopathy-1 is caused by autosomal dominant mutation in the *FZD4* gene which encodes the frizzled-4. Mutation in the *FZD4* gene is associated with a form of retinopathy of prematurity;; Familial exudative vitreoretinopathy-2 is caused by X-linked mutations in the *NDP* gene which encodes the Norrie disease (pseudoglioma). Mutations in the *NDP* gene are associated with Norrie disease;; Familial exudative vitreoretinopathy-3 has been mapped, but the gene has not been characterized;; Familial exudative vitreoretinopathy-4 can be caused by either autosomal dominant mutations or autosomal recessive mutations in the *LRP5* gene which encodes the low density lipoprotein receptor-related protein 5;; and Familial exudative vitreoretinopathy-5 is caused by autosomal dominant mutations in the *TSPAN12* gene which encodes the tetraspanin 12.

Retinopathy of prematurity or Retrolental fibroplasia: A disorder characterized by scarring of retinal blood vessels and retinal detachment. This condition can lead to blindness. The risk factors associated with retinopathy of prematurity include mainly non-genomic factors, such as preterm and low birth weight babies, oxygen toxicity and relative hypoxia. Retinopathy of prematurity can be caused by autosomal dominant mutations in the *FZD4* gene which encodes the frizzled-4.

3.07.05.4 Degeneration of Macula and Posterior Pole

Age-related macular degeneration: A disorder characterized by progressive degeneration of photoreceptors and underlying retinal pigment epithelium cells in the macula region of the retina. This condition is a common cause of blindness and visual impairment in older adults. Affected individuals showed extracellular deposits which are the hallmark of this condition. These deposits cause visual loss and choroidal neovascularization. There are several non-genomic factors contributing to this condition including: aging, hypertension, high fat intake and cholesterol levels, obesity, smoking and exposure to high-energy visible light. There are several genomic factors contributing to this condition including: family history, ethnicity (it is particularly more common among Caucasians than African Americans), and other genomic diseases including Retinitis pigmentosa and Stargardt's disease or Juvenile macular degeneration. There are several underlying causes of this disorder with at least 12 susceptibility autosomal genomic loci and at least 4 reduced risk autosomal genomic loci: Susceptibility to age-related macular degeneration-1 is associated with polymorphisms in the *HMCN1* gene which encodes the hemicentin;; Age-related macular degeneration-2 is associated with mutations in the *ABCR* gene which encodes the ATP-binding cassette (sub-family A member 4);; Age-related macular degeneration-3 is caused by mutations in the *FBLN5* gene which encodes the fibulin 5;; Susceptibility to age-related macular degeneration-4 is associated with polymorphisms in the *CFH* gene which encodes the complement factor H;; Age-related macular degeneration-5 is associated with mutations in the

ERCC6 gene which encodes the excision repair cross-complementing rodent repair deficiency (complementation group 6);; Age-related macular degeneration-6 is associated with mutations in the *RAX2* gene which encodes the retina and anterior neural fold homeobox 2;; Age-related macular degeneration-7 is associated with polymorphisms in the *HTRA1* gene;; Age-related macular degeneration-8 is associated with polymorphisms in the *ARMS2* gene which encodes the age-related maculopathy susceptibility 2;; Age-related macular degeneration-9 is associated with polymorphisms in the *C3* gene which encodes the complement component 3;; Age-related macular degeneration-10 is associated with polymorphisms in the *TLR4* gene which encodes the toll-like receptor 4;; Age-related macular degeneration-11 is associated with polymorphisms in the *CST3* gene which encodes the cystatin C;; and Age-related macular degeneration-12 is associated with polymorphisms in the *CX3CR1* gene which encodes the chemokine (C-X3-C motif) receptor 1. There is also a form of Age-related macular degeneration caused by mutations in the *MT-TL1* gene in the mitochondrial DNA. There are also haplotypes or genotypes associated with reduced risk for Age-related macular degeneration including polymorphisms: in the *CFB* gene which encodes the complement factor B, the *C2* gene which encodes the complement component 2, the *CFHR1* gene which encodes the complement factor H-related 1, or the *CFHR3* gene which encodes the complement factor H-related 3. Age-related macular degeneration is a common condition is estimated to affect around 10% of people 66-74 years of age and the prevalence rises to 30% in people 75-85 years of age.

3.07.05.5 Hereditary Retinal Dystrophy

Cone dystrophy: A disorder characterized by the loss of cone cells. These cells are the photoreceptors responsible for both central and color vision. Affected individuals exhibit vision loss, sensitivity to bright lights, photophobia, fine nystagmus and poor color vision. There are several underlying causes of this disorder with at least 5 autosomal genomic loci: Retinal cone dystrophy-1 can be caused by autosomal dominant inheritance, the locus has been identified, but the gene has not been characterized;; Cone dystrophy-3 is caused by autosomal dominant mutations in the *GUCA1A* gene which encodes the guanylate cyclase activator 1A;; Retinal cone dystrophy with supernormal rod electro-retinogram-3A is caused by mutations in the *PDE6H* gene which encodes the cone phosphodiesterase 6H;; Retinal cone dystrophy with supernormal rod electro-retinogram-3B is caused by autosomal recessive mutations in the *KCNV2* gene which encodes the potassium channel (subfamily V member 2);; and Cone dystrophy-4 is caused by autosomal recessive mutations in the *PDE6C* gene which encodes the cone phosphodiesterase 6C.

Cone rod dystrophies: Are a group of pigmentary retinopathies that lead to early impairment of vision. Affected individuals exhibit decline in color vision and visual acuity, photoaversion and decreased sensitivity in the central visual field. Later in this condition, patients show progressive loss in peripheral vision and night blindness. There are several underlying causes of this disorder with at least 16 autosomal genomic loci and 3 X-linked genomic loci: Cone-rod dystrophy-1 has been mapped, but the gene has not yet been characterized;; Cone-rod dystrophy-2 is caused by mutations in the *CRX* gene which encodes the cone-rod homeobox;; Cone-rod dystrophy-3 is caused by mutation in the *ABCA4* gene which encodes the ATP-binding cassette (sub-family A member 4);; Cone-rod dystrophy-4 has been mapped close to the *NF1* gene on 17q, but the gene causing this condition has not yet been characterized;; Cone-rod dystrophy-5 is caused by mutations in the *PITPNM3* gene which encodes the PITPNM family member 3;; Cone-rod dystrophy-6 is caused by mutations in the *GUCY2D* gene which encodes the membrane guanylate cyclase 2D (retina-specific);; Cone-rod dystrophy-7 is caused by mutations in the *RIMS1* gene which encodes the regulating synaptic membrane exocytosis 1;; Cone-rod dystrophy-8 has been mapped, but the gene has not yet been characterized;; Cone-rod dystrophy-9 is caused by mutations in the *ADAM9* gene which encodes the ADAM metallopeptidase domain 9;; Cone-rod dystrophy-10 is caused by mutations in the *SEMA4A* gene which encodes the semaphorin 4A;; Cone-rod dystrophy-11 is caused by mutations in the *RAX2* gene which encodes the retina and anterior neural fold homeobox 2;; Cone-rod dystrophy-12 is caused by mutations in the *PROM1* gene which encodes the prominin 1;; Cone-rod dystrophy-13 is caused by mutations in the *RPGRIP1* gene which encodes the retinitis pigmentosa GTPase regulator interacting protein 1;; Cone-rod dystrophy-14 is caused by mutations in the *GUCA1A* gene which encodes the guanylate cyclase activator 1A (retina);; Cone-rod dystrophy-15 is caused by mutations in the *CDHR1* gene which encodes the cadherin-related family member 1;; and Cone-rod dystrophy-16 is caused by mutations in the *C8orf37* gene which encodes the chromosome 8 open reading frame 37. The X-linked forms include: X-linked cone-rod dystrophy-1 is caused by mutations in an alternative terminal exon of the *RPGR* gene which encodes the retinitis

pigmentosa GTPase regulator;; X-linked cone-rod dystrophy-2 has been mapped, but the gene has not yet been characterized;; and X-linked cone-rod dystrophy-3 is caused by mutations in the *CACNA1F* gene which encodes the voltage-dependent calcium channel (L type subunit alpha 1F). Cone-rod dystrophy is estimated to affect 1 in 40,000 individuals. The four major forms associate with these conditions are Cone-rod dystrophy-2, 3, 6 and X-linked-1. Sometimes Cone-rod dystrophy is part of several syndromes, such as Bardet-Biedl syndrome and Spinocerebellar Ataxia Type 7.

Bestrophinopathy: A disorder characterized by loss of central vision, electro-retinogram anomalies and an absent electro-oculogram light rise. Bestrophinopathy can be caused by autosomal recessive mutations in the *BEST1* gene which encodes the transmembrane protein bestrophin-1. Bestrophinopathy is estimated to affect less than 1 in every 1,000,000 individuals.

Jalili syndrome: A disorder characterized by the combination of cone-rod dystrophy of the retina and amelogenesis imperfecta. Jalili syndrome is caused by autosomal recessive mutations in the *CNNM4* gene which encodes the cyclin M4, a metal transporter protein. This condition has been reported in a consanguineous Arab family living in the Gaza Strip. There are also a few cases reported in Lebanon and Kosovo. Jalili syndrome is estimated to affect less than 1 in every 1,000,000 individuals.

Retinitis pigmentosa: A group of disorders characterized by progressive retinal dystrophy. This condition often leads to incurable blindness. Affected individuals start manifesting night blindness in the 2nd decade of life. Years later, affected patients develop tunnel vision and slowly progressing decrease in central vision. Affected individuals become legally blind in the 4th-5th decade of life. Most of the forms of Retinitis pigmentosa are inherited; there are also *de novo* mutations, such as retinitis pigmentosa-3, which can be caused by *de novo* mutations in the *RPGR* gene which encodes the retinitis pigmentosa GTPase regulator. There are several underlying causes of this disorder with at least 65 loci. The patterns of inheritance include autosomal recessive, autosomal dominant, X-linked, Y-linked and mitochondrial: Patients with autosomal recessive inheritance account for 50-60% of all cases of retinitis pigmentosa: Retinitis pigmentosa-1 can be caused by biallelic mutations in the *RP1* gene which encodes the retinitis pigmentosa 1, an uncommon form reported among consanguineous Pakistani families;; Retinitis pigmentosa-4 is caused by biallelic mutations in the *RHO* gene which encodes the rhodopsin;; Retinitis pigmentosa-12 is caused by biallelic mutations in the *CRB1* gene which encodes the crumbs homolog 1;; Retinitis pigmentosa-14 is caused by biallelic mutations in the *TULP1* gene which encodes the tubby like protein 1;; Retinitis pigmentosa-19 is caused by biallelic mutations in the *ABCA4* gene which encodes the ATP-binding cassette (sub-family A member 4);; Retinitis pigmentosa-20 is caused by biallelic mutations in the *RPE65* gene which encodes the retinal pigment epithelium-specific protein 65kDa;; Retinitis pigmentosa-25 is caused by biallelic mutations in the *EYS* gene which encodes the eyes shut homolog;; Retinitis pigmentosa-26 is caused by biallelic mutations in the *CERKL* gene which encodes the ceramide kinase-like;; Retinitis pigmentosa-35 is caused by biallelic mutations in the *SEMA4A* gene which encodes the semaphorin 4A;; Retinitis pigmentosa-36 is caused by biallelic mutations in the *PRCD* gene which encodes the progressive rod-cone

degeneration;; Retinitis pigmentosa-37 is caused by biallelic mutations in the *NR2E3* gene which encodes the nuclear receptor subfamily 2;; Retinitis pigmentosa-38 is caused by biallelic mutations in the *MERTK* gene which encodes the c-mer proto-oncogene tyrosine kinase;; Retinitis pigmentosa-39 is caused by biallelic mutations in the *USH2* gene which encodes the Usher syndrome 2A;; Retinitis pigmentosa-40 is caused by biallelic mutations in the *PDE6B* gene which encodes the rod receptor phosphodiesterase 6B;; Retinitis pigmentosa-41 is caused by biallelic mutations in the *PROM1* gene which encodes the prominin 1;; Retinitis pigmentosa-43 is caused by biallelic mutations in the *PDE6A* gene which encodes the rod phosphodiesterase 6A;; Retinitis pigmentosa-44 is caused by biallelic mutations in the *RGR* gene which encodes the retinal G protein coupled receptor;; Retinitis pigmentosa-45 is caused by biallelic mutations in the *CNGB1* gene which encodes the cyclic nucleotide gated channel beta 1;; Retinitis pigmentosa-46 is caused by biallelic mutations in the *IDH3B* gene which encodes the isocitrate dehydrogenase 3 (NAD+) beta;; Retinitis pigmentosa-47 is caused by biallelic mutations in the *SAG* gene which encodes the S-antigen retina and pineal gland (arrestin);; Retinitis pigmentosa-49 is caused by biallelic mutations in the *CNGA1* gene which encodes the cyclic nucleotide gated channel alpha 1;; Retinitis pigmentosa-51 is caused by mutations in the *TTC8* gene which encodes the tetratricopeptide repeat domain 8;; Retinitis pigmentosa-53 is caused by biallelic mutations in the *RDH12* gene which encodes the retinol dehydrogenase 12 (all-trans/9-cis/11-cis);; Retinitis pigmentosa-54 is caused by biallelic mutations in the *C2orf71* gene which encodes the chromosome 2 open reading frame 71;; Retinitis pigmentosa-55 is caused by biallelic mutations in the *ARL6* gene which encodes the ADP-ribosylation factor-like 6;; Retinitis pigmentosa-56 is caused by biallelic mutations in the *IMPG2* gene which encodes the interphotoreceptor matrix proteoglycan 2;; Retinitis pigmentosa-57 is caused by biallelic mutations in the *PDE6G* gene which encodes the rod phosphodiesterase 6G;; Retinitis pigmentosa-58 is caused by biallelic mutations in the *ZNF513* gene which encodes the zinc finger protein 513;; Retinitis pigmentosa-59 is caused by biallelic mutations in the *DHDDS* gene which encodes the dehydrodolichyl diphosphate synthase;; Retinitis pigmentosa-61 is caused by biallelic mutations in the *CLRN1* gene which encodes the clarin 1;; Retinitis pigmentosa-62 is caused by biallelic mutations in the *MAK* gene which encodes the male germ cell-associated kinase;; Retinitis pigmentosa-64 is caused by biallelic mutations in the *C8orf37* gene which encodes the chromosome 8 open reading frame 37;; and Retinitis pigmentosa-65 is caused by biallelic mutations in the *CDHR1* gene which encodes the cadherin-related family (member 1). There are several autosomal recessive forms reported in consanguineous families which have been mapped, but whose genes have not yet been characterized, including Retinitis pigmentosa-22, 28, 29 and 32. Patients with autosomal dominant inheritance account for 30-40% of all cases of retinitis pigmentosa: Retinitis pigmentosa-1 is caused by monoallelic mutations in the *RP1* gene;; Retinitis pigmentosa-4 is caused by monoallelic mutations in the *RHO* gene. This form accounts for 25% of all the autosomal dominant forms of these conditions;; Retinitis pigmentosa-7 is caused by monoallelic mutations in the *PRPH2* gene which encodes the peripherin-2;; Retinitis pigmentosa-9 is caused by monoallelic mutations in the *RP9* gene which encodes the retinitis pigmentosa 9;; Retinitis pigmentosa-10 is caused by monoallelic mutations in the

IMPDH1 gene;; Retinitis pigmentosa-11 is caused by monoallelic mutations in the *PRPF31* gene which encodes the PRP31 pre-mRNA processing factor 31 homolog;; Retinitis pigmentosa-13 is caused by monoallelic mutations in the *PRPF8* gene which encodes the PRP8 pre-mRNA processing factor 8 homolog;; Retinitis pigmentosa-17 is caused by monoallelic mutations in the *CA4* gene which encodes the carbonic anhydrase IV;; Retinitis pigmentosa-18 is caused by monoallelic mutations in the *PRPF3* gene which encodes the PRP3 pre-mRNA processing factor 3 homolog;; Retinitis pigmentosa-27 is caused by monoallelic mutations in the *NRL* gene which encodes the neural retina leucine zipper;; Retinitis pigmentosa-30 is caused by monoallelic mutations in the *FSCN2* gene which encodes the fascin homolog 2;; Retinitis pigmentosa-31 is caused by monoallelic mutations in the *TOPORS* gene which encodes the topoisomerase I binding arginine/serine-rich E3 ubiquitin protein ligase;; Retinitis pigmentosa-33 is caused by monoallelic mutations in the *SNRNP200* gene which encodes the small nuclear ribonucleoprotein 200kDa;; Retinitis pigmentosa-35 is caused by monoallelic mutations in the *SEMA4A* gene;; Retinitis pigmentosa-37 is caused by monoallelic mutations in the *NR2E3* gene;; Retinitis pigmentosa-42 is caused by monoallelic mutations in the *KLHL7* gene which encodes the kelch-like family member 7;; Retinitis pigmentosa-44 is caused by monoallelic mutations in the *RGR* gene which encodes the retinal G protein coupled receptor;; Retinitis pigmentosa-48 is caused by monoallelic mutations in the *GUCA1B* gene which encodes the guanylate cyclase activator 1B (retina);; Retinitis pigmentosa-50 is caused by monoallelic mutations in the *BEST1* gene which encodes the bestrophin 1;; and Retinitis pigmentosa-60 is caused by monoallelic mutations in the *PRPF6* gene which encodes the PRP6 pre-mRNA processing factor 6 homolog. Other autosomal dominant forms include: Retinitis pigmentosa-63 whose locus has been mapped, but the gene has not yet been characterized. The autosomal dominant pericentral retinitis pigmentosa has not been mapped. Patients with X-linked forms of Retinitis pigmentosa account for 5-15% of all cases: Retinitis pigmentosa-2 is caused by X-linked mutations in the *RP2* gene which encodes the retinitis pigmentosa 2;; Retinitis pigmentosa-3 and Retinitis pigmentosa-15 are caused by X-linked mutations in the *RPGR* gene. Mutations in the *RPGR* gene account for approximately 70% of X-linked Retinitis pigmentosa. There are also childhood-onset and juvenile forms of retinitis pigmentosa which are more severe. This early onset condition could be autosomal recessive or dominant. The recessive form can be caused by mutations in the *SPATA7* gene which encodes the spermatogenesis associated 7, *LRAT* gene which encodes the lecithin retinol acyltransferase, and the *TULP1* gene which encodes the tubby like protein 1. The dominant form is caused by mutations in the *AIPL1* gene which encodes the aryl hydrocarbon receptor interacting protein-like 1. There are also other forms that have been mapped, but whose genes have not yet been characterized including Retinitis pigmentosa-6, 23, 24 and 34. There are also other minor forms of Retinitis pigmentosa including: Y-linked, caused by mutations in the *RPY* gene which encodes the Retinitis pigmentosa Y-linked;; Digenic such as Retinitis pigmentosa-7, caused by concomitant mutations in the *PRPH2* gene which encodes the peripherin 2, and the *ROM1* gene which encodes the retinal outer segment membrane protein 1;; Mitochondrial Retinitis pigmentosa-8 and Retinitis pigmentosa-21 caused by mutations in the *MT-TS2* gene. There are also other conditions which courses with retinitis

pigmentosa including abetalipoproteinemia, Alström syndrome, Refsum syndrome, Bardet-Biedl syndrome, Laurence-Moon syndrome, Usher syndrome, Cockayne syndrome and pallidal degeneration. Retinitis pigmentosa is the major cause of inherited blindness affecting 1 in every 3,000-5,000 individuals.

Fundus albipunctatus: A form of fleck retina disease with patients manifesting night blindness and presence of punctate white deposits in the retina. This disorder displays bilateral white dot-like lesions. Most of these lesions occur in midperiphery without macular involvement. There are several underlying causes of this disorder with at least 3 autosomal genomic loci: Fundus albipunctatus can be caused by autosomal dominant mutations in the *PRPH2* gene which encodes the peripherin 2, a tetraspanning membrane protein. This protein is involved in photoreceptor disc morphogenesis;; Fundus albipunctatus can be caused by autosomal recessive mutations in the *RDH5* gene which encodes the 11-cis retinol dehydrogenase. This protein is a member of the superfamily of short-chain alcohol dehydrogenases. Mutations in the *RDH5* gene are also associated with Bietti crystalline corneoretinopathy;; and Fundus albipunctatus and retinitis punctata albescens are caused by mutations in the RLBP1 gene which encodes the cellular retinaldehyde-binding protein. This protein, only present in the retina and pineal gland, has been associated with other conditions including Newfoundland rod-cone dystrophy and Bothnia retinal dystrophy.

Newfoundland rod-cone dystrophy: A disorder appearing during the first decade of life and with rapid progression characterized by severe rod-cone dystrophy. This condition leads to legal blindness by the 2nd-4th decades of life. Newfoundland rod-cone dystrophy is caused by autosomal recessive mutations in the *RLBP1* gene which encodes the retinaldehyde-binding protein-1. Some isolated groups of people, particularly in Canada, exhibit an increased prevalence of this disease.

Stargardt disease or Fundus flavimaculatus: A disorder characterized by juvenile macular degeneration causing a poor final visual outcome frequently to the point of legal blindness. Stargardt disease is one of the most frequent causes of inherited juvenile macular degeneration. There are several underlying causes of this disorder with at least 3 autosomal genomic loci: Stargardt disease-1 can be caused by autosomal recessive mutations in the *ABCA4* gene which encodes the ATP-binding cassette (sub-family A member 4), a transmembrane protein. In Europeans this pathogenic allele appears 2,400-3,000 years ago;; Stargardt disease-3 is caused by autosomal dominant mutations in the *ELOVL4* gene which encodes the elongation of very long chain fatty acids-4;; and Stargardt disease-4 is caused by autosomal recessive mutations in the *PROM1* gene which encodes the prominin 1, a transmembrane glycoprotein. Stargardt disease is estimated to affect 1 in every 8,000-10,000 individuals.

Usher syndrome: A disorder characterized by congenital hearing impairment and years later, patients develop progressive retinitis pigmentosa. In adults, Usher syndrome is one of the most frequent causes of combined deafness and blindness. Usher syndrome is classified into 3 clinical sub-types. Usher syndrome type 1 displays a profound deafness and vestibular impairment. Usher syndrome type 2 patients have hard of hearing, with normal vestibular function. Type 2 is the most common of the 3 forms of Usher syndrome. Usher syndrome type 3 patients have progressive hearing loss and approximately half have vestibular dysfunction. This condition is named after Charles Usher. There are several underlying causes of this disorder with at least 16 autosomal genomic loci: Usher syndrome-1B is caused by autosomal recessive mutations in the *MYO7A* gene which encodes the myosin VIIA;; Type-1B accounts for over half of all cases of Usher syndrome type I;; Usher syndrome-1C or the Acadian variety, is caused by autosomal recessive mutations in the *USH1C* gene which encodes the Usher syndrome 1C. This scaffold protein functions in the assembly of Usher protein complexes.;; Usher syndrome-1D is caused by autosomal recessive mutations in the *CDH23* gene which encodes the cadherin-related 23, a calcium dependent cell-cell adhesion glycoprotein;; Usher syndrome-1F is caused by autosomal recessive mutations in the *PCDH15* gene which encodes the protocadherin-15, a calcium dependent cell-cell adhesion glycoprotein;; Usher syndrome-1G is caused by autosomal recessive mutations in the *USH1G* gene which encodes the Usher syndrome 1G protein;; Usher syndrome-1E has been mapped, but the gene has not been characterized;; Usher syndrome-1H has been mapped, but the gene has not been characterized;; Usher syndrome-1IJ is caused by autosomal recessive mutations in the *CIB2* gene which encodes the calcium and integrin binding family protein (member 2);; and Usher syndrome-1K has been mapped, but the gene has not yet been characterized;; There are also a digenic form; Usher syndrome-1D/F which is characterize by concomitant monoallelic mutations in both the *CDH23* gene, and the *PCDH15* gene. Usher syndrome-2A is caused by autosomal recessive mutations in the *USH2A* gene which encodes the Usher syndrome 2A protein. This basement membrane protein is involved in the development and homeostasis of the inner ear and retina. Type-2A accounts for over

70% of cases of Usher syndrome type 2;; Usher syndrome-2C can be caused by autosomal recessive mutations in the *GPR98* gene which encodes the G protein-coupled receptor 98, or by biallelic digenic mutation in the *GPR98* and the *PDZD7* genes;; Usher syndrome-2D can be caused by autosomal recessive mutations in the *DFNB31* gene which encodes the deafness 31 protein;; Usher syndrome-3A is caused by autosomal recessive mutations in the *CLRN1* gene which encodes the clarin-1. This is an essential protein for the development and maintenance of the inner ear and retina. Mutations in the same gene can cause retinitis pigmentosa-61;; and Usher syndrome-3B is caused by autosomal recessive mutations in the *HARS* gene which encodes the histidyl-tRNA synthetase. Usher syndrome type 1 is estimated to af-fect 3-4 in every 100,000 individuals worldwide. Type 1 is more common in people of Ashkenazi Jewish and French-Acadian ancestry from Louisiana. Usher syndrome type 2 is estimated to affect 9-12 in every 100,000 individuals worldwide, but is often underdiagnosed.

Wagner syndrome: A disorder characterized by lesions of the vitreous and retina, and sometimes cataracts. Wagner syndrome can be caused by autosomal dominant mutations in the *VCAN* gene which encodes the versican or chondroitin sulfate proteoglycan-2. This form of proteoglycan is present in the vitreous body of the eye. Wagner syndrome is estimated to affect less than 1 in every 1,000,000 individuals. This condition has been reported mainly in people from Switzerland and Japan.

Table 3.07.05.1 Recognizable Disorders of Choroid and Retina

Disorder	OMIM	Gene	Biological Function	Product of Conception Alteration
Bietti crystalline corneoretinal dystrophy	210370	CYP4V2	Endoplasmic reticulum protein	AR
Pigmented paravenous chorioretinal atrophy	172870	CRB1	Cell surface receptor Enzyme-linked receptor	AD
Central areolar choriodal dystrophy	613105	PRPH2	Cytoskeletal protein	AD
Choroideremia	303100	CHM	Trafficking Rab	XR + de novo
Retinoschisis	300839	RS1	Extracellular protein of the retina	XR + de novo
Knobloch syndrome-1	267750	COL18A1	Extracellular matrix protein	AR
Knobloch syndrome-2	608454	ADAMTS18	Extracellular matrix protein	AR
Basal laminar drusen	126700	CFH	Complement factor	AD
Bradyopsia	608415	RGS9	G-protein signaling	AR
Bradyopsia	608415	RGS9BP	G-protein signaling	AR
Familial exudative vitreoretinopathy-1	133780	FZD4	Cell surface receptor G protein	AD
Familial exudative vitreoretinopathy-2	305390	NDP	Extracellular matrix protein	XR

Table 3.07.05.1. Continued.

Disorder	OMIM	Gene	Biological Function	Product of Conception Alteration
Familial exudative vitre-oretinopathy-4	601813	LRP5	Cell surface receptor Lipid receptor	AR or AD
Familial exudative vitre-oretinopathy-5	613310	TSPAN12	Transmembrane protein	AD
Familial exudative vitre-oretinopathy	604579	FZD4/LRP5	Cell surface receptor G protein/Lipid receptor	Dig
Snowflake vitreoretinal degeneration	193230	KCNJ13	Channelopathy Potassi-um	AD
Exudative retinopathy with bone marrow failure or Revesz syndrome	268130	TINF2	Nucleus protein-Telomere	AD + de no-vo
Retinopathy of prema-turity	133780	FZD4	Cell surface receptor G protein	AD
Age-related macular de-generation-1	603075	HMCN1	Organization of hemi-desmosomes	AD
Age-related macular de-generation-2	153800	ABCA4	ABC-transporter protein	AD
Age-related macular de-generation-3	608895	FBLN5	Extracellular matrix pro-tein	AD
Age-related macular de-generation-4	610698	CFH	Complement factor	AD
Age-related macular de-generation-5	613761	ERCC6	DNA repair Nucleotide excision repair	AD
Age-related macular de-generation-6	613757	RAX2	Nucleus transcription factor	AD
Age-related macular de-generation-7	610149	HTRA1	Serine proteases	AD
Age-related macular de-generation-8	613778	PLEKHA1	Plasma membrane pro-tein	AD
Age-related macular de-generation-9	611378	C3	Complement factor	AD
Age-related macular de-generation-10	611488	TLR4	Cell surface receptor Other/ungrouped	AD
Age-related macular de-generation-11	611953	CST3	Cysteine proteinase in-hibitors	AD
Age-related macular de-generation-12	613784	CX3CR1	Cell surface receptor G protein	AD
Cone dystrophy-3	602093	GUCA1A	Guanylate cyclase acti-vator	AD
Cone dystrophy-4	613093	PDE6C	Phosphodiesterase	AR
Retinal cone dystrophy-4	610478	CACNA2D4	Channelopathy Calcium	AR

The Concise Encyclopedia of Genomic Diseases

Table 3.07.05.1. Continued.

Disorder	OMIM	Gene	Biological Function	Product of Conception Alteration
Retinal cone dystrophy with supernormal rod electroretinogram or Retinal cone dystrophy-3A	610024	PDE6H	Phosphodiesterase	AR
Retinal cone dystrophy with supernormal rod electroretinogram or Retinal cone dystrophy-3B	610356	KCNV2	Channelopathy Potassium	AR
Cone-rod dystrophy-2	120970	CRX	Transcription factor	AD + de novo
Cone-rod dystrophy-3	604116	ABCA4	ABC-transporter protein	AR
Cone-rod dystrophy-5	600977	PITPNM3	Membrane-associated phosphatidylinositol transfer	AD
Cone-rod dystrophy-6	601777	GUCY2D	Cell surface receptor Enzyme-linked receptor	AR
Cone-rod dystrophy-7	603649	RIMS1	Regulates synaptic vesicle exocytosis	AD
Cone-rod dystrophy-9	612775	ADAM9	Variety of biological processes	AR
Cone-rod dystrophy-10	610283	SEMA4A	Axon guidance	AR
Cone-rod dystrophy-11	610381	RAX2	Nucleus transcription factor	AD
Cone-rod dystrophy-12	612657	PROM1	Transmembrane glycoprotein	AD
Cone-rod dystrophy-13	608194	RPGRIP1	Ciliary proteins, connecting cilia	AR
Cone-rod dystrophy-14	602093	GUCA1A	Guanylate cyclase activator	AD
Cone-rod dystrophy-15	613660	CDHR1	Cell-cell adhesion	AR
Cone-rod dystrophy-16	614500	C8orf37	Primary cilium of the outer segment	AR
X-linked cone-rod dystrophy-1	304020	RPGR	Ciliary proteins, connecting cilia	XR
X-linked cone-rod dystrophy-3	300476	CACNA1F	Channelopathy Calcium	XR
Bestrophinopathy	611809	BEST1	Channelopathy Chloride	AR
Vitreoretinochoroidopathy or Microcornea, rod-cone dystrophy, cataract, and posterior staphyloma	193220	BEST1	Channelopathy Chloride	AD

Table 3.07.05.1. Continued.

Disorder	OMIM	Gene	Biological Function	Product of Conception Alteration
Best disease or vitelliform macular dystrophy	153700	*BEST1*	Channelopathy Chloride	AD + de novo
Jalili syndrome	217080	*CNNM4*	Metal ion transport	AR
Juvenile retinitis pigmentosa	268000	*SPATA7*	Expressed in retina	AR
Juvenile retinitis pigmentosa	613341	*LRAT*	Endoplasmic reticulum protein	AD
Juvenile retinitis pigmentosa	613341	*LRAT*	Endoplasmic reticulum protein	AR
Juvenile retinitis pigmentosa	602280	*TULP1*	Ciliary proteins, connecting cilia	AR
Juvenile retinitis pigmentosa	604393	*AIPL1*	Aryl-hydrocarbon receptor-interacting protein	AD
Retinitis pigmentosa-1	180100	*RP1*	Ciliary proteins, connecting cilia	AR
Retinitis pigmentosa-4	613731	*RHO*	Transmembrane protein	AR
Retinitis pigmentosa-12	600105	*CRB1*	Cell surface receptor Enzyme-linked receptor	AR
Retinitis pigmentosa-14	600132	*TULP1*	Ciliary proteins, connecting cilia	AR
Retinitis pigmentosa-19	601718	*ABCA4*	ABC-transporter protein	AR
Retinitis pigmentosa-20	613794	*RPE65*	Isomerase activity	AR
Retinitis pigmentosa-25	602772	*EYS*	Extracellular region	AR
Retinitis pigmentosa-26	608380	*CERKL*	Ceramide kinase	AR
Retinitis pigmentosa-35	610282	*SEMA4A*	Axon guidance	AR
Retinitis pigmentosa-36	610599	*PRCD*	Progressive rod-cone degeneration	AR
Retinitis pigmentosa-37	611131	*NR2E3*	Transcription factor Zinc finger	AR
Retinitis pigmentosa-38	268000	*MERTK*	Tyrosine kinase	AR
Retinitis pigmentosa-39	268000	*USH2A*	Basement membrane	AR
Retinitis pigmentosa-40	613801	*PDE6B*	Phosphodiesterase	AR
Retinitis pigmentosa-41	612095	*PROM1*	Transmembrane glycoprotein	AR
Retinitis pigmentosa-43	613810	*PDE6A*	Phosphodiesterase	AR
Retinitis pigmentosa-44	613769	*RGR*	Retinal G protein coupled receptor	AR
Retinitis pigmentosa-45	268000	*CNGB1*	Channelopathy, cyclic nucleotide-gated cation channel	AR
Retinitis pigmentosa-46	604526	*IDH3B*	Mitochondrial protein	AR
Retinitis pigmentosa-47	613758	*SAG*	S-antigen, retina and pineal gland	AR

Table 3.07.05.1. Continued.

Disorder	OMIM	Gene	Biological Function	Product of Conception Alteration
Retinitis pigmentosa-49	613756	CNGA1	Channelopathy, cyclic nucleotide-gated cation channel	AR
Retinitis pigmentosa-51	613464	TTC8	Ciliary proteins Basal body	AR
Retinitis pigmentosa-53	612712	RDH12	Retinol dehydrogenase	AR
Retinitis pigmentosa-54	613428	C2orf71	Primary cilium of the outer segment	AR
Retinitis pigmentosa-55	613575	ARL6	Ciliary proteins Basal body	AR
Retinitis pigmentosa-56	613581	IMPG2	Interphotoreceptor matrix proteoglycan	AR
Retinitis pigmentosa-57	613582	PDE6G	Phosphodiesterase	AR
Retinitis pigmentosa-58	613617	ZNF513	Transcription factor Zinc finger	AR
Retinitis pigmentosa-59	613861	DHDDS	Dehydrodolichyl diphosphate synthase	AR
Retinitis pigmentosa-61	614180	CLRN1	Endoplasmic reticulum protein	AR
Retinitis pigmentosa-62	614181	MAK	Serine/threonine protein kinase	AR
Retinitis pigmentosa-64	614500	C8orf37	Primary cilium of the outer segment	AR
Retinitis pigmentosa-65	613660	CDHR1	Cell-cell adhesion	AR
Retinitis pigmentosa-66	615233	RBP3	Retinol binding protein	AR
Retinitis pigmentosa-1	180100	RP1	Ciliary proteins, connecting cilia	AD
Retinitis pigmentosa-4	613731	RHO	Transmembrane protein	AD
Retinitis pigmentosa-7	608133	PRPH2	Cytoskeletal Intermediate filaments	AD
Retinitis pigmentosa-9	180104	RP9	Nuclear pre-mRNA splicing	AD
Retinitis pigmentosa-10	180105	IMPDH1	Regulate cell growth	AD
Retinitis pigmentosa-11	600138	PRPF31	Nuclear spliceosomes	AD
Retinitis pigmentosa-13	600059	PRPF8	Nuclear spliceosomes	AD
Retinitis pigmentosa-17	600852	CA4	Carbonic anhydrase	AD
Retinitis pigmentosa-18	601414	PRPF3	Nuclear spliceosomes	AD
Retinitis pigmentosa-27	613750	NRL	Neural retina leucine zipper	AD
Retinitis pigmentosa-30	607921	FSCN2	Cytoskeletal protein	AD
Retinitis pigmentosa-31	609923	TOPORS	Nuclear protein zinc finger domain	AD

Table 3.07.05.1. Continued.

Disorder	OMIM	Gene	Biological Function	Product of Conception Alteration
Retinitis pigmentosa-33	610359	SNRNP200	Nuclear ribonucleopro-tein	AD
Retinitis pigmentosa-35	610282	SEMA4A	Axon guidance	AD
Retinitis pigmentosa-37	611131	NR2E3	Transcription factor Zinc finger	AD
Retinitis pigmentosa-42	612943	KLHL7	Protein degradation	AD
Retinitis pigmentosa-44	613769	RGR	Retinal G protein cou-pled receptor	AD
Retinitis pigmentosa-48	613827	GUCA1B	Guanylate cyclase acti-vator	AD
Retinitis pigmentosa-50 or concentric retinitis pigmentosa	613194	BEST1	Channelopathy Chloride	AD
Retinitis pigmentosa-60	613983	PRPF6	Nuclear spliceosomes	AD
Retinitis pigmentosa-7	608133	PRPH2/ROM1	Cytoskeletal Intermedi-ate filaments/Integral membrane protein	Dig
X-linked retinitis pigmen-tosa-2	312600	RP2	Beta-tubulin folding	XR
X-linked retinitis pigmen-tosa-3	300029	RPGR	Ciliary proteins, connect-ing cilia	XR + de no-vo or XsD
X-linked retinitis pigmen-tosa with recurrent res-piratory infections	300455	RPGR	Ciliary proteins, connect-ing cilia	XR
Retinitis pigmentosa, late-onset	268000	CRX	Transcription factor	AD
NARP syndrome or neu-ropathy, ataxia and reti-nitis pigmentosa	516060	MT-ATP6	Mitochondrial protein	Mit
Fundus albipunctatus	136880	RDH5	Retinol dehydrogenase	AR
Fundus albipunctatus	136880	PRPH2	Cytoskeletal Intermedi-ate filaments	AD
Fundus albipunctatus	136880	RLBP1	Retinaldehyde binding protein	AR
Bothnia retinal dystrophy	607475	RLBP1	Retinaldehyde binding protein	AR
Newfoundland rod-cone dystrophy	607476	RLBP1	Retinaldehyde binding protein	AR
Retinal macular dystro-phy 2	608051	PROM1	Transmembrane glyco-protein	AD
Retinitis punctata al-bescens	136880	PRPH2	Cytoskeletal protein	AD
Retinitis punctata al-bescens	136880	RLBP1	Retinaldehyde binding protein	AD

The Concise Encyclopedia of Genomic Diseases

Table 3.07.05.1. Continued.

Disorder	OMIM	Gene	Biological Function	Product of Conception Alteration
Retinitis punctata albescens	136880	PRPH2	Cytoskeletal protein	AD
Retinitis punctata albescens	136880	RHO	Transmembrane protein	AD
Doyne honeycomb degeneration of retina	126600	EFEMP1	Extracellular matrix protein	AD
Late-onset retinal degeneration	605670	C1QTNF5	Cell adhesion	AD
Early-onset severe Retinal dystrophy	613341	LRAT	Endoplasmic reticulum protein	AR
Enhanced S-cone syndrome	268100	NR2E3	Transcription factor Zinc finger	AR
Sorsby fundus dystrophy	136900	TIMP3	Extracellular matrix degradation	AD
Sveinsson choreoretinal atrophy	108985	TEAD1	Nuclear transcriptional enhancer factor	AD
Stargardt disease-1	248200	ABCA4	ABC-transporter protein	AR
Stargardt disease-3	600110	ELOVL4	Biosynthesis of fatty acids	AD
Stargardt disease-4	603786	PROM1	Transmembrane glycoprotein	AR
Usher syndrome-1B	276900	MYO7A	Cytoskeletal protein	AR
Usher syndrome-1C	276904	USH1C	Nuclear localization protein	AR
Usher syndrome-1D	601067	CDH23	Cytoskeletal protein	AR
Usher syndrome-1D/F	601067	PCDH15/ CDH23	Calcium-dependent adhesion/Cytoskeleton	Dig
Usher syndrome-1F	602083	PCDH15	Calcium-dependent adhesion	AR
Usher syndrome-1G	606943	USH1G	Homodimerization activity	AR
Usher syndrome-1J	614869	CIB2	Calcium and integrin binding member	AR
Usher syndrome-2A	276901	USH2A	Basement membrane	AR
Usher syndrome-2C	605472	GPR98	Cell surface receptor G protein	AR
Usher syndrome-2C	605472	GPR98/ PDZD7	Cell surface receptor G protein/Ciliary protein	AR
Usher syndrome-2D	611383	DFNB31	Cytoskeletal protein	AR
Usher syndrome-3A	276902	CLRN1	Endoplasmic reticulum protein	AR
Usher syndrome-3B	614504	HARS	Histidyl-tRNA synthetase	AR
Wagner syndrome	143200	VCAN	Extracellular matrix protein	AD

3.07.06 Glaucoma

Glaucoma: A disorder characterized by optic nerve damage leading to blindness if left untreated. Glaucoma is usually associated with increased intraocular pressure, but there are also other forms with normal or low intraocular tension. There are two main categories of glaucoma: open angle and closed angle. The angle refers to the space between the iris and cornea, through which fluid must flow to the trabecular meshwork. Open angle is a chronic form with slow progression. Closed angle could appear abruptly and is often painful. Eastern Asian individuals are prone to developing close angle glaucoma due to their shallower anterior chamber. Females are three times more likely than males to develop acute close angle glaucoma. Glaucoma is a major cause of blindness, particularly among African Americans, who are 3 times more likely to develop open angle glaucoma than Caucasians. Primary congenital glaucoma or Buphthalmos is the most common type of childhood glaucoma. The prevalence of Primary congenital glaucoma varies among ethnic groups: it is 1 in 1,250 in the Romani Gypsy people of Slovakia, 1 in 2,500 in the Middle Easterns, 1 in 3,300 in the Indian state of Andhra Pradesh, and 1 in 5,000-22,000 among people of western countries. Affected children exhibit tearing, photophobia, early onset increase of intraocular pressure and corneal diameter, enlarged globe, Haab striae, corneal edema and optic nerve head cupping. There are several underlying causes of this disorder with at least 19 autosomal genomic loci: Primary congenital glaucoma-3A is caused by autosomal recessive mutations in the *CYP1B1* gene which encodes the cytochrome P450 (family 1B member1). Sometimes mutations in the *MYOC* gene which encodes the myocilin may also contribute to the phenotype via digenic inheritance;; Primary congenital glaucoma-3B has been mapped, but the gene has not been characterized; Primary congenital glaucoma-3C has been mapped, but the gene has not been characterized;; and Primary congenital glaucoma-3D is caused by mutations in the *LTBP2* gene which encodes the latent transforming growth factor beta binding protein 2. Adult-onset forms include: Primary open angle glaucoma-1A is caused by autosomal dominant mutations in the *MYOC* gene;; Primary open angle glaucoma-1B has been mapped, but the gene has not been characterized;; Primary open angle glaucoma-1C has been mapped, but the gene has not been characterized;; Primary open angle glaucoma-1D has been mapped, but the gene has not been characterized;; Primary open angle glaucoma-1E is caused by mutations in the *OPTN* gene which encodes the optineurin;; Primary open angle glaucoma-1F is caused by autosomal dominant mutations in the *ASB10* gene which encodes the ankyrin repeat and SOCS box containing protein 10;; Primary open angle glaucoma-1G is caused by mutations in the *WDR36* gene which encodes the WD repeat domain 36;; Primary open angle glaucoma-1H has been mapped, but the gene has not been characterized;; Primary open angle glaucoma-1I has been mapped, but the gene has not been characterized;; Primary open angle glaucoma-1J has been mapped, but the gene has not been characterized;; Primary open angle glaucoma-1K has been mapped, but the gene has not been characterized;; Primary open angle glaucoma-1L has been mapped, but the gene has not been characterized;; Primary open angle glaucoma-1M has been mapped, but the gene has not been characterized;; Primary open angle glaucoma-1N has been mapped, but the gene has not been characterized;; Primary open angle glaucoma-1O is caused by mutations in the *NTF4* gene which encodes the neurotrophin 4;; and Primary open angle glaucoma-1P is caused by autosomal dominant mutations (300-kb duplication) in the TBK1 gene which encodes the TANK-binding kinase 1, a serine/threonine-protein kinase. Nail-patella syndrome has open angle glaucoma feature, it is caused by mutations in the *LMX1B* gene which encodes the LIM homeobox transcription factor 1 (beta). Susceptibility to normal tension glaucoma is associated with a particular intronic polymorphism of the *OPA1* gene which encodes the optic atrophy 1 and mutations in the *OPTN* gene. This form accounts for roughly one-third of all primary open angle glaucoma cases.

3.07.07 Disorders of Optic Nerve and Visual Pathways

Opticoacoustic nerve atrophy with dementia or Jensen syndrome: A disorder characterized by sensorineural hearing loss with onset during infancy. During adolescence patients exhibit progressive optic nerve atrophy with loss of vision. In adulthood patients manifest progressive dementia. Jensen syndrome is caused by X-linked recessive mutations in the *TIMM8A* gene which encodes the translocase of inner mitochondrial membrane 8 (homolog A). The same gene is also mutated in Mohr–Tranebjærg syndrome or Deafness-dystonia syndrome. Jensen syndrome is estimated to affect less than 1 in every 1,000,000 individuals.

Leber hereditary optic neuropathy: A disorder characterized by abrupt start of bilateral central vision loss. This condition leads to central scotoma in young adults. Sometimes this condition is also associated with motor disorders including dysto-

nia, postural tremor and cerebellar ataxia. There are several underlying causes of this disorder with at least 11 mitochondrial encoded genes associated with Leber hereditary optic neuropathy including: *MT-ND6, MT-ND4, MT-ND1, MT-CYB, MT-CO3, MT-ND5, MT-ND2, MT-COI, MT-ND5, MMT-ATP6* and *MT-ND4L* genes. Leber hereditary optic neuropathy is a maternally-inherited disease. The estimated prevalence of genotype associated with this condition is 1 in 9,000 among Northern Europeans, but the prevalence of this condition is around 1 in 50,000 individuals. The 3 most common genotypes associated with this condition among Europeans are mutations in: *MT-ND4*, G11778A; *MT-ND1*, G3460A; and *MT-ND6*, T14484C. Males are more commonly affected with Leber hereditary optic neuropathy than females.

Optic atrophy: A disorder characterized by visual impairment of insidious onset in the first decade of life. Affected individuals exhibit moderate to severe loss of visual acuity, color vision deficits, temporal optic disc pallor, and centrocecal scotoma of variable density. There are several underlying causes of this disorder with at least 6 autosomal genomic loci and 1 X-linked genomic locus: Optic atrophy-1, or Kjer optic atrophy, is caused by autosomal dominant mutations in the *OPA1* gene which encodes the optic atrophy 1;; Optic atrophy-2 has been mapped to chromosome X, but the gene has not been characterized;; Optic atrophy-3 with chorea and spastic paraplegia, is caused by autosomal recessive mutations in the *OPA3* gene which encodes the optic atrophy 3;; Optic atrophy-4 has been mapped, but the gene has not been characterized;; Optic atrophy-5 has been mapped, but the gene has not been characterized;; Optic atrophy-6 has been mapped, but the gene has not been characterized;; and Optic atrophy-7 is caused by mutations in the *TMEM126A* gene which encodes the transmembrane protein 126A. Optic atrophy is estimated to affect 1 in every 35,000 among individuals from North England.

Wolfram syndrome or DIDMOAD syndrome (Diabetes Insipidus, Diabetes Mellitus, Optic Atrophy, and Deafness): This disease is discussed under endocrine, nutritional and metabolic diseases in the subchapter "Diabetes Mellitus" (3.04.02.1).

3.07.08 Disorders of Ocular Muscles, Binocular Movement, Accommodation and Refraction

Chronic progressive external ophthalmoplegia: A disorder appearing during the 2nd- 4th decade of life characterized by weakness of the eye muscles. Affected individuals exhibit bilateral ptosis and weakness or paralysis of the muscles that move the eye. Affected individuals could show weakness of the skeletal muscles, especially during exercise. Progressive external ophthalmoplegia is caused by mitochondrial structural impairment affecting the oxidative phosphorylation. This condition can be associated with mutations in either nuclear or mitochondrial encoded genes: the nuclear forms include at least 3 autosomal genomic loci: mutations in the *POLG* gene which encodes the polymerase (DNA directed) gamma, or *SLC25A4* gene which encodes the solute carrier (family 25 member A4), a mitochondrial carrier adenine nucleotide translocator, or the *C10orf2* gene which encodes the chromosome 10 open reading frame 2, a hexameric DNA helicase. The mitochondrial forms include mutations in the *MT-TL1* gene or large deletions of mitochondrial DNA.

Horizontal gaze palsy with progressive scoliosis: A disorder characterized by progressive external ophthalmoplegia and severe scoliosis. Horizontal gaze palsy with progressive scoliosis is caused by autosomal recessive mutations in the *ROBO3* gene which encodes the roundabout homolog protein 3.

Congenital fibrosis of the extraocular muscles: A disorder characterized by bilateral blepharoptosis and ophthalmoplegia with the eyes fixed in an infraducted position about 20-30 degrees below the horizontal midline. There are several underlying causes of this disorder with at least 4 autosomal genomic loci: Congenital fibrosis of extraocular muscles-1 is caused by autosomal dominant mutations in the *KIF21A* gene which encodes the kinesin family member 21A;; Congenital fibrosis of extraocular muscles-2 is caused by autosomal recessive mutations in the *PHOX2A* gene which encodes the paired-like homeobox 2a. Affected individuals with type 2 exhibit bilateral ptosis with eyes fixed in an exotropic position;; Congenital fibrosis of extraocular muscles-3 generally has more variable clinical features: patients may display unilateral eye involvement, or may be able to raise their eyes above midline, or may not have blepharoptosis;; Congenital fibrosis of extraocular muscles-3A with or without extraocular involvement is caused by mutations in the *TUBB3* gene which encodes the class III beta-tubulin;; Congenital fibrosis of extraocular muscles-3B is caused by mutations in the *KIF21A* gene;; and Congenital fibrosis of extraocular muscles-3C has been mapped, but the gene has not been characterized. Congenital fibrosis of extraocular muscles is estimated to affect 1 in every 230,000 individuals.

Kearns–Sayre syndrome or Oculocraniosomatic neuromuscular disease with

ragged red fibers: A severe form of chronic progressive external ophthalmoplegia characterized by myopathy of early onset. Affected individuals may also exhibit diabetes mellitus, growth hormone deficiency, hypoparathyroidism, cerebellar ataxia, proximal muscle weakness and deafness. Kearns–Sayre syndrome is caused by large deletions in mitochondrial DNA. These deletions of DNA vary from 1.3-8 kb. The most common form includes a 4.9 kb deletion in mitochondrial DNA. Kearns-Sayre syndrome is estimated to affect 1-3 in every 100,000 individuals.

Duane retraction syndrome: A congenital form of strabismus characterized by horizontal eye movement limitation, globe retraction with palpebral fissure narrowing in attempted adduction. Affected individuals show paradoxical innervation of rectus lateral muscle. There are several underlying causes of this disorder with at least 2 autosomal genomic loci: Duane retraction syndrome-1 has been mapped, but the gene has not been characterized;; and Duane retraction syndrome-2 is caused by autosomal dominant mutations in the *CHN1* gene which encodes the chimerin 1. Duane retraction syndrome accounts for 1-5% of all strabismus cases. Isolated Duane retraction syndrome is estimated to affect 1 in every 1,000 people worldwide.

3.07.09 Visual Disturbances and Blindness

Leber congenital amaurosis: A disorder characterized by vision loss, nystagmus and severe retinal dysfunction. This condition affects the development of photoreceptor cells. Affected children exhibit profound vision loss and pendular nystagmus. Other features include hypermetropia, photodysphoria, oculo-digital sign, keratoconus and cataracts. This condition is named after Theodor Leber. There are several underlying causes of this disorder with at least 16 autosomal genomic loci: Leber congenital amaurosis-1 is caused by biallelic mutations in the *GUCY2D* gene which encodes the retinal guanylate cyclase. In Finnish people this pathogenic allele appears 3,000-3,750 years ago;; Leber congenital amaurosis-2 is caused by autosomal recessive mutations in the *RPE65* gene which encodes the retinal pigment epithelium-specific protein 65kDa;; Leber congenital amaurosis-3 is caused by autosomal recessive mutations in the *SPATA7* gene which encodes the spermatogenesis associated 7;; Leber congenital amaurosis-4 is caused by autosomal recessive mutations in the *AIPL1* gene which encodes the aryl hydrocarbon receptor interacting protein-like 1;; Leber congenital amaurosis-5 is caused by autosomal recessive mutations in the *LCA5* gene which encodes the Leber congenital amaurosis 5;; Leber congenital amaurosis-6 is caused by autosomal recessive mutations in the *RPGRIP1* gene which encodes the retinitis pigmentosa GTPase regulator interacting protein 1;; Leber congenital amaurosis-7 is caused by autosomal recessive or autosomal dominant mutations in the *CRX* gene which encodes the cone-rod homeobox;; Leber congenital amaurosis-8 is caused by autosomal recessive mutations in the *CRB1* gene which encodes the crumbs homolog 1;; Leber congenital amaurosis-9 has been mapped, but the gene has not been characterized;; Leber congenital amaurosis-10 is caused by autosomal recessive mutations in the *CEP290* gene which encodes the centrosomal protein 290kDa. This form may account for over 20% of cases of all cases of Leber congenital amaurosis. Mutations in the *CEP290* gene have been associated with Joubert syndrome;; Leber congenital amaurosis-11 is caused by autosomal recessive mutations in the *IMPDH1* gene which encodes the IMP (inosine 5'-monophosphate) dehydrogenase 1;; Leber congenital amaurosis-12 is caused by autosomal recessive mutations in the *RD3* gene which encodes the retinal degeneration 3;; Leber congenital amaurosis-13 is caused by autosomal recessive mutations in the *RDH12* gene which encodes the retinol dehydrogenase 12 (all-trans/9-cis/11-cis);; Leber congenital amaurosis-14 is caused by autosomal recessive mutations in the *LRAT* gene which encodes the lecithin retinol acyltransferase (phosphatidylcholine--retinol O-acyltransferase);; Leber congenital amaurosis-15 is caused by autosomal recessive mutations in the *TULP1* gene which encodes the tubby like protein 1;; and Leber congenital amaurosis-16 is caused by autosomal recessive mutations in the *KCNJ13* gene which encodes the inward rectifier potassium channel 13. Leber congenital amaurosis is estimated to affect 1-3 in every 100,000 individuals.

3.07.09.1 Colour Vision Deficiencies

Achromatopsia: A disorder characterized by complete inability to discriminate between colors. Affected individuals also manifest photophobia, reduced visual acuity and nystagmus. This condition affects the rod photoreceptor function. There are several underlying causes of this disorder with at least 4 autosomal genomic loci: Achromatopsia-1, or Achromatopsia-3, is caused by autosomal recessive mutations in the *CNGB3* gene which encodes the cyclic nucleotide gated channel (beta 3);; Achromatopsia-2 is caused by autosomal recessive mutations in the *CNGA3* gene which encodes the cyclic nucleotide gated channel (alpha 3);; Achromatopsia-4 is caused by autosomal recessive mutations in the *GNAT2* gene which encodes the guanine nucleotide binding protein (G pro-

tein) alpha transducing activity polypeptide 2;; and Achromatopsia-5 is caused by autosomal recessive mutations in the *PDE6C* gene which encodes the cGMP phosphodiesterase 6C. Achromatopsia is estimated to affect 1 in every 30,000 individuals. This condition is more prevalent among Moroccan, Iraqi and Iranian Jews. There is a high incidence of Achromatopsia-3 among Pingelapese islanders (Federated States of Micronesia).

Blue cone monochromacy: A disorder characterized by: inability to distinguish between colors and only an ability to perceive variations in brightness. There are several underlying causes of this disorder with at least 2 X-linked genomic loci: Blue cone monochromacy can be caused by X-linked recessive mutations in the *OPN1LW* gene which encodes the long-wave-sensitive opsin 1 (cone pigments), a red cone photoreceptor pigment. The encoded protein produces an alteration in the red visual pigment;; and Blue cone monochromatism can be caused by X-linked recessive mutations in the *OPN1MW* gene which encodes the medium-wave-sensitive opsin 1 (cone pigments), a green cone photoreceptor pigment. The encoded protein produces an alteration in the green visual photoreceptor pigment. Blue cone monochromatism is estimated to affect 1 in 100,000 individuals. The last 2 conditions are subtelomeric syndromes (X-linked).

Protan colorblindness or Protanopia: A disorder characterized by severely defective color vision based on the use of only 2 types of photoreceptors, blue plus green. Protan colorblindness is caused by X-linked recessive mutations in the *OPN1LW* gene which encodes the long-wave-sensitive opsin 1 (cone pigments), a red cone photoreceptor pigment. This subtelomeric condition affects 1% of all males.

Deutan colorblindness or Deuteranopia: A disorder characterized by inability or decreased ability to see color, or perceive color differences. This condition was named after John Dalton in 1798. Deutan colorblindness is caused by X-linked recessive mutations in the *OPN1MW* gene which encodes the medium-wave-sensitive opsin 1 (cone pigments), a green cone photoreceptor pigment. In people of northern European ancestry, colorblindness is estimated to affect around 8% of males and 0.5% of females. This condition is a subtelomeric syndrome.

Tritan colorblindness or Tritanopia: A disorder characterized by a selective deficiency of blue spectral sensitivity. Tritan colorblindness is caused by autosomal dominant mutations in the *OPN1SW* gene which encodes the short-wave-sensitive opsin 1 (cone pig-ments), a blue cone photoreceptor pigment. This condition affects 2 in 1,000 individuals.

3.07.09.2 Night Blindness

Congenital stationary night blindness: A non-progressive retinal disorder characterized by impaired night vision, decreased visual acuity, nystagmus, myopia and strabismus. There are different ways of classifying these conditions: in accordance with severity as complete or incomplete forms, and in accordance with genomic factors. These conditions also have different patterns of inheritance, including X-linked, autosomal recessive and autosomal dominant. There are several underlying causes of this disorder with at least 2 X-linked genomic loci and 10 autosomal genomic loci: X-linked complete congenital stationary night blindness-1 is caused by X-linked mutations in the *NYX* gene which encodes the nyctalopin. This protein is involved in retinal synapse formation or synaptic transmission;; X-linked incomplete congenital stationary night blindness-2A is caused by X-linked mutations in the *CACNA1F* gene which encodes the voltage-gated calcium channel (L type subunit alpha 1F). The prevalence of this condition is unknown, but it appears to be more common in people of Dutch-German Mennonite ancestry;; Complete congenital stationary night blindness-1B is caused by autosomal recessive mutations in the *GRM6* gene which encodes the glutamate receptor metabotropic 6;; Complete congenital stationary night blindness-1C is caused by autosomal recessive mutations in the *TRPM1* gene which encodes the transient receptor potential cation channel (subfamily M member 1);; Complete congenital stationary night blindness-1D is caused by autosomal recessive mutations in the *SLC24A1* gene which encodes the solute carrier (family 24 member A1), a sodium/potassium/calcium exchanger;; Complete congenital stationary night blindness-1E is caused by autosomal recessive mutations in the *GPR179* gene which encodes the G protein-coupled receptor 179;; Complete congenital stationary night blindness-1 is caused by autosomal dominant mutations in the *RHO* gene which encodes the rhodopsin. This transmembrane protein when photoexcited, initiates the visual transduction cascade;; Complete congenital stationary night blindness-2 is caused by autosomal dominant mutations in the *PDE6B* gene which encodes the rod cGMP-phosphodiesterase (beta-subunit);; Complete congenital stationary night blindness-3 is caused by autosomal dominant mutations in the *GNAT1* gene which encodes the guanine nucleotide binding protein (G protein) alpha transducing activity polypeptide 1;; Incomplete congenital stationary night blindness-2B is caused by autosomal recessive mutations in the *CABP4* gene which encodes the calcium binding protein 4;; and Congenital stationary night blindness, Oguchi type is a form in which all other visual functions are normal: Oguchi disease-1 is caused by autosomal recessive mutations in the *SAG* gene which encodes the S-antigen of retina and pineal gland;; and Oguchi disease-2 is caused by autosomal recessive mutations in the *GRK1* gene which encodes the rhodopsin kinase. Oguchi diseases are more frequent in individuals of Japanese ancestry.

Table 3.07.09.1 Recognizable Visual Disturbances and Blindness

Disorder	OMIM	Gene	Biological Function	Product of Conception Alteration
Leber congenital amaurosis-1	204000	GUCY2D	Cell surface receptor Enzyme-linked receptor	AR
Leber congenital amaurosis-2	204100	RPE65	Isomerase activity	AR
Leber congenital amaurosis-3	604232	SPATA7	Expressed in retina	AR
Leber congenital amaurosis-4	604393	AIPL1	Aryl-hydrocarbon receptor-interacting protein	AR
Leber congenital amaurosis-5	604537	LCA5	Ciliary proteins, connecting cilia	AR
Leber congenital amaurosis-6	605446	RPGRIP1	Ciliary proteins, connecting cilia	AR
Leber congenital amaurosis-7	602225	CRX	Transcription factor	AR or AD
Leber congenital amaurosis-8	604210	CRB1	Cell surface receptor Enzyme-linked receptor	AR
Leber congenital amaurosis-10	611755	CEP290	Ciliary proteins, Nephrocystin	AR
Leber congenital amaurosis-11	146690	IMPDH1	Regulate cell growth	AR
Leber congenital amaurosis-12	610612	RD3	Retinal protein	AR
Leber congenital amaurosis-13	612712	RDH12	Retinol dehydrogenase	AR
Leber congenital amaurosis-14	613341	LRAT	Endoplasmic reticulum protein	AR
Leber congenital amaurosis-15	613843	TULP1	Tubby like protein	AR
Leber congenital amaurosis-16	614186	KCNJ13	Channelopathy Potassium	AR
Achromatopsia-2	216900	CNGA3	Channelopathy, cyclic nucleotide-gated cation channel	AR
Achromatopsia-1 and 3	262300	CNGB3	Channelopathy, cyclic nucleotide-gated cation channel	AR
Achromatopsia-4	139340	GNAT2	Cell surface receptor G protein	AR
Achromatopsia-5	613093	PDE6C	Phosphodiesterase	AR
Blue cone monochromacy	303700	OPN1LW	Visual pigment	XR
Blue cone monochromacy	303700	OPN1MW	Visual pigment	XR

Table 3.07.09.1. Continued.

Disorder	OMIM	Gene	Biological Function	Product of Conception Alteration
Protan colorblindness or Protanopia	303900	OPN1LW	Visual pigment	XR
Deutan colorblindness or Deuteranopia	303800	OPN1MW	Visual pigment	XR
Tritan colorblindness or Tritanopia	190900	OPN1SW	Visual pigment	AD
X-linked complete congenital stationary night blindness-1	310500	NYX	Extracellular matrix protein	X
X-linked incomplete congenital stationary night blindness-2A	300071	CACNA1F	Channelopathy Calcium	XR
Complete congenital stationary night blindness-1B	257270	GRM6	Cell surface receptor G protein	AR
Complete congenital stationary night blindness-1C	613216	TRPM1	Channelopathy, transient receptor potential channels	AR
Complete congenital stationary night blindness-1E	613830	SLC24A1	Solute carrier	AR
Complete congenital stationary night blindness-1D	614565	GPR179	Cell surface receptor G protein	AR
Complete congenital stationary night blindness-1	610445	RHO	Transmembrane protein	AD
Complete congenital stationary night blindness-1	163500	PDE6B	Phosphodiesterase	AD
Complete congenital stationary night blindness-3	610444	GNAT1	Cell surface receptor G protein	AD
Incomplete congenital stationary night blindness-2B	610427	CABP4	Calcium-binding protein	AR
Oguchi disease-1	258100	SAG	Desensitization of G-protein-coupled receptors	AR
Oguchi disease-2	613411	GRK1	Cell surface receptor G protein	AR

3.07.10 Other Disorders of Eye and Adnexa

Idiopathic congenital nystagmus: A disorder characterized by conjugate, spontaneous and involuntary oscillations of the eyes. These oscillations are often symmetric and horizontal. Affected infants may also have other associated features including mildly decreased visual acuity, strabismus, astigmatism and occasional head nodding. There are several underlying causes of this disorder with at least 3 X-linked genomic loci and 4 autosomal genomic loci: X-linked congenital nystagmus-1 is caused by mutations in the *FRMD7* gene which encodes the FERM domain containing protein 7. This is the most common form. Mutations in the same gene are also associated with Infantile periodic alternating nystagmus;; X-linked congenital nystagmus-5 has been mapped, but the gene has not yet been characterized;; X-linked congenital nystagmus-6 is caused by mutations in the *GPR143* gene which encodes the G protein-coupled receptor 143;; There are several autosomal dominant forms which have been mapped, including Congenital nystagmus-2, Congenital nystagmus-3, Congenital nystagmus-4 and Congenital nystagmus-7, but whose genes have not been characterized;; and there are also rare forms of autosomal recessive Congenital nystagmus. There are several conditions which have congenital nystagmus as a feature including ocular diseases such as Albinism, Achromatopsia and Leber congenital amaurosis. Congenital nystagmus is associated with at least 3 X-linked disorders: Nettleship-Falls ocular albinism, Complete congenital stationary night blindness-1 and Blue-cone monochromatism. Congenital nystagmus is estimated to affect 1 in every 1,000 children.

3.08.0 Subchapter VIII

Diseases of the Ear and Mastoid Process

3.08.1 Disorders of Ear

Deafness: A common disorder characterized by profound or severe hearing loss. There are many causes of deafness; genomics is one of the most frequent factors, leading to deafness in 1 in 700-1,000 children. Deafness can be classified in many different forms: including: according to being syndromic or nonsyndromic; according to the form of hearing impairment (sensorineural, conductive and/or mixed hearing loss); according to the pattern of inheritance (autosomal recessive, autoso-

mal dominant, X-Linked, and/or mitochondrial). Patients with autosomal recessive deafness account for 85% of all the nonsyndromic cases. Patients with autosomal dominant deafness account for between 10-15% of nonsyndromic cases. There are also minor forms, accounting for around 1% including those inherited as an X-linked trait, mitochondrial or digenic, and those due to *de novo* mutations. There are loci that can exhibit several patterns of inheritance. For example, mutations in the *GJB2* gene could follow autosomal dominant, recessive or digenic patterns of inheritance. There are several underlying causes of this disorder with at least 91 loci associated with autosomal recessive nonsyndromic sensorineural deafness. Thirty-two of these loci have been identified at the gene level. The majority of cases of childhood perceptive deafness are caused by mutations in the *GJB2* gene. Autosomal recessive deafness-1A is caused by biallelic mutations in the *GJB2* gene which encodes the gap junction protein (beta 2). In Europeans at least one pathogenic allele appears 14,000-17,000 years ago;; Autosomal recessive deafness-1B is caused by biallelic mutations in the *GJB6* gene which encodes the gap junction protein beta 6;; Autosomal recessive deafness-3 is caused by biallelic mutations in the *MYO15A* gene which encodes the myosin XVA;; Autosomal recessive deafness-6 is caused by biallelic mutations in the *TMIE* gene which encodes the transmembrane inner ear;; Autosomal recessive deafness-7 is caused by biallelic mutations in the *TMC1* gene which encodes the transmembrane channel-like 1;; Autosomal recessive deafness-10 is caused by biallelic mutations in the *TMPRSS3* gene which encodes the transmembrane protease serine 3;; Autosomal recessive deafness-12 is caused by biallelic mutations in the *CDH23* gene which encodes the cadherin-related 23;; Autosomal recessive deafness-

16 is caused by biallelic mutations in the *STRC* gene which encodes the stereocilin. This protein is associated with the hair bundle of the sensory hair cells in the inner ear;; Autosomal recessive deafness-18 is caused by biallelic mutations in the *USH1C* gene which encodes the Usher syndrome protein 1C;; Autosomal recessive deafness-21 is caused by biallelic mutations in the *TECTA* gene which encodes the tectorin alpha;; Autosomal recessive deafness-22 is caused by biallelic mutations in the *OTOA* gene which encodes the otoancorin. This protein is involved in the attachment of the inner ear acellular gels to the apical surface of the underlying nonsensory cells;; Autosomal recessive deafness-23 is caused by biallelic mutations in the *PCDH15* gene which encodes the protocadherin-related 15;; Autosomal recessive deafness-24 is caused by biallelic mutations in the *RDX* gene which encodes the radixin. This cytoskeletal protein is involved in linking actin to the plasma membrane;; Autosomal recessive deafness-25 is caused by biallelic mutations in the *GRXCR1* gene which encodes the glutaredoxin cysteine rich 1;; Autosomal recessive deafness-28 is caused by biallelic mutations in the *TRIOBP* gene which encodes the TRIO and F-actin binding protein;; Autosomal recessive deafness-30 is caused by biallelic mutations in the *MYO3A* gene which encodes the myosin IIIA;; Autosomal recessive deafness-31 is caused by biallelic mutations in the *DFNB31* gene which encodes the deafness 31 protein;; Autosomal recessive deafness-35 is caused by biallelic mutations in the *ESRRB* gene which encodes the estrogen-related receptor beta;; Autosomal recessive deafness-37 is caused by biallelic mutations in the *MYO6* gene which encodes the myosin VI;; Autosomal recessive deafness-39 is caused by biallelic mutations in the *HGF* gene which encodes the hepatocyte growth factor;; Autosomal re-

cessive deafness-49 is caused by biallelic mutations in the *MARVELD2* gene which encodes the MARVEL domain containing protein 2;; Autosomal recessive deafness-53 is caused by biallelic mutations in the *COL11A2* gene which encodes the alpha 2 type XI collagen;; Autosomal recessive deafness-59 is caused by biallelic mutations in the *DFNB59* gene which encodes the deafness 59 protein;; Autosomal recessive deafness-63 is caused by biallelic mutations in the *LRTOMT* gene which encodes the leucine rich transmembrane and 0-methyltransferase domain containing protein;; Autosomal recessive deafness-67 is caused by biallelic mutations in the *LHFPL5* gene which encodes the lipoma HMGIC fusion partner-like protein 5;; Autosomal recessive deafness-77 is caused by biallelic mutations in the *LOXHD1* gene which encodes the lipoxygenase homology domains 1;; Autosomal recessive deafness-79 is caused by biallelic mutations in the *TPRN* gene which encodes the taperin;; Autosomal recessive deafness-84 is caused by biallelic mutations in the *PTPRQ* gene which encodes the protein tyrosine phosphatase receptor (type Q);; and Autosomal recessive deafness-91 is caused by biallelic mutations in the *SERPINB6* gene which encodes the serpin peptidase inhibitor (clade B member 6). There are several underlying causes of this disorder with at least 55 loci associated with autosomal dominant nonsyndromic sensorineural deafness. 27 of these loci have been identified at the gene level. Autosomal dominant deafness-1 is caused by monoallelic mutations in the *DIAPH1* gene which encodes the diaphanous homolog 1;; Autosomal dominant deafness-2A is caused by monoallelic mutations in the *KCNQ4* gene which encodes the potassium voltage-gated channel (subfamily KQT member 4);; Autosomal dominant deafness-2B is caused by monoallelic mutations in the *GJB3* gene which encodes the gap junction protein (beta 3);; Autosomal dominant deafness-3A is caused by monoallelic mutations in the *GJB2* gene which encodes the gap junction protein (beta 2);; Autosomal dominant deafness-3B is caused by monoallelic mutations in the *GJB6* gene which encodes the gap junction protein (beta 6);; Autosomal dominant deafness-4 is caused by monoallelic mutations in the *MYH14* gene which encodes the non-muscle myosin heavy chain 14;; Autosomal dominant deafness-5 is caused by monoallelic mutations in the *DFNA5* gene which encodes the deafness 5 protein;; Autosomal dominant deafness-6 is caused by monoallelic mutations in the *WFS1* gene which encodes the Wolfram syndrome protein 1;; Autosomal dominant deafness-8/12 is caused by monoallelic mutations in the *TECTA* gene which encodes the tectorin alpha;; Autosomal dominant deafness-9 is caused by monoallelic mutations in the *COCH* gene which encodes the coagulation factor C homolog cochlin;; Autosomal dominant deafness-10 is caused by monoallelic mutations in the *EYA4* gene which encodes the eyes absent homolog 4;; Autosomal dominant deafness-11 is caused by monoallelic mutations in the *MYO7A* gene which encodes the myosin VIIA;; Autosomal dominant deafness-13 is caused by monoallelic mutations in the *COL11A2* gene which encodes the alpha 2 type XI collagen;; Autosomal dominant deafness-14 is caused by monoallelic mutations in the *WFS1* gene;; Autosomal dominant deafness-15 is caused by monoallelic mutations in the *POU4F3* gene which encodes the POU class 4 (homeobox 3);; Autosomal dominant deafness-17 is caused by monoallelic mutations in the *MYH9* gene which encodes the myosin heavy chain, non-muscle 9;; Autosomal dominant deafness-20/26 is caused by monoallelic mutations in the *ACTG1* gene which encodes the actin gamma 1;; Autosomal dominant deafness-

23 is caused by monoallelic mutations in the *SIX1* gene which encodes the SIX homeobox 1;; Autosomal dominant deafness-25 is caused by monoallelic mutations in the *SLC17A8* gene which encodes the solute carrier (family 17 member A8), a sodium-dependent inorganic phosphate cotransporter;; Autosomal dominant deafness-28 is caused by monoallelic mutations in the *GRHL2* gene which encodes the grainyhead-like 2;; Autosomal dominant deafness-36 is caused by monoallelic mutations in the *TMC1* gene which encodes the transmembrane channel-like 1;; Autosomal dominant deafness-36 with dentinogenesis is caused by monoallelic mutations in the *DSPP* gene which encodes the dentin sialophosphoprotein;; Autosomal dominant deafness-38 is caused by monoallelic mutations in the *WFS1* gene;; Autosomal dominant deafness-40 is caused by monoallelic mutations in the *CRYM* gene which encodes the crystallin mu;; Autosomal dominant deafness-44 is caused by monoallelic mutations in the *CCDC50* gene which encodes the coiled-coil domain containing protein 50;; Autosomal dominant deafness-48 is caused by monoallelic mutations in the *MYO1A* gene which encodes the myosin IA;; and Autosomal dominant deafness-50 is caused by monoallelic mutations in the *MIR96* gene (microRNA 96), a non protein coding RNA. X-linked recessive deafness-1 is caused by mutations in the *PRPS1* gene which encodes the phosphoribosyl pyrophosphate synthetase 1;; X-linked recessive deafness-2 is caused by mutations in the *POU3F4* gene which encodes the POU class 3 homeobox 4;; and X-linked recessive deafness-4 is caused by mutations in the *SMPX* gene which encodes the small muscle protein X-linked. Digenic nonsyndromic deafness can be caused by concomitant mutations in the *GJB2* and *GJB3* genes;; Digenic nonsyndromic deafness can be caused by concomitant mutations in

the *GJB2* and *GJB6* genes. This condition can also be classified according to the acquisition of language, as prelingual or postlingual deafness. Syndromic deafness is hearing loss that is associated with other signs and symptoms. The most common dominant syndromic forms of hearing impairment include Waardenburg syndrome and Stickler syndrome. The most common recessive syndromic forms of hearing impairment are Usher syndrome, Pendred syndrome and Large vestibular aqueduct syndrome. Syndromic deafness forms include: ABCD syndrome or Albinism, black lock, cell migration disorder of the neurocytes of the gut, and sensorineural deafness;; Abruzzo-Erickson syndrome or Cleft palate, coloboma and deafness;; Albinism-deafness syndrome;; Alport syndrome or Alport deafness-nephropathy;; Ataxia, deafness, and mental retardation syndrome;; Atherosclerosis, deafness, diabetes, epilepsy and nephropathy;; Bilateral microtia, deafness and cleft palate;; Björnstad syndrome or Deafness, pili torti and hypogonadism;; Cardiomyopathy and maternally inherited deafness;; Cardiospondylocarpofacial syndrome or Mitral regurgitation, deafness and skeletal anomalies;; Cataract, ataxia and deafness;; Cataract, deafness and hypogonadism;; Central nervous system calcification, deafness, tubular acidosis and anemia;; Cerebellar ataxia, deafness and narcolepsy syndrome;; Charcot-Marie-Tooth disease type 1E, or Charcot-Marie-Tooth disease and deafness;; Choroideremia, deafness and obesity;; Conductive deafness and malformed external ear;; Conductive deafness, ptosis and skeletal anomalies;; Crandall syndrome or Alopecia, deafness and hypogonadism;; De Hauwere syndrome or Iris dysplasia, hypertelorism and deafness;; Deafness, enamel hypoplasia and nail defects;; Deafness and mental retardations, Martin-Probst type;; Deafness and oligodontia;; Deafness and onychodystro-phy;; Down syndrome;; Ectodermal dysplasia and sensorineural deafness;; Ehlers-Danlos syndrome, kyphoscoliotic and deafness type;; Familial steroid-resistant nephrotic syndrome with sensorineural deafness;; Fine-Lubinsky syndrome or Brachycephaly, deafness, cataract and mental retardation;; Gemignani syndrome or Spinocerebellar ataxia, amyotrophy and deafness;; Generalized resistance to thyroid hormone or Deafness and thyroid hormone resistance;; Growth delay due to insulin-like growth factor I deficiency or Growth delay, deafness and mental retardation;; Gusher syndrome or X-linked deafness mixed with perilymphatic gusher;; Hirschsprung disease, deafness and polydactyly;; Hypoparathyroidism, deafness and renal disease;; Jervell and Lange-Nielsen syndrome or Long QT interval and deafness;; Knuckle pads, leuconychia, sensorineural deafness and palmoplantar hyperkeratosis syndrome;; Large vestibular aqueduct syndrome;; Lethal ataxia with deafness and optic atrophy;; Lipodystrophy, intellectual deficit and deafness;; Lowe-Kohn-Cohen syndrome or Deafness, nephritis and ano-rectal malformation;; MEDNIK syndrome or Mental retardation, enteropathy, deafness, peripheral neuropathy, ichthyosis and keratodermia;; Maternally inherited diabetes and deafness;; Metaphyseal dysostosis, intellectual deficit and conductive deafness;; Microcephaly, deafness and intellectual deficit;; Mitochondrial myopathy, lactic acidosis and deafness;; Mohr–Tranebjærg syndrome or Deafness, dystonia and optic neuronopathy syndrome;; Muckle-Wells syndrome or Urticaria, deafness and amyloidosis;; Nathalie syndrome or Deafness, cataracts and skeletal anomalies;; Neutropenia, monocytopenia and deafness;; Olivopontocerebellar atrophy and deafness;; Pendred syndrome or Goiter and deafness;; Perrault syndrome or 46, XX gonadal dysgenesis and deafness;; Phocomelia, ectrodactyly, deafness and sinus arrhythmia;; Postlingual nonsyndromic genetic deafness;; Prelingual nonsyndromic genetic deafness;; Renal caliceal diverticuli and deafness;; Retinitis pigmentosa, intellectual deficit, deafness and hypogenitalism;; Richards-Rundle syndrome or Ketoaciduria, mental retardation, ataxia and deafness;; Schizophrenia, intellectual deficit, deafness and retinitis;; SeSAME syndrome or Seizures, sensorineural deafness, ataxia, intellectual deficit, electrolyte imbalance;; Sensorineural deafness with dilated cardiomyopathy;; Spastic paraparesis and deafness;; Spondyloepiphyseal dysplasia, MacDermot type or Spondyloepiphyseal dysplasia, myopia, sensorineural deafness;; Stickler syndrome;; Thiamine-responsive megaloblastic anemia syndrome or Thiamine-responsive megaloblastic anemia with diabetes mellitus and sensorineural deafness;; Thickened earlobes and conductive deafness;; Treft-Sanborn-Carey syndrome or Optic atrophy, ophthalmoplegia, ptosis, deafness and myopathy;; Usher syndrome or Retinitis pigmentosa and deafness;; Waardenburg syndrome;; Wolfram syndrome or Diabetes Insipidus, Diabetes Mellitus, Optic Atrophy and Deafness;; Woodhouse-Sakati syndrome or Diabetes, hypogonadism, deafness and mental retardation;; X-linked hereditary sensory and autonomic neuropathy with deafness;; X-linked spinocerebellar ataxia type 3 or X-linked ataxia and deafness syndrome.

Mohr–Tranebjærg syndrome: A deafness-dystonia syndrome and sometimes affected boys also exhibit cortical deterioration of vision and mental deterioration. Mohr–Tranebjærg syndrome is caused by X-linked recessive mutation in the *TIMM8A* gene which encodes the translocase of inner mitochondrial membrane 8 (homolog A). Mohr–Tranebjærg syndrome

is estimated to affect less than 1 in every 1,000,000 individuals.

Mitochondrial progressive myopathy with congenital cataract, hearing loss and developmental delay: A disorder characterized by congenital cataract, muscular hypotonia, sensorineural hearing loss and developmental delay. Mitochondrial progressive myopathy with congenital cataract, hearing loss and developmental delay is caused by autosomal recessive mutations in the *GFER* gene which encodes the growth factor (augmenter of liver regeneration).

LAMM syndrome (deafness with Labyrinthine Aplasia, Microtia and Microdontia): A disorder characterized by profound congenital sensorineural deafness, type I microtia with shortening of auricles above the crura of the antihelix, and microdontia with widely spaced teeth. LAMM syndrome is caused by autosomal recessive mutations in the *FGF3* gene which encodes the fibroblast growth factor 3. LAMM syndrome is estimated to affect less than 1 in every 1,000,000 individuals. This condition has been reported in a few countries including Turkey, Saudi Arabia, Albania and Italy.

Enlarged vestibular aqueduct: A disorder characterized by fluctuating and sometimes progressive sensorineural hearing loss caused by damage to sensory cells inside the cochlea. There are several underlying causes of this disorder with at least 3 autosomal genomic loci: Enlarged vestibular aqueduct is caused by autosomal recessive mutations in the *SLC26A4* gene which encodes the solute carrier (family 26 member A4), a sodium-independent chloride/iodide transporter. Mutations in the same gene also cause Pendred syndrome;; Enlarged vestibular aqueduct is caused by autosomal recessive mutations in the *FOXI1* gene which encodes the forkhead box I1. Mutations in the same gene also cause Pendred syndrome;; and Enlarged vestibular aqueduct can be caused by digenic monoallelic mutations in the *SLC26A4* gene and *FOXI1* gene, or in the *SLC26A4* gene and the *KCNJ10* gene which encodes the potassium inwardly-rectifying channel (subfamily J member 10).

Table 3.08.1.1 Recognizable Disorders of Ear

Disorder	OMIM	Gene	Biological Function	Product of Conception Alteration
Autosomal dominant deafness-1	124900	DIAPH1	Actin polymerization	AD
Autosomal dominant deafness-2A	600101	KCNQ4	Channelopathy Potassium	AD
Autosomal dominant deafness-2B	612644	GJB3	Channelopathy Connexin	AD
Autosomal dominant deafness-3A	601544	GJB2	Channelopathy Connexin	AD + de novo
Autosomal dominant deafness-3B	612643	GJB6	Channelopathy Connexin	AD
Autosomal dominant deafness-4A	600652	MYH14	Cystoskeletal protein	AD + de novo
Autosomal dominant deafness-4B	614614	CEACAM16	Cell adhesion molecule	AD
Autosomal dominant deafness-5	600994	DFNA5	Sensory perception of sound	AD
Autosomal dominant deafness-6/14/38	600965	WFS1	Endoplasmic reticulum protein	AD
Autosomal dominant deafness-8/12	601543	TECTA	Extracellular matrix protein	AD
Autosomal dominant deafness-9	601369	COCH	Extracellular matrix protein	AD
Autosomal dominant deafness-10	601316	EYA4	Transcriptional activator, eye development	AD
Autosomal dominant deafness-11	601317	MYO7A	Cystoskeletal protein	AD
Autosomal dominant deafness-13	601868	COL11A2	Extracellular matrix protein	AD
Autosomal dominant deafness-15	602459	POU4F3	Transcription factor Helix-turn-helix domains	AD
Autosomal dominant deafness-17	603622	MYH9	Cystoskeletal protein	AD
Autosomal dominant deafness-20/26	604717	ACTG1	Cystoskeletal protein	AD

The Concise Encyclopedia of Genomic Diseases

Table 3.08.1.1. Continued.

Disorder	OMIM	Gene	Biological Function	Product of Conception Alteration
Autosomal dominant deafness-22 and autosomal dominant deafness-22, with hypertrophic cardiomyopathy	606346	*MYO6*	Cystoskeletal protein	AD
Autosomal dominant deafness-23	605192	*SIX1*	Nucleus transcription factor	AD
Autosomal dominant deafness-25	605583	*SLC17A8*	Solute carrier	AD + de novo
Autosomal dominant deafness-28	608641	*GRHL2*	Transcription factor	AD
Autosomal dominant deafness-36	606705	*TMC1*	Transmembrane protein	AD
Autosomal dominant deafness-39 or dentinogenesis imperfecta and deafness	605594	*DSPP*	Extracellular matrix protein	AD
Autosomal dominant deafness-40	123740	*CRYM*	Crystallin	AD + de novo
Autosomal dominant deafness-44	607453	*CCDC50*	Coiled-coil domain-containing protein	AD
Autosomal dominant deafness-48	607841	*MYO1A*	Cystoskeletal protein	AD
Autosomal dominant deafness-50	613074	*MIR96*	Small regulator RNA disease	AD
Autosomal recessive deafness-1A	220290	*GJB2*	Channelopathy Connexin	AR
Autosomal recessive deafness-1B	612645	*GJB6*	Channelopathy Connexin	AR
Autosomal recessive deafness-2	600060	*MYO7A*	Cystoskeletal protein	AR
Autosomal recessive deafness-3	600316	*MYO15A*	Cystoskeletal protein	AR
Autosomal recessive deafness-4	600791	*SLC26A4*	Solute carrier	AR
Autosomal recessive deafness-6	600971	*TMIE*	Transmembrane protein	AR
Autosomal recessive deafness-7/11	600974	*TMC1*	Transmembrane protein	AR
Autosomal recessive deafness-8/10	601072	*TMPRSS3*	Transmembrane protease	AR
Autosomal recessive deafness-9 and autosomal recessive auditory neuropathy-1	601071	*OTOF*	Vesicle membrane fusion	AR
Autosomal recessive deafness-12	601386	*CDH23*	Cystoskeletal protein	AR
Autosomal recessive deafness-15/72/95	601869	*GIPC3*	Survival of spiral ganglion in the ear	AR

The Concise Encyclopedia of Genomic Diseases

Table 3.08.1.1. Continued.

Disorder	OMIM	Gene	Biological Function	Product of Conception Alteration
Autosomal recessive deafness-16	603720	*STRC*	Stereocilin	AR
Autosomal recessive deafness-18	602092	*USH1C*	Nuclear localization	AR
Autosomal recessive deafness-21	603629	*TECTA*	Extracellular matrix protein	AR
Autosomal recessive deafness-22	607039	*OTOA*	Attachment of the inner ear acellular gels	AR
Autosomal recessive deafness-23	609533	*PCDH15*	Calcium-dependent adhesion	AR
Autosomal recessive deafness-24	611022	*RDX*	Cystoskeletal protein	AR
Autosomal recessive deafness-25	613285	*GRXCR1*	Glutathionylation	AR
Autosomal recessive deafness-28	609823	*TRIOBP*	Cytoskeleton organization	AR
Autosomal recessive deafness-30	607101	*MYO3A*	Cystoskeletal protein	AR
Autosomal recessive deafness-31	607084	*DFNB31*	Cystoskeletal protein	AR
Autosomal recessive deafness-35	608565	*ESRRB*	Nuclear receptor	AR
Autosomal recessive deafness-37	607821	*MYO6*	Cystoskeletal protein	AR
Autosomal recessive deafness-39	608265	*HGF*	Growth factor	AR
Autosomal recessive deafness-42	609646	*ILDR1*	Immunoglobulin-like receptor	AR
Autosomal recessive deafness-48	609439	*CIB2*	Calcium and integrin binding member	AR
Autosomal recessive deafness-49	610153	*MAR-VELD2*	Membrane protein at the tight junction	AR
Autosomal recessive deafness-49	610153	*MAR-VELD2*	Membrane protein at the tight junction	AR
Autosomal recessive deafness-53	609706	*COL11A2*	Extracellular matrix protein	AR
Autosomal recessive deafness-59	610220	*DFNB59*	Sensory perception of sound	AR
Autosomal recessive deafness-61	613865	*SLC26A5*	Solute carrier	AR
Autosomal recessive deafness-63	611451	*LRTOMT*	Methyltransferase	AR
Autosomal recessive deafness-67	610265	*LHFPL5*	Membrane protein	AR
Autosomal recessive deafness-70	614934	*PNPT1*	Polyribonucleotide nucleotidyltransferase	AR
Autosomal recessive deafness-74	613718	*MSRB3*	Methionine sulfoxide reductase	AR

Table 3.08.1.1. Continued.

Disorder	OMIM	Gene	Biological Function	Product of Conception Alteration
Autosomal recessive deafness-77	613079	LOXHD1	Sensory perception of sound	AR
Autosomal recessive deafness-79	613307	TPRN	Sensory epithelial protein	AR
Autosomal recessive deafness-82 or Chudley-McCullough syndrome	604213	GPSM2	G-protein signaling modulator	AR
Autosomal recessive deafness-84A	613391	PTPRQ	Tyrosine phosphatase	AR
Autosomal recessive deafness-84B	614944	OTOGL	Otogelin-like	AR
Autosomal recessive deafness-91	613453	SERPINB6	Serine protease inhibitor	AR
Autosomal recessive deafness-93	614899	CABP2	Calcium binding protein	AR
Autosomal recessive deafness-98	614861	TSPEAR	Thrombospondin-type laminin G domain	AR
X-linked deafness-1	304500	PRPS1	Phosphoribosylation	XR
X-linked deafness-2	304400	POU3F4	Transcription factor Helix-turn-helix domains	XR
X-linked deafness-4	300066	SMPX	Maintenance of inner ear cells subjected to mechanical stress	XR
Digenic deafness GJB2/GJB3	220290	GJB2/ GJB3	Channelopathy Connexin	Dig
Digenic deafness GJB2/GJB6	220290	GJB2/ GJB6	Channelopathy Connexin	Dig
Digenic deafness TECTA/KCNQ4	600101	TECTA/ KCNQ4	Extracellular matrix protein/Channelopathy Potassium	Dig
Deafness with inner ear agenesis	610706	FGF3	Mitogenic and cell survival activities	AR
Deafness without vestibular involvement	609006	ESPN	Actin-bundling protein	AR
Bart-Pumphrey syndrome	149200	GJB2	Channelopathy Connexin	AD
Craniofacial-deafness-hand syndrome	122880	PAX3	Transcription factor Helix-turn-helix domains	AD

Table 3.08.1.1. Continued.

Disorder	OMIM	Gene	Biological Function	Product of Conception Alteration
Cutaneous hyperpigmentation with hypertrichosis, hepatosple-nomegaly, heart anomalies, hearing loss, and hypogonadism	612391	*SLC29A3*	Solute carrier	AR
Keratoderma, palmoplantar, with Deafness	148350	*GJB2*	Channelopathy Connexin	AD
Macrothrombocytopenia and progressive sensorineural deafness	600208	*MYH9*	Cystoskeletal protein	AD + de novo
Mohr-Tranebjaerg syndrome	304700	*TIMM8A*	Mitochondrial protein	XR + de novo
Myopathy with congenital cataract, hearing loss, and developmental delay	613076	*GFER*	Hepatic regenerative stimulation substance	AR
N syndrome	310465	*POLA1*	DNA polymerase	XR
SeSAME syndrome	612780	*KCNJ10*	Channelopathy Potassium	AR
Woodhouse-Sakati syndrome	241080	*DCAF17*	Nuclear transmembrane protein	AR
Congenital deafness with inner ear agenesis, microtia, and microdontia	610706	*FGF3*	Mitogenic and cell survival activities	AR
Enlarged vestibular aqueduct	600791	*SLC26A4*	Solute carrier	AR
Enlarged vestibular aqueduct	600791	*FOXI1*	Transcription factor Helix-turn-helix domains	AR
Digenic enlarged vestibular aqueduct SLC26A4/KCNJ10	605646	*SLC26A4/ KCNJ10*	Solute carrier/Channelopathy Potassium	Dig
Hystrix-like ichthyosis with deafness	602540	*GJB2*	Channelopathy Connexin	AD
Keratitis-ichthyosis-deafness syndrome	148210	*GJB2*	Channelopathy Connexin	AD
Low-frequency sensorineural hearing loss	600965	*WFS1*	Endoplasmic reticulum protein	AD
Autosomal recessive auditory neuropathy-1	601071	*OTOF*	Vesicle membrane fusion	AR
Sensorineural deafness with mild renal dysfunction	602522	*BSND*	Chloride channels, subunit	AR

3.09.0 Subchapter IX

Diseases of the Circulatory System

Table 3.09.0 Diseases of the Circulatory System

3.09.00-9	Diseases of the Circulatory System	
	3.09.1	Acute Rheumatic Fever
	3.09.2	Hypertensive Diseases
	3.09.3	Ischemic Heart Diseases
	3.09.4	Pulmonary Heart Disease and Diseases of Pulmonary Circulation
	3.09.5	Other forms of Heart Disease
	3.09.6	Cerebrovascular Diseases
	3.09.7	Diseases of Arteries, Arterioles and Capillaries
	3.09.8	Diseases of Veins, Lymphatic Lessels and Lymph nodes (not elsewhere classified)
	3.09.9	Other and Unspecified Disorders of the Circulatory System

3.09.1 Acute Rheumatic Fever

Rheumatic fever: An inflammatory disease which occurs after Streptococcus pyogenes infection. This condition affects children aged 5-17. There are several genomic factors which increase the predisposition to rheumatic fever, including familial aggregation and monozygotic twin's studies.

3.09.2 Hypertensive Diseases

Essential hypertension: A chronic disorder in which the blood pressure in the arteries is elevated, increasing the risk of cerebral, cardiac and renal events. Hypertension is a major risk factor for myocardial infarction, stroke, heart failure, peripheral arterial disease, aneurysms of the arteries and chronic kidney disease. Prevalence of essential hypertension increases with age. Patients with the essential form

account for 90-95% of all cases of hypertension. Essential hypertension seems to be the consequence of an interaction between environmental and genomic factors. In this condition, it is difficult to separate the effects of heritability and environmental factors. There are several environmental factors which influence blood pressure including salt intake, exercise, weight, alcohol intake, stress and others. The essential form tends to follow a familial pattern of inheritance. In the United States, about 1 in 4 adult persons have this condition. Hypertension is more common in African and Native Americans than Caucasians. Primary or essential hypertension has been associated with numerous genes with small effects on blood pressure, but the genomic basis of hypertension is not well understood. There are several underlying causes of this disorder with at least 20 loci which have been associated with this condition, but few genes have been characterized. Among the genes associated with this condition include: *NR3C2, ECE1, AGT,*

ADD1, PTGIS, AGTR1, PNMT, KCNMB1, CYP3A5, NOS3, GNB3, NOS2A and *RETN*. Polymorphism in the *ADD1* gene which encodes the adducin-1 has been associated with salt-sensitive essential hypertension. Variation in the *KCNMB1* gene which encodes the potassium large conductance calcium-activated channel (subfamily M, beta member 1) has been associated with resistance to diastolic hypertension. A polymorphism in the *CYP3A5* gene which encodes the cytochrome P450 (family 3A polypeptide 5) has been associated with salt-sensitivity essential hypertension. A mutation in the *NOS3* gene which encodes the nitric oxide synthase 3 (endothelial cell) has been associated with resistance to conventional therapy for essential hypertension and pregnancy-induced hypertension.

Secondary hypertension: A form of hypertension with an identifiable cause. Secondary hypertension accounts for 5-10% of cases of hypertension. The most common cause of secondary hypertension is renal disease. Other common causes are endocrine conditions; including Cushing syndrome, Conn syndrome or Hyperaldosteronism, Acromegaly, Hyperthyroidism, Hypothyroidism, Hyperparathyroidism and Pheochromocytoma. Other mainly environmental causes of secondary hypertension include illegal drugs consumption, obesity, sleep apnea, pregnancy and coarctation of the aorta. Renal diseases account for 60–70% of identifiable cause of hypertension in children. Several forms of secondary hypertension have a strong genomic basis, including renal disease, Pheochromocytoma, Hyperthyroidism, and others.

Liddle syndrome and pseudoaldosteronism: An early onset disorder characterized by severe hypertension concomitant with low plasma renin activity, metabolic alkalosis, hypokalemia and hypoaldosteronism. Liddle syndrome involves abnormal kidney function with excessive reabsorption of sodium and loss of potassium from the renal tubule. Liddle syndrome can be caused by autosomal dominant mutations (resulting in gain-of-function) in either the *SCNN1B* or the *SCNN1G* genes leading to constitutive activation of the renal epithelial sodium channel. These genes encode beta or the gamma subunits of nonvoltage-gated sodium channel. Liddle syndrome is estimated to affect less than 1 in every 1,000,000 individuals.

3.09.3 Ischemic Heart Diseases

Acute myocardial infarction or heart attack: A disorder characterized by coronary artery occlusion, resulting in the interruption of blood supply to a part of the heart. Acute myocardial infarction causes heart cells to die. Non-genomic factors associated with this condition include hypertension, tobacco smoking, diabetes, obesity, stress, alcohol consumption and others. Genomics factors associated with this condition may include: Hypercholesterolemia, Hyperhomocysteinemia, and gender (males are at a higher risk than females), and there is a higher risk for first-degree male relatives who suffered a coronary vascular event at or before age 55, or before age 65 for first-degree female relatives. In the United States, coronary heart disease is responsible for 1 out of 5 deaths.

3.09.4 Pulmonary Heart Disease and Diseases of Pulmonary Circulation

Dursun syndrome: A disorder characterized by pulmonary arterial hypertension, cardiac abnormalities such as atrial septal defect, leukopenia including intermittent neutropenia, lymphopenia, monocytosis and anemia. Dursun syndrome is caused by autosomal recessive mutations in the *G6PC3* gene which encodes the third subunit of glucose-6-phosphatase. Dursun syndrome is estimated to affect less than 1 in every 1,000,000 individuals.

Pulmonary venoocclusive disease: A disorder characterized by both intra- and extrapulmonary veins affected. Pulmonary venoocclusive disease can be caused by mutations in the *BMPR2* gene which encodes the bone morphogenetic protein receptor type 2 (serine/threonine kinase), a member of the TGF beta receptor superfamily.

Pulmonary arterial hypertension: A disorder characterized by elevated pulmonary arterial resistance leading to right heart failure. Pulmonary arterial hypertension can be idiopathic, familial or secondary, including collagen vascular disease, chronic obstructive pulmonary disease and sickle cell anemia. There are several underlying causes of this disorder with at least 3 autosomal genomic loci: Familial pulmonary arterial hypertension is caused by autosomal dominant mutations in the *BMPR2* gene;; Primary pulmonary hypertension can be caused by mutation in the *SMAD9* gene which encodes the SMAD family member 9;; and Primary pulmonary hypertension may also be associated with Hereditary hemorrhagic telangiectasia type 2 which is caused by mutations in the *ACVRL1* gene which encodes the activin A receptor (type II-like 1). There is also a possible autosomal recessive form of primary pulmonary hypertension.

3.09.5 Other forms of Heart Disease

3.09.5.1 Nonrheumatic Mitral Valve Disorders

Familial mitral valve prolapse: A common disorder characterized by systolic displacement or billowing of the mitral leaflets into the left atrium. This condition is often accompanied by mitral regurgitation. Mitral valve prolapse affect young adults and more frequently occurs in females than in males. Mitral valve prolapse shows an age-dependent penetrance. There are several underlying causes of this disorder with at least 3 autosomal genomic loci, following an autosomal dominant inheritance, but the genes responsible have not been characterized.

3.09.5.2 Cardiomyopathy

Dilated cardiomyopathy: A group of disorders frequently appearing during the 2nd or 3rd decade of life characterized by cardiac dilatation and reduced systolic function. Dilated cardiomyopathy can be caused by environmental and genomic factors. 20-50% of all dilated cardiomyopathy cases are caused by genomic or familial factors. Affected individuals suffer from heart failure, arrhythmia and are at risk of premature death. The risk of premature death is particularly manifested in the 2nd or 3rd decade of life. There are 3 forms: type 1 or monoallelic, type 2 or biallelic, and type 3 or X-linked. There are several underlying causes of this disorder with at least 38 autosomal genomic loci and 2 X-linked genomic loci: Most familial forms show autosomal dominant patterns of inheritance. There are also other patterns of inheritance including autosomal recessive, X-linked and Mitochondrial. Dilated cardiomyopathy-1A is caused by autosomal dominant mutations in the *LMNA* gene which encodes the lamin A/C. Mutations in this gene are also associated with Emery-Dreifuss muscular dystrophy and Hutchinson-Gilford progeria syndrome;; Dilated cardiomyopathy-1C with or without left ventricular noncompaction, is caused by autosomal dominant mutations in the *LDB3* gene which encodes the LIM domain binding 3;; Dilated cardiomyopathy-1D is caused by autosomal dominant mutations in the *TNNT2* gene which encodes the troponin T type 2 (cardiac);; Dilated cardiomyopathy-1E is caused by autosomal dominant mutations in the *SCN5A* gene which encodes the voltage-gated sodium channel (type V alpha subunit);; Dilated cardiomyopathy-1G is caused by autosomal dominant mutations in the *TTN* gene which encodes the titin;; Dilated cardiomyopathy-1I is caused by autosomal dominant mutations in the *DES* gene which encodes the desmin;; Dilated cardiomyopathy-1J is caused by autosomal dominant mutations in the *EYA4* gene which encodes the eyes absent homolog 4;; Dilated cardiomyopathy-1L is caused by autosomal dominant mutations in the *SGCD* gene which encodes the delta-sarcoglycan;; Dilated cardiomyopathy-1M is caused by autosomal dominant mutations in the *CSRP3* gene which encodes the cysteine and glycine-rich protein 3;; Dilated cardiomyopathy-1N is caused by autosomal dominant mutations in the *TCAP* gene which encodes the titin-cap;; Dilated cardiomyopathy-1O is caused by autosomal dominant mutations in the *ABCC9* gene which encodes the ATP-binding cassette (sub-family C member 9);; Dilated cardiomyopathy-1P is caused by autosomal dominant mutations in the *PLN* gene which encodes the phospholamban;; Dilated cardiomyopathy-1R is caused by autosomal dominant mutations in the *ACTC1* gene which encodes the alpha cardiac muscle actin 1;; Dilated cardi-

omyopathy-1S is caused by autosomal dominant mutations in the *MYH7* gene which encodes the cardiac muscle beta-myosin heavy chain 7;; Dilated cardiomyopathy-1T is caused by autosomal dominant mutations in the *TMPO* gene which encodes the thymopoietin;; Dilated cardiomyopathy-1U is caused by autosomal dominant mutations in the *PSEN1* gene which encodes the presenilin 1;; Dilated cardiomyopathy-1V is caused by autosomal dominant mutations in the *PSEN2* gene which encodes the presenilin 2;; Dilated cardiomyopathy-1W is caused by autosomal dominant mutations in the *VCL* gene which encodes the vinculin;; Dilated cardiomyopathy-1X is caused by autosomal dominant mutations in the *FKTN* gene which encodes the fukutin;; Dilated cardiomyopathy-1Y is caused by autosomal dominant mutations in the *TPM1* gene which encodes the tropomyosin 1 (alpha);; Dilated cardiomyopathy-1Z is caused by autosomal dominant mutations in the *TNNC1* gene which encodes the troponin C type 1 (slow);; Dilated cardiomyopathy-1AA is caused by autosomal dominant mutations in the *ACTN2* gene which encodes the alpha-actinin-2;; Dilated cardiomyopathy-1BB is caused by autosomal dominant mutations in the *DSG2* gene which encodes the desmoglein 2;; Dilated cardiomyopathy-1CC is caused by autosomal dominant mutations in the *NEXN* gene which encodes the nexilin (F actin binding protein);; Dilated cardiomyopathy-1DD is caused by autosomal dominant mutations in the *RBM20* gene which encodes the RNA binding motif protein 20;; Dilated cardiomyopathy-1EE is caused by autosomal dominant mutations in the *MYH6* gene which encodes the cardiac muscle myosin alpha heavy chain 6;; Dilated cardiomyopathy-1FF is caused by autosomal dominant mutations in the *TNNI3* gene which encodes the troponin I type 3 (cardiac);; Dilated cardiomyopathy-

1GG is caused by autosomal dominant mutations in the *SDHA* gene which encodes the succinate dehydrogenase complex subunit A flavoprotein;; Dilated cardiomyopathy-1HH is caused by autosomal dominant mutations in the *BAG3* gene which encodes the BCL2-associated athanogene 3;; Dilated cardiomyopathy-1II is caused by autosomal dominant mutations in the *CRYAB* gene which encodes the crystallin (alpha B);; Dilated cardiomyopathy-1JJ is caused by autosomal dominant mutations in the *LAMA4* gene which encodes the laminin (alpha 4);; and Dilated cardiomyopathy-1KK is caused by autosomal dominant mutations in the *MYPN* gene which encodes the sarcomeric protein myopalladin. There are several additional loci for familial or autosomal dominant dilated cardiomyopathy that have been mapped, but whose genes have not been characterized including: Dilated cardiomyopathy-1B;; Dilated cardiomyopathy-1H;; Dilated cardiomyopathy-1K;; Dilated cardiomyopathy-1Q;; Dilated cardiomyopathy-2A is caused by autosomal recessive mutations in the *TNNI3* gene;; Dilated cardiomyopathy-2B is caused by autosomal recessive mutations in the *GATAD1* gene which encodes the GATA zinc finger domain containing protein 1, a transcription factor;; Dilated cardiomyopathy-3A is caused by X-linked recessive mutations in the *TAZ* gene which encodes the tafazzin. This gene maps on chromosome Xq28, the subtelomeric region;; Dilated cardiomyopathy-3B is caused by X-linked recessive mutations in the *DMD* gene which encodes the dystrophin. This rod-like cytoskeletal protein is found at the inner surface of muscle fibers;; Dilated cardiomyopathy with or without left ventricular noncompaction is caused by autosomal dominant mutations in the *LDB3* gene which encodes the LIM domain binding 3;; and "Dilated cardiomyopathy with woolly hair and keratoderma" or Carvajal syndrome is caused by autosomal recessive mutations in the *DSP* gene which encodes the desmoplakin. This protein is an essential component of functional desmosomes that anchors intermediate filaments to desmosomal plaques. There are also mitochondrial deletions and mutations which presumably cause Dilated cardiomyopathy by altering myocardial ATP generation. Dilated cardiomyopathy is estimated to affect 1 in every 2,500 individuals.

Hypertrophic cardiomyopathy: A group of disorders characterized by an asymmetrical hypertrophy of the left ventricle, which regularly involves the interventricular septum. In the 2nd decade of life, patients suffer from dyspnea, syncope, collapse, palpitations and chest pain. This condition is a leading cause of sudden cardiac death in young athletes. About 50% of all cases are caused by genomic or familial factors. These conditions can either be inherited following autosomal dominant patterns or occur as a *de novo* event. When this condition occur *de novo,* it is associated with old paternal age. Genomic factors include a 25% rate of affectance among first-degree relatives of patients with this condition. Hypertrophic cardiomyopathy is slightly more common in males than in females. There are also hormonal, environmental, and modifying genomic factors such as those related with polymorphism in the *ACE* gene which encodes the angiotensin converting enzyme. This gene can contribute to the severity of this condition. Most of the mutated genes associated with this condition encode for sarcomeric proteins including beta-myosin heavy chain, myosin light chain, cardiac troponins T and I, cardiac protein C, alpha-tropomyosin, cardiac actin and titin. There are several underlying causes of this disorder with at least 25 autosomal genomic loci, but most cases are type 1 or 4: Hypertrophic cardiomyopathy-1 is caused by monoallelic mutations in the *MYH7* gene which encodes the cardiac muscle beta-myosin heavy chain 7. Type 1 is the most common form accounting for 45% of all cases of familial disease;; Hypertrophic cardiomyopathy-2 is caused by monoallelic mutations in the *TNNT2* gene which encodes the troponin T type 2 (cardiac);; Hypertrophic cardiomyopathy-3 is caused by monoallelic mutations in the *TPM1* gene which encodes the tropomyosin 1 (alpha);; Hypertrophic cardiomyopathy-4 is caused by monoallelic mutations in the *MYBPC3* gene which encodes the cardiac myosin binding protein C. Type 4 is the second most common form accounting for 35% of all cases of familial disease. In inhabitants of Indian subcontinent this pathogenic allele appears 33,000 years ago;; Hypertrophic cardiomyopathy-6 is caused by monoallelic mutations in the *PRKAG2* gene which encodes the AMP-activated protein kinase (subunit gamma-2);; Hypertrophic cardiomyopathy-7 is caused by monoallelic mutations in the *TNNI3* gene which encodes the troponin I type 3 (cardiac);; Hypertrophic cardiomyopathy-8 is caused by monoallelic or biallelic mutations in the *MYL3* gene which encodes the ventricular/slow twitch myosin alkali light chain;; Hypertrophic cardiomyopathy-9 is caused by monoallelic mutations in the *TTN* gene which encodes the titin;; Hypertrophic cardiomyopathy-10 is caused by monoallelic mutations in the *MYL2* gene which encodes the slow cardiac myosin regulatory light chain 2;; Hypertrophic cardiomyopathy-11 is caused by monoallelic mutations in the *ACTC1* gene which encodes the alpha cardiac muscle actin 1;; Hypertrophic cardiomyopathy-12 is caused by monoallelic mutations in the *CSRP3* gene which encodes the cysteine and glycine-rich protein 3;; Hypertrophic cardiomyopathy-13 is caused by monoallelic mutations in the *TNNC1* gene which encodes the troponin C type 1 (slow);;

Hypertrophic cardiomyopathy-14 is caused by monoallelic mutations in the *MYH6* gene which encodes the myosin heavy chain 6;; Hypertrophic cardiomyopathy-15 is caused by monoallelic mutations in the *VCL* gene which encodes the vinculin;; Hypertrophic cardiomyopathy-16 is caused by monoallelic mutations in the *MYOZ2* gene which encodes the myozenin 2;; Hypertrophic cardiomyopathy-17 is caused by monoallelic mutations in the *JPH2* gene which encodes the junctophilin 2;; Hypertrophic cardiomyopathy-18 is caused by monoallelic mutations in the *PLN* gene which encodes the phospholamban;; Hypertrophic cardiomyopathy-19 is caused by monoallelic mutations in the *CALR3* gene which encodes the calreticulin 3;; Hypertrophic cardiomyopathy-20 is caused by monoallelic mutations in the *NEXN* gene which encodes the nexilin (F actin binding protein);; Hypertrophic cardiomyopathy-21 has been mapped, but the gene has not been characterized;; and Hypertrophic cardiomyopathy-22 is caused by monoallelic mutations in the *MYPN* gene which encodes the sarcomeric protein myopalladin. There is also hypertrophic cardiomyopathy which has been associated with mutations in the *MYLK2, CAV3* or *SLC25A4* genes. These genes encode for cardiac myosin light-peptide kinase, caveolin-3 and the adenine nucleotide translocator, respectively. There is also hypertrophic cardiomyopathy which has been associated with mitochondrial tRNAs mutations in the *MT-TG* and *MT-TI* genes. These genes encode for mitochondrial tRNA-glycine and mitochondrial tRNA-isoleucine, respectively. Hypertrophic cardiomyopathy is estimated to affect 1 in every 500-2,000 individuals.

Familial restrictive cardiomyopathy: A group of disorders characterized by heart muscle stiffness and impairment in relaxation after each contraction. Familial restrictive cardiomyopathy can appear anytime from childhood to adulthood. Affected children exhibit dyspnea, pneumonia, syncope, ascites, inability to walk and heart murmurs. Adults exhibit shortness of breath, fatigue, and a reduced ability to exercise. There are several underlying causes of this disorder with at least 4 autosomal genomic loci: Familial restrictive cardiomyopathy-1 is caused by autosomal dominant mutations in the *TNNI3* gene which encodes the cardiac muscle isoform of troponin I;; Familial restrictive cardiomyopathy-2, has been mapped, but the gene has not been characterized;; Familial restrictive cardiomyopathy-3 is caused by autosomal dominant mutations in the *TNNT2* gene which encodes the troponin T type 2 (cardiac);; and Familial restrictive cardiomyopathy-4 is caused by autosomal dominant mutations in the *MYPN* gene which encodes the sarcomeric protein myopalladin. Familial restrictive cardiomyopathy is estimated to affect 1-9 in every 100,000 individuals.

3.09.5.3 Other Cardiomyopathy

Arrhythmogenic right ventricular dysplasia: A group of disorders characterized by different types of arrhythmias with a left branch block pattern. Affected individuals exhibit syncopal attacks and sudden death due to ventricular fibrillation. This is a leading genomic cause of juvenile sudden death. There are several underlying causes of this disorder with at least 11 autosomal genomic loci: Arrhythmogenic right ventricular dysplasia-1 is caused by autosomal dominant mutations in the *TGFB3* gene which encodes the transforming growth factor beta 3;; Arrhythmogenic right ventricular dysplasia-2 is caused by autosomal dominant mutations in the *RYR2* gene which encodes the

ryanodine receptor 2 (cardiac);; Arrhythmogenic right ventricular dysplasia-3 has been mapped, but the gene has not been characterized;; Arrhythmogenic right ventricular dysplasia-4 has been mapped, but the gene has not been characterized;; Arrhythmogenic right ventricular dysplasia-5 is caused by autosomal dominant mutations in the *TMEM43* gene which encodes the transmembrane protein 43;; Arrhythmogenic right ventricular dysplasia-6 has been mapped, but the gene has not been characterized;; Arrhythmogenic right ventricular dysplasia-8 is caused by autosomal dominant mutations in the *DSP* gene which encodes the desmoplakin. This protein is an essential component of functional desmosomes that anchors intermediate filaments to desmosomal plaques;; Arrhythmogenic right ventricular dysplasia-9 is caused by autosomal dominant mutations in the *PKP2* gene which encodes the plakophilin 2;; Arrhythmogenic right ventricular dysplasia-10 is caused by autosomal dominant mutations in the *DSG2* gene which encodes the desmoglein 2;; Arrhythmogenic right ventricular dysplasia-11 is caused by autosomal dominant mutations in the *DSC2* gene which encodes the desmocollin 2;; and Arrhythmogenic right ventricular dysplasia-12 is caused by autosomal dominant mutations in the *JUP* gene. Arrhythmogenic right ventricular dysplasia is estimated to affect 8-10 in every 10,000 individuals.

Left ventricular noncompaction: A disorder characterized by numerous prominent trabeculations and deep intertrabecular recesses in hypertrophied and hypokinetic segments of the left ventricle. This condition leads to systolic and diastolic dysfunction. This congenital cardiomyopathy may occur in isolation or in association with congenital heart disease. There are several underlying causes of this disorder with at least 6 autosomal genomic loci and 1 X-linked genomic locus: Left ventricular noncompaction-1 can be caused by autosomal dominant mutations in the *DTNA* gene which encodes the alpha-dystrobrevin. This is a component of the dystrophin-associated protein complex;; Left ventricular noncompaction-2 has been mapped, but the gene has not been characterized;; Left ventricular noncompaction-3 is caused by mutations in the *LDB3* gene which encodes the LIM domain binding protein 3;; Left ventricular noncompaction-4 is caused by mutations in the *ACTC1* gene which encodes the alpha cardiac muscle actin 1;; Left ventricular noncompaction-5 is caused by mutations in the *MYH7* gene which encodes the cardiac muscle beta-myosin heavy chain 7;; Left ventricular noncompaction-6 is caused by mutations in the *TNNT2* gene which encodes the troponin T type 2 (cardiac);; and there is also an X-linked form of Left ventricular noncompaction which can occur as part of Barth syndrome or 3-methylglutaconic aciduria type II. The last condition is caused by X-linked recessive mutations in the *TAZ* gene which encodes the tafazzin. This gene maps on chromosome Xq28, the subtelomeric region. The prevalence of left ventricular noncompaction is unknown.

3.09.5.4 Pre-excitation Syndrome

Wolff–Parkinson–White syndrome: A pre-excitation disorder characterized by an abnormal accessory electrical conduction pathway between the atria and the ventricles. This syndrome while remaining asymptomatic in most people, carries a risk of sudden cardiac death. These individuals could exhibit an atrioventricular reentrant tachycardia that can trigger ventricular fibrillation. Wolff-Parkinson-White syndrome is estimated to affect 1 in every 350-500 individuals. Wolff-Parkinson-White syndrome is the 2nd most common cause of paroxysmal supraventricular tachycardia worldwide. Wolff-Parkinson-White syndrome is the 1st cause of paroxysmal supraventricular tachycardia accounting for 70% of all the cases among Chinese. 70% of patients with Wolff-Parkinson-White syndrome are males, mostly at a young age. Most Wolff-Parkinson-White syndrome cases are sporadic. There are also other genomic factors associated with this condition including: first-degree relatives have 11 times more risk of suffering this condition than the average people; there is also a familial form associated with DNA mutations. Wolff-Parkinson-White syndrome can be caused by autosomal dominant mutations in the *PRKAG2* gene which encodes the gamma-2 regulatory subunit of AMP-activated protein kinase.

3.09.5.5 Other Specified Conduction Disorders

Congenital long QT syndrome: A group of disorders characterized by a prolongation of the QT interval leading to polymorphic ventricular arrhythmias. Affected individuals manifest high risk of developing life-threatening arrhythmias. These conditions are associated with mutations in various genes encoding ion channel subunits. Most of these conditions belong to the channelopathy group. There are several underlying causes of this disorder with at least 13 autosomal genomic loci, following mainly an autosomal dominant pattern of inheritance: Long QT syndrome-1 can be caused by autosomal dominant or monoallelic mutations in the *KCNQ1* gene which encodes the voltage-gated potassium channel (KQT-like 1). This gene maps on the subtelomeric region of chromosome

11p. Type 1 form is the most common accounting for at least 50% of all the Long QT syndrome cases. Jervell and Lange-Nielsen syndrome is a more severe form caused by autosomal recessive or biallelic mutations in the same gene;; Long QT syndrome-2 is caused by monoallelic mutations in the *KCNH2* gene which encodes the potassium voltage-gated channel (subfamily H member 2);; Long QT syndrome-3 is caused by monoallelic mutations in the *SCN5A* gene which encodes the voltage-gated sodium channel (type V alpha subunit);; Long QT syndrome-4 is caused by monoallelic mutations in the ANK2 gene which encodes the ankyrin 2;; Long QT syndrome-5 is caused by monoallelic mutations in the *KCNE1* gene which encodes the potassium voltage-gated channel (Isk-related family member 1);; Long QT syndrome-6 is caused by monoallelic mutations in the *KCNE2* gene which encodes the potassium voltage-gated channel (Isk-related family member 2);; Long QT syndrome-7 or Andersen cardiodysrhythmic periodic paralysis, is caused by monoallelic mutations in the *KCNJ2* gene which encodes the potassium inwardly-rectifying channel (subfamily J member 2);; Long QT syndrome-8 is caused by monoallelic mutations in the *CACNA1C* gene which encodes the voltage-dependent calcium channel (L type subunit alpha 1C);; Long QT syndrome-9 is caused by monoallelic mutations in the *CAV3* gene which encodes the caveolin 3;; Long QT syndrome-10 is caused by monoallelic mutations in the *SCN4B* gene which encodes the voltage-gated sodium channel (beta subunit type IV);; Long QT syndrome-11 is caused by monoallelic mutations in the *AKAP9* gene which encodes the A kinase (PRKA) anchor protein (yotiao) 9;; Long QT syndrome-12 is caused by monoallelic mutations in the *SNTA1* gene which encodes the alpha 1 syntrophin;; and Long QT syndrome-13 is caused by monoallelic

mutations in the *KCNJ5* gene which encodes the potassium inwardly-rectifying channel (subfamily J member 5). There is also Long QT syndrome associated with digenic inheritance. Long QT syndromes are estimated to affect 1 in every 2,500 live births.

Jervell and Lange-Nielsen syndrome: A disorder characterized by congenital deafness, prolongation of the QT interval, syncopal attacks due to ventricular arrhythmias, and a high risk of sudden death. Half of all individuals affected with this condition die by the age of 15 if untreated, due to ventricular arrhythmias. There are several underlying causes of this disorder with at least 2 autosomal genomic loci: "Jervell and Lange-Nielsen syndrome-1" is caused by autosomal recessive mutations in the *KCNQ1* gene which encodes the voltage-gated potassium channel (KQT-like 1);; and "Jervell and Lange-Nielsen syndrome-2" is caused by autosomal recessive mutations in the *KCNE1* gene which encodes the potassium voltage-gated channel (Isk-related family member 1). In England, Wales and Ireland, Jervell and Lange-Nielsen syndrome is estimated to affect 1.6-6 in every 1,000,000 children aged 4-15 years.

Timothy syndrome: A multi-system disorder characterized by lethal arrhythmias due to QT prolongation, webbing of fingers and toes, congenital heart disease, cognitive abnormalities, autism, intermittent hypoglycemia and characteristic facial features. Timothy syndrome or Long QT syndrome-8 is caused by de novo mutations in the *CACNA1C* gene which encodes the voltage-dependent calcium channel (L type subunit alpha 1C).

Romano-Ward syndrome: A major form of long QT syndrome that is not associated with deafness. Romano-Ward syndrome overlaps with the previously described Long QT syndrome. If untreated, patients exhibit irregular heartbeats leading to fainting, seizures or sudden death. Romano-Ward syndrome is caused by autosomal dominant mutations in the *ANK2, KCNE1, KCNE2, KCNH2, KCNQ1* and *SCN5A* genes. The encoded proteins include either form of channels that transport positively-charged ions, such as potassium and sodium, and ankyrin-2 which inserts ion channels into the cell membrane. Romano-Ward syndrome is estimated to affect 1 in every 7,000 individuals worldwide.

3.09.5.6 Paroxysmal Tachycardia

Catecholaminergic polymorphic ventricular tachycardia: An arrhythmogenic disorder characterized by adrenergically-induced ventricular tachycardia. Affected individuals exhibit syncope and sudden death of childhood onset. There are several underlying causes of this disorder with at least 4 autosomal genomic loci: Catecholaminergic polymorphic ventricular tachycardia-1 is caused by autosomal dominant mutations in the *RYR2* gene which encodes the cardiac ryanodine receptor;; Catecholaminergic polymorphic ventricular tachycardia-2 is caused by autosomal recessive mutations in the *CASQ2* gene which encodes the calsequestrin 2 (cardiac muscle). This sarcoplasmic reticulum protein binds and stores calcium for muscle function;; Catecholaminergic polymorphic ventricular tachycardia-3 has been mapped, but the gene has not been characterized;; and Catecholaminergic polymorphic ventricular tachycardia-4 is caused by autosomal dominant mutations in the *CALM1* gene which encodes the calmodulin 1 (delta phosphorylase kinase), a calcium-binding protein. Among Europeans, Catecholaminergic polymorphic ventricu-

lar tachycardia is estimated to affect 1 in every 10,000 individuals.

3.09.5.7 Atrial Fibrillation and Flutter

Familial atrial fibrillation: A group of disorders characterized by uncoordinated electrical activity in the atrial chambers, causing heartbeat to become fast and irregular. This abnormal heart rhythm can lead to dizziness, chest pain, palpitations, syncope and shortness of breath, and increases the risk of thromboembolic stroke. Atrial fibrillation is the most common type of sustained arrhythmia. In the United States, 1 in 100 individuals are affected by this uncoordinated electrical activity in the atrial chambers. The risk of developing this irregular heart rhythm increases rapidly with age. For example, 2.3% of individuals between the ages of 40 and 60 years are affected, and up to 5.9% over the age of 65 are also affected. Up to 30% of all cases of atrial fibrillation may be caused by a genomic or familial predisposition form. There are several underlying causes of this disorder with at least 12 autosomal genomic loci, following autosomal dominant patterns of inheritance. Most of these loci have been mapped, but the genes have not yet been characterized. Familial atrial fibrillation-1 has been mapped, but the gene has not been characterized;; Familial atrial fibrillation-2 has been mapped, but the gene has not been characterized;; Familial atrial fibrillation-3 is caused by mutations in the *KCNQ1* gene which encodes the voltage-gated potassium channel (KQT-like 1);; Familial atrial fibrillation-4 is caused by mutations in the *KCNE2* gene;; Familial atrial fibrillation-5 has been mapped, but the gene has not been characterized;; Familial atrial fibrillation-6 is caused by mutations in the *NPPA* gene which encodes the natriuretic peptide A;;

Familial atrial fibrillation-7 is caused by mutations in the *KCNA5* gene which encodes the potassium voltage-gated channel (shaker-related subfamily member 5);; Familial atrial fibrillation-8 has been mapped, but the gene has not been characterized;; Familial atrial fibrillation-9 is caused by mutations in the *KCNJ2* gene which encodes the potassium inwardly-rectifying channel (subfamily J member 2);; Familial atrial fibrillation-10 is caused by mutations in the *SCN5A* gene which encodes the voltage-gated sodium channel (type V alpha subunit);; Familial atrial fibrillation-11 is caused by mutations in the *GJA5* gene which encodes the gap junction protein alpha 5;; and Familial atrial fibrillation-12 is caused by mutations in the *ABCC9* gene which encodes the ATP-binding cassette (sub-family C member 9). The encoded proteins include either form of channels that transport positively-charged ions, such as potassium and sodium, and gap junction protein.

3.09.5.8 Other Cardiac Arrhythmias

Familial ventricular fibrillation: A disorder characterized by syncopal attacks and sudden cardiac arrest, without cardiac or noncardiac causes to account for the attacks. In the industrialized world, ventricular fibrillation is a leading cause of death. Most of these episodes are caused by either cardiac factors such as secondary to myocardial infarction, or noncardiac factors. In 5-12% of Familial ventricular fibrillation cases, these events can be of genomic or familial origin. There are several underlying causes of this disorder with at least 2 autosomal genomic loci: Familial ventricular fibrillation-1 can be caused by autosomal dominant mutations in the *SCN5A* gene which encodes the cardiac sodium channel;; and Familial ventricular fibrillation-2 can be caused by autosomal dominant mutations in the *DPP6* gene which encodes the dipeptidyl-peptidase 6.

Sick sinus syndrome or Sinus node dysfunction: A group of disorders encompassing a variety of conditions caused by sinus node dysfunction. Affected individuals can exhibit syncope, presyncope, dizziness and fatigue. When patients become symptomatic, electrocardiogram can display sinus bradycardia, sinus arrest, sino-atrial block and bradycardia-tachycardia syndrome. There are several underlying causes of this disorder with at least 3 autosomal genomic loci: Sick sinus syndrome-1 can be caused by autosomal recessive mutations in the *SCN5A* gene which encodes the voltage-gated sodium channel (type V alpha subunit);; Sick sinus syndrome-2 can be caused by autosomal dominant mutations in the *HCN4* gene which encodes the hyperpolarization activated cyclic nucleotide-gated potassium channel 4;; and Susceptibility to sick sinus syndrome-3 is influenced by variations in the *MYH6* gene which encodes the cardiac muscle myosin heavy chain 6 alpha. There are several conditions associated with Sick sinus syndrome, such as Amyloidosis, Sarcoidosis, Chagas disease and Cardiomyopathies. Atrial standstill can be caused by mutations in the *SCN5A* gene in combination with mutations in the *GJA5* gene which encodes the gap junction protein (alpha 5). Sick sinus syndrome is estimated to affect 3 in every 5,000 people over the age of 50.

Progressive familial heart block or Lev-Lenègre disease: A bundle branch disorder in the His-Purkinje system that may progress to complete heart block. When patients become symptomatic, they can exhibit palpitations or occasionally, symptoms of hemodynamic disorder such as dizziness, syncope, heart failure and sudden cardiac death. There are several underlying causes of this disorder with at least 3 autosomal genomic loci: Progressive familial heart block type IA is caused by autosomal dominant mutations in the *SCN5A* gene which encodes the voltage-gated sodium channel (type V alpha subunit);; Progressive familial heart block type IB is caused by autosomal dominant mutations in the *TRPM4* gene which encodes the a Ca^{2+}-activated nonselective cation channel;; and Progressive familial heart block type II is an autosomal dominant disorder that has been mapped, but the gene has not yet been characterized. Progressive familial heart block type II occurs in South African families of Dutch descent. Progressive familial heart block is estimated to affect 4-6 in every 1,000,000 individuals.

Brugada syndrome: A group of disorders characterized by an ST-segment elevation in the right precordial electrocardiogram leads. These disorders cause an increased risk of sudden cardiac death. This condition is the major cause of sudden death in young men without known underlying cardiac disease. The syndrome usually manifests during adulthood, but occurs in infants and children. Brugada syndrome accounts for up to 40-60% of all cases of ventricular fibrillation previously classified as idiopathic. This condition is named after Pedro Brugada and Josep Brugada. There are several underlying causes of this disorder with at least 8 autosomal genomic loci, following dominant patterns of inheritance: Brugada syndrome-1 is caused by autosomal dominant mutations in the

SCN5A gene which encodes the voltage-gated sodium channel (type V alpha subunit). Type 1 is a common form accounting for at least 20% of all cases;; Brugada syndrome-2 is caused by autosomal dominant mutations in the *GPD1L* gene which encodes the glycerol-3-phosphate dehydrogenase like peptide;; Brugada syndrome-3 is caused by autosomal dominant mutations in the *CACNA1C* gene which encodes the calcium channel (L-type);; Brugada syndrome-4 is caused by autosomal dominant mutations in the *CACNB2* gene which encodes the calcium channel (L-type). The phenotypes of Brugada syndrome type 3 and 4 include a relatively shortened QT interval on ECG; Brugada syndrome-5 is caused by autosomal dominant mutations in the *SCN1B* gene which encodes the voltage-gated sodium channel (beta subunit type I);; Brugada syndrome-6 is caused by autosomal dominant mutations in the *KCNE3* gene which encodes the potassium voltage-gated channel (Isk-related family member 3);; Brugada syndrome-7 is caused by autosomal dominant mutations in the *SCN3B* gene which encodes the voltage-gated sodium channel (beta subunit type III);; and Brugada syndrome-8 is caused by autosomal dominant mutations in the *HCN4* gene which encodes the hyperpolarization activated cyclic nucleotide-gated potassium channel 4. The exact prevalence of Brugada syndrome is unknown, although it is estimated to affect 1 in every 2,000 people worldwide. In south East Asian people, there is a higher prevalence for Brugada syndrome. In these countries this condition is referred to as Sudden unexplained nocturnal death syndrome.

The Concise Encyclopedia of Genomic Diseases

Table 3.09.5.1 Other Recognizable forms of Heart Disease

Disorder	OMIM	Gene	Biological Function	Product of Conception Alteration
Dilated Cardiomyopathy-1A	115200	LMNA	Nuclear Lamina Protein	AD + de novo
Dilated Cardiomyopathy-1C	601493	LDB3	Protein-protein interaction	AD
Dilated cardiomyopathy-1D	601494	TNNT2	Cytoskeletal Microfilaments	AD + de novo
Dilated Cardiomyopathy-1E	601154	SCN5A	Channelopathy Sodium	AD + de novo
Dilated Cardiomyopathy-1G	604145	TTN	Cytoskeletal Microfilaments	AD
Dilated Cardiomyopathy-1I	604765	DES	Cytoskeletal Intermediate filaments	AD
Dilated Cardiomyopathy-1J	605362	EYA4	Transcriptional activator	AD
Dilated Cardiomyopathy-1L	606685	SGCD	Linkage of the actin cytoskeleton to the extracellular matrix	AD
Dilated Cardiomyopathy-1M	607482	CSRP3	Development and cellular differentiation	AD
Dilated Cardiomyopathy-1N	607487	TCAP	Sarcomere assembly	AD
Dilated Cardiomyopathy-1O	608569	ABCC9	ABC-transporter protein	AD
Dilated Cardiomyopathy-1P	609909	PLN	Inhibitor of sarcoplasmic reticulum Ca(2+)-ATPase	AD
Dilated Cardiomyopathy-1R	613424	ACTC1	Cytoskeletal Microfilaments	AD
Dilated Cardiomyopathy-1S	613426	MYH7	Cytoskeletal Microfilaments	AD
Dilated cardiomyopathy-1T	613740	TMPO	Nucleus protein	AD
Dilated cardiomyopathy-1U	613694	PSEN1	Regulate APP processing	AD
Dilated cardiomyopathy-1V	613697	PSEN2	Regulate APP processing	AD
Dilated Cardiomyopathy-1W	611407	VCL	Cytoskeletal protein	AD
Dilated Cardiomyopathy-1X	611615	FKTN	Golgi-cis protein	AD
Dilated Cardiomyopathy-1Y	611878	TPM1	Cytoskeletal Microfilaments	AD
Dilated Cardiomyopathy-1Z	611879	TNNC1	Cytoskeletal Microfilaments	AD
Dilated cardiomyopathy-1AA	612158	ACTN2	Cytoskeletal Microfilaments	AD

Table 3.09.5.1. Continued.

Disorder	OMIM	Gene	Biological Function	Product of Conception Alteration
Dilated Cardiomyopathy-1BB	612877	DSG2	Cell adhesion	AD
Dilated cardiomyopathy-1CC	613122	NEXN	Cell adhesion and migration	AD
Dilated cardiomyopathy-1DD	613172	RBM20	Nuclear RNA binding	AD
Dilated Cardiomyopathy-1EE	613252	MYH6	Cytoskeletal Microfilaments	AD
Dilated Cardiomyopathy-1FF	613286	TNNI3	Cytoskeletal Microfilaments	AD + de novo
Dilated Cardiomyopathy-1GG	613642	SDHA	Mitochondrial protein	AD
Dilated cardiomyopathy-1HH	613881	BAG3	Cytoprotective protein	AD
Dilated cardiomyopathy-1II	615184	CRYAB	Crystallin	AD
Dilated cardiomyopathy-1JJ	615235	LAMA4	Extracellular matrix protein	AD
Dilated cardiomyopathy-1KK	615248	MYPN	Sarcomeric protein Myopalladin	AD
Dilated Cardiomyopathy-2A	611880	TNNI3	Cytoskeletal Microfilaments	AR
Dilated Cardiomyopathy-2B	614672	GATAD1	GATA zinc finger	AR
Dilated Cardiomyopathy-3A	302060	TAZ	Mitochondrial energy protein	XR
Dilated Cardiomyopathy-3B	302045	DMD	Cytoskeletal-Membrane	XR
Dilated cardiomyopathy with or without left ventricular noncompaction	601493	LDB3	Protein-protein interaction	AD
Dilated cardiomyopathy with woolly hair and keratoderma or Carvajal syndrome	605676	DSP	Cytoskeletal Desmoplakin	AD + AD = AR
Hypertrophic cardiomyopathy-1	192600	MYH7	Cytoskeletal Microfilaments	AD
Hypertrophic cardiomyopathy-2	115195	TNNT2	Cytoskeletal Microfilaments	AD
Hypertrophic cardiomyopathy-3	115196	TPM1	Cytoskeletal Microfilaments	AD
Hypertrophic cardiomyopathy-4	115197	MYBPC3	Cytosol sarcomere	AD
Hypertrophic cardiomyopathy-6, with WPW	600858	PRKAG2	Protein kinase, AMP-activated	AD
Hypertrophic cardiomyopathy-7	613690	TNNI3	Cytoskeletal Microfilaments	AD
Hypertrophic cardiomyopathy-8	608751	MYL3	Cytoskeletal Microfilaments	AD or AR

Table 3.09.5.1. Continued.

Disorder	OMIM	Gene	Biological Function	Product of Conception Alteration
Hypertrophic cardiomyopathy-9	188840	TTN	Cytoskeletal Micro-filaments	AD
Hypertrophic cardiomyopathy-10	608758	MYL2	Cytoskeletal Micro-filaments	AD
Hypertrophic cardiomyopathy-11	612098	ACTC1	Cytoskeletal Micro-filaments	AD
Hypertrophic cardiomyopathy-12	612124	CSRP3	Development and cellular differentia-tion	AD
Hypertrophic cardiomyopathy-13	613243	TNNC1	Cytoskeletal Micro-filaments	AD
Hypertrophic cardiomyopathy-14	613251	MYH6	Cytoskeletal Micro-filaments	AD
Hypertrophic cardiomyopathy-15	613255	VCL	Cytoskeletal protein	AD
Hypertrophic cardiomyopathy-16	613838	MYOZ2	Sarcomeric protein Myozenin	AD
Hypertrophic cardiomyopathy-17	613873	JPH2	Junctophilin	AD
Hypertrophic cardiomyopathy-18	613874	PLN	Inhibitor of sarco-plasmic reticulum Ca(2+)-ATPase	AD
Hypertrophic cardiomyopathy-19	613875	CALR3	Calreticulin	AD
Hypertrophic cardiomyopathy-20	613876	NEXN	Cell adhesion and migration	AD
Hypertrophic cardiomyopathy-22	615248	MYPN	Sarcomeric protein Myopalladin	AD
Hypertrophic cardiomyopathy, digenic	606566	MYLK2/MYH7	Cytoskeletal Micro-filaments	Dig
Hypertrophic cardiomyopathy	192600	SLC25A4	Solute carrier	AD
Hypertrophic cardiomyopathy	192600	CAV3	Trafficking Vesicle fusion	AD
Familial restrictive cardiomyopa-thy-1	115210	TNNI3	Cytoskeletal Micro-filaments	AD + de novo
Familial restrictive cardiomyopa-thy-3	612422	TNNT2	Cytoskeletal Micro-filaments	AD
Familial restrictive cardiomyopa-thy-4	615248	MYPN	Sarcomeric protein Myopalladin	AD
Arrhythmogenic right ventricular dysplasia-1	107970	TGFB3	Cytokine	AD
Arrhythmogenic right ventricular dysplasia-2	600996	RYR2	Channelopathy Cal-cium	AD

Table 3.09.5.1. Continued.

Disorder	OMIM	Gene	Biological Function	Product of Conception Alteration
Arrhythmogenic right ventricular dysplasia-5	604400	TMEM43	Transmembrane protein	AD
Arrhythmogenic right ventricular dysplasia-8	607450	DSP	Cytoskeletal Desmoplakin	AD
Arrhythmogenic right ventricular dysplasia-9	609040	PKP2	Cytoskeletal Plakophilin	AD
Arrhythmogenic right ventricular dysplasia-10	610193	DSG2	Cell adhesion	AD
Arrhythmogenic right ventricular dysplasia-11	610476	DSC2	Cell adhesion and desmosome formation	AD
Arrhythmogenic right ventricular dysplasia-12	611528	JUP	Cytoskeletal Catenin	AD
Left ventricular noncompaction-1, with or without congenital heart defects	604169	DTNA	Membrane protein	AD
Left ventricular noncompaction-3, with or without dilated cardiomyopathy	601493	LDB3	Protein-protein interaction	AD
Left ventricular noncompaction-4	613424	ACTC1	Cytoskeletal Microfilaments	AD
Left ventricular noncompaction-5	613426	MYH7	Cytoskeletal Microfilaments	AD
Left ventricular noncompaction-6	601494	TNNT2	Cytoskeletal Microfilaments	AD + de novo
Barth syndrome or 3-methylglutaconic aciduria type II	302060	TAZ	Mitochondrial energy protein	XR
Wolff-Parkinson-White syndrome or ventricular familial preexcitation syndrome	194200	PRKAG2	Protein kinase, AMP-activated	AD
Long QT syndrome-1	192500	KCNQ1	Channelopathy Potassium	AD + de novo
Long QT syndrome-2	152427	KCNH2	Channelopathy Potassium	AD + de novo
Long QT syndrome-1/2, digenic	152427	KCNH2/KCNQ1	Channelopathy Potassium	AD
Long QT syndrome-3	603830	SCN5A	Channelopathy Sodium	AD
Long QT syndrome-4	600919	ANK2	Cytoskeletal-Membrane	AD
Long QT syndrome-5	176261	KCNE1	Channelopathy Potassium	AD
Long QT syndrome-6	603796	KCNE2	Channelopathy Potassium	AD

Table 3.09.5.1. Continued.

Disorder	OMIM	Gene	Biological Function	Product of Conception Alteration
Long QT syndrome-3/6, digenic	600163	KCNE2/ SCN5A	Channelopathy Sodium/Potassium	AD
Long QT syndrome-7 or Andersen cardiodysrhythmic periodic paralysis	170390	KCNJ2	Channelopathy Potassium	AD
Long QT syndrome-8 or Timothy syndrome	601005	CACNA1C	Channelopathy Calcium	AD + 99%
Long QT syndrome-9	611818	CAV3	Trafficking Vesicle fusion	AD
Long QT syndrome-10	611819	SCN4B	Channelopathy Sodium	AD
Long QT syndrome-11	611820	AKAP9	Centrosome/Golgi apparatus	AD
Long QT syndrome-12	612955	SNTA1	Peripheral membrane protein associated with dystrophin	AD
Long QT syndrome-13	613485	KCNJ5	Channelopathy Potassium	AD
Jervell and Lange-Nielsen syndrome-1	220400	KCNQ1	Channelopathy Potassium	AD + AD = AR
Jervell and Lange-Nielsen syndrome-2	612347	KCNE1	Channelopathy Potassium	AD + AD = AR
Romano-Ward syndrome		KCNE1	Channelopathy Potassium	AD
Romano-Ward syndrome		KCNE2	Channelopathy Potassium	AD
Romano-Ward syndrome		KCNH2	Channelopathy Potassium	AD
Romano-Ward syndrome		KCNQ1	Channelopathy Potassium	AD
Romano-Ward syndrome	.	SCN5A	Channelopathy Sodium	AD
Romano-Ward syndrome		ANK2	Cytoskeletal-Membrane	AD
Short QT syndrome-1	609620	KCNH2	Channelopathy Potassium	AD
Short QT syndrome-2	609621	KCNQ1	Channelopathy Potassium	AD
Short QT syndrome-3	609622	KCNJ2	Channelopathy Potassium	AD
Catecholaminergic polymorphic ventricular tachycardia-1	604772	RYR2	Channelopathy Calcium	AD
Catecholaminergic polymorphic ventricular tachycardia-2	611938	CASQ2	Sarcoplasmic reticulum protein	AR

Table 3.09.5.1. Continued.

Disorder	OMIM	Gene	Biological Function	Product of Conception Alteration
Familial atrial fibrillation-3	607554	KCNQ1	Channelopathy Potassium	AD
Familial atrial fibrillation-4	611493	KCNE2	Channelopathy Potassium	AD
Familial atrial fibrillation-6	612201	NPPA	Natriuretic peptide	AD
Familial atrial fibrillation-7	612240	KCNA5	Channelopathy Potassium	AD
Familial atrial fibrillation-9	613980	KCNJ2	Channelopathy Potassium	AD
Familial atrial fibrillation-10	614022	SCN5A	Channelopathy Sodium	AD
Familial atrial fibrillation-11	614049	GJA5	Channelopathy Connexin	AD
Familial atrial fibrillation-12	614050	ABCC9	ABC-transporter protein	AD
Familial ventricular fibrillation-1	603829	SCN5A	Channelopathy Sodium	AD
Familial ventricular fibrillation-2	612956	DPP6	Membrane protein	AD
Idiopathic ventricular tachycardia	192605	GNAI2	Regulation of adenylate cyclase	AD
Sick sinus syndrome-1	608567	SCN5A	Channelopathy Sodium	AR
Sick sinus syndrome-2	163800	HCN4	Hyperpolarization-activated cyclic nucleotide-gated potassium channels	AD
Sick sinus syndrome-3	614090	MYH6	Cytoskeletal Microfilaments	
Atrial standstill	108770	SCN5A/ GJA5	Channelopathy Sodium/Connexin	Dig
Progressive familial heart block type IA	113900	SCN5A	Channelopathy Sodium	AD
Progressive familial heart block type IB	604559	TRPM4	Channelopathy, transient receptor potential channels	AD
Brugada syndrome-1	601144	SCN5A	Channelopathy Sodium	AD
Brugada syndrome-2	611777	GPD1L	Cytoplasm associated with the plasma membrane	AD
Brugada syndrome-3	611875	CACNA1C	Channelopathy Calcium	AD
Brugada syndrome-4	611876	CACNB2	Channelopathy Calcium	AD

The Concise Encyclopedia of Genomic Diseases

Table 3.09.5.1. Continued.

Disorder	OMIM	Gene	Biological Function	Product of Conception Alteration
Brugada syndrome-5	612838	SCN1B	Channelopathy Sodium	AD
Brugada syndrome-6	613119	KCNE3	Channelopathy Potassium	AD
Brugada syndrome-7	613120	SCN3B	Channelopathy Sodium	AD
Brugada syndrome-8	613123	HCN4	Hyperpolarization-activated cyclic nucleotide-gated potassium channels	AD

Table 3.09.5.2 Heart Disease in Different Ethnic Groups

Disease	Frequency of condition	People with higher prevalence (if available)
Wolff-Parkinson-White syndrome	2-3 in 1,000	Eastern Asians
Familial atrial fibrillation	1 in 1,000	Unknown
Hypertrophic cardiomyopathy	5-20 in 10,000	Unknown
Arrhythmogenic right ventricular dysplasia	8-10 in 10,000	Unknown
Brugada syndrome	5 in 10,000	Southeastern Asians
Dilated cardiomyopathy	4 in 10,000	Unknown
Long QT syndrome	4 in 10,000	Unknown
Sick sinus syndrome	2 in 10,000	Unknown
Catecholaminergic polymorphic ventricular tachycardia	1 in 10,000	Unknown
Familial restrictive cardiomyopathy	1-9 in 100,000	Unknown
Progressive familial heart block	4-6 in 1,000,000	Afrikaners
Jervell and Lange-Nielsen syndrome	1-6 in 1,000,000	Unknown

3.09.6 Cerebrovascular Diseases

Brain small vessel disease with Axenfeld-Rieger anomaly: A disorder characterized by retinal arteriolar tortuosity with retinal hemorrhage, hypopigmentation of the fundus, infantile hemiparesis, migraine with aura and leukoencephalopathy. Axenfeld-Rieger anomaly includes anterior chamber ocular malformations with early-onset of congenital cataract, congenital glaucoma, microcornea, peripheral opacities and unilateral amblyopia. Brain small vessel disease with Axenfeld-Rieger anomaly can be caused by autosomal dominant mutations in the *COL4A1* gene which encodes the alpha 1 type IV collagen. Mutations in the same gene also cause Porencephaly which is a more severe phenotype.

Retinal vasculopathy with cerebral leukodystrophy: An adult-onset disorder affecting the microvessels of the brain and resulting in central nervous system degeneration. Affected individuals display progressive loss of vision, stroke, motor impairment and cognitive decline. Affected individuals often live 5-10 years after the onset of signs and symptoms. Patients can also exhibit Raynaud's phenomenon, micronodular cirrhosis and glomerular dysfunction. Retinal vasculopathy with cerebral leukodystrophy is be caused by autosomal dominant mutations in the *TREX1* gene which encodes the three prime repair exonuclease 1.

CADASIL syndrome (Cerebral Autosomal Dominant Arteriopathy with Subcortical Infarcts and Leucoencephalopathy): This disease is discussed under mental and behavioral disorders in the subchapter "Organic Mental Disorders (including symptomatic)" (3.05.1).

Cerebral amyloid angiopathy (Dutch, Italian, Iowa, Flemish and Arctic variants): A progressive disorder characterized by spontaneous cerebral hemorrhage, ischemic lesions and progressive dementia. Cerebral amyloid angiopathy can be caused by autosomal dominant mutations in the *APP* gene which encodes the amyloid precursor protein. Mutations in the *APP* gene can also cause autosomal dominant Alzheimer disease-1. There are also similar phenotypes of genomic origin including: Cerebral amyloid angiopathy, Icelandic type, caused by mutations in the *CST3* gene which encodes the cystatin C, and the so-called British and Danish types, both caused by mutations in the *ITM2B*

gene which encodes the integral membrane protein 2B.

3.09.7 Diseases of Arteries, Arterioles and Capillaries

Carotid intimal medial thickness: A disorder strongly associated with atherosclerosis and decreased carotid artery intimal medial thickness. Carotid intimal medial thickness is associated with polymorphism in the *PPARG* gene which encodes the peroxisome proliferator-activated receptor gamma. In general terms, 35-45% of the variability in intimal medial thickness is explained by genomic factors.

Familial thoracic aortic aneurysm: A disorder characterized by cystic medial necrosis due to degeneration and fragmentation of elastic fibers, loss of smooth muscle cells and an accumulation of basophilic ground substance. On the contrary, degeneration leading to abdominal aortic aneurysm is typically caused by a combination of factors including age, atherosclerosis, hypertension, and infectious, inflammatory or autoimmune processes. Connective tissue disorders such as Marfan syndrome and vascular (type IV) Ehlers-Danlos syndrome cause medial necrosis and thoracic aortic aneurysm/dissection. Thoracic aortic aneurysm with dissection can occur as a manifestation of the Loeys-Dietz syndrome. The last condition is caused by mutations in the *TGFBR1* gene which encodes the transforming growth factor beta receptor 1, or the *TGFBR2* gene which encodes the transforming growth factor beta receptor 2. Moreover, most cases of medial necrosis occur in the absence of a clearly identifiable syndrome. There are several underlying causes of this disorder with at least 5 autosomal genomic loci: Familial thoracic aortic aneurysm-1 has been mapped, but the gene has not been characterized;; Familial thoracic aortic aneurysm-2 has been mapped, but the gene has not been characterized;; Familial thoracic aortic aneurysm-4 can be caused by mutations in the *MYH11* gene which encodes the myosin heavy chain 11;; Familial thoracic aortic aneurysm-6 can be caused by mutations in the *ACTA2* gene which encodes the aortic smooth muscle alpha 2 actin;; and Familial thoracic aortic aneurysm-7 can be caused by mutations in the *MYLK* gene which encodes the myosin light chain kinase.

Loeys-Dietz syndrome: A disorder with vascular features similar to those of Marfan syndrome. Patients exhibit a combination of aortic aneurysms, hypertelorism, cleft palate and/or bifid uvula, and generalized arterial tortuosity. Other manifestations include intellectual deficit, congenital heart disease, structural brain abnormalities, craniosynostosis, exotropy, micrognathia and retrognathia, translucent skin and joint hyperlaxity. This condition is named after Harry C. Dietz and Bart L. Loeys. There are several underlying causes of this disorder with at least 4 autosomal genomic loci: Loeys-Dietz syndrome-1A and 2A are both caused by monoallelic mutations in the *TGFBR1* gene which encodes the transforming growth factor beta receptor 1;; Loeys-Dietz syndrome-1B and 2B are both caused by monoallelic mutations in the *TGFBR2* gene which encodes the transforming growth factor beta receptor 2. Loeys-Dietz syndrome-1A and 1B are clinically indistinguishable, as are Loeys-Dietz syndrome-2A and 2B. Loeys-Dietz syndrome is inherited following an autosomal dominant pattern in 25% of cases and occurs de novo in 75% of cases;; Loeys-Dietz syndrome-3 is associated with early-onset osteoarthritis. Type 3 is caused by mutations in the *SMAD3* gene which encodes the SMAD family member 3;; and Loeys-Dietz syndrome-4 is caused

by mutations in the *TGFB2* gene which encodes the transforming growth factor beta 2.

Primary erythermalgia: A disorder characterized by intermittent episodes of symmetrical red, warm, painful burning extremities. These attacks are provoked by exercise, long periods of standing and exposure to warmth. This condition usually appears during early childhood and adolescence. This condition may occur either by genomic/familial factors, or due to unknown factors. Familial primary erythermalgia can be caused by autosomal dominant mutations in the *SCN9A* gene which encodes the voltage-gated sodium channel (alpha subunit). This channel is located predominantly in dorsal root ganglia and sympathetic ganglia neurons. Gain-of-function mutations in the *SCN9A* gene result in hyperexcitability of pain signaling neurons. Familial primary erythermalgia is estimated to affect less than 1 in every 1,000,000 individuals.

Hereditary hemorrhagic telangiectasia or Rendu-Osler-Weber disease: A vascular dysplasia leading to telangiectasia and arteriovenous malformations of skin, mucosa and viscera. Affected individuals can also exhibit both mucosal bleeding such as nosebleeds and gastrointestinal; and visceral shunting causing bleeding in the lung, liver and brain. This condition is named after Sir William Osler, Henri Jules Louis Marie Rendu and Frederick Parkes Weber. There are several underlying causes of this disorder with at least 4 autosomal genomic loci: Hereditary hemorrhagic telangiectasia-1 is caused by autosomal dominant mutations in the *ENG* gene which encodes the endoglin. This gene maps on the subtelomeric long arm of chromosome 9;; Hereditary hemorrhagic telangiectasia-2 is caused by autosomal dominant mutations in the *ACVRL1* gene which encodes the activin A receptor (type II-like 1). In French people at least one pathogenic allele appears 300 years ago. The *ENG* and *ACVRL1* genes are involved in the signaling pathway of the transforming growth factor-beta;; Hereditary hemorrhagic telangiectasia-3 has been mapped, but the gene has not been characterized;; and Hereditary hemorrhagic telangiectasia-4 has been mapped, but the gene has not been characterized. There is also juvenile polyposis/Hereditary hemorrhagic telangiectasia- syndrome that is caused by autosomal dominant mutations in the *SMAD4* gene which encodes the SMAD family member 4. In the United States and Europe, Hereditary hemorrhagic telangiectasia is estimated to affect 1 in every 5,000-8,000 individuals. The highest prevalence of this condition has been reported in the Afro-Caribbean people of the Netherlands Antilles, where it affects 1 in every 1,331 individuals.

3.09.8 Diseases of Veins, Lymphatic Lessels and Lymph nodes (not elsewhere classified)

Deep vein thrombosis: A disorder characterized by blood clot in a deep vein in the lower leg and thigh. Affected individuals exhibit pain, swelling, redness, warmness and engorged superficial veins. The most severe complication of deep vein thrombosis is a life-threatening pulmonary embolism. There are several non-genomic risk factors associated with this condition including: older age, cancer, trauma, surgery, immobilization, oral contraceptives, pregnancy, the postnatal period, inflammatory diseases, autoimmune diseases, nephrotic syndrome, obesity and antiphospholipid syndrome. Genomic factors that increase the risk of deep vein thrombosis include gain-of-function autosomal dominant or monoallelic mutations in either the

F2 gene which encodes the coagulation Factor II, or the *F5* gene which encodes the coagulation Factor V Leiden. Among individuals with European heritage, the G20210A monoallelic mutation in the *F2* gene occur in 1-3% of all individuals; Among individuals with European heritage, the biallelic mutation in the *F2* gene occurs in 1 in every 4,400-40,000 individuals. Mutations in Factor II increase the risk of deep vein thrombosis by 2-3 times. In Europeans this pathogenic allele appears 23,720 years ago. Among individuals with European heritage, the monoallelic Factor V Leiden mutation in the *F5* gene occurs in 3-5% of all individuals. Among individuals with European heritage, the biallelic mutation in the *F5* gene occurs in 1 in every 1,600-4,400 individuals. In Europeans this pathogenic allele appears 21,340 years ago. The Factor V Leiden monoallelic mutation increases the risk of deep vein thrombosis by 3-8 times, while the biallelic mutation in the *F5* gene increases the risk of deep vein thrombosis by up to 80 times. Other genomic factors that increase the risk of deep vein thrombosis include loss-of-function mutations in the *PROC* gene which encodes the protein C (inactivator of coagulation factors Va and VIIIa), or *PROS1* gene which encodes the protein S (alpha), or the *SERPINC1* gene which encodes the serpin peptidase inhibitor clade C member 1 (antithrombin). These mutations produce deficiencies of protein C, protein S and antithrombin which encodes anti-clotting proteins. Among individuals of European heritage, these forms have a prevalence of about 1 in every 5,000 individuals for protein C deficiency, 1 in every 500 individuals for protein S deficiency, and 1 in every 300 individuals for antithrombin deficiency. There is a higher prevalence of deficiences in protein S, protein C and antithrombin in individuals from China, Japan and Thailand. These loss-of-function forms increase the risk of deep vein thrombosis by about 10 times. Having an ancestral allele non-O blood type approximately doubles the risk of deep vein thrombosis. Other gene mutations associated with this condition affect the *SERPIND1, HRG, PLAT, HBB, FGA, FGB* and *FGG* genes.

Lymphangioleiomyomatosis: A disorder characterized by cystic lung changes, enlargement of the abdominal and pelvic lymphatics, and angiomyolipomas. Female patients can exhibit shortness of breath, lung collapse, coughing and chest pain. This condition starts manifesting during the 3rd-4th decade of life. This is a progressive condition which causes recurrent pneumothorax and chylous effusions leading to respiratory failure with a mean survival time of 10-20 years. Lymphangioleiomyomatosis can occur in association with tuberous sclerosis complex due to mutations in either the *TSC1* gene which encodes the tuberous sclerosis 1, a tumor suppressor, or the *TSC2* gene which encodes the tuberous sclerosis 2. Sporadic lymphangioleiomyomatosis typically results from 2 somatic mutations in the *TSC2* gene, although a small proportion of Sporadic lymphangioleiomyomatosis cases are caused by germline mutations in the *TSC1* gene. The *TSC1* and *TSC2* genes are both subtelomeric. *TSC1* is located on the long arm of chromosome 9, and *TSC2* is located on the short arm of chromosome 16. Sporadic lymphangioleiomyomatosis occurs mostly in women, and is estimated to affect 1 in every 1,000,000 individuals.

Hereditary lymphedema: This disease is discussed under congenital malformations, deformations and chromosomal abnormalities in the subchapter "Other Congenital Malformations of Skin" (3.17.9.3).

3.09.9 Other and Unspecified Disorders of the Circulatory System

Orthostatic intolerance: A disorder characterized by the development of adrenergic symptoms appearing during upright standing and relieved by recumbency. Most patients with orthostatic intolerance are women between the ages of 20-50 years. Orthostatic intolerance can be caused by disorders of the autonomic nervous system such as Chronic fatigue syndrome and Familial dysautonomia. Orthostatic intolerance can be caused by mutations in the *SLC6A2* gene which encodes the solute carrier (family 6 member A2), a sodium-dependent norepinephrine transporter.

3.10.0 Subchapter X

Diseases of the Respiratory System

Table 3.10.0 Diseases of the Respiratory System

3.10.00-4	Diseases of the Respiratory System	
	3.10.1	Influenza and Pneumonia
	3.10.2	Chronic Lower Respiratory Diseases
	3.10.3	Other Respiratory Diseases principally affecting the Interstitium
	3.10.4	Other Diseases of Pleura

3.10.1 Influenza and Pneumonia

Influenza or Flu: An infectious disease caused by RNA viruses of the family Orthomyxoviridae. Affected individuals exhibit general discomfort with fever, chills, sore throat, muscle pains, severe headache, coughing, nausea, vomiting and fatigue. Influenza spreads around the world in seasonal epidemics, resulting in 250,000-500,000 deaths yearly, but up to millions in some pandemic years. Susceptibility to severe influenza virus infection has been associated with variations in the *IFITM3* gene which encodes the interferon induced transmembrane protein 3.

3.10.2 Chronic Lower Respiratory Diseases

Emphysema: A progressive disorder of the lungs that primarily causes shortness of breath. The main risk for emphysema includes environmental factors, mainly cigarette smoking. Tobacco use significantly increases the risk of emphysema at an earlier age. Genomic factors include alpha-1-antitrypsin deficiency. The most common manifestation of this deficiency is emphysema, which becomes evident by the 3rd-4th decade of life. 15% of patients with this deficiency exhibit liver disease, resulting in cirrhosis and liver failure. Alpha-1-antitrypsin deficiency is caused by autosomal recessive PiZ allele in the *SERPINA1* gene which encodes the serpin peptidase inhibitor. Among people of European ancestry, the carrier frequency for Alpha-1-antitrypsin deficiency is estimated to be 1 in 19.5-29.5 individuals, which means the frequency of this condition is about 1 in 1,500-3,500 individuals or $1/(39)^2$ - $1/(59)^2$. Alpha-1 antitrypsin deficiency is uncommon in people of Asian descent.

Asthma: A disorder characterized by coughing, wheezing and dyspnea. Asthma is the most common chronic disease affecting children and young adults. Asthma affects around 300 million people and causes approximately 250,000 deaths per year worldwide. Asthma is essentially attributed to the interactions between environmental and genomic factors. The strongest risk factor for developing asthma is a history of atopic disease. Other environmental factors include exposure to indoor allergens and Western styles of housing. In the United States, Asthma prevalence is higher among Puerto Ricans, African Americans, Filipinos, Irish Americans and Native Hawaiians, than among Caucasians, Mexican Americans and Korean Americans. Asthma is a complex genomic disorder with over 100 genes being associated with this condition. Among the genes associated include: Asthma and asthma-related traits-1 are associated with mutations in the *PTGDR* gene which encodes the prostaglandin D2 receptor; Asthma and asthma-related traits-2 are associated with mutations in the *NPSR1* gene which encodes the neuropeptide S receptor 1; Asthma and asthma-related traits-3 have been mapped, but the genes have not been characterized; Asthma and asthma-related traits-4 have been mapped, but the genes have not been characterized; Asthma and asthma-related traits-5 are associated with variations in the *IRAK3* gene which encode the interleukin-1 receptor-associated kinase 3; Asthma and asthma-related traits-6 are associated with variations in the *ORMDL3* gene which encodes the ORM1-like protein 3; Asthma and asthma-related traits-7 are associated with polymorphisms in the *CHI3L1* gene which encodes the chitinase 3-like 1 (cartilage glycoprotein-39); and Asthma and asthma-related traits-8 have been mapped, but the genes have not been characterized. Other genes polymorphisms associated with Susceptibility to asthma include the *HNMT* gene which encodes the histamine N-methyltransferase, and the *ADRB2* gene which encodes the adrenoceptor beta 2.

Aspirin-induced asthma: A disorder characterized by bronchoconstriction, associated with aspirin use. Affected individuals exhibit greater cysteinyl leukotriene production and greater airway hyperresponsiveness. Aspirin-induced asthma can be caused by polymorphisms in the *TBX21* gene which encodes the T-box 21, the *PTGER2* gene which encodes the prostaglandin E receptors, or the *LTC4S* gene which encodes the leukotriene C4 synthetase. There seem to be autosomal dominant and recessive forms of Aspirin-induced asthma.

Bronchiectasis with or without elevated sweat chloride: A disorder with clinical features similar to those of Cystic fibrosis. Bronchiectasis with or without elevated sweat chloride patients also exhibit pancreatic exocrine dysfunction and infertility. There are several underlying causes of this disorder with at least 3 autosomal genomic loci: Bronchiectasis with or without elevated sweat chloride-1 is caused by mutations in the *SCNN1B* gene which encodes the beta subunit of the epithelial sodium channel;; Bronchiectasis with or without elevated sweat chloride-2 is caused by mutations in the *SCNN1A* gene which encodes the alpha subunit of the epithelial sodium channel;; and Bronchiectasis with or without elevated sweat chloride-3 is caused by mutations in the *SCNN1G* gene which encodes the gamma subunit of the epithelial sodium channel.

3.10.3 Other Respiratory Diseases principally affecting the Interstitium

Pulmonary alveolar microlithiasis: A disorder characterized by the deposition of calcium phosphate microliths throughout the lungs. Most patients are asymptomatic; for symptomatic patients, symptoms usually occur between the ages of 20-30 years. Around one-third of the cases are believed to be of genomic or familial origin. Pulmonary alveolar microlithiasis can be caused by autosomal recessive mutations in the *SLC34A2* gene which encodes the solute carrier (family 34 member A2), a sodium-phosphate cotransporter. This protein is involved in phosphate homeostasis in several organs. There are a few hundreds cases reported for this condition worldwide, predominantly among people from Turkey.

Pulmonary surfactant metabolism dysfunction: A disorder appearing in full-term neonates or infants characterized by severe respiratory insufficiency or failure. The surfactant (90% lipids and 10% proteins) is secreted by the alveolar cells of the lung and maintains the stability of pulmonary tissue by decreasing the surface tension of fluids that coat the lung. There are several underlying causes of this disorder with at least 4 autosomal genomic loci and 1 X-linked genomic locus: Pulmonary surfactant metabolism dysfunction-1 can be caused by autosomal recessive mutations in the *SFTPB* gene which encodes the surfactant protein B. This amphipathic protein is essential for lung function and homeostasis after birth;; Pulmonary surfactant metabolism dysfunction-2 can be caused by autosomal recessive mutations in the *SFTPC* gene which encodes the surfactant protein C;; Pulmonary surfactant metabolism dysfunction-3 can be caused by autosomal recessive mutations in the *ABCA3* gene which encodes the ABC transporter family;; Pulmonary surfactant metabolism dysfunction-4 can be caused by X-linked mutations in the *CSF2RA* gene which encodes the colony stimulating factor 2 (receptor alpha);; and Pulmonary surfactant metabolism dysfunction-5 can be caused by autosomal recessive mutations in the *CSF2RB* gene which encodes the colony stimulating factor 2 (receptor beta). Pulmonary surfactant metabolism dysfunction-1 is estimated to affect 1 in every 1,500,000 births.

Idiopathic pulmonary fibrosis: A chronic progressive interstitial lung disease characterized by fibrosis. This condition appears in the 6th-7th decade of life. Affected individuals exhibit exercise-induced breathlessness, dry coughing and clubbing associated with pulmonary heart disease. This condition has been related to both environmental factors such as cigarette smoking and exposure to silica, and genomic factors. There are several underlying causes of this disorder with at least 5 autosomal genomic loci. Idiopathic pulmonary fibrosis is caused by autosomal dominant mutations in the *SFTPA2* gene which encodes the pulmonary surfactant protein A2;; Susceptibility to Idiopathic pulmonary fibrosis can be conferred by polymorphisms in the *SFTPA1* gene which encodes the pulmonary surfactant protein A1, or by a promoter mutations in the *MUC5B* gene which encodes the mucin 5B oligomeric mucus/gel-forming;; and Pulmonary fibrosis can also be a feature in patients with mutations in the *TERT* gene which encodes the telomerase reverse transcriptase, or the *TERC* gene which encodes the telomerase RNA component. Idiopathic pulmonary fibrosis is estimated to affect 1 in every 5,000 males and 1 in every 7,700 females.

3.10.4 Other Diseases of Pleura

Primary spontaneous pneumothorax: A disorder characterized by an abnormal collection of air in the pleural space impairing breathing. This condition occurs in the absence of known lung diseases. The risk factors associated with this condition include male sex, smoking and a family history of pneumothorax. 1 in every 9 persons with a spontaneous pneumothorax have family member who has experienced a pneumothorax before.

Birt-Hogg-Dube syndrome: A disorder characterized by spontaneous pneumothorax, as well as fibrofolliculomas of the skin and increased risk of renal and colonic tumors. Primary spontaneous pneumothorax can be caused by autosomal dominant mutations in the *FLCN* gene which encodes the folliculin. Birt-Hogg-Dube syndrome is estimated to affect 1 in every 200,000 individuals.

Secondary spontaneous pneumothorax: A disorder occurring in the setting of a variety of lung diseases. Chronic obstructive pulmonary disease is the most common cause of pneumothorax accounting for 70% of all secondary cases. Several genomic disorders have been linked to pneumothorax, predominantly Marfan syndrome and Ehlers-Danlos syndrome. Other conditions also associated with pneumothorax include Alpha-1-antitrypsin deficiency, Acute severe asthma, Cystic fibrosis, Homocystinuria, and others.

3.11.0 Subchapter XI

Diseases of the Digestive System

Table 3.11.0 Diseases of the Digestive System

3.11.00-6	Diseases of the Digestive System	
	3.11.1	Diseases of Oral Cavity, Salivary Glands and Jaws
	3.11.2	Noninfective Enteritis and Colitis
	3.11.3	Other Diseases of Intestines
	3.11.4	Diseases of Liver
	3.11.5	Disorders of Gallbladder, Biliary Tract and Pancreas
	3.11.6	Other Diseases of the Digestive System

Table 3.11.0.1 Diseases of the Digestive System in Different Ethnic Groups

Disease	Frequency of condition	People with higher prevalence (if available)
Acute alcohol sensitivity	1-430 in 1,000	Eastern Asians
Celiac disease	1 in 100	Unknown
Inflammatory bowel disease	3-5 in 1,000	Ashkenazi Jews
Intrahepatic cholestasis of pregnancy	4-10 in 1,000 Pregnancy	Chileans, Bolivians and Scandinavians
Hypodontia or tooth agenesis	>1 in 1,000	Unknown
Cherubism	<1 in 10,000	Unknown
Amelogenesis imperfect	1 in 10,000	Swedish (North)
Dentinogenesis imperfecta	1-2 in 10,000	Unknown
Congenital diarrhea-5 or Congenital tufting enteropathy	1-2 in 100,000	Western Europeans
Progressive familial intrahepatic cholestasis	1-2 in 100,000	Unknown
Tooth agenesis and ectodermal dysplasia	6 in 100,000	Unknown
Inflammatory bowel disease	1-9 in 100,000	Unknown
Nance-Horan syndrome	1-9 in 1,000,000	Unknown
Tropical calcific pancreatitis and Hereditary pancreatitis	1-9 in 1,000,000	Unknown
Congenital defects of bile acid synthesis	1-9 in 1,000,000	Unknown
Hepatic venoocclusive disease with immunodeficiency	<1 in 1,000,000	Lebanese
Nance-Horan syndrome	<1 in 1,000,000	Unknown
Congenital chloride diarrhea	<1 in 1,000,000	Finnish
Congenital secretory sodium diarrhea	<1 in 1,000,000	Unknown
Congenital malabsorptive diarrhea	<1 in 1,000,000	Unknown
North American Indian childhood cirrhosis	<1 in 1,000,000	Ojibway-Cree Abitibi people (Quebec)
Enterokinase deficiency	<1 in 1,000,000	Unknown

3.11.1 Diseases of Oral Cavity, Salivary Glands and Jaws

3.11.1.1 Disorders of Tooth Development and Eruption

Aggressive periodontitis: A disorder characterized by severe and protracted gingival infections, leading to tooth loss. There are several underlying causes of this disorder with at least 3 autosomal genomic loci: Aggressive periodontitis can be caused by autosomal recessive mutations in the *CTSC* gene which encodes the cathepsin C. Another form of aggressive periodontitis has been mapped, but the gene has not been characterized. There is also susceptibility to periodontitis associated with HLA-A9 alleles.

Tooth agenesis: A group of disorders characterized by lack of one or a few permanent teeth, without any systemic disorders. Tooth agenesis can be either sporadic or familial in nature. Tooth agenesis is a common condition affecting 1 in 5 individuals. The most common form of tooth agenesis include the 3rd molars or wisdom teeth, followed by either upper lateral incisors or lower 2nd premolars; agenesis involving 1st and 2nd molars is infrequent. According to the severity of tooth agenesis this condition can be classified as: oligodontia is the more severe phenotype is defined as agenesis of 6 or more permanent teeth, and hypodontia, defined as agenesis of less than 6 teeth. The number of absent permanent teeth in both cases excludes 3rd molars or wisdom teeth. There are both environmental and genomic factors that can cause failure of tooth development. Environmental factors include low birth weight and perinatal factors, irradiation and chemotherapy particularly in children treated for malignant diseases. There are

several underlying causes of this disorder with at least 6 autosomal genomic loci and 1 X-linked genomic locus: Selective tooth agenesis-1 is caused by autosomal dominant mutations in the *MSX1* gene which encodes the msh homeobox 1;; Selective tooth agenesis-2 has been mapped, but the gene has not been characterized;; Selective tooth agenesis-3 is caused by autosomal dominant mutations in the *PAX9* gene which encodes the paired box 9;; Selective tooth agenesis-4 is caused by autosomal dominant mutations in the *WNT10A* gene which encodes the wingless-type MMTV integration site family (member 10A). This secreted signaling protein is involved in several developmental processes and in oncogenesis. Type 4 is the most common form accounting for the majority of isolated hypodontia patients;; Selective tooth agenesis-5 has been mapped, but the gene has not been characterized;; Selective tooth agenesis-6 is caused by autosomal recessive mutations in the *LTBP3* gene which encodes the latent transforming growth factor beta binding protein 3;; and X-Linked selective tooth agenesis-1 is caused by X-linked mutations in the *EDA* gene which encodes the ectodysplasin A. This protein is involved in cell-cell signaling during the development of ectodermal organs.

Amelogenesis imperfecta: A disorder characterized by impairment in the formation of the enamel which is the external layer of teeth. The people affected with amelogenesis imperfecta exhibit teeth with abnormal color: yellow, brown or grey. The teeth have hypersensitivity to temperature changes and a higher risk for dental cavities. There are several underlying causes of this disorder with at least 7 autosomal genomic loci and 2 X-linked genomic loci: Amelogenesis imperfecta type IB is caused by autosomal dominant mutations in the *ENAM* gene which encodes the enamelin. This protein is involved in the mineralization and structural organization of enamel. This form is the most common cause of Amelogenesis imperfecta;; Amelogenesis imperfecta type IC is caused by autosomal recessive mutations in the *ENAM* gene;; Amelogenesis imperfecta type IE or Hypoplastic amelogenesis imperfect, is caused by X-linked mutations in the *AMELX* gene which encodes the X-linked amelogenin. This protein is involved in biomineralization during tooth enamel development. There is another loci in chromosome X associated with this condition, but the gene has not yet been characterized;; Amelogenesis imperfecta type 2A1, or Pigmented hypomaturation-type amelogenesis imperfect, is caused by autosomal recessive mutations in the *KLK4* gene which encodes the kallikrein-related peptidase 4, a serine protease;; Amelogenesis imperfecta type 2A2 is caused by autosomal recessive mutations in the *MMP20* gene which encodes the matrix metalloproteinase 20. This enzyme is involved in the breakdown of extracellular matrix;; Amelogenesis imperfecta type 2A3 is caused by mutations in the *WDR72* gene which encodes the WD repeat-containing protein domain 72;; Amelogenesis imperfecta type 2A4 is caused by mutations in the *C4orf26* gene which encodes the chromosome 4 open reading frame 26;; Amelogenesis imperfecta type 3 or hypocalcified amelogenesis imperfecta is caused by autosomal dominant mutations in the *FAM83H* gene which encodes the family with sequence similarity 83 (member H);; and Amelogenesis imperfecta type 4 or Amelogenesis imperfecta of the hypomaturation-hypoplastic type with taurodontism can be caused by mutations in the *DLX3* gene which encodes the distal-less homeobox 3. Amelogenesis imperfecta is estimated to affect 1 in every 14,000 people in the United States. The largest preva-lence of this condition has been reported in the county of Vasterbotten in northern Sweden. In this area 1 in 700 people are affected with Amelogenesis imperfecta.

Dentinogenesis imperfecta: A disorder characterized by discolored and translucent teeth. The teeth may be gray to yellowish brown. This condition can affect both primary teeth and permanent teeth, making them prone to rapid wear, breakage and loss. Dentinogenesis imperfecta type I occurs in people that have features of osteogenesis imperfecta in which bones are brittle and easily broken. Dentinogenesis imperfecta II and III are caused by autosomal dominant mutations in the *DSPP* gene which encodes the dentin sialophosphoprotein. This protein is involved in the biomineralization process of dentin Dentinogenesis imperfecta is estimated to affect 1 in every 6,000-8,000 individuals.

Nance-Horan syndrome: A disorder characterized by the association of congenital cataracts with microcornea, dental anomalies and facial dysmorphism. Affected children also exhibit nystagmus, strabismus, and supernumerary incisors. Affected individuals have characteristic facial features including long face, prognathism, a large nose with a high nasal bridge and large often protruding ears. Other features of patients with this condition include intellectual impairment, mild or moderate motor delay and autistic features. Nance-Horan syndrome is caused by X-linked semi-dominant mutations in the *NHS* gene which encodes the Nance-Horan syndrome protein (congenital cataracts and dental anomalies). This protein is involved in the development of the eyes, teeth and brain. These mutations typically result in a truncated protein. The encoded protein contains four conserved nuclear localization signals. Males are more severely affected with this condition than females. Nance-Horan syndrome is estimated to affect less than 1 in every 1,000,000 individuals.

3.11.1.2 Other Disorders of Gingiva and Edentulous Alveolar Ridge

Gingival fibromatosis: An overgrowth disorder characterized by a benign, slowly progressing and nonhemorrhagic fibrous enlargement of maxillary and mandibular keratinized gingiva. There are several underlying causes of this disorder with at least 5 autosomal genomic loci: Gingival fibromatosis-1 is caused by autosomal dominant mutations in the *SOS1* gene which encodes the son of sevenless homolog 1 protein, a guanine nucleotide exchange factor. There are at least 4 additional loci which cause this condition, but the genes have not yet been characterized.

3.11.1.3 Diseases of Jaws

Cherubism: A benign fibro-osseous disorder appearing during childhood. This condition is limited to the lower half of the face, the maxilla and particularly the mandible. Affected children have a bilateral painless swelling of the jaws beginning around the 3rd or 4th year of life and progressing until the late teens. Affected individuals manifest an exaggerated inflammation, especially the submandibular lymph nodes. These children also exhibit abnormal dentition, multicystic bone tumors and eyes-to-heaven appearance. Cherubism can be a feature of other conditions such as Noonan syndrome and Noonan like syndrome, Ramon syndrome associated with gingival fibromatosis and Neurofibromatosis type 1. Cherubism can be caused by autosomal dominant mutations in the *SH3BP2* gene which encodes the SH3-domain binding protein 2, a cytoplasmic adaptor protein. This gene maps on the subtelomeric region of chromosome 4p16.3. Cherubism is estimated to affect less than 1 in every 10,000 individuals.

Table 3.11.1.1 Recognizable Diseases of Oral Cavity, Salivary Glands and Jaws

Disorder	OMIM	Gene	Biological Function	Product of Conception Alteration
Aggressive periodontitis	170650	*CTSC*	Lysosomal storage protein	AR
Selective tooth agenesis-1	106600	*MSX1*	Transcription factor Helix-turn-helix domains	AD
Selective tooth agenesis-3	604625	*PAX9*	Transcription factor Helix-turn-helix domains	AD
Selective tooth agenesis-4	150400	*WNT10A*	Several developmental processes	AR or AD
Selective tooth agenesis-6	613097	*LTBP3*	Extracellular matrix protein	AR
X-Linked selective tooth agenesis-1	313500	*EDA*	Ectodysplasin A	XD or XR
Amelogenesis imperfecta-1B	104500	*ENAM*	Matrix of developing teeth	AD

Table 3.11.1.1. Continued.

Disorder	OMIM	Gene	Biological Function	Product of Conception Alteration
Amelogenesis imperfecta-1C	204650	ENAM	Matrix of developing teeth	AR
Amelogenesis imperfecta-1E or hypoplastic amelogenesis imperfecta	301200	AMELX	Extracellular matrix protein	X
Amelogenesis imperfecta-2A1	204700	KLK4	Enamel matrix serine proteinase	AR
Amelogenesis imperfecta-2A2	612529	MMP20	Extracellular matrix protein	AR
Amelogenesis imperfecta-2A3	613211	WDR72	WD repeat domain	AR
Amelogenesis imperfecta-2A4	614832	C4orf26	Extracellular matrix protein	AR
Amelogenesis imperfecta-3 or hypocalcified amelogenesis imperfecta	130900	FAM83H	Development and calcification of tooth enamel	AD
Amelogenesis imperfecta-4 or amelogenesis imperfecta of the hypomaturation-hypoplastic type with taurodontism	104510	DLX3	Nuclear transcription factor	AD
Dentinogenesis imperfecta II and III	125490	DSPP	Extracellular matrix of the tooth	AD
Nance-Horan syndrome	302350	NHS	Nuclear localization	XsD
Gingival fibromatosis-1	135300	SOS1	Regulate RAS proteins	AD
Cherubism	118400	SH3BP2	Binds to tyrosine kinases	AD
Jalili syndrome	217080	CNNM4	Metal ion transport	AR
Haim-Munk syndrome	245010	CTSC	Lysosomal storage protein	AR
Primary failure of tooth eruption	125350	PTH1R	Cell surface receptor G protein	AD
Oligodontia-colorectal cancer syndrome	608615	AXIN2	Regulation of G-protein signaling	AD
Solitary median maxillary central incisor	147250	SHH	Patterning the early embryo	AD + de novo

3.11.2 Noninfective Enteritis and Colitis

Inflammatory bowel disease: A disorder characterized by a chronic relapsing intestinal inflammation. The major types of Inflammatory bowel disease are Crohn disease and Ulcerative colitis. In approximately 10% of cases confined to the rectum and colon, definitive classification of Crohn disease or Ulcerative colitis cannot be made and are designated "indeterminate colitis." Affected individuals exhibit pain, vomiting, diarrhea and other socially unacceptable symptoms. This condition is infrequently fatal on its own. In ulcerative colitis, the inflammation is continuous and limited to rectal and colonic mucosal layers; without fistulas and granulomas. On the contrary, Crohn disease may implicate any part of the gastrointestinal tract, but most often the terminal ileum and colon. Bowel inflammation is discontinuous and transmural; it may contain granulomas or be associated with perianal or intestinal fistulas. There are both genomic and environmental factors that seem to be important in the etiology of this disease. Crohn disease and ulcerative colitis are autoimmune diseases with higher prevalence in individuals with other autoimmune conditions, particularly Ankylosing spondylitis, Psoriasis, Sclerosing cholangitis and Multiple sclerosis. Inflammatory bowel disease has strong evidence of genomic factors from twin studies, familial risk data and segregation analysis. Ulcerative colitis and Crohn disease are complex genomic traits as inheritance does not follow any simple Mendelian models. People with an affected relative have a 10-fold increase in the risk of having Inflammatory bowel disease. Genomics factors include the fact that first-degree relatives have 1.6-4.5% lifetime risk for ulcerative colitis. Genomics factors for Crohn disease include the fact that first-degree relatives have 5.2-7.8% lifetime risk for the disease.. In other words, if a first-degree relative is affected, the risk is 20-35 times larger. About 10% of persons with Crohn disease have 1 or more close relatives with granulomatous disease of the bowel. There are several underlying causes of this disorder with at least 28 susceptibility loci. Crohn disease-associated growth failure can be caused by a promoter polymorphism in the *IL6* gene which encodes the

interleukin 6;; Inflammatory bowel disease-1 can be caused by mutations in the *NOD2* gene which encodes the nucleotide-binding oligomerization domain containing protein 2;; Inflammatory bowel disease-10 can be caused by mutations in the *ATG16L1* gene which encodes the autophagy related 16-like 1;; Inflammatory bowel disease-11 can be caused by mutations in the *MUC3A* gene which encodes the mucin 3A, a cell surface associated protein;; Inflammatory bowel disease-12 can be caused by mutations in the *MST1* gene which encodes the macrophage stimulating 1 (hepatocyte growth factor-like);; Inflammatory bowel disease-13 can be caused by mutations in the *ABCB1* gene which encodes the ATP-binding cassette (sub-family B member 1);; Inflammatory bowel disease-14 can be caused by mutations in the *IRF5* gene which encodes the interferon regulatory factor 5;; Inflammatory bowel disease-16 can be caused by mutations in the *TNFSF15* gene which encodes the tumor necrosis factor (ligand) superfamily (member 15);; Inflammatory bowel disease-17 can be caused by mutations in the *IL23R* gene which encodes the interleukin 23 receptor;; Inflammatory bowel disease-19 can be caused by mutations in the *IRGM* gene which encodes the immunity-related GTPase;; Early onset inflammatory bowel disease-25 is caused by biallelic mutations in the *IL10RB* gene which encodes the interleukin 10 receptor (beta);; and Early onset inflammatory bowel disease-28 is caused by biallelic mutations in the *IL10RA* gene which encodes the interleukin 10 receptor (alpha). The prevalence of this common inflammatory bowel disease is 1 in every 200-400 individuals in the United States and Europe.

3.11.3 Other Diseases of Intestines

Chronic intestinal pseudo-obstruction: A gastrointestinal motility disorder characterized by recurring episodes similar to mechanical obstruction without any mechanical lesions. Affected children exhibit abdominal pain, diarrhea and/or intractable constipation, distension/fullness, nausea/vomiting, and malabsorption of nutrients causing weight loss and/or failure to thrive. Chronic intestinal pseudo-obstruction affects 1.5-4 times more males than females. This condition may be present at birth. The prevalence of this condition is unknown. Chronic intestinal pseudo-obstruction can be caused by genomic or non-genomics factors including postoperatively, postinfectious or post abdominal radiation, or drugs. The genomic form includes myopathic and neuropathic categories. This condition may be a primary or secondary disorder due to muscular, neurologic, metabolic or endocrine disorders. Neuronal intestinal pseudoobstruction can be caused by mutations or duplication in the *FLNA* gene which encodes the filamin A. This gene maps on the subtelomeric long arm of chromosome X. Some primary forms of Chronic intestinal pseudo-obstruction are caused by defects of enteric neuronal cells e.g., Hirschsprung disease and autosomal recessive visceral neuropathy. For the latter form, the gene has not mapped.

3.11.3.1 Congenital Diarrhea

Congenital secretory chloride diarrhea: A severe form of chronic diarrhea characterized by excretion of large amounts of watery stool. Affected children are often born prematurely. Affected individuals exhibit dehydration, hypokalemia and metabolic alkalosis. The electrolyte disorder has similar features than those of the renal disorder Bartter syndrome, but that chloride diarrhea is not associated with calcium level abnormalities. There are several underlying causes of this disorder with at least 6 autosomal genomic loci: Congenital secretory chloride diarrhea-1 is caused by autosomal recessive mutations in the *SLC26A3* gene which encodes the solute carrier protein (family 26 member A3). The encoded gene transports chloride ions across the cell membrane in exchange for bicarbonate ions. In Finnish people this pathogenic allele appears 400 years ago;; Congenital diarrhea-2 or Microvillus inclusion disease is caused by mutations in the *MYO5B* gene which encodes the myosin VB;; Congenital secretory sodium diarrhea-3 is caused by autosomal recessive mutations in the *SPINT2* gene which encodes the serine peptidase inhibitor (Kunitz type 2);; Congenital malabsorptive congenital diarrhea-4 is caused by mutations in the *NEUROG3* gene which encodes the neurogenin 3;; Congenital diarrhea-5 or congenital tufting enteropathy is caused by mutations in the *EPCAM* gene which encodes the epithelial cell adhesion molecule;; and Congenital diarrhea-6 or Early-onset chronic diarrhea, is caused by mutations in the *GUCY2C* gene which encodes the guanylate cyclase 2C.

Congenital malabsorptive diarrhea-4 or Enteric anendocrinosis: A disorder characterized by a generalized malabsorption and absence of enteroendocrine cells. Affected neonates exhibit during the first weeks of life vomiting, diarrhea, dehydration, and a severe hyperchloremic metabolic acidosis after the ingestion of standard cow's milk-based formula. Affected individuals tend to develop a condition similar to diabetes type 1 during childhood. Congenital malabsorptive diarrhea-4 is caused by loss-of-function autosomal recessive mutations in the NEUROG3 gene which encodes the neurogenin 3.

This protein has been implicated in endocrine enteric and pancreatic cell development.

Congenital diarrhea-5 or Congenital tufting enteropathy: A congenital enteropathy characterized by early-onset severe intractable diarrhea occasionally causing irreversible intestinal failure. These infants manifest watery diarrhea within the first days after birth. In affected individuals, diarrhea persists in spite of bowel rest and parenteral nutrition. Congenital diarrhea-5 or Congenital tufting enteropathy is caused by autosomal recessive mutations in the *EPCAM* gene which encodes the epithelial cell adhesion molecule. In Western Europe, Congenital diarrhea-5 is estimated to affect 1 in every 50,000-100,000 live births. The prevalence seems to be higher in areas with a high degree of consanguinity and in patients of Arabic origin.

Table 3.11.2-3.1 Recognizable Noninfective Enteritis and Colitis and Other Diseases of Intestines

Disorder	OMIM	Gene	Biological Function	Product of Conception Alteration
Crohn disease-associated growth failure	266600	*IL6*	Interleukin	
Inflammatory bowel disease-1	266600	*NOD2*	Nucleotide Oligomerization Domain receptors	
Inflammatory bowel disease-10	611081	*ATG16L1*	Autophagy, targeted to lysosomes	
Inflammatory bowel disease-11	191390	*MUC3A*	Secreted and membrane bound protein	
Inflammatory bowel disease-12	612241	*MST1*	Hepatocyte growth factor	
Inflammatory bowel disease-13	612244	*ABCB1*	ABC-transporter protein	
Inflammatory bowel disease-14	612245	*IRF5*	Transcription factor Helix-turn-helix domains	
Inflammatory bowel disease-16	612259	*TNFSF15*	Tumor necrosis factor ligand	
Inflammatory bowel disease-17	612261	*IL23R*	Cell surface receptor Other/ungrouped	
Inflammatory bowel disease-19	612278	*IRGM*	Immunity-related GTPase	
Early onset inflammatory bowel disease-25	612567	*IL10RB*	Cell surface receptor Other/ungrouped	AR
Early onset inflammatory bowel disease-28	613148	*IL10RA*	Cell surface receptor Other/ungrouped	AR
Neuronal intestinal pseudoobstruction	300048	*FLNA*	Cytoskeletal disease Microfilaments	XR
Congenital secretory chloride diarrhea-1	214700	*SLC26A3*	Solute carrier	AR
Congenital secretory chloride diarrhea-2 or microvillus inclusion disease	251850	*MYO5B*	Myosin	AR

Table 3.11.2-3.1. Continued.

Disorder	OMIM	Gene	Biological Function	Product of Conception Alteration
Congenital secretory sodium diarrhea-3	270420	SPINT2	Serine peptidase inhibitor	AR
Congenital malabsorptive congenital diarrhea-4	610370	NEUROG3	Transcription factor Basic helix-loop-helix	AR
Congenital diarrhea-5 or congenital tufting enteropathy	613217	EPCAM	Cell adhesion molecule	AR
Congenital diarrhea-6 or early-onset chronic diarrhea	614616	GUCY2C	Guanylate cyclase	AD

3.11.4 Diseases of Liver

Acute infantile liver failure: A transient disorder of hepatic function. These infants exhibit elevated liver enzymes, hyperbilirubinemia, jaundice, vomiting, coagulopathy and hyperlactatemia. This condition is consistent with a defect in mitochondrial respiratory function. Acute infantile liver failure can be caused by autosomal recessive mutations in the *TRMU* gene which encodes the tRNA 5-methylaminomethyl-2-thiouridylate methyltransferase. This protein is involved in mitochondrial protein translation. There is also a more severe, permanent disorder with some overlapping features that is associated with mitochondrial DNA depletion. For example, Mitochondrial DNA depletion syndrome-3 is caused by autosomal recessive mutations in the nuclear-encoded *DGUOK* gene which encodes the deoxyguanosine kinase.

Hereditary North American Indian childhood cirrhosis: A disorder characterized by severe intrahepatic cholestasis. Affected individuals exhibit first a transient neonatal jaundice. This disease progresses into periportal fibrosis and cirrhosis during a period fluctuating from childhood to adolescence. North American Indian childhood cirrhosis is caused by autosomal recessive mutations in the *CIRH1A* gene which encodes the cirhin, a WD40-repeat-containing protein that is localized to the nucleolus. This condition has been described among Ojibway-Cree children from First Nations communities in the Abitibi region of northwestern Quebec, Canada. In the latter group, the carrier frequency for North American Indian childhood cirrhosis trait is estimated to be 1 in 11 individuals, which means the frequency of this condition is about 1 in 484 individuals or $1/(22)^2$. Affected individuals require hepatic transplantation in childhood or early adulthood.

Reynolds syndrome: A disorder characterized by the association of scleroderma with primary biliary cirrhosis. Scleroderma is primarily limited on CREST syndrome. Affected individuals exhibit calcinosis cutis, Raynaud's phenomenon, esophageal dysmotility and telangiectasis. Symptoms of scleroderma are combined with primary biliary cirrhosis in 3-17% of all patients. Reynolds syndrome can be caused by autosomal dominant mutations in the *LBR* gene which encodes the lamin B receptor. This protein anchors the lamina and the heterochromatin to the membrane.

Hepatic venoocclusive disease with immunodeficiency: A disorder characterized by the association of hepatic venoocclusive disease, combined T and B cell immunodeficiency with absent lymph node germinal centers, absent tissue plasma cells and severe hypogammaglobulinemia. Hepatic venoocclusive disease with immunodeficiency is caused by autosomal recessive mutations in the *SP110* gene which encodes the SP110 nuclear body protein. The encoded protein may have a role in the regulation of gene transcription. In Lebanese individuals, the carrier frequency for Hepatic venoocclusive disease with immunodeficiency trait is estimated to be 1 in 25 individuals, which means the frequency of this condition is about 1 in 2,500 individuals or $1/(50)^2$. This condition has also been reported among Australians. The mortality rate of this condition is around 85% if the syndrome remains unrecognized and untreated.

3.11.5 Disorders of Gallbladder, Biliary Tract and Pancreas

Congenital defects of bile acid synthesis: A disorder of neonatal onset characterized by a primary failure to synthesize bile acids. These children exhibit progressive cholestatic liver disease and malabsorption of lipids and lipid-soluble vitamins from the gastrointestinal tract. Affected infants show failure to thrive and secondary coagulopathy. Affected adults show mainly neurological disease. There are several underlying causes of this disorder with at least 4 autosomal genomic loci: Congeni-

tal defects of bile acid synthesis-1 can be caused by autosomal recessive mutations in the *HSD3B7* gene which encodes the 3-beta-hydroxy-delta-5-C27-steroid oxidoreductase;; Congenital defects of bile acid synthesis-2 is caused by autosomal recessive mutations in the *AKR1D1* gene which encodes the delta(4)-3-oxosteroid 5-beta-reductase;; Congenital defects of bile acid synthesis-3 is caused by autosomal recessive mutations in the *CYP7B1* gene which encodes the 7-alpha hydroxylase;; and Congenital defects of bile acid synthesis-4 is caused by autosomal recessive mutations in the *AMACR* gene which encodes the alpha-methylacyl-CoA racemase. "Congenital defects of bile acid synthesis" is estimated to affect 1-9 in every 1,000,000 individuals.

Susceptibility to gallbladder disease: A disorder that has been reported more commonly among Females, in their Forties, that are Fertile and Fat or the 4 F's of gallbladder disease. Affected individuals exhibit nausea, vomiting, or pain in the abdomen, back or just under the right arm. Others conditions associated with gallbladder disease include hypothyroidism, rapid weight loss, alcohol intake, diabetes and chronic inflammatory bowel disease. Genomics factors associated with this condition include ethnicity. For example, in the United States, this condition is more common among Pima Indians and Mexican-Americans than any other ethnic groups. There are also native people in Chile and Peru that are especially prone to developing gallstones. Family history of gallbladder disease is also associated with susceptibility to gallbladder disease. There are several underlying causes of this disorder with at least 4 autosomal genomic loci: Low phospholipid-associated cholelithiasis or Gallbladder disease-1 is caused by mutations in the *ABCB4* gene which encodes the ATP-binding cassette (subfamily B member 4). Low phospholipid-associated cholelithiasis is a disorder characterized by symptomatic and recurring cholelithiasis, due to low biliary phospholipid concentration. In affected individuals, bile has a low phospholipid content that increased lithogenicity resulting in cholestasis;; Gallbladder disease-2 and 3 have been mapped, but the genes have not yet been characterized;; and Susceptibility to gallbladder disease-4 can be influenced by variations in the *ABCG8* gene which encodes the ATP-binding cassette (sub-family G member 8). This protein promotes excretion of cholesterol and sterols into bile, and also facilitates transport of sterols back into the intestinal lumen. These variations involve a gain-of-function mechanism that increases efficiency of cholesterol transport into the bile lumen. This causes cholesterol hypersaturation of bile and promotes the formation of gallstones. Among Germans, the risk of gallbladder disease is 2.2 times higher than average for monoallelic carriers and 7.1 times higher than average for biallelic carriers of *ABCG8* H19 allele. In people of German ancestry, the carrier frequency for heterozygous *ABCG8* H19 trait is estimated to be 1 in 9 individuals, which means the frequency of homozygous is about 1 in 320 individuals or $1/(18)^2$.

Intrahepatic cholestasis of pregnancy: A cholestatic disorder characterized by pruritus. Affected individuals manifest symptoms during the third trimester of pregnancy. Affected individuals exhibit elevated serum aminotransferases and bile acid levels and spontaneous relief after delivery. Intrahepatic cholestasis of pregnancy is associated with fetal complications, including placental insufficiency, premature labor, fetal distress and intrauterine death. This condition recurs in 45-70% of subsequent pregnancies. In North America and in most regions of Central and Western Europe, intrahepatic cholestasis of pregnancy affects 0.4-1% of pregnancies. There is a higher prevalence of intrahepatic cholestasis of pregnancy in individuals of Chile and Bolivia, affecting around 5-15% of pregnancies. Among Scandinavia and the Baltic states there is an intermediate prevalence; around 1-2% of pregnancies are associated with this condition. There are hormonal, environmental and genomic factors which may contribute to this condition. There are several underlying causes of this disorder with at least 2 autosomal genomic loci: Intrahepatic cholestasis of pregnancy-1 can be caused by autosomal dominant mutations in the *ATP8B1* gene which encodes the ATPase (class I type 8B). Mutation in the same gene can also cause progressive familial intrahepatic cholestasis-1 and benign recurrent intrahepatic cholestasis-1;; and Intrahepatic cholestasis of pregnancy-3 can be caused by mutations in the *ABCB4* gene which encodes the ATP-binding cassette (sub-family B member 4).

Progressive familial intrahepatic cholestasis: A severe hepatocellular neonatal disorder which affects bile formation. Affected neonates exhibit signs of cholestasis with recurrent or permanent jaundice associated with hepatomegaly and severe pruritus. There are several underlying causes of this disorder with at least 3 autosomal genomic loci: Progressive familial intrahepatic cholestasis-1 is caused by autosomal recessive mutations in the *ATP8B1* gene which encodes the ATPase (class I type 8B);; Progressive familial intrahepatic cholestasis-2 or Byler syndrome is caused by autosomal recessive mutations in the *ABCB11* gene which encodes the ATP-binding cassette (sub-family B member 11), a liver-specific transporter;; and Progressive familial intrahepatic cholestasis-3 is caused by autosomal recessive mutations in the *ABCB4* gene which encodes the ATP-binding cassette (subfamily B member 4), a class III multidrug resistance P-glycoprotein. Progressive familial intrahepatic cholestasis-1 and 2 are associated with mildly elevated or normal serum levels of gamma-glutamyltransferase, whereas Progressive familial intrahepatic cholestasis-3 is associated with high serum gamma-glutamyltransferase levels and liver histology that shows portal inflammation and ductular proliferation in an early stage. Progressive familial intrahepatic cholestasis types are estimated to affect 1 in every 50,000-100,000 live births. Type 2 represents half of all the progressive familial intrahepatic cholestasis cases.

Benign recurrent intrahepatic cholestasis: A disorder characterized by intermittent periods of cholestasis without extrahepatic bile duct obstruction. Affected individuals start exhibiting elevation of serum bile acids, followed by cholestatic jaundice that last weeks to months. There are several underlying causes of this disorder with at least 2 autosomal genomic loci: Benign recurrent intrahepatic cholestasis-1 is caused by either autosomal dominant or recessive mutations in the *ATP8B1* gene;; and Benign recurrent intrahepatic cholestasis-2 is caused by mutations in the *ABCB11* gene. A founder mutation in the *ATP8B1* gene has been reported in the Faroe Islands.

Tropical calcific pancreatitis: A disorder with similar clinical features to those of alcoholic pancreatitis, except that tropical calcific pancreatitis appears at a younger age. Patients with this condition also displays pancreatic calcification which remarkably increases the incidence of pancreatic cancer and Insulin-dependent but ketosis-resistant diabetes mellitus. This condition is more prevalent among tropical developing countries. Tropical calcific pancreatitis can be caused by mutations in the *SPINK1* gene which encodes the pancreatic secretory trypsin inhibitor.

Hereditary chronic pancreatitis: A disorder appearing during childhood with similar clinical features to those of alcoholic chronic pancreatitis. Hereditary chronic pancreatitis is milder, and has an earlier onset and a slower progression than alcoholic chronic pancreatitis. Affected individuals exhibit chronic abdominal pain, nausea, vomiting, maldigestion, impairment of endocrine and exocrine function of the pancreas, diabetes, pseudocysts, bile duct and duodenal obstruction, and pancreatic cancer. Hereditary chronic pancreatitis can be caused by mutations in the *PRSS1* gene which encodes the serine protease 1 (trypsinogen). The encoded protein is a member of the trypsin family of serine proteases. This mutated gene plays a role due to an increase autocatalytic conversion of trypsinogen to active trypsin. Other genes mutations and/or polymorphisms that cause Hereditary chronic pancreatitis include: the *PRSS2* gene which encodes the anionic trypsinogen; the *SPINK1* gene; the *CFTR* gene which encodes the cystic fibrosis transmembrane conductance regulator; and mutations in the *CTRC* gene which encodes the chymotrypsin C.

Table 3.11.5.1 Recognizable Disorders of Gallbladder, Biliary Tract and Pancreas

Disorder	OMIM	Gene	Biological Function	Product of Conception Alteration
Congenital defects of bile acid synthesis-1	607765	HSD3B7	Endoplasmic reticulum protein	AR
Congenital defects of bile acid synthesis-2	235555	AKR1D1	Reductase	AR
Congenital defects of bile acid synthesis-3	613812	CYP7B1	Hydroxylase	AR
Congenital defects of bile acid synthesis-4	214950	AMACR	Racemase	AR
Gallbladder disease-1 or low phospholipid-associated cholelithiasis	600803	ABCB4	ABC-transporter protein	AR
Gallbladder disease-4	611465	ABCG8	ABC-transporter protein	
Intrahepatic cholestasis of pregnancy-1	243300	ATP8B1	ATPase protein	AD
Intrahepatic cholestasis of pregnancy-3	147480	ABCB4	ABC-transporter protein	AD
Progressive familial intrahepatic cholestasis-1	211600	ATP8B1	ATPase protein	AR
Progressive familial intrahepatic cholestasis-2	601847	ABCB11	ABC-transporter protein	AR
Progressive familial intrahepatic cholestasis-3	602347	ABCB4	ABC-transporter protein	AR
Benign recurrent intrahepatic cholestasis-1	243300	ATP8B1	ATPase protein	AD
Benign recurrent intrahepatic cholestasis-2	605479	ABCB11	ABC-transporter protein	AR
Tropical calcific pancreatitis	608189	SPINK1	Serine peptidase inhibitor	AD
Susceptibility to chronic Pancreatitis	601405	CTRC	Peptidase	AD
Hereditary Pancreatitis	276000	PRSS1	Serine protease	AD
Protection against hereditary pancreatitis	601564	PRSS2	Serine protease	AD
Hereditary pancreatitis	167800	SPINK1	Serine peptidase inhibitor	AD
Idiopathic pancreatitis	167800	CFTR	Channelopathy Chloride	AD

3.11.6 Other Diseases of the Digestive System

Celiac disease: A chronic immune-mediated enteropathy triggered by consumption gluten, a protein present in some types of grain. In predisposed individuals, the ingestion of gluten-containing food such as wheat rye and barley proteins induces a flat jejunal mucosa with infiltration of lymphocytes. Affected individuals exhibit maldigestion and malabsorption of most nutrients and vitamins. The main manifestations of this condition include: diarrhea, gas and bloating, stomach pain, weight loss, anemia, edema, and bone or joint pains. Celiac disease is the most common food intolerance with prevalence close to 1% in both Europe and North America. Celiac disease can be diagnosed either in early childhood at around 2 years of age, or as a latent form which may be diagnosed at around 40 years of age. Now it is considered to be more often diagnosed in adults than in children. Celiac disease is influenced by both environmental and genomic factors. Celiac disease is associated with either the HLA-DQ2 or the HLA-DQ8 haplotype. Nevertheless, 20% of all healthy people carry these haplotype. Patients usually have a normal life expectancy.

Primary bile acid malabsorption: An intestinal disorder associated with chronic watery diarrhea, excess fecal bile acids and steatorrhea. There are 3 forms of bile acid malabsorption: Primary bile acid malabsorption type 1 is due to ileal dysfunction resulting from Crohn disease or ileal resection;; Primary bile acid malabsorption type 2 can be caused by autosomal recessive mutations in the *SLC10A2* gene which encodes the sodium/bile acid co-transporter (family 10 member A2);; and Primary bile acid malabsorption type 3 is secondary to other conditions, including cholecystectomy, post-vagotomy, celiac disease and pancreatic insufficiency.

Diarrhea, failure to thrive and hypoproteinemic edema: A disorder caused by autosomal recessive mutations in the *TMPRSS15* gene which encodes the serine transmembrane protease 15, a proenterokinase.

Renal-hepatic-pancreatic dysplasia: A disorder characterized by enlarged polycystic kidneys with gross renal dysplasia, enlarged nodular, cystic pancreas, and portal fibrosis with bile duct proliferation. Sometimes patients display hypoplasia of the spleen, cardiac anomalies and other situs abnormalities. Renal-hepatic-pancreatic dysplasia can be caused by autosomal recessive mutations in the *NPHP3* gene which encodes the nephronophthisis 3. The encoded protein interacts with nephrocystin. This protein is required for normal ciliary development and it functions in renal tubular development. Mutations in the same gene result in nephronophthisis-3 and in the more severe disorder the Meckel syndrome-7.

3.12.0 Subchapter XII

Diseases of the Skin and Subcutaneous Tissue

Table 3.12.0 Diseases of the Skin and Subcutaneous Tissue

3.12.00-7	Diseases of the Skin and Subcutaneous Tissue	
	3.12.1	Pigmentation of Hair, Eye, and Skin
	3.12.2	Dermatitis and Eczema
	3.12.3	Papulosquamous Disorders
	3.12.4	Urticaria and Erythema
	3.12.5	Radiation-related Disorders of the Skin and Subcutaneous Tissue
	3.12.6	Disorders of Skin Appendages
	3.12.7	Other Disorders of the Skin and Subcutaneous Tissue

3.12.1 Pigmentation of Hair, Eye and Skin

Pigmentation of hair, eye and skin: A trait that is amongst the most evident examples of human phenotypic difference. The variation in pigmentation is thought to be caused by biochemical differences that affect the number of melanosomes produced, the size and shape of the melanosomes, and the type of melanin synthesized. Pigmentation of hair, eye and skin is attributable to the number, type, and cellular distribution of melanosomes. These are subcellular compartments produced by melanocytes that synthesize and store the light-absorbing polymer melanin. The biological function of skin pigmentation appears to be absorbed and protected from ultraviolet radiation. Individuals with lighter skin pigmentation originally inhabited locations distant from the equator. Most of the dark pigmented people originally lived within 20 degrees of the equator. The geographic distribution of human skin pigmentation reflects a history of adaptation to latitude-dependent levels of ultraviolet radiation. There is a preponderant variation in human eye and hair color among individuals of European ancestry. Most other non-European people fixed for brown eyes and black hair. There are at least variations in 11 autosomal gene/loci which influences skin, hair and/or eye pigmentation: Skin, hair, and/or eye pigmentation-1, or Blond/brown hair and blue/nonblue eyes, is influenced by variations in the *OCA2* gene which encodes the oculocutaneous albinism II protein, a small molecule that transports tyrosine - a precursor of melanin. There are also noncoding variants in the *HERC2* gene 200 kb downstream of *OCA2* gene that are thought to affect OCA2 expression;; Skin, hair and/or eye pigmentation-2 is associated by variations in the *MC1R* gene which encodes the melanocortin 1 receptor (alpha melanocyte stimulating hormone receptor). This 7 transmembrane G-protein-coupled receptor controls melanogenesis. This phenotype is principally characterized by red hair and fair skin;; Skin, hair and/or eye pigmentation-3 encompasses pigment variations influenced by the *TYR* gene which encodes the tyrosinase;; Skin, hair and/or eye pigmentation-4 includes pigment variations influenced by the *SLC24A5* gene which encodes the solute carrier (family 24 member A5);; Skin, hair and/or eye pigmentation-5 involves pigment variations influenced by the *SLC45A2* gene which encodes the solute carrier (family 45 member A2);; Skin, hair and/or eye pigmentation-6 encompasses pigment variations influenced by the *SLC24A4* gene which encodes the solute carrier (family 24 member A4), a sodium/potassium/calcium exchanger;; Skin, hair and/or eye pigmentation-7 includes pigment variations influenced by DNA sequence variation that is thought to affect expression of the *KITLG* gene which encodes the KIT ligand;; Skin, hair and/or eye pigmentation-8 has been associated with SNPs at chromosome 6p25.3;; Skin, hair and/or eye pigmentation-9 involves pigment variations influenced by polymorphisms in the 3-prime-untranslated region of the *ASIP* gene which encodes the agouti signaling protein. There are ASIP haplotype that has been associated with susceptibility to cutaneous malignant melanoma and basal cell carcinoma.;; Skin, hair and/or eye pigmentation-10 association comprises variations in the *TPCN2* gene which encodes the two pore segment channel 2;; and Skin, hair and/or eye pigmentation-11 is associated with polymorphisms near the *TYRP1* gene which encodes the tyrosinase-related protein 1.

3.12.2 Dermatitis and Eczema

Atopy or Atopic syndrome: A predisposition to develop certain allergic hypersensitivity reactions. Asthma, hay fever, and eczema are encompassed under the general term atopic hypersensitivity. Atopy is caused by both genomic and environmental factors. Atopy is a predisposition to an excessive IgE reaction. Studies in first-

degree family display the risk of developing atopic dermatitis or atopy. In general, the risk rises by a factor of 2 with each first-degree family member affected with this condition. There are several mutations in genes that are associated to Atopy including *SELP, SPINK5, HAVCR1, PLA2G7, MS4A2, PHF11, IL4R* and *IL21R* among others.

3.12.3 Papulosquamous Disorders

Psoriatic arthritis: A chronic inflammatory skin disorder characterized by a form of inflammatory arthritis. Psoriatic arthritis seems to start manifesting about 10 years after the first signs of psoriasis. The strongest genomic factors associated with this condition include a concordance rate of 50-70% among monozygotic twins. Psoriatic arthritis is more common among Caucasians than African or Asian Americans. Susceptibility to psoriatic arthritis is determined by multiple genes. There are strong associations between psoriasis/psoriatic arthritis and HLA-Cw*0602. There are strong associations between psoriatic arthritis and sequence variants of the *NOD2, TNFA*, and the *LTA* genes.

3.12.4 Urticaria and Erythema

Familial cold autoinflammatory syndrome: A disorder characterized by recurrent attacks of a maculopapular rash. Affected individuals display a rash that is associated with conjunctivitis, arthralgias, myalgias, fever and chills. Patients display swelling of the extremities which is elicited by exposure to cold. There are also overlapping and more severe syndromes including: Muckle-Wells syndrome and CINCA syndrome. Muckle-Wells syndrome is caused by mutations in the *NLRP3* gene which encodes the NLR family pyrin domain containing protein 3.

Other features of Muckle-Wells syndrome include a high frequency of amyloidosis and late-onset sensorineural deafness. Chronic neurologic cutaneous and articular syndrome or CINCA syndrome is another overlapping condition which shows earlier onset and a more severe phenotype. There are several underlying causes of this disorder with at least 3 autosomal genomic loci: Familial cold autoinflammatory syndrome-1 is caused by autosomal dominant mutations in the *NLRP3* gene which encodes the NLR family pyrin domain containing protein 3;; Familial cold autoinflammatory syndrome-2 is caused by mutations in the *NLRP12* gene which encodes the NLR family pyrin domain containing protein 12;; and Familial cold autoinflammatory syndrome-3 is caused by mutations in the *PLCG2* gene which encodes the gamma 2 phospholipase C (phosphatidylinositol-specific). In the United States and Europe, Familial cold autoinflammatory syndrome is estimated to affect 1 in every 1,000,000 individuals

CINCA syndrome or Chronic Infantile Neurological, Cutaneous, and Articular: This disease is discussed under diseases of the nervous system in the subchapter "Inflammatory Diseases of the Central Nervous System" (3.06.01).

Muckle-Wells syndrome: This disease is discussed under endocrine, nutritional and metabolic diseases in the subchapter "Amyloidosis" (3.04.18.1).

3.12.5 Radiation-related Disorders of the Skin and Subcutaneous Tissue

Disseminated superficial actinic porokeratosis: A disorder characterized by skin lesions in sun-exposed areas. This condition usually appears during the 2nd decade of life. There are several underlying causes of this disorder with at least 4 autosomal genomic loci: Disseminated superficial actinic porokeratosis-1 is caused by autosomal recessive mutations in the *SART3* gene which encodes the squamous cell carcinoma antigen (recognized by T cells 3). Disseminated superficial actinic porokeratosis-2, 3 and 4 have been mapped, but the genes have not yet been identified.

3.12.6 Disorders of Skin Appendages

3.12.6.1 Nail Disorders

Nonsyndromic congenital nail disorder: A group of nail disorders unrelated to iron deficiency or trauma. These conditions seem to follow an autosomal dominant pattern of inheritance. There are several underlying causes of this disorder with at least 10 clinical nonsyndromic congenital nail disorder types; in 4 cases the gene has been identified, and in two additional cases the loci have been mapped. Nonsyndromic congenital nail disorder-1 or Trachyonychia is a form of nail dystrophy characterized by excessive longitudinal striations and numerous superficial pits on the nails, which have a distinctive rough sandpaper-like appearance;; Nonsyndromic congenital nail disorder-2 or Koilonychia;; Nonsyndromic congenital nail disorder-3 or Leukonychia is caused by mutations in the *PLCD1* gene which encodes the phospholipase C delta 1;; Nonsyndromic congenital nail disorder-4 or Anonychia/hyponychia is caused by mutations in the *RSPO4* gene which encodes the R-spondin 4. This protein is involved in activation of Wnt/beta-catenin signaling pathways;; Nonsyndromic congenital nail disorder-5 with partial onycholysis with scleronychia;; Nonsyndromic congenital nail disorder-6 with anonychia of thumbs with onychodystrophy;; Nonsyndromic congenital nail disorder-7 or Onychodystrophy has been mapped, but the gene has not been characterized;; Nonsyndromic congenital nail disorder-8 or Toenail dystrophy is caused by mutations in the *COL7A1* gene which encodes the alpha 1 type VII collagen;; Nonsyndromic congenital nail disorder-9 or Onychodystrophy has been mapped, but the gene has not been characterized;; and Nonsyndromic congenital nail disorder-10 or Onychodystrophy is caused by mutations in the *FZD6* gene which encodes the frizzled family receptor 6.

3.12.6.2 Alopecia

Alopecia universalis congenita: A severe hair follicle disorder characterized by rapid loss of all the hair-bearing areas of the body. Alopecia universalis congenital is caused by biallelic mutations in the *HR* gene which encodes the hair growth associated protein. This form of alopecia universalis has been described among members of a consanguineous Pakistani family. Mutations in the *HR* gene also caused Atrichia with papular lesions.

Hypotrichosis simplex: A disorder characterized by progressive hair follicle miniaturization. Affected individuals usually show normal hair at birth, but hair loss is diffused and progressive, and frequently begins during early childhood and progresses with age. This condition can be divided into 2 forms: the scalp-limited form and the generalized form, in which all body hair is affected. Localized autosomal recessive hypotrichosis is characterized by fragile hairs that break easily, leaving short and sparse scalp hairs. The localized form affects the trunk, extremities, scalp, and the eyebrows and eyelash-

es, whereas the beard, pubic and axillary hairs are largely spared. There are several underlying causes of this disorder with at least 10 autosomal genomic loci: Hypotrichosis simplex-1 can be caused by monoallelic mutations in the *APCDD1* gene which encodes the adenomatosis polyposis coli down-regulated protein 1;; Hypotrichosis simplex-2 and Hypotrichosis simplex of the scalp-1 can be caused by monoallelic mutations in the *CDSN* gene which encodes the corneodesmosin;; Hypotrichosis simplex-3 and Hypotrichosis simplex of the scalp-2 can be caused by monoallelic mutations in the *KRT74* gene which encodes the keratin74, an intermediate filament protein;; Hypotrichosis simplex-4 can be caused by monoallelic mutations in the *HR* gene which encodes the hair growth associated protein;; Hypotrichosis simplex-5 has been mapped but the gene has not been characterized;; Hypotrichosis simplex-6 or autosomal recessive localized hypotrichosis-1 can be caused by biallelic mutations in the *DSG4* gene which encodes the desmoglein 4. This protein plays a role in cell-cell adhesion in epithelial cells;; Hypotrichosis simplex-7 or Woolly hair with or without hypotrichosis-2 or Autosomal recessive localized hypotrichosis-2 can be caused by biallelic mutations in the *LIPH* gene which encodes the lipase H. This membrane-bound enzyme catalyzes the production of 2-acyl lysophosphatidic acid;; Hypotrichosis simplex-8 or Woolly hair with or without hypotrichosis-1 or Autosomal recessive localized hypotrichosis-3 can be caused by biallelic mutations in the *LPAR6* gene which encodes the lysophosphatidic acid receptor 6;; Hypotrichosis simplex-9 has been mapped but the gene has not been characterized;; and Hypotrichosis simplex-10 has been mapped but the gene has not been characterized.

3.12.6.3 Follicular Cysts of Skin and Subcutaneous Tissue

Steatocystoma multiplex: A disorder characterized by hundreds to two thousand asymptomatic dermal cysts. These cysts start to appear during the 1st or 2nd decade of life. These dermal cysts appear on the sternal region, upper back, axilla and proximal parts of the extremities. Steatocystoma multiplex is caused by either inherited or *de novo* monoallelic mutations in the *KRT17* gene which encodes the keratin17, an intermediate filament protein.

3.12.7 Other Disorders of the Skin and Subcutaneous Tissue

Vitiligo-associated multiple autoimmune: An autoimmune disorder characterized by melanocyte loss. This condition results in patchy depigmentation of skin and hair. This condition is associated with a higher risk of other autoimmune diseases. There are several underlying causes of this disorder with at least 6 autosomal genomic loci: Vitiligo-associated multiple autoimmune disease susceptibility-1 is caused by variants in the *NLRP1* gene which encodes the NLR family, pyrin domain containing protein 1;; Vitiligo-associated multiple autoimmune disease susceptibility-2 is caused by mutations in the *FOXD3* gene which encodes the forkhead box D3;; and Susceptibility to vitiligo-associated multiple autoimmune disease can be cause by 4 additional loci, but the genes have not yet been characterized.

3.12.7.1 Other Disorders of Pigmentation

Familial progressive hyperpigmentation: A disorder characterized by hyperpigmented patches in the skin. This condi-

tion appears during early infancy and increase in size and number with age. There are at least 2 autosomal dominant forms of this condition. Familial progressive hyperpigmentation-1 has been mapped, but the gene has not been characterized;; and Familial progressive hyperpigmentation-2 is caused by autosomal dominant gain-of-function mutations in the *KITLG* gene which encodes the KIT ligand.

Dyschromatosis symmetrica hereditaria or Acropigmentation of Dohi: A genodermatosis characterized by the presence of hyperpigmented and hypopigmented macules on the face and dorsal aspects of the extremities. This condition appears during infancy or early childhood. Dyschromatosis symmetrica hereditaria is caused by autosomal dominant mutations in the *ADAR* gene which encodes the double-stranded RNA-specific adenosine deaminase. Mutations in the same gene cause Aicardi-Goutieres syndrome 6. Dyschromatosis symmetrica hereditaria is estimated to affect less than 1 in every 1,000,000 individuals.

Dowling-Degos disease or Reticulate acropigmentation of Kitamura: A disorder characterized by postpubertal reticulate pattern of hyperpigmentation that is progressive and disfiguring, particularly in the flexures and great skin folds. Other features include hyperkeratotic dark brown papules, pitted perioral acneiform scars, and genital and perianal reticulated pigmented lesions. These areas of hyperpigmentation do not darken with exposure to sunlight and cause no health problems. Dowling-Degos disease is caused by monoallelic loss-of-function mutations in the *KRT5* gene which encodes the keratin 5, an intermediate filament protein.

Griscelli syndrome: A disorder characterized by pigmentary dilution of the skin and hair. Affected individuals exhibit large clumps of pigment in hair shafts, and an accumulation of melanosomes in melanocytes. This condition is also associated with immunologic abnormalities, hemophagocytic syndrome and severe neurologic impairment. There are several underlying causes of this disorder with at least 3 autosomal genomic loci: Griscelli syndrome type 1 with primary neurologic impairment and without immunologic impairment is also refered to as Elejalde syndrome. Type 1 is caused by biallelic mutations in the *MYO5A* gene which encodes the myosin VA;; Griscelli syndrome type 2 or "Griscelli syndrome with immune impairment" is caused by biallelic mutations in the *RAB27A* gene which encodes the RAB27A member (RAS oncogene family);; and Griscelli syndrome type 3 is characterized by hypomelanosis with no immunologic or neurologic manifestations. Type 3 can be caused by biallelic mutations in the *MLPH* gene which encodes the melanophilin.

Piebaldism: A melanocyte development disorder characterized by congenital absence of melanocytes in affected areas of the skin and hair. Affected children exhibit a white forelock of triangular shape, often with unpigmented patches on the forehead, and scattered normal pigmented and hypopigmented macules. Affected individuals exhibit absence of melanocytes in eyebrows and eyelashes. Patients display irregularly shaped white patches on the face, trunk and limbs, typically in a symmetrical distribution. Piebaldism can be caused by autosomal dominant mutations either in the *KIT* gene which encodes the v-kit Hardy-Zuckerman 4 feline sarcoma viral oncogene homolog, or in the *SNAI2* gene which encodes the snail family zinc finger 2, a transcription factor.

ABCD syndrome (Albinism, Black lock, Cell migration disorder of the neurocytes of the gut and sensorineural Deafness): A disorder characterized by albinism, absence of black lock and retinal depigmentation, severe Hirschsprung disease and sensorineural deafness. ABCD syndrome is caused by biallelic mutations in the *EDNRB* gene which encodes the Endothelin B receptor. Mutation in the same gene causes Waardenburg-Shah syndrome or type 4A Waardenburg syndrome. ABCD syndrome has been reported among children of consanguineous Kurdish families.

3.12.7.2 Pyoderma Gangrenosum

PAPA syndrome (Pyogenic sterile Arthritis, Pyoderma gangrenosum and Acne): An autoinflammatory disorder appearing during childhood, predominantly affecting the joints and skin. Affected individuals exhibit intermittent and migratory arthritis with characteristic accumulation of sterile pyogenic material within the joint space if left untreated. Recurrent sterile arthritis frequently appears after minor trauma, but can also appear spontaneously. PAPA syndrome is caused by autosomal dominant mutations in the *PSTPIP1* gene which encodes the PSTPIP1 protein. This protein binds pyrin/marenostrin, the protein encoded by the *MEFV* gene. Familial Mediterranean fever is caused by mutations in the *MEFV* gene. PAPA syndrome is estimated to affect less than 1 in every 1,000,000 individuals.

3.12.7.3 Lupus Erythematosus

Systemic lupus erythematosus: A chronic autoinflammatory multisystemic disorder of connective tissue. This condition especially affects females in child-bearing years, with symptoms starting to appear between ages 15-35. Systemic lupus erythematosus principally affects the skin, joints, kidneys and serosal membranes. Systemic lupus erythematosus is a type III hypersensitivity reaction caused by antibody-immune complex formation. There are both environmental triggers and genomic susceptibilities associated with this condition. The most important inclusion criteria to consider Systemic lupus erythematosus of genomic origin include: Twin studies show that the correlation for Systemic lupus erythematosus in monozygous twins is 24%, whereas for dizygous twins it is 2%. The familial risk in first-degree family members of affected individuals is 100 times higher than average people. Systemic lupus erythematosus is more prevalent in certain ethnic groups. For example, African-American ancestries display a higher prevalence than Caucasians. In the United States, Systemic lupus erythematosus is estimated to affect 53 in every 100,000 individuals. The highest prevalence has been described among those of Afro-Caribbean ancestry, affecting 159 in 100,000 individuals. Multiple genes are involved in the causation of systemic lupus erythematosus, including *PTPN22, FCGR2B, TREX1, BANK1, C4A*, HLA-DR2, HLA-DR3 and *DNASE1* genes. Polymorphisms in the HLA loci explain that this condition is 3-4 times more frequent in African-Americans than in Caucasians. An autosomal recessive form of systemic lupus erythematosus is caused by biallelic mutations in the *DNASE1L3* gene which encodes the deoxyribonuclease I-like 3. The most important exclusion criteria to exclude the genomic origins of Systemic lupus erythematosus include: this condition is 9 times more prevalent in women than in men, although most of the causative gene have been mapped on autosomal chromosome, and the fact that in the United States, the incidence of this condition has increased several-fold in the last 50 years.

Childhood systemic lupus erythematosus: A condition starts to manifest between the ages of 3 and 15. This form is 4 times more common in females than in males.

Chilblain lupus: A chronic cutaneous form of systemic lupus erythematosus. Affected individuals exhibit painful bluish-red papular or nodular lesions of the skin in acral locations. There are several underlying causes of this disorder with at least 2 autosomal genomic loci: Chilblain lupus-1 is caused by monoallelic mutations in the *TREX1* gene which encodes the three prime repair exonuclease 1;; and Chilblain lupus-2 is caused by monoallelic mutations in the *SAMHD1* gene which encodes the SAM domain and HD domain 1. Mutations in the *TREX1* and *SAMHD1* genes also cause Aicardi-Goutieres syndrome type 1 and 5, respectively.

3.13.0 Subchapter XIII

Diseases of the Musculoskeletal System and Connective Tissue

Table 3.13.0 Diseases of the Musculoskeletal System and Connective Tissue

3.13.00-9		Diseases of the Musculoskeletal System and Connective Tissue
	3.13.1	Inflammatory Polyarthropathies
	3.13.2	Arthrosis
	3.13.3	Other Joint Disorders
	3.13.4	Systemic Connective Tissue Disorders
	3.13.5	Spondylopathies
	3.13.6	Other Dorsopathies
	3.13.7	Disorders of Muscles
	3.13.8	Soft Tissue Disorders
	3.13.9	Osteopathies and Chondropathies

Table 3.13.01 Diseases of the Musculoskeletal System and Connective Tissue in Different Ethnic Groups

Disease	Frequency of condition	People with higher prevalence (if available)
Paget disease of bone	>1 in 1,000	Unknown
Lysyl hydroxylase 3 deficiency	1 in 100,000	Unknown
Generalized lipodystrophy	1-9 in 1,000,000	Unknown
Familial thrombotic thrombocytopenic purpura	2-11 in 1,000,000	Unknown
Chondrocalcinosis	<1 in 1,000,000	Unknown
Hyperphosphatemic tumoral calcinosis	<1 in 1,000,000	Druze people and African Americans
Normophosphatemic familial tumoral calcinosis	<1 in 1,000,000	Yemenite Jews
Wrinkly skin syndrome and Cutis laxa	<1 in 1,000,000	Unknown
Fibrodysplasia ossificans progressiva	<1 in 1,000,000	Unknown
Hyaline fibromatosis	<1 in 1,000,000	Unknown
Sclerosteosis, hyperostosis corticalis generalisata or Van Buchem disease	<1 in 1,000,000	Afrikaners
Juvenile Paget disease	<1 in 1,000,000	Unknown

3.13.1 Inflammatory Polyarthropathies

3.13.1.1 Seropositive Rheumatoid Arthritis

Rheumatoid arthritis: A chronic inflammatory disorder that may affect many tissues and organs, but primarily the joints. Rheumatoid arthritis affects around 1% of people, most frequently appearing during the 4th-5th decades of life. There are both environmental triggers and genomic susceptibilities associated with this condition. The most important inclusion criteria to consider Rheumatoid arthritis of genomic origin include: Twin studies show that the correlation of this condition in monozygous twin is 12-15%, while in dizygous twins it is 4%. The familial risk in first-degree family members of affected individuals is 1.6 times higher than in average people. Rheumatoid arthritis is more prevalent among certain ethnic groups, particularly Native Americans. Multiple genes and haplotypes are involved in the causation of Rheumatoid arthritis, including HLA-DRB1 (most specifically DR0401 and 0404), *SLC22A4, PTPN8, MHC2TA, IRF5* and *NFKBIL1* genes. The most important exclusion criteria to exclude the genomic origins of Rheumatoid arthritis include: this condition is 3 times more prevalent in women than in men, while most of the genes have been mapped on autosomal chromosome. Rheumatoid arthritis is 3 times more common in smokers than in non-smokers.

3.13.1.2 Crystal Arthropathies

Chondrocalcinosis: A chronic articular disorder characterized by intermittent attacks of arthritis. Affected individuals display calcium pyrophosphate crystals in synovial fluid, cartilage and periarticular soft tissue. There are at least 2 familial forms of chondrocalcinosis associated with autosomal dominant inheritance. Chondrocalcinosis-1 with early-onset osteoarthritis has been mapped, but the gene has not been characterized; and Chondrocalcinosis-2 is caused by monoallelic mutations in the *ANKH* gene which encodes the progressive ankylosis protein homolog. There are also chondrocalcinosis associated with metabolic disorders such as Hemochromatosis, Hyperparathyroidism, Hypothyroidism and Wilson disease.

Hyperphosphatemic familial tumoral calcinosis: A metabolic disorder characterized by increased renal absorption of phosphate. Affected individuals suffer from a progressive deposition of basic calcium phosphate crystals in periarticular spaces, soft tissues and sometimes bones. There are several underlying causes of this disorder with at least 3 autosomal genomic loci: Hyperphosphatemic familial tumoral calcinosis can be caused by autosomal recessive loss-of-function mutations in the *GALNT3* gene which encodes the UDP-N-acetyl-alpha-D-galactosamine:polypeptide N-acetylgalacto-saminyltransferase 3 (GalNAc-T3), or *FGF23* gene which encodes the fibroblast growth factor 23, or the *KL* gene which encodes the klotho, a type-I membrane protein that is related to beta-glucosidases. Most cases are caused by mutations in the *GALNT3* gene. In Druze people this pathogenic allele appears 88-200 years ago. The prevalence of this condition is unknown, but it seems more prevalent among individuals from the Middle East and Africa.

Normophosphatemic familial tumoral calcinosis: A disorder appearing during the 1st year of life characterized by reddish-to-hyperpigmented skin lesions. Years later patients develop calcified nodules, distributed primarily over the extremities. Affected individuals manifest other features including severe conjunctivitis and gingivitis. Normophosphatemic familial tumoral calcinosis can be caused by autosomal recessive mutations in the *SAMD9* gene which encodes the sterile alpha motif domain-containing protein 9. This condition has been reported among Yemenite Jews.

3.13.2 Arthrosis

Osteoarthritis: A degenerative disorder of the joints characterized by degradation of the hyaline articular cartilage and remodeling of the subchondral bone with sclerosis. Affected individuals exhibit pain, and joint stiffness that frequently causes an significant disability and requires joint replacement. Osteoarthritis is the most common form of arthritis, affecting around 9% of the people in the United States. There are several syndromes associated with early onset osteoarthritis, including Marfan syndrome, Alkaptonuria, Ehlers-Danlos Syndrome, Hemochromatosis and Wilson disease. There are several underlying causes of this disorder with at least 6 autosomal genomic loci: Susceptibility to osteoarthritis-1 is associated with variations in the *FRZB* gene which encodes the frizzled-related protein;; Susceptibility to osteoarthritis-2 is associated with variations in the *MATN3* gene which encodes the matrilin 3. This protein is involved in the formation of filamentous networks in the extracellular matrices of various tissues;; Susceptibility to osteoarthritis-3 is associated with variations in the *ASPN* gene which encodes the asporin;; Susceptibility to osteoarthritis-5 is associated with variations in the *GDF5* gene which encodes the growth differentiation factor 5;; and Susceptibility to osteoarthri-

tis-4 and 6 have been mapped, but the genes have not been characterized.

3.13.3 Other Joint Disorders

Trismus-pseudocamptodactyly syndrome: A disorder characterized by inability to open the mouth entirely leading to problems in mastication. Affected individuals exhibit other features including short finger-flexor tendons resulting in camptodactyly and in foot deformity. Trismus-pseudocampto-dactyly syndrome can be caused by autosomal dominant mutations in the *MYH8* gene which encodes the myosin heavy chain 8. This condition has been reported among members of large Dutch kindred.

3.13.4 Systemic Connective Tissue Disorders

Lysyl hydroxylase 3 deficiency: A connective tissue disorder characterized by craniofacial, skeletal, cutaneous and vascular features. Lysyl hydroxylase 3 deficiency is caused by autosomal recessive mutations in the *PLOD3* gene which encodes the 2-oxoglutarate 5-dioxygenase 3 procollagen-lysine. This rough endoplasmic reticulum enzyme (cofactors iron and ascorbate) catalyzes the hydroxylation of lysyl residues in collagen-like peptides.

Familial thrombotic thrombocytopenic purpura or Schulman-Upshaw syndrome: A congenital hemolytic anemia characterized by fragmentation of erythrocytes, thrombocytopenia, fever, diffuse and nonfocal neurologic features, and decreased renal function. Familial thrombotic thrombocytopenic purpura is caused by autosomal recessive mutations in the *ADAMTS13* gene which encodes the von Willebrand factor-cleaving protease.

Systemic lupus erythematosus: This disease is discussed under diseases of the skin and subcutaneous tissue in the subchapter "Lupus Erythematosus" (3.12.7.3).

Syndromic multisystem autoimmune disease: A disorder characterized by infiltration of autoimmune inflammatory cell in the lungs, liver and intestine. Affected individuals also exhibit failure to thrive, organomegaly, developmental delay and dysmorphic features. Affected individuals have a larger prevalence of other autoimmune disorders including hypothyroidism, autoimmune hepatitis and type 1 diabetes mellitus. Syndromic multisystem autoimmune disease is caused by autosomal recessive mutations in the *ITCH* gene which encodes the itchy E3 ubiquitin protein ligase. This condition has been reported among members of a consanguineous Old Order Amish family.

3.13.5 Spondylopathies

Spondyloarthropathy: A joint disorder characterized by vertebral column inflammation. The most important inclusion criteria to consider Ankylosing spondylitis of genomic origin include: Twin studies show that the correlation of this condition in monozygous twin is 75% whereas in dizygous twins, it is 12.5%. There are several underlying causes of this disorder with at least 3 autosomal genomic loci: Susceptibility to ankylosing spondylitis can be conferred by variations in the HLA-B27 allele. There are two additional susceptibility loci associated with spondyloarthropathy, but the genes have not yet been characterized. The prevalence of ankylosing spondylitis among Europeans is between 0.1-1.4%.

3.13.6 Other Dorsopathies

Spinal disc herniation: A disorder affecting the spine, usually due to trauma or lifting injuries. Affected individuals exhibit lower back pain and frequently leg pain as well. There are also genomics factors associated with Spinal disc herniation including mutations in genes coding for proteins involved in the regulation of the extracellular matrix, such as the *MMP2* gene and the *THBS2* gene. In 90% of cases the area affected is the lumbar spine. This is a common condition affecting about 5% of all individuals. Males have a slightly higher incidence than females.

Degeneration of the intervertebral disc: A disorder characterized by severe constant chronic pain. Multiple genes are involved in the causation of degeneration of the intervertebral disc, including *COL9A2*, *COL11A1*, *THBS2*, *ASPN*, *CILP* and the *COL9A3*.

Ossification of the posterior longitudinal ligament of the spine: A disorder that can cause spinal-cord compression leading to tetraparesis. There are several underlying causes of this disorder with at least 3 autosomal genomic loci: Susceptibility to abnormal ossification of the posterior longitudinal ligament of the spine can be caused by single-nucleotide polymorphisms in the *COL6A1* and *ENPP1* genes. *ENPP1* gene encodes nucleotide pyrophosphatase which produce inorganic pyrophosphate, is a major inhibitor of calcification and mineralization. There is also a male-specific association to abnormal ossification due to *COL11A2* haplotype. Ossification of the posterior longitudinal ligament of the spine is a common disorder among East Asians affecting between 1.9-4.3% of these individuals.

3.13.7 Disorders of Muscles

Fibrodysplasia ossificans progressiva: A severely disabling connective tissue disorder characterized by progressive ossification of skeletal muscle, fascia, tendons and ligaments. Affected individuals exhibit other features including short thumbs, fifth finger clinodactyly, malformed cervical vertebrae, short broad femoral necks, deafness, scalp baldness and mild mental retardation. Children with this condition appear normal at birth excluding the congenital malformations of the great toes. During the 1st decade of life, patients suffer from soft tissue swellings leading to heterotopic bone, making movement impossible. Most patients are wheelchair-bound by the end of the 2nd decade of life. The median lifespan is about 40 years of age. Fibrodysplasia ossificans progressiva is either inherited or caused by *de novo* monoallelic mutations in the *ACVR1* gene which encodes the Activin A receptor type 1. Fibrodysplasia ossificans progressiva is estimated to affect 1 in every 2,000,000 worldwide.

Progressive osseous heteroplasia: A disorder appearing during infancy characterized by dermal ossification, followed by increasing and extensive bone formation in deep muscle and fascia. Patients with this condition have a collection of physical findings referred to as Albright hereditary osteodystrophy. Progressive osseous heteroplasia is caused by monoallelic paternal mutations in the GNAS gene which encodes the GNAS complex locus. This locus exhibits a highly complex imprinted expression pattern. This mutation occurs on the paternal allele resulting in loss-of-function of the Gs-alpha isoform of the GNAS.

Muscle weakness associated with depletion of mtDNA in skeletal muscle or Mi- tochondrial DNA depletion syndrome-2: A disorder appearing during childhood characterized mainly by muscle weakness associated with depletion of mtDNA in skeletal muscle. Mitochondrial DNA depletion syndrome-2 is caused by autosomal recessive mutations in the *TK2* gene which encodes the nuclear mitochondrial thymidine kinase.

3.13.8 Soft Tissue Disorders

Hyaline fibromatosis syndrome: A fibromatosis disorder apparent either at birth or within the first six months of life. Affected individuals manifest severe chronic pain, progressive joint contractures, skin abnormalities, and extensive deposition of hyaline material in many tissues including: skeletal muscle, cardiac muscle, spleen, thyroid, adrenal glands, skin, gastrointestinal tract and lymph nodes. Patients with this condition exhibit other features including intractable diarrhea and increased susceptibility to infection. Hyaline fibromatosis syndrome is caused by autosomal recessive mutations in the *ANTXR2* gene which encodes the anthrax toxin receptor 2. This protein may be involved in extracellular matrix adhesion. Hyaline fibromatosis syndrome is estimated to affect less than 1 in every 1,000,000 individuals.

3.13.9 Osteopathies and Chondropathies

Osteoporosis: A disorder of bones that leads to a higher risk of fracture. This disorder may be classified in 3 groups; primary type 1, primary type 2, or secondary. Primary type 1 osteoporosis or Postmenopausal osteoporosis is more common in women after menopause. Primary type 2 osteoporosis or senile osteoporosis affects patients after 75 years of age. Secondary osteoporosis has equal prevalence among males and females. This condition arises following certain medical problems, chronic diseases, and prolonged use of glucocorticoids. There are environmental triggers and genomic susceptibilities associated with this condition. The most important inclusion criteria for this condition to be of genomic origin include: Twin studies show that the heritability of bone mass is about 90% in the lumbar spine and 70% in the femoral neck. Ethnic studies show that osteoporosis affects people from all ethnic groups, particularly those of European or East Asian ancestry. Patients with a family history of fracture or osteoporosis are at a higher risk; the heritability factors of the fracture as well as low bone mineral density range from 25-80%. Multiple genes are involved in the causation of osteoporosis, including *PDLIM4, CALCR, COL1A2, LRP5, VDR* and *COL1A1*. There are several non-genomic factors associated with the development of osteoporosis including: malnutrition, underweight/inactivity, vitamin D deficiency, excessive alcohol consumption, tobacco smoking and lack of endurance training. Osteoporosis affects 1 out of 3 women and 1 out of 12 men over the age of 50 worldwide. This condition is 4 times more common in women than in men, although most of the genes have been mapped on autosomal chromosome.

3.13.9.1 Disorders of Bone Density and Structure

Hyperostosis corticalis generalisata or van Buchem disease: An endosteal hyperostosis disorder characterized by a symmetrically enlarged thickness of bones. The bones more commonly affected include the jawbone, skull, ribs, diaphysis of long bones, as well as tubular bones of the hands and feet. Consequently,

patients exhibit hearing loss, visual problems, neurologic pain, facial nerve palsy and blindness. Affected individuals display bone anomalies appearing during the 1st decade of life and progressing with age. There are several underlying causes of this disorder with at least 2 autosomal genomic loci: van Buchem disease is caused by autosomal recessive mutations in the *SOST* gene. The mutations include a 52-kb deletion approximately 35 kb downstream of the *SOST* gene. This deletion removes a SOST-specific regulatory element;; and van Buchem disease features that can be caused by autosomal dominant mutations in the *LRP5* gene. This form has been reported among a Sardinian family, with osteosclerosis of the skull and enlarged mandible.

Melorheostosis with osteopoikilosis: A mesenchymal dysplasia characterized by hyperostosis of the cortex of tubular bones. In affected individuals, lesions are typically asymmetric and involve only 1 extremity or correspond to a particular sclerotomal distribution. Patients may also manifest abnormalities of adjacent soft tissue, including joint contractures, sclerodermatous skin lesions, muscle atrophy or hemangiomas. Sometimes these individuals are asymptomatic, but pain and stiffness, with limitation of motion may occur. Melorheostosis with osteopoikilosis may be caused by loss-of-function mutations in the *LEMD3* gene which encodes the LEM domain (containing 3 an inner nuclear membrane protein). Melorheostosis may sometimes be a feature of Buschke-Ollendorff syndrome which is a benign disorder caused by mutations in the *LEMD3* gene. While germline or somatic LEMD3 mutations had been hypothesized to cause isolated melorheostosis, several studies have not been able to prove it.

Paget disease of bone: A chronic disorder characterized by enlarged and misshapen bones. Affected individuals exhibit pain, fractures and arthritis in the joints adjacent to the affected bones. Around 10-40% of cases of Paget disease of bone are inherited or familial. There are several underlying causes of this disorder with at least 4 autosomal genomic loci: Paget disease of bone-2 is caused by mutations in the *TNFRSF11A* gene which encodes the tumor necrosis factor receptor superfamily member. This is an essential protein in osteoclast formation;; Paget disease of bone-3 is caused by mutations in the *SQSTM1* gene which encodes the sequestosome 1. This protein is associated with the tumor necrosis factor receptor superfamily pathway;; and Paget disease of bone-1 and 4 have been mapped, but the genes have not been characterized.

Primary hypertrophic osteoathropathy: A disorder characterized by digital clubbing, pachydermia and periostosis. Sometimes these patients manifest other features including delayed closure of the fontanels and congenital heart disease. There are several underlying causes of this disorder with at least 3 autosomal genomic loci: Autosomal recessive primary hypertrophic osteoathropathy-1 is caused by biallelic mutations in the *HPGD* gene which encodes the hydroxyprostaglandin dehydrogenase 15-(NAD). Biallelic mutations in the same gene cause Cranioosteoarthropathy;; Autosomal recessive primary hypertrophic osteoathropathy-2 is caused by biallelic mutations in the *SLCO2A1* gene which encodes the solute carrier organic anion transporter (family 2 member A1);; and there is also other Autosomal dominant form of Primary hypertrophic osteoarthropathy, but the gene has not yet been characterized. Primary hypertrophic osteoathropathy occurs 7 times more frequently in men than in women.

Familial expansile osteolysis: A bone dysplasia characterized by focal skeletal changes of peripheral distribution. This condition appears during the 2nd decade of life. Affected individuals exhibit severe, painful, disabling bone deformity leading to pathologic fracture. Familial expansile osteolysis can be caused by autosomal dominant mutations in the *TNFRSF11A* gene which encodes the tumor necrosis factor receptor superfamily member 11a (NFKB activator). This protein is involved in osteoclast formation.

Caffey disease or Infantile cortical hyperostosis: A disorder characterized by acute inflammatory reaction affecting one or several areas of the body. Affected infants exhibit severe pain resulting in either pseudoparalysis or true localized palsies. Affected individuals exhibit other features including irritability, tenderness, hyperaesthesia, soft tissue swelling and redness. Caffey disease appears after 5 months of age and habitually resolves spontaneously by 2 years of age. There is also a lethal form of prenatal cortical hyperostosis. Caffey disease is caused by autosomal dominant mutations in the *COL1A1* gene which encodes the alpha-1 type 1 collagen.

3.13.9.2 Chondropathies

Familial osteochondritis dissecans: A skeletal disorder characterized by multiple osteochondritic lesions due to separation of cartilage and subchondral bone. This condition mainly affects the knees, hips and/or elbows. Affected individuals exhibit other features including early-onset osteoarthritis and disproportionate short stature. Familial osteochondritis dissecans is caused by autosomal dominant mutations in the *ACAN* gene which encodes the aggrecan. This protein is a major component of cartilage extracellular matrix. The prevalence of Familial osteochondritis dissecans is unknown. Sporadic osteochondritis dissecans is more prevalent; it is estimated to occur in the knee in 15-29 in every 100,000 individuals.

3.14.0 Subchapter XIV

Diseases of the Genitourinary System

Table 3.14.0 Diseases of the Genitourinary System

3.14.00-8	Diseases of the Genitourinary System	
3.14.00-4	Diseases of the Genitourinary System: Urinary System	
	3.14.1	Glomerular Diseases
	3.14.2	Renal Tubulo-interstitial Diseases
	3.14.3	Urolithiasis
	3.14.4	Other Disorders of Urinary System
3.14.5-8	Diseases of the Genitourinary System: Pelvis, Genitals and Breasts	
	3.14.6	Diseases of Male Genital Organs
	3.14.7	Sex Reversion
	3.14.8	Noninflammatory Disorders of Female Genital Tract

Table 3.14.01 Diseases of the Genitourinary System in Different Ethnic Groups

Disease	Frequency of condition	People with higher prevalence (if available)
Ovarian hyperstimulation and dysgenesis syndrome	5-200 in 1,000 pregnancy	Yoruba people and Cândido Godói people (Brazil)
Sertoli-cell-only syndrome	4 in 10,000 Males	Worldwide
Globozoospermia	1 in 1,000 in male infertile patients	Unknown
Ovarian dysgenesis with normal XX karyotype	1-10 in 100,000 Females	Finnish
Azoospermia and Spermatogenic failure	1-5 in 10,000	Worldwide
Gitelman syndrome	2-5 in 100,000	Unknown
Nephronophthisis	1 in 100,000	Unknown
Nephrotic syndrome	1-2 in 100,000	Finnish
Focal segmental glomerulosclerosis	1-2 in 100,000	Finnish
Nephronophthisis	1-2 in 100,000	Unknown
Senior–Løken syndrome	1-2 in 100,000	Unknown
Nephrogenic diabetes insipidus	1-9 in 1,000,000	Unknown
Distal renal tubular acidosis	<1 in 1,000,000	Inhabitants of Antioquia (Colombia) and Turkish
SERKAL syndrome	<1 in 1,000,000	Unknown
Renal tubular acidosis with ocular abnormalities	<1 in 1,000,000	Unknown
Pierson syndrome	<1 in 1,000,000	Unknown
Frasier syndrome	<1 in 1,000,000	Unknown
Glomerulopathy with fibronectin deposits	<1 in 1,000,000	Unknown
Urofacial syndrome or Ochoa syndrome	<1 in 1,000,000	Unknown
Hypophosphatemic nephrolithiasis/ osteoporosis	<1 in 1,000,000	Unknown

3.14.00-4 Diseases of the Genitourinary System: Urinary System

3.14.1 Glomerular Diseases

Frasier syndrome: A disorder characterized by the association of progressive glomerular nephropathy, male pseudohermaphroditism and high risk of developing gonadoblastoma. Female patients exhibit primary amenorrhea, while males show female external genitalia and streak gonads. During childhood, patients manifest proteinuria and nephrotic syndrome. In adolescence or adulthood, patients develop end-stage renal disease. In female patients, Frasier syndrome is caused by monoallelic de novo mutations in the *WT1* gene which encodes the Wilms tumor 1 protein, a transcription factor essential to the development of the urogenital system. Mutations in the same gene also cause several conditions, including Type 1 Wilms tumor, Denys-Drash syndrome, Meacham syndrome and type 4 Nephrotic syndrome. Frasier syndrome is estimated to affect less than 1 in every 1,000,000 individuals.

Denys-Drash syndrome: A disorder characterized by the association of pseudohermaphroditism, Wilms tumor, hypertension and degenerative renal disease. Denys-Drash syndrome is caused by monoallelic de novo mutations in the *WT1* gene.

Benign familial hematuria: A disorder characterized by nonprogressive isolated microscopic hematuria that does not result in renal failure. Benign familial hematuria is caused by autosomal dominant mutations in either the *COL4A3* gene which encodes the alpha-3 type 4 collagen, or the *COL4A4* gene which encodes the alpha-4 type 4 collagen.

Nephrotic syndrome: A nonspecific renal disorder that leaks large amounts of protein from the blood into the urine. Affected individuals exhibit proteinuria, hypoalbuminemia, hyperlipidemia and edema. Around 20% of affected individuals have a genomic steroid-resistant form that progresses to end-stage renal failure. There are several underlying causes of this disorder with at least 8 autosomal genomic loci: Nephrotic syndrome-1 is caused by autosomal recessive mutations in the *NPHS1* gene which encodes the congenital nephrosis protein 1, Finnish type (nephrin). This cell adhesion molecule functions in the glomerular filtration barrier in the kidney.;; Nephrotic syndrome-2 is caused by autosomal recessive mutations in the *NPHS2* gene which encodes the idiopathic nephrosis protein 2 steroid-resistant (podocin). This protein plays a role in the regulation of glomerular permeability, and acts as a linker between the plasma membrane and the cytoskeleton. This is a steroid-resistant form progressing to end-stage renal failure in the 1st or 2nd decade of life;; Nephrotic syndrome-3 is caused by autosomal recessive mutations in the *PLCE* gene which encodes the plectin 1, an intermediate filament binding protein (500kDa);; Nephrotic syndrome-4 is caused by autosomal dominant mutations in the *WT1* gene;; Nephrotic syndrome-5 with or without ocular abnormalities is caused by autosomal recessive mutations in the *LAMB2* gene which encodes the laminin beta 2;; Nephrotic syndrome-6 is caused by autosomal recessive mutations in the *PTPRO* gene which encodes the protein tyrosine phosphatase receptor type O;; Nephrotic syndrome-7 is caused by autosomal recessive mutations in the *DGKE* gene which encodes the diacylglycerol kinase (epsilon 64kDa);; and Nephrotic syndrome-8 is caused by autosomal recessive mutations in the *ARHGDIA* gene which encodes the Rho GDP dissociation inhibitor alpha.

Nephrotic syndrome type 1 or Finnish congenital nephrosis: A disorder appearing during fetal life characterized by protein loss. In 80% of these cases, the affected children are born prematurely and have low birth weight. Other features occurring during these pregnancies include larger placental volume and a 10-fold increase in alpha-fetoprotein concentration in the amniotic fluid. Nephrotic syndrome type 1 is caused by autosomal recessive mutations in the *NPHS1* gene which encodes the congenital nephrosis protein 1, Finnish type (nephrin). Several loss-of-function mutations have been identified, and 2 of them account for 90% of all cases in the people from Finland. In this group, the carrier frequency for Nephrotic syndrome type 1 trait is estimated to be 1 in 45 individuals, which means the frequency of this condition is about 1 in 8,200 individuals or $1/(90)^2$. This condition has been observed in various ethnic groups worldwide.

Focal segmental glomerulosclerosis: A disorder characterized by proteinuria, exhibiting nephrotic syndrome which progresses to impairment of renal function. Focal segmental glomerulosclerosis is a common cause of end-stage renal disease. There are several underlying causes of this disorder with at least 5 autosomal genomic loci: Focal segmental glomerulosclerosis-1 is caused by mutations in the *ACTN4* gene which encodes the alpha actinin 4. This protein is involved in binding actin to the membrane;; Focal segmental glomerulosclerosis-2 is caused by mutations in the *TRPC6* gene which encodes the transient receptor potential cation channel (subfamily C member 6);; Focal segmental glomerulosclerosis-3 is caused by mutations in the *CD2AP* gene which encodes the CD2-

associated protein;; Focal segmental glomerulosclerosis-4 is caused by mutations in the *APOL1* gene which encodes the apolipoprotein L 1;; and Focal segmental glomerulosclerosis-5 is caused by mutations in the *INF2* gene which encodes the inverted formin FH2 and WH2 domain containing protein. Focal segmental glomerulosclerosis is estimated to affect 1 in every 70,000 individuals in the United States.

Familial steroid-resistant nephrotic syndrome: An early onset disorder characterized by a nephrotic syndrome. This disorder usually progresses to end-stage renal failure. There are several underlying causes of this disorder with at least 3 autosomal genomic loci: Familial idiopathic steroid-resistant nephrotic is caused by either autosomal recessive or dominant inheritance. The autosomal recessive form is caused by biallelic mutations in the *NPHS2* gene;; and the autosomal dominant form is caused by monoallelic mutations in either the *ACTN4* gene or the *TRPC6* gene.

Pierson syndrome: A disorder characterized by the association of congenital nephrotic syndrome and ocular anomalies with microcoria. Affected individuals also exhibit psychomotor delay, marked muscle hypotonia and movement disorders. Pierson syndrome is caused by autosomal recessive mutations in the *LAMB2* gene which encodes the laminin beta 2. This protein is primarily expressed in the glomerular basement membrane, at the neuromuscular junctions, and in the intraocular muscles, lens and retina. The disease usually progresses towards renal failure during the 1st few months of life. Pierson syndrome is estimated to affect less than 1 in every 1,000,000 individuals.

Lipoprotein glomerulopathy: A disorder characterized by proteinuria usually progressing to kidney failure. Affected individuals also exhibit a unique lipoprotein thrombi formation that distend and occlude the glomerular capillary lumina. Lipoprotein glomerulopathy can be caused by monoallelic haplotype 2 or 3 in the *APOE* gene which encodes the apolipoprotein E. Lipoprotein glomerulopathy mainly affects people of Japanese and Chinese ancestry, and has infrequently been reported among Caucasians.

Glomerulopathy with fibronectin deposits: A disorder characterized by the association of proteinuria, microscopic hematuria, type IV renal tubular acidosis and hypertension. Affected individuals progress to end-stage renal failure in the 2nd-6th decade of life. There are several underlying causes of this disorder with at least 2 autosomal genomic loci: Glomerulopathy with fibronectin deposits-1 has been mapped, but the gene has not been characterized;; and Glomerulopathy with fibronectin deposits-2 is caused by autosomal dominant mutations in the *FN1* gene which encodes the fibronectin 1. This protein is involved in cell adhesion and migration processes. This form accounts for around 40% of all cases. Glomerulopathy with fibronectin deposits is estimated to affect less than 1 in every 1,000,000 individuals.

Table 3.14.1.1 Recognizable Glomerular Diseases

Disorder	OMIM	Gene	Biological Function	Product of Conception Alteration
Frasier syndrome	136680	WT1	Transcription factor Zinc finger	AD + 99%
Denys-Drash syndrome	194080	WT1	Transcription factor Zinc finger	AD + 99%
Benign familial hematuria	141200	COL4A3	Extracellular matrix protein	AD
Benign familial hematuria	141200	COL4A4	Extracellular matrix protein	AD
Nephrotic syndrome-1	256300	NPHS1	Channelopathy slit diaphragm	AR
Nephrotic syndrome-2	600995	NPHS2	Channelopathy slit diaphragm	AR
Nephrotic syndrome-3	610725	PLCE	Cytoskeletal Plectin	AR
Nephrotic syndrome-4	256370	WT1	Transcription factor Zinc finger	AD
Nephrotic syndrome-5	614199	LAMB2	Extracellular matrix protein	AR
Nephrotic syndrome-6	614196	PTPRO	Tyrosine phosphatase receptor	AR
Nephrotic syndrome-7	615008	DGKE	Diacylglycerol kinase	AR
Nephrotic syndrome-8	615244	ARHGDIA	Rho GDP dissociation inhibitor	AR
Focal segmental glomerulosclerosis-1	603278	ACTN4	Cytoskeletal Microfilaments	AR

Table 3.14.1.1. Continued.

Disorder	OMIM	Gene	Biological Function	Product of Conception Alteration
Focal segmental glomerulosclerosis-2	603965	TRPC6	Channelopathy, transient receptor potential channels	AD
Focal segmental glomerulosclerosis-3	607832	CD2AP	Cell surface receptor Other/ungrouped	AR
Focal segmental glomerulosclerosis-4	612551	APOL1	Cholesterol transport	
Focal segmental glomerulosclerosis-5	613237	INF2	Polymerization and depolymerization of actin filaments	AD
Focal segmental glomerulosclerosis, digenic	602716	NPHS1/NPHS2	Channelopathy slit diaphragm	Dig
Susceptibility to nondiabetic end-stage renal disease	612551	APOL1	Cholesterol transport	
Familial idiopathic steroid-resistant nephrotic	600995	NPHS2	Channelopathy slit diaphragm	AR or AD
Familial idiopathic steroid-resistant nephrotic	603278	ACTN4	Cytoskeletal Microfilaments	AD
Familial idiopathic steroid-resistant nephrotic	603965	TRPC6	Channelopathy, transient receptor potential channels	AD
Pierson syndrome	609049	LAMB2	Extracellular matrix protein	AR
Lipoprotein glomerulopathy	611771	APOE	Triglyceride-rich lipoprotein	AD
Glomerulopathy with fibronectin deposits 1	601894	FN1	Cell surface and extracellular matrix protein	AD

3.14.2 Renal Tubulo-interstitial Diseases

Nephronophthisis: A chronic tubulointerstitial nephropathy which often progresses to end-stage renal failure before the age of 15. There are 3 forms: infantile, juvenile and adolescent. Juvenile is the most common form of nephronophthisis, accounting for 15% of all cases of childhood end-stage renal failure. Infantile nephronophthisis develops to end-stage renal failure before age 5. In the juvenile form, affected children (at 2 years of age) exhibit failure to thrive, polyuria, polydipsia and progressive deterioration of renal function. There are several underlying causes of this disorder with at least 15 autosomal genomic loci: Familial juvenile nephronophthisis-1 is caused by autosomal recessive loss-of-function either due to mutations or due to deletion of the *NPHP1* gene which encodes the nephrocystin 1, a protein that appears to function in the control of cell division, as well as in cell-cell and cell-matrix adhesion signaling; Infantile nephronophthisis-2 is caused by autosomal recessive mutations in the *INVS* gene which encodes the inversin; Nephronophthisis-3 is caused by autosomal recessive mutations in the *NPHP3* gene which encodes the nephronophthisis 3; Juvenile nephronophthisis-4 is caused by mutations in the *NPHP4* gene which encodes the nephrocystin 4 protein; Nephronophthisis-7 is caused by autosomal recessive mutations in the *GLIS2* gene which encodes the GLIS family zinc finger 2; Nephronophthisis-8 is caused by autosomal recessive mutations in the *RPGRIP1L* gene which encodes the RPGRIP1-like; Nephronophthisis-9 is caused by autosomal recessive mutations in the *NEK8* gene which encodes the NIMA-related kinase 8; Nephronophthisis-11 is caused by autosomal recessive mutations in the *TMEM67* gene which encodes the transmembrane protein 67; Nephronophthisis-12 is caused by autosomal recessive mutations in the *TTC21B* gene which encodes the tetratricopeptide repeat domain 21B; Nephronophthisis-13 is caused by autosomal recessive mutations in the *WDR19* gene which encodes the WD repeat domain 19; Nephronophthisis-14 is caused by autosomal recessive mutations in the *ZNF423* gene which encodes the zinc finger protein 423; and Nephronophthisis-15 is caused by autosomal recessive mutations in the *CEP164* gene which encodes the centrosomal protein 164kDa. Several loci have not been characterized yet at the gene level. Nephronophthisis is estimated to affect 1 in every 100,000 individuals.

3.14.2.1 Obstructive and Reflux Uropathy

Urofacial syndrome or Ochoa syndrome: A disorder characterized by the association of severe voiding dysfunction, and facial grimace when trying to smile or cry. Affected individuals manifest other features including incontinence, hydronephrosis, urinary tract infection and moderate to severe constipation. Urofacial syndrome is caused by autosomal recessive mutations in the *HPSE2* gene which encodes the heparanase 2. Urofacial syndrome is estimated to affect less than 1 in every 1,000,000 individuals.

Vesicoureteral reflux: A disorder characterized by a retrograde movement of urine from the bladder into ureters and sometimes into the kidneys. This condition is a risk factor for urinary tract infections. Vesicoureteral reflux is a common condition affecting 1 in 100 children. Genomic factors associated with this condition include: a higher prevalence among first-degree relatives and the fact that vesicoureteral reflux is rare in certain ethnicities such as African Americans. There are several underlying causes of this disorder with at least 6 autosomal genomic loci: Vesicoureteral reflux-1 has been mapped, but the gene has not been characterized; Vesicoureteral reflux-2 is caused by mutations in the *ROBO2* gene which encodes the roundabout axon guidance receptor homolog 2; Vesicoureteral reflux-3 is caused by mutations in the *SOX17* gene which encodes the SRY (sex determining region Y)-box 17, a transcription factor; Vesicoureteral reflux-4 has been mapped, but the gene has not been characterized; Vesicoureteral reflux-5 has been mapped, but the gene has not been characterized; and Vesicoureteral reflux-6 has been mapped, but the gene has not been characterized. A possible Vesicoureteral reflux-X-linked form has been reported.

3.14.3 Urolithiasis

Hypophosphatemic nephrolithiasis/osteoporosis-1: A disorder characterized by hypophosphatemia due to decreased renal phosphate absorption. Affected individuals also manifest recurrent urolithiasis and bone demineralization. There are several underlying causes of this disorder with at least 2 autosomal genomic loci: Hypophosphatemic nephrolithiasis/osteoporosis-1 is caused by monoallelic mutations in the *SLC34A1* gene which encodes the solute carrier (family 34 member A1), a sodium phosphate transporter;; and Hypophosphatemic nephrolithiasis/osteoporosis-2 is caused by monoallelic mutations in the *SLC9A3R1* gene which encodes the solute carrier (family 9 member A3 regulator 1), a cation proton antiporter 3. Hypophosphatemic nephrolithiasis/osteoporosis is estimated to affect less than 1 in every 1,000,000 individuals.

Hypercalciuria and recurrent calcium oxalate stones: A disorder characterized by hypercalciuria. Affected individuals exhibit primary intestinal over absorption of dietary calcium. There are several underlying causes of this disorder with at least 2 autosomal genomic loci: Absorptive hypercalciuria-1 has been mapped, but the gene has not been characterized;; and Absorptive hypercalciuria-2 is caused by mutations in the *ADCY10* gene which encodes the soluble adenylyl cyclase 10.

3.14.4 Other Disorders of Urinary System

Renal tubular acidosis: A disorder characterized by an elevation in urinary pH regardless of the presence of serum acidosis. Affected individuals exhibit vomiting, dehydration, failure to thrive and/or growth impairment. There are several underlying causes of this disorder with at least 5 autosomal genomic loci: Renal tubular acidosis with progressive sensorineural deafness is caused by autosomal recessive mutations in the *ATP6V1B1* gene which encodes the lysosomal transporting H+ ATPase (V1 subunit B1);; Distal renal tubular acidosis with preserved hearing or late-onset sensorineural hearing loss is caused by autosomal recessive mutations in the *ATP6V0A4* gene which encodes the lysosomal transporting H+ ATPase (V0 subunit A4);; Autosomal recessive distal renal tubular acidosis with hemolytic anemia is caused by mutations in the *SLC4A1* gene which encodes the solute carrier (family 4 member A1), an anion exchange protein;; Autosomal dominant distal renal tubular acidosis is caused by monoallelic mutations in the *SLC4A1* gene;; and Autosomal recessive osteopetrosis with renal tubular acidosis is caused by mutations in the *CA2* gene which encodes the carbonic anhydrase II.

Autosomal recessive proximal renal tubular acidosis with ocular abnormalities: A disorder characterized by severe hyperchloremic acidosis, due to reduction of maximum tubular capacity for bicarbonate reabsorption. Affected individuals exhibit mental retardation, growth retardation, short stature, cataract, corneal opacities, glaucoma, nystagmus, and defects in the enamel of the permanent teeth. "Autosomal recessive proximal renal tubular acidosis with ocular abnormalities" is caused by biallelic mutations in the *SLC4A4* gene which encodes the solute carrier (family 4 member A4), a sodium bicarbonate cotransporter.

Nephrogenic diabetes insipidus: A disorder characterized by polyuria with polydipsia. Affected individuals exhibit acute hypernatremic dehydration, recurrent attacks of fever, constipation and sometimes neurological sequelae. Nephrogenic diabetes insipidus results by the inability of the renal collecting ducts to absorb water in response to antidiuretic hormone. There are several underlying causes of this disorder with at least 1 X-linked genomic locus and 1 autosomal genomic locus: Nephrogenic diabetes insipidus-1 is caused by X-linked loss-of-function mutations in the *AVPR2* gene which encodes the vasopressin V2 receptor. This gene maps on the long subtelomeric arm on chromosome Xq28;; and Nephrogenic diabetes insipidus-2 is caused by either autosomal recessive or autosomal dominant mutations in the *AQP2* gene which encodes the aquaporin 2. This protein is involved in the transportation of water in the renal tubules. Nephrogenic diabetes insipidus is estimated to affect 1-2 in every 1,000,000 individuals. 90% of these patients are males, since most cases are X-linked.

Pseudohypoaldosteronism: A disorder characterized by renal salt wasting and high concentrations of sodium in sweat, stool and saliva. This disorder is especially life threatening in the neonatal period. There are several underlying causes of this disorder with at least 9 autosomal genomic loci: Pseudohypoaldosteronism type I can be caused by autosomal recessive mutations in any one of 3 genes which encode subunits of the epithelial sodium channel: *SCNN1A*, *SCNN1B* and *SCNN1G*, encoding the alpha, beta or the gamma subunit respectively;; Pseudohypoaldosteronism type I can be caused by autosomal dominant mutations in the *NR3C2* gene which encodes the nuclear receptor subfamily 3 (group C member 2). This encoded protein exhibits mineralocorticoid receptor function;; Pseudohypoaldosteronism type IIA or Gordon hyperkalemia-hypertension syndrome has been mapped, but the gene has not been characterized;; Pseudohypoaldosteronism type IIB can be caused by autosomal dominant mutations in the *WNK4* gene which encodes the WNK lysine deficient protein kinase 4;; Pseudohypoaldosteronism type IIC can be caused by autosomal dominant mutations in the *WNK1* gene which encodes the WNK lysine deficient protein kinase 1;; Pseudohypoaldosteronism type IID can be caused by autosomal dominant mutations in the *KLHL3* gene which encodes the kelch-like family member 3;; and Pseudohypoaldosteronism type IIE can be caused by autosomal dominant mutations in the *CUL3* gene which encodes the cullin 3. Gitelman syndrome is another example of primary renal tubular salt wasting. Pseudohypoaldosteronism type I is estimated to affect 1 in every 80,000 newborns.

Bartter syndrome: A life-threatening disorder affecting the thick ascending limb of the loop of Henle. Affected neonates exhibit hypokalemia, alkalosis and normal to low blood pressure. There are several underlying causes of this disorder with at least 5 autosomal genomic loci: Antenatal Bartter syndrome type 1 is caused by autosomal recessive mutations in the *SLC12A1* gene which encodes the solute carrier (family 12 member A1), a sodium-potassium-chloride cotransporter;; Antenatal Bartter syndrome type 2 is caused by loss-of-function mutations in the *KCNJ1* gene which encodes the potassium inwardly-rectifying channel (subfamily J member 1);; Bartter syndrome type 3 or classic Bartter is caused by autosomal recessive mutations in the *CLCNKB* gene which encodes the kidney voltage-sensitive chloride channel B;; Infantile Bartter syndrome with sensorineural deafness or Bartter syndrome type 4A is caused by mutations in the *BSND* gene which encodes the Bartter syndrome with sensorineural deafness, an essential beta subunit for CLC chloride channels;; and Infantile Bartter syndrome with sensorineural deafness or Bartter syndrome type 4B is caused by simultaneous mutations in both the *CLCNKA* gene, and the *CLCNKB* gene which encodes the kidney voltage-sensitive chloride channel subunit A and B, respectively. Bartter syndrome is estimated to affect 1 in every 1,000,000 individuals.

Gitelman syndrome: A milder phenotype, and numerically more predominant than Bartter syndrome. Gitelman syndrome is a disorder characterized by affecting the distal convoluted tubule of the kidneys. Affected individuals exhibit hypochloremic metabolic alkalosis, hypokalemia, hypocalciuria and occasionally hypomagnesemia. Gitelman syndrome is caused by autosomal recessive loss-of-function mutations in the *SLC12A3* gene which encodes the solute carrier (family 12 member A3), a thiazide-sensitive sodium-chloride cotransporter. Gitelman syndrome is estimated to affect 1 in every 40,000 individuals.

Dent disease: A renal disorder appearing during early childhood characterized by proximal tubule dysfunction. These conditions predominantly affect male patients exhibiting low-molecular-weight proteinuria associated with hypercalciuria.

Other features include rickets or osteomalacia, bone pain, difficulty in walking, nephrolithiasis, nephrocalcinosis and progressive renal failure. 30-80% of affected males progress to end-stage renal failure between the 3rd and 5th decades of life. The most severely affected patients are males, whereas female carriers show a milder form. There are two loci associated with this condition, both located on the chromosome X. Dent disease-1 is caused by X-linked recessive mutations in the *CLCN5* gene which encodes the voltage-sensitive chloride channel 5, an electrogenic Cl-/H+ exchanger. Mutations in the same gene also cause: X-linked recessive nephrolithiasis, X-linked recessive hypophosphatemic rickets and low molecular weight proteinuria;; and Dent disease-2 is caused by X-linked recessive mutations in the *OCRL* gene which encodes the phosphatidylinositol bisphosphate 5-phosphatase. Mutations in the same gene are also associated with Lowe Syndrome. Dent disease type 2 is associated with extra-renal features including: mild intellectual impairment, hypotonia and sub-clinical cataract. The prevalence of the disease is unknown, but it has been reported in at least 250 families worldwide.

The Concise Encyclopedia of Genomic Diseases

Table 3.14.4.1 Other Recognizable Disorders of Urinary System

Disorder	OMIM	Gene	Biological Function	Product of Conception Alteration
Renal tubular acidosis with progressive sensorineural deafness	267300	ATP6V1B1	ATPase protein	AR
Distal renal tubular acidosis with preserved hearing or late-onset sensorineural hearing loss	267300	ATP6V1B1	ATPase protein	AR
Autosomal recessive distal renal tubular acidosis with hemolytic anemia	611590	SLC4A1	Solute carrier	AR
Autosomal dominant distal renal tubular acidosis	179800	SLC4A1	Solute carrier	AD
Autosomal recessive osteopetrosis-3, with renal tubular acidosis	259730	CA2	Carbonic anhydrase	AR
Autosomal recessive proximal renal tubular acidosis with ocular abnormalities	604278	SLC4A4	Solute carrier	AR
Nephrogenic diabetes insipidus-1	304800	AVPR2	Cell surface receptor G protein	XR
Nephrogenic diabetes insipidus-2	125800	AQP2	Channelopathy Porin	AR or AD
Pseudohypoaldosteronism type I	264350	SCNN1A	Channelopathy Sodium	AR
Pseudohypoaldosteronism type I	264350	SCNN1B	Channelopathy Sodium	AR
Pseudohypoaldosteronism type I	264350	SCNN1G	Channelopathy Sodium	AR
Pseudohypoaldosteronism type I, digenic	264350	SCNN1B/ SCNN1G	Channelopathy Sodium	Dig
Autosomal dominant pseudohypoaldosteronism type I	177735	NR3C2	Transcription factor Zinc finger	AD + 20%
Pseudohypoaldosteronism type IIB	145260	WNK4	Serine/threonine-protein kinase	AD
Pseudohypoaldosteronism type IIC	145260	WNK1	Serine/threonine-protein kinase	AD
Pseudohypoaldosteronism type IID	614495	KLHL3	Kelch-like member	AR or AD
Pseudohypoaldosteronism type IIE	614496	CUL3	Cullin	AD
Antenatal Bartter syndrome-1	601678	SLC12A1	Solute carrier	AR
Antenatal Bartter syndrome-2	241200	KCNJ1	Channelopathy Potassium	AR

Table 3.14.4.1. Continued.

Disorder	OMIM	Gene	Biological Function	Product of Conception Alteration
Bartter syndrome-3 or classic Bartter	607364	CLCNKB	Channelopathy Chloride	AR
Bartter syndrome-4A or Infantile Bartter syndrome with sensori-neural deafness	602522	BSND	Channelopathy Chloride	AR
Bartter syndrome-4B or Infantile Bartter syndrome with sensori-neural deafness	613090	CLCNKA/ CLCNKB	Channelopathy Chloride	Dig
Gitelman syndrome	263800	SLC12A3	Solute carrier	AR
Dent disease-1	300009	CLCN5	Channelopathy Chloride	XR + de novo
Dent disease-2	300555	OCRL	Trans-Golgi network	XR
Hypophosphatemic rickets	300554	CLCN5	Channelopathy Chloride	XR
Nephrolithiasis-1	310468	CLCN5	Channelopathy Chloride	XR
Low molecular weight proteinuria with hypercalciuric nephrocalcin-osis	308990	CLCN5	Channelopathy Chloride	XR

3.14.5 Diseases of the Genitourinary System: Pelvis, Genitals and Breasts

3.14.6 Diseases of Male Genital Organs

Testicular microlithiasis: An asymptomatic, non-progressive condition characterized by the deposition of calcium phosphate microliths within the seminiferous tubules. Testicular microlithiasis can be conferred in some cases by mutations in the *SLC34A2* gene which encodes the solute carrier (family 34 member A2), a sodium phosphate transporter. Testicular microlithiasis is a common condition affecting 0.6-9% of males. This condition is associated with subfertility. There are also patients suffering from symptomatic tes-ticular microlithiasis, which is associated with an 8-10 fold increase in the risk, for testicular germ cell tumor.

Spermatogenic failure: A disorder characterized by a severe deficiency of spermatogenesis, causing male infertility. Patients suffering from spermatogenic failure exhibit meiotically arrested spermatocytes which accumulate in the tubules. These spermatocytes degenerate, and are easily distinguishable due to their partially condensed chromosomes. Spermatogenic failure can be caused due to underlying endocrinologic-genomic disorders including hypogonadotropic hypogonadism or ciliary dyskinesias. There are several underlying causes of this disorder with at least 2 Y-linked genomic loci, 2 X-linked genomic loci, and 11 autosomal genomic loci. In Spermatogenic failure there are 3 main forms of product of conception alterations: Y-linked, X-linked, and autosomal form: Y-linked Spermatogenic failure-1, or Sertoli cell-only syndrome, is caused by *de novo* deletions in the azoospermia factor region on the long arm of the Y chromosome. This region includes the *USP9Y, DBY* and the *UTY* genes. These genes encode for ubiquitin-specific protease 9, the DEAD/H box 3, and the ubiquitously transcribed tetratricopeptide repeat, respectively. These product of conception alterations are defined as classic Y-linked syndrome plus 99% (Y + 99%);; Y-linked spermatogenic failure-2, or Nonobstructive spermatogenic failure, is caused by interstitial deletions on the Y chromosome. Y-linked Spermatogenic failure-1 is estimated to affect 1 in every 2,500 males. There are 2 forms of X-linked spermatogenic failure: Type 1 is a Sertoli cell-only syndrome;; Type 2 is associated with defects in meiosis. The gene/locus has not been characterized. There are at least 9 autosomal forms of this condition, including: Spermatogenic failure-1 is caused by an autosomal recessive form due to defects in meiosis. The gene has not been characterized;; Spermatogenic failure-2 is associated with rearrangements on chromosome 1;; Spermatogenic failure-3, or Nonobstructive azoospermia, is caused by monoallelic mutations in the *SLC26A8* gene which encodes the solute carrier (family 26 member A8);; Spermatogenic failure-4, or Azoospermia due to perturbations of meiosis, is caused by monoallelic mutations in the *SYCP3* gene which encodes the synaptonemal complex protein 3;; Spermatogenic failure-5, or "Male infertility due to large-headed, multiflagellar and polyploid sperm", is caused by autosomal recessive mutations in the *AURKC* gene which encodes the aurora kinase C;; Spermatogenic failure-6, or Acrosome malformation resulting in globozoospermia, is caused by autosomal recessive

mutations in the *SPATA16* gene which encodes the spermatogenesis associated 16;; Spermatogenic failure-7 is caused by autosomal recessive mutations in the *CATSPER1* gene which encodes the cation channel sperm associated 1;; Spermatogenic failure-8 is caused by monoallelic mutations in the *NR5A1* gene which encodes the nuclear receptor (subfamily 5 group A) member 1;; Spermatogenic failure-9, or Globozoospermia, is caused by autosomal recessive mutations in the *DPY19L2* gene which encodes the dpy-19-like 2;; Spermatogenic failure-10, or Spermatogenic failure with defective sperm annulus, is caused by monoallelic mutations in the *SEPT12* gene which encodes the septin 12;; and Spermatogenic failure-11 is caused by monoallelic mutations in the *KLHL10* gene which encodes the kelch-like family member 10.

3.14.7 Sex Reversion: Most examples of sex reversal syndrome are discussed under congenital malformations, deformations and chromosomal abnormalities in the subchapter "Indeterminate Sex and Pseudohermaphroditism" (3.17.6.5).

SERKAL syndrome (SEx Reversion, Kidneys, Adrenal and Lung dysgenesis): A disorder characterized by female to male sex reversal and developmental anomalies of the kidneys, adrenal glands and lungs. SERKAL syndrome can be lethal due to renal agenesis. SERKAL syndrome is caused by autosomal recessive (loss-of-function) mutations in the *WNT4* gene which encodes the wingless-type MMTV integration site family member 4. This secreted signaling protein is involved in several developmental processes and in oncogenesis. SERKAL syndrome is estimated to affect less than 1 in every 1,000,000 individuals.

Meacham syndrome: A disorder characterized by the association of pseudohermaphroditism, cardiac and pulmonary malformations. Meacham syndrome is caused by monoallelic mutations in the *WT1* gene which encodes the Wilms tumor 1 protein, a transcriploin factor.

3.14.8 Noninflammatory Disorders of Female Genital Tract

Ovarian hyperstimulation syndrome: A condition associated with dizygotic twinning. Females with this condition exhibit abdominal distention due to massive ovarian enlargement. Affected individuals ex-hibit other features including capillary leak with fluid sequestration making this condition potentially life-threatening. The main current cause of Ovarian hyperstimulation syndrome is due to ovarian stimulation treatments for in vitro fertilization. Dizygotic twinning exhibit large ethnic variations, especially among people from the city of Cândido Godói (Brazil), Yoruba people (Nigeria), and people from the archipelago of Åland and Åboland (southwest Finland). Ovarian hyperstimulation syndrome can be caused by autosomal recessive mutations in the *FSHR* gene which encodes the follicle-stimulating hormone receptor.

Table 3.14.5-8.1 Recognizable Diseases of the Genitourinary System: Pelvis, Genitals and Breasts

Disorder	OMIM	Gene	Biological Function	Product of Conception Alteration
Testicular microlithiasis	610441	*SLC34A2*	Solute carrier	
Y-linked Spermatogenic failure-1 or Sertoli cell-only syndrome	400042	*USP9Y/DBY /UTY*	Azoospermia factor	Y + 99%
Y-linked Spermatogenic failure-1 or Sertoli cell-only syndrome	415000	*USP9Y*	Ubiquitin-specific pro-teases	Y + 99%
Spermatogenic failure-3 or Nonobstructive azoospermia	606766	*SLC26A8*	Solute carrier	AD
Spermatogenic failure-4 or Azoospermia due to perturbations of meiosis	270960	*SYCP3*	Nucleus synaptonemal complex	AD
Spermatogenic failure-5 or Male infertility due to large-headed, multiflagellar, polyploid sperm	243060	*AURKC*	Aurora kinase	AR
Spermatogenic failure-6 or Acrosome malformation resulting in globozoospermia	102530	*SPATA16*	Golgi apparatus protein	AR
Spermatogenic failure-7	612997	*CATSPER1*	Cation channel, sperm associated	AR
Spermatogenic failure-8	613957	*NR5A1*	Nuclear receptor	AD

Table 3.14.5-8.1. Continued.

Disorder	OMIM	Gene	Biological Function	Product of Conception Alteration
Spermatogenic failure-9 or globozoospermia	613958	DPY19L2	Dpy-19-like 2	AR
Spermatogenic failure-10 or spermatogenic failure with defective sperm annulus	614822	SEPT12	Septin	AD
Spermatogenic failure-11	615081	KLHL10	Kelch-like member	AD
SERKAL syndrome	611812	WNT4	Several developmental processes	AR
Meacham syndrome	608978	WT1	Transcription factor Zinc finger	AD + de novo
Ovarian hyperstimulation syndrome	608115	FSHR	Cell surface receptor G protein	AD
Ovarian response to FSH stimulation	276400	FSHR	Cell surface receptor G protein	AR

3.15.0 Subchapter XV

Pregnancy, Childbirth and the Puerperium

Table 3.15.0 Pregnancy, Childbirth and the Puerperium

3.15.00-3	Pregnancy, Childbirth and the Puerperium	
	3.15.1	Pregnancy with Abortive Outcome
	3.15.2	Skewed X-inactivation
	3.15.3	Other Maternal Disorders Predominantly Related to Pregnancy

3.15.1 Pregnancy with Abortive Outcome

Hydatidiform mole: A benign gestational trophoblastic proliferation that appears during an abnormal fertilization. Affected individuals exhibit during the second trimester vomiting, metrorrhagia, and abnormal increase in the size of the uterus. On the basis of genomic origin, hydatidiform mole can be divided into 3 types:

partial moles, complete homozygous and complete heterozygous. Partial moles are triploid, and the additional sets of chromosomes are mostly paternally derived. Partial moles account for 25% of all cases. Complete moles are diploid but all the chromosomes are paternally derived. There are two forms of complete hydatidiform: either, homozygous or heterozygous. The homozygous form arises from duplication of a haploid sperm. This form accounts for 90% of all complete moles. The heterozygous form arises due to the fertilization of an anucleate egg by 2 sperms. Hydatidiform mole is estimated to affect 1 in every 1,500 pregnancies in the United States. There is a larger prevalence in Eastern Asian individuals, affecting 1 in every 250 pregnancies. There are several underlying causes of this disorder with at least 2 autosomal genomic loci: Recurrent hydatidiform moles and reproductive wastage can be caused by autosomal recessive mutations in the *NLRP7* gene which encodes the NLR family pyrin domain containing protein 7;; and Recurrent biparental complete hydatidiform mole-2 is caused by autosomal recessive mutations in the *KHDC3L* gene which encodes the KH domain containing protein 3-like subcortical maternal complex member. This protein is specifically expressed in the oocytes.

Cocoon syndrome: A fetal encasement disorder characterized by pregnancy ended in the 1st or 2nd trimester. Affected fetuses exhibit multiple malformations including abnormal cyst in the cranial region, a large defect of the craniofacial area and an omphalocele, as well as immotile and hypoplastic limbs. Affected individuals manifest other features including microcephaly, defect in the diaphragm, tetralogy of Fallot, defects in both lungs and a horseshoe kidney. Cocoon syndrome is caused by autosomal recessive mutations in the *CHUK* gene which encodes the conserved helix-loop-helix ubiquitous kinase.

3.15.2 Skewed X-inactivation

Familial skewed X-inactivation: A female condition characterized by preferential use at a level greater than 95% of the paternal X chromosome. There are several underlying causes of this disorder with at least 2 X-linked genomic loci: Familial skewed X-inactivation-1 is caused by X-linked mutations in the *XIST* gene which encodes the X inactive specific transcript (non-protein coding). This gene is essential for the initiation and spread of X-inactivation. This gene was the first non-coding gene identified within the X inactivation center (XIC);; and Familial skewed X-inactivation-2 has been mapped to chromosome Xq25-q26, but the gene has not been characterized.

3.15.3 Other Maternal Disorders Predominantly Related to Pregnancy

Gestational diabetes mellitus: A condition that affects pregnant women exhibiting high blood glucose levels, without previously diagnosed diabetes. There are several complications in babies born to mothers with gestational diabetes, including being large for gestational age, low blood sugar, jaundice, and proneness to developing childhood obesity and type 2 diabetes later in life. There are several complications in women with gestational diabetes, including higher incidence of pre-eclampsia and increased risk of developing type 2 diabetes mellitus after pregnancy. Gestational diabetes affects 3-10% of pregnancies. There are several ethnic groups with high prevalence of gestational diabetes including: African-Americans,

Hispanic-Americans, Native Americans and Pacific Islanders. Gestational diabetes has been shown to be caused by mutations in the *GCK* gene which encodes the glu-cokinase. About 50% of women who are carriers for those mutations may develop gestational diabetes.

3.16.0 Subchapter XVI

Conditions Originating in the Perinatal Period

Table 3.16.0 Conditions Originating in the Perinatal Period

3.16.00-5	Conditions Originating in the Perinatal Period	
	3.16.1	Disorders Related to Length of Gestation and Fetal Growth
	3.16.2	Respiratory and Cardiovascular Disorders Specific to the Perinatal Period
	3.16.3	Hemorrhagic and Hematological Disorders of Fetus and Newborn
	3.16.4	Transitory Endocrine and Metabolic Disorders Specific to Fetus and Newborn

3.16.1 Disorders Related to Length of Gestation and Fetal Growth

Severe growth retardation and developmental delay, coarse facies and early death: A disorder characterized by severe multiple congenital anomalies and death by age 3 years. Affected infants also exhibit intrauterine growth retardation, severe failure to thrive, heart defects, severe developmental delay, microcephaly, lissencephaly, seizures and Dandy-Walker malformation. "Severe growth retardation and developmental delay, coarse facies and early death" can be caused by autosomal recessive mutations in the *FTO* gene which encodes the fat mass and obesity associated protein. This condition has been reported in a consanguineous Palestinian Arab family.

3.16.2 Respiratory and Cardiovascular Disorders Specific to the Perinatal Period

Choreoathetosis, congenital hypothyroidism and neonatal respiratory distress: A disorder characterized by neonatal respiratory distress with recurrent pulmonary infections requiring mechanical ventilation. Affected individuals exhibit other features including: mildly increased serum TSH, muscular hypotonia, delayed development and staggering gait. Years later, patients develop choreiform hyperkinesia. "Choreoathetosis, congenital hypothyroidism and neonatal respiratory distress" is caused by monoallelic mutations in the *NKX2-1* gene which encodes the NK2 homeobox 1. Mutations in the same gene also cause Benign hereditary chorea. Choreoathetosis, congenital hypothyroidism and neonatal respiratory distress is estimated to affect less than 1 in every 1,000,000 individuals.

Congenital alveolar capillary dysplasia with misalignment of pulmonary veins: A disorder characterized by lack of formation and ingrowth of alveolar capillaries. Affected neonates exhibit thickened alveolar walls, reduced number of alveoli, medial muscular thickening of small pulmonary arterioles with muscularization of the intraacinar arterioles and aberrant situated pulmonary veins. "Congenital alveolar capillary dysplasia with misalignment of pulmonary veins" is associated with persistent pulmonary hypertension of the neonate, respiratory distress, and multiple congenital malformations affecting the cardiovascular, musculoskeletal, genitourinary and gastrointestinal systems. Congenital alveolar capillary dysplasia with misalignment of pulmonary veins is caused by autosomal dominant mutations in the *FOXF1* gene which encodes the transcription factor forkhead box F1. Congenital alveolar capillary dysplasia with misalignment of pulmonary veins is estimated to affect less than 1 in every 1,000,000 individuals.

3.16.3 Hemorrhagic and Hematological Disorders of Fetus and Newborn

ABO blood group: A trait characterized by A, B, AB and O antigen in the red cells. Besides the importance in transfusion medicine, the ABO blood group is also associated with susceptibility to other conditions including infectious disease, pancreatic cancer and von Willebrand disease. For example, individuals with blood group AB are relatively resistant to cholera, due to E1 Tor biotype of V. cholerae 01. On the contrary, individuals with blood group O are at higher risk of contracting cholera. Individuals with blood group O are resistant to severe malaria due to Plasmodium falciparum, but non-O blood groups are sensitive. ABO blood groups are caused by autosomal codominant polymorphisms in the *ABO* gene which encodes the glycosyltransferase. This enzyme catalyzes the transfer of carbohydrates to the H antigen, forming the antigenic structure of the ABO blood groups. The *ABO* gene maps on band 9q34. There are large distributions of these ancestral alleles, each of them with frequencies of 2-70% worldwide. For instance, type A and O has the highest prevalence worldwide. Type B is more common in people of Asian ancestry, but less common among those with Western European ancestry.

H-Bombay blood group: A condition characterized by lack on the red blood cell of antigen A, B and H. The H antigen serves as a precursor for producing A and B antigens. Moreover, individuals with H-Bombay blood group produce isoantibodies to antigen H as well as to both A and B antigens. Therefore, H-Bombay blood group can receive blood only from other Bombay donors. H-Bombay blood group is caused by autosomal recessive mutations either in the *FUT1* or the *FUT2* gene. These genes encode fucosyltransferase 1 and 2, respectively. H-Bombay blood group is common in tribal areas of India. In this group, the carrier frequency for H-Bombay blood group trait is estimated to be 1 in every 5-8 individuals, which means the frequency of this condition is about 1 in 100-256 individuals or $1/(10)^2$ - $1/(16)^2$. H-Bombay blood group is estimated to affect around 4 in every 1,000,000 individuals in the United States.

3.16.3.1 Hemolytic Disease of Fetus and Newborn

Rh blood group: A trait that is clinically the 2nd most important after the ABO blood group. Hemolytic disease of the newborn appears in an RhD-positive newborn carried by an RhD-negative woman. They become sensitized by transplacental passage of RhD-positive red blood cells during a previous pregnancy. These newborns exhibit anemia, enlargement of the liver and spleen, jaundice, dyspnea and severe edema. Rh blood group negative is caused by autosomal recessive deletion of the *RHD* gene which encodes the D antigen. These loci contain a second gene that encodes both the RhC and RhE antigens on a single polypeptide. Rh blood group negative is more prevalent among Caucasians, especially among Basque people, of whom around 60% are Rh blood group negative. Rh blood group negative affects around 1% of Africans and South East Asians.

3.16.3.2 Neonatal Jaundice

Transient familial neonatal hyperbilirubinemia or Lucey-Driscoll syndrome: A disorder characterized by neonatal unconjugated hyperbilirubinemia due to inability to metabolize bilirubin. Affected neonates at days 2-4 of life display a significant increase in bilirubin levels. Sometimes, patients could even die of kernicterus. Transient familial neonatal hyperbilirubinemia is caused by autosomal recessive mutations in the *UGT1A1* gene which encodes the uridine diphosphate-glucuronosyltransferase. This condition can be associated with a common SNP in the exon 1 of the *UGT1A1* gene, resulting in a substitution gly71-to-arg. Mutations in the same gene also cause Crigler-Najjar syndrome and Gilbert's syndrome. Neonatal hyperbilirubinemia can also be caused by maternal steroids passed on through breast milk to the newborn. Transient familial neonatal hyperbilirubinemia seems to have an unusually high frequency

among Yemenite Jews. This condition is particularly common among Japanese, Koreans and Chinese.

3.16.4 Transitory Endocrine and Metabolic Disorders Specific to Fetus and Newborn

Neonatal diabetes mellitus: A disorder appearing within the 1st months of life characterized by hyperglycemia, dehydration, ketoacidosis, failure to thrive and coma. Transient neonatal diabetes infants develop diabetes in the first few weeks of life but it is resolved at a median age of 3 months, with possible recurrence to a permanent diabetes state during adolescence or as adulthood. Permanent neonatal diabetes is caused by pancreatic dysfunction due to insulin secretory failure. In most occasions, very early onset diabetes mellitus appears to be unrelated to autoimmunity. There are several underlying causes of this disorder with at least 8 autosomal genomic loci: Transient neonatal diabetes-1, or Transient neonatal diabetes with hypomethylation at 6q24, can be caused by biallelic mutations in the *ZFP57* gene which encodes the ZFP57 zinc finger protein, this form can involve imprinting;; Transient neonatal diabetes-2 is caused by either monoallelic or biallelic mutations in the *ABCC8* gene which encodes the ATP-binding cassette (sub-family member 8). This is another example of subtelomeric gene located on chromosome 11p15.1;; Transient neonatal diabetes-3 is caused by mutations in the *KCNJ11* gene which encodes the potassium inwardly-rectifying channel (subfamily J member 11);; Permanent neonatal diabetes mellitus can be caused by biallelic mutations in the *GCK* gene which encodes the glucokinase;; Permanent neonatal diabetes mellitus can be caused by concomitant monoallelic mutations in the *KCNJ11* gene and the *INS* gene which encodes the insulin;; Permanent neonatal diabetes mellitus can be caused by either monoallelic or biallelic mutations in the *ABCC8* gene;; Permanent neonatal diabetes mellitus can also be caused by pancreatic agenesis, which results in exocrine pancreatic deficiency as well as permanent neonatal-onset diabetes mellitus;; Permanent neonatal diabetes mellitus can be caused by monoallelic mutations in the *PDX1* gene which encodes the pancreatic and duodenal homeobox 1;; Pancreatic agenesis associated with congenital cardiac defects can be caused by mutations in the *GATA6* gene which encodes the GATA binding protein 6, a transcription factor;; and Pancreatic agenesis associated with cerebellar agenesis can be caused by mutations in the *PTF1A* gene which encodes the pancreas specific transcription factor, 1a. Transient and Permanent neonatal diabetes is estimated to affect 1 in every 100,000-500,000 live births.

3.17.0 Subchapter XVII

Congenital Malformations, Deformations and Chromosomal Abnormalities

Table 3.17.0 Malformations, Deformations and Chromosomal Abnormalities

3.17.00-9		Congenital Malformations and Deformations
	3.17.1	Nervous System
	3.17.2	Eye, Ear, Face and Neck
	3.17.3	Circulatory System
	3.17.4	Respiratory System
	3.17.5	Digestive System
	3.17.6	Genital Organs
	3.17.7	Urinary System
	3.17.8	Musculoskeletal System
	3.17.9	Other Disorders (not elsewhere classified)
3.17.10		Chromosomal Abnormalities (not elsewhere classified)

The primary goal of the subchapter on Malformations, Deformations and Chromosomal Abnormalities is to provide a sense of the prevalence of these diseases. The first section will address altogether the set of people where variations in DNA correlate with a disease. Table 3.17.0.1 displays the diseases, the worldwide frequencies of conditions, and the set of people with higher prevalence (if available). This section illustrates common and rare diseases. With regard to the table, only Chromosome 16 trisomy, Chromosome X monosomy or Turner syndrome, Brachydactyly, Chromosome 21 trisomy or Down syndrome, Fetal alcohol syndrome, Opitz-Kaveggia syndrome or FG syndrome, Adult polycystic kidney disease, Unilateral renal agenesis, Cleft palate, Bicuspid aortic valve, Hypospadias, Madelung deformity, Triple X syndrome, Klinefelter syndrome, Chromosome X/Y trisomy or XYY syndrome, Ichthyosis vulgaris, Holoprosencephaly, Polydactyly and Dizygotic twins are considered common in the nationalities and/or ethnic groups described. The rest of these diseases are considered to be rare, occuring in less than 1 in every 2,000 individuals.

Table 3.17.0.1 Malformations, Deformations and Chromosomal Abnormalities in Different Ethnic Groups

Disease	Frequency of condition	People with higher prevalence (if available)
Chromosome 16 trisomy	3-10 in 100 Pregnancies	Worldwide
Chromosome X monosomy or Turner syndrome	3-10 in 100 Pregnancies	Worldwide
Brachydactyly	2 in 100	Unknown
Chromosome 21 trisomy or Down syndrome	10-14 in 10,000	Worldwide
Fetal alcohol syndrome	1-2 in 1,000	
Opitz-Kaveggia syndrome or FG syndrome	2 in 1,000	Worldwide
Dizygotic twins	6-20 in 1,000 deliveries	Yoruba people
Adult polycystic kidney disease	1-2,5 in 1,000	Worldwide
Unilateral renal agenesis	1-2 in 1,000	Worldwide
Cleft palate	10-15 in 10,000	Native Americans
Bicuspid aortic valve	>1 in 1,000	Worldwide
Hypospadias	1 in 1,000	Worldwide
Madelung deformity	1 in 1,000	Worldwide
Triple X syndrome	1 in 1,000 Females	Worldwide
Klinefelter syndrome karyotype 47, XXY	1 in 1,000 Males	Worldwide
Chromosome X/Y trisomy or XYY syndrome	1 in 1,000 Males	Worldwide
Ichthyosis vulgaris	4 in 1,000	Worldwide
Holoprosencephaly	4 in 1,000 Pregnancies	Worldwide
Polydactyly	5-19 in 10,000	Worldwide
Monozygotic twins	3 in 1,000 deliveries	Worldwide
Postaxial polydactyly	8 in 10,000	Africans
Chromosome 22q11.2 microdeletion Syndrome or Velocardiofacial syndrome or DiGeorge Syndrome	2-5 in 10,000	Worldwide
Fragile X syndrome, Martin–Bell syndrome or Escalante syndrome	2-5 in 10,000	Worldwide
Chromosome X monosomy or Turner syndrome	2-5 in 10,000 Females	Worldwide
Congenital diaphragmatic hernia	2-4 in 10,000	Worldwide
Marfan syndrome	2-3 in 10,000	Worldwide
Neurofibromatosis type 1 or von Recklinghausen disease	2-4 in 10,000	Israelis (North African Origin)
Perinatally lethal renal diseases	2-3 in 10,000	Worldwide
Transposition of great arteries	2-3 in 10,000	Worldwide
Cleft lip/palate-ectodermal dysplasia syndrome	6-9 in 10,000	Worldwide
Blepharophimosis-ptosis-epicanthus inversus syndrome	2 in 10,000	Unknown

The Concise Encyclopedia of Genomic Diseases

Table 3.17.0.1. Continued.

Disease	Frequency of condition	People with higher prevalence (if available)
Chromosome 18 trisomy or Edwards syndrome	2 in 10,000	Worldwide
Chromosome 1p36 deletion syndrome	2 in 10,000	Worldwide
Complete atrioventricular canal	2 in 10,000	Worldwide
Ehlers–Danlos syndrome	2 in 10,000	Worldwide
Hirschsprung disease or Aganglionic megacolon	2 in 10,000	Worldwide
Mayer-Rokitansky-Küster-Hauser syndrome	2 in 10,000	Worldwide
Partial atrioventricular canal	2 in 10,000	Worldwide
Scaphocephaly	2 in 10,000	Worldwide
Double-outlet right ventricle	3 in 10,000	Worldwide
Preaxial polydactyly	3 in 10,000	American Indians
Noonan syndrome	4-10 in 10,000	Worldwide
Dextro-looped transposition of great arteries	2-3 in 10,000	Worldwide
Craniosynostosis	4-5 in 10,000	Worldwide
Congenital bilateral absence of the vas deferens	1-5 in 10,000	Unknown
Stickler syndrome	1-5 in 10,000	Worldwide
Bilateral renal agenesis or aplasia, Potter syndrome	16-25 in 100,000	Worldwide
Partial atrioventricular septal defect with heterotaxy syndrome	5 in 10,000	Worldwide
Atrial septal defect	6 in 10,000	Worldwide
Congenital cataracts	1-3 in 10,000	Worldwide
X-linked Ichthyosis	1-3 in 10,000	Worldwide
Tetralogy of Fallot	4 in 10,000	Worldwide
Tuberous sclerosis	10-16 in 100,000	Worldwide
Hypoplastic left heart syndrome	1-2 in 10,000	Worldwide
Microphthalmia	1-2 in 10,000	Worldwide
Supravalvular aortic stenosis	1-2 in 10,000	Worldwide
CHARGE syndrome	1 in 10,000	Worldwide
Chromosome 13 trisomy 13 or Patau syndrome	1 in 10,000	Worldwide
Congenital contractural arachnodactyly	1 in 10,000	Worldwide
Distal arthrogryposis	1 in 10,000	Worldwide
Ehlers-Danlos syndromes types III or Hypermobility type	1 in 10,000	Worldwide
Holoprosencephaly	1 in 10,000	Worldwide
Multiple epiphyseal dysplasia	1 in 10,000	Worldwide
Septooptic dysplasia	1 in 10,000	Unknown

Table 3.17.0.1. Continued.

Disease	Frequency of condition	People with higher prevalence (if available)
Primary congenital glaucoma	1-16 in 20,000	Romani Gypsies (Slovakia), Middle Easterns and inhabitants of Andhra Pradesh (India)
Angelman syndrome	5-10 in 100,000	Worldwide
Williams-Beuren syndrome	5-13 in 100,000	Worldwide
Cornelia de Lange syndrome 1	3-10 in 100,000	Worldwide
Persistent truncus arteriosus	3-10 in 100,000	Worldwide
Chromosome 17q24.3-q25.1 deletion syndrome or Pierre Robin syndrome	3-12 in 100,000	Worldwide
Premature ovarian failure	3-4 in 100,000	Worldwide
Swyer syndrome	3-4 in 100,000	Worldwide
Ehlers-Danlos syndromes types I and II or Classic type	2-5 in 100,000	Worldwide
Smith-Lemli-Opitz syndrome	2-5 in 100,000	Europeans
Thanatophoric dysplasia	2-5 in 100,000	Worldwide
Cri du chat syndrome or Lejeune syndrome	2-6 in 100,000	Worldwide
Trigonocephaly	2-6 in 100,000	Worldwide
Chromosome 8 trisomy	2-4 in 100,000	Worldwide
Primary microcephaly	2-4 in 100,000	Pakistanis (North)
Saethre-Chotzen Syndrome	2-4 in 100,000	Worldwide
Autosomal recessive polycystic kidney disease or Polycystic kidney and hepatic disease	2-3 in 100,000	Afrikaners and Finnish
Branchiootorenal syndrome	2-3 in 100,000	Worldwide
Epidermolysis bullosa	2-3 in 100,000	Worldwide
Epidermolysis bullosa simplex	2-3 in 100,000	Unknown
Incontinentia pigmenti	2-3 in 100,000	Worldwide
von Hippel-Lindau syndrome	2-3 in 100,000	Worldwide
Retinoschisis	4-20 in 100,000	Worldwide
Acrokeratosis verruciformis of Hopf and Darier disease	1-9 in 100,000	Worldwide
Currarino syndrome	1-9 in 100,000	Worldwide
Hereditary lymphedema	1-9 in 100,000	Worldwide
Isolated ectopia lentis	1-9 in 100,000	Worldwide
Opitz G syndrome type 1	1-9 in 100,000	Worldwide
VATER syndrome or VACTERL association	1-9 in 100,000	Worldwide
46, XY sex reversal or Swyer syndrome	3 in 100,000	Worldwide
Dandy–Walker syndrome or Dandy–Walker complex or Dandy–Walker Malformation	3 in 100,000	Worldwide

The Concise Encyclopedia of Genomic Diseases

Table 3.17.0.1. Continued.

Disease	Frequency of condition	People with higher prevalence (if available)
Hydrocephalus with stenosis of aqueduct of Sylvius or Bickers-Adams syndrome	3 in 100,000	Unknown
Hypochondroplasia	3 in 100,000	Worldwide
Kabuki syndrome	3 in 100,000	Worldwide
Prader–Willi syndrome	4-10 in 100,000	Worldwide
Beckwith-Wiedemann syndrome	7 in 100,000	Worldwide
Sotos syndrome	7-20 in 100,000	Worldwide
XX male syndrome or de la Chapelle syndrome or 46, XX sex reversal-1	1-5 in 100,000	Worldwide
Cataract	1-6 in 100,000	Worldwide
Peutz-Jeghers syndrome	1-6 in 100,000	Worldwide
Autosomal dominant osteopetrosis	5 in 100,000	Worldwide
Autosomal recessive osteopetrosis	5 in 100,000	Unknown
Brachycephaly	5 in 100,000	Worldwide
Buschke-Ollendorff syndrome	5 in 100,000	Unknown
chromosome 17p11.2 microduplication syndrome or Potocki-Lupski syndrome	5 in 100,000	Worldwide
Osteogenesis imperfecta	5 in 100,000	Worldwide
Stüve-Wiedemann syndrome	5 in 100,000	Worldwide
Heterotaxia or situs ambiguus	6 in 100,000	Worldwide
Hypohidrotic ectodermal dysplasia	6 in 100,000	Worldwide
Primary ciliary dyskinesia	6 in 100,000	Worldwide
Trigonocephaly, Jackson-Weiss, Pfeiffer, and Kallmann syndrome	6 in 100,000	Worldwide
Parietal foramina	4-6 in 100,000	Unknown
Polycystic kidney and hepatic disease	1-3 in 100,000	Worldwide
Alport syndrome	2 in 100,000	Worldwide
Apert syndrome	2 in 100,000	Worldwide
Chromosome 4p deletion syndrome or Wolf–Hirschhorn syndrome or Pitt-Rogers-Danks syndrome	2 in 100,000	Worldwide
Crouzon disease	2 in 100,000	Worldwide
Darier disease	2 in 100,000	Worldwide
Hailey-Hailey disease or Benign chronic familial pemphigus	2 in 100,000	Worldwide
Multiple exostoses	2 in 100,000	Worldwide
Nail patella syndrome	2 in 100,000	Unknown
Pseudoachondroplasia	2 in 100,000	Worldwide
Treacher Collins syndrome	2 in 100,000	Worldwide
Wolf–Hirschhorn syndrome or Pitt-Rogers-Danks syndrome	2 in 100,000	Worldwide

The Concise Encyclopedia of Genomic Diseases

Table 3.17.0.1. Continued.

Disease	Frequency of condition	People with higher prevalence (if available)
Aarskog-Scott syndrome	4 in 100,000	Unknown
Achondroplasia	4 in 100,000	Worldwide
Chromosome 17p13.1 deletion syndrome	4 in 100,000	Worldwide
Keratosis palmoplantaris striata 1	4 in 100,000	Worldwide
Neurofibromatosis type 2	4 in 100,000	Worldwide
Pseudoxanthoma elasticum or Grönblad–Strandberg syndrome	4 in 100,000	Afrikaners
Smith–Magenis Syndrome	4 in 100,000	Worldwide
Split-hand/foot malformation or Lobster-claw deformity	4 in 100,000	Worldwide
Achondrogenesis	1-2 in 100,000	Unknown
Alagille syndrome	1-2 in 100,000	Worldwide
Cat eye syndrome or Schmid Fraccaro Syndrome	1-2 in 100,000	Worldwide
Chromosome 22p trisomy or Cat Eye Syndrome or Schmid Fraccaro Syndrome	1-2 in 100,000	Worldwide
Isolated aniridia	1-2 in 100,000	Unknown
Joubert syndrome	1-2 in 100,000	Worldwide
Opitz G/BBB syndrome	1-2 in 100,000	Worldwide
Schizencephaly	1-2 in 100,000	Worldwide
Silver-Russell syndrome	1-2 in 100,000	Worldwide
Van der Woude synsyndromedrome 2	1-2 in 100,000	Worldwide
Zellweger syndrome	1-2 in 100,000	Worldwide
Caudal regression syndrome	1-2 in 100,000 Pregnancies	Unknown
Miller–Dieker syndrome or Miller–Dieker lissencephaly syndrome	<1 in 100,000	Unknown
X-linked lissencephaly	<1 in 100,000	Unknown
Absence of ulna and fibula with severe limb deficiency or Al-Awadi/Raas-Rothschild/Schinzel phocomelia syndrome	1 in 100,000	Unknown
Caudal duplication anomaly	1 in 100,000	Unknown
Chromosome chimera 46, XX/46, XY true hermaphrodite	1 in 100,000	Worldwide
Craniofrontonasal dysplasia	1 in 100,000	Worldwide
Glomerulocystic kidney disease, hyperuricemia and isosthenuria	1 in 100,000	Worldwide
Holt-Oram syndrome	1 in 100,000	Unknown
Keratosis linearis with ichthyosis congenita and sclerosing keratoderma	1 in 100,000	Worldwide

The Concise Encyclopedia of Genomic Diseases

Table 3.17.0.1. Continued.

Disease	Frequency of condition	People with higher prevalence (if available)
Larsen syndrome	1 in 100,000	Inhabitants of Réunion Island (France)
Medullary cystic kidney disease	1 in 100,000	Unknown
Pfeiffer syndrome	1 in 100,000	Worldwide
Polycystic liver disease	1 in 100,000	Unknown
Pycnodysostosis	1 in 100,000	Unknown
Rhizomelic chondrodysplasia punctata	1 in 100,000	Worldwide
Spondyloepiphyseal dysplasia congenita	1 in 100,000	Unknown
Thrombocytopenia-absent radius syndrome	1 in 100,000	Worldwide
Rubinstein-Taybi syndrome	8-10 in 1,000,000	Worldwide
Bardet-Biedl syndrome	1-9 in 1,000,000	Bedouins (Kuwait) and inhabitants of Newfoundland (Canada)
Congenital Clubfoot	1-9 in 1,000,000	Hawaiians and Maori people (New Zealand)
Congenital ichthyosis	1-9 in 1,000,000	Unknown
Greig cephalopolysyndactyly syndrome	1-9 in 1,000,000	Unknown
Klippel-Feil syndrome, Spondylocostal dysostosis, and microphthalmia	1-9 in 1,000,000	Worldwide
Male pseudohermaphroditism with gynecomastia	1-9 in 1,000,000	Unknown
Meckel–Gruber Syndrome	1-9 in 1,000,000	Finnish and Tatar people (Russian Federation)
Nonbullous congenital ichthyosiform erythroderma	1-9 in 1,000,000	Unknown
Asphyxiating thoracic dystrophy	2-10 in 1,000,000	Unknown
Ehlers-Danlos syndromes types IV or Vascular type or Ecchymotic type	4-10 in 1,000,000	Worldwide
Ellis-van Creveld syndrome	5-16 in 1,000,000	Old Order Amish (Pennsylvania) and Aboriginal People (West Australia)
Orofaciodigital syndrome I or Papillon-Leage and Psaume syndrome	4 in 1,000,000	Unknown
Sjögren-Larsson syndrome	4 in 1,000,000	Swedish
Townes-Brocks syndrome	4 in 1,000,000	Unknown
Xeroderma pigmentosum	4 in 1,000,000	Japanese, North Africans and the Middle Easterns
Diastrophic dysplasia	2 in 1,000,000	Worldwide
Ichthyosis bullosa of Siemens	2 in 1,000,000	Unknown
Lamellar ichthyosis	2 in 1,000,000	Unknown
Campomelic dysplasia	3 in 1,000,000	Unknown

Table 3.17.0.1. Continued.

Disease	Frequency of condition	People with higher prevalence (if available)
Popliteal pterygium syndrome	3 in 1,000,000	Unknown
Ichthyosis	3-5 in 1,000,000	Norwegians
Axenfeld-Rieger syndrome	5 in 1,000,000	Unknown
Birt-Hogg-Dube syndrome	5 in 1,000,000	Unknown
Cockayne syndrome	5 in 1,000,000	Unknown
Cowden syndrome	5 in 1,000,000	Worldwide
Epidermolytic hyperkeratosis or Bullous congenital ichthyosiform erythroderma	5 in 1,000,000	Unknown
Iridogoniodysgenesis	5 in 1,000,000	Unknown
Netherton syndrome	5 in 1,000,000	Unknown
Persistent Mullerian duct syndrome	5 in 1,000,000	Unknown
Spondylocostal dysostosis	5 in 1,000,000	Puerto Ricans
Spondyloepiphyseal dysplasia tarda	5-6 in 1,000,000	Unknown
Muenke syndrome	8 in 1,000,000	Worldwide
Papillon-Lefèvre syndrome or Palmoplantar keratoderma with periodontitis	1-4 in 1,000,000	Unknown
Craniometaphyseal dysplasia	1 in 1,000,000	Worldwide
Crisponi syndrome and Cold-induced sweating syndrome	1 in 1,000,000	Sardinians (Italy), Norwegians, Israelis and Canadians
Dyskeratosis congenita or Zinsser-Cole-Engman syndrome	1 in 1,000,000	Unknown
Senior–Løken syndrome	1 in 1,000,000	Unknown
Opitz trigonocephaly syndrome	1-2 in 1,000,000	Unknown
Spondyloepiphyseal dysplasia tarda with progressive arthropathy	1-2 in 1,000,000	Unknown
3M syndrome	<1 in 1,000,000	Unknown
6q25 microdeletion syndrome	<1 in 1,000,000	Unknown
9q34 deletion syndrome or Kleefstra syndrome	<1 in 1,000,000	Unknown
Acheiropodia	<1 in 1,000,000	Unknown
Acrocallosal syndrome	<1 in 1,000,000	Unknown
Acrocapitofemoral dysplasia	<1 in 1,000,000	Unknown
Acrocapitofemoral dysplasia and Brachydactyly	<1 in 1,000,000	Belgians and Dutch
Acromesomelic chrondrodysplasia and Brachydactyly	<1 in 1,000,000	Unknown
Acromesomelic dysplasia, Hunter-Thompson type	<1 in 1,000,000	Unknown
Acromesomelic dysplasia, Maroteaux type	<1 in 1,000,000	Unknown

Table 3.17.0.1. Continued.

Disease	Frequency of condition	People with higher prevalence (if available)
Adducted thumbs - arthrogryposis, Dundar type	<1 in 1,000,000	Turkish and Austrians
Adrenal insufficiency, congenital, with 46, XY sex reversal, partial or complete	<1 in 1,000,000	Unknown
ADULT syndrome	<1 in 1,000,000	Unknown
Agenesis of the corpus callosum with mental retardation, ocular coloboma and micrognathia, or Graham-Cox syndrome	<1 in 1,000,000	Unknown
Agnathia-otocephaly complex	<1 in 1,000,000	Unknown
Alström syndrome	<1 in 1,000,000	French Acadians (Nova Scotia and Louisiana)
Amish lethal microcephaly	<1 in 1,000,000	Amish
Andermann syndrome or Charlevoix disease	<1 in 1,000,000	French Canadians (Saguenay-Lac-Saint-Jean)
Anhidrotic and hypohidrotic ectodermal dysplasia	<1 in 1,000,000	Unknown
Anterior segment mesenchymal dysgenesis and polar cataract	<1 in 1,000,000	Inhabitants of Newfoundland (Canada) and Australians
Antley-Bixler syndrome	<1 in 1,000,000	Unknown
Antley-Bixler-like syndrome - ambiguous genitalia - disordered steroidogenesis	<1 in 1,000,000	Unknown
Arterial tortuosity syndrome	<1 in 1,000,000	Unknown
Arthrogryposis, renal dysfunction and cholestasis	<1 in 1,000,000	Unknown
Arts syndrome	<1 in 1,000,000	Unknown
Atelosteogenesis type I	<1 in 1,000,000	Unknown
Atelosteogenesis type II	<1 in 1,000,000	Unknown
Atelosteogenesis type III	<1 in 1,000,000	Unknown
Athabaskan brainstem dysgenesis syndrome	<1 in 1,000,000	Navajo and Apache (Athabaskan)
Baller-Gerold syndrome	<1 in 1,000,000	Unknown
Beare-Stevenson cutis gyrata syndrome	<1 in 1,000,000	Unknown
Bifid nose with or without anorectal and renal anomalies	<1 in 1,000,000	Egyptians, Afghans and Pakistanis
Bilateral frontoparietal polymicrogyria	<1 in 1,000,000	Palestinians
Björnstad syndrome	<1 in 1,000,000	Finnish
Bloom syndrome	<1 in 1,000,000	Ashkenazi Jews
Bohring-Opitz syndrome or C-like syndrome	<1 in 1,000,000	Unknown
Boomerang dysplasia	<1 in 1,000,000	Unknown

The Concise Encyclopedia of Genomic Diseases

Table 3.17.0.1. Continued.

Disease	Frequency of condition	People with higher prevalence (if available)
Borjeson-Forssman-Lehmann syndrome	<1 in 1,000,000	Unknown
Bosley-Salih-Alorainy syndrome or	<1 in 1,000,000	Saudi Arabians and Turkish
Bowen-Conradi syndrome	<1 in 1,000,000	Hutterite people (Canada and the United States)
Brachiootic syndrome and Deafness	<1 in 1,000,000	Unknown
Brachydacytly-mental retardation syndrome	<1 in 1,000,000	Unknown
Branchio-oculo-facial syndrome	<1 in 1,000,000	Unknown
Brittle cornea syndrome	<1 in 1,000,000	Tunisians, Syrians and Palestinians
Bruck syndrome	<1 in 1,000,000	Unknown
Camptodactyly-arthropathy-coxa vara-pericarditis syndrome	<1 in 1,000,000	Unknown
Camurati-Engelmann disease	<1 in 1,000,000	Unknown
Cardiofaciocutaneous Syndrome	<1 in 1,000,000	Unknown
Carpenter syndrome	<1 in 1,000,000	Unknown
Cartilage-hair hypoplasia	<1 in 1,000,000	Finnish and Amish
CATSHL syndrome	<1 in 1,000,000	Unknown
CEDNIK syndrome	<1 in 1,000,000	Unknown
Cenani-Lenz syndactyly syndrome	<1 in 1,000,000	Unknown
Cerebellar ataxia and mental retardation with or without quadrupedal locomotion	<1 in 1,000,000	Iraqis
Cerebrooculofacioskeletal syndrome	<1 in 1,000,000	Finnish and the indigenous people of Manitoba (Canada)
Char syndrome	<1 in 1,000,000	Unknown
CHILD syndrome	<1 in 1,000,000	Unknown
Chromosome 11p11.2 deletion syndrome or Potocki-Shaffer syndrome	<1 in 1,000,000	Unknown
Chromosome 7p21.1 deletion syndrome	<1 in 1,000,000	Unknown
Cleidocranial dysplasia	<1 in 1,000,000	South Africans
Cohen syndrome	<1 in 1,000,000	Finnish
Cold-induced sweating syndrome	<1 in 1,000,000	Norwegians, Israelis and Canadians
Congenita anonychia	<1 in 1,000,000	Unknown
Congenital cataracts, facial dysmorphism, and neuropathy	<1 in 1,000,000	Unknown
Coppock-like cataract	<1 in 1,000,000	Unknown
Cornea plana	<1 in 1,000,000	Finnish

The Concise Encyclopedia of Genomic Diseases

Table 3.17.0.1. Continued.

Disease	Frequency of condition	People with higher prevalence (if available)
Cortical dysplasia-focal epilepsy syndrome and Autism susceptibility	<1 in 1,000,000	Old Order Amish (Pennsylvania)
Costello syndrome	<1 in 1,000,000	Unknown
Cousin syndrome or Pelviscapular dysplasia	<1 in 1,000,000	Maghrebians
Cranioectodermal dysplasia	<1 in 1,000,000	Unknown
Craniofacial-deafness-hand syndrome	<1 in 1,000,000	Unknown
Craniolenticulosutural dysplasia or Boyadjiev-Jabs syndrome	<1 in 1,000,000	Saudi Arabians
Cutis laxa	<1 in 1,000,000	Unknown
Denys-Drash syndrome	<1 in 1,000,000	Unknown
Dermopathy and mandibuloacral dysplasia	<1 in 1,000,000	Unknown
Desbuquois dysplasia	<1 in 1,000,000	Unknown
Desmosterolosis	<1 in 1,000,000	Unknown
Dolicospondylic dysplasia or 3-M syndrome	<1 in 1,000,000	Unknown
Donnai-Barrow syndrome	<1 in 1,000,000	Unknown
Dowling-Degos disease	<1 in 1,000,000	Unknown
Dyggve-Melchior-Clausen disease	<1 in 1,000,000	Unknown
Anhidrotic ectodermal dysplasia with T-cell immunodeficiency	<1 in 1,000,000	Unknown
Ectodermal dysplasia, ectrodactyly and macular dystrophy	<1 in 1,000,000	Unknown
Ehlers-Danlos syndrome, musculocontractural type	<1 in 1,000,000	Unknown
Ehlers-Danlos syndromes types VIIA and VIIB or Arthrochalasia type	<1 in 1,000,000	Unknown
Ehlers-Danlos syndromes types VIIC or Dermatosparaxis type	<1 in 1,000,000	Unknown
Eiken syndrome	<1 in 1,000,000	Turkish
Endocrine-cerebroosteodysplasia	<1 in 1,000,000	Old Order Amish
Epidermolysis bullosa dystrophica	<1 in 1,000,000	Unknown
Erythrokeratodermia variabilis	<1 in 1,000,000	Unknown
Exocrine pancreatic insufficiency, dyserythropoietic anemia, and calvarial hyperostosis	<1 in 1,000,000	Arabs
Feingold syndrome	<1 in 1,000,000	Unknown
Focal dermal hypoplasia or Goltz-Gorlin syndrome	<1 in 1,000,000	Unknown
Frank-ter Haar syndrome	<1 in 1,000,000	Unknown
Fraser syndrome	<1 in 1,000,000	Unknown

Table 3.17.0.1. Continued.

Disease	Frequency of condition	People with higher prevalence (if available)
Frontometaphyseal dysplasia	<1 in 1,000,000	Unknown
Frontonasal dysplasia	<1 in 1,000,000	Unknown
Fuhrmann syndrome	<1 in 1,000,000	Unknown
Geleophysic dysplasia	<1 in 1,000,000	Unknown
Geroderma osteodysplasticum or Walt Disney dwarfism	<1 in 1,000,000	Unknown
Ghosal hematodiaphyseal dysplasia syndrome	<1 in 1,000,000	Unknown
Glomuvenous malformations	<1 in 1,000,000	Unknown
Gnathodiaphyseal dysplasia and muscular dystrophy	<1 in 1,000,000	Japanese
Goldston syndrome or Renal-hepatic-pancreatic dysplasia	<1 in 1,000,000	Unknown
Guttmacher syndrome	<1 in 1,000,000	Unknown
Haim-Munk syndrome	<1 in 1,000,000	Inhabitants of Cochin (India)
Hallermann-Streiff syndrome or François dyscephalic syndrome	<1 in 1,000,000	Unknown
HEM skeletal dysplasia (Hydrops-Ectopic calcification-Moth-eaten skeletal dysplasia) or Greenberg dysplasia	<1 in 1,000,000	Unknown
Hennekam lymphangiectasia-lymphedema syndrome	<1 in 1,000,000	Unknown
Hydrolethalus syndrome	<1 in 1,000,000	Finnish
Hypergonadotropic hypogonadism and Male pseudohermaphroditism	<1 in 1,000,000	Unknown
Hypoparathyroidism, sensorineural deafness and renal disease or Barakat syndrome	<1 in 1,000,000	Unknown
Hypotrichosis and recurrent skin vesicles	<1 in 1,000,000	Unknown
Hypotrichosis simplex	<1 in 1,000,000	Unknown
Hypotrichosis, lymphedema and telangiectasia syndrome	<1 in 1,000,000	Unknown
Ichthyosis follicularis, atrichia, and photophobia syndrome	<1 in 1,000,000	Unknown
Ichthyosis histrix, Curth-Macklin type	<1 in 1,000,000	Unknown
Ichthyosis prematurity syndrome	<1 in 1,000,000	Unknown
Ichthyosis with hypotrichosis	<1 in 1,000,000	Unknown
Ichthyosis, leukocyte vacuoles, alopecia, and sclerosing cholangitis	<1 in 1,000,000	Unknown
Infantile spinal muscular atrophy 2	<1 in 1,000,000	Unknown
Isolated congenital vertical talus	<1 in 1,000,000	Unknown
Jackson-Weiss syndrome	<1 in 1,000,000	Unknown

The Concise Encyclopedia of Genomic Diseases

Table 3.17.0.1. Continued.

Disease	Frequency of condition	People with higher prevalence (if available)
Johanson-Blizzard syndrome	<1 in 1,000,000	Unknown
Junctional epidermolysis bullosa	<1 in 1,000,000	Unknown
Juvenile cataract with microcornea and glucosuria	<1 in 1,000,000	Unknown
KBG syndrome	<1 in 1,000,000	Turkish
Kenny-Caffey syndrome	<1 in 1,000,000	Unknown
Keratitis (and hystrix-like) ichthyosis with deafness syndrome	<1 in 1,000,000	Unknown
Keutel syndrome	<1 in 1,000,000	Unknown
Kindler syndrome	<1 in 1,000,000	Unknown
Langer mesomelic dysplasia	<1 in 1,000,000	Worldwide
Laryngoonychocutaneous syndrome and epidermolysis bullosa	<1 in 1,000,000	Indian subcontinent
Lathosterolosis	<1 in 1,000,000	Unknown
Legius syndrome	<1 in 1,000,000	Unknown
LEOPARD syndrome	<1 in 1,000,000	Unknown
Lethal acantholytic epidermolysis	<1 in 1,000,000	Unknown
Lethal Arthrogryposis with anterior horn cell disease or Vuopala disease	<1 in 1,000,000	Finnish
Lethal congenital contracture syndrome-1	<1 in 1,000,000	Finnish (North-East)
Lethal congenital contracture syndrome-3	<1 in 1,000,000	Israeli Bedouins
Lethal restrictive dermopathy or Lethal tight skin contracture syndrome	<1 in 1,000,000	Unknown
Lhermitte-Duclos disease	<1 in 1,000,000	Unknown
Lissencephaly and Subcortical laminal heteropia	<1 in 1,000,000	Unknown
Lymphedema, distichiasis, renal disease and diabetes mellitus syndrome	<1 in 1,000,000	Unknown
Macrocephaly, alopecia, cutis laxa and scoliosis	<1 in 1,000,000	Israeli-Arabs
Macrocephaly, macrosomia and facial dysmorphism syndrome	<1 in 1,000,000	Unknown
Macrocephaly/autism syndrome	<1 in 1,000,000	Unknown
Marshall syndrome	<1 in 1,000,000	Unknown
Martsolf syndrome	<1 in 1,000,000	Unknown
McKusick-Kaufman syndrome and Bardet-Biedl syndrome	<1 in 1,000,000	Old Order Amish
Megalencephaly-polymicrogyria-polydactyly-hydrocephalus syndrome	<1 in 1,000,000	Unknown
Meleda disease	<1 in 1,000,000	Inhabitants of Mljet island (Croatia)

Table 3.17.0.1. Continued.

Disease	Frequency of condition	People with higher prevalence (if available)
Melnick-Needles syndrome	<1 in 1,000,000	Unknown
Melorheostosis and osteopoikilosis	<1 in 1,000,000	Unknown
Mental retardation, stereotypic movements, epilepsy and/or cerebral malformations, or 5q14.3 microdeletion syndrome	<1 in 1,000,000	Unknown
Metachondromatosis	<1 in 1,000,000	Unknown
Metaphyseal anadysplasia	<1 in 1,000,000	Unknown
Metaphyseal chondrodysplasia, Murk Jansen type	<1 in 1,000,000	Unknown
Metatropic dysplasia	<1 in 1,000,000	Unknown
Microcephalic osteodysplastic primordial dwarfism	<1 in 1,000,000	Unknown
Microcephaly and digital abnormalities with normal intelligence	<1 in 1,000,000	Unknown
Microcephaly, seizures and developmental delay	<1 in 1,000,000	Unknown
Microtia, hearing impairment, and cleft palate	<1 in 1,000,000	Unknown
Microvillus inclusion disease	<1 in 1,000,000	Unknown
Midface retraction syndrome, Schinzel-Giedion type	<1 in 1,000,000	Unknown
Miller syndrome	<1 in 1,000,000	Unknown
Mirror-image polydactyly or Laurin-Sandrow Syndrome	<1 in 1,000,000	Unknown
Mowat-Wilson syndrome	<1 in 1,000,000	Unknown
MULIBREY nanism	<1 in 1,000,000	Finnish
Multiple cutaneous and mucosal venous malformations	<1 in 1,000,000	Unknown
Multiple synostoses syndrome	<1 in 1,000,000	Unknown
Naegeli-Franceschetti-Jadassohn syndrome	<1 in 1,000,000	Unknown
Naxos disease	<1 in 1,000,000	Inhabitants of Greek islands
Neurodegeneration with brain iron accumulation, and cataract syndrome	<1 in 1,000,000	Unknown
Nevo syndrome	<1 in 1,000,000	Middle Easterns
Oculoauricular syndrome, Schorderet type	<1 in 1,000,000	Swiss
Oculodentodigital dysplasia	<1 in 1,000,000	Unknown
Oculo-oto-radial and Duane-radial ray syndrome	<1 in 1,000,000	Unknown
Odontoonychodermal dysplasia	<1 in 1,000,000	Unknown

Table 3.17.0.1. Continued.

Disease	Frequency of condition	People with higher prevalence (if available)
Omodysplasia 1	<1 in 1,000,000	Unknown
Osteoglophonic dysplasia	<1 in 1,000,000	Unknown
Osteopathia striata with cranial sclerosis	<1 in 1,000,000	Unknown
Otopalatodigital syndrome	<1 in 1,000,000	Unknown
Otospondylomegaepiphyseal dysplasia	<1 in 1,000,000	Unknown
Pachyonychia congenita	<1 in 1,000,000	Unknown
Pachyonychia congenita, Jackson-Lawler type	<1 in 1,000,000	Unknown
Pachyonychia congenita, Jadassohn-Lewandowsky type	<1 in 1,000,000	Unknown
Pallister ulnar-mammary syndrome	<1 in 1,000,000	Unknown
Palmoplantar hyperkeratosis and true hermaphroditism	<1 in 1,000,000	Unknown
Papillo-renal syndrome	<1 in 1,000,000	Unknown
Partial agenesis of the pancreas	<1 in 1,000,000	Unknown
PCWH syndrome or Neurologic Variant of Waardenburg-Shah syndrome	<1 in 1,000,000	Unknown
Perrault syndrome	<1 in 1,000,000	Unknown
Peters-plus syndrome	<1 in 1,000,000	Unknown
Phelan-McDermid Syndrome or 22q13 deletion syndrome	<1 in 1,000,000	Unknown
Phocomelia	<1 in 1,000,000	Unknown
Pitt-Hopkins syndrome	<1 in 1,000,000	Unknown
Pitt-Hopkins-like syndrome-1	<1 in 1,000,000	Old Order Amish (Pennsylvania)
Polymicrogyria with optic nerve hypoplasia	<1 in 1,000,000	Unknown
Pontocerebellar hypoplasia	<1 in 1,000,000	Unknown
Radioulnar synostosis with amegakaryocytic thrombocytopenia	<1 in 1,000,000	Unknown
Raine syndrome	<1 in 1,000,000	Middle Easterns
RAPADILINO syndrome	<1 in 1,000,000	Unknown
Renal tubular dysgenesis	<1 in 1,000,000	Unknown
Roberts syndrome or Pseudothalidomide syndrome	<1 in 1,000,000	Unknown
Robinow syndrome	<1 in 1,000,000	Turkish, Omanis, Pakistanis, and Brazilians
Rothmund-Thomson syndrome	<1 in 1,000,000	Unknown
Schimke immuno-osseous dysplasia or Spondyloepiphyseal dysplasia - nephrotic syndrome	<1 in 1,000,000	Unknown

Table 3.17.0.1. Continued.

Disease	Frequency of condition	People with higher prevalence (if available)
Schinzel-Giedion midface retraction syndrome	<1 in 1,000,000	Unknown
Schneckenbecken dysplasia or Chondrodysplasia with snail-like pelvis	<1 in 1,000,000	Unknown
Schopf-Schulz-Passarge syndrome	<1 in 1,000,000	Unknown
Sclerosteosis	<1 in 1,000,000	Afrikaners
Seckel syndrome or Microcephalic primordial dwarfism	<1 in 1,000,000	Unknown
Short rib-polydactyly syndrome 2 digenic phenotype DYNC2H1/NEK1	<1 in 1,000,000	Unknown
Shprintzen-Goldberg syndrome	<1 in 1,000,000	Unknown
Simpson-Golabi-Behmel syndrome	<1 in 1,000,000	Unknown
Skin fragility-woolly hair syndrome	<1 in 1,000,000	Unknown
Small patella syndrome or Coxo-podo-patellar syndrome	<1 in 1,000,000	Unknown
Smith-McCort dysplasia	<1 in 1,000,000	Inhabitants of Madeira Island and Guam
Spondyloepimetaphyseal dysplasia with joint laxity, type 2	<1 in 1,000,000	Americans
Spondyloepimetaphyseal dysplasia, Pakistani type	<1 in 1,000,000	Pakistanis and Turkish
Spondyloepimetaphyseal dysplasia, Strudwick type	<1 in 1,000,000	Unknown
Spondyloepiphyseal dysplasia, Kimberley type	<1 in 1,000,000	Unknown
Spondyloepiphyseal dysplasia, Maroteaux type	<1 in 1,000,000	Unknown
Spondylo-megaepiphyseal-metaphyseal dysplasia	<1 in 1,000,000	Unknown
Spondylometaepiphyseal dysplasia, short limb-hand type	<1 in 1,000,000	Unknown
STAR syndrome or Syndactyly - telecanthus - anogenital and renal malformations	<1 in 1,000,000	Unknown
Terminal osseous dysplasia	<1 in 1,000,000	Unknown
Tetra-amelia or Zimmer phocomelia	<1 in 1,000,000	Arabs and Turkish
Torg-Winchester syndrome	<1 in 1,000,000	Unknown
Trichodontoosseous syndrome	<1 in 1,000,000	Unknown
Trichorhinophalangeal syndromes	<1 in 1,000,000	Unknown
Ulnar-mammary syndrome	<1 in 1,000,000	Unknown
Uniparental disomy	<1 in 1,000,000	Unknown
UV-sensitive syndrome	<1 in 1,000,000	Unknown
Warburg Micro syndrome	<1 in 1,000,000	Unknown

The Concise Encyclopedia of Genomic Diseases

Table 3.17.0.1. Continued.

Disease	Frequency of condition	People with higher prevalence (if available)
Warsaw breakage syndrome	<1 in 1,000,000	Unknown
Weaver syndrome	<1 in 1,000,000	Unknown
Weill-Marchesani syndrome	<1 in 1,000,000	Unknown
Weissenbacher-Zweymuller syndrome	<1 in 1,000,000	Unknown
Whistling face syndrome or Freeman-Sheldon syndrome	<1 in 1,000,000	Unknown
Wolcott-Rallison syndrome	<1 in 1,000,000	Unknown

3.17.00-9 Congenital Malformations and Deformations

3.17.1 Nervous System

3.17.1.1 Microcephaly

Microcephaly: A disorder defined by an occipitofrontal circumference that is more than 2 standard deviations smaller than average for the person's age and sex. There are many environmental causes of congenital microcephaly, including fetal alcohol syndrome, diabetes, radiation, and infection by the varicella zoster virus, rubella or CMV. Moreover, microcephaly is the only proven congenital malformation, found in the surviving children of the Hiroshima and Nagasaki bombings. Most patients with microcephaly exhibit usually moderate mental retardation, but do not have systematized neurological defects or seizures. There are several underlying causes of this disorder with at least 10 autosomal genomic loci: Primary microcephaly-1 can be caused by autosomal recessive mutations in the *MCPH1* gene which encodes the microcephalin 1, a DNA damage response protein;; "Primary microcephaly-2 with or without cortical malformations" is caused by mutations in the *WDR62* gene which encodes the WD repeat domain 62. This protein plays a role in cerebral cortical development;; Primary microcephaly-3 is caused by mutations in the *CDK5RAP2* gene which encodes the CDK5 regulatory subunit (associated protein 2);; Primary microcephaly-4 is caused by mutations in the *CASC5* gene which encodes the cancer susceptibility candidate 5, a protein phosphatase;; Primary microcephaly-5 is caused by mutations in the *ASPM* gene which encodes the asp (abnormal spindle) homolog microcephaly associated);; Primary microcephaly-6 is caused by mutations in the *CENPJ* gene which encodes the cen-

tromere protein J, a protein involved in centrosome integrity;; Primary microcephaly-7 is caused by mutations in the *STIL* gene which encodes the SCL/TAL1 interrupting locus;; Primary microcephaly-8 is caused by mutations in the *CEP135* gene which encodes the centrosomal protein 135kDa;; Primary microcephaly-9 is caused by mutations in the *CEP152* gene which encodes the centrosomal protein 152kDa;; and Primary microcephaly-10 is caused by mutations in the *ZNF335* gene which encodes the zinc finger protein 335. Primary microcephaly is estimated to affect 1 in every 25,000-50,000 births.

Amish lethal microcephaly: A disorder characterized by extreme microcephaly and early death occuring within the 1st year. Affected fetuses manifest a visible microcephaly on ultrasounds taken on the 22nd week of pregnancy. Patients with this condition have a very poor prognosis; the average life span of affected infants is 5-6 months. Affected children have high urinary levels of alpha-ketoglutaric acid. Amish lethal microcephaly is caused by biallelic mutations in the *SLC25A19* gene which encodes the solute carrier (family 25 member A19), a mitochondrial thiamine pyrophosphate carrier. Biologically, this is a membrane transport disease. This disease is ethnic specific, reported only in the Old Order Amish of Lancaster County Pennsylvania. In the latter group, the carrier frequency for the Amish lethal microcephaly trait is estimated to be 1 in 11 individuals, which means the frequency of this condition is about 1 in 500 individuals or $1/(22)^2$.

Bowen-Conradi syndrome: A congenital malformations disorder characterized by microcephaly, low birth weight, mild joint restriction, a prominent nose, micrognathia, fifth finger clinodactyly and 'rocker-bottom' feet. Patients with this condition have a very poor prognosis, with all reported infants dying within the first few months of life. Bowen-Conradi syndrome is caused by biallelic mutations in the *EMG1* gene which encodes the EMG1 N1-specific pseudouridine methyltransferase (nucleolar protein homolog). This essential protein is involved in ribosome biogenesis. This disease is ethnic specific; there are around 40 cases that have been reported in the literature, all of which involved the Hutterite people. This disease is ethnic specific, reported only in the Hutterite people (Canada and the United States). In the Hutterite people, the carrier frequency for Bowen-Conradi syndrome trait is estimated to be 1 in 9 individuals, which means the frequency of this condition is about 1 in 355 individuals or $1/(18)^2$.

Microcephaly - seizures - developmental delay: A disorder characterized by microcephaly, early-onset and intractable seizures, developmental delay, and variable behavioral problems, especially hyperactivity. Microcephaly - seizures - developmental delay is caused by autosomal recessive mutations in the *PNKP* gene which encodes the polynucleotide kinase 3'-phosphatase, a DNA repair protein. Microcephaly - seizures - developmental delay is estimated to affect less than 1 in every 1,000,000 individuals.

3.17.1.2 Congenital Hydrocephalus

Hydrocephalus with stenosis of aqueduct of Sylvius or Bickers-Adams syndrome: A disorder characterized by the association of hydrocephaly, severe intellectual deficit, spasticity and adducted thumbs. Affected individuals display pyramidal tract signs and numerous cerebral malformations including: dilation of the cerebral ventricles, stenosis of the aqueduct of Sylvius and agenesis of the corpus callosum. Patients with this condition have a very poor prognosis. Hydrocephalus with stenosis of aqueduct of Sylvius is caused by X-linked mutations in the *L1CAM* gene which encodes the L1 cell adhesion molecule. This axonal glycoprotein plays an important role in nervous system development, including neuronal migration and differentiation. This condition is X-linked recessive in most cases but occur *de novo* in 30% of cases. This gene maps on the subtelomeric long arm on chromosome X. There are 2 similar conditions caused by mutations in the *L1CAM* gene including: MASA syndrome ("Mental retardation, Aphasia, Spasticity of the lower limbs with hyperreflexia and Adducted thumbs") and Spastic paraplegia associated with intellectual deficit. In comparison with Hydrocephalus with stenosis of aqueduct of Sylvius, the last 2 diseases are associated with prolonged survival. Bickers-Adams syndrome is the commonest form of congenital hydrocephalus. Hydrocephalus with stenosis of aqueduct of Sylvius is estimated to affect 1 in every 30,000 male live births.

Dandy–Walker syndrome or Dandy–Walker complex or Dandy–Walker Malformation: A group of disorders that do not represent a single entity, but rather several abnormalities of brain development which coexist. Patients with this condition exhibit congenital brain malformations involving the cerebellum and the fluid filled spaces around it. A key feature of this condition is the partial or even complete absence of the part of the brain located between the two cerebellar hemispheres. This condition is associated with 3 signs: hydrocephalus, partial or complete absence of the cerebellar vermis, and posterior fossa cyst contiguous with the fourth ventricle. Patients manifest motor deficits such as delayed motor development, hypotonia and ataxia. Half of these patients manifest mental retardation and some have hydrocephalus. A locus responsible for this condition has been mapped on 3q22-q24. Dandy–Walker syndrome displays a 1-2% recurrence risk suggesting that Mendelian inheritance is doubtful. Dandy–Walker syndrome is estimated to affect 1 in every 30,000 live births.

3.17.1.3 Other Congenital Malformations of Brain

3.17.1.3.1 Congenital Malformations of Corpus Callosum

Andermann syndrome or Charlevoix disease: An early-onset disorder characterized by corpus callosum agenesis and neuronopathy. Patients with this condition have developmental milestone delay, severe sensory-motor polyneuropathy with areflexia, and a variable degree of agenesis of the corpus callosum, amyotrophy, hypotonia and cognitive impairment. Andermann syndrome is caused by autosomal recessive protein-truncating mutations in the *SLC12A6* gene which encodes the solute carrier (family 12 member A6), a potassium/chloride transporter. Biologically this is a Membrane transport disease. This disorder has been reported infrequently worldwide. This condition has been reported mainly in the Saguenay-Lac-St-Jean region of the province of Quebec (Canada), predominantly due to the founder effect. In the latter group, the carrier frequency for Andermann syndrome trait is estimated to be 1 in 23 individuals, which means the frequency of this condition is about 1 in 2,117 individuals or $1/(46)^2$.

Acrocallosal syndrome: A polymalformative syndrome characterized by agenesis of corpus callosum, distal anomalies of limbs, minor craniofacial anomalies and intellectual deficit. Acrocallosal syndrome can be caused by autosomal recessive mutations in the *KIF7* gene which encodes the kinesin family member 7. Acrocallosal syndrome is estimated to affect less than 1 in every 1,000,000 individuals, but fewer than 50 cases worldwide have been published.

Proud syndrome: A disorder characterized by agenesis of the corpus callosum with abnormal genitalia and severe mental retardation. Patients with this condition also manifest seizures and spasticity. Male patients are more severely affected, whereas females may be unaffected or have a milder phenotype. Proud syndrome is caused by X-linked recessive mutations in the *ARX* gene which encodes the aristaless related homeobox.

Agenesis of the corpus callosum with mental retardation, ocular coloboma and micrognathia, or Graham-Cox syndrome: A disorder characterized by agenesis of the corpus callosum, coloboma of the iris and optic nerve, high forehead, severe retrognathia and intellectual deficit. "Agenesis of the corpus callosum with mental retardation, ocular coloboma and micrognathia" is caused by X-linked recessive mutations in the *IGBP1* gene which encodes the immunoglobulin (CD79A) binding protein 1. "Agenesis of the corpus callosum with mental retardation, ocular coloboma and micrognathia" is estimated to affect less than 1 in every 1,000,000 individuals.

3.17.1.3.2 Holoprosencephaly

Holoprosencephaly: A complex brain malformation resulting from incomplete cleavage of the prosencephalon occurring between the 18th and 28th day of gestation.

Patients with this condition are affected in both the forebrain and face, which results in neurological manifestations and facial anomalies of variable severity. The etiology of this condition is very heterogeneous; it could occur due to environmental factors such as maternal diabetes or hypocholesterolemia during gestation; or due to chromosomal abnormalities such as chromosome 13 trisomy, and other known syndromes such as Smith-Lemli-Opitz syndrome, CHARGE syndrome, etc... The genomic complexity underlying this condition is yet to be clarified. This condition displays 3 degrees of severity based on their anatomical characteristics: lobar, semi-lobar and alobar. In lobar holoprosencephaly, the ventricles are separated, but there is incomplete frontal cortical separation. Semilobar holoprosencephaly is the most common of this form in neonates who survive gestation. These newborns display partial cortical separations with rudimentary cerebral hemispheres and a single ventricle. In alobar holoprosencephaly, there is a single ventricle and no interhemispheric fissure. The olfactory bulbs and tracts and the corpus callosum are usually absent. Holoprosencephaly is the most common structural malformation of the human forebrain. There are several underlying causes of this disorder with at least 11 autosomal genomic loci: Holoprosencephaly-1 has been mapped, but the gene has not been characterized;; Holoprosencephaly-2 is caused by mutations in the *SIX3* gene which encodes the SIX homeobox 3;; Holoprosencephaly-3 is caused by mutations in the *SHH* gene which encodes the sonic hedgehog protein. This protein has been implicated in patterning the early embryo (ventral neural tube, the anterior-posterior limb axis and the ventral somites);; Holoprosencephaly-4 is caused by mutations in the *TGIF1* gene which encodes the TGFB-induced factor homeobox 1;; Holoprosencephaly-5

is caused by mutations in the *ZIC2* gene which encodes the Zic family member 2;; Holoprosencephaly-6 has been mapped, but the gene has not been characterized;; Holoprosencephaly-7 is caused by mutations in the *PTCH1* gene which encodes the patched 1. This protein is the receptor for sonic hedgehog;; Holoprosencephaly-8 has been mapped, but the gene has not been characterized;; Holoprosencephaly-9 is caused by mutations in the *GLI2* gene which encodes the GLI family zinc finger 2. This protein plays a role during embryogenesis;; Holoprosencephaly-10 has been mapped, but the gene has not been characterized;; and Holoprosencephaly-11 is caused by mutations in the *CDON* gene which encodes the oncogene regulated cell adhesion associated protein. There are 4 major loci associated with this condition. Type 2-3-4 and 5 are the most common and the 4 causative genes are the *SIX3*, the *SHH*, the *TGIF1*, and the *ZIC2*, respectively. These conditions can either be inherited following an autosomal dominant pattern or occur *de novo*. Mutations in the *SHH*, *SIX3*, and *TGIF1* genes are inherited in more than 70% of cases, whereas mutations in the *ZIC2* gene occur *de novo* in 70% of cases. Molecular analyses of the 4 main genes are routinely performed with a mutation detection rate of 25%. Holoprosencephaly is estimated to affect up to 1 in every 250 fetuses during gestations, but occurs in only 1 in every 8,000 live and still births. This condition has similar distributions in countries across the world.

3.17.1.3.3 Other Reduction Deformities of Brain

Lissencephaly: A brain formation disorder caused by defective neuronal migration during the 9th - 13th weeks of gestation. Patients with this condition display an absence of development of brain folds and grooves. Children with lissencephaly are severely neurologically impaired and often die within several months of birth. There are non-genomic causes such as viral infections during the first trimester. There are also a number of genomic causes of lissencephaly, including *de novo* or inherited form. There are around 20 different types of lissencephaly which make up the spectrum. Other causes which have not yet been identified are likely as well. There are several underlying causes of this disorder with at least 4 autosomal genomic loci and 2 X-linked genomic loci: Lissencephaly-1 and Subcortical band heterotopia are caused by monoallelic mutations in the *PAFAH1B1* gene which encodes the platelet-activating factor acetylhydrolase 1b (regulatory subunit 1). Lissencephaly-1 is caused by de novo monoallelic mutations in the *PAFAH1B1* gene;; Lissencephaly-2 or Norman-Roberts syndrome is caused by autosomal recessive mutations in the *RELN* gene which encodes the reelin. This extracellular matrix protein is involved in controlling cell-cell interactions critical for cell positioning and neuronal migration during brain development. This disorder is caused by a disruption of the reelin-signaling pathway;; Lissencephaly-3 is caused by de novo monoallelic mutations in the *TUBA1A* gene which encodes the tubulin alpha 1a. The alpha and beta tubulins represent the major components of microtubules;; Lissencephaly-4 is caused by autosomal recessive mutations in the *NDE1* gene which encodes the nudE neurodevelopment protein 1. This centrosomal protein plays an essential role in microtubule organization, mitosis and neuronal migration;; X-linked Lissencephaly-1 is caused by semi-dominant mutations in the *DCX* gene which encodes the doublecortin;; and X-linked Lissencephaly-2 is caused by mutations in the *ARX* gene which encodes the aristaless-related homeobox. X-linked lissencephaly is esti-

mated to affect 1 in every 85,000 individuals.

Miller-Dieker lissencephaly syndrome: A disorder caused by a contiguous gene microdeletion syndrome involving the subtelomeric area on chromosome 17p13 and including the *PAFAH1B1* gene, and the *YWHAE* gene which encodes the tyrosine 3-monooxygenase/tryptophan 5-monooxyge-nase activation protein (epsilon polypeptide). Miller-Dieker lissencephaly syndrome is caused by *de novo* mutations in 85% of cases, and by autosomal dominant mutations in the remaining cases. Miller-Dieker lissencephaly syndrome can be caused by subtelomeric 17p13 haploinsufficiency. Miller-Dieker lissencephaly syndrome is estimated to affect 1 in every 100,000 individuals.

X-linked lissencephaly with abnormal genitalia: A severe neurological disorder characterized by lissencephaly with posterior-to-anterior gradient and only a moderate increase in the thickness of the cortex. Patients with this condition also exhibit absent corpus callosum, neonatal-onset severe epilepsy, hypothalamic dysfunction with defective temperature regulation, and ambiguous genitalia with micropenis and cryptorchidism. X-linked lissencephaly with abnormal genitalia is caused by X-linked mutations in the *ARX* gene which encodes the aristaless-related homeobox. This condition is only manifested in genotypic males.

Polymicrogyria: A cerebral cortical malformation characterized by excessive cortical folding and shallow sulci. This condition can be caused by both genomic and environmental factors. For example, non-genomic forms can be caused by cytomegalovirus intrauterine infections or due to defects in placenta perfusion.

Bilateral symmetrical perisylvian polymicrogyria: A disorder characterized by mild mental retardation, epilepsy and pseudobulbar palsy. Affected individuals manifest other features including difficulties with speech learning and feeding. This is the most frequent form of genomic polymicrogyria, but its prevalence is unknown. This disease maps on the subtelomeric region on chromosome X (Xq27.2-q28), but the gene has not been characterized.

Bilateral frontoparietal polymicrogyria: A disorder characterized by developmental delay and a nonprogressive cerebellar ataxia. Bilateral frontoparietal polymicrogyria can be caused by autosomal recessive mutations in the *GPR56* gene which encodes the G protein-coupled receptor 56. This protein is involved in the regulation of brain cortical patterning.

Symmetric or asymmetric polymicrogyria: A disorder characterized by cognitive impairment, congenital contralateral hemiparesis and focal seizures. "Symmetric or asymmetric polymicrogyria" is caused by monoallelic mutations in the *TUBB2B* gene which encodes the class IIb beta-tubulin. This condition can either be inherited following an autosomal dominant pattern or occur *de novo*.

Bilateral occipital polymicrogyria: A seizure disorder characterized by complex partial seizures with visual hallucinations and secondary generalization. In the reported cases, this condition could appear anywhere between birth and 24 years. Affected individuals manifest other features including salivation, constriction of the larynx, auditory buzzing and schizophrenic-type psychotic disorder. A Bilateral occipital polymicrogyria locus has been mapped, but the gene has not been characterized. This condition has been reported in a large consanguineous Moroccan family.

Polymicrogyria with optic nerve hypoplasia: A disorder characterized by severe developmental delay, hypotonia, seizures, optic nerve hypoplasia and extensive polymicrogyria. Polymicrogyria with optic nerve hypoplasia is caused by autosomal recessive mutations in the *TUBA8* gene which encodes the tubulin alpha 8. This condition has been reported in consanguineous Pakistani families.

Polymicrogyria with seizures: A disorder characterized by microcephaly, moderate to severe mental retardation, poor speech, dysarthria and seizures. Polymicrogyria with seizures is caused by autosomal recessive mutations in the *RTTN* gene which encodes the rotatin, a protein that might be involved in the maintenance of normal ciliary structure. This condition has been reported in a consanguineous Turkish family.

Megalencephaly-polymicrogyria-polydactyly-hydrocephalus syndrome: A disorder characterized by little intellectual progress, and completely lacking motor and speech development. Affected individuals manifest a distinctive facial aspect with frontal bossing, low nasal bridge and large eyes, but no cutaneous abnormalities. There are several underlying causes of this disorder with at least 2 autosomal genomic loci: Megalencephaly-polymicrogyria-polydactyly-hydrocephalus syndrome can be caused by monoallelic mutations in the *PIK3R2* gene which encodes the phosphoinositide-3-kinase, regulatory subunit 2 (beta), or the *AKT3* gene which encodes the v-akt murine thymoma viral oncogene homolog 3 (gamma protein kinase B). This condition occurs as a *de novo* mutation, particularly affecting a specific CpG dinucleotide in the *PIK3R2* gene. Megalencephaly-polymicrogyria-polydactyly-hydrocephalus syndrome is estimated to affect less than 1 in every 1,000,000 individuals.

Band-like calcification with simplified gyration and polymicrogyria: A disorder characterized by congenital microcephaly, intracranial calcifications and severe developmental delay. "Band-like calcification with simplified gyration and polymicrogyria" is caused by autosomal recessive mutations in the *OCLN* gene which encodes the occluding, a tight junction protein. This condition has been reported in consanguineous families from Egypt and Turkey.

Joubert syndrome: A disorder characterized by absence or underdevelopment of the cerebellar vermis and a malformed brain stem. Affected individuals display hypotonia, cerebellar ataxia, dysregulated breathing patterns and developmental delay. Patients manifest other features including hyperpnea, sleep apnea, abnormal eye and tongue movements. Joubert syndrome is one of the many genomic syndromes associated with syndromic retinitis pigmentosa. Some patients have a mild form with minimal motor disability and good mental development, while others may have severe motor disability and moderate mental retardation. This syndrome is named after Marie Joubert. Typically, this condition follows an autosomal recessive pattern of inheritance, except Joubert syndrome-10 which is X-linked recessive. Most proteins produced from the genes causing this condition are known or suspected to play roles in cilia structures. There are several underlying causes of this disorder with at least 19 autosomal genomic loci and 1 X-linked genomic locus: Joubert syndrome-1 can be caused by autosomal recessive mutations in the *INPP5E* gene which encodes the inositol polyphosphate-5-phosphatase. This gene maps on the subtelomeric long arm on chromosome 9;; Joubert syndrome-2 is caused by autosomal recessive mutations in the *TMEM216* gene which encodes the transmembrane protein 216;; Joubert syndrome-3 is caused by autosomal recessive mutations in the *AHI1* gene which encodes the Abelson helper integration site 1;; Joubert syndrome-4 is caused by autosomal recessive mutations in the *NPHP1* gene which encodes the nephronophthisis 1;; Joubert syndrome-5 is caused by autosomal recessive mutations in the *CEP290* gene which encodes the centrosomal protein 290kDa;; Joubert syndrome-6 is caused by autosomal recessive mutations in the *TMEM67* gene which encodes the transmembrane protein 67;; Joubert syndrome-7 is caused by autosomal recessive mutations in the *RPGRIP1L* gene which encodes the RPGRIP1-like;; Joubert syndrome-8 is caused by autosomal recessive mutations in the *ARL13B* gene which encodes the ADP-ribosylation factor-like 13B;; Joubert syndrome-9 is caused by autosomal recessive mutations in the *CC2D2A* gene which encodes the coiled-coil and C2 domain containing protein 2A;; Joubert syndrome-10 is caused by X-linked recessive mutations in the *OFD1* gene which encodes the oral-facial-digital syndrome 1, a centrosomal protein;; Joubert syndrome-11 is caused by autosomal recessive mutations in the *TTC21B* gene which encodes the tetratricopeptide repeat domain 21B;; Joubert syndrome-12 is caused by autosomal recessive mutations in the *KIF7* gene which encodes the kinesin family member 7;; Joubert syndrome-13 is caused by autosomal recessive mutations in the *TCTN1* gene which encodes the tectonic family member 1;; Joubert syndrome-14 is caused by autosomal recessive mutations in the *TMEM237* gene which encodes the transmembrane protein 237;; Joubert syndrome-15 is caused by autosomal recessive mutations in the *CEP41* gene which encodes the centrosomal protein 41 kDa;; Joubert syndrome-16 is caused by autosomal recessive mutations in the *TMEM138* gene which encodes the transmembrane protein 138;; Joubert syndrome-17 is caused by autosomal recessive mutations in the *C5orf42* gene which encodes the chromosome 5 open reading frame 42;; Joubert syndrome-18 is caused by autosomal recessive mutations in the *TCTN3* gene which encodes the tectonic family member 3;; Joubert syndrome-19 is caused by autosomal recessive or autosomal dominant mutations in the *ZNF423* gene which encodes the zinc finger protein 423;; and Joubert syndrome-20 is caused by autosomal recessive mutations in the *TMEM231* gene which encodes the transmembrane protein 231. Joubert syndrome is estimated to affect 1 in every 80,000-100,000 newborns.

COACH syndrome (Cerebellar vermis hypoplasia, Oligophrenia, congenital Ataxia, Coloboma and Hepatic fibrosis): A disorder characterized by mental retardation, ataxia due to cerebellar hypoplasia and hepatic fibrosis. COACH syndrome is a variant of Joubert syndrome with congenital liver fibrosis. Affected individuals manifest other features including congenital hepatic fibrosis, portal hypertension, coloboma and renal cysts. There are several underlying causes of this disorder with at least 3 autosomal genomic loci: COACH syndrome is most commonly caused by autosomal recessive mutations in the *TMEM67* gene which encodes the transmembrane protein 67. COACH syndrome can be caused by autosomal recessive mutations in the *CC2D2A* gene which encodes the coiled-coil and C2 domain containing protein 2A, or the *RPGRIP1L*

gene which encodes the RPGRIP1-like protein.

Pitt-Hopkins syndrome: A disorder characterized by mental retardation, wide mouth and distinctive facial features, and intermittent hyperventilation followed by apnea. Pitt-Hopkins syndrome is mainly caused by *de novo* monoallelic mutations in the *TCF4* gene which encodes the transcription factor 4. Haploinsufficiency in this gene causes this condition. Pitt-Hopkins syndrome is estimated to affect less than 1 in every 1,000,000 individuals. There are 2 conditions with similar clinical features including: Pitt-Hopkins-like syndrome-1, caused by autosomal recessive mutations in the *CNTNAP2* gene which encodes the contactin associated protein-like 2. Pitt-Hopkins-like syndrome-1 is estimated to affect less than 1 in every 1,000,000 individuals. This condition is an ethnic specific disorder affecting people from the Old Order Amish of Lancaster, Pennsylvania;; and Pitt-Hopkins-like syndrome-2, caused by autosomal recessive mutations in the *NRXN1* gene which encodes the neurexin 1.

Pontocerebellar hypoplasias: Are a group of congenital brain morphogenesis disorders associated with diverse etiologies, such as genomics and environmental. There are several underlying causes of this disorder with 8 clinical forms. Type 1 form is a severe disorder, with death usually occuring early in life. Patients with type 1 display central and peripheral motor dysfunction associated with anterior horn cell degeneration similar to infantile spinal muscular atrophy. Patients with type 2 manifest a progressive microcephaly from birth combined with extrapyramidal dyskinesias. Patients with type 3 display hypotonia, hyperreflexia, microcephaly, optic atrophy and seizures. Patients with type 4 manifest hypertonia, olivopontocerebellar hypoplasia, joint contractures and early death. Patients with type 5 or "Severe Olivopontocerebellar hypoplasia and degeneration" display cerebellar hypoplasia apparent in the 2nd trimester and experience seizures. Patients with type 6 exhibit an uncharacteristically small cerebellum and brainstem, associated with severe developmental delay. Other features of these patients include mitochondrial respiratory chain defects. Patients with type 7 manifest delayed psychomotor development, hypotonia, breathing abnormalities and gonadal abnormalities. Patients with type 8 display severe psychomotor retardation, abnormal movements, hypotonia, spasticity and variable visual defects. There are several underlying causes of this disorder with at least 10 autosomal genomic loci, following an autosomal recessive pattern of inheritance: Pontocerebellar hypoplasia-1A can be caused by autosomal recessive mutations in the *VRK1* gene which encodes the vaccinia related kinase 1;; Pontocerebellar hypoplasia-1B is caused by mutations in the *EXOSC3* gene which encodes the exosome component 3;; Pontocerebellar hypoplasia-2A is caused by mutations in the *TSEN54* gene which encodes the tRNA splicing endonuclease 54 homolog;; Pontocerebellar hypoplasia-2B is caused by mutations in the *TSEN2* gene which encodes the tRNA splicing endonuclease 2 homolog;; Pontocerebellar hypoplasia-2C is caused by mutations in the *TSEN34* gene which encodes the tRNA splicing endonuclease 34 homolog;; Pontocerebellar hypoplasia-2D is caused by mutations in the *SEPSECS* gene which encodes the Sep (O-phosphoserine) tRNA:Sec (selenocysteine) tRNA synthase;; Pontocerebellar hypoplasia-4 is caused by mutations in the *TSEN54* gene;; Pontocerebellar hypoplasia-6 is caused by mutations in the *RARS2* gene which encodes the mitochondrial arginyl-tRNA synthetase 2;; and Pontocerebellar hypo-

plasia-8 is caused by mutations in the *CHMP1A* gene which encodes the charged multivesicular body protein 1A. This protein is involved in the sorting of cell-surface receptors into multivesicular endosomes. Pontocerebellar hypoplasia-3 has been mapped, but the gene has not been characterized. Pontocerebellar hypoplasia-5 and 7 have not been mapped. Most forms of Pontocerebellar hypoplasia appear to affect less than 1 in every 1,000,000 individuals.

Warburg Micro syndrome: A disorder characterized by microcephaly, microphthalmia, microcornea, congenital cataracts, optic atrophy, corpus callosum hypoplasia, cortical dysplasia, severe mental retardation and spastic diplegia. Other features include hypogonadism, short stature, minor digital abnormalities, cardiomyopathy, heart failure and mild facial dysmorphism. There are several underlying causes of this disorder with at least 3 autosomal genomic loci: Warburg Micro syndrome-1 can be caused by autosomal recessive mutations in the *RAB3GAP1* gene which encodes the RAB3 GTPase activating protein subunit 1 (catalytic). The encoded protein regulated exocytosis of neurotransmitters and hormones;; Warburg Micro syndrome-2 can be caused by autosomal recessive mutations in the *RAB3GAP2* gene which encodes the RAB3 GTPase activating protein subunit 2 (non-catalytic);; and Warburg Micro syndrome-3 can be caused by autosomal recessive mutations in the *RAB18* gene which encodes the small GTPase RAB18, a RAS oncogene family member. Warburg Micro syndrome is estimated to affect less than 1 in every 1,000,000 individuals.

3.17.1.3.4 Septooptic Dysplasia

Septooptic dysplasia: A clinically heterogeneous disorder characterized by the classical triad of optic nerve hypoplasia, pituitary hormone abnormalities and midline brain defects. Severity of this condition varies and only 30% of patients manifest the complete clinical triad with most patients having associated features. Septooptic dysplasia can be caused by either monoallelic or biallelic mutations in the *HESX1* gene which encodes the HESX homeobox 1. Mutations in the *HESX1* gene can also cause Combined pituitary hormone deficiency-5 without associated optic nerve hypoplasia or defects of midline brain structures. Less than 1% of Septooptic dysplasia patients have mutations in the *HESX1* gene. There is an additional form of Septooptic dysplasia associated with cardiomyopathy and exercise intolerance caused by mutations in the *MT-CYB* gene. This mitochondrial gene encodes the cytochrome b. Septooptic dysplasia is estimated to affect 1 in every 10,000 live births. The disorder is more common in infants born to younger mothers.

3.17.1.3.5 Congenital Cerebral Cysts

Porencephaly: A disorder characterized by a circumscribed intracerebral cavity of variable size that may be bordered by abnormal polymicrogyric grey matter. The disease is usually unilateral and starts manifesting symptoms during infancy. The clinical manifestations of Porencephaly depend on the location and the size of the lesion. Affected individuals exhibit seizures, hemiplegia and intellectual disability, although the severity is variable. There are several underlying causes of this disorder with at least 2 autosomal genomic loci: Porencephaly-1 is caused by monoallelic mutations in the *COL4A1* gene which encodes the alpha 1 type IV collagen;; and

Porencephaly-2 is caused by monoallelic mutations in the *COL4A2* gene which encodes the alpha 2 type IV collagen. Most cases are sporadic, and the prevalence of this condition is unknown.

Schizencephaly: A grey matter malformation of the brain characterized by a full-thickness cleft within the cerebral hemispheres. Affected individuals exhibit abnormal continuity of histologic grey matter tissue extending from the ependyma lining of the cerebral ventricles to the pial surface of the cerebral hemisphere surface. There are 2 types of schizencephaly, classified based on the size of the area involved and the separation of the cleft lips. Patients with type I are often almost normal but they may experience seizures and spasticity. Type I Schizencephaly consists of a fused cleft. Patients with type II frequently manifest mental retardation, seizures, hypotonia, spasticity, inability to walk or speak, and blindness. In type II Schizencephaly, there is a large defect - a holohemispheric cleft in the cerebral cortex filled with fluid and lined by polymicrogyric gray matter. The cause of the grey matter malformation is not known, but it could be caused by either genomic factors such as predisposition due to a gene mutation, or a physical insult such as infection, infarction, hemorrhage or toxin. Sometimes identical twins are discordant for this condition. For example, in a case reported, one twin was diagnosed at 7 months of pregnancy with this condition while the other was not affected. There are several underlying causes of this disorder with at least 6 autosomal genomic loci: Schizencephaly can be caused in some patients by mutations in the *EMX2* gene which encodes the empty spiracles homeobox 2, the *SIX3* gene which encodes the SIX homeobox 3, or the *SHH* gene which encodes the sonic hedgehog protein.

Schizencephaly is estimated to affect 1 in every 67,500 live births in California.

3.17.1.3.6 Other Specified Congenital Malformations of Brain

Focal cortical dysplasia of Taylor: A cerebral developmental malformation characterized by patients displaying an intractable epilepsy usually requiring surgery. Focal cortical dysplasia of Taylor or Focal cortical dysplasia type II can be caused by mutations in the *TSC1* gene which encodes the tuber-ous sclerosis 1, a tumor suppressor protein that is thought to play a role in the stabiliza-tion of tuberin. This gene maps on the sub-telomeric long arm on chromosome 9. Sometimes in this condition the alteration in the DNA is pro-duce postzygotically. The same gene is mutated in Tuberous sclerosis-1.

Periventricular nodular heterotopia: A brain malformation disorder characterized by abnormal neuronal migration. Patients with this condition have a subset of neu-rons which fails to migrate into the devel-oping cerebral cortex and remains as nod-ules that line the ventricular surface. This disorder is far more frequently diagnosed in females, since it is associated with pre-natal lethality in males (similar to Rett syndrome). These female patients exhibit average intelligence to borderline mental retardation, epilepsy of variable severity and extra-central nervous
system signs, especially cardiovascular de-fects or coagulopathy. There are several un-derlying causes of this disorder with at least 2 X-linked genomic loci and 3 auto-somal genomic loci: Periventricular nodu-lar het-erotopia-1 or X-linked periventricular heter-otopia is caused by X-linked dominant mon-oallelic mutations in the *FLNA* gene which encodes the fila-min A. This protein is involved in remod-eling the cytoskeleton to produce changes in cell shape and migration. This gene maps on the subtelomeric long arm on chromosome Xq28. This is an example of de novo alteration in this subtelomeric gene. Types 2, 3 and 5 are autosomal forms (with similar occurrences in males and females) and have a later seizure onset and fewer nodules than the X-linked dom-inant form. Periventricular nodular hetero-topia-2 is caused by autosomal recessive mutations in the *ARFGEF2* gene which encodes the ADP-ribosylation factor gua-nine nucleotide-exchange factor 2 (bre-feldin A-inhibited);; Periventricular nodu-lar heterotopia-3 has been associated with anomalies of 5p;; Periventricular nodular heterotopia-4 associated with Ehlers-Danlos syndrome is also caused by X-linked mutations in the *FLNA* gene;; and Periventricular nodular het-erotopia-5 has been associated with dele-tions of chromo-some 5q. Prevalence for this condition is unknown and difficult to assess because individuals with the mild phenotype may never seek medical evaluation.

Lhermitte-Duclos disease: A disorder characterized by hamartomatous over-growth of hypertrophic ganglion cells which substitutes the granular cell layer and Purkinje cells of the cerebellum. Af-fected individuals show an abnormal de-velopment, enlargement of the cerebellum, and an increased intracranial pressure. Pa-tients with this condition exhibit headache, nausea, cerebellar dysfunction, occlusive hydrocephalus, ataxia, visual disturbances and other cranial nerve palsies. This condi-tion appears most commonly in the 3rd–4th decades of life. Lhermitte-Duclos disease can be caused by monoallelic mutations in the *PTEN* gene which encodes the phos-phatase and tensin homolog, a tumor sup-pressor. Lhermitte-Duclos disease can co-exist with Cowden disease. Lhermitte-

Duclos disease is estimated to affect less than 1 in every 1,000,000 individuals.

3.17.1.3.7 Neural Tube Defects

Neural tube defects: Are common disorders affecting 1 in every 1,000 individuals of Caucasian American ancestry. Neural tube defects are the 2nd most common type of birth defect after Congenital heart defects. The most common forms of Neural tube defects are Spina bifida and Anencephaly. Multifactorial etiology has been described for these complex traits encompassing both genomic and environmental components. The estimated recurrence risk of spina bifida in siblings of affected children is 4%. There are several underlying causes of this disorder with at least 6 autosomal genomic loci and 1 X-linked genomic loci: Polymorphic variants in genes have been associated with increased risk of neural tube defects, including SNPs in the *MTHFR* gene which encodes the methylenetetrahydrofolate reductase, and mutations in the *VANGL1* gene which encodes the VANGL planar cell polarity protein 1, and the *VANGL2* gene which encodes the VANGL planar cell polarity protein 2. Neural tube defects can be caused by mutations in the *FUZ* gene which encodes the fuzzy planar cell polarity protein, and the *DACT1* gene which encodes the dapper antagonist of beta-catenin homolog 1. There is an X-linked form of either Anencephaly or Spina bifida, but the gene has not been mapped or characterized. It is inferred that this condition follows X-linked recessive patterns of inheritance.

Table 3.17.1 Recognizable Congenital Malformations and Deformations of Nervous System

Disorder	OMIM	Gene	Biological Function	Product of Conception Alteration
Primary microcephaly-1	251200	MCPH1	Nuclear DNA G2/M checkpoint	AR
Primary microcephaly-2, with or without cortical malformations	604317	WDR62	Neurogenesis	AR
Primary microcephaly-3	604804	CDK5RAP2	Neuronal differentiation	AR
Primary microcephaly-4	604321	CASC5	Cancer susceptibility candidate 5	AR
Primary microcephaly-5	608716	ASPM	Mitotic spindle protein	AR
Primary microcephaly-5, with or without simplified gyral pattern	608716	ASPM	Mitotic spindle protein	AR
Primary microcephaly-6	608393	CENPJ	Centrosome integrity protein	AR
Primary microcephaly-7	612703	STIL	Cytoplasmic protein	AR
Primary microcephaly-8	614673	CEP135	Centrosomal protein	AR
Primary microcephaly-9	614852	CEP152	Centrosomal protein	AR

Table 3.17.1. Continued.

Disorder	OMIM	Gene	Biological Function	Product of Conception Alteration
Primary microcephaly-10	615095	ZNF335	Transcription factor Zinc finger	AR
Amish lethal microcephaly	607196	SLC25A19	Solute carrier	AR
Bowen-Conradi syndrome	211180	EMG1	Ribosome biogenesis	AR
Microcephaly, seizures, and developmental delay	613402	PNKP	DNA repair	AR
Microcephaly and digital abnormalities with normal intelligence	602585	MYCN	Transcription factor Basic domains	AD
Hydrocephalus with stenosis of aqueduct of Sylvius or Bickers-Adams syndrome	307000	L1CAM	Axonal glycoprotein	XR + 30%
Andermann syndrome or Charlevoix disease	218000	SLC12A6	Solute carrier	AR
Acrocallosal syndrome	200990	KIF7	Cilia-associated protein	AR
Proud syndrome	300004	ARX	Transcription factor Helix-turn-helix domains	XR
Agenesis of the corpus callosum with mental retardation, ocular coloboma, and micrognathia or Graham-Cox syndrome	300472	IGBP1	Surface phosphoprotein	XR
Partial agenesis of corpus callosum	304100	L1CAM	Axonal glycoprotein	XR
Holoprosencephaly-2	157170	SIX3	Nuclear transcription factor	AD + de novo
Holoprosencephaly-3	142945	SHH	Patterning the early embryo	AD + 50%
Holoprosencephaly-4	142946	TGIF1	Nuclear transcription regulator	AD + de novo
Holoprosencephaly-5	609637	ZIC2	Zinc finger protein	AD + 70%
Holoprosencephaly-7	610828	PTCH1	Cell surface receptor Other/ungrouped	AD
Holoprosencephaly-9	610829	GLI2	Transcription factor Zinc finger	AD + de novo

The Concise Encyclopedia of Genomic Diseases

Table 3.17.1. Continued.

Disorder	OMIM	Gene	Biological Function	Product of Conception Alteration
Holoprosencephaly-11	614226	CDON	Cell adhesion associated protein	AD + de novo
Lissencephaly-1 and subcortical band heterotopia	607432	PAFAH1B1	Variety of biologic and pathologic processes	AD + 99%
Lissencephaly-2 or Norman-Roberts syndrome	257320	RELN	Extracellular matrix protein (control cell-cell interactions)	AR
Lissencephaly-3	611603	TUBA1A	Cytoskeletal Microtubules	AD + 90%
Lissencephaly-4	614019	NDE1	Centrosomal protein	AR
X-linked Lissencephaly-1	300067	DCX	Cytoplasmic protein (bind microtubules)	XsD
X-linked Lissencephaly-2	300215	ARX	Transcription factor Helix-turn-helix domains	XR
Miller-Dieker lissencephaly syndrome	247200	PAFAH1B1/ YWHAE	Variety of biologic and pathologic processes	AD + 85%
Lissencephaly syndrome, Norman-Roberts type	257320	RELN	Extracellular matrix protein (control cell-cell interactions)	AR
X-linked lissencephaly with abnormal genitalia	300215	ARX	Transcription factor Helix-turn-helix domains	XR
Bilateral frontoparietal polymicrogyria	606854	GPR56	Cell surface receptor G protein	AR
Symmetric or asymmetric polymicrogyria	610031	TUBB2B	Cytoskeletal Microtubules	AD + de novo
Polymicrogyria with optic nerve hypoplasia	613180	TUBA8	Cytoskeletal Microtubules	AR
Polymicrogyria with seizures	614833	RTTN	Maintenance of normal ciliary structure	AR
Megalencephaly-polymicrogyria-polydactyly-hydrocephalus syndrome	603157	PIK3R2	Phosphoinositide-3-kinase, regulatory subunit	AD + de novo

Table 3.17.1. Continued.

Disorder	OMIM	Gene	Biological Function	Product of Conception Alteration
Megalencephaly-polymicrogyria-polydactyly-hydrocephalus syndrome	611223	AKT3	Protein kinase	AD + de novo
Band-like calcification with simplified gyration and polymicrogyria		OCLN	Tight junction protein	
Joubert syndrome-1	213300	INPP5E	Ciliary proteins, primary cilia	AR
Joubert syndrome-2	608091	TMEM216	Ciliary proteins other	AR
Joubert syndrome-3	608629	AHI1	Ciliary proteins Basal body	
Joubert syndrome-4	609583	NPHP1	Channelopathy slit diaphragm	AR
Joubert syndrome-5	610188	CEP290	Ciliary proteins Nephrocystin	AR
Joubert syndrome-6	610688	TMEM67	Ciliary proteins, primary cilia	AR
Joubert syndrome-7	611560	RPGRIP1L	Ciliary proteins Nephrocystin	AR
Joubert syndrome-8	612291	ARL13B	Ciliary proteins, primary cilia	AR
Joubert syndrome-9	612285	CC2D2A	Ciliary proteins Basal body	AR
Joubert syndrome-10	300804	OFD1	Ciliary proteins Basal body	XR
Joubert syndrome-11	613820	TTC21B	Cilium axoneme	AR
Joubert syndrome-12	200990	KIF7	Cilia-associated protein	AR
Joubert syndrome-13	614173	TCTN1	Tectonic family protein	AR
Joubert syndrome-14	614424	TMEM237	WNT signaling	AR
Joubert syndrome-15	614464	CEP41	Centrosomal protein	AR
Joubert syndrome-16	614465	TMEM138	Ciliogenesis	AR
Joubert syndrome-17	614615	C5orf42	Transmembrane protein	AR
Joubert syndrome-18	614815	TCTN3	Tectonic family protein	AR
Joubert syndrome-19	614844	ZNF423	Transcription factor Zinc finger	AR or AD
Joubert syndrome-20	614970	TMEM231	Transmembrane protein	AR

Table 3.17.1. Continued.

Disorder	OMIM	Gene	Biological Function	Product of Conception Alteration
COACH syndrome	216360	CC2D2A	Ciliary proteins Basal body	AR
COACH syndrome	216360	RPGRIP1L	Ciliary proteins Nephrocystin	AR
COACH syndrome	216360	TMEM67	Ciliary proteins, primary cilia	AR
Pitt-Hopkins syndrome	610954	TCF4	Transcription factor	AD + 90%
Pitt-Hopkins-like syndrome-1	610042	CNTNAP2	Cell adhesion molecule	AR
Pitt-Hopkins-like syndrome-2	614325	NRXN1	cell adhesion molecules and receptors	AR
Pontocerebellar hypoplasia-1A	607596	VRK1	Serine/threonine protein kinase	AR
Pontocerebellar hypoplasia-1B	614678	EXOSC3	Exosome component	AR
Pontocerebellar hypoplasia-2A	277470	TSEN54	Nuclear tRNA splicing endonuclease	AR
Pontocerebellar hypoplasia-2B	612389	TSEN2	Nuclear tRNA splicing endonuclease	AR
Pontocerebellar hypoplasia-2C	612390	TSEN34	Nuclear tRNA splicing endonuclease	AR
Pontocerebellar hypoplasia-2D	613811	SEPSECS	Phosphoserine selenocysteine tRNA synthase	AR
Pontocerebellar hypoplasia-4	225753	TSEN54	Nuclear tRNA splicing endonuclease	AR
Pontocerebellar hypoplasia-6	611523	RARS2	Mitochondrial protein	AR
Pontocerebellar hypoplasia-8	614961	CHMP1A	Multivesicular body protein	AR
Warburg Micro syndrome-1	600118	RAB3GAP1	Rab GTPase activating protein	AR
Warburg Micro syndrome-2	614225	RAB3GAP2	Rab GTPase activating protein	AR
Warburg Micro syndrome-3	614222	RAB18	GTP-binding protein	AR
Septooptic dysplasia or De Morsier syndrome	182230	HESX1	Transcriptional repressor	AR or AD
Exercise intolerance, cardiomyopathy, and septooptic dysplasia	516020	MT-CYB	Mitochondrial protein	Mit
Porencephaly-1	175780	COL4A1	Extracellular matrix protein	AD

The Concise Encyclopedia of Genomic Diseases

Table 3.17.1. Continued.

Disorder	OMIM	Gene	Biological Function	Product of Conception Alteration
Porencephaly-2	614483	COL4A2	Extracellular matrix protein	AD
Schizencephaly	269160	EMX2	Transcription factor	AD + de novo
Schizencephaly	269160	SHH	Patterning the early embryo	AD
Schizensephaly	269160	SIX3	Nuclear transcription factor	AD
Focal cortical dysplasia of Taylor	607341	TSC1	Tumor suppressor	AD
Cortical dysplasia-focal epilepsy syndrome	610042	CNTNAP2	Cell adhesion molecule	AR
Periventricular nodular heterotopia-1 or X-linked periventricular heterotopia	300049	FLNA	Cytoskeletal Microfilaments	XD + de novo
Periventricular nodular heterotopia-2	608097	ARFGEF2	Transcription factor Zinc finger	AR
Periventricular nodular heterotopia-4 associated with Ehlers-Danlos syndrome	300537	FLNA	Cytoskeletal Microfilaments	XD
Lhermitte-Duclos syndrome	158350	PTEN	Tumor suppressor	AD
Susceptibility to Neural tube defects	158105	CCL2	Secreted-Chemokine	
Folate-sensitive neural tube defects	172460	MTHFD1	Dehydrogenase	
Folate-sensitive neural tube defects	607093	MTHFR	Cytosol reductase	
Susceptibility to Neural tube defects	610132	VANGL1	Integral membrane protein	
Folate-sensitive neural tube defects	156570	MTR	Methyltransferase	
Folate-sensitive neural tube defects	602568	MTRR	Mitochondrial protein	

3.17.2 Eye, Ear, Face and Neck

3.17.2.1 Congenital Malformations of Eyelid, Lacrimal Apparatus and Orbit

Blepharophimosis-ptosis-epicanthus inversus syndrome: A condition characterized by dysplasia of the eyelids either with premature ovarian failure or without. Blepharophimosis-ptosis-epicanthus inversus syndrome is caused by monoallelic mutations in the *FOXL2* gene which encodes the forkhead box L2, a transcription factor. This condition is inherited following an autosomal dominant pattern of inheritance in 50% of the cases and occurs as a *de novo* event in the other 50%. There is evidence of advanced maternal age effect in de novo mutations. About a hundred articles of Blepharophimosis-ptosis-epicanthus inversus syndrome have been written worldwide. Although this condition has not been evaluated accurately, Blepharophimosis-ptosis-epicanthus inversus syndrome is estimated to affect less than 1 in every 5,000 individuals. About 12% of individuals with this condition do not have an identified *FOXL2* gene mutation.

Autosomal dominant aplasia of lacrimal and salivary glands: A disorder characterized by irritable eyes, epiphora and xerostomia. Affected individuals manifest an increased risk of dental erosion, dental caries, periodontal disease and oral infections. Autosomal dominant aplasia of lacrimal and salivary glands is caused by monoallelic loss-of-function mutations in the *FGF10* gene which encodes the fibroblast growth factor 10. Lacrimoauriculodentodigital syndrome is an allelic disorder with a more severe phenotype.

3.17.2.2 Anophthalmos, Microphthalmos and Macrophthalmos

Anterior segment mesenchymal dysgenesis: A heterogeneous group of complex diseases, encompassing aniridia, Axenfeld and Rieger anomalies, iridogoniodysgenesis, Peters anomaly, and posterior embryotoxon. Anterior segment mesenchymal dysgenesis can be caused by monoallelic mutations either in the *PITX3* gene which encodes the paired-like homeodomain 3, or the *FOXE3* gene which encodes the forkhead box E3.

Microphthalmia: A developmental disorder of the eye that literally means small eye. Microphthalmia can be either unilateral or bilateral. People with microphthalmia may also have a condition called coloboma. Colobomas are missing pieces of tissue in structures that form the eye. Sometimes Microphthalmia is associated with environmental factors such as Fetal alcohol syndrome or infections during pregnancy. Genomic causes of Microphthalmia include chromosomal abnormalities (Patau syndrome and Wolf-Hirschhorn Syndrome) or monogenomic Mendelian disorders. The latter may be autosomal dominant, autosomal recessive or X linked. There are dozens of genes that have been implicated in microphthalmia including many transcription and regulatory factors. There are several forms of microphthalmia including: 8 types of isolated forms, 4 types associated with cataract form, 7 types associated with coloboma form, and 12 types associated with syndromic form. There are several forms of microphthalmia for which the genes have not been characterized. Isolated microphthalmia-2 can be caused by mutations in the *VSX2* gene which encodes the visual system homeobox 2;; Isolated microphthalmia-3 can be caused by mutations in the *RAX* gene which encodes the retina and anterior neural fold homeobox;; Isolated microphthalmia-4 can be caused by mutations in the *GDF6* gene which encodes the growth differentiation factor 6;; Isolated microphthalmia-5 can be caused by mutations in the *MFRP* gene which encodes the membrane frizzled-related protein;; Isolated microphthalmia-7 can be caused by mutations in the *GDF3* gene which encodes the growth differentiation factor 3;; Microphthalmia with cataract-2 can be caused by mutations in the *SIX6* gene which encodes the SIX homeobox 6;; Microphthalmia with cataract-4 can be caused by mutations in the *CRYBA4* gene which encodes the crystallin beta A4;; Microphthalmia with coloboma-3 can be caused by mutations in the *VSX2* gene;; Microphthalmia with coloboma-5 can be caused by mutations in the *SHH* gene;; Microphthalmia with coloboma-6 can be caused by mutations in the *GDF3* gene;; Microphthalmia with coloboma-7 can be caused by mutations in the *ABCB6* gene which encodes the ATP-binding cassette (sub-family B member 6);; Microphthalmia with coloboma-8 can be caused by mutations in the *STRA6* gene which encodes the protein stimulated by retinoic acid 6;; Syndromic microphthalmia-2 or "Microphthalmia, congenital cataract, persistent primary teeth, dental radiculomegaly and atrial septal defect" can be caused by mutations in the *BCOR* gene which encodes the BCL6 corepressor;; Syndromic microphthalmia-3 or "Anophthalmia or microphthalmia with or without defects of the optic nerve, optic chiasm, optic tract and extraocular abnormalities" can be caused by mutations in the *SOX2* gene which encodes the SIX homeobox 2;; Syndromic microphthalmia-5 or "Microphthalmia with early-onset retinal dystrophy and pituitary dysfunction" can be caused by mutations in the *OTX2* gene

which encodes the orthodenticle homeobox 2;; Syndromic microphthalmia-6 or "Microphthalmia with brain and digital anomalies" can be caused by mutations in the *BMP4* gene which encodes the bone morphogenetic protein 4;; Syndromic microphthalmia-7 or "Unilateral or bilateral microphthalmia and linear skin defects" can be caused by X-linked dominant mutations in the *HCCS* gene which encodes the holocytochrome c synthase;; Syndromic microphthalmia-9 or "Anophthalmia, bilateral pulmonary agenesis, overriding aorta with high ventricular septal defect, nodular vestige of the main pulmonary artery and eventration of the left hemidiaphragm" can be caused by mutations in the *STRA6* gene;; and Syndromic microphthalmia-11 or "Microphthalmia associated with cleft lip and palate and agenesis of the corpus callosum" is caused by biallelic mutations in the *VAX1* gene which encodes the ventral anterior homeobox 1. Microphthalmia is estimated to affect 1 in every 7,100 live births. This condition is responsible for blindness in 3-11% of children.

3.17.2.3 Congenital Lens Malformations

Cataracts: A group of disorders characterized by clouding that develops in the crystalline lens of the eye or in its envelope, obstructing the passage of light. There are several non-genomic causes of cataract, including long-term exposure to ultraviolet light, exposure to ionizing radiation, secondary effects of diseases such as diabetes, hypertension and advanced age, or trauma. Genomic factors are often a cause of congenital cataracts, and positive family history may also play a role in predisposing someone to cataracts at an earlier age, a phenomenon of "anticipation" in presenile cataracts. There are Chromosomal disorders associated with cataract including, 1q21.1 deletion syndrome, Cri-du-chat syndrome, Patau syndrome, Schmid-Fraccaro syndrome, Edwards syndrome and Turner syndrome. There are single-gene disorders associated with cataract including Alport syndrome, Conradi syndrome, Myotonic dystrophy, Basal-cell nevus syndrome, Ichthyosis, Fabry disease, Galactosemia or galactosemic cataract, Homocystinuria, Hypothyroidism, Mucopolysaccharidoses and Wilson disease. There are dozens of genes that have been implicated in cataract. There are several forms of cataracts for which the genes have not been characterized. There are several forms of congenital bilateral cataracts including: Cerulean (Patients show bluish and white opacifications in the superficial layers of the lens nucleus), Nuclear (involves the central or nuclear part of the lens), Zonular (involves the Zonules of Zinn part of the lens), Coppock-like (family named), Crystalline aculeiform, Lamellar, Posterior polar, and syndromic forms. Congenital cataract can be caused by biallelic mutations in the *CRYAA* gene which encodes the crystallin alpha A;; Congenital cataract can be caused by X-linked dominant monoallelic mutations in the *NHS* gene which encodes the Nance-Horan syndrome;; Congenital cerulean cataract -2 can be caused by monoallelic mutations in the *CRYBB2* gene which encodes the crystallin beta B2;; Congenital cerulean cataract -3 can be caused by monoallelic mutations in the *CRYGD* gene which encodes the crystallin gamma D;; Autosomal recessive congenital cataract-2 is caused by biallelic mutations in the *FYCO1* gene which encodes the FYVE and coiled-coil domain containing protein 1;; Autosomal recessive congenital cataract-4 is caused by biallelic mutations in the *TDRD7* gene which encodes the tudor domain containing protein 7;; and Autosomal recessive congenital cataract-5 is caused by biallelic mutations in the *AGK* gene which encodes the acylglycerol kinase. "Congenital cataracts, facial dysmorphism and neuropathy" can be caused by biallelic mutations in the *CTDP1* gene which encodes the CTD (carboxy-terminal domain RNA polymerase II polypeptide A) phosphatase subunit 1;; Congenital nuclear cataract-2 can be caused by biallelic mutations in the *CRYBB3* gene which encodes the crystallin beta B3;; Congenital nuclear cataract-3 can be caused by biallelic mutations in the *CRYBB1* gene which encodes the crystallin beta B1;; Congenital or juvenile-onset cataract can be caused by monoallelic mutations in the *BFSP2* gene which encodes the beaded filament structural protein 2 phakinin;; Congenital zonular cataract with sutural opacities can be caused by monoallelic mutations in the *CRYBA1* gene which encodes the crystallin beta A1;; Coppock-like cataract can be caused by monoallelic mutations in the *CRYBB2* gene which encodes the crystallin beta B2;; Coppock-like cataract can be caused by monoallelic mutations in the *CRYGC* gene which encodes the crystallin gamma C;; Crystalline aculeiform cataract can be caused by monoallelic mutations in the *CRYGD* gene crystallin gamma D;; Juvenile cataract with microcornea and glucosuria can be caused by monoallelic mutations in the *SLC16A12* gene which encodes the solute carrier (family 16 member A12), a monocarboxylic acid transporter;; Lamellar cataract-2 can be caused by monoallelic mutations in the *CRYBA4* gene which encodes the crystallin beta A4;; Lamellar cataract, Marner type can be caused by monoallelic mutations in the *HSF4* gene which encodes the heat shock transcription factor 4;; Nance-Horan syndrome can be caused by either X-linked dominant or *de novo* monoallelic mutations in the *NHS* gene;; Nonnuclear polymorphic congenital cataract can be caused by monoallelic mutations in the

CRYGD gene;; Nuclear cataract can be caused by monoallelic mutations in the *CRYAA* gene;; Polymorphic and lamellar cataract can be caused by monoallelic mutations in the *MIP* gene which encodes the major intrinsic protein of lens fiber;; Posterior polar cataract-1 can be caused by monoallelic mutations in the *EPHA2* gene which encodes the EPH receptor A2;; Posterior polar cataract-2 can be caused by monoallelic mutations in the *CRYAB* gene;; Posterior polar cataract-3 can be caused by monoallelic mutations in the *CHMP4B* gene which encodes the charged multivesicular body protein 4B. This protein is involved in the sorting of cell-surface receptors into multivesicular endosomes;; Posterior polar cataract-4 can be caused by monoallelic mutations in the *PITX3* gene which encodes the paired-like homeodomain 3;; Pulverulent zonular cataract-1 can be caused by monoallelic mutations in the *GJA8* gene which encodes the gap junction protein (alpha 8);; Pulverulent zonular cataract-3 can be caused by monoallelic mutations in the *GJA3* gene which encodes the gap junction protein (alpha 3);; Sutural cataract with punctate and cerulean opacities can be caused by monoallelic mutations in the *CRYBB2* gene which encodes the crystallin beta B2;; and Zonular central nuclear cataract can be caused by monoallelic mutations in the *CRYAA* gene. Congenital cataracts are estimated to affect 1-3 in every 10,000 children worldwide. This condition is responsible for blindness in 5-20% of children.

Ectopia lentis: A disorder characterized by an anomalous extending of the zonular fibers that leads to lens dislocation, and occasionally acute or chronic visual impairment. There are several underlying causes of this disorder with at least 2 autosomal genomic loci: Familial ectopia lentis can be caused by monoallelic mutations in

the *FBN1* gene which encodes the fibrillin 1, a glycoprotein which helps in providing the structural support in elastic and nonelastic connective tissue throughout the body;; and Isolated ectopia lentis is caused by biallelic mutations in the *ADAMTSL4* gene which encodes the ADAMTS-like 4, a disintegrin and metalloproteinase with thrombospondin motifs. This protein is involved in diverse biological functions including cellular adhesion, angiogenesis, and patterning of the developing nervous system.

Congenital primary aphakia: A developmental disorder characterized by absence of the lens. Congenital primary aphakia is caused by biallelic mutations in the *FOXE3* gene which encodes the forkhead box E3, a transcription factor.

3.17.2.4 Congenital Malformations of Anterior Segment of Eye

Aniridia: A congenital ocular malformation disorder characterized by the complete or partial absence of the iris. This condition can be isolated or occur as part of a syndrome. Isolated aniridia can occur in association with a range of other ocular anomalies including cataracts, glaucoma (usually occurring during adolescence), corneal pannus, optic nerve hypoplasia, absence of macular reflex, ectopia lentis, nystagmus, and photophobia, all of which generally result in poor vision. Syndromic aniridia is associated with other non-ophthalmological anomalies, for instance: Wilms tumor, Aniridia, Genitourinary anomalies, and mental Retardation syndrome in the case of the WAGR syndrome. Aniridia is caused by monoallelic mutations in the *PAX6* gene which encodes the paired box 6, a transcriptional regulator. This protein is involved in oculogenesis and other developmental processes. Mutations in this gene can either be inherited or occur *de novo*. Isolated aniridia is estimated to affect 1 in every 64,000-96,000 individuals.

Peters anomaly: A disorder characterized by central corneal leukoma, and absence of the posterior corneal stroma and Descemet membrane. Affected individuals manifest a variable degree of iris and lenticular attachments to the central aspect of the posterior cornea. Peters anomaly can occur as an isolated ocular abnormality or in association with other ocular defects; including the Cataract-microcornea syndrome. Peters anomaly is a feature of the Krause-Kivlin syndrome and the Peters-plus syndrome. There are several underlying causes of this disorder with at least 4 autosomal genomic loci: Peters anomaly can be caused by mutations in the *PAX6* gene, the *PITX2* gene which encodes the paired-like homeodomain 2, the *CYP1B1* gene which encodes the cytochrome P450 (subfamily 1B polypeptide 1), or the *FOXC1* gene which encodes the forkhead box C1, a transcription factor.

Peters-plus syndrome: A disorder characterized by the association of Peters' congenital glaucoma with dwarfism, intellectual deficit, disorders of the ear, and cleft palate. Peters' congenital glaucoma results from a genomic developmental disorder at the end of the 3rd week of uterine life when the mesencephalic neural crests migrate. Peters-plus syndrome is caused by biallelic mutations in *B3GALTL* gene which encodes the beta 1,3-galactosyltransferase-like.

Autosomal recessive cornea plana: A disorder characterized by reduced corneal curvature leading in most cases to hyperopia, hazy corneal limbus and arcus lipoides at an early age. Autosomal recessive cor-

nea plana can be caused by biallelic mutations in the *KERA* gene which encodes the keratocan, a sulfate proteoglycan that is involved in corneal transparency. There is an autosomal dominant form that is mild, whereas this autosomal recessive form is severe and frequently associated with additional ocular manifestations.

Axenfeld-Rieger syndrome: A disorder characterized by ocular and non-ocular features. Patients with this condition exhibit ocular abnormalities mainly affecting the iris, cornea and the chamber angle. There are several underlying causes of this disorder with at least 3 autosomal genomic loci: Axenfeld-Rieger syndrome-1 can be caused by monoallelic mutations in the *PITX2* gene which encodes the paired-like homeodomain 2, a transcription factor;; Axenfeld-Rieger syndrome-2 has been mapped, but the gene has not been characterized;; and Axenfeld-Rieger syndrome-3 can be caused by monoallelic mutations in the *FOXC1* gene which encodes the forkhead box C1, a transcription factor. A large number of different mutations have been identified in these genes, but there is no clear genotype-phenotype relationship. Axenfeld-Rieger syndrome is estimated to affect 1 in every 200,000 individuals.

Iridogoniodysgenesis: A disorder characterized by malformations of the iridocorneal angle of the anterior chamber of the eye. Iridogoniodysgenesis is the result of abnormal migration or terminal induction of neural crest cells. These cells lead to formation of most of the anterior segment structures of the eye. There are several underlying causes of this disorder with at least 2 autosomal genomic loci: Iridogoniodysgenesis-1 can be caused by monoallelic mutations in the FOXC1 gene which encodes the forkhead box C1, a transcription factor;; and Iridogoniodysgenesis-2 can be caused by monoallelic mutations in the PITX2 gene which encodes the paired-like homeodomain 2, a transcription factor. Iridogoniodysgenesis is estimated to affect 1 in every 200,000 individuals in the United States.

3.17.2.5 Congenital Malformations of Posterior Segment of Eye

Ocular coloboma: A congenital abnormality caused by defective closure of the embryonic fissure of the optic cup. The defect is usually located in the inferonasal part of the iris. Ocular coloboma can be caused by monoallelic mutations in the PAX6 gene which encodes the paired box 6, a transcriptional regulator. This protein is involved in oculogenesis and other developmental processes. There is another loci associated with ocular developmental anomalies that occurs from disruption of the GDF6 gene which encodes the growth differentiation factor 6, and the GDF3 gene which encodes the growth differentiation factor 3. Coloboma is a prime feature of the CHARGE syndrome. Coloboma is estimated to affect 1 in every 10,000 individuals.

Coloboma of the optic nerve: A disorder caused by mutations in the PAX6 gene. Coloboma of the optic nerve is estimated to affect 1 in every 2,000,000 individuals in the United States.

3.17.2.6 Congenital Malformations of Ear

Microtia, hearing impairment and cleft palate: A disorder characterized by bilateral microtia, partial cleft palate, and mixed symmetric severe to profound hearing impairment. "Microtia, hearing im-

Papillorenal syndrome: A disorder characterized by optic nerve congenital cleft leading to variable visual disorders, and renal anomalies such as hypoplasia (i.e. insufficient kidney growth). Papillorenal syndrome is caused by autosomal dominant mutations in the PAX2 gene which encodes the paired box 2, a transcription factor. About half of all patients with this condition do not have an identified PAX2 gene mutation.

Norrie disease: A vitreoretinal dysplasia disorder characterized by the association of microphthalmia, hypoplasic irides, synechiae, glaucoma and cataracts. This condition only affects males whereas female carriers have a normal phenotype; but in rare cases, female carriers may display some symptoms of the disorder. Patients with this condition exhibit bilateral leucocoria appearing during the 1st weeks of life. Affected individuals sometimes manifest bilateral perceptive deafness occurring between the ages of 20 and 30 years. Patients also frequently display psychomotor retardation, along with many systemic abnormalities including pulmonary, cardiac, genitourinary, skeletal and gastrointestinal. Norrie disease is caused by X-linked recessive mutations in the NDP gene which encodes the Norrie disease (pseudoglioma) protein, a secreted protein with a cystein-knot motif that activates the Wnt/beta-catenin pathway.

pairment and cleft palate" is caused by biallelic point mutations in the *HOXA2* gene which encodes the homeobox A2, a transcription factor. This condition has been described in a consanguineous Iranian family.

3.17.2.7 Congenital Malformations of Face and Neck

Branchio-oculo-facial syndrome: A disorder characterized by growth retardation resulting in low birth weight and hemangiomatous bilateral branchial clefts. Affected individuals sometimes exhibit linear skin lesions behind the ears, obstructed nasolacrimal ducts, a broad nasal bridge with a flattened nasal tip, congenital strabismus and a protruding upper lip. About 50 cases have been reported so far. Branchio-oculo-facial syndrome is caused by monoallelic mutations in the *TFAP2A* gene which encodes the transcription factor AP-2 (alpha). The mode of transmission can be either inherited following an autosomal dominant pattern or occur *de novo*. Branchio-oculo-facial syndrome is estimated to affect less than 1 in every 1,000,000 individuals.

Table 3.17.2 Recognizable Congenital Malformations and Deformations of Eye, Ear, Face and Neck

Disorder	OMIM	Gene	Biological Function	Product of Conception Alteration
Blepharophimosis, epicanthus inversus, and ptosis	110100	*FOXL2*	Transcription factor Helix-turn-helix domains	AD + 50%
Autosomal dominant aplasia of lacrimal and salivary glands	180920	*FGF10*	Mitogenic and cell survival activities	AD
Anterior segment mesenchymal dysgenesis	107250	*PITX3*	Transcription factor Helix-turn-helix domains	AD
Anterior segment mesenchymal dysgenesis	107250	*FOXE3*	Transcription factor Helix-turn-helix domains	AD
Isolated microphthalmia-2	610093	*VSX2*	Nuclear transcription factor	AR
Isolated microphthalmia-3	611038	*RAX*	Transcription factor	AR
Isolated microphthalmia-4	613094	*GDF6*	Cell surface receptor Enzyme-linked receptor	AD
Isolated microphthalmia-5	611040	*MFRP*	Membrane protein	AR
Isolated microphthalmia-7	613704	*GDF3*	Cell surface receptor Enzyme-linked receptor	AD
Microphthalmia with cataract-2	212550	*SIX6*	Nuclear transcription factor	AD
Microphthalmia with cataract-4	610426	*CRYBA4*	Crystallin	AD
Microphthalmia with coloboma-3	610092	*VSX2*	Nuclear transcription factor	AR
Microphthalmia with coloboma-5	611638	*SHH*	Patterning the early embryo	AD
Microphthalmia with coloboma-6	613703	*GDF3*	Cell surface receptor Enzyme-linked receptor	AD

Table 3.17.2. Continued.

Disorder	OMIM	Gene	Biological Function	Product of Conception Alteration
Microphthalmia with coloboma-7	614497	ABCB6	ABC-transporter protein	AD
Microphthalmia with coloboma-8	601186	STRA6	Membrane protein	AR
Syndromic microphthalmia-2	300166	BCOR	Zinc finger transcription repressor	XD
Syndromic microphthalmia-3	206900	SOX2	Transcription factor β-Scaffold factors	AD + de novo
Syndromic microphthalmia-5	610125	OTX2	Transcription factor	AD + de novo
Syndromic microphthalmia-6	607932	BMP4	Cell surface receptor Enzyme-linked receptor	AD + 99%
Syndromic microphthalmia-7	309801	HCCS	Links a heme group to the apoprotein of cytochrome c	XD
Syndromic microphthalmia-9	601186	STRA6	Membrane protein	AR
Syndromic microphthalmia-11	614402	VAX1	Ventral anterior homeobox protein	AR
Congenital cataract	123580	CRYAA	Crystallin	AR
Congenital cataract	302200	NHS	Development of the eyes, teeth and brain	XR + de novo
Congenital cerulean cataract-2	601547	CRYBB2	Crystallin	AD
Congenital cerulean cataract-3	608983	CRYGD	Crystallin	AD
Autosomal recessive congenital cataract-2	610019	FYCO1	Transport of autophagic vesicles	AR
Autosomal recessive congenital cataract-4	613887	TDRD7	Determining the fate of mRNAs	AR
Autosomal recessive congenital cataract-5	614691	AGK	Acylglycerol kinase	AR
Congenital cataracts, facial dysmorphism, and neuropathy	604168	CTDP1	Transcription initiation factor	AR
Congenital nuclear cataract-2	609741	CRYBB3	Crystallin	AR
Congenital nuclear cataract-3	611544	CRYBB1	Crystallin	AR
Congenital or juvenile-onset cataract	604219	BFSP2	Cytoskeletal Intermediate filaments	AD
Congenital zonular cataract with sutural opacities	600881	CRYBA1	Crystallin	AD
Coppock-like cataract	604307	CRYBB2	Crystallin	AD

Table 3.17.2. Continued.

Disorder	OMIM	Gene	Biological Function	Product of Conception Alteration
Coppock-like cataract	604307	*CRYGC*	Crystallin	AD
Crystalline aculeiform cataract	115700	*CRYGD*	Crystallin	AD
Juvenile cataract with microcornea and glucosuria	612018	*SLC16A12*	Solute carrier	AD
Lamellar cataract 2	610425	*CRYBA4*	Crystallin	AD
Lamellar cataract, Marner type	116800	*HSF4*	Transcription factor	AD + de novo
Nance-Horan syndrome	302350	*NHS*	Development of the eyes, teeth and brain	XD + de novo
Nonnuclear polymorphic congenital cataract	601286	*CRYGD*	Crystallin	AD
Nuclear cataract	123580	*CRYAA*	Crystallin	AD
Polymorphic and lamellar cataract	604219	*MIP*	Channel proteins	AD
Posterior polar cataract-1	613020	*EPHA2*	Tyrosine kinase	AD
Posterior polar cataract-2	613763	*CRYAB*	Crystallin	AD
Posterior polar cataract-3	605387	*CHMP4B*	Endosomal Sorting Complex Required for Transport	AD
Posterior polar cataract-4	610623	*PITX3*	Transcription factor Helix-turn-helix domains	AD
Pulverulent zonular cataract-1	116200	*GJA8*	Channelopathy Connexin	AD
Pulverulent zonular cataract-3	601885	*GJA3*	Channelopathy Connexin	AD
Sutural cataract with punctate and cerulean opacities	607133	*CRYBB2*	Crystallin	AD
Zonular central nuclear cataract	123580	*CRYAA*	Crystallin	AD
Cataract-1 or Cataract-microcornea syndrome	116150	*GJA8*	Channelopathy Connexin	AD
Cataract with late-onset corneal dystrophy	604219	*PAX6*	Transcription factor Helix-turn-helix domains	AD
Familial ectopia lentis	129600	*FBN1*	Cytoskeletal Microfilaments	AD
Isolated ectopia lentis	225100	*ADAMTSL 4*	Extracellular matrix protein	AR
Congenital primary aphakia	610256	*FOXE3*	Transcription factor Helix-turn-helix domains	AR
Aniridia	106210	*PAX6*	Transcription factor Helix-turn-helix domains	AD + de novo

The Concise Encyclopedia of Genomic Diseases

Table 3.17.2. Continued.

Disorder	OMIM	Gene	Biological Function	Product of Conception Alteration
Peters anomaly	604229	CYP1B1	Endoplasmic reticulum protein	AR
Peters anomaly	604229	PAX6	Transcription factor Helix-turn-helix domains	AR
Peters anomaly	604229	PITX2	Transcription factor Helix-turn-helix domains	AR
Peters-plus syndrome	261540	B3GALTL	Membrane protein	AR
Autosomal recessive cornea plana	217300	KERA	Keratocan	AD + AD = AR
Axenfeld-Rieger syndrome-1	180500	PITX2	Transcription factor Helix-turn-helix domains	AD + de novo
Axenfeld-Rieger syndrome-3	602482	FOXC1	Transcription factor Helix-turn-helix domains	AD + de novo
Iridogoniodysgenesis-1	601631	FOXC1	Transcription factor Helix-turn-helix domains	AD
Iridogoniodysgenesis-2	137600	PITX2	Transcription factor Helix-turn-helix domains	AD
Iris hypoplasia and glaucoma	601631	FOXC1	Transcription factor Helix-turn-helix domains	AD
Ocular coloboma	120200	PAX6	Transcription factor Helix-turn-helix domains	AD
Coloboma of optic nerve and morning glory disc anomaly	120430	PAX6	Transcription factor Helix-turn-helix domains	
Papillorenal syndrome	120330	PAX2	Transcription factor Helix-turn-helix domains	AD
Norrie disease	310600	NDP	Secreted protein that activates the Wnt/beta-catenin pathway	XR
Microtia, hearing impairment, and cleft palate	612290	HOXA2	Transcription factor Helix-turn-helix domains	AR
Branchiooculofacial syndrome	113620	TFAP2A	Nuclear transcription factor	AD + de novo

3.17.3 Circulatory System

3.17.3.1 Congenital Malformations of Cardiac Chambers and Connections

Double outlet right ventricle: A disorder characterized by cardiac anomaly in which both the aorta and pulmonary trunk originate, predominantly or exclusively, from the right ventricle. This condition is associated with a wide spectrum of malformations. Some patients have clinical features similar to those of Tetralogy of Fallot and others resemble Transposition with ventricular septal defect. Double outlet right ventricle on occasion can have an identifiable genomic cause. Double outlet right ventricle can be caused by mutations in the *CFC1* gene which encodes the cripto FRL-1 cryptic (family 1), or the *GDF1* gene which encodes the growth differentiation factor 1. Double-outlet right ventricle can also be caused by mutations in the *ZFPM2* gene which encodes the zinc finger protein FOG (family member 2). Double outlet right ventricle is estimated to affect 1 in every 3,000 live births. This condition is responsible for 3% or more of all congenital heart defects.

Dextro-looped transposition of the great arteries: A disorder characterized by atrioventricular concordance and ventriculoarterial discordance. The exact etiology remains unknown. Several non-genomic risk factors have been shown to be implicated including: gestational diabetes mellitus, maternal use of antiepileptic drugs, and maternal exposure to rodenticides and herbicides. There are several underlying causes of this disorder with at least 3 autosomal genomic loci: Dextro-looped transposition of the great arteries-1 can be caused by mutations in the *MED13L* gene which encodes the mediator complex sub-unit 13-like;; Dextro-looped transposition of the great arteries-2 can be caused by mutations in the *CFC1* gene which encodes the cripto FRL-1 cryptic (family 1);; and Dextro-looped transposition of the great arteries-3 can be caused by monoallelic mutations in the *GDF1* gene which encodes the growth differentiation factor 1. Dextro-looped transposition of the great arteries is estimated to affect 1 in every 3,500-5,000 live births, with a male-to-female ratio of 1.5-3.2 to 1.

Conotruncal heart malformations: A group of disorders characterized by cardiac outflow tract defects, such as Tetralogy of Fallot, Pulmonary atresia, Double-outlet right ventricle, Persistent truncus arteriosus and Aortic arch anomalies. These conditions account for between one fourth and one third of all nonsyndromic congenital heart defects. 30% of cases of isolated Conotruncal anomalies are associated with microdeletions of 22q11.2. Studies of families with children with cardiac malformations found that conotruncal malformations carry a higher recurrence risk than any other cardiac defects. Dextro-looped transposition of the great arteries and Double-outlet right ventricle can also be caused by mutations in the *CFC1* gene, which plays a role in the etiology of this condition. Conotruncal anomaly face syndrome can be caused by mutations in the *TBX1* gene which encodes the T-box 1, a transcription factor. The *TBX1* gene maps on the 22q11.2. Persistent truncus arteriosus can be caused by biallelic mutations in the *NKX2-6* gene which encodes the NK2 homeobox 6. This form has been identified in a consanguineous Kuwaiti family;; Interrupted aortic arch and persistent truncus arteriosus can be caused by mutations in the *NKX2-5* gene which encodes the NK2 homeobox 5, a transcription factor with heart formation and development functions;; and Persistent truncus arteriosus can be caused by mutations in the *GATA6* gene which encodes the GATA binding protein 6, a transcription factor.

Yorifuji-Okuno syndrome: A disorder characterized by partial pancreatic agenesis, diabetes mellitus and heart anomalies. Yorifuji-Okuno syndrome can be caused by autosomal dominant mutations in the *GATA6* gene. This condition has been reported in Japan.

3.17.3.2 Congenital Malformations of Cardiac Septa

Atrial septal defect or Interauricular communication: A disorder characterized by a communication between the atrial chambers of the heart. About one third of Atrial septal defect children have an associated genomic syndrome, such as Down's syndrome, Noonan syndrome, Alagille syndrome, Holt-Oram syndrome and Ellis-van Creveld syndrome. Most defects appear to occur sporadically as a result of spontaneous genomic mutations, but hereditary forms have been reported. Atrial septal defect occasionally displays a familial pattern of inheritance, suggesting that a Mendelian form may sometimes be responsible for this condition. There are several underlying causes of this disorder with at least 9 autosomal genomic loci: Atrial septal defect-1 has been mapped, but the gene has not been identified;; Atrial septal defect-2 can be caused by mutations in the *GATA4* gene which encodes the GATA binding protein 4, a transcription factor;; Atrial septal defect-3 is not associated with other cardiac abnormalities and is caused by mutations in the *MYH6* gene which encodes the cardiac muscle myosin heavy chain 6 alpha;; Atrial septal defect-4 can be caused by mutations in the *TBX20* gene which encodes the T-box 20, a transcription factor;; Atrial

septal defect-5 is not associated with other cardiac abnormalities and is caused by mutations in the *ACTC1* gene which encodes the cardiac muscle actin alpha 1;; Atrial septal defect-6 associated with aneurysm of the interatrial septum and cardiac arrhythmias is caused by mutations in the *TLL1* gene which encodes the tolloid-like 1;; Atrial septal defect-7 is often associated with atrioventricular conduction defects and is caused by mutations in the *NKX2-5* gene which encodes the NK2 homeobox 5, a transcription factor;; Atrial septal defect-8 is associated with other cardiac anomalies and is caused by mutations in the *CITED2* gene which encodes the Cbp/p300-interacting transactivator with Glu/Asp-rich carboxy-terminal domain 2;; and Atrial septal defect-9 can be caused by mutations in the *GATA6* gene which encodes the GATA binding protein 6, a transcription factor. Atrial septal defect is estimated to affect 1 in every 1,500 live births, with a female-to-male ratio of 2-4 to 1. This condition is responsible for 6-8% or more of all congenital heart defects.

Atrioventricular septal defect: A disorder characterized by a defect in the atrioventricular septum of the heart. This condition is caused by an inadequate fusion of the endocardial cushions with the mid portion of the atrial septum and the muscular portion of the ventricular septum. Atrioventricular septal defect occurs either as part of a malformation syndrome or as a nonsyndromic form. The syndromic form occurs in patients with Down syndrome and Heterotaxy syndromes. There are several underlying causes of this disorder with at least 5 autosomal genomic loci: Atrioventricular septal defect-1 has been mapped, but the gene has not been identified;; Atrioventricular septal defect-2 is caused by mutations in the *CRELD1* gene which encodes the cysteine-rich with

EGF-like domains 1;; Atrioventricular septal defect-3 is caused by mutations in the *GJA1* gene which encodes the gap junction protein (alpha 1);; Atrioventricular septal defect-4 is caused by mutations in the *GATA4* gene;; and Atrioventricular septal defect-5 is caused by mutations in the *GATA6* gene.

Complete atrioventricular canal: A disorder characterized by a defect in the ostium primum atrial septum, and a variable deficiency of the ventricular septum inflow. Complete atrioventricular canal is estimated to affect 1 in every 5,000 live births. Both sexes are similarly affected and this condition is more prominently associated with Down syndrome.

Partial atrioventricular canal: A disorder characterized by defective fusion of the endocardial cushions with the atrial septum primum. Partial atrioventricular canal is estimated to affect 1 in every 5,000 live births. This condition is responsible for 4% of all congenital heart defects.

Tetralogy of Fallot: A disorder characterized by the combination of interventricular communication, override of the ventricular septum by the aortic root, obstruction of the right ventricular outflow tract and right ventricular hypertrophy. This condition is named after Étienne-Louis Arthur Fallot. Tetralogy of Fallot is thought to be cause by environmental or genomic factors or a combination. The etiology of this condition is multifactorial including untreated Maternal diabetes, Phenylketonuria and intake of retinoic acid. Several chromosomal anomalies can cause this condition including trisomies 21, 18, and 13, and especially microdeletions of chromosome 22. There are several underlying causes of this disorder with at least 7 autosomal genomic loci: Tetralogy of Fallot can be caused by monoallelic mutations in the *JAG1* gene which encodes the jagged 1, in the *NKX2-5* gene which encodes the NK2 homeobox 5, or in the *GATA4* gene which encodes the GATA binding protein 4, a transcription factor. Sporadic Tetralogy of Fallot form can be caused by mutations in the *ZFPM2* gene which encodes the zinc finger protein FOG family (member 2), in the *GDF1* gene which encodes the growth differentiation factor 1, in the *TBX1* gene which encodes the T-box 1, or in the *GATA6* gene which encodes the GATA binding protein 6, a transcription factor. The heritability for Tetralogy of Fallot is about 54%. The recurrence risk for Fallot tetralogy in siblings is between 1-3%. Tetralogy of Fallot is estimated to affect 1 in every 2,500 live births.

Hypoplastic left heart syndrome: A disorder characterized by abnormal development of the left-sided cardiac structures, resulting in obstruction to blood flow from the left ventricular outflow tract. Affected individuals exhibit underdevelopment of the left ventricle, aorta and aortic arch, as well as mitral atresia or stenosis. Hypoplastic left heart syndrome is compatible with multifactorial inheritance, the recurrence risk among later-born siblings is about 2%. There are several underlying causes of this disorder with at least 2 autosomal genomic loci: Hypoplastic left heart syndrome-1 can be caused by mutations in the *GJA1* gene which encodes the gap junction protein (alpha 1);; and Hypoplastic left heart syndrome-2 can be caused by mutations in the *NKX2-5* gene which encodes the NK2 homeobox 5. Hypoplastic left heart syndrome is estimated to affect 1-2 in every 6,250 live births.

3.17.3.3 Congenital Malformations of Aortic and Mitral Valves

Bicuspid aortic valve: A disorder that is the most common congenital condition of the aortic valve where 2 of the aortic valvular leaflets fuse during development. This condition results in a valve that is bicuspid instead of the normal tricuspid configuration. There are several underlying causes of this disorder with at least 2 autosomal genomic loci: Aortic valve disease-1 is caused by monoallelic mutations in the *NOTCH1* gene which encodes the notch 1, a protein involved in a variety of developmental processes. This gene maps on the subtelomeric long arm on chromosome 9;; and Aortic valve disease-2 is caused by monoallelic mutations in the *SMAD6* gene which encodes the SMAD family member 6. Bicuspid aortic valve is estimated to affect 1-2% of all people, affecting males twice as often as females. The incidence of bicuspid aortic valve can be as high as 10% in families affected with the valve problem.

Supravalvular aortic stenosis: A disorder that is usually localized and characterized by a diffuse narrowing of the ascending aorta or other arteries such as coronary arteries and branch pulmonary arteries. This narrowing of the aorta or pulmonary branches may impede blood flow, causing heart murmur and ventricular hypertrophy. Supravalvular aortic stenosis can be caused by monoallelic mutations in the *ELN* gene which encodes the elastin, a protein that is one of the two components of elastic fibers. The disease can it occur de novo or be transmitted as an autosomal dominant pattern; the proportion of cases caused by de novo mutations is unknown. Supravalvular aortic stenosis can be part of the Williams-Beuren syndrome due to a contiguous gene deletion syndrome that includes hemizygous deletion of the *ELN* gene on chromosome 7q11.23. Supravalvular aortic stenosis is estimated to affect 1 in every 7,500 births.

3.17.3.4 Other Congenital Malformations of Heart

Meacham-Winn-Culler syndrome or Rhabdomyomatous dysplasia - cardiopathy - genital anomalies: A multiple malformation syndrome characterized by congenital diaphragmatic abnormalities, genital defects and cardiac malformations. In some patients, Meacham-Winn-Culler syndrome can be caused by monoallelic mutations in the *WT1* gene which encodes the Wilms tumor 1 protein, a transcription factor. Less than 15 patients have been reported with this condition worldwide and there is no sex predilection.

Char syndrome: A disorder characterized by the triad of patent ductus arteriosus, facial dysmorphism and hand anomalies. Char syndrome is caused by monoallelic mutations in the *TFAP2B* gene which encodes the transcription factor AP-2 beta (activating enhancer binding protein 2 beta). This condition can be inherited or occur de novo; the proportion of cases caused by de novo mutations is unknown. Char syndrome is estimated to affect less than 1 in every 1,000,000 individuals.

3.17.3.5 Congenital Malformations of Peripheral Vascular System

Parkes Weber syndrome: A disorder characterized by cutaneous flushing with underlying multiple micro-arteriovenous fistulas. In some instances Parkes Weber syndrome can be caused by mutations in the *RASA1* gene which encodes the Ras GTPase-activating protein.

Arterial tortuosity syndrome: A connective tissue disorder characterized by tortuosity and elongation of the large and medium-sized arteries and a predisposition towards aneurysm formation. Patients with this condition also exhibit stenosis of the pulmonary arteries and vascular dissection. Arterial tortuosity syndrome is caused by autosomal recessive mutations in the *SLC2A10* gene which encodes the solute carrier (family 2 member A10), a facilitative glucose transporter. Arterial tortuosity syndrome is estimated to affect less than 1 in every 1,000,000 individuals.

Glomuvenous malformations: Are vascular malformations characterized by the presence of small, multifocal bluish-purple venous lesions involving the skin. These lesions are painful on palpation and display a cobble-stone appearance. Glomuvenous malformations are caused by monoallelic mutations in the *GLMN* gene which encodes the glomulin, a FKBP associated protein. The encoded protein is specifically expressed in vascular smooth muscle cells and is essential for normal development of the vasculature. Prevalence of this condition is unknown.

Multiple cutaneous and mucosal venous malformations: Are venous malformations charac-terized by lesions that tend to be multifocal and small. Malformations in affected individuals are comprised of grossly dilated vascular spaces lined by a single continuous layer of endothelial cells. Some patients with this condition exhibit venous malformation located in internal organs, and some have additional anomalies, including cardiac malformations. "Multiple cutaneous and mucosal venous malformations" can be caused by monoallelic mutations in the *TEK* gene which encodes the epithelial-specific tyrosine kinase receptor. There is also another familial or autosomal dominant form of venous malformation called the Blue rubber bleb nevus, but the gene has not been mapped. Blue rubber bleb nevus is a bladderlike variety of hemangioma appearing predominantly on the trunk and upper arms. "Cerebral cavernous malformations" is a similar condition without cutaneous lesions, and has been discussed under Neoplasms in the subchapter "Neoplasms of Uncertain or Unknown Behavior" (3.02.11).

Table 3.17.3 Recognizable Congenital Malformations and Deformations of Circulatory System

Disorder	OMIM	Gene	Biological Function	Product of Conception Alteration
Double-outlet right ventricle	217095	*GDF1*	Cell surface receptor Enzyme-linked receptor	
Double-outlet right ventricle	217095	*CFC1*	Intercellular signaling pathways	
Double-outlet right ventricle	603693	*ZFPM2*	zinc finger protein	
Dextro-looped transposition of the great arteries-1	608808	*MED13L*	RNA polymerase 2 (initiation of transcription)	
Dextro-looped transposition of the great arteries-2	613853	*CFC1*	Intercellular signaling pathways	
Dextro-looped transposition of the great arteries-3	613854	*GDF1*	Cell surface receptor Enzyme-linked receptor	AD
Conotruncal anomaly face syndrome	217095	*TBX1*	Nuclear transcription factor	AD + de novo

Table 3.17.3. Continued.

Disorder	OMIM	Gene	Biological Function	Product of Conception Alteration
Persistent truncus arteriosus	217095	NKX2-6	Transcription factor	AR
Interrupted aortic arch and persistent truncus arteriosus	600584	NKX2-5	Transcription factor	
Persistent truncus arteriosus	601656	GATA6	Transcription factor Zinc finger	
Yorifuji-Okuno syndrome	11	GATA6	Transcription factor Zinc finger	AD + de novo
Susceptibility to cardiac conduction defect	115080	AKAP10	Mitochondrial protein	AD
Atrial septal defect-2	607941	GATA4	Transcription factor Zinc finger	AD
Atrial septal defect-3	614089	MYH6	Cytoskeletal Microfilaments	AD
Atrial septal defect-4	611363	TBX20	Nuclear transcription factor	AD
Atrial septal defect-5	612794	ACTC1	Cytoskeletal Microfilaments	AD
Atrial septal defect-6	613087	TLL1	Peptidase protease	
Atrial septal defect-7	108900	NKX2-5	Transcription factor	
Atrial septal defect-8	614433	CITED2	Interacting transactivator protein	
Atrial septal defect-9	614475	GATA6	Transcription factor Zinc finger	AD
Atrioventricular septal defect-2	606217	CRELD1	Cell adhesion molecule	
Atrioventricular septal defect-3	600309	GJA1	Channelopathy Connexin	
Atrioventricular septal defect-4	614430	GATA4	Transcription factor Zinc finger	AD
Atrioventricular septal defect-5	614474	GATA6	Transcription factor Zinc finger	AD
Pancreatic agenesis and congenital heart defects	600001	GATA6	Transcription factor Zinc finger	AD
Atrial septal defect with atrioventricular conduction defects	108900	NKX2-5	Transcription factor	
Tetralogy of Fallot	187500	GDF1	Cell surface receptor Enzyme-linked receptor	AD
Tetralogy of Fallot	187500	JAG1	Signalling through notch 1	
Tetralogy of Fallot	187500	ZFPM2	Transcription factor	
Tetralogy of Fallot	187500	GDF1	Cell surface receptor Enzyme-linked receptor	
Tetrology of Fallot	187500	NKX2-5	Transcription factor	AD

Table 3.17.3. Continued.

Disorder	OMIM	Gene	Biological Function	Product of Conception Alteration
Tetralogy of Fallot	187500	GDF1	Cell surface receptor Enzyme-linked receptor	AD
Tetralogy of Fallot	187500	JAG1	Signalling through notch 1	
Tetralogy of Fallot	187500	ZFPM2	Transcription factor	
Tetralogy of Fallot	187500	GDF1	Cell surface receptor Enzyme-linked receptor	
Tetrology of Fallot	187500	NKX2-5	Transcription factor	AD
Tetrology of Fallot	187500	GATA4	Transcription factor Zinc finger	AD
Tetrology of Fallot	187500	TBX1	Nuclear transcription factor	
Bicuspid aortic valve-1	109730	NOTCH1	Notch signaling pathway	AD
Bicuspid aortic valve-2	614823	SMAD6	Signal transducers and transcriptional modulators	AD
Hypoplastic left heart syndrome-1	241550	GJA1	Channelopathy Connexin	
Hypoplastic left heart syndrome-2	614435	NKX2-5	Transcription factor	
Meacham-Winn-Culler syndrome or Rhabdomyomatous dysplasia - cardiopathy - genital anomalies	608978	WT1	Transcription factor Zinc finger	AD + de novo
Char syndrome	169100	TFAP2B	Nuclear transcription factor	AD + de novo
Supravalvular aortic stenosis	185500	ELN	Extracellular matrix protein	AD + de novo
Parkes Weber sindrome	608355	RASA1	GTPase-activating protein	
Arterial tortuosity syndrome	208050	SLC2A10	Solute carrier	AR
Glomuvenous malformations	138000	GLMN	Intracellular protein	AD
Multiple cutaneous and mucosal venous malformations	600195	TEK	Tyrosine kinase	AD
Capillary malformation-arteriovenous malformation	608354	RASA1	GTPase-activating protein	AD + de novo

3.17.4 Respiratory System

Laryngoonychocutaneous syndrome: A disorder characterized by hoarseness, dystrophic changes in the nails, chronic bleeding, and crusted lesions of the skin of the face. In addition, the teeth are deformed. Laryngoonychocutaneous syndrome is caused by biallelic mutations in the *LAMA3* gene which encodes the alpha-3 subunit of laminin 5. Laminins are basement membrane components involved in the organization of cells into tissues during embryonic development by interacting with other extra-cellular matrix components. This condition has been reported in people of Punjabi Mus-lim ancestry living in Lahore, Pakistan.

3.17.5 Digestive System

3.17.5.1 Cleft Lip and Cleft Palate

Cleft lip/palate: A common facial developmental disorder affecting 1 in every 700 children. This condition is thought to be caused by environmental or genomic factors, or a combination. The clefting mainly occurs during the 6th–8th weeks of pregnancy. There are many environmental influences which may also cause or interact with genomic factors to produce orofacial clefting. This condition has been linked to maternal hypoxia (due to smoking), maternal alcohol abuse and maternal hypertension. There are syndromic forms of Cleft lip and cleft palate, such as those associated with patients with Van der Woude syndrome, Siderius X-linked mental retardation, Stickler's Syn-drome, Loeys-Dietz syndrome, Patau Syn-drome and Treacher Collins Syndrome. There are several underlying causes of this disorder with at least 13 autosomal genomic loci; very few genes have been characterized: Orofacial cleft-5 is caused by mutations in

the *MSX1* gene which encodes the msh homeobox 1, a transcription repressor;; Orofacial cleft-7 is caused by mutations in the *PVRL1* gene which encodes the poliovirus receptor-related 1;; Orofacial cleft-8 is caused by mutations in the *TP63* gene which encodes the tumor protein p63, a transcription factor;; Orofacial cleft-10 is caused by mutations in the *SUMO1* gene which encodes the SMT3 suppressor of mif two 3 (homolog 1);; Orofacial cleft-11 is caused by mutations in the *BMP4* gene which encodes the bone morphogenetic protein 4. The recurrence risk for this condition varies from 4-17%. For example, in subsequently born children, it is 4% if one child has it, 4% if one parent has it, 17% if one parent and one child have it, and 9% if two children have it. There is evidence of advanced maternal age effect in de novo cases. This trait seems to be dominant with low penetrance. There is a large prevalence of clefting in different ethnic backgrounds. The highest prevalence rate of clefting has been reported for Native Americans with 3.74 in every 1,000 children and the lowest was reported for Africans at 0.18-1.67 in every 1,000 children.

Cleft lips: A disorder characterized by the association with cleft palate in two-thirds of cases. Males are 1.6 times more frequently affected than females. The lip or the lip and palate together fail to close in approximately 1 in every 1,000 live births. The recurrence risk in a subsequent baby is 2-8%.

Cleft lip/palate-ectodermal dysplasia syn-drome or Zlotogora-Ogur syn-drome: A disorder characterized by cleft lip/palate and ectodermal dysplasia syndrome with limb involvement. Affected individuals manifest syndactyly, ectodermal dysplasia and cleft/lip palate. Cleft lip/palate-ectodermal dysplasia syndrome is caused by autosomal recessive mutations in the *PVRL1* gene which encodes the poliovirus receptor-related 1, a cell-cell adhesion molecule/ herpesvirus receptor. The prevalence for this condition is unknown, but males and females are equally affected. This syndrome is particularly common on Margarita Island, Venezuela.

3.17.5.2 Congenital Malformations of Lips (not elsewhere classified)

Van der Woude syndrome: A craniofacial disorder characterized by the association of pits or sinuses on the lower lip with cleft lip and/or cleft palate. This condition represents the most frequent form of syndromic cleft lip and palate, affecting around 1 in every 60,000 individuals. Patients with this condition sometimes exhibit hypodontia. Van der Woude syndrome is caused by monoallelic or autosomal dominant mutations in the *IRF6* gene which encodes the interferon regulatory factor 6. The syndrome exhibits a high penetrance at around 80-97%.

3.17.5.3 Congenital Absence, Atresia and Stenosis of Large Intestine

Caudal duplication anomaly: A disorder characterized by duplication to various extents of the embryonic cloaca and notochord. Patients with this condition exhibit duplications of different organs in the caudal region. For example, these patients could have a duplication of the colon, bladder, distal spine, ureter and genital organs. Caudal duplication anomaly encompasses a spectrum of anomalies associated with conjoined twinning. Caudal duplication anomaly is associated with hypermethylation in the *AXIN1* gene promoter. This gene encodes the axin 1 protein. Caudal duplication anomaly is estimated to affect less than 1 in every 100,000 individuals.

Currarino syndrome: A disorder characterized by partial sacral agenesis associated with a presacral mass and ano-rectal malformation. Sometimes patients with this condition have presacral teratoma. Currarino syndrome is caused by monoallelic mutations in the *MNX1* gene which encodes the motor neuron and pancreas homeobox 1, a transcription factor. Currarino syndrome occurs as *de novo* in 15-30% of cases and is inherited with an autosomal dominant pattern of inheritance with reduced penetrance in the remaining cases. Females are more frequently diagnosed, since this condition is often associated with gynecological and urinary tract problems. Currarino syndrome is estimated to affect 1-9 in every 100,000 individuals.

3.17.5.4 Other Congenital Malformations of Intestine

Hirschsprung disease or Aganglionic megacolon: A disorder characterized by the failure of the neural crest cells to migrate completely during fetal development of the intestine. The affected segment of the colon fails to relax, causing an obstruction. Hirschsprung disease is estimated to affect 1 in every 5,000 births. Caucasian children are more commonly affected than other ethnic groups. Hirschsprung disease occurs as an isolated trait in 70% of cases. Hirschsprung disease occurs in association with chromosomal anomaly in 12% of cases, and occurs with additional congenital anomalies in 18% of cases. The most common conditions associated with Hirschsprung disease include: Down syndrome, Waardenburg-Shah syndrome,

Mowat-Wilson syndrome, Goldberg-Shpritzen megacolon syndrome and Congenital central hypoventilation syndrome. There are several underlying causes of this disorder with at least 9 autosomal genomic loci: Susceptibility to the development of Hirschsprung disease-1 is caused by mutations in *RET* gene which encodes the ret proto-oncogene;; Hirschsprung disease-2 is associated with variations in the *EDNRB* gene which encodes the endothelin receptor type B;; Hirschsprung disease-3 is associated with variations in the *GDNF* gene which encodes the glial cell derived neurotrophic factor;; Hirschsprung disease-4 is associated with variations in the *EDN3* gene which encodes the endothelin 3;; Hirschsprung disease-5 has been mapped, but the gene has not been characterized;; Hirschsprung disease-6 has been mapped, but the gene has not been characterized;; Hirschsprung disease-7 has been mapped, but the gene has not been characterized;; Hirschsprung disease-8 has been mapped, but the gene has not been characterized;; and Hirschsprung disease-9 has been mapped, but the gene has not been characterized. This condition does not strictly follow a Mendelian mode of inheritance. The most important exclusion criteria for a genomic origin of Hirschsprung disease include the fact that this condition is 3.5-4 times more often in males than in females, while most of the genes are mapped on autosomal chromosome.

Shprintzen-Goldberg syndrome: A disorder characterized by a marfanoid habitus, with craniofacial, skeletal and Hirschsprung disease. Other features of these patients include cardiovascular abnormalities and learning disabilities. Shprintzen-Goldberg craniosynostosis syndrome can be caused by *de novo* monoallelic mutations in the *SKI* gene which encodes the v-ski sarcoma viral oncogene homolog, a nuclear proto-oncogene protein that func-tions as a repressor of TGF-beta signaling. This gene maps on the subtelomeric short arm on chromosome 1. Shprintzen-Goldberg syndrome is estimated to affect less than 1 in every 1,000,000 individuals.

PCWH syndrome or Neurologic Variant of Waardenburg-Shah syndrome: A disorder characterized by the combination of 4 distinct syndromes including: Peripheral demyelinating neuropathy, Central dysmyelination, Waardenburg syndrome and Hirschsprung disease. The neurologic variant of Waardenburg-Shah syndrome seems to be caused by *de novo* monoallelic mutations in the *SOX10* gene which encodes the SRY (sex determining region Y)-box 10, a transcription factor. Neurologic variant of Waardenburg-Shah syndrome is estimated to affect less than 1 in every 1,000,000 individuals.

Mowat-Wilson syndrome: A multiple congenital anomaly syndrome characterized by a distinctive facial phenotype, moderate-to-severe intellectual deficiency and epilepsy. Affected individuals also manifest variable congenital malformations including Hirschsprung disease, genitourinary anomalies, agenesis of the corpus callosum, congenital heart defects and eye anomalies. Mowat-Wilson syndrome is caused by *de novo* monoallelic mutations in the *ZEB2* gene which encodes the Zinc finger E-box-binding homeobox 2. This condition can occur either by point mutations or by deletions in the *ZEB2* gene. Mowat-Wilson syndrome is estimated to affect less than 1 in every 1,000,000 individuals.

3.17.5.5 Congenital Malformations of Gallbladder, Bile Ducts and Liver

Polycystic liver disease: A disorder characterized by the appearance of numerous cysts spread throughout the liver. Females are predominantly affected and have larger numbers of cysts than affected males. These cysts appear after the age of 40 years. Most polycystic liver disease cases are inherited as an autosomal dominant trait, but a few sporadic cases have been reported. Patients suffering from the autosomal dominant polycystic kidney disease also exhibit multiple liver cysts. There are several underlying causes of this disorder with at least 2 autosomal genomic loci: Isolated polycystic liver disease can be caused by monoallelic mutations in the *PRKCSH* gene which encodes the protein kinase C substrate 80K-H. This gene encodes the noncatalytic beta subunit of glucosidase II, an N-linked glycan-processing enzyme in the endoplasmic reticulum;; and Autosomal dominant polycystic liver disease can be caused by monoallelic mutations in the *SEC63* gene which encodes the SEC63 homolog protein. This encoded protein is part of an essential component of the protein translocation apparatus of the endoplasmic reticulum membrane. Autosomal dominant polycystic liver disease is estimated to affect 1 in every 100,000 individuals.

Alagille syndrome: A disorder characterized by a paucity of intrahepatic bile ducts, in association with 5 main clinical abnormalities: characteristic facial phenotype, cholestasis, skeletal abnormalities, cardiac disease and ocular abnormalities. This condition is named after Daniel Alagille. There are several underlying causes of this disorder with at least 2 autosomal genomic loci: Alagille syndrome-1 can be caused by monoallelic mutations in the *JAG1* gene which encodes the protein jagged 1. This protein is the ligand for the receptor notch 1. This condition occurs due to a microdeletion of the 20p12 loci corre-

sponding to the *JAG1* gene. This encoded protein is involved in signaling between adjacent cells during embryonic development;; and Alagille syndrome-2 is caused by monoallelic mutations in the *NOTCH2* gene which encodes the notch 2 protein. This condition can either be inherited following an autosomal dominant pattern or occur de novo; the proportion of cases caused by de novo mutations is unknown. Alagille syndrome is estimated to affect 1 in every 70,000 births.

3.17.5.6 Other Congenital Malformations of Digestive System

Partial agenesis of the pancreas: A disorder characterized by the congenital absence of a critical mass of pancreatic tissue. The severity of the condition depends on the amount of functional pancreatic tissue present. Pancreatic agenesis is commonly associated with other malformations, particularly pancreaticobiliary duct anomalies, pancreatitis, hyperglycemia and polysplenia. There are several underlying causes of this disorder with at least 3 autosomal genomic loci: Congenital pancreatic agenesis is caused by biallelic mutations in the *PDX1* gene which encodes the pancreatic and duodenal homeobox 1;; Pancreatic and cerebellar agenesis or "Permanent neonatal diabetes mellitus with cerebellar agenesis" can be caused by mutations in the *PTF1A* gene which encodes the pancreas specific transcription factor 1a;; and "Pancreatic agenesis associated with congenital cardiac defects" can be caused by mutations in the *GATA6* gene which encodes the GATA binding protein 6, a transcription factor. Agenesis of the dorsal pancreas has been consistent with either autosomal dominant or X-linked dominant inheritance, but the gene has not been mapped. Partial agenesis of the pancreas is estimated to affect less than 1 in　　every　　1,000,000　　individuals.

Table 3.17.5 Recognizable Congenital Malformations and Deformations of Digestive System

Disorder	OMIM	Gene	Biological Function	Product of Conception Alteration
Orofacial cleft-5	608874	MSX1	Transcription factor Helix-turn-helix domains	
Orofacial cleft-6	608864	IRF6	Transcription factor Helix-turn-helix domains	
Orofacial cleft-7	225060	PVRL1	Adhesion protein	
Orofacial cleft-8	129400	TP63	Transcription factor	AD + de novo
Orofacial cleft-10	613705	SUMO1	Variety of cellular processes	
Orofacial cleft-11	600625	BMP4	Cell surface receptor Enzyme-linked receptor	
Cleft lip/palate-ectodermal dysplasia syndrome or Zlotogora-Ogur syndrome	225060	PVRL1	Adhesion protein	AR
Van der Woude syndrome-1	119300	IRF6	Transcription factor Helix-turn-helix domains	AD
Van der Woude syndrome-2	606713	WDR65	Protein-protein complexes	AD
Isolated cleft palate	119540	UBB	Mark cellular proteins for degradation	AD
Rapp-Hodgkin syndrome	129400	TP63	Transcription factor	AD
Cleft palate and mental retardation	119540	SATB2	Transcription regulation and chromatin remodeling	AD + de novo
Cleft palate with ankyloglossia	303400	TBX22	Nuclear transcription factor	XR
Caudal duplication anomaly	607864	AXIN1	Regulation of G-protein signaling	Hypermethylation
Currarino syndrome	176450	MNX1	Transcription factor Helix-turn-helix domains	AD + 25%
Hirschsprung disease-1	142623	RET	Proto-oncogene	
Hirschsprung disease-2	600155	EDNRB	Cell surface receptor G protein	
Hirschsprung disease-3	142623	GDNF	Growth factor	
Hirschsprung disease-4	613712	EDN3	Endothelin	
Shprintzen-Goldberg craniosynostosis syndrome	182212	SKI	Repressor of TGF-beta signaling	AD + 99%

Table 3.17.5. Continued.

Disorder	OMIM	Gene	Biological Function	Product of Conception Alteration
Neurologic variant of Waardenburg-Shah syndrome or PCWH syndrome	609136	SOX10	Transcription factor β-Scaffold factors	AD + 99%
Mowat-Wilson syndrome	235730	ZEB2	Transcription factor Helix-turn-helix domains	AD + 99%
Polycystic liver disease	174050	PRKCSH	Endoplasmic reticulum protein	AD
Polycystic liver disease	174050	SEC63	Post-translational protein translocation into the ER	AD
Alagille syndrome-1	118450	JAG1	Signalling through notch 1	AD + de novo
Alagille syndrome-2	610205	NOTCH2	Notch signaling pathway	AD + de novo
Congenital pancreatic agenesis	260370	PDX1	Transcription factor Helix-turn-helix domains	AR
Pancreatic and cerebellar agenesis	609069	PTF1A	Pancreas specific transcription factor	AR
Pancreatic agenesis associated with congenital cardiac defects	600001	GATA6	Transcription factor Zinc finger	AD

3.17.6 Genital Organs

3.17.6.1 Congenital Malformations of Uterus and Cervix

Mayer-Rokitansky-Küster-Hauser syndrome: A disorder characterized by congenital aplasia of the uterus and of the upper two third part of the vagina. Female patients display both a normal development of secondary sexual characteristics and 46, XX karyotype. Other features may include unilateral renal agenesis, skeletal abnormalities, and to a lesser extent hearing loss or heart defects. This condition could be of genomic origin, but there are several genes that have been tested to corroborate this possibility and no single factor has yet been identified as being responsible. There are several candidate genes which have been excluded including the *WNT4, TCF2,* and *LHX1* genes. In familial cases, the syndrome appears to be transmitted through autosomal dominant patterns of inheritance limited to females. This disorder can also be transmitted through normal males. Mayer-Rokitansky-Küster-Hauser syndrome is the second most common cause of primary amenorrhea after Turner syndrome. Mayer-Rokitansky-Küster-Hauser syndrome is estimated to affect 1 in every 4,500 women.

Hand-foot-uterus syndrome: A disorder characterized by small feet with remarkably short great toes and abnormal thumbs. Females with the disorder have duplication of the genital tract, including longitudinal vaginal septum. Affected individuals sometimes exhibit strabismus, hypoplastic thenar eminences, fifth finger clinodactyly, malformed thumbs, longitudinal vaginal septum, uterus bicornis bicollis and bilateral ureterovesical reflux. Hand-foot-uterus syndrome is caused by mutations in the *HOXA13* gene which encodes the homeobox A13, a transcription factor.

3.17.6.2 Other Congenital Malformations of Female Genitalia

McKusick-Kaufman syndrome: A disorder characterized by hydrometrocolpos, post-axial polydactyly, and to a lesser extent, cardiac defects. Affected individuals exhibit hydrometrocolpos due to an imperforate hymen or vaginal atresia. McKusick-Kaufman syndrome is caused by autosomal recessive mutations in the *MKKS* gene which encodes the McKusick-Kaufman syndrome. The protein encoded shares similarity to members of the chaperonin family. The encoded protein may have a role in protein folding, processing and assembly. Mutations in the same gene cause Bardet-Biedl syndrome-6. McKusick-Kaufman syndrome has mostly been reported in individuals of the Old Order Amish ancestry. In this community, the carrier frequency for the McKusick-Kaufman syndrome trait is estimated to be 1 in 50 individuals, which means the frequency of this condition is about 1 in 10,000 individuals or $1/(100)^2$. The incidence of this condition in other ethnic groups is unknown.

3.17.6.3 Congenital Malformations of Male Genital Organs

Cryptorchidism: A disorder characterized by a failure of testicular descent. Multifactorial etiology has been proposed involving endocrine, environmental and hereditary factors. This is the most common birth defect regarding male genitalia, affecting around 1% of males. Undescended testes are associated with augmented risk of testicular tumors. In most full-term infant boys that have this condition but no other genital abnormalities, a cause cannot be established. There are several congenital malformation syndromes that have been associated with much higher rates of cryptorchidism including: Prader-Willi syndrome, Noonan syndrome and cloacal exstrophy. Several environmental risk factors are associated with this condition including: maternal exposure to regular alcohol consumption during pregnancy, being a twin, gestational diabetes and caffeine use during gestation. There are few cases of familial isolated cryptorchidism. Cryptorchidism occurs in 4% of males if fathers were affected and 6-10% if brothers were affected. Heritability in first-degree male relatives has been estimated at 67%. Bilateral cryptorchidism was associated with a higher recurrence risk for brothers. Isolated cryptorchidism in few instances may be caused by autosomal dominant mutations either in the *INSL3* gene which encodes the insulin-like-3, or the *RXFP2* gene which encodes the relaxin/insulin-like family peptide receptor 2, a G protein-coupled receptor.

Hypospadias: A disorder characterized by defect of the urethra in the male that involves an abnormally placed urinary meatus. In most cases, the cause of this birth defect cannot be established. Several environmental risk factors are related with this condition including maternal use of hormones such as progesterone and diethylstilbestrol. There may also be an increased risk for this condition in infant males born to women of an advanced age. The incidence of hypospadias has been increasing in recent decades worldwide. Hypospadias is estimated to affect 1 in every 1,000 births. Familial forms hypospadias account for 7-10% of cases. There are several underlying causes of this disorder with at least 3 X-linked genomic loci and 1 autosomal genomic locus and: X-linked isolat-

ed hypospadias-1 can be caused by mutations in the *AR* gene which encodes the androgen receptor;; X-linked isolated hypospadias-2 can be caused by mutations in the *MAMLD1* gene which encodes the mastermind-like domain containing protein 1;; Isolated hypospadias-3 has been mapped to autosomal chromosome, but the gene has not been characterized;; and X-linked isolated hypospadias-4 can be caused by mutations in the *DGKK* gene which encodes the diacylglycerol kinase kappa.

3.17.6.4 Other Congenital Malformations of Male Genital Organs

Congenital bilateral absence of the vas deferens: A condition leading to male infertility. It accounts for 6-8% of cases of obstructive azoospermia and affects about 1 in every 1,000 males. This condition is also found in 98% of males with Cystic fibrosis. Infertile patients produce small volumes of acidic sperm. In 42% of infertile male patients, this condition was caused by monoallelic mutations in the *CFTR* gene which encodes the cystic fibrosis transmembrane conductance regulator (ATP-binding cassette sub-family C member 7). On occasion, the mechanism might be caused by alternative splicing of exon 9 in the *CFTR* gene.

Persistent Müllerian duct syndrome: A disorder characterized by the presence of a uterus and sometimes other müllerian duct derivatives in 46, XY karyotype or genomically male individuals. There are several underlying causes of this disorder with at least 2 autosomal genomic loci: Persistent müllerian duct syndrome type I can be produced by autosomal recessive mutations in the AMH gene which encodes the anti-mullerian hormone;; and

Persistent müllerian duct syndrome type II can be produced by autosomal recessive mutations in the AMHR2 gene which encodes the anti-mullerian hormone receptor. The exact prevalence of this condition is unknown.

3.17.6.5 Indeterminate Sex and Pseudohermaphroditism

Frasier syndrome: A disorder characterized by pseudohermaphroditism and progressive glomerulopathy. Male chromosomal pattern, or 46, XY karyotype, individuals have female external genitalia. Affected individuals do not have any functional gonads, but instead have streak gonads. These underdeveloped gonads often become cancerous. Affected females usually have normal genitalia and gonads. These individuals develop progressive renal disease during adolescence. Frasier syndrome is caused by monoallelic mutations in the WT1 gene. The encoded protein exhibit a zinc-finger DNA-binding function. Most cases are not familial or inherited, but are results of a de novo mutation.

Leydig cell hypoplasia: A disorder characterized by complete 46, XY male pseudohermaphroditism, low testosterone and high LH levels, and total lack of responsiveness to the LH/CG challenge. Leydig cell hypoplasia types I and II are caused by inactivating autosomal recessive mutations in the LHCGR gene which encodes the luteinizing hormone/choriogonadotropin receptor. Mutations in the same gene in females are associated with luteinizing hormone resistance.

Male pseudohermaphroditism, due to 17-beta-hydroxysteroid dehydrogenase isozyme 3 deficiency: A disorder characterized by incomplete differentiation of the male genitalia in 46, XY males. "Male pseudohermaphroditism, due to 17-beta-hydroxysteroid dehydrogenase isozyme 3 deficiency" is caused by autosomal recessive mutations in the HSD17B3 gene which encodes the 17-beta-hydroxysteroid dehydrogenase. This enzyme catalyzes the conversion of androstenedione to testosterone in the testis. Male pseudohermaphroditism, due to 17-beta-hydroxysteroid dehydrogenase isozyme 3 deficiency is estimated to affect 1 in every 147,000 individuals in the Netherlands.

Male pseudohermaphroditism, due to Steroid 5-alpha-reductase 2 deficiency: A disorder similar to the previous condition. It is caused by autosomal recessive mutations in the SRD5A2 gene which encodes the steroid 5-alpha-reductase 2.

46, XX sex reversal-1 or 46, XX true hermaphroditism: A disorder characterized by a phenotypically standard male who has a female genotype. There are several underlying causes of this disorder with at least 1 Y-linked genomic locus and 1 X-linked genomic locus: 46, XX sex reversal-1 or 46, XX true hermaphroditism is caused by translocation of a subtelomeric segment of the Y chromosome containing the SRY gene to the X chromosome. These products of conception alterations are defined as chromosome XY recombination (XY + Rec). This gene maps on the subtelomeric short arm on chromosome Y;; and Dosage-sensitive 46, XX sex reversal-2 is caused by X-linked duplication in a regulatory region upstream of the NR0B1 gene, which encodes the nuclear receptor (subfamily 0 group B 1).

46, XY sex reversal or Swyer syndrome: A disorder characterized by complete gonadal dysgenesis. Individuals with 46, XY karyotype show externally female features with streak gonads. Affected individuals manifest primary amenorrhea, tall stature, and hypoestrogenized vagina and cervix. If these patients are left untreated, they will not experience puberty. 46, XY sex reversal is a form of "pure gonadal dysgenesis" as opposed to Turner syndrome. A fetus whose cells do not produce a functional sex-determining region Y protein will develop as a female. There are several underlying causes of this disorder with at least 1 Y-linked genomic locus, 1 X-linked genomic locus, and 6 autosomal genomic loci: 46, XY sex reversal-1 is caused by point mutations or deletions in the *SRY* gene which encodes the sex determining region Y. These product of conception alterations are defined as classic Y-linked syndrome plus 99 % (Y + 99%);; 46, XY sex reversal-2 is caused by autosomal monoallelic duplication in a regulatory region upstream of the *SOX9* gene which encodes the SRY (sex determining region Y)-box 9;; 46, XY sex reversal-3 is caused by autosomal recessive mutations in the *NR5A1* gene which encodes the nuclear receptor (subfamily 5 group A 1);; 46, XY sex reversal-4 is caused by autosomal recessive deletion on chromosome 9p24.3;; 46, XY sex reversal-5 is caused by autosomal recessive mutations in the *CBX2* gene which encodes the chromobox homolog 2;; 46, XY sex reversal-6 is caused by autosomal recessive mutations in the *MAP3K1* gene which encodes the mitogen-activated protein kinase kinase kinase 1 (E3 ubiquitin protein ligase);; 46, XY sex reversal-7 is caused by autosomal recessive mutations in the *DHH* gene which encodes the desert hedgehog;; and 46, XY sex reversal-8 is caused by autosomal recessive mutations in the *AKR1C2* gene which encodes the aldo-keto reductase (family 1 member C2), with a possible contribution from the contiguous *AKR1C4* gene which encodes the aldo-keto reductase (family 1 member C4). 46, XY sex reversal is estimated to affect 1 in every 30,000 individuals.

Table 3.17.6 Recognizable Congenital Malformations and Deformations of Genital Organs

Disorder	OMIM	Gene	Biological Function	Product of Conception Alteration
Hand-foot-uterus syndrome	140000	HOXA13	Transcription factor Helix-turn-helix domains	AD
McKusick-Kaufman syndrome	236700	MKKS	Ciliary proteins Basal body	AR
Bilateral cryptorchidism	219050	RXFP2	Cell surface receptor G protein	AD
Idiopathic cryptorchidism	219050	INSL3	Insulin-like 3 (Leydig cell)	AD
X-linked isolated hypospadias-1	300633	AR	Transcription factor Zinc finger	XR
X-linked isolated hypospadias-2	300758	MAMLD1	Transcriptional co-activator	XR
X-linked isolated hypospadias-4	300856	DGKK	Diacylglycerol kinase	XR
Congenital bilateral absence of vas deferens	277180	CFTR	Channelopathy Chloride	AD

Table 3.17.6. Continued.

Disorder	OMIM	Gene	Biological Function	Product of Conception Alteration
Persistent Müllerian duct syndrome-1	261550	AMH	Cell surface receptor Other/ungrouped	AR
Persistent Müllerian duct syndrome-2	261550	AMHR2	Cell surface receptor Enzyme-linked receptor	AR
Frasier syndrome	136680	WT1	Transcription factor Zinc finger	AD + de novo
Leydig cell hypoplasia with hypergonadotropic hypogonadism	238320	LHCGR	Cell surface receptor G protein	AR
Leydig cell hypoplasia with pseudohermaphroditism	238320	LHCGR	Cell surface receptor G protein	AR
Female luteinizing hormone resistance	238320	LHCGR	Cell surface receptor G protein	AR
Male pseudohermaphroditism, due to 17-beta-hydroxysteroid dehydrogenase isozyme 3 deficiency	300438	HSD17B10	Mitochondrial protein	XR + de novo
Male pseudohermaphroditism, due to Steroid 5-alpha-reductase 2 deficiency	264600	SRD5A2	Endoplasmic reticulum protein	AR
Male pseudohermaphroditism, due to defective LH	152780	LHB	Luteinizing hormone beta polypeptide	
Male pseudohermaphroditism with gynecomastia	264300	HSD17B3	Endoplasmic reticulum protein	AR
46, XX sex reversal-1 or 46, XX true hermaphroditism	400045	SRY	Transcription factor Zinc finger	XY + Rec
Dosage-sensitive 46, XX sex reversal-2	300018	NR0B1	Nuclear receptor	XR
46, XY sex reversal-1	400044	SRY	Transcription factor Zinc finger	Y + 99%
46, XY sex reversal-2	278850	SOX9	Transcription factor β-Scaffold factors	AD + de novo
46, XY sex reversal-3	612965	NR5A1	Transcription factor β-Scaffold factors	AR
46, XY sex reversal-5	613080	CBX2	Chromatin remodeling protein	AR
46, XY sex reversal-6	613762	MAP3K1	Mitogen-activated protein kinase	AR
46, XY sex reversal-7	233420	DHH	Cell surface protein	AR
46, XY sex reversal-8	614279	AKR1C2	Aldo-keto reductase	AR
46, XY partial gonadal dysgenesis, with minifascicular neuropathy	607080	DHH	Cell surface protein	AR

3.17.7 Urinary System

3.17.7.1 Renal Agenesis and other Reduction Defects of Kidney

Perinatally lethal renal diseases: Are a group of sometimes fatal disorders including severe bilateral renal dysplasia, unilateral renal agenesis with contralateral dysplasia and severe obstructive uropathy. This absence of functional kidneys causes oligohydramnios leading to further malformations. Patients with this condition exhibit the typical "Potter facies" including wide-set eyes, flattened nose, receding chin, and large and low-set ears deficient in cartilage. This condition is thought to be cause by environmental or genomic factors or a combination. This condition is more common in infants born to one or more parents with a kidney adysplasia. In some patients, Renal adysplasia can be caused by mutations in the *PAX2* gene which encodes the paired box 2, the *RET* gene which encodes the ret proto-oncogene, or the *UPK3A* gene which encodes the uroplakin 3A. Perinatally lethal renal diseases are estimated to affect 1 in every 4,000-6,400 births with an apparent male preponderance. The twin correlation for Bilateral renal agenesis has been reported in 1 out of 6 twin monozygous pregnancies. There is also a similar disorder, the Unilateral renal agenesis, which generally does not exhibit any major health consequences as long as the unaffected kidney is healthy. Unilateral renal agenesis is a common disorder, affecting 1 in 750 live births. Unilateral renal agenesis is associated with an increased incidence of mullerian duct abnormalities and hypertension. Asymptomatic renal malformations, most often caused by Unilateral renal agenesis, runs in families. For example, the frequency of Renal agenesis in the fa-milial form is 4.5% whereas it affects 0.3% of all adults.

Renal tubular dysgenesis: A severe disorder of renal tubular development characterized by persistent fetal anuria and perinatal death, probably due to pulmonary hypoplasia from early-onset oligohydramnios. There are several underlying causes of this disorder with at least 4 autosomal genomic loci: Renal tubular dysgenesis can be caused by autosomal recessive mutations in the *REN* gene which encodes the renin, the *AGT* gene which encodes the angiotensinogen (serpin peptidase inhibitor clade A member 8), the *ACE* gene which encodes the angiotensin I converting enzyme (peptidyl-dipeptidase A) 1, and the *AGTR1* gene which encodes the angiotensin II receptor type 1.

3.17.7.2 Cystic Kidney Disease

Familial juvenile hyperuricemic nephropathy: A disorder characterized by elevated serum uric acid concentrations due to a low fractional excretion of uric acid. Patients with this condition exhibit interstitial nephropathy, defective urinary concentrating ability and progression to end-stage renal failure. Affected individuals could die from renal failure at a relatively early age. There are several underlying causes of this disorder with at least 4 autosomal genomic loci: Familial juvenile hyperuricemic nephropathy-1 is caused by monoallelic mutations in the *UMOD* gene which encodes the uromodulin. This protein is a constitutive inhibitor of calcium crystallization in renal fluids;; Familial juvenile hyperuricemic nephropathy-2 is caused by mutations in the *REN* gene which encodes the renin. This aspartyl protease cleaves angiotensinogen to form angiotensin I;; Familial juvenile hyperuricemic nephropathy-3 has been mapped, but the gene has not been characterized. There is also a form of Medullary cystic kidney disease that is caused by mutations in the *UMOD* gene. The last condition is a form of glomerulocystic kidney disease with hyperuricemia and isosthenuria. There is also an atypical form of familial juvenile hyperuricemic nephropathy associated with renal cysts and diabetes that is caused by mutations in the *HNF1B* gene which encodes the HNF1 homeobox B.

Goldston syndrome or Renal-hepatic-pancreatic dysplasia: A disorder characterized by association of polycystic kidneys and Dandy Walker malformation with or without hepatic fibrosis. Goldston syndrome is caused by autosomal recessive mutations in the *NPHP3* gene which encodes the nephronophthisis 3 protein. Mutations in this gene also result in Nephronophthisis-3 and in the more severe Meckel syndrome-7. Goldston syndrome is estimated to affect less than 1 in every 1,000,000 individuals.

Autosomal recessive polycystic kidney disease or Polycystic kidney and hepatic disease: A disorder characterized by the development of cysts affecting the collecting ducts. Liver involvement is detectable in approximately 45% of infants and is often the major feature in older patients. Half of all affected neonates die of pulmonary hypoplasia, which is caused by oligohydramnios from severe intrauterine kidney disease. Autosomal recessive polycystic kidney disease is caused by biallelic mutations in the *PKHD1* gene which encodes the polycystic kidney and hepatic disease 1 protein. This condition may affect around 1 in every 40,000 live children worldwide. In Afrikaners, the carrier frequency for Autosomal recessive polycystic kidney trait is estimated to be 1 in 49 individuals, which means the frequency of this condition is about 1 in 9,600 individuals or

$1/(98)^2$. There is a similar prevalence for this condition in Finland.

Autosomal dominant polycystic kidney disease: A systemic disorder that predominantly affects the kidneys, but may affect other organs including the liver, pancreas, brain and arterial blood vessels. Affected individuals could display renal cysts, liver cysts and intracranial aneurysm. About half of all individuals with this condition will develop end stage kidney disease and require dialysis or kidney transplantation. Progression to end stage kidney disease frequently occurs in the 4th-6th decades of life. There are several underlying causes of this disorder with at least 3 autosomal genomic loci: Polycystic kidney disease-1 is caused by monoallelic mutations in the *PKD1* gene which encodes the polycystic kidney disease 1 protein. This integral membrane protein plays a role in renal tubular development. This protein modulates G-protein-coupled signal-transduction pathways. This gene maps on the subtelomeric short arm on chromosome 16. Type 1 is thought to be responsible in 85% of patients;; Polycystic kidney disease-2 is caused by monoallelic mutations in the *PKD2* gene which encodes the polycystic kidney disease 2 protein;; and for Polycystic kidney disease-3, the gene have not been characterized. The known genes encode members of the polycystin protein family. Polycystic kidney disease can either be inherited or occur de novo. These conditions belong to the ciliopathies disorder. Other known ciliopathies include Primary ciliary dyskinesia, Bardet-Biedl syndrome, Polycystic liver disease, Nephronophthisis, Alström syndrome, Meckel-Gruber syndrome and some forms of retinal degeneration. Polycystic kidney disease is estimated to affect 1 in every 400-1,000 people worldwide.

Autosomal dominant medullary cystic kidney disease: A chronic tubulointerstitial nephropathy characterized by formation of renal cysts at the corticomedullary junction. In the 3rd decade of life, patients with this condition lose the ability to produce concentrated urine, generating polyuria and stable low urinary osmolality. Years later, patients exhibit anemia, metabolic acidosis and uremia, which reflect the progressive renal insufficiency. Affected individuals manifest end-stage renal disease occurring in the 6th decade of life. There are several underlying causes of this disorder with at least 2 autosomal genomic loci: Medullary cystic kidney disease type 1 is associated with end-stage renal disease at the mean age of 62 years. Type 1 has been mapped, but the gene has not been characterized;; and Medullary cystic kidney disease type 2 is caused by monoallelic mutations in the *UMOD* gene which encodes the uromodulin. Type 2 is characterized by manifesting an earlier age for end-stage renal disease, around the 3rd decade of life that is often associated with hyperuricemia and gout. Medullary cystic kidney disease is estimated to affect 1 in every 100,000 individuals.

Senior–Løken syndrome: An oculo-renal disorder characterized by the association of retinal dystrophy with nephronophthisis. Affected individuals show an early-onset severe visual loss, then manifest photophobia, nystagmus and hyperopia. During childhood, patients exhibit polyuria, polydipsia, secondary eneuresis and anemia. The progression of the disease involves acute or chronic renal insufficiency leading to end-stage kidney disease. There are several underlying causes of this disorder with at least 7 autosomal genomic loci: Senior–Løken syndrome-1 is caused by autosomal recessive mutations in the *NPHP1* gene which encodes the nephronophthisis 1 protein;; Senior–

Løken syndrome-4 is caused by autosomal recessive mutations in the *NPHP4* gene which encodes the nephronophthisis 4 protein;; Senior–Løken syndrome-5 is caused by autosomal recessive mutations in the *IQCB1* gene which encodes the IQ motif containing protein B1. This nephrocystin protein interacts with calmodulin and the retinitis pigmentosa GTPase regulator protein;; Senior–Løken syndrome-6 is caused by autosomal recessive mutations in the *CEP290* gene which encodes the centrosomal protein 290kDa;; and Senior–Løken syndrome-7 is caused by autosomal recessive mutations in the *SDCCAG8* gene which encodes the serologically defined colon cancer antigen 8. The cause of Senior–Løken syndrome involves an adverse effect in the protein formation mechanism of the cilia. This condition belongs to the ciliopathy group. Senior–Løken syndrome is estimated to affect 1 in every 1,000,000 individuals worldwide.

Meckel–Gruber Syndrome: A lethal ciliopathic disorder characterized by renal cystic dysplasia, central nervous system malformations, polydactyly, hepatic developmental defects, and pulmonary hypoplasia due to oligohydramnios. The malfunctions of this ciliary protein cause this lethal disorder. This condition is named after Johann Meckel and Georg Gruber. There are several underlying causes of this disorder with at least 10 autosomal genomic loci: Meckel syndrome-1 is caused by biallelic mutations in the *MKS1* gene which encodes the Meckel syndrome type 1 protein, a component of the flagellar apparatus basal body proteome;; Meckel syndrome-2 is caused by biallelic mutations in the *TMEM216* gene which encodes the transmembrane protein 216;; Meckel syndrome-3 is caused by biallelic mutations in the *TMEM67* gene which encodes the transmembrane protein 67;;

Meckel syndrome-4 is caused by biallelic mutations in the *CEP290* gene which encodes the centrosomal protein 290kDa;; Meckel syndrome-5 is caused by biallelic mutations in the *RPGRIP1L* gene which encodes the RPGRIP1-like protein;; Meckel syndrome-6 is caused by biallelic mutations in the *CC2D2A* gene which encodes the coiled-coil and C2 domain containing protein 2A;; Meckel syndrome-7 is caused by biallelic mutations in the *NPHP3* gene which encodes the nephronophthisis 3 protein;; Meckel syndrome-8 is caused by biallelic mutations in the *TCTN2* gene which encodes the tectonic family member 2;; Meckel syndrome-9 is caused by biallelic mutations in the *B9D1* gene which encodes the B9 protein domain 1;; and Meckel syndrome-10 is caused by biallelic mutations in the *B9D2* gene which encodes the B9 protein domain 2. The incidence for this condition is not precise; it varies from 0.02-0.7 in 10,000 births worldwide. In Finland, the carrier frequency for Meckel syndrome trait is estimated to be 1 in 47.5 individuals, which means the frequency of this condition is about 1 in 9,000 individuals or $1/(95)^2$. Meckel syndrome is estimated to account for 5% of all neural tube defects in Finland. This condition also showed a relatively high prevalence among Tatar people (Russian Federation).

Table 3.17.7 Recognizable Congenital Malformations and Deformations of Urinary System

Disorder	OMIM	Gene	Biological Function	Product of Conception Alteration
Renal adysplasia	167409	PAX2	Transcription factor Helix-turn-helix domains	AD + de novo
Renal adysplasia	164761	RET	Proto-oncogene	AD
Renal adysplasia	611559	UPK3A	Transmembrane protein	AD + de novo
Urogenital adysplasia	611559	UPK3A	Transmembrane protein	AD
Renal tubular dysgenesis	267430	REN	Aspartyl protease	AR
Renal tubular dysgenesis	267430	AGTR1	Cell surface receptor G protein	AR
Renal tubular dysgenesis	267430	ACE	Integral membrane protein	AR
Renal tubular dysgenesis	267430	AGT	Plasma and extracellular spaces protein	AR
Renal adysplasia	120330	PAX2	Transcription factor Helix-turn-helix domains	AD
Familial juvenile hyperuricemic nephropathy-1	162000	UMOD	Uromodulin	AD
Familial juvenile hyperuricemic nephropathy-2	613092	REN	Aspartyl protease	AD
Renal cysts and diabetes syndrome or Maturity-onset diabetes of the young-5	137920	HNF1B	Transcription factor	AD

Table 3.17.7. Continued.

Disorder	OMIM	Gene	Biological Function	Product of Conception Alteration
Goldston syndrome or Renal-hepatic-pancreatic dysplasia	208540	NPHP3	Channelopathy slit diaphragm	AR
Autosomal recessive polycystic kidney disease or Polycystic kidney and hepatic disease	263200	PKHD1	Channelopathy, transient receptor potential channels	AR
Adult polycystic kidney disease-1	173900	PKD1	Channelopathy, transient receptor potential channels	AD + de novo
Adult polycystic kidney disease-2	613095	PKD2	Channelopathy, transient receptor potential channels	AD + de novo
Medullary cystic kidney disease-2	603860	UMOD	Uromodulin	AD
Glomerulocystic kidney disease with hyperuricemia and isosthenuria	609886	UMOD	Uromodulin	AD
Senior–Løken syndrome-1	266900	NPHP1	Channelopathy slit diaphragm	AR
Senior–Løken syndrome-4	606996	NPHP4	Channelopathy slit diaphragm	AR
Senior–Løken syndrome-5	609254	IQCB1	Ciliary proteins Nephrocystin	AR
Senior–Løken syndrome-6	610189	CEP290	Ciliary proteins Nephrocystin	AR
Senior–Løken syndrome-7	613524	SDCCAG8	Ciliary proteins other	AR
Meckel syndrome-1	249000	MKS1	Ciliary proteins Basal body	AR
Meckel syndrome-2	603194	TMEM216	Ciliary proteins other	AR
Meckel syndrome-3	607361	TMEM67	Ciliary proteins, primary cilia	AR
Meckel syndrome-4	611134	CEP290	Ciliary proteins Nephrocystin	AR
Meckel syndrome-5	611561	RPGRIP1L	Ciliary proteins Nephrocystin	AR
Meckel syndrome-6	612284	CC2D2A	Ciliary proteins Basal body	AR
Meckel syndrome-7	267010	NPHP3	Channelopathy slit diaphragm	AR
Meckel syndrome-8	613885	TCTN2	Tectonic family protein	AR
Meckel syndrome-9	614209	B9D1	B9 protein domain	AR
Meckel syndrome-10	614175	B9D2	B9 protein domain	AR

www.ingramcontent.com/pod-product-compliance
Lightning Source LLC
Chambersburg PA
CBHW081212220326
41598CB00037B/6755